Nicolaus Steno

Nicolaus Steno

Biography and Original Papers
of a 17th Century Scientist

Second Edition

Edited and Translated
by Troels Kardel and Paul Maquet

From

Vilhelm Maar, ed., *Nicolai Stenonis Opera Philosophica*, Copenhagen 1910.
Gustav Scherz, ed., *Steno. Geological Papers*, Odense 1969.
Gustav Scherz, *Niels Stensen, eine Biographie*, vol. 1, Leipzig 1986.

 Springer

Troels Kardel   Paul Maquet
Gammel Holte   Aywaille
Denmark   Belgium

Additional material to this book can be downloaded from http://extras.springer.com

ISBN 978-3-662-58553-5   ISBN 978-3-662-55047-2   (eBook)
https://doi.org/10.1007/978-3-662-55047-2

1st edition: © Springer-Verlag Berlin Heidelberg 2013
2nd edition: © Springer-Verlag GmbH Germany, part of Springer Nature 2018
Sofcover re-print of the Hardcover 2nd edition 2018

Endpaper illustrations: The title page and a text page from the *Elementorum Myologiæ Specimen* in Stensen's characteristic handwriting, NKS 4019, the Royal Library, Copenhagen, Denmark (REX), see also Fig. 4.1, p. 158

Printed on acid-free paper

This Springer imprint is published by the registered company Springer-Verlag GmbH, DE
part of Springer Nature
The registered company address is: Heidelberger Platz 3, 14197 Berlin, Germany

The production of this book is supported by
The Lundbeck Foundation
and
GEUS, the Geological Survey for Denmark and Greenland

Book 1
Gustav Scherz: Niels Stensen – Eine Biographie.
Bearbeitet von Harriet M. Hansen
Herausgegeben von Franz Peter Sonntag.
St. Benno-Verlag GmbH, Leipzig. Band I, 1987.
English translation by Paul Maquet
by agreement with
Stiftung zur Herausgabe der Niels-Stensen-Biographie von Gustav Scherz

Book 2
NICOLAI STENONIS OPERA PHILOSOPHICA [OPH] I-II
Edited by Vilhelm Maar, Copenhagen, Vilhelm Tryde Publ., 1910
Translated into English by
Paul Maquet, OPH 1, 2, 3, 4, 5, 6, 7, 8, 9, 10, 11, 12, 13, 14, 15,
16, 18, 19, 20, 21, 22, 24, 25, 26, 28, 31, 32, 33
OPH 22, 31a and 31b by Paul Maquet in cooperation
with S. Emmanuel Collins
OPH 17 by Margaret Tallmadge May
OPH 23 and 27 by Alex J. Pollock who also translated DE THERMIS
OPH 29, 30 and ORNAMENTS by Maureen Rhode

e.Transcripts[1]
The scientific works of Stensen in original languages transcribed by
August Ziggelaar, Paul Maquet and Troels Kardel

---

[1]Additional material to this book, the transcripts of the original editions, can be downloaded from
http://extras.springer.com

Preface to the First Edition

Boni enim interpretis est, non ipsa modo verba seorsum considerare, sed & cum antecedentibus simul & conseqventibus conferre. (Steno, OPH I, 66)
A good interpreter considers not only the words as such but also confers them with what precedes and what follows. <474>

Vilhelm Maar's scholarly edition from 1910, the *Nicolai Stenonis Opera Philosophica* (OPH), is after one hundred years still the foundation for the study of Niels Stensen's scientific works in their original languages. Most works were written in Latin, one is in French and two small works were published later in Italian. Gustav Scherz's *Niels Stensen, eine Biographie*, for 25 years the most exhaustive existing biography, provides the documentation that lies behind the author's contribution on Stensen in the *Dictionary of Scientific Biography*.[2]

Both these editions came out in small numbers, the biography from a mainly religious publisher in the former German Democratic Republic. It would be of little service just to make re-editions or to provide access on the Web, since only few readers today master the Latin sufficiently enough to study Maar's edition of Stensen; or read German well enough to study Scherz with benefit; or read or even can get hold of the first and only complete translation of OPH in modern language, the richly illustrated Italian edition, *Niccolò Stenone—Opere Scientifiche*, edited by Luciana Casella and Enrico Coturri in 1986.

---

[2]DSB 13, 30–35.

We present here what was missing, English translations of all Stensen's printed scientific works and of the first volume in Scherz's biography on Stensen. In ten chapters it deals with genealogy, youth, education, his life and achievements as a scientist and his religious conversion. A brief summary of Scherz's volume 2 deals with Stensen as a Catholic bishop in Northern Germany between 1677 and 1686 is found on <p. 3–4>. That volume and Stensen's theological writings await a new team to take over.

33 published works in science and additional texts were collected and edited in the OPH by Vilhelm Maar in 1910. Stensen's first published work in science from 1661, De Thermis—*On hot springs*, was still undiscovered when Vilhelm Maar numbered the works.

Stensen's active life in research lasted altogether from 1660 to 1674. His science falls into two main categories, the anatomical works and the geological works. The *geological* works between 1667 and 1670 are OPH 23, 27, 29, 30.

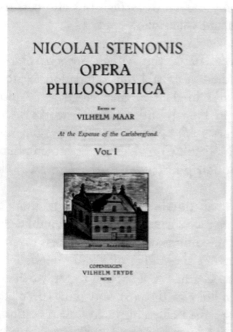

Vilhelm Maar (1871–1940) completed medical studies at Copenhagen University in 1896. After studies at learned centres in Europe, mainly in London, Maar became an assistant to his teacher of physiology, Prof. Christian Bohr. Based on studies of the respiration, Maar became Doctor of Medicine in 1902. Alongside the academic career, Maar practiced medicine. Maar was Professor of the History of Medicine in Copenhagen between 1911 and 1933 and published several bibliophile editions, the first of which was the Danish translation of the *Prodromus on solids* in 1902 in cooperation with August Krogh, the zoo physiologist and later Nobel laureate

Thematically, the *anatomical* works can be subgrouped as follows, with the time and place of origin indicated.

1. *Works on the glands and lymphatics* comprise OPH 1, 2, 3, 4, 5, 6, 7, 8, 11, 12, 13 from the Netherlands 1661–1663 and the second part of OPH 15 from Copenhagen in 1664. Many pages were spent on a dispute with Gerard Blaes and his allies on priority for the discovery of the excretory duct of the parotid gland.
2. *Works on the heart and muscles*, OPH 9, 10 and 14 from the Netherlands in 1662–1663, the first part of OPH 15 from Copenhagen in 1664, the main work, *Specimen of elements of myology* OPH 22, from Florence in 1667, and OPH 32 from Copenhagen in 1673. Annotated English translations of OPH 14 and 22 were published by the Transactions of the American Philosophical Society in 1994.
3. *Works on the brain*, OPH 22, the *Discourse of anatomy of the brain*, from Paris in 1665 and OPH 28 from Innsbruck in 1669.
4. *Works on the circulation of the blood*, OPH 11 from Leiden in 1662 and OPH 33 from Copenhagen in 1673.
5. *Works on foetal malformations*. OPH 20 is from Paris in 1665 and OPH 28 from Innsbruck in 1669.
6. *Comparative works on reproductive organs and foetal growth*. OPH 16 and 17 are from Copenhagen in 1664 and OPH 19, 20, 21, 24, 25 and 26 from Florence in 1668.
7. An *essay on scientific endeavour* is the *Prooemium*, OPH 31 from Copenhagen in 1673, which contains Steno's most famous words, *Beautiful is what we see, more beautiful what we know, but by far the most beautiful is what we do not know* <857>.

There are thematic overlaps. Thus, the main geological work, the *Prodromus on solids*, OPH 27, contains an original description of the compartments of the water space in the body with a conclusion that the formation of stones, such as bladder and gallstones, takes place by accretion in the so-called external water space of the body. Stensen's interest in stone formation in living beings was revealed in the Chaos-manuscript as a medical student, a prompter for his later interests.

English translations of the scientific works remained few and scattered when Gustav Scherz edited *Steno—Geological Papers* (GP) in 1969 with new translations from Latin by Alex J. Pollock of 2 of the 33 works in OPH and of the dissertation *on hot springs*. The GP further contains two letters in

Italian on Grottos, OPH 29 and 30, translated by Maureen Rhode.[3] These five geological papers with an excerpt of the notes by Scherz are reprinted here with kind permission by the publisher, the University Press of Southern Denmark.

Of 29 anatomical works, only a few were accessible in English when we published two works in *Steno on Muscles* and one work, the *Prooemium*, in *Steno—Life, Science, Philosophy* in 1994. These translations were based on preliminary work by the late Sister M. Emmanuel Collins, OSF, in Rochester, Minnesota. We thank the editors of the *Transactions of the American Philosophical Society* and the *Acta Historica Scientiarum Naturalium et Medicinalium* for permission to republish these translations.

In continuation, Paul Maquet, assisted by August Ziggelaar, S.J., D.Phil., translated all remaining scientific works into English with the exception of OPH 17, which was uniquely translated by Margaret Tallmadge May in 1950. The *Journal of History of Medicine and Allied Sciences* granted the permission to republish Dr. May's translation.

The scholarly annotations by Vilhelm Maar in OPH are to be found in the additional material to the present edition (Part III) which can be downloaded from http://extras.springer.com, together with the anatomical works by Stensen in their original languages transcribed from OPH.

Those translations completed, Paul Maquet took up the task of translating Gustav Scherz's biography of Stensen from the German.

---

[3]We were unable to trace Alex J. Pollock and Maureen Rohde (or Rhode?) to give them acknowledgement for their translations which is hereby presented.

Gustav Scherz

NIELS STENSEN

Eine Biographie

BAND I

1638 – 1677

Im Auftrag der
»Stiftung zur Herausgabe der
Niels-Stensen-Biographie von Gustav Scherz«
bearbeitet von Harriet M. Hansen

Herausgegeben von Franz Peter Sonntag

1987
St. Benno-Verlag GmbH Leipzig

Fr. Gustav Scherz, C.ss.R, D.Phil. & theol. & h.c. (1895–1971), is the second great scholar on Stensen of the twentieth century. Among numerous books and papers, Scherz co-edited Stensen's *Opera theologica* (OTH) and edited his correspondence, the *Epistolae* (EP), both works having been published in quarto two-volume sets like the *Opera philosophica*. See the Gustav Scherz biography by Ivar Hoel in <p. xvii>

Dr. August Ziggelaar, of Copenhagen, now Nijmegen in the Netherlands, also assisted in this translation and revised Scherz's many quotations from Stensen's student journal, the Chaos-manuscript, based on his own English edition (N.S. *Chaos*, ed. Z).

Texts and illustrations were edited by Troels Kardel. Alan Cutler, Ph.D., Washington D.C., geologist and the author of a widely distributed scientific biography on Stensen,[4] kindly undertook the revision of geological issues and

---

[4]Recent English biographies: Hans Kermit, *Niels Stensen. The scientist who was beatified*; Alan Cutler, *The seashell on the mountaintop*.

helped with the language editing. Harriet M. Hansen, M.A., historian and the editor of the German edition of the Stensen-biography, contributed with several notes. Health problems prevented her from taking part in the editing. We are most grateful for the assistance and inspiration we have received over the years of cooperation from August Ziggelaar, Alan Cutler and Harriet M. Hansen and for inputs from several other specialists. We would also like to thank the Committee for Publication of Gustav Scherz's Biography on Niels Stensen for permission to make this English edition as granted on 3 November 2009 by Jørgen Nybo Rasmussen.

We are grateful to the Uffizi Gallery, Florence, Italy, for providing a new digital photograph of Stensen's portrait for this edition. Scans and photographs from books were kindly provided by the Royal Library, Copenhagen, Denmark; the Biblioteca Nazionale Centrale and the Archivio di Stato, Florence, Italy; the Library Company, Philadelphia, Pennsylvania, USA; the Historische Bibliothek St. Anna, Schwerin, Germany; and the University Library in Tromsø, Norway. Other illustrations are from books in the possession of the editors. The graphical ornaments are reproduced from Vilhelm Maar's edition, if not from the *Elementorum Myologiae*, org. edition.

It has been our aim to keep the texts as complete as possible. Annexes were added within areas of new research, that is on the muscles—myology—and on geology. Texts were supplied with comments mainly as footnotes marked by letters or added in earlier endnotes after *. In the words of Gustav Scherz, when he himself edited A.D. Jørgensen's classical biography of "Nils Stensen" from 1884, notes and annexes were made "in order that old mistakes should not be given prolonged life, and not deny the certain knowledge that recent research has come to". It is our hope that the access to Niels Stensen's scientific works and biography in translation will enhance further studies on an important figure in European science and erudition in the seventeenth century.

October 2012

Troels Kardel, M.D.
Doct. Med. Copenhagen
Gammel Holte, Denmark

Paul Maquet, M.D.
Doct. h.c. Univ. Paris XII
Doct. med. h.c. R.W.T.H. Aachen
Aywaille, Belgium

Preface to the Second Edition

The interest in 'Nicolaus Steno—Biography and Original Papers of a 17th Century Scientist' allows us by the discretion of *Springer Nature* to present a second edition with additions and revisions.

Numbering of the chapters in three layers is introduced partly to ease navigation, partly to eliminate Roman chapter numbers. The first digit denotes 1 for Book 1 = Steno's biography, or 2 for Book 2 = Steno's scientific papers. Digits in the second layer indicate the chapter numbers. In Book 2, the chapter numbers follow the traditional OPH numbers from *N.S. Opera Philosophica* with few exceptions in the high numbers as mentioned in Contents.

In Book 1, the annexes on Stensen's myology and geology after Scherz are rewritten. Thus, new research enables in vivo confirmation in humans by advanced ultrasound technique of Stensen's controversial theory of muscular mechanics <198>. There are also referred to two newly recovered letters. < >

Three works are added in Book 2: To follow up on Chapter 2.32, the *Preface to Anatomical Demonstrations in the Copenhagen Theatre in the Year 1673*, we appended Thomas Bartholin's grandiose invitation to these demonstrations, Chapter 2.31. Also added is, Holger Jacobæus's manuscript, Chapter 2.33, that describes Stensen's demonstrations over a week following the Preface-lecture. Jacobæus' text offers a unique ringside description of a seventeenth-century anatomist at work in the dawn of physiology. See Gustav Scherz's comments on this work, <349>. These works are testimonies of the high esteem shown by Stensen's former teacher and by his student. No less than ten works published by Thomas Bartholin <350> indicates that the *Royal Anatomist* received a recognition in his home that is not often being mentioned.

The *Ornaments: Monuments, Signs, Arguments*, a sermon with geological imagery from the author's transition from scientist to priest, is added as Chapter 2.36.[5]

The translation of the Specimen of Elements of Myology, Chapter 2.22, is now given unabridged.

There are given excerpts from two earlier unrecorded letters providing information on Stensen's arrival in Tuscany in 1666 and his thoughts of returning from Copenhagen to Tuscany in 1673.

The electronic transcripts include the published scientific texts by Stensen in original languages with the extensive notes by Vilhelm Maar and Gustav Scherz. The Web address, http://extras.springer.com opens up the transcripts when entering the books ISBN number with hyphens.

To help the reader navigate from Book 1 to Book 2, numbers like <999> refer to the pagination in this book.

What follows the sign [*] is an addition by the editors.

Despite renowned editions, Gustav Scherz declared himself a non-specialist in geology <261, note 65>. Scherz built, as he writes, "on the interpretations of others, in recognition that a thorough and embracing scientific evaluation of Stensen's (scientific) works, of what he owed to his predecessors and contemporaries, remains to be written". Likewise, with the help of specialists we have, to the best of our abilities from medical schooling and in continuation of our research on Stensen's and his contemporaries' writings, intended to provide readers access to all Stensen's printed scientific papers and the almost forgotten wealth of information and sources brought together by Gustav Scherz and Vilhelm Maar.

We wish to remind that a new team is needed for translating the second volume, the religious part, of Gustav Scherz, *Niels Stensen, Eine Biographie*, and, not to forget, translate the *N.S. Opera Theologica* and his extensive correspondence in the *Epistolae*, to improve coverage of the spiritual life and work of a remarkable figure in European science and church.

---

[5]The 'Ornaments' were published in Latin by the Danish Royal Academy of Science and Letter edited by Vilhelm Maar in 1910; by Gustav Scherz in 'N.S. Opera Theologica', 1944; and edited by Scherz in *Steno - Geological Papers* in translation 1969. The sermon introduces the ethical aspect of vanity: glitter to the ruling classes costs health and lives for miners, pearl-divers, etc. Already in his student notes, the Chaos-manuscript, Stensen mentioned the health risk for guilders working in the family's goldsmith workshop.

The science–faith relationship and the philosophical context of Stensen's writings have recently been addressed in several editions and a well-documented heavy-reader.[6]

Over 25 years the editors have received numerous inputs, suggestions and valuable critique from scholars in many countries. In the second edition August Ziggelaar contributes by translations; Peter A. Huijing contributes on Stensen's myology; Alan Cutler and Gary D. Rosenberg on Stensen's geology; and Arne Friemuth Petersen on Stensen's method in science. We wish to express our sincere gratitude to all new and earlier contributors, most of all to Harriet M. Hansen, the editor of the German edition of Scherz's Steno Biography, who for health reasons was unable to co-edit this book.

Gammel Holte, Denmark   Troels Kardel
Aywaille, Belgium   Paul Maquet
April 2018

---

[6]Gian Battista Vai, 'The scientific revolution and Nicholas Steno's twofold conversion', in G.A. Rosenberg, ed., *The Revolution in Geology from the Renaissance to Enlightenment*, 2009, p. 187 ff.

Stefano Miniati, *Nicholas Steno's Challenge for Truth. Reconciling science and faith*, Milan, FrancoAngeli 2009.

Frank Sobiech, 'Ethos, Bioethics, and Sexual Ethics in Work and Reception of the Anatomist Niels Stensen (1638–1686)', *Springer—Philosophy and Medicine*, v. 117, 2016.

Raphaële Andrault and Mogens Laerke, eds., *Steno and the Philosophers*. Leiden, Brill 2018.

Rovere, Maxime, *Le Clan Spinoza: Amsterdam 1677 L'invention de la liberté*, un roman / une recherche. Paris, Flammarion 2017.

Short Biography of Gustav Scherz by Ivar Hoel

I

The first time I met Fr. Gustav Scherz was in July 1965. A small incident during that initial meeting in a glimpse gave an introduction both to his personality and to his profound interest. Coming from Norway, I had become engaged to be married to a Danish girl, and as he was a priest in her parish and had been a good friend of her family for decades, he was to be the marrying priest. As part of the marriage preparations, I was interrogated on my knowledge of Niels Stensen. What did I know about him? I sat there in his humble attic office and was forced to admit that I had never heard of him. It was as if I was not able to pass an examination and did not deserve to marry. Some remarks of his made me venture to say that Niels Stensen seemed to resemble Descartes. He lit up in a great smile, protested that the two were certainly not of the same opinion, turned around and took an offprint from the shelf and gave it to me as compulsory reading for our next meeting: a paper he had written on "Steno and Descartes"—as well as several others of his Steno publications.

II

Gustav Scherz was born on 17 February 1895 in Vienna as the second of seven children in an Old Viennese middle-class family. His father was a government official in the Ministry of the Interior. As a boy, he wanted to join the army, and his parents tried to get him admitted to a cadet school. The application was declined—luckily, Fr. Scherz later said—because of his myopia, so he carried on in the normal school system. However, since he did not receive good grades in spite of his abilities, his mother found it better if he were to go to boarding school. At the age of 12, he started in a Viennese boarding school run by the Redemptorist congregation, but without any purpose of becoming a priest. His interest in spiritual matters grew, however, during the school years, and in 1913, he entered the Redemptorist congregation as a novice. He was ordained a priest on 31 July 1919.

His health had suffered during the war years due to food shortage, and he contracted a serious lung disease. His work as a priest started in 1920–1921 when he was both patient and chaplain in the sanatorium of Enzenbach near Graz. In March 1922, his superiors sent him to Denmark—to him the land of milk and honey—for a period of recreation with the Redemptorists in Copenhagen. Thanks to the good living conditions he found there, and thanks to his disciplined energy, he regained his health. He would, for example, go for a summer morning swim in the cold waters of Øresund irrespective of the weather. At 70, he could easily compete with me, almost 50 years his junior, when it came to mountain walking. Upon arrival in Denmark, he at once set out to learn Danish, of course successfully, again thanks to his dedication and discipline and his flair for languages. He once remarked to me that he did not consider that he knew a language unless he could deliver a sermon in it. In the many years that he enjoyed being summer replacement for the vicar in Riva del Garda in Italy, he gave sermons in Italian, German, French and English—and in the Nordic countries in Danish.

He stayed for the rest of his life in Copenhagen, still in close contact with his family in Vienna. In 1933, he obtained Danish citizenship. He worked as a priest, as primary school founder, as headmaster and teacher, as congregation vice-provincial, as editor and as a researcher. 49 years in the same house was unusual in the congregation. It was in charge of three parishes in Denmark and circulated its priests among them. This can be explained by the fact that over the years his scientific research took more and more of his time and it required easy access to libraries and archives in the capital city.

Gustav Scherz was an affable and courteous man, his eyes smiling, his Viennese charm intact. He was generous, kind, helpful. He was modest on his own behalf, but not on behalf of his beloved church. He could at times be brusque, especially when questions to the Catholic Church or faith were raised that he considered irreverent. He was an old-school catholic, with no great understanding of the changes introduced by the Second Vatican Council. Yet he was looked upon with loving respect also by young people who did not see eye to eye with him in matters Catholic. "He really is some guy!" one of them said.

On Sunday 28 March 1971, after having said mass in the parish home for elderly, alive and well as ever, he went on his scooter to visit a hospital patient. He was involved in a traffic accident and died the next day. He is buried in the Roman Catholic department at Sundby Cemetery in Copenhagen, next to his fellow priests. His library of books on Steno and on his times, especially on baroque science, was acquired by Odense University Library (now the University Library of Southern Denmark), where it is catalogued as "The Steno Collection". His manuscripts and papers (51 capsules) as well as letters (39 capsules) were donated by his family in 1976 to the Royal Library in Copenhagen, where it is available to Steno researchers under the call numbers "NKS 4963 4°" and "Tilg 621", respectively.

III

During the years 1932–1939, Fr. Gustav Scherz was the editor of the Danish Catholic weekly journal, *Nordisk Ugeblad for katholske Kristne*. As editor, he was efficient, wrote well and had a great outlook, also on the contemporary situation in Europe. His anti-Hitlerian stance was so explicit that he was warned by the German embassy. Becoming an editor turned out to have a decisive impact on the rest of his life, for it was in that capacity he was first made aware of Niels Stensen. For the 1936 Steno jubilee (250 years after his death) and later for the 1938 jubilee (300 years after his birth), he edited illustrious special editions of the journal. In the early summer of 1936, he was preparing the first of these editions. My mother-in-law has recounted how he, sitting in her mother's allotment garden and joyfully reading Steno texts, beamed on her and her husband and said "Jeg er blevet lun på Niels Stensen—I've got a crush on Niels Stensen!". She had known and respected him since he first came to Denmark, but admitted that such a remark made her think there might be some subtleties of the Danish language that he still was not acquainted with.

That "crush" became his lifelong dedication and overriding project. It is because of his work, disciplined, dedicated, devoted and demanding, that he has been named "the Father of the Steno Renaissance". He was not only demanding of himself but also of others, at times exceeding the limits of their capabilities. He was the Steno researcher–editor–promoter–writer–correspondent–fund-raiser, working from the attic-office hub of the world-wide Steno network that he created. At home, he also had a network, a circle of ladies committed to helping him with typing, proofreading and suchlike clerical chores. Most of the Steno texts and letters that we have available today were published for the first time by him. A substantial part of them were texts that he was the first to extricate from unrecognized obscurity.

For his accomplishments, he was created doctor of philosophy from the University of Copenhagen in 1956 for his work "Vom Wege Niels Stensens". He was awarded doctorates honoris causa in theology from the University of Münster and in natural science from the University of Fribourg. During his many research journeys through all countries that might have any connection with Steno, he was unrelenting in his chase for Steno-related documents or any other new information on him. "Everything that he undertook or was assigned he did thoroughly" said his fellow priest and biographer, Fr. Klar. Dr. Scherz's Steno fact-finding research is an ample illustration of that. Over the years, he searched through 67 archives and libraries in nine countries on two continents. "Detective work" he himself rightly called it. No minute detail bearing on Steno was considered too small to go unnoticed and unrecorded. He was not only interested in texts; he was also very keen on acquiring the best available pictures for illustrations in his many and different Steno publications.

IV

His death came as a shock. He was mourned, and his achievements praised, by scientists and churchmen from throughout the world. He himself had certainly not considered his work finished, for there was one issue that was still unconcluded, even after decades of work: the cause for canonization of Niels Stensen as a saint in the Roman Catholic Church. The first initiatives had been taken in 1936 and 1938, with Fr. Scherz in many of the roles, from inspirator, writer and publisher to odd-job man, a man with connections both in the many different dioceses (Osnabrück, Paderborn, Hildesheim, Münster, Cologne and Copenhagen were all involved in a joint committee) and in scientific circles. The local preliminary informatory process was opened in Osnabrück in 1959, and by 1964, the historical process was concluded and a case for sainthood filed in Rome.

SACRA CONGREGATIO PRO CAUSIS SANCTORUM

OFFICIUM HISTORICUM

38

OSNABRUGEN.

BEATIFICATIONIS et CANONIZATIONIS

SERVI DEI

NICOLAI STENONIS

EPISCOPI TITIOPOLITANI

(† 1686)

POSITIO

SUPER INTRODUCTIONE CAUSAE ET SUPER VIRTUTIBUS

EX OFFICIO CONCINNATA

ROMAE MCMLXXIV

Cover of the *Positio*, the Vatican, Rome 1974, 4°

A large part of Fr. Scherz's investigations into the life and work of Steno had been done for the purpose of promoting that cause. All the material that was sent to Rome had been assembled by him. However, during the latter part of the 1960s, Fr. Scherz became somewhat impatient. "Why does it take so long in Rome", he wondered, "everything is done, all the information has been assembled, it is all there—why is nothing happening?" In 1968, a Vatican postulator to further the case was named, and Fr. Scherz started to work with him. To him, it was the Steno cause, not his personal interests that mattered. Yet it was not to be until three years after his death in 1971 that the 1107-page "Positio", a summary of the documentation to prove the virtues of the candidate for sainthood, appeared (see Figure). "Gustav Scherz did more than anybody else for the Causa Stenone" this weighty Vatican volume says, and rightly so. It took 14 more years for the canonization process to come to a conclusion. On 23 October 1988, Pope John Paul II declared Niels Stensen beatified in the church of St. Peter's. Many of the thousands of Danes and Germans assembled in Rome on that festive day have in their gratefulness surely given Fr. Scherz his due share of their thanks. On that day, his work was completed.

Ivar A.L. Hoel
Civ. Eng. Phys., library director (retired)

Literature

Fr. Klar, *Niels Stensen-forskeren Redemptoristpater Gustav Scherz*, København 1972.

Contents

List of Illustrations

Niels Stensen's portrait, digital photo kindly provided by the Uffizi Gallery, Florence, Italy.

Graphical elements and some sketches are from **NICOLAI STENONIS OPERA PHILOSOPHICA** (OPH)

Biblioteca Nacionale Centrale, Firenze (FBN)
Part of cover, p. xxxiv and Figs. 1.1.8, 1.6.4

Firenze, Archivio di Stato (FAS)
p. xxxviii

Gustav Scherz, Niels Stensen eine Biographie / Steno as Geologist
Figs. 1.1.1, 1.2.6, 1.5.1, 1.6.1, 1.7.1
The sketches displaying Stensen's travel routes (Figs. 1.2.6, 1.6.1, 1.7.1) were drawn by the Cartographer Axel Grevsen

Journal of Biomechanics
Fig. 1.4.4

Nationalmuseet, Copenhagen
p. xxviii

Photo by courtesy of J. Wilhelm I. Holst, Oslo
Figs. 1.1.2, 1.1.3

Steno—Geological Papers (GP)
Figures in Chap. 2.19

The Library Company, Philadelphia, Pennsylvania
p. xxxiii

The Royal Library, Copenhagen (REX)
p. xxxvi, xxxvii, 548–556
Figs. 1.1.6, 1.2.4, 1.2.7, 1.2.8, 1.3.2, 1.4.1, 1.4.3, 1.6.2, 1.6.3, 1.6.5, 1.9.2, 1.9.3, 1.9.4, 2.15.1, 2.22.2, 2.22.3, 2.22.7, 2.22.12, 2.31.1. Back cover, Detail from the frontispiece in OPH 15

Tromsø Universitetsbibliotek (TUB)
p. xxxi, Fig. 1.9.1

Wikimedia
p. xxxii

The editor's collection
p. xxxii, Figs. 1.1.4, 1.1.5, 1.1.7, 1.2.1, 1.2.2, 1.2.3, 1.2.5, 1.3.1, 1.4.2, Figures in Chaps. 2.11, 2.12, 2.13, 2.21

Colour Plates

Niels Stensen/Nicolaus Stenonis (or as indicated in the portrait Stenonius), 1638–1686. No signature, ascribed to the court painter Justus Sustermans. The portrait was kindly provided by the Uffizi Gallery, Florence, Italy

Signature from a letter, 1670

Stensen's letter seal—replica

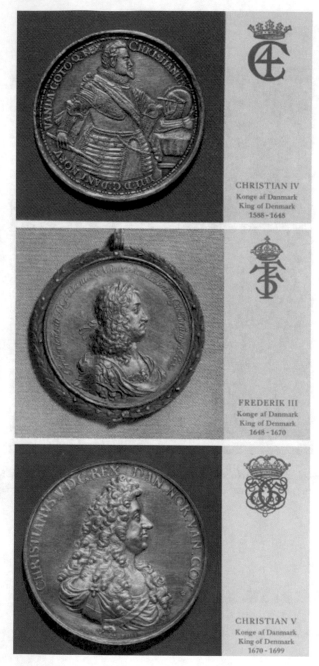

CHRISTIAN IV
Konge af Danmark
King of Denmark
1588 - 1648

FREDERIK III
Konge af Danmark
King of Denmark
1648 - 1670

CHRISTIAN V
Konge af Danmark
King of Denmark
1670 - 1699

Monarchs of Denmark in Stensen's lifetime. Source Nationalmuseet, Copenhagen

Fig. 1.1.2 Silverware dated 1637 marked Sten Pedersen's workshop, chalice in the church of Valle, Norway, photograph by courtesy of J. Wilhelm I. Holst, Oslo

Fig. 1.2.1 Barge drawn by a horse, Jacobæus, 16

Fig. 1.2.2 The Athenaeum at Oudezijds Voorburgwal is among the oldest buildings of Amsterdam University. The former cloister is still in use for doctoral defences, see "Agnietenkapel, een gebouw met een rijke historie", Universiteit van Amsterdam, 2007. Photograph 2010, TK <71>

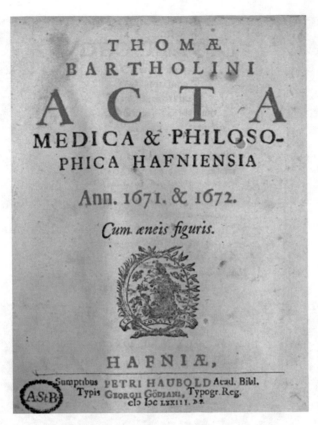

Fig. 1.9.1 Title page of *Acta Hafniensia*, TUB

Ferdinand II (1610–1670), source: Wikimedia Commons

Cosimo III (1642–1723), *Source* Wikimedia Commons

The altar of grace in the SS Annunziata in Florence, photograph TK

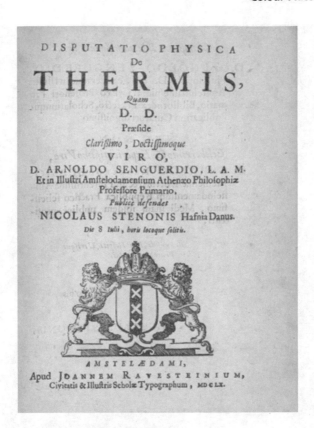

Stensen's first work, the Library Company, Philadelphia, Penn. <413>

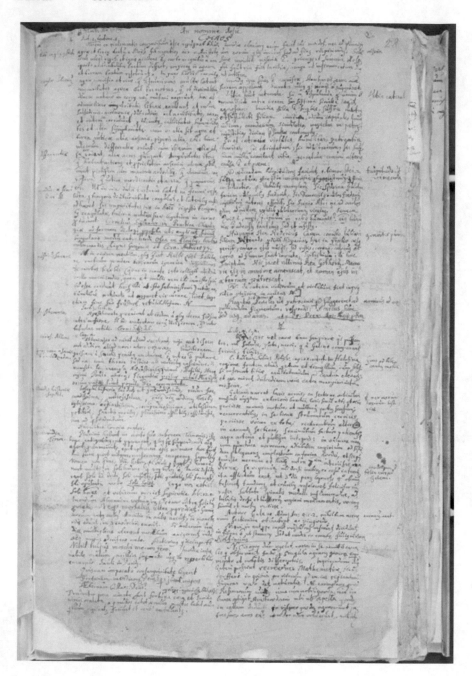

Fig. 1.1.8 Stensen's student notes, the *Chaos-manuscript*, e FBN: Gal. 291, first folio page of the manuscript with the headline, *In nomine Jesu CHAOS* <61>

CANIS CARCHARIÆ
DISSECTVM CAPVT.

ON dubito , quin Lectori faftidiofa_ fuerit longa , & non interrupta , rerum mufculos fpectantium expofitio; quocirca , cum in varietate oblectationem animus inueniat , materiam_ priori fubiungendam iudicaui, quę varijs, nec cohęrentibus inter fe, obferuationibus recenfendis occafionem præberet . Nec commodiorem aliam defiderare potui , quam quæ mihi fefe obtulit , Caput Canis Carchariæ diffecanti . Scilicet Sereniffimus Magnus Etruriæ Dux , cùm Liburno ipfi fignificatum effet , ad aliquot milliarium a Portu diftantiam captum infolitæ magnitudinis pifcem effe , caput illius Florentiam apportari iuffit , mihiq; tradi diffecandum . Licuit mihi in eo nonnulla videre, animalium partibus cognofcendis lucem haud obfcuram afferentia , quæ varijs alijs obferuationibus iluftrata hìc exponere volui .

The introduction to OPH 23 (This figure is part of Chap. 2.23)

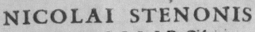

NICOLAI STENONIS

DE SOLIDO

INTRA SOLIDVM NATVRALITER CONTENTO

DISSERTATIONIS PRODROMVS.

A D

SERENISSIMVM

FERDINANDVM II.

MAGNVM ETRVRIÆ DVCEM.

FLORENTIÆ

Ex Typographia fub figno STELLÆ MDCLXIX.

SVPERIORVM PERMISSV.

REX (This figure is part of Chap. 2.27)

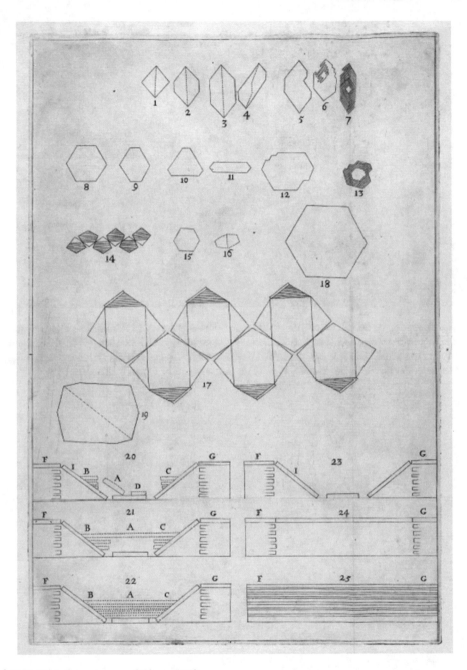

REX (This figure is part of Chap. 2.27)

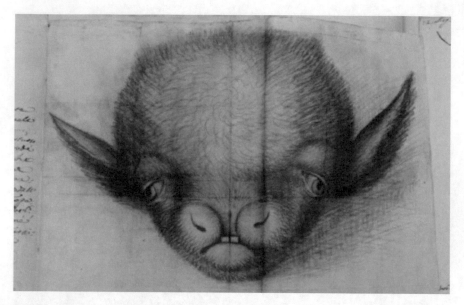

The illustration of the calf with hydrocephalus dissected by Stensen in Innsbruck is drawn in red chalk. No other such drawing exists from Stensen's hand. It may well have been provided by a court artist at Innsbruck since it is found attached to the letter by Anna de' Medici, the Archduchess of Austria to Ferdinand II, her brother, FAS. (This figure is part of Chap. 2.28)

Steno's Scientific Life in a Nutshell (1638–1677)

Niels Stensen was born in 1638 in Copenhagen. His father was a goldsmith, but many members of the family had been or were Lutheran clergymen (Chap. 1.1). His father died in 1645, and the workshop was carried on by the successive husbands of his mother. In 1656, Stensen graduated from the School of Our Lady and began studying science at Copenhagen University under the famous Thomas Bartholin. From Stensen's time as a student here, we have a most interesting collection of his notes and abstracts, the so-called *Chaos*-manuscript, depicting three months of intensive study in the year 1659.

Stensen went abroad to conclude his studies (Chap. 1.2): first to Amsterdam, where he made his first discovery, the parotid salivary duct, and then to Leiden, where he spent three fruitful years under François de le Boë Sylvius and van Horne and among friends like Jan Swammerdam and Reinier de Graaf. A dispute about his first discovery intensified his glandular research, and in 1661, he published his *On glands of the mouth and salivary ducts*, followed in 1662 by *Observationes anatomicae* with the former reprinted and three more papers on glands. He became deeply involved in the discussion of Descartes' philosophy and became friends with Spinoza. His anatomical research, however, soon made him realize the weaknesses of their philosophical systems.

Stensen returned to Denmark in the spring of 1664, where he published one of his principal works, the *On muscles and glands*. In spite of his success, he failed to obtain a professorship at the university and left again for Europe with a small legacy after his deceased mother.

First, he went to Paris (Chap. 1.3), where he stayed from November 1664 to September 1665. Here, he lived among the best scientists and earned for himself a reputation as a most distinguished anatomist. His famous *Discourse of anatomy of the brain* was held at Thevenot's house for a gathering of scholars. In the autumn of 1665, he continued his travels, probably staying a few months in Montpellier, where he met such English scientists as William Croone, John Ray and Martin Lister. In March 1666, Stensen finally reached Pisa, where he was received by the Grand Duke of Tuscany, Ferdinand of Medici, who became his benevolent patron (Chap. 1.4).

During the next years, Stensen settled in Tuscany, recognizing it as his second home country. Here, he worked among the members of the Cimento Academy, notably Francesco Redi and Vincenzio Viviani, in close contact with the Grand Duke, Ferdinando II, and Prince Leopold. During the summer of 1666, he concluded his treatise on muscle movements. The head of a large shark, brought to him for dissection, gave the research a new direction: the shark's teeth led to the investigation of fossil shark-teeth and further into the field of geology. In April 1667, his *Specimen of elements of myology* was published, dealing with three treatises, the first and main part with myology, thereafter palaeontology and embryology. He then seems to have concentrated on geology, but this was interrupted by a major religious crisis, which in November 1667 led to his conversion to the Catholic Church (Chap. 1.5).

In the summer of 1668, he seems to have received a summons from the Danish king to return to Denmark and hastily he put together a short introduction to his results in geology, the now classic *Prodromus on Solids*, which was written before the end of August (Chap. 1.6).

In November 1668, he set out on his long journey (Chap. 1.7) to Rome, Naples, Bologna, Venice, Innsbruck, Vienna, the mines of Hungary and Bohemia and finally to Amsterdam, but not to Denmark. In Holland, he found his conversion to be a subject of long discussions, and when he returned to Florence due to news about the Grand Duke's illness and death, he took up a long correspondence particularly with Johannes Sylvius, a reformed clergyman in Amsterdam, about religious matters (Chap. 1.8). Although at this time he is still known to have pursued scientific research, in which he was encouraged by the new Grand Duke, Cosimo III, his lifelong friend, he stood, in fact, on the threshold of a life devoted to God.

In 1672, he was summoned to Copenhagen, where he stayed from July 1672 to July 1674 (Chap. 1.9). Here, he held his famous opening speech on the occasion of a public dissection of a female body, stating that beautiful is that which we see; more beautiful that which we know; by far the most beautiful that which we do not know. The conditions for the convert in orthodox Lutheran Denmark, however, were unfavourable, and in 1674, Stensen decided to accept Cosimo's offer to become the tutor of the Crown Prince of Tuscany and returned to Florence (Chap. 1.10).

Steno as Bishop (1677–1686): Summary of "Gustav Scherz, Niels Stensen, Vol. 2" [not included in this book]

This quiet happy life ended abruptly in the spring of 1677, when Johann Friedrich, the Duke of Hannover, asked for Stensen as successor to the late Vicar Apostolic Valerio Maccioni (Chap. 2.0). The decision was made by Stensen's superiors, and in May 1677, he left his beloved Florence to spend the summer in Rome. On September 13, he was appointed bishop of Titiopolis, and on September 19, he was consecrated bishop.

The new Vicar Apostolic arrived in Hannover in November 1677 (Chap. 2.1). At the court of Johann Friedrich, he met not only the Duke's family, including Ernst August, the Prince Bishop of Osnabruck, later to become the Elector of Hannover, and his wife Sophia, but also the polymath and mathematician G.W. Leibniz, the Duke's librarian and counsellor. In Hannover, Stensen continued his theological writings and several treatises appeared between 1678 and 1680. His household, into which he received numerous converts, was formed on the basis of the bishop's highly idealistic notions. His way of life was ascetic, though never imprudently so.

The Duke of Hannover died suddenly in January 1680, leaving his dukedom to his protestant successor Ernst August. Ferdinand von Fürstenberg, the Prince Bishop of Münster and Paderborn, then asked for Stensen as an auxiliary bishop; Rome agreed, and in the summer of 1680, Stensen left Hannover for Münster.

The main duties of the auxiliary bishop were to administer the sacrament of confirmation, to consecrate altars and to ordain priests; for three years, Stensen led a very active life in the vast area of the diocese (Chap. 2.2). Best known is his careful visitation of the deanery of Vechta in 1682. He also wrote *Parochorum hoc age*—a booklet about the duties of a parish priest. To increase his modest income, he was appointed dean of St. Ludgeri in Münster, but since he felt unable to fulfil the duties of both bishop and dean, he resigned from his deanery a year later. Stensen's high ideals about the responsibility and sincerity of every member of the clergy clashed with the more easy ways of the time, and the ordination of unworthy priests was for him cause of acute suffering.

Stensen was looked upon as too severe, and when Ferdinand died in June 1683, he found himself in open conflict with the chapter of Münster (Chap. 2.3), which now possessed all power. When a new Prince Bishop was found in the pious, but weak Maximilian Heinrich, Elector Prince of Cologne, Archbishop of Cologne and Prince Bishop of Hildesheim and Liège, and this election was made possible by promises of money and offices, Stensen refused to accept such Simonistic machinations and left Münster on 1 September 1683. By applying directly to the Roman Curia, Stensen ensured, not the resignation of Maximilian Heinrich, but the Pope's refusal to confirm his election.

Stensen went to Hamburg, where he found an abode with the Tuscan envoy, his study-friend from Leiden, Th. Kerckring (Chap. 2.4). Hamburg's Magistrate was in a crisis, and the Catholic community was divided by dissension. As a measure to stabilize the mission, Stensen privately suggested a replacement of the two oldest missionaries, but brought down resentment on his own head when this leaked out. Stensen always had some reservations about his usefulness as bishop and Vicar Apostolic, and in Hamburg, he was convinced that he had no means of bettering the evils he plainly saw. At last, in the summer of 1685, he persuaded Rome to grant him a period of solitude and study in Italy, but then, the Catholic Duke of Mecklenburg issued permission for Stensen to work in Schwerin. What was needed there was a church independent of the palace to secure the Catholic faith in the region under protestant successors of the present Duke. Stensen felt obliged to avail himself of the opportunity, and after a delay of several months, he finally arrived in Schwerin in December 1685.

The small Catholic community stood in great need of instruction. Father Steffani, the only priest, was suffering from ill health and wanted to retire; Stensen, working as an ordinary priest, took up the work at his side, carefully avoiding any suggestion of supervision. In the spring of 1686, Stensen was suggested for the post of auxiliary bishop of Trier. He feared a repetition of the difficulties in Münster and longed for a time of reflections in Italy, but the decision was placed in the hands of the Pope. Stensen never lived to hear it.

In November, Father Steffani died peacefully. Shortly after, Stensen fell mortally ill. After four days of severe pain, he died on 25 (st. v.) November 1686. By instructions of the Grand Duke Cosimo, his body was transferred to Florence, where it was laid to rest in the crypt of S. Lorenzo. In 1953, the earthly remains were transferred from the crypt and placed in an antique sarcophage in a separate chapel facing the high alter of S. Lorenzo.

Niels Stensen was beatified by his Holiness Pope John Paul II in Rome on 23 October 1988.

Part I

Biography of Nicolaus Steno by Gustav
Scherz, Translated by Paul Maquet

1.1 Childhood and Youth in Copenhagen 1638–1659

Copenhagen, Coat of Arms

1.1.1 Genealogy and Family

Times were stormy when Niels Stensen was born in 1638. Twenty years previously the defenestration of Prague had started the war that lasted for 30 years in the heart of Europe. That brought mercenaries of the Emperor's Marshal Wallenstein even to the northern part of Denmark on the peninsula of Jutland in a victorious push of the Catholic League. The counterthrust of the Swedish King Gustav Adolf and the rough bravery of his army indeed drove away the imperial troops. The young Niels attended school for the first time in 1649, the year after the Peace of Westphalia. His youth and adulthood were marked by war after war, not only far away but also in the Baltic

© Springer-Verlag GmbH Germany, part of Springer Nature 2018
T. Kardel and P. Maquet, *Nicolaus Steno*,
https://doi.org/10.1007/978-3-662-55047-2_1

Sea at the border of his country with Sweden for the *dominium Maris Baltici*, for the predominance of the Baltic Sea to the South and East of Denmark.

Since the conquest of England and Norway by Danish King Canute the Great in the Viking period, since the time of the Valdemars in the High Middle Ages and the crusade of Danish knights in Estonia, since the Kalmar Union under Queen Margrethe, who could mount all the three Nordic crowns into the Danish coat of arms, the Land at the Sound (Øresund) and the Belts had had supremacy at sea. Even after the powerless century of the reformation, politically the sun seemed to shine again through clouds at the beginning of the long reign, 1588–1648, of Christian IV. However, the manhood of this popular king did not fulfil what his youth had promised. After the defeat at the battle of Lutter am Bahrenberg in 1626, things went downhill rapidly for the "Ruler of the Baltic". The Peace of Brömsebro in 1645 cost Denmark not only half of its Sound Dues, the most important revenue of the country, but also sovereignty over Gotland Island in the Baltic and of the province of Halland, thus disconnecting the line to Norway. In the Peace of Roskilde in 1658, Scania and Blekinge, which had been major parts of Denmark since ancient times, and provinces in Norway were also lost to Sweden. They were not to be recovered despite all efforts by subsequent Danish kings Frederik III and Christian V.[1]

The Stensen family, on the father's side,[2] came from Scania with its venerable cathedral of Lund, which in the first half of the seventeenth century was still a Danish province. However, far from the fruitful ground of Scania at The Sound, so rich in pastures, fields and fishing which make such an overwhelming impression on the visiting traveller, the family originated from the barren northern district, Göngeherred. The undulatory chain of hills that continues into the heights of Småland where deep forests are interspersed with lakes, moors and rivers, was the home of farmers who, fond of liberty, had always given the strongest resistance to foreign intruders.[3] Here, in the rural communities of Kviinge and Gryt, the forefathers of our Niels had worked for three generations as parsons and preachers. In the seventeenth

---

[1]An introduction to the history of Denmark is *Danmarks historie*, edited by Aksel E. Christensen, H. P. Clausen, Svend Ellehøj and Søren Mørch, Copenhagen 1977 ff. See vol. 2, 2 (1559–1648), 1982; vol. 3 (1648–1730), 1989.

[2]When Stensen speaks about *paese* or *mia patria* he thinks of his homeland Denmark to which he felt himself linked by the king, *il mio re* (EP I, 316, 252, 280).

[3]For more information on Scania and its history in general see Johannesson, *Skåne, Halland og Blekinge*. The Gönge gunsmiths were skilled craftsmen. Even in Copenhagen, the capital, people from this region had a reputation of being skilled in weaponry. It was a countryside into which one transferred deeds of heroes that were completed in the borderland (see K. Fabricius, *Skaanes overgang*, 43. On the "Gönge chieftains" of the sixteenth and seventeenth centuries, see DBL, 3rd ed., 5, 437 and 11, 494 f).

century, the status of Lutheran preachers in Scania was generally bad: ignorance, greediness and drunkenness were far from rare. In this regard, however, the Stensen forefathers seem to have been excellent pastors of souls.[4]

We meet the oldest ancestor, Niels Pedersen, for the first time on August 1, 1546 as the representative of Kviinge. He was still present on April 2, 1569 when his son Peter was ordained as its chaplain.[5] The childhood and youth of these forefathers still occurred in Catholic times and during the stormy and violent years 1528–1530 when the new faith arrived in Malmö and from there gradually spread to the countryside. As a consequence of the fact that the family brought up student sons in Kviinge the then frequent and tragic-comic custom that a new parson had to marry the widow of the deceased colleague if he wished to take over the job could be avoided. The grandfather of Stensen, Peter Nielsen, married Beate Pedersdatter. We meet him 16 years after his ordination at the Landsting in Lund, where in 1584 in the name of his parish he signed the act of allegiance to Christian IV. He died in Kviinge on June 2, 1599, after about 30 years of service.[6]

The sons of Peter Nielsen, Claus, the successor of his father in Kviinge, Niels and Sten, the father of our Niels, then appear in the full light of history. Born at the beginning of November 1577, Claus was 22 years old when he took over the parish from his father, which he administered for 43 years and 3 months until his death on October 6, 1642.[7] A large decayed tombstone under the tower on the Southern side of the Kviinge church still reminds us of "the honest, highly educated man and his dear wife". The growing tension between Sweden and Denmark was already throwing deep shadows over the communities in Göngeherred. During the Kalmar war with Sweden, the parsonage in Kviinge burst into flames on St. Nicholas day in 1612.[8]

The visit protocols of two bishops testify the pastoral zeal of Claus Pedersen. Thus the bishop of Lund, Paul Mortensen Aastrup, after his visit on January 19, 1618, (attended by Claus Bille from Vanaas, to whose noble family the pastors of Kviinge felt particularly close) presents him as a model

---

[4]The accounts of visits, if such visits took place at all, give a gloomy impression of these municipalities and their pastors. See Kornerup, *Til Lunde stifts kirkehistorie*, 28 ff. A perennial plaint was the tenacious adherence to Catholic practice like the ringing of bells for the deceased, pilgrimages and the worship of saints, Kornerup, 34 ff.

[5]LLA, original Stiftskisten, Skånebrevsfort. Div. eccl. 188. De ældste danske Arkivregistraturer 4, 350. Lovén, *De Gothungia*, 1745, 52. LLA, Tyge Asmussens Ordin. längd. Cf. in the following: EP II, Add. 1, 898–906 with further details on Stensen's family. Dates are in old style.

[6]Cawallin, 4, 433. LUB Sommelius Saml. Nr. 19.

[7]Cawallin, 4, 433 f. Niels Pedersen left an autobiography also telling about other members of the family, see below.

[8]Cawallin, 4, 323.

and exemplary preacher. He takes first place as far as erudition and way of life are concerned. His successor, Bishop Dr. Mads Jensen Medelfar, expresses the same opinion after his visit on February 3, 1623.[9]

A booklet preserved from the library of Claus Pedersen allows an interesting conclusion concerning the erudite links of the remote parish leader. It is a transcript of the precious Scania chronicle by Henrik Vejger or Smith (c. 1495–1563) from Malmö.[10] The latter, after completing studies in Rostock and Leipzig, came home in 1519. He joined the Mayor of Malmö, Hans Mikkelsen, and in 1523 went with him, in banishment in the retinue of King Christian II. Smith probably returned to his native town soon before the Reformation in 1536 and stayed there until his death. He endorsed the new doctrine and assisted in the first Danish translation of the Bible in Wittenberg. He displayed a profound interest in humanism and classics, as shown by his book *Hortulus Synonymorum, ad usum Danorum concinatus* (1520), and was even a skilful practitioner of medicine whose medical writings and herbals written in Danish were highly appreciated.[11] His interests in history found their expression in the above-mentioned chronicle. The transcript originates from Johannes Petersen who claims to be its original and legitimate proprietor[12] and who, according to his name, could well be counted among the ancestors of Stensen. Johannes Petersen continued the records of Vejger and therewith passed on much otherwise unknown information on the history of the Lund bishopric. There may also be a connection between Claus Pedersen and Malmö, which was then in its heyday with 9000 inhabitants and the second largest city of Denmark but was later to suffer badly from warfare and epidemics.

Claus Pedersen married Apollone Knudsdatter, the daughter of a preacher. In the span of their 41-year-long union, she gave birth to five children. She died on August 10, 1642. Regarding the two sons' history not much more is

---

[9] "... *qui si qvis alius inter Göthinos pastores officium debito modo facit, hic optimus et doctissimus vir palmam tenet, sive doctrinam sive mores respicias.*" Thus in 1618. In 1623: "*Pastor vir bonus bonis ornatus donis, catechumeni bene informati, omnia in Dei gloriam.*" Kornerup, Visitatsbog, 103 f.

[10] In LUB. He is also well known under the name Henrik Smid (Smith). Excerpts from his chronicle can be found in Samlingar til Skånes Historia 1871, 1 ff. Cf. Rørdam, *Monumenta historiae Danicae* I, 571–662.

[11] DBL, 3rd ed. 13, 527. FL 7, 399–403. See also the epilogue by A.E. Brade in Smith, *Lægebog* I–VI, 3–37, 1976.

[12] Johann Winslöw's words on the title page are: *Memoriale hoc chronologicum e possessione viri Dn. Claudij Petri past. quondam Quidingens, in Gothungia, avi mei materni, transijt in possessionem parentis mei charissimi, cujus dono meum tandem esse coepit. A. C. 1659 die 6. Maij.* Also: *Hic liber a Joanne Petri scriptus est, estque meus is possessor legittimus sine omnj amphibologia.* Could it be that Johannes Pedersen was the brother of Claus, Niels and Sten?

recalled than that they studied in Copenhagen.[13] Of the daughters, Karine, Anna and Birgitte, however, who all had children and enlarged a family of parsons, the first, who died in 1673, deserves our particular attention. She successively married the preacher Peder Nielsen, who died in 1630, and the dean Jakob Hansen in Winslöf. She had at least 11 children[14] and, through the first son from her second marriage, she became the grandmother of Jakob (Benignus) Winsløw (1669–1760), the famous anatomist and convert who followed the tracks of his genial granduncle.[15] In his autobiography, he gives a lively description of the deeply religious education to which his father, then parson at Our Lady's church in Odense, submitted him and his 12 sisters and brothers: common daily prayers, regular participation in services and religious teaching, dutifulness, frugality, care for the poor.[16] Our Stensen certainly received a similar upbringing.

The life of Niels, like that of his brother Claus Pedersen, appears as clear and demanding as recorded in his own *Vitae curriculum Nicolai Petri Quidingii*.[17] Born in Kviinge on October, 6 1581, Niels arrived in 1590 in the old school of Wä, which he later would endow with monetary gifts. On April 18, 1602, he was sent from the Latin school to the University of Copenhagen. He was accommodated in the Royal Foundation for impecunious students, the so-called Kommunitet, where the cost and lodging of students in theology were provided.[18] After the university years we find Niels Pedersen as a private teacher in the castles of the Bille family with whom the son of a parson from Kviinge felt attached by tradition.[19] Anders Bille, royal councillor and commandant in Helsingborg and Herredsvad, was particularly well disposed toward him. He helped him to obtain the office of preacher in Melby and Tjörnarp, and his noble protectors seem to have played a role in his marriage too. After his betrothal to Maren Eriksdatter on Epiphany day in 1611 in Rygaard (Zealand), Mrs. Anne Bille organized their wedding some weeks later, on February 24, 1611.

---

[13]Cawallin, 4, 431. Kbh Univ. Matr. I, 84.

[14]Cawallin, 4, 323 f.

[15]DBL 2nd ed. 26, 104.

[16]Winslow, *L'autobiographie*, 5 ff.

[17]Written down by the church verger Peter Theilsson, LUB MS, Biogr. Saml. Records Sommelius Nr. 19, ser. B, 1 fol. 164. See also Cawallin, 4, 33 ff.

[18]Rørdam, *Kommunitetets matrikel*, 353.

[19]Until July 28, 1608 he taught Peder Bille, the son of Oluf B., at Vallø or at the family's estate in Jutland Urup and Herringsholm. From August 8, 1608 he served Eske Bille from Ellinge, at that time captain in Elsinore, and taught the family's six children. He was highly praised by the nobleman and Mrs. Hedwig Giøe Falk.

For 34 years Niels Pedersen held office in Melby and Tjörnarp; from July 9, 1625 also as dean in the Göngeherred. He lived through all the horrors of the Kalmar war. His parsonage erupted in flames on that same December 6, 1612 as that of his brother in Kviinge. Rebuilt, the house was again the prey of flames on Easter Monday, April 20, 1617 and, for the third time, on May 31, 1624, while people were busy working in the fields. The "Confession book"[20] which Niels Pedersen set up gives an idea of his pastoral zeal. Above all, the Episcopal records of visits highly praise the parson of Melby and they also praise his wife.[21] A letter of April 29, 1632 shows Niels Pedersen in friendly relation with his bishop, Dr. Mads Jensen Medelfar, whom he calls his protector, while he also thanks the wife of the latter for good deeds rendered. His request not to let his remarks concerning a parson reach the court, shows a cautious gentleman. The elegant and easy hand-writing as well as the agreeable Danish and Latin style tell of the erudite who writes with the same ease in both languages. The letter seal shows a kind of Hermes figure over a globe with the subscription: *Varie pingit amicos*. N.P.[22] These words make those of the epitaph of Niels Pedersen more understandable, which calls him *praesulis adjutor Lundani*.

The dean of Melby died on November 8, 1644 and was buried in his parish church where initially an epitaph written by no less than bishop Winstrup, the successor of Mads Jensen Medelfar, was erected. There the deceased is described as a good shepherd and an example for his herd, and his flawless, worthy life is pointed out. Although childless, he did, however, give life to many "mystical clemency children".[23] Niels Pedersen died a wealthy man whose heritage gave rise to two lawsuits. In one of them there appeared as witnesses one Søren Schultz from Copenhagen, as well as the

---

[20]Ahnfelt, *En skriftebok från Göinge härad på 1600-talet*.

[21]Kornerup, *Visitatsbog*, 101. The parish priest had to deal with crude manners as shown by the anger of a farmer in Hedensø who had to be cooled down in a prison. One Sunday he had unhinged the church door in Kjørnerup and placed it so that Niels Pedersen could not enter the church and carry out his worship (KRA, *Skånske tegnelser* 5, 356b, Marquard, *Kancelliets brevbøger 1627–1629*, 619). Cf. also the argument between the ministers of the deanery, on which the bishop had himself to arbitrate in Ignaberga on June 5, 1628. On that occasion he strongly reprimanded the bad manners of the ministers (Kornerup, *Visitatsbog*, 106).

[22]KRA, Danske Kancelli, B 160. Indlæg til reg. og tegn. mm., 1632, April 29 and May 16. Cf. also the relation with Sivert Grubbe (1566–1636), sheriff and travel companion of King Christian IV. Grubbe heard on February 1, 1629 the sermon in Melby and afterwards was invited to dinner by the minister. In January 1630 he himself received Mr. Niels from Melby as visitor at Hovdala. On February 7 the rural dean came to Hovdala to celebrate the engagement of the church advocate Willum Jensen and Mette Gregersdatter. He also visited in March and on June 22, the latter time with the bishop Medelfar (Grubbe, *Dagbog*, pp. 78–81).

[23]Passed on in Cawallin, 4, 324 and EP II, 900.

Superintendent of Scania, Dr. Peder Winstrup.[24] The second lawsuit was of immediate interest to the nephew of the parson of Melby, Niels Stensen. On June 30, 1655 his guardian, the revenue officer Jørgen Carstensen, demands the implementation of a judgment of May 6, 1652 which obliged Christopher Mogensen from Lund to return an illegally appropriated part of the heritage.[25] As bailiff of the cathedral of Lund, Christopher Mogensen who had married Anne Sophie Clausdatter, a niece of Niels Pedersen, was in an eminent position since he had to administer half the fortune of the Chapter and, in the following decades, he also became bailiff of the university. His children, following a good family tradition, gradually occupied many parsonages of Scania.[26] The court of justice granted the claim. Thus in his last year in Latin school Niels had certainly obtained payment of the heritage from his uncle Niels.

Thus on his father's side the forefathers of our Niels constituted a healthy, intellectually and morally prominent lineage of Lutheran ministers. They were zealous and practical people who, by their way of life and their work, acquired the trust and respect of their fellow citizens and had connections with the leading classes of society. It was certainly above all a religious attitude that led generations of sons to the service of the church. However, it must not be ignored that this hereditary pastorate was also a consequence of the structure of seventeenth society, in which the main purpose of Latin schools and universities was the upbringing of preachers, while later in life the parishes offered the students almost the only possibility of subsistence. In the national political circles of the time the sympathies of these families were certainly on the side of the old kingdom and its hereditary royal house.

Stensen's origins on his mother's side are more obscure. His mother's name, Anne Nielsdatter, means that her father was called Niels; there was also a sister, Ermgard Nielsdatter, Mrs. Jeremias Figenschug. The Figenschugs were saddlers and strap makers who had immigrated from Germany in 1600 and soon became at home in Copenhagen. A Hans Figenschug had his own house in Østergade before 1614. He belonged to the St. Nikolaj Parish and was buried there in an imposing burial place on September 21, 1629. His son Jeremias took over his father's house and workshop and became guild

---

[24]January 9, 1651. KRA, *Herredags dombog* t. 50, 1651, fol. 220r–229v.

[25]KRA, *Herredags Dombog* t. 53. 1655, fol. 341r–346r.

[26]On his son, the preacher Niels Christoffersen Londin, Tryde, cf. Cawallin, 4, 139 f. On the sisters Anna Elisabeth and Hedwig, see Cawallin, 4, 91 and 451.

master.[27] From his marriage with Ermgard he had one daughter and two sons. He died in 1649 but Ermgard was still alive in 1654 and the young Niels may have visited this family several times. His cousin Nicolaus became saddler too and was guild master from 1680. He was buried in St. Nikolaj on April 30, 1682.

The sepulchre of Anne herself in the St. Nikolaj church tells us the surprising fact that she was married four times.[28] Her first husband was Henrik Kur who was buried in St. Nikolaj on August 16, 1632. The tombstone and burial are those of a wealthy man.[29] The last three spouses were all goldsmiths and they were all buried in the same grave in St. Nikolaj.

So far most tracks indicate a connection of Anne with town people and craftsmen in Copenhagen, particularly with such people in the parish of St. Nikolaj. Some tracks lead elsewhere. Indeed, in the personal history one can find reliable collections of the country judge J. Bircherod (1693–1737) concerning his family history, a small indication of the origin of Niels Stensen from the family of Karen Lethsdatter.[30] Karen Leth is known as the daughter of the mayor of Odense Jakob Leth († 1622) and sister of Inger Lethsdatter married to the bishop of Lund, the above-mentioned Dr. Mads Jensen Medelfar. Before her second marriage with Bertel Wichmand, Karen had been married to a man whose Christian name was Niels. We do not know his last name. Thus there is a possibility that Stensen was related to Karen Leth. As descendants of her first marriage, Jakob Nielsen and the sisters Maren and Bodil are only briefly mentioned. There is no mention anywhere of an Anne (or Ermgard) Nielsdatter.[31]

In conclusion: the forefathers of Niels Stensen belonged to the educated middle class and constituted a parsons' lineage in the then Danish province of Scania. Claus and Niels, brothers of his father Sten Pedersen, were distinguished Lutheran preachers with a good education. They had a connection with the bishops of Lund and also with noble families. They certainly professed the orthodox Lutheran faith of the time. Bishop Mads Jensen Medelfar

---

[27]Concerning the following see V. Hermansen, esp. 78 f. Figenschuh's family tombs in the church of St. Nikolaj, see *Danmarks kirker I*, 1, 1610 and O. Nielsen in PHT 1880, 221.

[28]O. Nielsen in PHT 1880, 211 f, *Danmarks kirker I*, 1, 599.

[29]Otherwise nothing is known on Henrik Kur. The name Kur (Cur, Chur, Kuur) is found in Copenhagen in 17th century for a glazier, Christoffer Kur, mentioned between 1656 and 1668. Cf. Kbh. Dipl.s Register.

[30]REX, Kallsk. Saml. Nr. 130, fol. 398. In such a case, Jakob B. Winsløw (1669–1760), who later went in Stensen's footprints, is related to Stensen on both his father's and his mother's side. Cf. genealogical table and EP II, 901.

[31]Anne Nielsdatter possibly also had relatives in Scania, because she paid liabilities for Mr. Laurids Hansen from Scania who was buried on January 13, 1645. O. Nielsen in PHT 1880, 220.

(1579–1637) as well as his successor Peder Winstrup (1605–1679) undoubtedly stood alongside the two Zealand bishops, Hans Poulsen Resen and Jesper Brochmand, being strong advocates of absolute Lutheranism. The faith of these ancestors, however, never went into the harshness and rigidity of Calvinism as the result of the spiritual welfare of their upbringing. The same is true of the mother's family of craftsmen from the middle class that also included parsons.

1.1.2   Childhood Years

Sten Pedersen, the third son of Peter Nielsen and father of our Niels, may have been born about 1585 (Fig. 1.1.1). He spent his youth in Scania and may have been educated either in Malmö or Lund.[32] We meet him for the first time in Copenhagen in 1620 as a distinguished goldsmith incorporated in the Danish Company at a Christmas drink in the year 1620. Originally, this guild from the Middle Ages served the interests of the profession on a religious basis and was dedicated to the Holy Trinity. It gathered its members in the guild house in the Kompagnistræde, mostly on the occasion of church feast days for dances, armed parades, bird shootings and similar activities. The members of the guilds were "honest people", as the wealthy townsmen were then called. Even kings and queens could be "brothers and sisters". Christian IV did not disdain a drink for Christmas in the circle of the guilds and, at Christmas 1620, among others, the imperial Field Marshal and Royal Count Heinrich Holck from Eskildstrup was admitted to the guild.[33]

The same year, the king had lent the considerable sum of 300 rigsdaler to Sten Pedersen, probably to transform his two "shops" which were previously designated in the ground book of 1625 as a court at the corner of Købmagergade and Klareboderne. This was a very considerable building inside the town over a ground of about 8×10 m provided with a ground wall of 17 window panels and it was not too far from a piece of land outside the walls in the vicinity of the so-called Jarmer-Redoubt, that also belonged to Sten Pedersen.[34] His deliveries to the court testify a goldsmith that enjoyed

---

[32]Also written as Steen, Stenn, Stehen, Sthen, Szten and thus an interesting example of the indifference as to the orthography of names at the time. (Cf. EP II, 904 f).

[33]N.P. Jensen, *Medlemmerne i Hellig Trefoldigheds laug*, 43—Kbh. Dipl. 6, 55.

[34]The sum of 300 Daler, for which already in 1621 he regularly paid in tax to the king and which he apparently had to pay back in 1626, may indicate that he worked upwards from a small beginning (*Kong Christian IV's egenhændige breve*, 1, 231, Marquard, *Kancelliets brevbøger 1624–1626*, 586. Schlegel, *Samling* 2, p. 46 and 69. Also: Kbh. Dipl. 5, 93. On the houses: 1, 610; 3, 48; 6, p. 217 and 225). *A memory plaque marks the site.

the favour of the king. Soon the workshop would make a traveller's wine pub with two silver bottles, a silver inkstand, a kettle, a silver lamp, a golden chalice or four silver pillars for the canopy of Corfitz Ulfeldt. To win rich appointments of thousands of daler, one needed feast occasions such as weddings, which often, particularly in war time, were followed by a low tide in the State's money chest and thereby belated payment.[35] The poor financial state of the kingdom perhaps stimulated Sten Pedersen to open a wine-cellar in 1625 where "wine, spirits, ale and others" were poured, which certainly attracted many guests and where the mood often became very joyful, if one thinks of the expressions of conviviality of the times.[36] Sten Pedersen stood in the first rank of his corporation. This is testified by the 4 years 1623–1627 during which the dignity of an old master was bestowed on him. His nice and careful written reports and accounts from this period in the so-called "Tegnebog" offer a lively picture of his works and duties inside and outside the guild assemblies (Figs. 1.1.1 and 1.1.2).[37]

Sten Pedersen had two children from his first marriage, a son and a daughter, who were already in the middle of their lives when our Niels still lay in his cradle. The son, Johann, came from the cathedral school in Roskilde on November 12, 1632[38] to university in Copenhagen and his private preceptor was Professor Jakob Matthiesen the court preacher and later bishop of Aarhus married to Anne Bartholin. Their son Holger would become the fervent pupil of Niels Stensen. Living in the home of this influential man, Johann spent 3 years of study at the university and his preceptor showed him the example of a spotless life and of undisputed proficiency. Johann also held a public dissertation. We meet him in 1636 at Leiden University during the study tour so customary in the circles of townsmen and nobles.[39] Soon after his return home he passed the theological professional test on January 26, 1639 and was declared well-acquainted *in fidei articulis ac lectione biblica.*

---

[35]Bering Liisberg, *Christian den Fjerde og guldsmedene.* Nyrop, *Dansk guldsmedekunst,* 43 f.

[36]Stichman came into an argument with the guild of wine merchants. He held that wine had been traded in the same place for more than 30 years (KRA, Da. Kanc. B 57, Sjæl. Tegn. 1654–1656 28.8. 1656, fol. 572).

[37]EP II, 901, KSTDA, Guldsmed Laug Nr. 3, Guldsmedenes Laugs Theignebog, wdj Stehen Pedersen Oldermandts Tid. Cf. EPII, 901. The stamp of Sten Pedersen in the year 1639, an intertwining S and P, can be found in a chalice in Holmens Church and in Ulse Church (Bøje, *Danske guld og sølvmærker,* I, 84).

[38]Kbh. Univ. Matr. 113. KRA, Danske Kancelli, B 160, Indlæg til reg. og tegn. mm. 5.11. 1641, from the university notary Peder Spormand with the most important dates of the life of Johann Stensen.

[39]Bricka in PHT, 2, 1881, 124: "Johannes Stenius Hafniensis 22, lit. stud."; the short remark in the protocol not only reveals his humanistic studies but also his year of birth 1614.

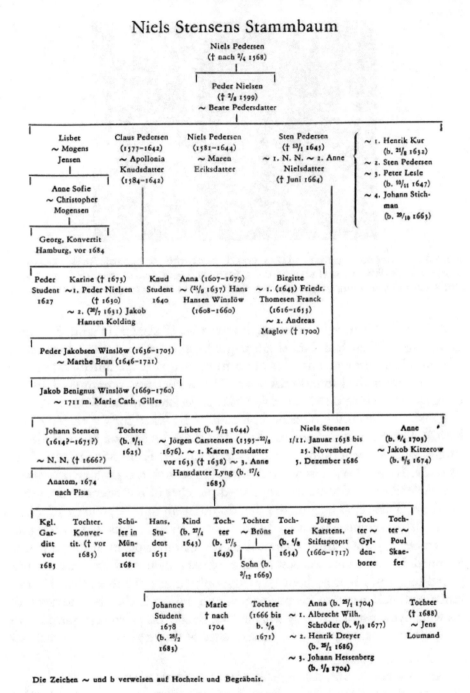

Fig. 1.1.1 Niels Stensen's genealogical table, Scherz, *Biographie* I, 18

Fig. 1.1.2 Silverware dated 1637 marked with the stamp of Sten Pedersen's workshop. Chalice in the church of Valle, Norway, photo by courtesy of J. Wilhelm I. Holst, Oslo (For a colour version, please see the plates at the beginning of this book.)

A stay in Elsinore intervened disastrously. Through Johann Stensen's knowledge of French and other languages he in some way aroused the anger of a drunken customs officer. It came to a duel with the outcome that the customs officer died four weeks later. The verdict was pronounced in the presence of the king on October 4, 1641. It declared Johann Stensen to be a free man but unworthy of clerical office until he asked for the mercy of the king.[40] This he obtained and, on May 15, 1643, Johann could conclude his studies with the acquisition of the grade of master. From May 15, 1644 till April 29, 1646, he acted as rector of the Latin school in Nykøbing on Falster. On November 1, 1648 he was appointed preacher of Balkager and Snarestad in Scania. There are some matters that do not seem to have gone well for him as his many complaints even to the king reveal.[41] His last writing of November 17, 1669 from Snarestad to the commission in Scania is a true jeremiad. He served the smallest parishes of the province for 22 years and was forced to spend his small heritage as the only means of subsistence. He had much bad luck, many of his cattle died, and, immediately after the death of his wife, five servants ran away with horses. Finally his parsonage burnt down completely on January 17, as a consequence of the carelessness of somebody

[40]KRA, Danske Kancelli, B 54, Sjæl. Reg. t. 21, 1641–1648, fol. 111 v, 112 r.

[41]Upon his withdrawal the local priest felt compelled on May 8, 1646 to report a complaint to the bishop over Mr. Johann. We do not know its contents. (Barfod, *Den falsterske gejstligheds personalhistorie I*, 64, KRA, Da. Kanc. B 50, koncepter og indlæg til Skånske registre 1656–1660, May 15, 1656. Landshøvding i Malmøhus Lens Arkiv, Skrivelser fra embedsmænd 1665).

else. At his age he ought to have had something to live from to bring up his children. He died in 1675. It seems that then Niels Stensen had to help two of his sons.[42]

Of Lisbeth Stensdatter, the half-sister of Niels, who got married in the year of his birth, we hardly know any more than that she had a small daughter buried in St. Nikolaj on April the 17, 1643 and that she followed her child to the grave on December 8, 1644.[43] Her husband, Jørgen Carstensen originated from a distinguished family of townsmen and was born in Sønderborg on November 15, 1595.[44] To Sten Pedersen he must have been a precious son-in-law. On June 14, 1625, he was appointed secretary in the royal revenue chamber, in the department of "Danske Kancelli" that dealt with the finances of the kingdom. He soon proved to be a competent clerk readily employed by the government. In 1633 he was assigned the superintendence of the accounts of the Sound customs. The court also entrusted him with many missions in the country and abroad.[45] As proved by his appointment in the revenue chamber, Jørgen Carstensen must have been a clever business man who successfully dealt with legal issues. We know of several lawsuits with townsmen, noble men and even the court.

On January 1, 1661, Carstensen was finally appointed secretary of the register (properly on February 16, 1661) with a yearly salary of 350 rigsdaler. The contract books of the revenue chamber, which he held until 1671, have been preserved. The grounds and houses of which he was the proprietor testify the wealth that Jørgen Carstensen progressively acquired. From July 9, 1633 he was the proprietor of a courtyard in Admiralgade. In 1668, his two houses were evaluated at 400 rigsdaler. Already in 1633, a garden outside the city gate Nørreport belonged to him.[46]

Sten Pedersen may have been about 50 years old—judging from his brother's age—when, well after 1632, he contracted a second marriage with Anne Nielsdatter, a marriage which resulted in our Niels.[47] It was also Anne's second marriage and, as mentioned she would still contract a third and a fourth spouse. At the time this was nothing unusual, partly because the frequent early deaths left many young widows and widowers, and partly for

---

[42]Carlquist, *Lunds stifts herdaminne*, Ser. II, vol. 5, 381 f.

[43]EP II, 901, *Danmarks Kirker* I, 1, 594.

[44]J.C. Bloch, *Den Fyenske gejstligheds historie I*, 652 ff.

[45]Cf. in reference the sources, EP II, 903 ff.

[46]Matr. No. 210 in Laxegade. See Ramsing, *Københavns ejendomme 1377–1728 I*, 112. Kbh. Dipl. 2, 810 and 825; 6, 227.

[47]EP II, 900 ff. Anne's first husband was buried in 1632.

social reasons since a single woman could not easily carry on a business like the gold jewelers. The bishop of Lund at the time, Mads Jensen Medelfar may have played a role as an intermediary in this wedding. As indicated by the name he was born in Middelfart. He had married Inger Leth, the sister of this Karen Leth (who may have been Anne's mother) and was, between 1612 and 1620, also parson at St. Nikolaj and royal confessor in Copenhagen. In Scania he kept excellent relations with the family of Sten Pedersen.[48] From this marriage first sister Anne was born, the playmate of the childhood years of our Niels. She would survive her brother for many years (she died in 1703).[49] Niels Stensen himself was born on New Year's Day 1638 (Jul. Calendar, which corresponds to January the 11, in the Greg. Calendar).[50]

Stensen mentions only few details from his early childhood at his father's home; however, they are of great importance for his whole later life. A disease during 3 years from 1641–1644 deeply impressed him. We do not know what kind of illness it was. The expression *morbus satis difficilis* is quite a general one that can be translated in different ways (burdensome, severe, troublesome, or dangerous). In any case it prevented the child from taking part in games and entertainments with companions since he always had to be under the protection of his parents or adult friends. Moreover, it also appears that the upbringing of our Niels must have been very quiet. The parents were of a mature age. It was a family of few children at a time when ten children and more was not uncommon in a townsman's family and in the parish homes. Stensen ascribes to this circumstance a quite decisive influence on his later choice of friends: "Therefore, I had got used to listen to what older people said, particularly about religious subjects, rather than to the light talk of younger ones", and he continued: "Also during my travels, I avoided as

---

[48]On Medelfar see DBL 2nd ed., 15, 430 and FL 5, 349–352.

[49]EP I, 905.

[50]Stensen's birthday cannot be ascertained from any birth register. If such a register did exist—the official command to keep Church Registers was given in 1644—it would have burned in the fire of Copenhagen on June 5, 1795. Cf. G. Bang, *Kirkebogstudier*. Efforts to confirm the date from official documents (Universities of Copenhagen and Leiden, conversion, ordination as priest and bishop) were all in vain (EP II, 904). More important are three contemporary Danish records that speak for January 1. Thus the register to Petri Johannis Resentii Bibliotheca (Hafnia 1685), which says: *Nicolaus Stenonius Hafniensis … natus Hafniae d. 1. Janr. 1638, mortuus Sverini 1686, d. 25. Nov.* Peder Resen (1623–1688) was a professor trained in law and burgomeister. Thanks to his historical interest we have 39 volumes with manuscripts of his Atlas Danicus (DBL 3rd ed. 12, 146–148). Also his coauthor on the library-catalogue, Jens Birkerod (1658–1708), the later bishop in Odense covers the above date with authority. Further J. Th. Bircherod (1693–1737) in his already mentioned family history: *Af hendes* (Karen Letsdatters) *Familie var dend laerde Nicolaus Stenonius … Episcopus Titiopolitanus og Hannoveranus. Hand var fød i Kiøbenhavn 1638, den 1. Jan. Og var een Guldsmeds søn, var liden og kleinlig og altiid svag …* The Cimbria literata by J. Mollerus supports January 1 (2,867). (Cf. M. Børup, *Hvornaar blev Stensen født [When was Stensen born]?*—Børup held on January 10/20).

much as possible the discourse of futile and periculous people and I always attempted to get in friendly communication with the people who somehow stood out by their way of life or their erudition."[51] In this he succeeded, as shown by the number of acquaintances with spiritually and morally prominent people.

It would, however, be a mistake to conclude from his disease in childhood that he had a lifelong ailing and frail constitution. Apart from the health and even sturdiness and longevity of the families from which the boy came, his research was carried out with an iron will; his distant travels, his trekking in the Alps and in the Apennines, as well as the energetic visits of the later bishop indicate altogether a strong constitution, which certainly did not fall victim of a deficiency disease.

Another trait also reveals that the young boy was already then an alert observer. Deeply involved in his study of the tear apparatus of the eye, in 1662 Stensen wrote to Thomas Bartholin to ask why tobacco dust makes the eyes clearer, among other matters: "For the limpidity of the humour lining the eyelids inside is not always such that sometimes intermingled thicker and more viscous particles do not adhere to the tunics of the eye, the thinner ones being removed. This is what occurred rather often to my father of blessed memory, who, after an excessive outflow of humour, quickly restored their pristine acuity to the eyes by bringing near them ginger root moistened with saliva."[52] Since Sten Pedersen was buried on January 13, 1645, we are dealing here with an observation and a remembrance of a 6 or 7-year-old boy. As the later notes of the student show, he had certainly not been sitting idle in the goldsmith workshop of his father. Moreover, when later in his research and work as anatomist, besides his visual acuity, his sure hands, his art of preparations and his drawing talent stood out, the basis of such skills had certainly already been laid down in the workshop at home in Klareboderne.

Later, Stensen particularly and strongly pointed out a third remembrance from his childhood when describing his conversion and religious crises and naming the reasons that kept him in his old faith. It was the paternal home and the strict Christian-Lutheran upbringing to which he had been subjected. He ascribes to "his parents and first teachers"[53] his firm confessional conviction and summarizes the end of the faith crisis which he went through in

---

[51]OTH I, 394.
[52]OTH I, 109 = EP I, 152 f.
[53]OTH I, 387.

Holland in the circle of Cartesianism in the following words: "However, I continued to hold fast to the paternal faith."[54] There one sees a Christian family in which the child grew up, imbued with Lutheran pietism like many of the townsmen families in Copenhagen, and with strong moral idealism. Little Niels himself felt something of the *maxima sollicitudo* that he later, as a bishop, recommended to others concerning the religious upbringing of small children who "still believe everything from their first best teacher and, once used to rules, keep to them for a long time".[55]

The family had its chapel in Sankt Nikolaj, the church then at the sea-shore, which was dedicated to the sailors and in which Niels had surely been baptized. The street Klareboderne ("the stalls of St. Clare") in the Eastern quarter belonged to its parish. It was the church of the wealthy townsmen. A new church had just been opened under Christian IV (in about 1638 with a tower from 1581, the top of which was destroyed by a violent storm in 1628 and was not rebuilt until in 1666; the church burned down in 1795). The inside was second to none in splendour. Little Niels will have looked up to the powerful buttresses which anchored the building in the filled-up shore ground. For the first time he saw human bones and skulls when he went along the charnel house at the northern part of the church or admired the big gold-plated statue of Saint Nicolaus below the Eastern gable of the church.[56]

Not only the parish church but also the whole city and the new image it presented will often have been looked at with surprise by an alert boy. In the middle of the seventeenth century Copenhagen was a fortified town of about 25,000–30,000 inhabitants[57] and still the capital and residence city for Denmark, Norway and Iceland with two million people altogether. Nearly half of the citizens of the capital were somehow in the service of the court and the government. As seen with modern eyes, proportions were small, the walls had fallen in disrepair, and in the 3 m wide streets many jutting-out buildings hindered traffic. Booths and fences on the sidewalks were not rare. At night one had to have a light-boy and could, however, get lost in the narrow streets with the risk of a more than doubtful shower-bath spilled from a window. However, it was also already the Copenhagen with its towers and buildings that Christian IV led to modernity and that the modern-day visitor finds pretty and picturesque. Church and kingdom and nobility and rising

---

[54]OTH I, 126.

[55]OTH II, 372.

[56]*Danmarks Kirker* I, 1, 459 ff.

[57]Cf. *Fridericia*, 219–263. Marquard, *Kjøbenhavns borgere*.

townsmen had their buildings here. However, most places on which the amazed eyes of the gold-smith's son fell have since disappeared.

Very close to the parental home and a series of artisan booths the great court of the nobleman Tage Thott was situated, which in 1661 was evaluated as being worth 1700 rigsdaler[58]—six times that of Sten Pedersen's house. On the opposite side of Købmagergade, at the corner of Løvstræde, lay the tradesmen's hall in which the two-and-a-half years older Peder Schumacher, later Chancellor Griffenfeld, grew up.[59] The narrow street Klareboderne was in itself full of historical reminiscences. It led to a place where the pious Queen Christine in the beginning of the sixteenth century had established a large court as an addition to a cloister for the Clarisses, "her particular and chosen children". The kings of the Reformation had then put their hands on the imposing complex and the cloister became a bell foundry and then the royal mint, whereas the big pretty cloister church which still stood in about 1570 was demolished soon afterwards.[60] However, the consecrated monuments had taken such a place in the heart of the people that at the beginning of the seventeenth century it was envisaged to rebuild them. In a letter of the apostolic vicar to the Great Duke Cosimo III, a child remembrance of these narrow streets by Stensen emerges when he adds to clarify the date: "On the St. Clare's day on whose intercession I expected particular help since in my country I was born in a street which still today bears her name."[61]

Only some 300 m away from the parental home, at the Northern end of the Købmagergade, facing the student college "Regensen", the Round Tower loomed and, connected to it, the Trinitatis church.[62] According to Christian Longomontanus, the enthusiastic pupil of Tycho Brahe, the stately tower with its interior spiral gangway was completed in 1642—the year when Galileo died—to be solemnly consecrated in 1656 together with the church for a triple goal: church for the spiritual needs of the students, a library for the university (under the church roof) and, in the tower, an observatory. At the opposite end of the Købmagergade, along St. Nikolaj, the son of the gold-smith certainly often went for a stroll with his father to the royal castle, the seat of the government and administration. The king himself kept to the charming castle Rosenborg most of the time, completed around 1625. With the ladies of the house little Niels may have gone along Holmens church to

---

[58]Kbh. Dipl. 1, 755. Cf. Scherz, *Niels Stensens København*.
[59]Fabricius, *Griffenfeld*, 23.
[60]Bruun, 1, 360 f, Nielsen, *Kjøbenhavns historie*, 1, 170 f.
[61]EP II, 800.
[62]*Danmarks kirker*, I, 2, 251 ff.

the stock exchange, this "stone legend about the first lucky years of Christian IV's reign" in 20 booths of which one could buy what the land and foreign countries had to offer in clothes, jewelry, books and all sorts of goods. Since 1610 there was also the New Town Hall at the Old Market, which Sten Pedersen often visited. He was indeed one of the valuers of the town.[63]

The years 1644–1645 opened painful gaps in the Stensen family. On November 8, 1644 the dean Niels Pedersen died in Melby. On December 8, Lisbeth, the wife of the revenue secretary Carstensen was buried in St. Nikolaj after having lost a probably new-born child in May of the previous year. Now, on January 13, 1645, Sten Pedersen, the father of Niels, was also brought to his grave in St. Nikolaj.[64] Since Claus Pedersen, the parson of Kviinge had already died in October 1642, the family had lost the three Pedersen brothers who had worked themselves forwards to such honourable positions.

After the death of her husband, it was up to Anne Nielsdatter to continue the goldsmith shop. As long as she remained unmarried, however, the law required the appointment of a guardian and the execution of an inheritance partition as well as the administration of the fortune of the children under proper tutelage. Since mother Anne carried on the business at least until 1646 under her own name and Niels also remained in the peace of his parents' home, a guardian was appointed.[65] He was the business associate Jørgen Carstensen who, soon after the death of Lisbeth, had concluded his third marriage with Anna Hansdatter Lyng from Ystad in Scania. On May 4, 1646, Carstensen deposited the very considerable sum of 1402 rigsdaler, the part of the paternal heritage of Niels, in the court in Købmagergade.[66] In the same year, mother Anne had her third marriage with the goldsmith Peter Lesle, who is otherwise completely unknown. Lesle was never mentioned by Niels

[63]Kbh. Dipl. 6, 339 f. In 1631 Sten Pedersen worked with taxing at the shops. The financial need in 1638 and the following years several times caused the Stænderforsamling to be called and to make requests to the citizens (*Danmarks historie* 2, 2, 532 ff).

[64]EP II, 900. Cf., moreover, V. Hermansen, *Niels Stensens Mutter*. Hermansen proves that Sten Petersen was buried on January 13, 1645, not on January 13, 1644 (p. 76).

[65]KLA, *Københavns Pantebog* 1645–52, fol. 12. Anne signed bills to be recovered after the death of Sten Pedersen (KRA, Rentekammeret 216, 276 Afregninger, XIII, position 9, Sten Pedersen guldsmed og hustru Anna). On January 13, 1645 she paid the bill for the tomb of Laurits Hansen from Snægod in Scania (O. Nielsen in PHT 1880, 220). She conducted a lawsuit against Mrs. Ingeborg Ulfstand on February 22, 1646 because of a money claim of her husband as of May, 20, 1640 (KRA, Kongens Retteting. *Stævningsbog* 1643–1646, fol. 799v–800r).

[66]KLA, *Københavns Pantebog* 1645–1652, fol. 12: Anno 1646. *mandagen dend 4. Maij, Anna sl. Steen Pederszen gulldszmidtz efterleffversche, wed hindis fullmechtig Madz hannszen, giorde hinndes sönn Nicolaj Steensens formönder Jörgen Cartsennssens renndteschriffer, forskrifvning udi hinndes gaard som hunn sielff paa Kiöbmangergade iboer, for 1402 rdl. Som er bete Nicolaj hanns fæderne arffuepart.*

and, the very next year he was buried along side his two predecessors on November the 10, 1647.[67]

We do not know what kind of education Niels enjoyed before going to the Latin school during his earliest childhood and before his exam of maturity. Here, however, a letter from Stensen to Cosimo III at the end of 1686 throws some light on the matter. The then apostolic vicar in Hamburg met a friend of his youth, Jakob Henrik Paulli (1637–1704), who first was professor of anatomy, later became royal historiographer and finally worked as a diplomat.[68] Stensen introduces him to the Grand Duke with these words: "He was the son of the first physician of His Majesty. And, as children, we were brought up together in Copenhagen."[69] Of Jakob Henrik Paulli we know that in 1655 he came *ex privata institutione* to the Copenhagen University,[70] i.e. that he was taught by a private preceptor or in a private school under the direction of his father himself. Niels has taken part in such kind of education.[71] When in 1661 he dedicated his first larger paper to six particular professors, he named Simon Paulli first and 2 years later he acknowledged him as "a teacher whom I must respect as a father".[72] The only two poems from Stensen's hand that we know of also reveal deep ties with the Paulli family. The first one consists of the distich of a mourning poem which Jakob Henrik Paulli had encouraged him to write in 1657 for the burial of Regina Kallenbach, a member of the family. "A satisfied life is the best life", Stensen says in the poem, "it does not know death. When it goes to heavens, to the fatherland, who could call such a journey a death voyage?"[73]

When in Strasbourg Simon Paulli's second son, also Simon, married the widow of a printer on August 28, 1661, Stensen wished his friend good luck by sending a wedding poem with these lines: "Every love, you new husband, is a yoke[74]; it forces two hearts together which were still recently separated." The difference of the tie of love is then explained with "stories of the olden

---

[67]EP II, 902.

[68]DBL, 3rd ed. 12, 373 f. He was knighted under the name Rosenschildt.

[69]EP II, 887, cf. Chaps. 26 and 27.

[70]*Kbh. Univ. Matr.*, 255.

[71]Close to the home of Stensen the well-known teacher, David Schulmeister (Reich), gave lessons in his impressive yard (Kyrre, *Byens skole*, 31 ff.). From more private teaching in his childhood Blondel knows that: *Comme ils* (the parents) *étoint fort riches, ils lui donnèrent un précepteur* ... , 1575, a remark that can go back to J.B. Winsløw and Anne Kitzerow.

[72]OPH I, 77—*praeceptori parentis loco*, OPH I, 196.

[73]*Indice*, p. 290.

[74]*Omnis amor, Nove Sponse, jugum est* ..., a metaphor with double meaning: a yoke is a burden and can as well be a help to lift a burden.

time" like that of Theseus and Persephone to end up with the prayer that a lucky star might lead their hearts together.[75]

The influence of Paulli on the boy cannot be overemphasized. Born in Rostock, Simon Paulli (1603–1680), after studies in his native town, in Leiden and in Wittenberg, was professor of medicine in Rostock. In 1639, he came to Copenhagen as professor of medicine, surgery and botany.[76] He enjoyed the favour of Christian IV. His projects of reform, however, met the opposition in university circles. It was an event which raised the interest of the king as well as of simple townsmen, when, in 1644–45, he succeeded in erecting the remarkable *Theatrum Anatomicum* on the place of Our Lady, in which he undertook to dissect a male and a female corpse. He resigned his professorship in 1648, became physician to the royal household in 1650, and physician in ordinary to Frederik III in 1656. Botany was his favourite discipline and above all made him famous. His botanical excursions with pupils were popular. A progressive mind that strongly engaged, for instance, in the education of surgeons and pharmacists, he was in advance of his time from many points of view. Obviously, a personality that must have influenced our Niels (Fig. 1.1.3).

The homage of Stensen to the marvel of nature, pronounced during his extensive dissection of a woman's corpse in the above-mentioned *Theatrum Anatomicum* in the year 1673 may have been inspired from one of these excursions of Paulli. It says: "Who looks at a meadow from a distance in the loveliest time of the year, experiences a most pleasant visual sensation from the mingling of delightful colours. But when then he stoops over the meadow itself to look more closely at the leaves and flowers of the individual plants, such a variety and beauty of shapes and colours unveil themselves that he is compelled to exclaim: from a distance they appear beautiful, but nearby they are far more beautiful! But if, indeed, he proceeds even further and examines in one single plant the inner structure of the different constituent parts, the goings and displacements of the fluids which perform all (vital) actions, in the transition from seed to the mature plant producing new seed, even if he becomes acquainted very little with all that as if it were through a mist, yet he

---

[75]Indice, 291.

[76]DBL 3rd ed. 11, 181 ff. Jul. Petersen, *Bartholinerne*, 28–46, C. Christensen, *Den danske botanik* I, 25–34. See also Paulli, *Flora Danica*, Kbh. 1971, with an introduction by J. Lange and V. Møller-Christensen.

EPIGRAMMA.

MULLERUS pro Matre fua implorabat O-
lympum;,
Longius ut Matri vivere poffe daret.
Mors, ut id audivit, qvid ait, MULLERE,
precaris,
Non fatis eft Medicos, tot mihi ferre moras?
Tu qvoq; vis precibus nobis contrarius effe?
Non fic. Et Parcas fila fecare jubet.

In
Honeftiffimam & Integerrimam
B. MATRONÆ
Vitam.

Ita fuis contenta bonis eft optima vita,
Ludibrióq; vices fortis & orbis habet.
Non annos optat Pylii, Pylii tamen annos,
Si fatum jubeat ferre, qvieta feret.
Non huic lenta nimis, non velox Parca videtur;
Huic matura placet mors, & acerba placet.
Qvid? qvod & ignoret mortem, fempérq; fruatur
Triftiáve foli, lætitiáve poli,
Egreffamq; folo patriæ polus inferat urbi.
Hoc qvis iter vellet dicere mortis iter?

NIC. STENONIS.

Fig. 1.1.3 The first of two poems by Niels Stensen. This is for the funeral of Regina Kallenbach in the church of Our Lady in Copenhagen on October 16, 1657. It is the first printed text from Stensen's hand, REX

sees enough to acknowledge that the pleasure experienced from what is known is nothing compared to that he would experience if he could know all that lies hidden."[77]

However, it was not only the teacher and scientist who may not have been without influence on an impressionable precocious boy, for instance for his choice of profession. Stensen certainly also learnt from the standard of life in the house in Læderstræde[78] of which well-known maxims were painted on the wall and door. One can indeed read on the window shutters: "O man, think of eternity! God's eye is on you!" On the wall of the dining room, Aeneas who carried his father from Troy in flames claims: "Respectable fear for the father!" Above the mirror was written: "Know thyself!" Above the entry to the kitchen: "Fat cooking is the mother of diseases!" Well in view on the table of the record-office there was a book decorated with laurels and the inscription: "One gains respect through knowledge and weapon!" Over the entry and exit, the psalmist word admonished: "Commit your way to the Lord; trust in him, and he will act." (Ps 37.5).[79]

The childhood days of our Niels saw the last glare of the government of Christian IV. As Swedish troupes under Torstensson invaded Jutland without any declaration of war in December 1643 and soon occupied the whole peninsula, the king, in the greatest threat for the kingdom, appeared as a clever and energetic ruler who inspired confidence in everybody. He sailed from Copenhagen with 40 vessels on July 1, 1644 to fight the Swedish navy in the battle on the Kolberger Heide. When he returned, he drove away a Dutch fleet in a battle in the Sound which could be observed from the towers of the city. When he landed on August 13, he was welcomed with endless shouts of joy. However, with the execution of the old Admiral Peter Galt on August 31—after a summary trial in front of the castle under the eyes of the king—a sad time began. After a decisive defeat of the Danish fleet to the Swedes in the autumn and two epidemics in Copenhagen at the turn of the year, the Peace of Brömsebro of August 13, 1645 sealed the termination of Danish supremacy, its *dominium Maris Baltici* was in decline. King Christian IV himself died on February the 29, 1648 in his palace, Rosenborg. Four months earlier, he had buried his eldest son and heir to the throne. Within the country, the authority of the king and the party of the nobles were

---

[77] <854–855> OPH II, 252.

[78] The residence of Simon Paulli as professor was in Nørregade in a house belonging to the university close to Domus Anatomica. According to the census of 1645 it must have been a large house, Kbh. Dipl. 6, 277. But in 1668 he had a house in Læderstræde. Kbh. Dipl. 2, 812.

[79] P.H. Resen has described the apartment in his *Inscriptiones Haffnienses* 262 ff.

engaged in a bitter fight. Therefore, Frederik III took over with more limited powers than hardly ever any other Danish ruler who had assumed succession. The deceased king had left his kingdom poor and weakened. Yet, despite all his character flaws and despite all the failures of his politics, people mourned for this king more than for any other king of the lineage of Oldenburg. They remembered his dedication to work and his sense of duty, his courage in the hours of danger and his masculine and stout appearance.[80] Deep regret undoubtedly reigned in the house of the goldsmith in Klareboderne whose family also lost a friend and protector.

1.1.3   At Our Lady's School

When Christian IV was carried on his bier under a black canopy to his grave by 32 nobles on November 18, 1648, this happened with the deployment of a mourning splendour that Copenhagen had never experienced. 132 pupils of the Latin school of Our Lady in long mourning dresses opened the endless march, followed by students, university professors, parsons and bishops of the country.[81] At the time our Niels already belonged to this school in which one began at 10 years of age in the lowest of eight classes[82] However, he was probably not among the singer boys of the procession who had to provide a very tiresome service. Most of them originated from the poorest layers and lived from welfare foundations and church service, on which generally the teachers also depended.[83] Only a third of all the schoolboys were children from "good people" who, for most part, were registered on the express

---

[80]DBL 3rd ed. 3, 303–311. Also *Danmarks Historie* 2, 2 and Bruun, I, 618 ff.

[81]Christian IV was not buried until eight months had passed after his death. Bruun, II, 2 ff.

[82]It is certain only that Stensen was matriculated from the School of Our Lady to the University of Copenhagen on November 27, 1656 (Kbh. Univ. Matr. 1, 265). Maturity, age, pensum and duration of school were then very flexible. Caspar Bartholin the elder, 15 years old with 3 years in Latin school in Malmø, wrote long discourses in Latin. His grandson had the same name. He obtained his university entrance qualification when aged 15 (Iacobæus, *En lærd Familie*, 46). Thomas Bartholin was 18 before that, but he had had private teachers (Garboe, *Thomas Bartholin* 1, 20). Peter Schumacher, the later Lord Chancellor, entered the Latin school when 4 years of age and obtained his university entrance qualification at the age of 12. Cf. the solemn piety typical for the spirit of the time of his father: "Wednesday, September 12, 1639 after the morning prayer I let my first born son call into my room. Then I knelt down for my God and, while the boy knelt by my side, I recommended him in a special prayer. After me the boy also prayed, whereupon I put my hand on his head and said the blessing to the Lord. At last we both shouted: Amen, Jesus, Amen. And then I send him to school" (Jørgensen, *Peter Schumacher Griffenfeld* 1, 6 f). Johann Monrad had had private lessons in Ribe and came to the School of Our Lady only in 1650 and received university entrance qualification after 6 years. (*Selvbiografi*, 25 f).

[83]Lorenzen, *Metropolitanskolen*, p. 50. Thus in 1650 there were delivered clothes to the 355 pupils.

condition not to be obliged to participate in burials, to attend on early preach and to take part in singing.

The School of Our Lady, the respectable Latin school of the city, on the southeast side of Our Lady Cathedral and not far from the paternal house of Niels Stensen, dated from the Catholic High Middle Ages. It was founded in 1209 by a relative of the great bishop and statesman Absalon. Originally it was graded after the cathedral schools of the episcopal cities. Its reputation grew with the development of Copenhagen into a political and cultural centre in the middle of the realm. From the middle of the seventeenth century it was called the Metropolitan school, particularly since, about the end of the Middle Ages, the *schola universalis*, the only university of the country, had been erected just opposite it at the place of Our Lady.[84]

Christianity and humanism provided this school with its decisive character. Above all, the religious element exerted an influence that we can hardly imagine today. The most distinguished duty of the school was not teaching but the *exercitia pietatis*.[85] Bishops and clerics kept the strictest control and could dismiss rectors and co-rectors, and the school records, etc., were submitted to their inspection. Johann Monrad, a schoolmate of Stensen, gave a lively description of the impression that the visitation of the bishop exerted on the boys: "When Dr. Jesper Brochmand came to the school, what an alarm the short, old man provoked, more than if the Roman emperor had arrived."[86] The religious teaching revolved around Luther's Catechism from the first to the eighth grade. It began with the *Catechismus minor*, a booklet of 64 pages printed in double columns with a Danish text facing the Latin one. On the back of the title page, the declaration of war of Luther against the papacy greeted the reader: *Pestis ero vivens, moriens tua mors ero, Papa, ipse Lutherus est, credite, Papicolae.* The different paragraphs of the catechism had to be learnt *de verbo ad verbum*, and to not leave any doubt about the seriousness of the matter, it was ordered that the family fathers should even deprive any recalcitrants of food. The highest grades used the *Catechismus integer*, the contents of which were a somewhat enlarged edition of the Minor.

---

[84]Dalberg and Plum, *Metropolitanskolen*, 2 ff., Lorenzen, *Metropolitanskolen*. Both sources are used in the following. The Metropolitanskole ended by a school-reform in 2010.

[85]Lorenzen, 35. The manuscript, *Liber scholae Hafniensis* by Jørgen Eilersen from the year 1666 is an important source on the school.

[86]Monrad's Selvbiografi, 29 f. Jesper Rasmussen Brochmand (1585–1632) published his *Universae theologicae systema* I and II in 1633, distinguished by lucidity and erudition (DBL, 3rd ed., 2, 534–538, FL 2, 79–88). On March 21, 1685 Stensen acknowledged Cosimo III from Hamburg for offering him an Otium in Livorno that he wanted to use to write a reply to this work, *al livro fondamentale de' theologhi Luterani della mia patria* (EP II, 766).

It was completed in the eighth grade by reading of texts from the Holy Scripture and the New Testament in Greek.

Even more than catechism and teaching, the way of life, religious exercises and church duties determined the behaviour of the Latin schoolboy. Erudite Scholars and clerics were also often similar in their outside appearance and change from one study to the other happened often and easily. Most schoolboys were indeed destined to a clerical status. Dignified earnestness like wearing a dress was considered a duty of the schoolboy.[87] Above all the diploma of the candidates for the final examination revealed how the schoolboy "had behaved with fear of God, application and with suitable and modest dress".[88] The teachers were called *collegae* for more than one reason. They were bachelors, lived in the school and formed, under the leadership of the rector who appointed them, a corporate entity with almost military discipline. The school rules and their oath enjoined not only their obedience to the rector and their dignified behaviour in front of the schoolboys; there was even a ban on drinking bouts and smoking. One was not allowed to leave the town without permission nor spend the night outside.

Most of the schoolboys were, as mentioned above, kinds of singer boys who carried out choir services. How that interfered with the daily teaching is shown in the instructions for the singing and music teacher who, as early as at 6 or 7 a.m., led the morning prayers, at 10 o'clock gave the indications for an eventual burial, at noon practised the songs of the burials, which then started at 1 p.m. On Sundays, the schoolboys performed choir services in different churches of the city led by their teacher.[89]

---

[87]The remark by the General superintendent J. Hildebrand in 1680 in a religious dispute in Celle: *Miraretur hoc episcopus, cum ipse a puero aliter sensisset*, was a surprise to the bishop (i.e. Stensen). From childhood he had learned that differently (OTH I, 448). During the discussion the Protestant teaching of the Eucharist was presented so that the transformation of the bread into the body of Christ took place not at the consecration but at the consumption of the species, and the unification of the body of Christ and the bread and of the blood of Christ and the wine would happen not in the hand of the dispenser but in the receiver's mouth. This had been a surprise to him.

[88]Still at the time of Ludvig Holberg, for 100 students of theology there was hardly one student of medicine or law (Olrik, *Borchs Kollegium*, 11). *Farffven paa deris Klæder schall verre Sort, eller udi det ringeste mörchagtig, fornemmelig udi de öffverste Classibus* (Liber scholae, fol. 24 v, see also fol. 25 and fol. 6.7)—Thus in the school of nobility at Sorø there were four to five hours in Latin and Greek daily and two to three hours worship preparations (Sorø 1, 344). The Latin school in Frederiksborg required that pupils prayed lying on their knees with a bare head, thereby setting an example for others to follow. Two to three times a year pupils of the highest and second-highest grade should have an evening meal after eight days of preparation by the rector or a teacher. In the chambers the older pupils surveyed the younger ones during morning and evening prayers (Kornerup, *Frederiksborg Statsskole*, 107).

[89]Liber scholae: *Pro infimae classis informatore leges quaedam singulares*. "One should not overlook that the school of Our Lady then, together with music at the university, the court and of the city, constituted higher music in Copenhagen and in part the whole country. The school contributed in many

The profane teaching in the Our Lady school, in agreement with the time, was governed by the ideal of humanist education, that the formal schooling of the mind, the *fari posse*, required the Latin eloquence. The specialists of the trivium were still always in honour: the grammar would procure the *sermo emendatus*, the dialectic the *sermo probabilis*, and the rhetoric the *sermo ornatus*. Latin conversation began early with the help of the humanist booklet of children's conversation (*Colloquia puerilia*). The only verses by the hand of Niels that have been preserved show that he also mastered the highest skill, the *imitationes* (writing of verse).[90] The old *Quadrivium* with its transmission of knowledge in geometry and arithmetic, in music and astronomy were at first in the background. Erudition, broader knowledge, was reserved for the continued study at mature age and above all for learned circles.[91]

It is not difficult to draw a more concrete picture of the subject matter and the school management from the subject matter or objectives that were used during Stensen's school years. The rector of Stensen's secondary school himself contributed to these objectives and exactly described the program in his *Liber scholae*.[92] If one begins in the lowest grades with the simplest grammar and calligraphy,[93] in the 4th and 5th grades one comes to deal with Greek, arithmetic and geometry and then logic and the introduction to philosophy in the 7th and 8th grades. In the 7th grade one then had to grasp and imitate the particular style of the Latin author and for the 8th grade the requirement was: *Scribant Latine indolem linguae, quae in auctoribus purioris Latinitatis observatur.*[94] One even learnt some Hebrew in the highest grades.

(Footnote 89 continued)

performances of operas by foreign companies" (Lorenzen, 49). The children choir often brightened festive events that then blossomed and was divided into three groups according to their quality.

[90]*Indice*, 290 f.

[91]Cf. the development of the school system in Denmark and the influence of the Jesuit school, Kornerup, *Hans Poulsen Resen*, 246 ff.

[92]The university tables of 1656 came into effect in the year of Stensen's entry to the university. They had already been commissioned on December 29, 1653 by King Frederik III. Jørgen Eilersen who had had the seat and rank of professor at the university from July 5, 1654 may have proposed the ordering already practised by himself. The program in mathematics and geography for the 8th grade is also verbatim taken up as the undertitle of his *Progymnasmatum mathematicorum enchiridion* that appeared in those years.

[93]Stensen had a clear and orderly handwriting that remained the same from his student notes until his deathbed. He understood how to cut a goose feather and to move it over the paper like his father and uncles (examples in EP II, p. VII–X). According to G.M. Moretti's opinion of a graphologist in Sten. Cath, 7, 1961, 36 ff.

[94]Liber Scholae, fol. 27 f.

Due to the flexible study program at that time, the lessons did not have as much importance as the spirit of the school and the personality of the teachers; we must above all consider that Stensen spent his school years in the blossom time of the Metropolitan school which then had prominent *Hører* (professors). Some years later it was obviously an urge for him to mention two of them among the "most highly deserving teachers" to whom he dedicated his paper *De glandulis oculorum.*[95]

In the first place, Stensen remembers his former rector, the mathematician and linguist (*mathematico et litteratori*) Jørgen Eilersen (Georgius Hilarius, 1616–1686), the son of the mayor in Vordingborg. Eilersen had initially studied medicine at Copenhagen university, to switch over to theology to please his father and finally to pursue the study of the classic languages in Leiden and in Paris and to listen to professors of philosophy and mathematics. Co-rector since 1641 and rector of the Metropolitan school since October the 15, 1646, he had also become calendariograph in 1645 with the exclusive right of printing calendars and almanacs. From 1672 and to his death he then worked as professor of mathematics at the university.[96]

Eilersen must have been an outstanding personality by his sense of duty and his fatherly goodness. Among the groups of young fellows whom he sent to the university, none was refused because of *in literis ruditatem.*[97] On the contrary, many of his students achieved the highest honours. Five of them became bishops. Upon being nominated as rector, he almost immediately put the school buildings in order, provided his two-storey residence west of the church of Our Lady with a garret and also improved the lodgings of the teachers. His own rooms, partly panelled, gleamed with bright green and vermilion colours and were warmed by large cosy tiled stoves. From the red room, obviously the working room and office of rector, one could enter directly into the highest grade for the master lesson. In the school buildings, the new rector introduced a particularly beneficent novelty that most of the other schools adopted more than 100 years later. He partitioned the large teaching room in which all the grades were taught by their teachers at the same time, separated from each other only by a passage. Since the school then was attended by 400–500 pupils, about 50 boys crowded in one classroom, somewhat fewer in the highest grades but many more in the two lowest ones.[98]

[95]OPH I, 77.
[96]FL II, 423–425. DBL 3rd ed. 4, 139.
[97]Universitetsprogram og Mindeskrift, 11.
[98]Dalberg and Plum, 61 ff. Lorenzen, 50.

Eilersen also conquered the hearts of his schoolboys. The preacher Jens Bergendal wrote a fervently enthusiastic Memorial about his former teacher and protector in his honour. The schoolboy praises his teacher's excellent method, his clear and intelligible explanations, his zeal for church visit and song.[99] The school leader also knew how to make the days of childhood and youth festive and to give the schoolboys light remembrances for life.[100]

The importance of such an educator personality at a time of big disasters, which then happened to Eilersen's school is easily understood. At the age of 14, Niels Stensen had already suffered a bad dysentery.[101] In the spring of 1654, a severe epidemic occurred which caused many victims in the schools. By the end of May, the university closed. The king and court left the city and with them many townsmen. One day, the school had to take care of 60 burials and altogether some 9000 people, about one third of the population, died in the course of that year. Eilersen writes about this: "In the year 1654, an awful and severe plague appeared here in Copenhagen, during which all who had relatives either here in the city or somewhere else left the school and went home; and many did never come back again, the poor alone remained ... Then 246 schoolboys, small as well as tall, died."[102] Such events prepared in the boy that sentiment in which he as bishop years later urged his listeners to be well prepared for death: "The first preparation is the consideration of the vanity of the world ... We are not here to stay. The life here on earth is an inn with summer weather for the ones and winter weather for the others, but the fatherland is heaven."[103]

Whilst the title of master of language which Stensen gives to his rector is meant to express his gratitude to the humanist Eilersen who taught in the two

---

[99]Bergendal, Sørgeligste ... *Copenhagen 1686*, cf. FL I, 352.

[100]The schoolmate of Stensen, the titular counsellor of state Monrad, describes: "When it came to Christmas time we were treated by Magister Jørgen, and there was the cheerful Christmas celebrations. Then we carried all the tables and benches away and played throughout the night" (Monrad, *Selvbiografi*, 30).

[101]Eilersen writes in Liber Scholae: "Anno 1652 appeared a severe blood-epidemics during the months of July, August and September, and during these months we could not read in the school because of the funerals. The highest number was 21 that in one day had to be taken hand of" (which means, buried). Regarding these epidemics (dysentery, cold fever) that killed 20 people daily in Copenhagen, cf. Mansa, 384 ff.

[102]Dalberg and Plum, 65. Monrad describes: "The great plague in Copenhagen 1654 occurred during my childhood ... At 8 a.m. one began ringing the bells for the dead, and all day passed with the dead, and still in the evening one rang the bell. There died so many hundreds of schoolchildren with whom I had daily been together" (*Selvbiografi*, 35 f). Mansa, 391 ff. The epidemic was brought into the city by Dutch ships. It started in April, its peak was in August and it finished early in 1655.

[103]OTH II, 350.

highest grades and induced his pupils to master the language of Rome,[104] the title of mathematician points to a deeper tie between Niels and his rector based on a common enthusiasm for this science and its application.[105] It is known that in 1662 Stensen, then in the middle of his successful anatomical observations, was stirred by such a deep longing for geometry as the first and proper field of research that he would rather have returned to it.[106] It can hardly be doubted that this passion for mathematics and physics, so characteristic for the time, resulted from the years in the secondary school and, since the school schedule did not leave much time for that, one can think of a more private, personal tie with Eilersen, as reported by other gifted schoolboys.[107] Nor should it be forgotten that Copenhagen, precisely at the time of Stensen's Latin school gave shelter to one of the most important mathematicians of the time, Nicolaus Mercator (Claus Kauffman, 1620–1687) from Holstein, whom his more recent biographer calls one of "the greatest pioneers who have opened the way to higher analysis." After his sojourn in Rostock and Leiden, since June 1, 1648, he studied at the university of Copenhagen where, in 1651 and 1653, he published several of his papers, among others, on cosmology and spherical astronomy. He was also close to the nobleman J. Gersdorff (the friend of Ole Borch).[108]

The second of the famous and outstanding men and highly deserving teachers to whom Stensen dedicated his above-mentioned paper was the recently mentioned Hører at Our Lady's School, Ole Borch (Borrichius, 1626–1696), to whom a lifelong friendship would bind him.[109] In 1644 Borch came as a student from the time-honoured cathedral school in Ribe in Jutland to the Copenhagen university and when Niels entered the Latin school, he was still engaged in his own studies. In 1650 when aged 24 he was appointed teacher at the school of Our Lady so that the young Stensen for 6 years was under his teaching influence. Borch found himself so well here that

---

[104]Monrad, 41f.

[105]Eilersen had then published the *Trigonometria plana* (Hafniae 1644) and as well the earlier mentioned *Enchiridion* (Hafniae 1656).

[106]Stensen writes on this to Thomas Bartholin: "When I published my few observations, I had decided to lay down the anatomical knife until more convenient times and to take up again the nearly cast away geometer's rod, so as not to appear to have wasted all time and labour if I deserted completely my research for which I spent many hours in the past and which I would have treated not as my primary, but as my unique [work], if strained circumstances at home had not so much convinced as forced me to prefer the useful to the pleasant." (OPH I, 115 = EP I, 158 f).

[107]Monrad, 29.

[108]Scherz, *Vom Wege*, 30 f; Hofmann, *Mercator*, 46.

[109]On him cf. BOD, I, Introduction, also E.F. Koch, *Oluf Borck*, 1866 Vilh. Andersen, *Tider og typer*, 199–233, 1907, *Mindeskrift for Oluf Borch*, 1926, DBL, 3rd ed. 2, 317–360.

Fig. 1.1.4 Ole Borch with pupils on a botany excursion outside the city walls, Jacobæus, 68

he declined a rectorship in the renowned boarding school Herlufsholm. The company of the youth replaced family and home for the young fellow. In 1655, he accepted the position of teacher for the sons of the royal court master Joachim Gersdorff, and in 1660 he became ordinary professor of philology at the Copenhagen university and extraordinary professor in botany and chemistry. Before that, however, Borch, accompanied by several former pupils during many months, undertook a 6-year study travel in order to take a thorough look at the world (Fig. 1.1.4).

Stensen attributes two titles to Borch, that of polyhistor and nature researcher.[110] He thus points out the aspects of the teacher's personality which particularly influenced on him. Borch displayed a wideness and diversity of knowledge which must have acted fruitfully in the training of the youth. He was first of all reputed to be one of the best Latinists of his time. "It should be a matter at heart", says the bright humanist in him, "to keep this queen of the languages, which was given to Europe as a gift of God and is still in power. Through many centuries it has been the guardian of all knowledge and the true servant of our knowledge. With its help we can unveil our sincere opinions to many widely scattered people. Through it alone we can differentiate ourselves from the barbarians. With it we command with authority, we comfort eloquently, and we embrace all in one. Therefore, we

---

[110]OPH I, 77.

alone deserve the name of erudites."[111] This enthusiasm entirely corresponds to the eloquent and exquisite Latin that Stensen uses in his writings.[112] He remained very far from the erudite passion of the time for an excess of classical quotations and comparisons but he knew the ancient authors and their words and images trickled unconstrained to his pen.[113] One traces classical schooling in the terse epigrammatic style in which Stensen readily clothes his thoughts.[114] The years of travel in Europe and the study and use of modern languages like Dutch, French, Italian and German then weakened the brightness of the classical discourse. He remarked on this when, as royal anatomist on October 5, 1672, he wrote from Copenhagen to the Nuncius Pallavicini that his Latin caused him some concern. He had been invited, unprepared, to dissect a human corpse in the anatomical theatre. It was something disquieting, among others, "because, by using several modern languages, I had neglected the practice of Latin" (Fig. 1.1.5).[115]

Borch was also an inspiring nature researcher! During experiments in chemistry he generated oxygen nearly 100 years before Scheele and Priestley without being really conscious of the importance of his discovery. He was so active as a botanist that Johann Monrad, the above-mentioned schoolfellow of Stensen, counted as one of his youth greatest joys, "when we went *herbatum* searching for plants and determined them, and, so doing, walked miles away".[116] Borch may also have encouraged his pupils to study medicine since he himself at that time completed his turn to physician. His open outlook and his preference of experience as the main road to knowledge, as well as his warm heart, were directed to the human needs by the bubonic

---

[111]In *Dissertationes anatomicae* II, 180, see Vilh. Andersen, 227. Just then he had his first schoolbook, the Parnassus in nuce (1654) published to introduce youths into Latin poetry. In this the literature's "Augustusaltar" was still rather neglected. In 1670 he fought in the Lingua Pharmacopoeorum the mean Latin of doctors and pharmacists and published a great work in 1675, *Olai Borichii Cogitationes de variis Latinae linguae aetatibus* ... according to which Latin lives through infancy, youth, adulthood, grey years and death and then resurrection, in which Cicero, "the man with the eternel genious" naturally represents youth (Vilh. Andersen, 221 ff.).

[112]Cf. the introduction to Prodromus de Solido or the Prooemium of 1673 (OPH II, 183 ff, 231 ff).

[113]Cf. the *Apes et fuci vespa iudice*, or *The bees and the drones* from Aesop's *Phaedrus*. The reference to Apelles and Alexander, the history of nature of Plinius Secundus, Galen and Hippocrates naturally goes in the direction of a coming anatomist. One also finds excerpts from Aristotle, the Symposion by Plato, Xenophon's Memorabilia of Socrates, from Seneca, Diogenes, Laertes, Lucian and Epicur, also from Plutarch, Tacitus, Horace and Sophocles, works with which Stensen was familiar (cf. OPH I, 360 ff, OTH II, 559 ff).

[114]Cf. his well known motto: *Pulchra sunt, quae videntur* ... (OPH II, 254), sentences of the priest (OTH II, 543), or an aphorism from Hesiod: *Malum consilium consultori pessimum* (EP II, 770).

[115]EP I, 274.

[116]Monrad, 30.

Fig. 1.1.5 Ole Borch (1626–1690), teacher and friend of Stensen

plague of 1654. Then Borch made his début as a physician and said of himself in his biography: "For the first time I understand what human medicines can do against so fierce enemies."[117] Many drinks were brewed there, in the laboratory in Østergade which the royal court master had put at disposal for the teacher of his own children.

Stensen did not face the polyhistorian uncritically. This appears truly in a notice in the chaos-manuscript from the university years. There he had been at Borch's where the most varied subjects were discussed, which he refers to and adds: "Another time when a conversation is made, one topic should be kept to and all reasons for and against what is propounded should be put forward and harmful leaping from one subject to another should be avoided, particularly in the case of Borch."[118]

The Latin school years between 1648–1656, of which we know nothing personal of Niels, were not a quiet time, neither at home nor in Europe. In Germany the Peace of Westphalia had finally been sealed but then followed the period of its application. England suffered the troubles of Cromwell's time and was at war for 2 years against the Low Countries. In Sweden, Christine, the daughter of Gustav Adolph, reigned, at first during a very quiet period, but in 1654 she abdicated the throne and devoted herself to the Catholic Church. Frederik III (1648–1670) took over as king of Denmark, a humiliated country which continued to feel threatened. From then on it was also a poor country. While our Niels heard of feast banquets for the crowning on November 23, 1648 and saw people brawl about the big roasted beefs and the wine casks on the castle place, he may have noticed the unrest in his hometown. However, the king, who preferred to retire to his library, had to wrestle with the noble council and the royal council which had acquired unusual power through the decision at his election. In the year 1651, on July 11, Copenhagen experienced the execution of the mendacious Dina Vinhofver and, on July 13, the secret escape of Corfitz Ulfeldt, the son-in-law of Christian IV, and wife Leonora Christine to Sweden.[119]

Several changes occurred in the goldsmith workshop of Klareboderne. Jørgen Carstensen had indeed been appointed as guardian and led the many year long lawsuit about the heritage of his ward after the dean Niels

[117]BOD I, xv. *Chaos*, ed. Z, 77.
[118]EP II, 908. *Chaos*, ed. Z, 77.
[119]Bruun, 2, 5–13.

Pedersen.[120] The parental home also existed further but mother Anne, after the death of her third husband, went to her fourth marriage with the goldsmith Johann Stichman in about 1650.

Stichman is attested as goldsmith in Copenhagen for the first time on April 27, 1651.[121] However, it appears from a lawsuit with Christopher Ulfeld of Linnstrup in Scania about a money debt, that he was already in relation with the nobleman on July 10, 1649.[122] He must have been a strong personality. As master of the goldsmith workshop of Sten Pedersen, he at first maintained the connection with the royal court.[123] In his accounts he readily used the German language. He thus writes: "Year 1663, on June 6, on the gracious order of Your Royal Majesty order to Fridrichsburg, I have polished everything anew and fixed the altar in the church, the pulpit as well as the small altar in Your Royal Majesty's cabinet. Besides, in the church in Fridrichsburg I kept two fellows to repair everything, for the cost of 160 Rthr. Johann Stichman."[124] Artisans originating from Germany were then no rarity in Copenhagen. Among 14 guild masters, 7 were German or Dutch. Stichman was a man with good connections as shown by his struggle in 1656 against the guild of the wine merchants who had acquired the monopoly for the retail of wine and brandy and now wanted to close the retail by Stichman; yes, they even seized his goods. On September 12, 1656 the goldsmith complained in a Danish letter to the king of this step against the small commerce in the cellar under his house although his late predecessors, his wife and he himself had kept the inn for more than 30 years. On August 23, the second bailiff had come to his door with many people and had broken into the cellar and violently demolished what was in their way, and had then

---

[120]Dean Niels Pedersen died in 1644 at a time of warfare with Sweden that prevented his family in Copenhagen from taking part in negotiations on the inheritance, now taken care of by the husband, Christopher Mogensen, of a niece Anne Sophie Clausdatter for which reason she was bestowed with a share of one fifth rather than one twelfth allotted. On November 18, 1645 Jørgen Carstensen appealed in the name of Stensen. In two decisions of November 18, 1645 and November 24, 1652 his standpoint was accepted. Since the opponent did not give in, Carsten brought it to the Herredag in Copenhagen on June 30, 1652 and ultimately won the case (KRA, *Herredagsdombog*, t. 53, 1655, fol. 341–346).

[121]EP I, 902. The name Stichman is also attested in Funen and in Holland (A.D. Jørgensen, *Nils Stensen*, 12).

[122]KRA, *Herredagsdombog*, t. 52, 1653, fol. 282–284. See also his demand for a share in the estate of Mrs. Maren Urup in Scania (KRA, Da. Kanc. B 51, Skaanske tegn, t. 8, 1643–1656, fol. 336).

[123]On November 25, 1652 he was paid 300 daler, on August 12, 1655 more than 742 daler, on June 2, 1656 200 daler, on and on December 24, 1656 he delivered silverware and received more than 370 daler for his deliveries to the royal silver chamber (Marquard, *Kongelige kammerregnskaber*, 8, 10, 24 f, 38, 46, 48 ff). There still exist cups coming from his workshop marked by his stamp, the intertwining letters I and S. Thus one can and seven goblets in Rosenborg, a plate in Gentofte Church and a pitcher with lid in the fishermen's guild in Flensburg (Bøje, *Danske guld og sølvsmedemærker*, I, 90 f.).

[124]KRA, Rentek. 216, 276. Afregninger XIII, 3, J, 174, Lage Jakob Kitzerow Guldsmed.

rushed out of the house as if they had been in an enemy country.[125] For the same matter, Stichman, on September 13, 1656 sent a letter in German to the Queen and described that the wine dealers had treated him "so mercilessly and unchristianly" and had plundered him. The accounts of the cellar mates show that the retail cannot have been very small: between May 12, 1654 and July 24, 1655 wine, brandy and beer were sold for an amount of 5829 daler. Thereupon, the king decided that one could make an exception for Stichman and allow him his wine retail.[126]

In Niels' last year in the Latin school, there also occurred the betrothal and wedding of his only sister, Anne Stensdatter, who would survive her brother for so long and would be buried in the family grave in St. Nikolaj as late as April the 6, 1703, after 30 years widowhood. She married the goldsmith Jakob Kitzerow, whom we meet for the first time on April 29, 1655 when he was introduced to the guild by Johann Stichman and Johan Welbom. He had then passed his first journeyman's examination and would produce his indenture and authentic letter, which he did on May 2. Thereafter he was instructed to make his master piece at Jakob Brun's. This seems to have been carried out quickly since he was introduced as goldsmith as early as July 1 to participate regularly from then on at the meetings of the guild.[127] Kitzerow is likely to have migrated from Germany—his accounts are always drafted in German, which he wrote easily and fluently. He was a travelling fellow who had found work in Klareboderne and was found fit to sue for the hand of the master's daughter.

The old goldsmith workshop in the Købmager quarter was now doubly protected and had received new blood. This workshop of the paternal house must have been some sort of a laboratory for the young Niels. There he was taught many technical skills and given knowledge in chemistry and physics, as his notes preserved from the last years at the university show us. There he displayed an active interest, for instance, in glass polishing. To this end he knew a better work tool than Descartes and explained it through a small drawing.[128] Another time he wrote: "For grinding lenses one can use what is available in our workshop, cut out in many concave segments; for concave lenses those small convex ones, etc."[129] The student knew the metals and their craftsmen's treatment. He gave an outstanding description of the

---

[125]KRA, Da. Kanc. B 57, Sjæl. Tegn. 1654–1656, fol. 572 v.

[126]*ibid.*, fol. 572 v, 627 r. Further Da. Kanc. B 54, Sjæl. Reg. 1651–1656, fol. 862 r.

[127]KSTDA, Guldsmedelauget nr. 2, *Klattebog* 1651–57.

[128]EP II, 918 f.

[129]Chaos, ed. Z, 290.

properties of quicksilver and noticed, among others: "Quicksilver is often harmful to gilders as their limbs become paralyzed and powerless, their nerves are weakened, often when they put their hand to their mouth, they drop onto their side with tremor. During the gilding process, gold remains in the pores of the silver …"[130] Another time he was interested in colours and writes: "On should enquire from goldsmiths about the nature of colours."[131] "Goldsmiths make a red dye either from burnt vitriol or from thin iron sheets which have been corroded with nitric acid and melted in fire, but when the dye is to be used, some fluid should be mixed in it; moreover from burnt vitriol also a black colour is obtained."[132]

The influence of the goldsmith workshop on young Niels can be quite specifically established from such notes. We, therefore, must not forget that, if he turned away without lingering from the pure humanistic and theological education of his paternal forefathers and brothers to the natural sciences—as his father had turned from the preacher profession to become artisan—this choice and its energetic realization undoubtedly also resulted from the fact that the political decline of Denmark in the middle of seventeenth century occurred at a time that Danish natural research blossomed, inspired by its simultaneous rise in Europe.

As early as the sixteenth century, the professor at Copenhagen university Peder Sørensen/Petrus Severinus had published in Denmark the ideas, sometimes mysticophantastic, sometimes scientific ingenious, of Paracelsus.[133] Sørensen, in his writing *Idea medicinae philosophicae* (Basel 1571), transmitted and explained the dark attempts of the Swiss author at knowing Nature and its basic lines and the coherence of existence through Platonic thinking. Sørensen earned the unlimited praise of Francis Bacon who greatly appreciated his lecture on *vitale principium in natura*. With that, Sørensen had sound ideas on the method and exclaimed to the youth with inflaming eloquence: "Go, my sons, sell your fields, sell your houses, your garments and your rings. Burn your books, buy strong shoes, come to the mountains, investigate the valleys and the wilderness, the shores of the sea and the deepest abysses of the earth. Observe the distinction of the animals, the differences of the plants, the orders of minerals, and the properties of all things and the way they come into existence. Carefully learn the astronomy and terrestrial philosophy of the peasants, and be not ashamed. Finally,

---

[130] Chaos, ed. Z, 86 f.

[131] Chaos, ed. Z, 268.

[132] Chaos, ed. Z, 298.

[133] DBL, 2nd ed., 23, 206–209. Bastholm in Sørensen, *Petrus Severinus*, 1–32.

purchase coals, build furnaces, be vigilant and tend to your preparations without weariness. For thus will you come to an understanding of bodies and their proprieties, and not otherwise."[134] An even brighter star was Sørensen's contemporary, the astronomer Tycho Brahe, whose paracelsic, alchemistic and astrological ideas still revealed his love for mysticism of the older natural science, but who recognized observation always more clearly and with decisive success to be the foundation of all natural science.[135]

In the seventeenth century there existed in Copenhagen a large treasury which attracted visitors even from abroad, the so-called *Museum Wormianum*, the remarkable collection of items of the physician, professor of medicine and researcher Ole Worm (1588–1654).[136] In his residence in Kannikestræde, Worm had been collecting minerals, plants, animals, skeletons and antics from all over the world since 1623. He had, moreover, written a catalogue which after his death appeared in a superb edition in Amsterdam, 1655. The king had purchased this collection for his art room in the castle. Niels will certainly have seen the collection and have studied several objects, like, for instance, the fossil shark teeth from Malta. We know nothing of any personal connections with Ole Worm. The fancy and superstitious explanations of the great collector may have kept him away. On the contrary, Niels came in relation with his son Willum Worm, who in 1673 was appointed inspector of the royal art room. He then showed Stensen petrified fish from Norway as we know them from the Ice Age and Stensen undertook to send some samples to Florence.[137]

As far as his choice of profession is concerned, the name of Thomas Bartholin looms above all, whose family dominated Copenhagen University for a century.[138] His father Caspar Bartholin the elder had acquired a European reputation a long time ago.[139] Originally the compiler of textbooks of philosophy, during studies in Basel and Padua he felt such a wide interest for medicine (however, he ended his short life as theologian of orthodox-pietistic observance) that he wrote his *Institutiones anatomicae* (Wittenberg 1611), the most used textbook of the century as a consequence of its pedagogic qualities. First of all, his son Thomas Bartholin (1616–1680)

---

[134]Chap. VI, *De principiis corporum*, 73. Translation from Shackelford, *Petrus Severinus*, 264.

[135]DBL, 3rd ed. 3, 429.

[136]DBL, 2nd ed. 26, 279 ff. Schepelern, *Museum Wormianum*.

[137]EP I, 278.

[138]Cf. esp. KU 1479–1979, 7, 22 ff, 34 ff.

[139]DBL, 3rd ed. 1, 475 f.

would work as an epoch-making discoverer in anatomy.[140] Gifted and covetous for honour, the young Thomas set out for a journey 1 year before the birth of Stensen and studied successfully for 9 years at the most famous universities of the time, in Leiden and Paris, in Montpellier, Padua and Basel. He became widely known through a new edition of the anatomy book of his father (1641) and at the same time he revealed the peculiarities of the working of his mind: open-mindedness to the new knowledge like the discovery of Harvey, in the method a multiple application to new research, even if he could also gather in his Centuria—letter collections—the most peculiar stories; above all, a literary passion and gift. Back home, in 1648 he was appointed professor of anatomy and displayed a bursting activity in the new *Theatrum Anatomicum* appreciated in the town as well as by the court. The bright period of anatomical creativity to which he had summoned the competent German surgeon Michael Lyser as an assistant also met, however, many adverse reactions from O. Rudbeck in Uppsala and J. Riolan in Paris. This occurred in the early 1650s at which time Stensen was still attending the Latin school. As early as 1656, after his appointment as Medicus primarius and dean of the university Bartholin made his last dissection, and 4 years later for reasons of health he was exempted from giving lectures (Fig. 1.1.6).

In Thomas Bartholin's work, *De lacteis thoracicis in homine brutisque nuperrime observatis* (Hafniae 1653), he demonstrated in man that chyle from the intestines flows to the *ductus thoracicus* and from there to the left subclavian vein and the heart, and thus that it does not flow straight on to the liver. This had been discovered in a dog by J. Pecquet and was thus a confirmation. In a second paper, which drew great attention, the *Vasa lymphatica nuper Hafniae in animantibus inventa et hepatis exsequiae* (Hafniae 1653), followed the presentation of a new important discovery, namely the lymphatic vessels in animals which then, in a third dissertation *Vasa lymphatica in homine nuper inventa* (Hafniae 1654), were detected in man also. The discoveries of these new vessels overruled the old Galenic theory of the liver being the producer of the blood. Stensen would later in his first writing from Leiden pay homage with warm enthusiasm for this research when he presented it as an instance of the necessity of pursuing true research by observation and deductions. He says: "Who during so many centuries, even

---

[140]DBL, 3rd ed. 1, 477–480, Garboe, *Thomas Bartholin*, I–II, Th. Bartholin, *Cista Medica*, 1982, esp. 15 ff. * New biography in Danish by Jesper Brandt Andersen, *Thomas Bartholin, Lægen og anatomen*, Copenhagen, FADL' Forlag 2017. Bartholin dedicated the *Cista Medica* (1662) to the Grand Duke Ferdinand of Tuscany. The dedication may have paved the way for visits by Stensen and Bartholin's sons in Tuscany.

Fig. 1.1.6   Cista Medica, Copenhagen 1662. In the *centre* Thomas Bartholin surrounded by professors of Copenhagen University, Henrik Fuiren, Ole Worm, Thomas Fincke, Caspar Bartholin, Simon Paulli and Johannes Rode. REX

most ready by natural inclination or through dreams, thought of the lymphatics before You, most famous Bartholin, presented to the eyes of everybody what You observed with your eyes, not those of the mind but those of the body? But, since then, who succeeded in revealing through his intelligence alone the matters which are still hidden? Who was able to demonstrate with trustworthy arguments from where the lymphatic vessels originate, in the liver or in the rest of the body?"[141]

On November 27, 1556, Niels Stensen was registered at Copenhagen University—his first step towards a scientific career.[142]

1.1.4 University Years 1656–1659

A Latin schoolboy entering Copenhagen University in the seventeenth century had to present himself to the dean with a certificate from a Latin school —usually delivered on St. Hans' (John's) day—or eventually from his private teacher. There also had to be a *testimonium eruditionis* concerning his knowledge and maturity as well as a *testimonium vitae* concerning faith and the moral way of life. Often an *examen stili*, a written Latin examination or the *examen artium* concerning other knowledge was requested.[143]

Then he went to the "deposition" which would remind him with drastic symbols the value and dignity of education and academic life. The schoolboy appeared disgustingly dressed like a fool in the court of the university: the peaked hood, the horns on the head, the big nose with blackened face, a hump on the back, as well as his behaviour and gesticulations would betray a blockhead and ruffian. Then the trustee who was also disguised approached the schoolboy and drove before him the deponents like a cattle horde with blows and invectives, with stick, pliers and knife, removing from them with any sort of comic gesticulations the buffoonery symbols. After these often quite tough rituals, the students put on their proper clothes and one of them humbly begged to be admitted as an academic citizen. Then the dean, as a token of joy, poured wine over their heads and, as a symbol of wisdom, put salt on the tongue of each one. All that should remind the schoolboys that, with toil and pains, they would rise from a lower to a higher life, the life of the educated and learned, which was also distinguished by some privileges like

---

[141]OPH I, 14 f.

[142]Kbh. Univ. Matr. I, 265.

[143]Rørdam, Kjbh. *Univ. Hist. 3*, 417 ff, Norvin, 1, 231 ff, that are used several times in the following. See also Stybe, *Universitet og Åndsliv*.

that of tax exemption and the right to have their own jurisdiction. On the day after the deposition, the students went to the rector and their names were recorded in the university register.

The Latinized name of Niels Stensen is Nicolaus Stenonis. It was recorded on Thursday, November 27, 1656. Copenhagen University was founded in 1479. Since the Reformation in 1536 it had had its seat in the court of the formerly Catholic Roskilde bishop to the north of the Church of Our Lady. The greatest academic events still took place there. Where the main entrance and the feast hall stand today, there was a building in Dutch Renaissance style. The upper and lower lecture halls in this building had been in use since 1601, the latter hall for the theological lectures.[144] To attend these lectures—around 20 a week—was mandatory and did not consist in dictations. They took place on Mondays and Tuesdays and as well on Thursdays and Fridays, whereas the Wednesday was kept for the repetition and disputations. One began early in the morning at 6 a.m., enjoyed a break from 10–12 o'clock and turned on from noon to 5 p.m. with lectures. In winter, because of the cold, the lectures were often delivered in the private lodgings of the professors. In those days one began to reduce the academic year with the teaching taking place from September 1 to July 31. The students tried to distinguish themselves from 'ordinary' townsmen by their clothes and haircut and demanded their right to carry a sword, which was often necessary. Regular visits to the students' church, the Trinity Church consecrated in 1656, and attendance at the evening meals were mandatory. There were harsh punishments for failures. Moreover, mandatory signs of fidelity and devotion to the king and royal house were more and more stressed during the time of Stensen.[145]

The 2–3 years spent by the students at the university usually began—when needed—by studies of language, philosophy and theology. In 1675 Peder Griffenfeld made the *examen philosophicum* mandatory. However, since philosophy teaching was then the obvious prerequisite for university studies all over in Europe and Denmark, for instance, possessed such an important scholastic as bishop Jesper Brochmand,[146] one can assume that Stensen acquired a good philosophical background either at the Latin school or at the beginning of his time at the university. This was, moreover, especially

---

[144]Concerning buildings of the university, see also: Københavns Universitet 1479–1979, vol. 4: Gods, bygninger, biblioteker, esp. 151–178. Of particular interest is moreover in vol. 1: Almindelig historie 1479 til 1788, and vol. 7: Det lægevidenskabelige fakultet.

[145]Thus in 1667 a direct loyalty oath to the king was added, the *Sacrae Regiae Majestati fideles et obsequentes sunto*, Norvin, 1, 238.

[146]DBL 3rd ed. 2, 534–538 see below.

intended for theologians. The advice that Caspar Bartholin the elder gave to his own sons (they became Stensen's teachers, Thomas and Erasmus B.) in his much read booklet *De studio medico* of 1628, which was still fairly valid for the next generations, required not a little linguistic knowledge: rhetoric and logic cannot be neglected, practical philosophy *perfunctorie percurrisse sufficiet*, and of the metaphysics it is true that, *nullum pene usum medico praestat*. Then follows advice concerning the specialties and the textbooks for physics, geometry, optics, astrology, botany, anatomy, etc.[147] The medical lectures of the professors dealt with the practice and theory of medicine and above all with the explanations of the ancient (Galen and Hippocrates) and more recent authors. Without any proper examination one could eventually obtain a *testimonium publicum* as the final testimony. The medical doctoral degree, a prerequisite for the *jus practicandi*, was however almost always acquired abroad.

In this working frame, the university studies of Stensen could have unfolded over 2–3 years if the political evolution had not offered the most unfavourable conditions imaginable. On June 1, 1657 Frederik III put his name on the declaration of war with Sweden. Soon the Swedish King Karl Gustav led his highly trained army from Poland against Jutland, which was conquered in July and August, whilst the disgraceful loss of the fortress Frederiksodde (later Fredericia) was a particularly bitterly resented warning. In the course of the extreme cold winter, the Swedish troops rushed over the frozen Belts into the southern island of Lolland towards Zealand and Copenhagen where the mood became increasingly gloomy. On February 26, 1658, Karl Gustav enforced the Peace Treaty of Roskilde, which brought to the lips of the Danish negotiator the words of Nero: "How I wish I'd never learned to write!" The remaining provinces of Denmark east of The Sound with Scania, the home land of the Stensen family on the paternal side, were lost with Bornholm and part of Norway.

The peace had hardly been concluded when the king of Sweden again invaded Denmark to capture what remained of the exhausted country. On August 7, 1658, Karl Gustav's fleet sailed into the harbour of Korsør. The year that followed, from Sunday, August 8, when, during the divine service, the rumour ran through the capital that Swedish ships were already anchoring outside Dragør south of the city, must have been deeply imprinted in the memory of the 20-year-old student. The resolute royal manifesto which inspired resistance, the rush for prayers in the churches, but even so the misery, the lack of food, the hard labour on the city walls and the flames that burnt

---

[147] *De studio medico inchoando, continuando et absolvendo consilium.* Hafniae 1628.

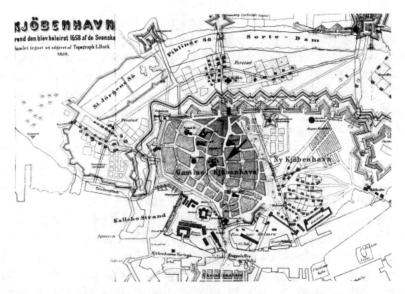

Fig. 1.1.7 Copenhagen and the walls before the siege in 1658, L. Both, 1859. The *arrow* indicates the corner of Købmagergade and Klareboderne, the birthplace of Niels Stensen

down all the suburbs! Niels will also have rejoiced at the successful general sortie on August 23, and the attack against a Swedish fleet that wanted to conquer Amager, the island and larder close to the city. If one lamented about the announcement of the fall of Kronborg Castle on September 6, then October 20 was again a day of jubilation when a Dutch relief-fleet broke the blockade in The Sound bringing in fresh supplies. The book of Anders Mathisøn Hiøring, later so popular, Leyrs-Krantz, describes almost every day during the hard winter, when women brought food to their men on the walls at Christmas time and spies and deserters spread alarming information on the planned main attack. From two months before the assault there were alarms almost every night. What must Niels have experienced during the main assault against Copenhagen between February 10 and 11, that night which became so bloody for the Swedish troops and happily decisive for Denmark (Fig. 1.1.7).[148]

The students of Copenhagen were directly drawn into the defence of the city from the beginning of the siege and with their professors declared to offer life and blood for their king and country. They formed a regiment of 4 companies, a total of 266 men, under the command of a colonel and were to receive the greatest respect from the enemy. Stensen served in the 1st

---

[148]Cedergreen Bech, *Københavns historie* 2, 146–148 (reference p. 318), Bruun, 2, 56 ff.

company.[149] As the siege drew out and misery affected particularly those students separated from their family, the youth by all means lost patience during the long watches on the walls. At the second summoning in the beginning of 1659, it appeared that many students had left the city and many suffered misery, above all the Norwegians. Only after an action of relief did more of them register.[150] One of them, the dear brother and fellow-student Niels Kristensen Krag was accused of "rebellion and opposition to his royal Majesty", and he was accused of having made other students disloyal to the service. However, 66 comrades, among them Niels Stensen, testified his innocence in writing. Krag would later become a parson and Anne, Stensen's sister, his sons' godmother.[151]

The goldsmith workshop in Klareboderne played a particularly active part in the historical events of the city and country. The determined courage with which the king Frederik III answered in the hours of danger the ordinary citizens urging him to fly: "I will die in my nest", had particularly inflamed the townsfolk who in the person of the mayor Hans Nansen possessed a prominent leader. The king and royal council had signed a letter of liberty for Copenhagen on August 10. Its privileges, some of which remain valid to this day, stimulated townsfolk to extreme resistance on the walls. In the first ranks of the defenders stood Johann Stichman, the stepfather of Niels Stensen, who had become a well-distinguished townsman of Copenhagen in the 1650s. He not only presided cleverly and firmly in his own house[152] but, since April 14, 1659, he belonged to the 32 "best men" who, together with the mayors, formed the town representation and advisory on revenues and expenses.[153] Stichman did not only support the ordinary charges of the siege. During the famous night of the main assault he commanded 409 townsmen of his Købmager quarter on the curtain-wall between Nørreport and the New-Nørreport bastion. He was one of nine captains awarded a golden chain by the king. After having surmounted the danger, one also wanted to take part in shaping the peace and, together with other townsmen, Stichman signed an open request of April 26, 1660, which surprised Frederik III considerably since the townsmen allowed themselves to put the question why

---

[149]Rørdam, *Danske og Norske studenter*, 159.

[150]How profoundly student life was affected by the war is evident from the number of students provided with meals by the Kommunitet. Before the war there were 144, in the beginning of the war the number had increased to 160. At the end of the war there were only 12. Norvin, 1, 301.

[151]Rørdam, *Danske og Norske studenter*, 58 ff, 130 ff, 202 ff, EP I, 179, note 2.

[152]Kbh. Dipl. 5, 402 ff, cf. EP II, 902.

[153]Cf. moreover Kbh. Dipl. I, 698–710. The names of the 32, see p. 704 f. Bruun, 2, 134 f.

peace negotiations were protracted so long. He was also a co-signer of the new absolutist constitution and was appointed royal master of the mint on March 11, 1662.[154]

It was against this bloody background that Niels Stensen completed his university studies. The famous anatomist mentioned before, Thomas Bartholin, became his private preceptor. The mentor in the person of a professor had to survey application and progress, moral behaviour and economic conditions. Bartholin was at the height of his fame as professor of anatomy, and although more learned than a researcher, he was in fact the ideal teacher for young Stensen. Whether Niels himself had chosen Thomas Bartholin as his preceptor or whether he had been assigned to him, the circumstances presume a quite regular tie between teacher and student. However, when Stensen began his university studies in 1656, Bartholin gradually withdrew from medical teaching.

Already during the disastrous plague in the spring and summer of 1654 Bartholin had left Copenhagen together with other professors. On November 4, while staying in Roskilde, he became elected rector for the rest of the office period of Ole Worm who had passed away during the plague while staying in the city. Bartholin sent in a program on February 1, 1655. After the death of the 96-year-old Thomas Fincke at the beginning of 1656, Bartholin was promoted to *Medicus primarius*. However, as early as February 9, 1656, he ended his medical activity with a dissection in the *Theatrum Anatomicum*. During the Swedish war he was not in Copenhagen at all from the summer 1657 until the summer 1658.[155] He was then, moreover, interested in that the son of Simon Paulli, Jakob Henrik Paulli, be appointed at the university.[156] When, however, Stensen, later, by far preferred the study in Copenhagen to the teaching of Gerard Blasius in Amsterdam,[157] it was very likely more due to the books of the professors,[158] their instructions for the

---

[154]EP II, 902. On January 1, 1659 men of the household able to carry arms were a servant Lucas Tellemand and two apprentices both named Hans. In October seven peoples stayed in the house being their quarter. Marquard, *Kongelige kammerregnskaber*, 55, 56, 121. Kbh. Dipl. 6, 395. Jørgensen, *Nils Stensen*, 13.

[155]Garboe, *Thomas Bartholin* 1, 209.

[156]*Ibid.*, 2, 14.

[157]He writes so in 1663: " … *jam tum enim integro, et quod excedit, triennio a Clariss. Viris D. Simone Paulli, Dn. Th. Bartholino, Dn. Backmeistero aliisque medicinae tum candidatis, tum studiosis ea in anatomicis, ut taceam caetera, videram, quae apud Blasium frustra quaererem*". OPH I, 147.

[158]Thomas Bartholin's edition of the *Anatomy* by his father, 3rd edition, 1655, in which he entered the discovery of the circulation of blood by Harvey and his own discovery on the lymphatics, was certainly the most important textbook for the students. Moreover the *Flora Danica* by Simon Paulli, 1648, was important for medicinal studies, reprinted in 1971.

Fig. 1.1.8 Stensen's student notes, the *Chaos-manuscript* upper half of first folio page of the manuscript, BNCF: Gal. 291, "Caos", http://teca.bncf.firenze.sbn.it/ImageViewer/servlet/ImageViewer?idr=BNCF0003561685#page/58/mode/2up

study of the medical literature and, above all, the stimulation in the private circles of friends and researchers that were eager for knowledge.

The faculty then comprised 17 professors of whom, in medicine, there were two "ordinary" and, at the time of Niels Stensen, one "extraordinary".[159] Thomas Bartholin was, as has been mentioned, first physician. The second professor of medicine, Christian Ostenfeld (1619–1671) meant as much as nothing to his faculty. He was rather a good jurist and an outstanding negotiator, but, in the university years of Stensen, no lectures by him are recorded in the catalogue.[160] Rasmus (or Erasmus) Bartholin (1625–1698), the brother of Thomas, who became third professor of medicine in 1657, was a mathematician that may have come nearer Stensen as a physicist as well as mineralogist. However, noticeably there is no written record that indicates closer ties between these two men (Fig. 1.1.8).[161]

---

[159]Slottved, 105 ff.

[160]DBL 3rd ed. I, 475 f. Møller-Christensen counts him as a faithful teacher who took lessons until June 1657 when Frederik III declared war against Sweden (Niels Stensen, 174). Ostenfeld was rector from 1657–1658 and prorector during the siege.

[161]DBL 3rd ed. I, 475 f. Møller-Christensen, *Niels Stensen*, 174, describes him as a conscientious teacher that presumably chaired the natural science education from 1657–1659. He was the only professor of medicine who taught in the winter of 1659 (p. 181).

Beyond the realm of suppositions, the valuable *Chaos-manuscript* is the first to lead us to secure historical ground. It consists of notes by Stensen from the time between March 8 and July 3, 1659. The manuscript came to notice when discovered by Gustav Scherz in Florence in 1946. It provides an uncommon lively insight into the scientific and human trends of young Stensen. Perhaps it was also his purpose: the student would here have wanted to demonstrate his maturity to his preceptor or his professor before an intended study journey abroad. The confidential religious thoughts would hardly have stood in the way. Even the regulations of the university required such a cast of mind.[162]

Of the books referred to here, let us first mention those by Danish authors, such as the works of his preceptor. A place, for instance, refers to nr. 96 of the 3rd Centuria (on worms and teeth) in Thomas Bartholin's *Historiarum anatomicarum centuriae*. The third volume had just been published in 1657.[163] At another time Stensen seems to have studied Bartholin's also recent *De lacteis thoracicis*, together with Pecquet's *Experimenta nova anatomica* on the blood circulation and chyle system,[164] both quite new discoveries, and he made notes on Bartholin's *De luce animalium* (1647).[165]

Stensen knows Peder Sørensen's *Idea medicinae philosophicae*[166] and Tycho Brahe's *Epistolae astronomicae*[167] and also Otto Sperling (1602–1681), the physician and botanist who ended up in jail because of Corfitz Ulfeldt,[168] and he knows Christian Nold (1626–1683), the professor of theology, with whom, later, he as a bishop had a controversy.[169] In several places, one has the impression of dealing with a conversation in a circle of friends, above all

---

[162]The manuscript has around 90 pages (fol. 28 r–79 v) in a volume containing manuscripts of the Accademia del Cimento (Signature: Post 32 III 17. Nowadays Gal. 291, web-accessible) at the Biblioteca Nazionale Centrale, Florence, under the covering title *Scritti di Niccolò Stenone*. Complete edition with Introduction, Notes and Commentary by August Ziggelaar, 1997, *Acta Historica Naturalium et Medicinalium*, vol. 44. Copenhagen, Munksgaard, 520 pp. Abbreviated here as ed. Z.

[163]Chaos fol. 57 v, col. 111, ed. Z, 276.

[164]Chaos fol. 66 rv, col. 146, ed. Z, 349 ff.

[165]Chaos fol. 35 r, on a loose sheet, ed. Z, 82.

[166]Chaos fol. 35 v, on a loose sheet, ed. Z, 83.

[167]Chaos fol. 36 r, col. 26, ed. Z, 86.

[168]Chaos fol. 42 r, col. 50, ed. Z, 145.

[169]Chaos fol. 54 r, col. 97, ed. Z, 250, EP II, 914, OTH I, 257 ff.

with Ole Borch, without being able of identifying the cited names like Hofman,[170] Pretzler and some others.[171]

Of course one often comes across excerpts from the medical literature of the time, which the student read on the explicit advice of his professor.[172] He made notes: "First, de Roy's Medicina must be rapidly skimmed through, then his Physica, and outline tables must be made briefly showing the plan of the whole work, and the same must be done with Sennert, then with Kyper and finally with other authors. Before noon nothing must be done except Medicine."[173] Henri de Roy or Henricus Regius (1598–1679) was professor of medicine and botanic in Utrecht and particularly appeared as defender of Harvey's theory of blood circulation. As a supporter of Descartes he made Utrecht a stronghold of Cartesianism. His manual *Medicina et praxis medica* (1655) like his *Fundamenta physices* (1646) are written in the mood of his master.[174]

The chemist and learned compilator Daniel Sennert (1572–1637) was professor of medicine in Prague. He had helped others in six plague epidemics and finally fell victim to the disease himself.[175] His books were widely distributed: the *Epitome universa Dan. Sennerti doctrinam summa fide complectens* in which, in well ordered chapters, organ after organ, disease after disease were dealt with. Further, the *Institutionum medicinae libri V* and the *Practicae medicinae libri VI*. In the *Anthropologia corporis humani contentorum.* ... Albert Kypers († 1655) which was published several times, and, in his *Collegium medicum*, Stensen met a Hippocratic who wanted to reconcile modern times with the old.[176] He retained these *veterum placita quae manifeste rationi et experientiae non adversabantur*.

As an introduction to surgery, Stensen used the booklet *Practica chirurgiae brevis et facilis* (of 1557) of the Belgian physician Cornelius Schylander, a kind of catechism with questions and answers.[177] Stensen also knew Werner Rolfinck (1599–1673), the distinguished professor and dissector in Iena,

---

[170]*Georg and Gottfried Hoffmann were brothers born in Lauban, Silesia and educated at the University of Leipzig. They worked as fortress engineers for the Danish king. See Ziggelaar in Chaos, ed. Z., 490.

[171]EP II, 907 f. Chaos-MS mentions Borch's first, and at that time his only published book, *De cabala characterali dissertatio* from 1649, in which he turned downsuperstition and a healing power of words and letters (Chaos, fol. 37 v, col. 32, ed. Z, 99 and note 169).

[172]Cf. fol. 61 r, col. 126, ed. Z, 299. "He" (N.194) is probably Ole Borch (AZ).

[173]Chaos fol. 36 v, col. 27, ed. Z, 89.

[174]Mouy, 14 ff, 73 ff.

[175]Chaos fol. 36 v, col. 27, ed. Z, 90 ff; fol. 37 rv, col. 30, ed. Z, 96, note 156.

[176]Chaos fol. 36 v, col. 27, 89; fol. 37 r, col. 31, ed. Z, 97.

[177]Chaos fol. 28 r, col. 1, ed. Z, 22; fol. 36 r, col. 25, ed. Z, 84.

who, through his textbooks *Epitome methodi cognoscendi et curandi particulares affectus* and *Dissertationes anatomicae*, exerted a great influence on the anatomy study of his time.[178] They were clear and practical manuals mixed with many unctuous considerations.

For the theory of diseases Stensen studied, beside Borel's or Bartholin's case reports, the Polish author Johann Jonstonus (1603–1675), namely the *Idea universae medicinae practicae libri VIII absoluta* with its short, clear diagnoses and medical prescriptions. Moreover, Ionston's *De naturae constantia*. He writes: "As shallowness and haste are of much harm,—and anyway I have most hope of making progress in my studies, if I commence on other books besides the one with reports, and on diseases it will be on Ionston's procedure, on medicaments in Schröder's, on other topics in other ways, and it is not going to be all jumbled and with imprecise titles, but, after reading Ionston, taking the others afterwards, or also while reading him, and writing down in loci communes everything that is not found there, referring to the same, etc., but in such a way that additions or changes are made to the questions occurring on qualities on loose sheets for insertion in the Marasius book, where the doses too are to be added. But Ionston must be read very attentively and Schröder included, but from Sennert nothing is to be included but symptoms and signs. But immediately after the dinner interpreters are to be explained and consulted."[179]

Johann Schröder (1600–1664), mentioned here, was the city physician of Frankfurt, highly appreciated for the theory of medicine, and a former military physician in the Swedish army.[180] His *Pharmacopoeia medicochymica sive thesaurus pharmalogicus* appeared in many editions until in the eighteenth century and owed its reputation to clarity, thoroughness and its practical presentation.

From the zeal with which Stensen studied pharmacy one can conclude that he, who later hardly practiced the medical profession, consciously had prepared himself for it. He also explains very precisely the theory of remedies of Pierre Morel (1593–1668), the book which was printed often in the seventeenth century, *Methodus praescribendi formulas remediorum*,[181] as well as Jean Renou's *Institutiones pharmaceuticae* and *Dispensatorium galenicochymicum*. Yes, in the Chaos-Manuscript one meets not only many notes about different

---

[178]Chaos fol. 33 v, col. 23–24, ed. Z, 78.
[179]Chaos fol. 58 r, col. 114, ed. Z, 281.
[180]Cf. ADB 32, 518.
[181]Chaos fol. 61 r, col. 126, ed Z, 299.

medicines but also whole tables to indicate how many authors are involved for one determined medicine.[182]

Particularly the last pages deal with practical medical matters. Thus for instance: "All symptoms should be observed and it should be seen in how many different diseases this symptom is found, how different are the accompanying symptoms in various people and diseases. If convulsions, which diseases it accompanies, which it follows upon, which it ends in, which other symptoms it is connected with, when it is cured, when not, etc."[183] There is a note about the diet.

One is amazed by the many excerpts from polyhistorical works of the time, natural science in general, as well as medical ones. However, one soon understands the student. These books, also that of Thomas Bartholin, were simply standard literature of the time and, despite several false conceptions and an abundance of funny stories, a collection of manifold knowledge that otherwise was not easily accessible. Athanasius Kircher's (1602–1680) *Magnes sive de arte magnetica* (Rome 1641) was extensively quoted,[184] being the first book on the foundations of magnetism and also dealing with its application in the second book, as well as the *magnetica catena*, the links between everything, magnetism in music, in love, and in the central magnet: God.

One meets the typical and longest example of this kind in the first pages of the Chaos-Manuscript. Columns 3–21 contain point after point abstracts from the main work of the personal physician of Louis XIV, Pierre Borel (1620–1689), whom Stensen would come to know personally during his sojourn in Paris. It refers to the *Historiarum et observationum medicophysicarum centuriae IV* (Paris 1656),[185] a book that properly presents himself in the subtitle when it promises *multa utilia sed et rara, stupenda ac inaudita*. The reader is prepared for a *scientia curiosa*, for *historiae prodigiosae*. Stensen only quotes these stories and traits with Borel's own expressions, among which many good and sharp-witted remarks such as the opinion of Borel on fossil snails, fishes and mussels, which led back to the deluge or originated from a lake that had withdrawn later.[186] Nor does that sentence in the book escape the reader through which the author became known in the history of medicine, namely the discovery of the cause of the grey cataract, the cloudy

---

[182]Chaos fol. 41 v, col. 47, ed. Z, 140; fol. 63 rv, col. 132–134, ed. Z, 322–324.

[183]Chaos fol. 56 r, col. 105, ed. Z, 265.

[184]Chaos fol. 38 r-45 v, col. 33–62, ed. Z, 113–169.

[185]DSB II, 303 f.

[186]Chaos fol. 30 r, col. 10, ed. Z, 46.

crystalline liquid.[187] Similarly the remark of Borel that bloody spots erupting over night in the face of a child can result either from insect stings or from the long needle of a witch or others, from bewitched food that a woman served by hate against her husband, or from the servant who called his master back to life by kisses and others.[188]

Strikingly, Stensen is very interested in iatrochemistry and iatrophysics; the representatives of a quantitative conception of nature. One thus finds a smart opinion about the above-mentioned Swiss philosopher of nature, Theophrastus Paracelsus (1493–1541), the precursor of iatrochemistry, a well known, controversial author. Although he studied little from books, threw all the writings of Galen and of other physicians on the pyre in Basel and got his knowledge from hangmen, barbers, gypsies and quacks, he qualified as physician. He dedicated himself above all to the occult philosophy and called for magic power. His theory of the divine original force that impregnates all nature and the microcosm (the human organism) in connection with the macrocosm (all the world) holds, his endeavour to find the "elixir of life", a universal medicine, the mystical darkness and confusion in many of his expressions and writings are as characteristic for the man as the fact that he was the first to introduce chemical remedies in the treatment of sick people. For instance, he recommended quicksilver as a specific means against syphilis.[189] Stensen releases a mature judgement on Paracelsus, which corresponds to our actual knowledge of this remarkable man. He notes after a conversation with Borch: "Paracelsus and many magicians were not devil's servants but nature researchers. They went astray, however, in that they have not published their discoveries and the reasons for that." And he speaks another time of "Aureolus Paracelsus, by most people more easily criticized than understood".[190] Stensen of course also read the work of the above-mentioned Peder Sørensen, the highly gifted and in the whole of Europe famous pupil of Paracelsus.

Stensen also read the work of Johann Bapt. Helmont (1577–1644), the iatrochemist, who was, on the one hand, a supporter of Paracelsus and a speculative philosopher and, on the other hand, an accurate researcher interested in experiments.[191] Stensen would later come into a close relationship with Helmont's most important successor, Franz Sylvius. In the

---

[187]Chaos fol. 31 v, col. 15, ed. Z, 56.

[188]Chaos fol. 29 rv, col. 7, ed. Z, 37–42.

[189]DSB X, 304–313.

[190]Chaos fol. 33 v, col. 23, ed. Z, 76 f; fol. 36 r, col. 25, ed. Z, 86.

[191]Chaos fol. 31 r, col. 14, ed. Z, 54.

Chaos-Manuscript one also finds the first iatrophysician, the famous professor in Padua, Santorio Santorio (1561–1636) and his work *De statica medicina*. The idea of Santorio, to determine the physiological and pathological changes in man by means of a balance (see the later remarks of Stensen about "imperceptible perspiration"[192]) already preoccupied the student who critically notices: "Santorio's *Statica*, cannot be quite accurate unless he made these trials with a naked person or someone who had only a shirt on, or in some other way he took the clothes into consideration. For I do not believe that it is possible to avoid some of the sweat clinging to the clothes, as there is always found dirt in them. So if some things turn up that contradict each other, it might be excused in this way."[193] It is certainly not an uncritical interest that appears from the fact that one often meets Hermann Conring (1608–1681), the polyhistorian from Helmstedt, who in his book *De hermetica medicina* turns against the supporters of Paracelsus and defends medicine with herbs. His book must also have been assiduously discussed in Copenhagen.[194]

This preoccupation with iatrochemist and iatrophysics authors already discloses a strong interest of young Stensen in chemistry and physics. First on the background of the Chaos-manuscript one understands his desire to devote himself entirely to "geometry" and his love for the exact sciences. Here he deals with metal alloys and their properties,[195] with the currents of heated air and the effects of thrust and blow.[196] He studies the properties of quicksilver,[197] he describes and draws snow crystals,[198] he interests himself in fire and smoke and he makes experiments with sulphur.[199]

Whether the technical dexterity of Stensen came from the goldsmith workshop of his father—he once measured also the weight and volume of gold[200]—or not, he must anyway have often worked there to provide himself with the necessary instruments or to have noticed how to make them. He built a hydraulic machine that carried out feats of all kinds.[201] With the help of a bladder he determined the volume of expired air[202] and he possessed

---

[192]"*insensibilis transpiratio*", OPH II, 118.

[193]Chaos fol. 56 v, col. 108, ed. Z, 270, probably Ole Borch's comment (AZ).

[194]Chaos fol. 41 v, col. 48, ed. Z, 141.

[195]Chaos fol. 33 v, col. 24, ed. Z, 80.

[196]Chaos fol. 36 r, col. 25, ed. Z, 84.

[197]Chaos fol. 36 r, col. 26, ed. Z, 86–87.

[198]Chaos fol. 36 v, col. 36, ed. Z, 120.

[199]Chaos fol. 38 v, col. 36, ed. Z, 120.

[200]Chaos fol. 60 r, col. 121, ed. Z, 290.

[201]Chaos fol. 39 r, col. 37, ed. Z, 122.

[202]Chaos fol. 47 rv, col. 70, ed. Z, 193.

thermoscopes.[203] He elaborately described and drew an instrument *ad pro-portionandum.*[204] He fixed the lenses of a telescope to a stick and with them he measured refraction.[205] To polish glass he used the tool of his father's workshop[206] and has also a microscope and a telescope.[207] He described and sketched a contrivance to drag glasses, which surpassed the machine of Descartes.[208] Optic research was of particular interest to him. He wanted to penetrate the essence of the colours, prepared red colour from vitriol, observed the refraction of the light in windows, colours, the optic of the eyes, *et cetera.*[209]

From the fact that Stensen mentions a microscope only once in his later writings, it should not be concluded that he did not know of or did not adequately appreciate it. The Chaos-Manuscript shows a student actively preoccupied with microscopes and his notes reveal an understanding of the relevant problems and difficulties.[210]

There is much to indicate that the student then joined a circle of Danish friends with the same interests. In a note, an astrolabe is mentioned with the three circles of the equator, the poles and the meridian and as the owner of this sun-dial is named Mathesius, presumably Henrik Mathesius († 1681). Then teacher of the future king Christian V, Mathesius had earlier under-taken year long travels with the children of the Marshal Anders Bille.[211] But, the optician, calendar- and instrument maker Thomas Rasmussen Walgesten (c. 1627–1681) from Gotland may also then have been back in Copenhagen after his study years in Leiden. It is well-known that later Walgesten was admired in Paris and Florence because of his *laterna magica*, the first one that was serviceable.[212]

One is not less amazed by the interest of the student Stensen for French research in physics. He knows Ismaël Boulliau (1605–1694), the courageous supporter of the Copernican theory, who as early as 1639 had published *De vero systemate mundi*, and he showed a great interest in his book *De natura*

[203]Chaos fol. 38 v, col. 36, ed. Z, 121; fol. 56 v, col. 107, ed. Z, 269.

[204]Chaos fol. 58 r, col. 113–114, ed. Z, 279–282.

[205]Chaos fol. 59 r, col. 118, ed. Z, 286.

[206]Chaos fol. 60 r, col. 121, ed. Z, 290.

[207]Chaos fol. 60 v, col. 124, ed. Z, 296–297.

[208]Chaos fol. 60 r, col. 122, ed. Z, 293.

[209]EP II, 917 f.

[210]Cf. EP II, 919, OP I, 131.

[211]Chaos fol. 44 f, col. 58, ed. Z, 161. DBL 3rd ed., 9, 468. Scherz, *Niels Stensen und Galilei*, 9 ff. As noted by AZ, 'Mathesius' may be Matthias Jacobsen.

[212]Chaos fol. 75 v, col. 84, ed. Z, 448. Scherz, *Niels Stensen und Galilei*, 9 f, 30.

lucis (Paris 1638) of which he noted some 40 sentences about light, its properties and colours, its velocity and refraction, and also the views of the ancients like Plato and Aristotle.[213] Elaborate abstracts are found about the fundamental problems of French physics of the time, the atomic theory and the problem of the empty space from the work of Pierre Gassendi (1592–1655) under the title *Animadversiones in decimum librum Diogenis Laertii, qui est de vita, moribus placitisque Epicuri I-III* (Lugduni 1649). Like Descartes before him, and later Christiaan Huygens and Isaac Newton, Gassendi developed ideas about the structure of matter and finally arrived in connection with the atomic theory of Epicure at a system of corpuscular physics according to which all matter consists of solid impenetrable mobile particles. When Gassendi attributed cold to a remarkable property of new atoms, Stensen commented critically: "But, dear Gassendi, that no cooling atoms enter into the constitution of ice, that I have proved and, if you like, you may easily do the experiment thus. Take in a very cold winter a glass vessel of any capacity, fill it for a third part with water, close the orifice with a threefold bladder, set in the middle of a court yard, it will freeze immediately; take then a balance and weigh the whole thing; then you will see that the weight has neither increased nor decreased a scruple, not even a grain, and that would altogether be impossible if those atoms had penetrated."[214]

The student had made notes for the problem of the empty space under the title: *Experimenta physico-mathematica de vacuo.* He had himself taken up experiments concerning this problem. He knew the experiment of Pascal at Puy-de-Dôme and the controversy of priority which developed about that, among others the control experiment carried out by Pascal's brother-in-law. In a 15 m suspended long tube, closed at the top, open at the bottom, air pressure maintained the water inside the tube until a height of elevation of about 10 m.[215]

In this *Chaos-manuscript* one seeks with particular close attention places from which one can conclude the views of the student concerning the Ptolemaic and Copernican systems, the most sensational problem around the middle of the seventeenth century. Then at Copenhagen University one kept to the intermediary system of Tycho Brahe, according to whom planets move around the sun at the Copernican distances and times of revolution, but the sun with its whole retinue still always moves around the earth in 365 days.

---

[213]Chaos fol. 69 rv, col. 156–159, ed. Z, 370–376. Scherz, *Niels Stensen und Galilei*, 13 f.

[214]Chaos fol. 72 v, col. 171, ed. Z, 415–416. Scherz, *Niels Stensen und Galilei*, 16 ff. Stensen here quoted Borch. (AZ).

[215]Chaos fol. 73 r, col. 173, ed. Z, 422. Scherz, *Niels Stensen und Galilei*, 17 ff.

Since this system did not contradict the conventional explanations of the Bible, almost everybody paid homage to it in the seventeenth century, for example, Longomontanus who had initiated the building of the Round Tower and its observatory but also the then professor of anatomy in Copenhagen Villum Lange (1624–1682), who we know had later connections with Stensen.[216] Certainly other influences were also already being brought to attention, like that of the Copernican Nicolaus Mercator who, at the beginning of the 1650s, when he was in Copenhagen, dedicated to Joachim Gersdorff, the patron of Ole Borch, a writing (*Cosmographia*), which describes the orbit revolution of the celestial bodies in the sense of Galileo.[217]

It is against that background that the notes in the Chaos-Manuscript must be read, in which Stensen let himself become inspired by Galileo. At the latest in the spring of 1659, the student read Galileo's *Sidereus Nuncius*, the messenger of the stars, which appeared in 1610 in Venice and agitated the world by the announcement of new observations in the sky.[218] Galileo here advices the reader to build a telescope and gives his instructions how to do so. After reading this Stensen noted the following practical remark for checking its performance: "Two circles or two squares of paper are to be made, one of them 400 times larger than the other, i.e. the diameter of the larger one is 20 times that of the smaller one; then both surfaces are fixed on the same wall, and they are to be viewed from a long distance, the smaller one through the telescope and the larger one with the eye, then they will appear of the same size. Galileo's Sidereus Nuncius."[219]

This excerpt reveals that Stensen knew Galileo and his work. It is astonishing that the student excerpted the work of a convinced and famous Copernican very approvingly. This is the *Geographia generalis in qua affectiones generales telluris explicantur. Auctore Bernhardo Varentio, Med. Dr.* (Amsterdam 1650). Bernhard Varen (1622–1650), the son of a preacher from northern Germany, had studied medicine by necessity but his passion was mathematics. His book remained quite unnoticed in Germany but in England nobody less than Isaac Newton published an improved edition as early as in 1672, and he delivered lectures about it. Later the work was counted among the books which mark the border between two epochs.[220] Stensen had thus studied the book already in 1659. Its praise of mathematics

---

[216]Scherz, *Niels Stensen und Galilei*, 25 f.

[217]Scherz, *Vom Wege*, 30, *Niels Stensen und Galilei*, 28 f.

[218]Scherz, *Niels Stensen und Galilei*, 6 ff.

[219]Chaos fol. 61 v, col. 127, ed. Z, 301 f.

[220]DSB XIII, 583 f, Scherz, *Niels Stensen und Galilei*, 31 ff.

in the Introduction may have pleased him. However, he became absorbed by the proofs from the old and the new time for the spherical shape of the Earth and the movement of the earth, for which Varen presents eight arguments in the sense of the Copernican conception.[221]

Stensen knows Willebrord Snellius (1580–1626) who had undertaken the measurement of the degree following the rigorous rules of triangulation. This measurement provided an accurate value of the meridian degree.[222] He also knew the physicist Cornelius Drebbel (1572–1631), the inventor of a thermoscope.[223] He mentioned the Jesuit Christoph Scheiner (1575–1650) who acquired the greatest merit for the foundation in optics, for the invention of optical instruments and for the physics of the sun, and noticed his *Rosa Ursina*.[224]

Yet, best of all, one can recognize the future pioneering researcher in these Chaos notes in the alert zeal with which the student shared the enthusiastic trend of his time for a mathematical foundation of science and above all for the most reliable methods of research. There he appeared as the best of his time. He knew Johannes Kepler from the booklet *De nive sexangula* and let himself be stimulated for his own researches on snow crystals.[225] Stensen made only few notes from Francis Bacon's *De augmentis and dignitate scientiae* (1605).[226] The powerful herald of the inductive approach to research who put his splendid eloquence and writing talent at its service. Bacon was also called its prophet but hardly engaged himself in research of importance, yet in many ways he was inspiring. His conviction that *tantum possumus quantum scimus* (we can do as much as we know) stimulated him for the project of a new frame for the sciences. This was first exposed in the above-mentioned writing which also bears the title *Instauratio magna*.

There is no doubt that Stensen knew of and studied Descartes and his works published then, above all his methods of obtaining certain knowledge. When the student speaks of exhalations passing through the skins and muscles, he says: "This should be investigated more carefully and systematically according to Descartes' method, or by considering directly what enters the pores of the blood, what its particles are, how they move, what is expelled

---

[221]Chaos fol. 55 v, col. 103, ed. Z, 261.

[222]Chaos fol. 55 v, ibid.

[223]Chaos fol. 39 r, col. 37, ed. Z, 122; (Scherz: fol. 57 v ?).

[224]Chaos fol. 55 v, col. 103, ed. Z, 262.

[225]Chaos fol. 39 v, col. 39, ed. Z, 125—Stensen was stimulated for his research on snow crystals by Hendrik de Roy (Henricus Regius), ed. Z, 472. (AZ).

[226]Chaos fol. 33 v, col. Col. 24, ed. Z, 80–81.

from there and how. Thus, by asking all similar questions that can be thought of and by collating them and doing experiments about it."[227] This is obviously the expression of the rules which Descartes postulates in his *Discours de la méthode* to acquire mathematically precise knowledge. One may suppose that the student also knows Descartes as the founder of analytic geometry and a physicist with merits from his theory of the refraction of the light, from the explanation of the rainbow, from the determination of the gravity of air. In any case, Stensen also knew the main work, the "Principles of Philosophy",[228] in which Descartes in the first book deals with the principles of human knowledge, in the second with the principles of the body, and in the last two with the visible world and the earth. The student also corrects some erroneous conceptions of Cartesian theory. The above-mentioned Christian Nold had published his *Leges distinguendi, seu de virtute et vitio distinctionis* (Franequerae 1657) in which he examined different concepts. Stensen was discontented with him: "In examining the teaching of others one should not proceed from one's own foundations but from those of the author. As Nold did who, when he will criticize Descartes' division of place, forms for himself his notion of space and examines Descartes in relation to it. Thus also words are to be interpreted not otherwise than the author took them, as Nold does when he understands position. ... as position of the parts in regard to the whole—e.g. that a bent person is in a different position from a standing one, etc."[229] Another time Stensen spoke of an observation of Descartes that vapours rises from earth more easily and more often than from water. The place is to be found in *Les météores*, in the second *Discours des vapeurs et des exhalaisons*.[230] The dioptrics of Descartes stimulated Stensen to do an observation at the window: "Sometimes light rays are not refracted but reflected in windows. Thus I have often myself seen my finger in windows as if in a mirror but not as clearly. Perhaps therefore Descartes' Dioptrique, p. 81 § 6."[231] Another text testifies of his criticism of Descartes: "There are people in Belgium who after an accurate examination of the Cartesian philosophy noted that there are certain things in it which cannot be approved; furthermore there is someone who has written against him with geometrical argumentation."[232]

---

[227]Chaos fol. 39 r, col. 37–38, ed. Z, 123.

[228]Scherz, *Vom Wege*, 50.

[229]Chaos fol. 54 r, col. 97, ed. Z, 250.

[230]Chaos fol. 55 v, col. 103, ed. Z, 286; Descartes, *Oeuvres*, 6, 239 f.

[231]Chaos fol. 59 r, col. 117, ed. Z, 286.

[232]Chaos fol. 75 v, col. 184, ed. Z, 447.

The methodological observations which Stensen arrived at in correspondence with the creative minds of his time are quite important. He notes, for instance: "In the natural sciences it is better not to be bound to any knowledge, but to refer everything that can be observed to certain titles and then, by one's own exertions, to elicit something from it, if nothing else at least some kind of certain knowledge."[233]

A statement of basic importance is: "One sins against the majesty of God by being unwilling to look into nature's own works and contenting oneself with reading others; in this way one forms and creates for oneself various fanciful notions and thus not only does one not enjoy the pleasure of looking into God's wonders but also wastes time that should be spent on necessities and to the benefit of one's neighbour and states many things which are unworthy of God. Such are those scholastics, such are most philosophers and those who devote their whole lives to be spent on explaining and defending, indeed scarcely on examining them, and one must not boldly and impetuously assign anything to art on the observing a single thing. From now on then I shall spend my time, not on musings, but solely in investigation, experience and recording of natural objects and reports of the ancients on the observation of such things, as well as in testing out these reports, if that be possible."[234]

Somewhere else Stensen opposes the then highly appreciated analogies with these words: "One should criticize the error of those who in the natural sciences want to demonstrate phenomena by means of similar things. Thus Peiresc in his biography, p. 36, where he will have it that all light is a small but very rare flame which burns when it is collected, just as steam is rarefied water which appears as liquid when collected; but if this were true, then it should follow that, just as steam is not taken away at the moment when the water out of which the steam comes, is taken away, thus light does not disappear either, when the sun of the candle from which it comes, is taken away; but that is wrong because steam is not spread instantly like light, etc. Thus they have said that the foetus is attached to the womb by the umbilical cord just as plants to the soil by their roots, thus that the mesenteric veins suck in nourishment from the intestines as roots do from the soil; thus Kircher [says] that like a magnet attracting iron in some manner sucks spirits out of it, thus man has a double magnet implanted in him, etc."[235]

---

[233]Chaos fol. 37 v, col. 32, ed. Z, 100.
[234]Chaos fol. 44 r, col. 58, ed. Z, 159 f.
[235]Chaos fol. 53 v—54 r, col. 96, ed. Z, 249.

Towards the end of the manuscript the clear, methodical remarks become particularly determined and critical, like for instance this statement excerpted from A. Auzout: "In physical science we know nothing beyond experiments and observations and that which is deduced from them according to metaphysical and mechanical principles."[236] About the medicine of his time a sarcastic opinion is issued: "I would fear that someone might define medicine as: The art of, standing with furrowed brow in front of the patient, uttering inanities and of using uncertain remedies in order to ease slightly the worries of the mind and so that restoration of good health may be awaited calmly, while nature does its work,—or death when fate takes its course."[237]

The records of the Chaos-Manuscript allow many insights into Stensen's studies and scientific interests during the time at university in Copenhagen, the circle within which he moved, and the books which he read. They thus reveal above all the deepest insight imaginable in the spiritual life of the 21-year-old fellow. These short notes and excerpts, the fugitive remarks and commentaries on the brim of the day, these confessions unfurl, as already said, before our eyes the image of a serious respectable striving for an unusually deep and warm life of the faith. The passages quoted above have already shown that faith and scientific research are for him one whole in complete harmony.

The opening title near the date of March 8, 1659 reveals the mind of the whole, "In nomine Jesu". This is not just a pious formula. After a long day entirely devoted to studies, on a day in which an experiment partly succeeded, partly failed, one finds: "The whole day for studies, though before dinner an experiment was tried which partly succeeded, say, for the immersed part. I should never be disturbed by any evil nor become angry because of insults against myself or my friends or relatives. God sees and foresees everything; and all that happens comes from him and for the glory of his name."[238]

Here is a student who works hard, is parsimonious with his time, reads thoroughly, and often notes the contents of a book point by point. On another day he wrote: "Almost the whole day my disturbed mind, being preoccupied with various reflections, could do nothing else than skim lightly over everything, then immediately to leave the causes of everything aside, but I pray, thee, o God, take this plague from me and grant me the power to free

---

[236]Chaos fol. 67 v, col. 152, ed. Z, 360.

[237]Chaos fol. 67 v, 68 r, col. 153, ed. Z, 363. Stensen quotes from Pecquet's book of 1654, *On milky vessels*, a letter by Samuel Sorbière. (AZ).

[238]Chaos fol. 30 v, col. 12, ed. Z, 50.

my soul from all distraction, to work on one thing alone and to make myself familiar with the tables of medicine alone."[239] Another time he wrote: "In the name of Jesus. Before noon after prayers first one of Sennert's *Tabulæ* must be read, first to understand it, then to take it in. Then I must look at Sennert's explanations, giving in the form of the position of the author together with the method of treatment, adding ampler explanations or narrative in Chaos or in the outline table, if space permits. If I have leisure, Duncan Lidell should be consulted on the same subject and then Kyper and finally the note book. When this has been done, I will revise the outline and expound it in teaching-form without looking at the books. (If anything else should appear in other writers worth being in the outline, it is to be entered in the book with the outline tables.) Immediately after dinner I must attentively go through one of the later tables starting out from Sphinx and I must familiarize myself with it, but without looking at either Sennert or the others. Stand by me, Jesus, by thy grace!"[240]

The entire work was inspired by the world-outlook of the student who writes: "Everywhere and nowhere we are at home while we are outside heaven"; or he notes: "Being right-minded one is never without God."[241] After an anatomic exposé we suddenly bump into this exclamation: "Otherwise done too little today. Guide me, God, to the glory of thy name, grant me to be able to work properly and persistently and to do something sensible."[242] The faith should now and later determine the way of life: "The Lord's name be blessed forever. Always and everywhere one ought to seek occasion to observe and something to imitate so that it conduces to a pious, seemly and prudent way of life, for the present or for the future. If ever God makes me a father of a family, then with my whole house I shall praise God on feast days and on the eve of them with prayers and hymn-singing and find my joy in it. Yea, each morning and evening but not with many nor long, yet adequate [prayers]. If my means allow, I shall always give two poors a meal and two others clothes, stockings and other necessary items, but no money."[243] Some lines in the middle of completely different notes appear like an evening conscience investigation: "I was inattentive during divine service and have done little good. Forgive me, God, and grant me grace to come to myself again. Let me always keep death in view, and always have in my mouth:

---

[239]Chaos fol. 37 r, col. 30, ed. Z, 95.

[240]Chaos fol. 31 rv, col. 30–31, ed. Z, 96 f.

[241]Chaos fol. 39 r, col. 26, ed. Z, 88 and 87.

[242]Chaos fol. 45 v, col. 38, ed. Z, 123.

[243]Chaos fol. 45 v, col. 63, ed. Z, 169.

M e m e n t o   m o r i."[244] Or we read: "Let us lay down at God's feet the soiled cloths of our sins. … Grant me, o God, to be able to impose upon myself abstinence from all sins and particularly from a hasty and ill-considered judgement or opinion on things about which I know either little or much, unless very exactly."[245]

If the mind of these pronouncements reminds us in some way of the "imitation of Christ" or the ascetic writers of the Middle Ages, one is even more surprised by the early Catholic part in many notes. Many of the thoughts and sayings are actually the fruit of reading, often literal quotations from a work of the court preacher Jeremias Drexel. The son of protestant parents from Augsburg he became a Catholic early on. He joined the Society of Jesus, worked as teacher in the colleges of Munich and Augsburg and finally became court preacher in the Bavarian capital. From 1620–1639 he published 21 edifying writings which enjoyed a comprehensive publication of 158 700 issues and many translations into foreign languages (one also into Danish: *Betænkninger om Evigheden*, København 1681).[246] The genuine piety of the author, his attractive lively language with many examples and sayings, the Latin epigrams which easily impressed, made him one of the loveliest ascetic writers of his time.[247] The book which Stensen obtained carried the title *Joseph Aegypti prorex descriptus* and in ten chapters with three paragraphs each dealt with the story of the Egyptian Joseph, aiming at moral consider-ations about humility and trust in God, purity and vice, et cetera. Stensen had read the book meditatively in segments and written thorough notes.[248]

These highly interesting and unique notes of the Chaos-Manuscript cover the period from March 18 until July 3, 1659. After the night of the main assault in February, the power of the Swedish siege was broken and the student could again become absorbed by his studies. Yet the war dragged about the city and on the countryside. The misery in Copenhagen was still profound—many students left the town. Not until after the death of the Swedish King Karl Gustav in the night of February 12–13, 1660, did peace negotiations make progress more quickly, and on March 1, they led to an armistice. Finally, on May 27, 1660, peace was concluded.[249]

---

[244]Chaos fol. 37 r, col. 30, ed. Z, 96.

[245]Chaos fol. 41 v, col. 48, ed. Z, 142.

[246]Bibliotheca Danica I, 450.

[247]L. Koch, *Jesuiten-Lexicon*, 455.

[248]Chaos fol. 28 ff, col. 3 ff, ed. Z, 26 ff.

[249]Bruun, 2, 143 ff.

Stensen, however, had then completed his studies at the university for a long time in order to undertake the study journey, usual also in his family and necessary for a student of medicine. On September 19, 1659 he was still in Copenhagen, in March 1660 he worked in Amsterdam.[250] He must have departed at the latest in the beginning of 1660. His journey must have been made possible partly and encouraged by three men to whom he dedicated his first important work in Holland: Thomas Bartholin who had given him a letter of commendation, the archbishop Hans Svane who may have helped him financially, and Otto Krag, the Danish resident in the Netherlands. Stensen calls the two latter his Maecenas and promoters.[251]

Stensen undertook the journey well-prepared as appears from the existing statements about his studies. The actual operation of the university may well have suffered much from the unfavourable conditions of the time, but his relations with enthusiastic teachers like Borch, Paulli and Thomas Bartholin and the exchange of scientific ideas that must have taken place among the younger generation, were important compensations. Above all, the new science of nature had provided the researcher to become with books, and he already knew many erudite men with whom he would become acquainted face to face thoroughly through their work. Yet none of the friends and patrons who then helped Stensen for his journey foresaw that this student would develop into one of the greatest natural scientists during the next few years.

---

[250]He testified the innocence of Niels Christensen Krag in September 1659. See above. On April 7, 1660 st.n., he dissected in Amsterdam, OPH I, 4 = EP I, 134.

[251]<425> OPH I, 11.

1.2   Years of Discovery in Holland 1660–1664

When at the age of 22 he undertook his first journey to a foreign country Niels Stensen had no idea that he would see his home country again only rarely and on brief occasions.

The student, we presume, did not travel directly to Holland but spent some time in Rostock. When later in Amsterdam he reproached Professor Blasius to have taught him nothing but chemical truisms, Stensen writes that he had seen during the previous 3 years of training with Simon Paulli, Thomas Bartholin and Bacmeister what he had looked for in vain with Blasius.[1] Johan Bacmeister the Younger (1624–1686) was professor of medicine and town physician in Rostock from 1654.[2] Unfortunately only sparse information on his life is available. It is not known whether he had been in Copenhagen, although this is unlikely considering the war. We must rather assume that Stensen stayed with him in Rostock during his journey.

A significant member of the Hanseatic League in the Middle Ages, Rostock had lost much of its importance from the time of the Reformation. Besides the sumptuous churches of St. Peter, St. Nicolaus, Our Lady and St. Jacob, there were several old hospitals and institutions of charity.[3] The university founded in 1419 is one of the oldest in Germany and had become a stronghold of Lutheranism. It was often visited by Danish students, and there was there also St. Olaf's burse. Stensen would certainly have been recommended by Simon Paulli to all his connections in Rostock.

---

[1] <536> OPH I, 147. Backmeister (Maar).
[2] Krey, *Andenken*, 17.
[3] On Stensen's travels, see *Reisen*, esp. p. 15 ff.

© Springer-Verlag GmbH Germany, part of Springer Nature 2018
T. Kardel and P. Maquet, *Nicolaus Steno*,
https://doi.org/10.1007/978-3-662-55047-2_2

The name of Stensen does not appear in the register of the high school and it is unlikely that he stayed in the city for more than one month. The postal route westwards to Lübeck went over Wismar which, however, was in the hands of the Swedes. Stensen, accordingly, might have preferred the road over Schwerin, the town in which years later he found his death. Lübeck, the city on the Trave river with some 31 000 inhabitants in 1650, had had a strong cultural influence on Copenhagen and Denmark in the Middle Ages. Besides three large medieval churches, the cathedral and the churches of St. Mary and St. Peter, there were several ancient social institutions such as the Holy Spirit Hospital, which ought to have been of particular interest to the medical student. Stensen might then have continued to Hamburg with the strong fortified walls which had kept the horrors of the Thirty Years' War away from the city. In this city of more than 60 000 inhabitants, the student might have made several friends. He would no longer have been able to meet the most important teacher of the old cathedral school, Joachim Jungius (1587–1637) who founded a Societas Ereunetica. While he was still in Copenhagen Stensen had read the *Geographica generalis* of his disciple, Bernhard Varen. Another possible host in Hamburg would have been the professor of law, Johann Müller. In a poem Stensen as a student in Copenhagen had addressed his "most respectful regards" to the mother of the professor at the time of her decease (Fig. 1.1.4).[4] Surely Stensen, like his pupil Holger Jakobsen, visited the big old churches and above all the cathedral, as well as the spinning house and the penitentiary, which were charity institutions along the Dutch model (Fig. 1.2.1).

The route to Amsterdam which Stensen took from Hamburg was the same one that Ole Borch chose 1 year later. Borch went over Blankenese and Buxtehude to Bremen, which had hardly half the inhabitants of Hamburg but attracted many travellers because of buildings such as the Cathedral of St. Peter and the town hall with the statue of Roland. Stensen might have been reminded of the battles for the city in the Thirty Years' War in which the Danish king had lost the "archbishopric" to the Swedes. Delmenhorst, also on the way, was at the time in Danish hands. Smaller towns like Löningen and Lingen hardly retained the attention of our traveller, contrarily to the first true Dutch city, Zwolle. Here, from the 91 m high tower of the Our Lady's church, he might have had a first view over the Dutch landscape with canals on which barges were drawn by horses, a way of transport Stensen himself in time would become accustomed to. It is very likely that, like Borch, he completed the journey from Hamburg to Amsterdam in 10 days.

---

[4] *Indice*, 290.

Fig. 1.2.1 Barge drawn by a horse, Jacobæus, 16 (For a colour version of this figure, please see the colour plates at the beginning of this book.)

1.2.1 In Amsterdam

The Amsterdam coat of arms

Dutchmen consider the seventeenth century as "De gouden eeuw", their golden century. Even before the recognition of an independent Holland at the Peace of Westphalia, the Netherlands had powerfully blossomed, above all through an impressive development of navigation and trade, even during wars with England and France. While the Dutch East India Company sent their fleet to East Asia, the Dutch West India Company controlled the coasts of

Africa and South America. Amsterdam became a centre of world trade and the biggest financial market. Dutch admirals dominated the seas, Dutch merchants brought tradesmen, nature researchers and collectors on to the colonies and back again. The intellectual life blossomed, great men were leaders in art, literature and science, not to mention the technical developments as they appeared, for instance, in ship-building, with cartographers, and in the development of optical instruments.[5]

Holland was closely connected to Denmark not only as a consequence of a politico-military alliance against Sweden. There were as many portraitists and architects, tradesmen and sailors who went north as there were Danish students who attended Dutch universities or sons of Danish merchants who were trained in the trade offices of the Amstel city.[6]

Stensen arrived around March 20, 1660 in this Dutch Venice, a major city with some 200 000 inhabitants. He was admitted in the house of the anatomist Gerard Blasius on the Verversgracht by recommendation of Thomas Bartholin. It seems that Blasius invited the student to accommodation in his home.[7] At first the young traveller must have visited the beautiful city with the curious eyes of the youth. He must have strolled through the rich canal streets of the Heerengracht, Keizersgracht and of the Singels, admiring the stock exchange, the superb city hall with the chimes and the churches of the city. He certainly heard the common saying: Amsterdam stands on Norway, i.e. on Norwegian wooden palings. Like an underground wall they maintained the whole magnificence above the water. Like his later pupil and friend Holger Jakobsen he was surely interested in the social and sanitary conditions and might have seen the Rasphuys, the men's workhouse, the Spinhuys, where women worked, the Weeshuys for poor boys and girls and Het Dolhuys for mentally disabled patients.[8] The intellectual life ought to have impressed the student. He came to the Amsterdam of Rembrandt and the still vigorous 70-year-old poet Vondel who worked as librarian not far from the Athenaeum, a city enjoying a free civilian administration and government and a quite unusual freedom from an ideological point of view. Amsterdam had a

---

[5]On the Netherlands in general during this period, see *Algemene geschiedenis der Nederlanden*, vol. 7–9, 1979–1980. On science also Hackman, *The Growth of Science in the Netherlands*, and Lindeboom, G.A., *Geschiedenis van de medische Wetenschap in Nederland*.

[6]*Holland-Danmark*, I-II.

[7]<536> OPH I, 147. *Eric Jorink has indicated Stensen's lodges in Amsterdam and Leiden often staying under the same roof as Ole Borch. Jorink, Modus politicus vivendi: Nicolaus Steno and the Dutch (Swammerdam, Spinoza and other friends). In Andrault and Laerke (eds.), *Steno and the Philosophers*, 2018, 13–44.

[8]Jakobsen, 9.

strong Jewish colony and Catholicism was still widely spread in the city and in the country. The century of the Enlightenment was already unfolding.

From the very first days Stensen was deeply involved in the study of anatomy to which he must have been incited by his personal intercourse with his host. Gerard Blaes (about 1625–1692), Latinized Blasius, was born under a lucky star. His father had served as an architect under the Danish King Christian IV. Blasius was physician and, on September 4, 1660, he became extraordinary Professor at the Athenaeum, the city university, which was situated in the former St. Agnes cloister church. He was also librarian in the "Buchkammer" above the lecture-halls.[9] The city hospital, "het gasthuis", was a stately complex with buildings, courts and gardens, with departments for men and women, soldiers and sailors, for casualties, foreigners and the poor who could find refuge here for 3 days. It offered favourable opportunities of work to an anatomist. Moreover, there was an anatomy room in one of the four butcher halls of the city, installed in the former church of St. Margaret. In this building, the Collegium Medicum held its meetings and, for a long time already, dissections of corpses.[10]

Immediately after Easter, which fell on March 28, Blasius delivered a 6-day course of lectures on the anatomy of the head and brain, which Stensen attended with interest. However, he was anxious to find an opportunity for his own research and soon asked permission to do so. "As I had wanted it, alone in the study room, on April 7, I dissected first the head of a sheep which I had bought. I had luck and found a passage which, to my knowledge, had not been described by anybody. After having removed the usual membranes, I considered to undertake a dissection of the brain but decided by chance to examine the vessels which lead through the mouth. While, to this end, I examined the course of the arteries and veins by introducing a probe, I felt that the point of my knife, no longer confined between the membranes, more freely moved in a wide cavity and, as I pushed the iron further, I even heard it clink against the teeth. Amazed by this discovery, I called my host to hear his opinion. He first accused me of forcible processing, then resorted to whims of nature and finally referred to Wharton."[11] Soon after, an examination of a dog's head carried out with the utmost care confirms the discovery.

---

[9]NNBW, 7, 138 f. Vugs, 10.

[10]Van der Vijver, 3, 50 ff.

[11]<421> OPH I, 4. Thomas Wharton (1614–1673), a popular London physician and rediscoverer of the ductus named after him. In his Adenographia he had published the latest results in research on the glands. The ductus parotideus had, moreover, been discovered in 1655 by Walter Needham (c. 1631–1691), who, however, had not published until in his Disquisitio Anatomica de Formatio Foetu (Londini 1667, new ed. 2009). DNB XIV, 164 f. See also Philipsen, Ductus parotideus stenonianus.

This discovery of the *ductus Stenonianus* was to determine the career of Steno as a researcher. The excretory duct of the parotid gland even today rightly carries his name as anatomical eponym all over the world, even though the young erudite would later consider this discovery to be of less importance because of his discoveries to follow.[12] Not in Amsterdam, however.

We have little information on the work and social circles of Stensen. We know that his housemate Christian Rudnick from Bütow in Prussia soon went to Leiden with him.[13] The acquaintance of Jan Blasius, a poet and lawyer who graduated in Leiden on May 28, 1660, is marked by friendliness. In the quarrel soon to occur with his brother Gerard about the discovery of the parotid duct, in a letter to Eyssonius, Jan Blasius stood on Stensen's side, against his brother, which, however, he later seemed to regret.[14]

The lifelong friendship with Jan Swammerdam (1637–1680), so rewarding both person to person and scientifically, is most likely to have begun here in Amsterdam. Jan was destined to the service of the church by his father, a wealthy pharmacist. He first worked in his father's curiosity collection in the house at "Montelbaantoren"—one of many such collections in the Amsterdam of this time—to prepare a catalogue. He completed Latin school in 1661, to start the study of medicine in Leiden, from which he acquired a doctorate in 1667 on a reputed work on breathing. Already as a boy, Swammerdam with true passion began investigations of the animal world which would make him a famous master in comparative anatomy, particularly in the field of entomology. He became known world-wide many years after his death when his *Biblia naturae* was published. For Stensethe congenial youth must have been a precious friend from the very beginning and particularly when Swammerdam began to study in Leiden in December 1661 (Fig. 1.2.2).[15]

Stensen must soon have decided to leave Amsterdam. The lectures in chemistry by Blasius which he attended with Christian Rudnick did not impress. He calls them *vulgares et nudas operationes chymicas*.[16] The academic life in Amsterdam was still mostly at its beginning. There were neither proper examinations nor graduations. Not until 1632 had the Athenaeum been promoted to a university. Blasius was above all a studious compilator who

---

[12]<419> OPH I, 1.

[13]<470> OPH I, 62, in which Rudnick is mentioned, cf. OPH I, 237. G.Scherz says that this house-friend of Stensen in Amsterdam was, however impossible it be, the source of this information.

[14]<421> OPH I, 4–5, EP I, 134, 135, 138, 148, 149.

[15]EP I, 7 f. DSB XIII, 168–175. Schierbeek, Jan Swammerdam, *Nordström*, Swammerdamiana.

[16]<536> OPH I, 147.

Fig. 1.2.2 The Athenaeum at Oudezijds Voorburgwal is among the oldest buildings of Amsterdam University. The former cloister is still in use for doctoral defences, see "Agnietenkapel, een gebouw met een rijke historie", Universiteit van Amsterdam, 2007. Photo 2010, TK (For a colour version of this figure, please see the colour plates at the beginning of this book.)

later made himself known particularly by his *Anatome animalium* (Amsterdam 1681), a precious outline of literature about 119 animal species, in which Stensen is frequently quoted.[17]

On the other hand, the *Professor primarius* Arnold Senguerd or Senckward (1610–1667), a native Amsterdamer, seems to have offered more to the students. He must have been a kind and diligent teacher. We know of his

---

[17]Cole, 150 f.

relations to Stensen from a final scholastic disputation of the student that was held in the Athenaeum on Burgwal on July 8 (Fig. 1.2.2).[18] Its title is *De thermis* and it is the 35th of 93 preserved dissertations. It displays a growing interest of the professor of philosophy for natural science. Almost at the same time, Jan Blasius disputed on the human skeleton and Christian Rudnick on *Miscellanea*.[19]

In *De thermis, On hot springs*, after an explanation of the words, the question of the origin of heat is discussed, whether it is by friction, burning sulphur or from an underground fire. Then the solid, fluid and gaseous components of these sources are discussed, particularly their metal contents. Of the two conclusive questions, the first, whether gold particles are essentially different from water was answered affirmatively, the second, whether iron and steel are essentially different, negatively.[20]

1.2.2   Research on Glands in Leiden

Leiden coat of arms

Stensen went to Leiden light-heartedly. He was registered at the university on July 27, 1660 shortly before the beginning of the lectures.[21] We do not know

---

[18]Scherz, *Stensen's First Dissertation*, 251 f, Blasius could not preside since he did not become professor until September 1660.

[19]See the survey of dissertations in Scherz, *Stensen's First Dissertation*. Jan Blasius is listed as no. 24, Stensen as no. 35, Rudnick as no. 50.

[20]<413>. Facsimile in Scherz, *Stensen's First Dissertation*. Printed with an English translation in GP, 49 ff. Italian translation: *Disputatio physica de Thermis*, 53 ff. Cf. Pfeiffer, *Stensens erste geologische Schrift*.

[21]Bricka, 202.

where he resided, perhaps in one of the *hospitia pro perigrinis* which Jakobsen enumerates.[22] Jakobsen himself was quartered in "Poolse Sabel" in the Houtstraat. The learned city at the "Alten Rhein" which appeared like a double chain of canals around the nice houses and streets were the scenery of his scientific activity for the next 3 years. Leiden was at the time much larger than it is today and truly a metropolis of the minds. The famous university drew many students from all over the world. In the year 1660 alone, there were 444 registrations.[23] Already in the first century of its existence, 750 students came in from Denmark.[24] The town with the beautiful city-hall, the old St. Pankratius church, the historical souvenirs and artistic traditions formed a beautiful background for first-rate scientific institutions such as the university, the library, the museum and the famous botanical garden.

Judging from the *Series lectionum* which took place from September on for the winter semester, the day of a Leiden medical student was quite busy.[25] At 7 a.m. a "demonstration" began in the botanical garden. At 8 o'clock a lecture on the *Epitome anatomicum* followed, preceding a lecture on mathematics by Professor J. Golius at 9 o'clock. Apparently there was no break. At 10 o'clock followed a medical lecture and another on methods to diagnose and cure diseases. In the early afternoon Professor Sylvius stood in the chair and the anatomical exercises began, which he shared with van Horne.

In these two men Stensen found ideal teachers. Franz de le Boë (Dubois), Latinized Sylvius (1614–1672) descended from a Huguenot family. He had studied at German universities and became medicine doctor in Basel in 1637. After several years in practice in Hanau and Amsterdam, he finally acted as professor of clinical medicine in Leiden from 1658. He soon gained the adoring respect of the higher school youth.[26] Besides his handsome, imposing appearance and oratorical gifts, above all his developed inclination for the new, his delight for dissection, his sharp gift of observation explained this enthusiasm. He claimed: *Nihil in medicina vel naturalium rerum cognitione admittendum pro vero, nisi quod verum esse ostenderit aut confirmaverit per sensus externos experientia.* (In medicine or in the natural sciences that nothing be considered as being true unless the experience has shown through the external senses or confirmed that it is true.) His daily teaching at the bedside

---

[22]Jakobsen, 10.

[23]Molhuysen, *Bronnen*, 3, 164.

[24]Cf. Bricka, 104.

[25]Molhuysen, *Bronnen*, 140.

[26]EP I, 6, NNBW 8, 1290–1294, DSB IV, 198 f, Baumann, François de le Boë Sylvius, Gotfredsen, *Medicinens Historie*, 216 ff.

of the patient, a new way of lecturing, made him popular. Further, Sylvius was a successful researcher, famous for his investigations of the brain (fissura cerebri lateralis Sylvii[27]), his classification of the glands in conglobate and conglomerate and his tubercle theory. Sylvius also ought to have interested the students as a leader of the iatrochemical school of his time. Already in Copenhagen, Stensen had shown an interest for this orientation. He saw in the professor an erudite man who, free from authoritative belief and even critical of Descartes, wanted to provide medicine with a scientific foundation based on anatomy, chemistry and practical observations. In physiology he followed the Helmont theory of fermentation and, for the rest, was of the opinion that the life processes were identical with the processes in a test tube and that base or acid reactions caused health or disease. Anyway, during his Leiden time, Stensen assiduously took part in the anatomical exercises of the professor (Fig. 1.2.3).[28]

Together with his disciple Sylvius stated the reality of the *ductus Stenonianus* in Leiden's *Theatrum Anatomicum* close to the hospital (Fig. 1.2.2).[29] Together they demonstrated the parotid glands of an ill woman.[30] The professor readily attended the dissections of Stensen, left him cadavers and invited others to witness the research of Stensen, such as the Amsterdam physician W. Piso.[31]

Stensen had another outstanding professor in Johannes van Horne (1621–1670) , the son of an Amsterdam patrician, who had taught anatomy, surgery and botany in Leiden since 1653 after having discovered and described the *Ductus thoracicus* in man in 1652 independently of Pecquet.[32] He gladly invited Stensen and presided his disputation.[33]

Leiden must have offered the young anatomist unusually good working conditions for the time and Stensen used them well, as one can ascertain in Ole Borch's diary. As a reward for his courageous behaviour in the defence of Copenhagen, Borch had obtained an extraordinary professorship in botany and chemistry in Copenhagen. Before taking office, he sat out on journey that lasted 6 full years during which he was in Amsterdam from November 1660

---

[27]*Beukers, *The Sylvian Fissure*, 53.

[28]Cf. OPH I, 22, 37. Robert Sibbald (1641–1722) studied anatomy and surgery with van Horne, botany with Adolph Vorstius and medicine with Sylvius in Leiden from the spring in 1660 until September 1661. Sibbald watched 23 dissections (Sibbald, 56 f.).

[29]<421> OPH I, 5. <534> OPH I, 146.

[30]<439> OPH I, 27.

[31]<470> OPH I, 62, <501> OPH I, 101 f.

[32]NNBW VII, 624.

[33]OPH I, 5, 55, 62, 102. Facsimile of the title pages NSVO 13, 19.

Fig. 1.2.3 Leiden's anatomical theatre, drawing by J.C. Woudanus, W. Swanenburgh (etching), A. Cloucq (publ.) 1610, reprint Universiteitsbibliotheek Leiden, 1975

to January 1661, and in Leiden from January 1661 on. Borch noticed that Stensen introduced the very first day of the year 1661 by dissecting a male corpse. On January 5–7, a live dog was dissected, on January 14, a pregnant dog. On January 27, the corpse of a woman, Helene, was available, on February 7, that of a Janicke Jansen who had died from syphilis. On February 24, the anatomy theatre obtained the beheaded victim of a murder for demonstration.[34] Most of the year was spent like that. For the times it was a flourishing activity, fundamentally different from that in Copenhagen.

Here was the richest material for anatomical research. Stensen was actually forced by a feud to concentrate on the investigation of the glandular system during the first 2 years in Leiden. When he had arrived, he had shown Sylvius and van Horne his discovery, the duct from the parotid gland and had pointed it out to the former in the dissection of a man in the hospital; van Horne publicly attributed the discovery to Stensen in the large *theatrum*

---

[34]On Borch's travels, see BOD = *Olai Borrichii Itinerarium 1660–1665*, ed. by H.D. Schepelern, 1983. BOD I, xii, ff. Dissections I, 61, 18, 71 f, 74 ff. *See also Jorink, 2018.

anatomicum and gave it his name.[35] That did not please Stensen's Amsterdam Professor Blasius. As Borch reports to Bartholin on March 20, Blasius, in private letters to different people, had claimed the discovery for himself and to the whole world in his *Medicina generalis*, a small handbook which was published at that time.[36] On April 22, in a letter to Bartholin, Stensen described the circumstances of the discovery and explained that he had to claim his right of priority for his own reputation. Blasius had called him a plagiarist. He only wanted to prove his claims through a disputation on which he was already working and for which he was preparing drawings on the advice of Bartholin.[37] Since at the end of May he still discovered several rows of glands so far unknown, in the oral cavity of an ox head, he also published the new discoveries together with the mentioned findings in the disputation under the title *Nicolai Stenonis de glandulis oris, et novis earundeem vasis observationes anatomicae.* The disputation was defended on July 6 and 9 and presided by van Horne.[38] It was reprinted in Leiden in 1662 by Jacob Chouër as the first of four treatises in *Nicolai Stenonis de glandulis oris et novis earundem vasis observationes anatomicae.*

The discussion is dedicated to Stensen's preceptor Thomas Bartholin and to the commander of Nyborg Castle, Otto Krag, as well as to the archbishop Hans Svane. They are called his Maecenas and promoters. It revealed to the world the 23-year-old young man as a mature researcher, not only thanks to the number of new discoveries which represented the main contents of the document but also because Professor Blasius had proven with his own words that he did not yet know the *Ductus Stenonianus* and, as far as the glands were concerned, he only plagiarized Wharton. Stensen recognized that the conventional name *Parotis* actually designates two different "glands", one of which pours saliva through the duct which he discovered, whereas the other like usual lymph nodes without an excretory duct belongs to the system of lymphatic vessels. Stensen, with Sylvius, placed the former among the *conglomerate glands* and the latter among the *conglobate glands.* Of those *conglomeratae* or proper glands he then found according to his own words: "Apart from the maxillaries and the tonsils which are described by the famous Wharton, I have observed others below the ears in the region of the cheeks

---

[35] <421> OPH I, 5.

[36] BEM III, no. 87, 375 f. EP I, 3.

[37] <419> OPH I, 3 ff, *Pionier*, 50 f. NSVO, 1 ff. Cf. Gosch II, 1, 150 ff.

[38] <425> OPH I, 11 ff. Cf. note p. 224. NSVO, 11 ff. On the research on lymphatics and glands before Stensen see f.i. Vugs, 110 f, Moe, *When Steno brought new esteem to glands.*

(*buccales*), under the tongue (*sublinguales*) and in the palate (*palatinae*)."[39] He described these comprehensively as far as situation, shape and colour and corrected a series of erroneous conceptions (Figs. 1.2.4 and 1.2.5).

However, he also achieved basic investigations on the function of these glands and, as an important result he determined "that the saliva consists of the liquid secreted by the glands of the mouth from arterial blood. This liquid is brought into the mouth by the excretory lymph vessels with the help of the *spiritus animales* (nerves) which flow in the glands and the distant muscles. However, the round or conglobate glands in the neighbourhood of the former pour back into the veins the lymph received from the external parts. In the veins, they become mixed with the blood flowing back to the heart."[40] This completely contradicted the theory of Bils and other contemporaries, that is that all aqueous secretions and liquids came from the thoracic duct. Because of his discovery, Stensen became involved in a new controversy, not with Bils himself, but with A. Deusing (1612–1666), a learned professor in practical anatomy but with little experience, in Groningen, according to whom saliva could also be secreted in another way, for example, in exaggerated secretion of saliva, the aqueous liquid of the brain would flow directly through the *os sphenoidale* into the cavity of the mouth in which, influenced by the nerves with the help of the glands it would separate the saliva.[41] As opposed to the Middelburg physician Anton Everaerts († 1679) who endorsed the point of view of Bils concerning the function of the lymph and pretended that the lymph was led directly to the *mammae* there to become changed into milk, Stensen decisively explained that *mammae* are glands and that milk appears in these special glands like the secretion of saliva as a filtration from arterial blood.[42]

The remarks on the method are still more explicit as they appear in the Foreword: "Reasoning should consider the observation and evaluate the matter from every point of view as far as possible so that a right presentation of the matter stamps itself upon the mind."[43] Here observation and conclusion meet in a beautiful harmony, and Stensen condemns the frequent demonstration through analogies. That was actually a reminder of the book of Athanasius Kircher which he had read in.

---

[39]<432> OPH I, 20.

[40]<462> OPH I, 49. §55. *Stensen shows concord with his teacher on the lymphatic vessels by reprinting a section from chapter VIII of Bartholin's *Vasa lymphatica*, see <481> The famous Bartholin himself saw this ... OPH I,72.

[41]NNBW, 383 f. Vugs, 13.

[42]Vugs, 120 f.

[43]<428> OPH I, 15.

DISPUTATIO ANATOMICA.
DE
Glandulis Oris, & nuper obſer-
vatis inde prodeuntibus Vaſis

PRIMA.

QVAM,
Divinâ Favente Gratiâ,
SUB PRÆSIDIO
Viri Clariſſimi
D. JOHANNIS van HORNE, Medicin. Doct.
Anatomiæ & Chirurgiæ Profeſſoris Celeberrimi,
Placido eruditorum examini ſubjicit
NICOLAUS STENONIS, Hafniâ-Danus.
Ad-diem 6. Iulii, loco horiſque ſolitis.

LUGDUNI BATAVORUM,
Apud JOHANNEM ELSEVIRIUM,
Academ. Typograph.
cIɔ Iɔc LXI.

Fig. 1.2.4 Stensen's anatomical dissertation on his discoveries on glands of the mouth was defended in Leiden on July 6, 1661. Note that this was after the death of the renowned Johannes Elsevier (1622–June 8, 1661) who printed the book, 4°, REX

NICOLAI STENONIS
OBSERVATIONES ANATOMICÆ,

QUIBUS

Varia Oris, Oculorum, & Narium
Vafa defcribuntur, novique falivæ, lacrymarum & muci fontes deteguntur,

ET

Novum Nobiliſſimi

B   I   L   S   I   I

De lymphæ motu & ufu
commentum

Examinatur & rejicitur.

LVGDVNI BATAVORVM
Apud JACOBUM CHOUËR
cIↄ Iↄ c LXII.

Fig. 1.2.5   Four dissertations on glands, 8°, 1662, facsimile ed. 1951

Copenhagen when he had criticized the thinking process which, amazed by the attractive force of the magnet, attempted at explaining everything by magnetism.[44] His attitude towards ancient times is harmonically balanced and as far from uncritical glorification as from deprecatory undervaluing. A homage immediately follows the evidence that the ancients knew very little of the glands: "However, I would not like that it be understood as if I wanted to belittle the credit of the ancients whom I have always respected and honoured since I am firmly convinced that so far nothing has been immediately discovered completely. These ignited the light but we should take care that it continues to burn and over the course of time always more clearly shines."[45] Stensen saw in Hippocrates the eminent and venerable (divine) genius.[46]

The mind of Stensen, who was deeply religious, turning to God already reveals itself in this work. The dedication begins with a great amazement about the work of the Creator. How admirable that a man in his little skull can scurry through the infinite universe and its different parts: "Soon he will make his way to the stars to depict to us the firm order of the fix star, the completely ascertained deviations of the planets and the totally erratic rovings of the comets. In the next moment he turns back to stroll about the space to paint the diverse and pretty colours or the imposing shapes of a volcanic eruption which can be seen again and again in these regions. When he comes down to the earth, he will portray the different, finely ordained works of the nature and the reproduction of art which are hardly second of these. Finally he will penetrate the inside of the earth and discover the hidden mysteries of the minerals. All these representations respond to a sign as if the macrocosmos laid hidden in the microcosmos."[47] Stensen recognizes with Socrates "in the structure of the living the work of art of a wise Creator who loves every living thing since yes all their organs and their structure reveal that they are created with deliberation." Painfully he perceives that many people conceive natural phenomena as miraculous signs only and despise the works of the world architect.[48]

The strong scientific concentration, however, did not make Stensen an eremite. We rather find him in interaction with friends and erudite men, above all corresponding with his fellow countrymen. One of the witnesses of

---

[44] <430> OPH I, 18.

[45] <431> OPH I, 19.

[46] <433> OPH I, 21.

[47] <426> OPH I, 13.

[48] <429> OPH I, 17.

his first successes was Matthias Jakobsen (1637–1688), medical student and son of the Aarhus bishop Jakob Matthiasen and nephew of Thomas Bartholin. Jakobsen who 1 year later, in 1664, would obtain the professorate in Copenhagen, which ought to have been given to his by far greater and more genial fellow countryman Stensen.[49] Another time we see Henrik a Møinichen (1631–1709) in Stensen's company.[50] He distinguished himself as anatomist and had also stayed in Leiden in 1661.

Besides the letters to his preceptor, we have only a poem of well wishes for Simon, son of Professor Paulli, written on July 26, 1661 in Leiden, to show Stensen's connection with his home town.[51] Of his correspondence with friends and relatives at home, only a reference to a letter to his fellow student Niels Krag exists.[52]

We know of a brilliant feast of promotion which Stensen was invited to. This was on March 15, 1661 after a solemn disputation of an Achatius Hager from Hamburg who defended a thesis on fevers where Borch was the opponent. Anybody who had a name and a status in Leiden seems to have been present: the professors of medicine Vorstius, Sylvius, van Horne, the mathematician Origanus and the physician Christian Marckgravius. Further, Joachim Merian from Frankfurt a.M., Wilhelm Schröder, a German from Franken, and Christian Lagermann from Hamburg, perhaps a relative of the lawyer Lukas Lagermann and if so of the rector J.F. Gronow. Heinrich Meibom (1638–1670) from Lübeck was also there. He distinguished himself by the discovery of the tarsal glands of the eyelids.[53] All in all this gives a good picture of the learned society in Leiden.[54]

Above all we often find Stensen at the side of his former teacher Ole Borch. Stensen spent Christmas 1660 in Amsterdam in the company of Borch and Walgesten.[55] On March 2, 1661, the friends visited the Rotterdam anatomist Ludwig Bils for whom a new method of embalming corpses had gained

---

[49]FL IV, 225 f. EP I, 151. He wrote a homage in the second ed. of *De glandulis* … , Leiden 1662, see OPH I, 225 f.

[50]<457> OPH I, 44, Ingerslev I, 508 f.

[51]*Indice* 290 f.

[52]EP I, 175 f.

[53]*Heinrich Meibom, later professor in Helmstedt, and Stensen remained in respectful contact as shown by their correspondence between 1669–1673 recovered by Niels W. Bruun, *Fund og Forskning*, 2008.

[54]BOD I, 51. It appears from the following remark that Stensen also met other fellows: *I was a fellow student with Steno, who became famous afterwards for his wrytings. He dissected in my chamber sometymes, and showed me there the ductus salivalis superior, he had discovered.* (Sibbald, 57).

[55]BOD I, 38 ff, on Walgesten <55>.

150 000 guldens. Bils pretended to be able to dissect without any blood-shed.[56] Stensen depicted him as a man whose "dexterous hands were pointed out by his dissections and whose character appeared in his conversations and writing."[57] Besides Borch and Stensen, two Danish students went to "De Engelsche Court" where Bils lived, and the physician Angilles introduced them. Bils pretended grandiloquently that his principles could renew the entire anatomy. In the presence of the friends he dissected a dog and opened it in one cut. However, this resulted in much bloodshed, as Stensen and Borch noticed, "because Bils had injured the *lobum hepatis*." After this visit, the friends strolled via Overschie to Schiedam where they spent the night at Angilles' home. The next day they went to Delft and looked over the "Prinsenhof" in which Wilhelm van Oranje was murdered in 1584, and visited his sumptuous tomb in the Nieuwe Kerk. From Delft they continued to Den Haag where, among others, they visited a Danish resident Charisius. On September 17, Stensen was in Amsterdam with Borch and, on the Rokin, not far from the stock exchange, the friends enjoyed "that Turkish trank coffij" in a restaurant. Then together with Gottfried Gersdorf they discussed chemical questions. The next day Borch visited different erudite men, attended a Baptist service, then a marionette theatre and of course Borri's. Not until October 7 did he return to Leiden.

Outside Leiden, Stensen must primarily have become acquainted with Amsterdam and its learned world but we have had no clear indication of the how and when of the first encounters. He came in relation with Paul Barbette, a very popular physician whose writings were much read and also translated[58] and who very highly appraised the young researcher. "Your Stensen", the physician writes to Thomas Bartholin on July 30, 1661, "means much to us here in Amsterdam. We admire his erudition, his zeal and his genius. One thing only is wanting in him: more restraint towards Blasius. But this young mind promises with his vehemence greater things when he has first cooled down."[59] His acquaintance with Willem Piso (1611–1678) was also very honourable and stimulating.[60] Born in Leiden and a graduated Dr. Med., Piso had lived in Amsterdam since 1648 as a respected physician.

---

[56]NNBW IV, 150 f, Vugs, 14 f.

[57]<453> OPH I, 42.

[58]EP I, 182, DBF V, 286. See *Stensen's letter to Barbette* dated June 22, 1664, st. V., <593> OPH I, 211 ff.

[59]BEM III, 196 f.

[60]NNBW 9, 805 f. Houtzager, *Willem Piso*. <514> OPH I, 118. See *Stensen's letter to Piso* dated April 24, 1664. <579> OPH I, 195 ff.

He had been in Brazil for years as the personal physician of Graf Johann Mauritius von Nassau. His two works on Brazilian medicine and the natural history of the Indian countries were certainly known by Stensen. Years later, Stensen showed his friend Viviani a publication of Nicolaus Witsen on ship-building, so he is very likely to have met and visited this mayor.[61] We also know of a connection with Johan Wandelman, Catholic parson (since 1663) of the St. Catherina church "outside the Utrechter Tor."[62]

The first and second semesters in Leiden must have slipped by quickly, mostly with scientific work to which even an Easter holiday—in 1661 Easter fell on April 17—was sacrificed.

1.2.3 A Holiday Journey to North Holland

Holland proper, the province with the two great cities of enterprise and culture, Amsterdam and Rotterdam, and with Den Haag, still "the prettiest village in Europe", lay within reach of a student in Leiden. In the summer 1661, a group of friends including Borch and Stensen, wanted to have a look around to the neighbouring provinces of Friesland, Drenthe, Overijssel and Gelderland. From fertile Dutch deep country with its world cities, the journey went to isolated areas with sand stretches, moraine walls from the ice age and moors, with various dialects spoken and proper historical remembrances (Fig. 1.2.6).

On the anniversary of Stensen's registration in Leiden, on Wednesday July 27, 1661, the company, to which a Franciscus from Silesia and a Uromantus from England belonged, proceeded on foot from Leiden to Amsterdam.[63] From other references we companions: know the names of two of the Danish Bartolus Herland, registered in Leiden 1658[64] and Jørgen Hasebard (†1670), son of the vicar of the church of Our Lady in Copenhagen and a university friend of Stensen.[65]

These friends spent a whole week in Amsterdam because of an illness of Walgesten and Hasebard. They visited learned people or looked at the town.

---

[61]EP I, 29. On Nicolaus Witsen see NNBW IV, 1473 ff. Cf. Reisen, 32 ff.

[62]Van de Pas, 21. Neercassel mentions Wandelman as *Stenoni amicissimus*, HARA, Neercassel-Korrespondenz. Neercassel to Terhoente, 5.3.1681.

[63]Ole Borch recorded the most important impressions from this trip in his diary (BOD I, 184 ff), only briefly mentioned by Stensen (EP I, 140), cf. Reisen, 32 ff.

[64]DBL 3rd ed. 6, 314.

[65]EP I, 143.

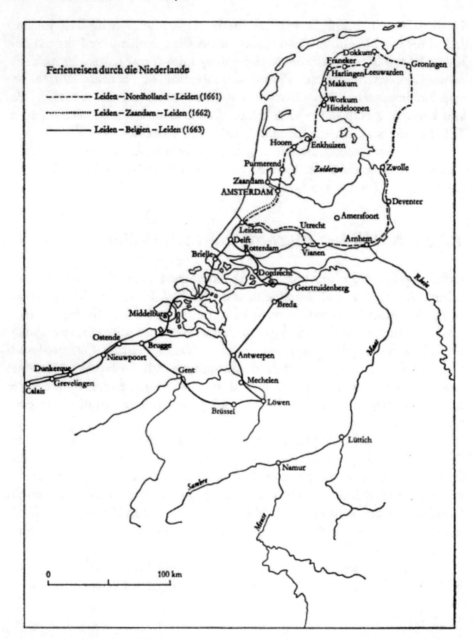

Fig. 1.2.6 Vacation in Holland, Scherz I, 79

Borch felt attracted by the chemists and visited the German Johann Glauber (1604–1670), one of the most competent practical chemists of the seventeenth century. Glauber lived in Amsterdam were his collected works were published in seven volumes as late as 1661.[66] Borch was already then an associate of the gifted Italian Giuseppe Francesco Borri who had arrived in the Amstel city during his far peregrinations which ultimately would lead him into the jail of the Inquisition in Rome.[67] Initial scepticism towards the knowledge and character of Borri soon developed into open admiration. Borch saw in him a noble man, a genius who explored the mysteries of chemistry and he had great confidence in the medical treatments of the Italian. Never did the open admission by Borri of God and nature being one and the same for him, or that the power of Christ could be explained by chemical forces, seem to have challenged the orthodox Borch, whereas the world around had already passably seen through the quack. Just then the Amsterdam goldsmiths had set a prize on the seizing of Borri who seems to have avoided the town for a while after having coaxed jewels and valuables out of these people.

On the Egelantiersgracht Stensen visited the well-known pedagogic writer and bigot Johann Amos Comenius (1592–1670) who so warmly fought for the popular school, the mother tongue and self-efficiency of children. Comenius showed the guest his books—he had written some 140 of them. Borch found his writings somewhat confusing but he praised the common sense of his host.[68] Comenius complained about the many letters which he received from all over the world, expected, with his chiliastic ideas of the century, the enlightenment which he held for closely approaching. A remarkable gleam which had appeared in the preceding night over the houses and narrow streets of Amsterdam, he explained as a miracle that confirmed his ideas.

For many visits, the presence of Stensen is expressively attested as when they met Major General Frederik von Buchwald who came back from the sources of Spa, or the Duke von Norburg and his aristocrats who came back home from Italy where the Duke had studied musical theory with the all-round talent Pater Athanasius Kircher. On August 2, they met 27-year-old Peter Schultz, probably a relative of Stensen, who had come to Holland a month before to learn political science in Leiden. On the same day the only attested meeting with Peder Schumacher took place. He was soon to become the Danish Lord Chancellor, named Griffenfeld, and was on his return to Copenhagen after long travels abroad. On August 3, Borch, Stensen and

---

[66]BOD I, 184. DSB V, 419 ff. Stillman, 386 ff. Partington 2, 341 ff.BOD I, 184. DSB V, 419 ff. Stillman, 386 ff. Partington 2, 341 ff.

[67]BOD I, 186 f. DBI 13, 4 ff. On Borri and Denmark, see Veibel, esp. 81 ff.

[68]BOD I, 185. Göttler, 128 ff. Thijsen-Schroute §393, 613 ff.

Hasebard began the proper summer journey. From Turm in the Y-Bay they went over the Sound to Buiksloot and spent the night in Purmerend. After a visit to the castle, the next day they walked through the blossoming landscape which earlier had been under water but now enlightened the eye with pretty flowers and neat small villages. They reached Hoorn, the birth town of Willem Schoutens who in 1616 had given this name to the most southern point of America and the harbour of which was then a sight, with some 14 naval vessels.

Through a landscape in which dikes and palings reminded the fight of men against the power of the sea, they came to Enkhuizen, a blossoming town which could send yearly 400 boats to catch herrings. Here also battle ships brought the sea power of Holland to the conscience of the tourists. Downtown they looked for the dyke inspector Stenberger who welcomed them and showed them a large collection of rarities. A ferry brought them to the narrowest part of the Zuiderzee and they strolled through Hinderloopen, Workum and Makkum to Harlingen which had been engulfed completely by the sea in the Middle Ages. Now they found a well fortified town, an out-standing harbour, and high walls with archaic guns. Ten naval vessels lay at anchor. The friends stayed at the "De vergülte Pfaw".

They reached Franeker, a well-known university town since 1585, on August 6. The Stock Exchange with 7 tables for 50 students hurt Borch's nose by the smell of the food. In contrast, the auditoria and the consistory pleased him. He looked with interest at the university church where a graduation was being held. In the *Theatrum Anatomicum* they found three complete skeletons and in the library many old books chained to the cup-boards. Their hostel "De Falch" gave the impression of an old cloister.

A 3-mile walk brought the travellers to the old capital city of Friesland, Leeuwarden. The tongue of the proudest and most freedom-loving Dutch stock sounded strange to them. They heard a French preacher, visited the painter Yelstein from Utrecht and visited the church of the town.

A walk of 3 miles brought the friends to Dokkum where the apostle of Friesen, Bonifacius, was murdered. The travellers visited the city hall and the school of the small town and, after dinner, took the night boat pulled by horse to Groningen where they staid in "Te Torlast" at the market place. In this the most important trade city of North Holland they visited the uni-versity, library and church, the simple *Theatrum Anatomicum*, the city hall and the walls. They met Eysonius and spoke with him of *De foetu Missipontano*. However, Stensen was surely disappointed as he did not see Deusing. Instead of the professor, he soon saw his script *Vindiciae hepatis redivivi*, against which he would direct his *Responsio* on November 28, 1661.[69]

From Groningen, the journey continued through the fairly barren and poor Drenteland to Zwolle. They spent the night 3 miles before the city in Drenten at a village inn. The next morning they visited the city of Zwolle with its gymnasium and the library, the old volumes of which caused the admiration of Borch, and then went on to Deventer, the university town where Johann Georg Greve (Graevius), the famous humanist, at the time was the rector.[70] The next year he became professor in Utrecht. Eight years later, on a quick journey through Utrecht, Stensen sent two letters to Greve, which presuppose friendly relations. So perhaps, they had met for the first time on this holiday travel.[71]

The next day, August 11, the journey went on over the flat wasteland of Gelderland. They left Zutphen on their left and, after a walk of 4 miles, reached Arnhem the big chimes of which deafened the travellers whereas their eyes enjoyed the charming situation. In a comfortable inn, "De goude arent", they ate together with Captain Suartzenburg, his daughter and the mayor of Bergen op Zoom. The innkeeper claimed that there was a gold mine under his house but Borch was more interested in a picture that represented a young woman riding an old man who had a stick entangled in his beard, or in the inscription on the window: "woe to the good when the wicked is the law-maker". From the window in their room they could see the Rhine bridge, which was built on 21 cutters. The next day, they passed it and, during a 7-h walk, they were subjected to an awful storm. The road went along gallows on which, the year before, a man had been hanged who, Borch tells us, changed to a werewolf. Over the Rhine and along Wijck bij Duurstede they came to Fartium (Waardeburg?) in the evening, where they spent the night in the "Schwan".

The next day they walked through the lowlands along the Rhine to Vianen, visited the objects of interest in the Earls' castle, crossed the Rhine at lunchtime and took the boat to Utrecht, the city where the seven provinces under Wilhelm the Taciturn had concluded the Union bound. Here stands the splendid church of St.-Martin, the nave of which crumbled some years later so that the choir and the transept have since then been separated from the tower. They climbed the 456 steps to the top of the tower and enjoyed the broad view over the country. The university founded in 1636 of course

---

[69]EP I, 140, cf. OP I, 58 ff.

[70]BOD I, 196, NNBW 4, 669 f.

[71]EP I, 213–215. See in Chap. 1.7.

awakened their particular interest but the churches of Our Lady and Catherina were visited too. In the library they spent some time over the books and Borch noticed here the sentence of the Genesis: "She will crush your head down". Finally they went back to the inn where a Dutch society, as Borch wrote, made itself noticed by a particularly rude language.

On August 14, the friends came back to Leiden in good shape after a walk of 8 miles through forest areas. Only Jørgen Hasebard came back ill but soon recovered under the care of Borch. On the whole holiday journey of 500–600 km we have no record of Stensen. We know about this journey through the diary of Borch with its precise indications about the country and people, about the course of the journey and all the inns visited, and which conveys a very lively picture of such a walk in the seventeenth century.

1.2.4   Two Fruitful Years of Work

After his return from the holiday journey in North Holland, Stensen turned to scientific work, soon spurred by the controversy with Blasius which now fully blazed. After the disputation on June 6 and 9, 1661, in which Stensen, in front of the learned world had refuted the claim of Blasius, now professor in Amsterdam, on the discovery of the *ductus parotideus*, the latter, on July 16 wrote a long letter to his friend Thomas Bartholin in Copenhagen. Here, Blasius accused his former disciple of the blackest ingratitude because he had abused his trust and published this discovery as his own.[72] Blasius admitted not to have demonstrated the duct to Stensen in the lectures due to the lack of time but to have done so on other occasions. His attention to the claim of Stensen having been aroused by the winter lectures of van Horne had made him protest in several letters as well as in the Foreword of his *Medicina generalis* but without success. Deeply offended by the remark of Stensen of not knowing facts rightly, he attempted to refute it by a reference to his disputation *De triplici coctione* of July 8, 1661. Blasius also complained about being misnamed Johannes, a name that Stensen, this *malicious animus*, had used twice instead of his proper name Gerard.

The answer of Bartholin in September was conciliatory and reserved.[73] However, Stensen produced his next great reply in the beginning of 1662 by the publication of his *Observationes anatomicae*.[74] In four parts it brought a

---

[72]BEM III, no. 43, 158 ff.
[73]BEM III, no. 44, 184 ff.
[74]OPH, no. II, IV, V, and VI (Fig. 1.2.5).

re-print of the earlier disputation on glands of the mouth with a correction of
the Christian name of Blasius, with the addition of the § 41 on the course of
the lymphatic vessels and their junction into the venous system, and a quite
new § 52 with the investigations on the course of the milk juice together. As
Part 4 it brings a reply to the formerly mentioned attack by Deusing, the
Response to vindications of rehabilitation of the liver a rejoinder on Deusing's
Vindicias hepatis redivivi of July 1661. It also contained two dissertations on
the lachrymal glands and the glands of the nose, new discoveries with which
he stressed his priority claim to the discovery in Amsterdam with the utmost
force. We shall come back to this but here we tackle the conclusion of the
debate immediately.

Blasius saw his reputation extremely compromised and let three of his
leading disciples (Keyser, M.D., Dr. Med. Leonaertz, and a student in theology, Croese) arrange for explanations (of April 22 and May 21, 1662)
according to which the professor would have shown them the disputed duct
before the arrival of Stensen. Blasius himself sent these and a copy of his letter
of July 16 to Bartholin and to a young friend, Nicolaus Hoboken, who had
just become professor in Utrecht. Ostensibly on the request of the latter, he
says in an accompanying script, that he himself had found Stensen *refutatione
omnino indignum* and had privately corrected the matter vis-à-vis Bartholin.
Hoboken published the whole in the shape of a letter to Blasius under the
title *Novus ductus salivalis Blasianus in lucem protractus*. Not satisfied with this,
he added to the script an extremely hateful note in an anagram in the style of
the time (like Matthias Jakobsen had accompanied the disputation of Stensen
formed as an anagram). By reading the name Stenonis backwards, Hoboken
formed the sentence: *Sinon est*, thus making reference to the traitor through
whom Troja fell to the Greek.[75]

Stensen felt deeply insulted. He replied at the beginning of 1663, in the
Prologue of the apologia, a folio-page which is today extremely rare, formulated
as a lawsuit in which *Blasius accusator, ego reus, and Hoboken judex*.[76] First a
choice of invectives of the plaintiff, an impressive file to which one truly had
only to add: "What a list of titles! Those are the honourable words of Blasius!
Who would have expected these from a professor, or from any respectable
gentleman, in a lawsuit not yet tried?"[77] Some quotations of Hoboken and
the Sinon anagram were also ironically dispatched: "Perceive how free of any

---

[75]Gosch II, 1, 157.
[76]<533> OPH I, 143 ff.
[77]<534> OPH I, 145.

bias the mind of the judge."[78] Then followed a short description of the history of the discovery: Blasius had had no idea of the finding when he was called in, neither had Sylvius and van Horne previously seen or heard of it when he showed them the duct 7 months after its discovery. Blasius may well have then claimed the honour of the discovery in private letters and in the *Medicina generalis* but he never demonstrated it. Since Blasius in his book has determined neither the origin of the duct the discovery of which he claims, nor its end, since he even assigns the gland from which it comes another unworthy function, he could neither know it nor have shown it to anybody else, even less have discovered it. Stensen ironically mentions the "great" good deeds to him of which Blasius boasts, that he actually was pushed to be accommodated by him by intrigue and, except for some chemical futilities, he had learned nothing from him. By the words of the *Medicina generalis* it was proven that Blasius could not have seen the duct and possessed an incomplete knowledge of it. If he had seen something, it was certainly not the Stensen duct. Here the author applies the method which he readily uses elsewhere. He confronts the statements of both parties and lets the reader himself recognize the contradiction. Stensen proposes to prove the falsity of the assertions by a demonstration to which he invites the professor. Sarcastically he comments on an offer of Blasius to strengthen by an oath his claim that he had demonstrated the *ductus parotideus* before the arrival of Stensen. So it is to be certified by an oath that Blasius should have demonstrated something that was never demonstrated by Blasius nor by anybody else because it simply does not exist. With his apology Stensen hopes to have unveiled the true Blasius. "Indeed, I will not make him odious by any bitterness of epithets. I content myself with writing down the history based on the facts so that it appears to everybody how worthy are for me the lustre of the Athenaeum of Amsterdam, the dignity of a professorial person, and above all my love for modesty, so that, although he showered me with abuse, I will not return the slightest on him."[79] One is amazed again and again by the assurance and considerable calm with which the student opposed a professor and radically refuted him.

Now this controversy seemed essentially at an end. Blasius had no desire to continue. The lines written by Stensen suggest deep irritation but he remained factual and no word had to be retrieved. The statement of Hoboken in his letter of June 26, 1663 to Thomas Bartholin[80] that Stensen had visited

---

[78]<534> OPH I, 146.
[79]<542> OPH I, 153.
[80]EP I, 5, Gosch II, 1, 161.

Blasius in Amsterdam and had been modest, may well be true. The next assertion, however, that he should have put down or retrieved the *Prologue of the apologia* sounds very unlikely. Anyway, Blasius undertook no further attacks but actually came to acknowledge the anatomical discoveries of Stensen. Thus in Blasius' compilatory work the *Anatome animalium* Stensen is among the most frequently quoted authors.[81]

This controversy did not hinder the essential and rich research work in which Stensen was deeply involved during these years. This is clearly revealed by a letter of September 12, 1661 to Thomas Bartholin on "Various new observations on eyes and nose".[82] He mentions here only briefly his discoveries on the lachrymal apparatus and announces the immediate publication of a script on the glands of the eyes. This appeared in 1662 as the third part of the above-mentioned *Observationes anatomicae* under the sub-title *Anatomical Observations on the Glands of the Eyes*.[83]

The dedication is very interesting. The publication is dedicated to three professors of Copenhagen and three of Leiden to whom Stensen obviously felt particularly obliged. Additional labels display his motives. He names as his "highly deserving teacher" in Copenhagen the general physician Dr. Simon Paulli, the mathematician Dr. Jørgen Eilersen, the polyhistor and nature researcher Dr. Ole Borch. The Dutch professors particularly bound to him in Leiden he names as the clinician Dr. Franz de le Boë, Sylvius, the anatomist Dr. Johannes van Horne and the mathematician and teacher of Arabic, Dr. Jacob Golius.

In the Preface, Stensen calls his designation of the discovered vessels "new". They actually had always existed and may have been recognized by others previously. With regard to the word of Stevin of a golden age of science, he says: "I have no doubt that, if we possessed all the scripts of the ancients, we could learn from them without sweat and blood many things which we admire as new and which have cost us much work." Yet to him, they were new and anyway he did not rely on others.

Stensen then pays homage to the "most brilliant of all mechanics" who, like men grease the axles or rollers of the shipyard under a ship, has done the same since the beginning of time of the livings. "But in the self-maintaining body of animals all that proceeds more skillfully or, I should say, more divinely. The parts are arranged so that the liquid concealed in the vicinity as

---

[81]Gosch II, 1, 164.
[82]<465> OPH I, 55 ff = EP I, 140 ff.
[83]<483> OPH I, 75 ff.

in a storeroom is secreted more scarcely or more copiously depending on the more or less intensive usage, without our noticing it. Then, after it has accomplished its task, it is carried once more to the parts through other paths."[84] This is how it is with the movements in the mouth, when swallowing food, so that the inside of the intestine as well as other organs are coated with viscous slime.

Then follows a description of the lachrymal glands (Borch had written about this work as early as on March 3, 1661),[85] how he observes them in animals, particularly in calves and which glands Wharton had called *innominata*. They lie in the lateral angle of the eye. Narrow pathways led out from these to the inner slime membrane in the eyelid as discovered on November 11, 1660. At the other glands in the medial angle of the eye, on June 11, 1661, he saw the two exits that lead out the lymph. The liquid of the tears which comes from the glands flows to the nose through the lachrymal points. Stensen then enumerates several hypothetical explanations given for the overflow of tears illustrating, "how little even the force of shrewd intelligence can do if it is not based on sufficient research." He finally concludes: "I mean, therefore, that tears are nothing else than a liquid which, flowing profusely, is aimed at keeping the eyes moist and that one should not look for them as abducted from other, hidden or distant ways, since the vessels which one can observe in glands of the eye show obvious and nearby ways."[86] Further it is explained that the flow of tears is provoked through the nerves.

To the dissertation on the tear apparatus, Stensen joined another one, *On vessels of the nose*.[87] Correctly understanding the remark of Hippocrates that in order to smell properly one must have dry nostrils—he meant only no overabundant liquid—, Stensen also held the moisture of the nose as necessary and found as its source at first the Eustachian tube which, as he thought, brought liquid from the ear, later the lachrymal canal and particularly the many glands which he discovered under the membrane of the nose and which Steiner had described in his work on the catarrh under the name of *Membrana pituitaria*. However, Stensen also looked after the ways of exit for

---

[84] <484> OPH I, 81.

[85] BEM III, no. 87, 374 ff.

[86] <490> OPH I, 86 ff. Cf. Pionier, 51 ff. Danish: NSVO, 113. English: Anatomical observations ... , ed. by Edv. Gotfredsen. * See also Thomas Bartholin on the source of tears, *Vasa lymphatica*, 1653, 3: "Water of the brain drips through the orbits of the eyes and wet the face by a large stream of tears."

[87] <495> OPH I, 91 ff.

the moisture and pointed out, besides the nostrils and the opening in the throat, the so-called Stenonian ducts or *ductus nasopalatini* which open in the palate closely behind the teeth.[88] As far as the moisture of the nose does not leave the body through the nostrils, it flows down in the mouth and from there on to the stomach and returns to the blood in its proper circulation.[89]

From the various facts which he sees, Stensen always strove to general insights and connections for which he had always an open eye, without forgetting experiment and observation. Thus, in a letter to Bartholin dated January 9, 1662 he says: "I am not far from believing that all the aqueous liquids which one finds in the different cavities are to be attributed to these glands."[90] He also mentions the sweat as a new discovery and, since he now had attributed to glands three body liquids the origin of which was not known previously, namely the tears, the slime of the nose and the sweat, and, moreover, also the analogous production of saliva, he proposes the hypothesis that all moistures found in the different cavities are secreted by conglomerate glands and not directly by the membranes.

1.2.5 Mathematics and Research on Muscles

At mid-year 1662, a confession of our so successful anatomist occurs which one reads only with amazement. It is a well-known passage in the letter of August 26, 1662 to Bartholin where Stensen explains that when he had completed his observation on the glands, he had been close to making the change from anatomy to geometry, "in order not completely to give up the study to which I once devoted so many hours and which I would have pursued as the most important and maybe the only, if the limited means of my family had not only urged me but rather enforced me to prefer the useful to the pleasant".[91] The research of Stensen at that time is in accordance with these words. The holiday journeys of 1662 and 1663 remind us how much he was interested in Dutch mechanics and technology. From the short, dry

---

[88]*This duct, present in rats, is absent in man.

[89]*In the fourth dissertation, <465> OPH 3, Stensen described a small gland concealed under the mucous membrane of the nose in sheep and dogs, now named the lateral gland (Steno's gland). It is well-developed in the rat and not present in man. It serves rapid humidification of inspired air to prevent the drying out of the mucous membrane and to ensure the function of smell and ciliary activity in the nose. (Moe and Bojsen-Møller, *Lateral nasal gland (Steno's gland)*; Moe, *When Steno brought new esteem to glands*, 87).

[90]<501>OPH I, 101.

[91]<511> OPH I, 115.

indications in the diary of Braems, we know of a longer evasion. Braem notes that he travelled from Amsterdam, where he had come to deliver a lecture, to Zaandam. In his company, besides Borch and the sons of Gersdorf, were also Walgesten and Stensen. In Zaandam there was the largest Dutch shipyard and the travellers saw some 40 finished ships, and moreover, they visited an oil press.[92]

In this connection, we must consider the abundance of anatomical research and problems that kept him very busy at the time, and in which the practical-technical mathematics also played a role. The letter of May 12, 1662 to Thomas Bartholin mentions the image of the sun in the eye and the effect of snuff on the eyes, which Bartholin had asked him a question about.[93] In the letter of August 26, 1662, Stensen tells of his observation that the vena cava (hollow vein) possesses a force of contraction, which still persists even a long time after the heart has stopped beating or, as Sylvius expressed it, that the vein in the vicinity of the heart dies last. He had observed this phenomenon on a sea snake of which he also studied the liver, the pancreas and other organs, and it was confirmed in several vivisected rabbits.[94,95] Notes about similar observations show how much materials Stensen opened, i.e. vivisected, to arrive at his results.[96] In the letter of March 5, 1663 he speaks of a microscope with the help of which he confirmed the pulmonary vesicles recently discovered by Malpighi, and he added observations of the aspect of the lungs of a new-born baby and of a swan.[97] He also writes of the fact already established by himself that the part of the vena cava lying in the thorax in mammals fills with blood during breathing in while the adjacent parts of the large veins in the neck and lower part of the body empty themselves (Fig. 1.2.7).[98]

---

[92]Corfitz Braem (1639–1688) from Elsinore was the son of Dutch parents. His father was customs officer, his brother was bishop in Aarhus and his sister was married to Rasmus Bartholin. He himself became Burgomeister in Aarhus. He took part in the defence of Copenhagen in 1659 but already on November 16, 1659 he departed for his first travel to Holland, in fact on a Danish diplomatic mission. He went on his second journey to Holland on October 1, 1661, which lasted until the autumn of 1663. DBL, 2nd ed., 3, 356, REX, Thott 1926, 4°. See also Reisen, 34 f. BOD, I, 168.

[93]<505> OPH I, 107 ff.

[94]<515> OPH I, 115.

[95]*"Finally the heart failed ... ", <514> OPHI, 117. Stensen reported three locations for contractions of the hearts of dying animals, in the vena cava, in the atrium, and in the ventricles. These phenomena found an explanation and descriptive terms, the sino-atrial and the atrio-ventricular blocking, from electro-physiological examinations in the twentieth century.

[96]<517> OPH I, 123 ff.

[97]<523> OPH I, 132.

[98]<525> OPH I, 133.

Thomæ Bartholini
EPISTOLARUM
MEDICINALIUM
Centuria IV.

Variis Obſervationibus
curioſis & utilibus re-
ferta.

HAFNIÆ,
Typis MATTHIÆ GODICCHENII.
Sumptibus PETRI HAUBOLD, Acad. Bibl.
ANNO cIɔ Iɔc LXVII.

Fig. 1.2.7 Letters from Stensen published in *Epistolarum medicinalium centuria*/Hundred medical letters of Thomas Bartholin, vols. 3 and 4, 1667

- Discovery of the external salivary duct 3, 86–95, OPH 1
- Observations in eyes and nose 3, 224–230, OPH 3
- Origin of sweat from glands 3, 262–266, OPH 7
- Why tobacco makes eyes clearer 4, 1–10, OPH 8
- Anatomical observations in birds and rabbits 4, 103–113, OPH 9
- On vesicles in the lung, etc. 4, 348–359, OPH 11
- New structure of the muscles and heart 4, 414–421, OPH 14

We can easily ascertain the enthusiastic dedication of young Stensen to pure and applied mathematics in his Copenhagen years but how could it suddenly blaze again in Leiden and nearly divert him from anatomy in which he had done epoch-making discoveries?

First the intellectual scientific atmosphere in Holland and above all in Leiden must have aroused once again his esteem for mathematics. His visit to the old mathematician Gregory of S. Vincent (1584–1667) during his holiday journey in May 1661 may have reminded him of other important mathematicians who were the pride of the Netherlands in the seventeenth century: the war engineer Simon Stevin (1548–1620) from Bruges, who also came to Leiden, Willibrord Snell (1580–1626) who, aged 20, had delivered lectures in Leiden, or Frans van Schooten (1615–1660), also a professor of mathematics in Leiden.[99] Stensen met again in Holland Walgesten and in Leiden, as he reports himself, he worked with the microscope. This brings to mind the Dutch physicist Christiaan Huygens (1629–1695), the discoverer of the pendulum clock and successful optician and astronomer,[100] and even more so the highly developed Dutch optics of the seventeenth century. It is particularly interesting that Stensen dedicated his *Observationes anatomicae* of 1662 not only to physicians like Sylvius and van Horne but also to the professor of mathematics and Arabic Jakob Golius. Born in Den Haag, that universal erudite, after oriental studies in Morocco, was active in Leiden as professor of Arabic and, from 1629, as professor of mathematics and astronomy.[101] As such Golius came into relation with the greatest of all "Dutch" mathematicians—at that time, it was said, the country of Rembrandt as well of Descartes.

Descartes had become the friend of Isaac Beeckman (1588–1637) in the camp of Breda and had collaborated with him in the Netherlands. In the impassionate discussions over the ideas of Descartes which then spread in the Netherlands, above all at the universities of Utrecht, Leiden and Amsterdam, Golius was called not only a true friend but also a gallant defender of the philosopher whose greatness of mind and beauty of the discoveries, particularly in his *Geometria*, he admired. Golius had put the family Huygens in relation with Descartes and, referring to this, might also have been a good guide for Stensen.[102] On this matter, one can think of the analytic geometry, as, for example, his understanding concerning the refraction of light, the

---

[99]On mathematics of this period see, for instance, Struik, *The land of Stevin and Huygens.*
[100]DSB VI, 597–615. Bell, Christiaan Huygens, 13 ff.
[101]NNBW 10, 287–289.
[102]Thijsen-Schroute, 74 ff, 76 ff. Descartes in DSB IV, 1, 51–65. TRE 8, 499–510.

rainbow, the gravity of air, the trajectory of thrown bodies or the expansion of water by freezing.

At the end of 1662 or beginning of 1663 anatomical research made Stensen more cautiously critical of the mathematical method of Descartes and its application in the domain of nature, but, on the other hand he felt himself attracted by the mathematics of Spinoza, which promised greater reliability. He got into an ideological crisis. The outcome was to the benefit of his own exact research.

As shown in his Chaos-Manuscript, Stensen shared the enthusiasm of his time for reliable knowledge, for experiment and practice. He wanted to follow the Descartes method when he says: "By method I mean the reliable and simple rules according to which anybody who follows them precisely will never take the false for the true or apply oneself to useless speculation, but by increasing his knowledge little by little arrives at a true understanding of that which he can grasp."[103] Stensen was certainly also impressed by the struggle for such universal mathematical rules which Descartes wanted to set as the basis of the complete explanation and domination of nature, organic as well as inorganic. All phenomena in nature should be explained by extension and movement and the whole understanding of nature become geometry. In the strife for reliability Stensen willingly accepted the methodical doubt and Descartes' picture of two mechanics, God who produced the machine of the world and gave it movement the laws of which we study, and man who is provided with thinking and uses the particular machine of the human body while the beasts on the whole are machines. He praised the Cartesian method as an appropriate help to discard prejudice and, when in the year 1662 the professor of philosophy Florentius Schuyl published in Leiden the Latin edition of *De homine* (Fig. 1.2.8) by Descartes, Stensen attended the discussion with lively interest, particularly in the physics and the anatomical lessons of that script.[104]

Soon his critical sense was stirred as he compared Descartes' anatomical constructions to explain the human body with the facts that his dissection knife and his observations together with sound deduction revealed to him, quoting almost exclusively the name of the late philosopher with irony to correct his views. When he quotes the opinion of Descartes according to whom tears proceed from the brain (end of 1661), he mentions the "highly

---

[103]The fourth rule, cf. Descartes, *Oeuvres complètes* X, 371. On Stensen and Descartes, see a.o. Faller, *Niels Stensen und Cartesianismus*, Rothschuh, *Descartes, Stensen und der Discours sur l'anatomie du cerveau*, Olden-Jørgensen, *Nicholas Steno and René Descartes*.

[104]OTH I, 388, <516> OPH I, 120.

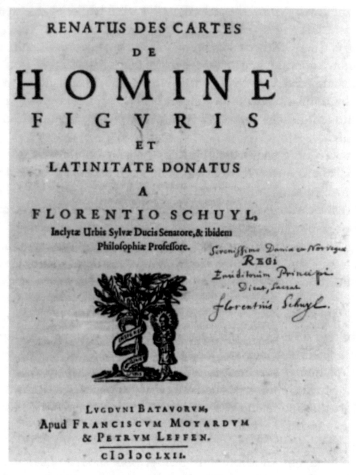

Fig. 1.2.8 Descartes' *De homine*—Treatise of Man—with a dedication to the most serene King of Denmark and Norway by the editor Florentius Schuyl. REX

gifted Cartesius" who derives the tears from the steams which flow out of the eyes in large quantity because of the magnitude of the optical nerves and the amount of small arteries. Yet Stensen adds right away: "But, although various fairly clever opinions seem to be pressed upon with great probability from these people, experience however demonstrates paths far different, more in agreement with the usual way of nature."[105] About the treatise *De homine*, Stensen notices (on August 26, 1662): "In this book there are some not inelegant figures. It is certain that they have proceeded from a clever brain.

---

[105]<489> OPH I, 85.

I doubt, however, whether such images can be seen in any brain … "[106]
Another time he writes (in a letter dated March 5, 1663): "Certainly, the more I open brains, either of other animals or of birds of various kinds, the less the structure of the brain of animals thought out by the noble Descartes, most ingenious and very appropriate otherwise to explain animal actions, seems to fit animals"[107]

The strongest doubts about Cartesianism occurred to the anatomist from vivi-sections. Descartes' concept of the animals as machines and automats actually led to the concept that they are impervious to pain. It has been said that "perhaps never in the history of philosophy has a speculation apart from reality led to so tragicomic situations as in the seventeenth century where Cartesian circles almost enjoyed inflicting pricks, blows and cuttings to animals in order to delight in their 'well imitated' reactions. The poor victims may wriggle, shout and squirm as much as they well wanted, the philosopher was not to be deterred from his aprioristic ideas."[108] In Stensen, heart and brain protested against this "idealistic" insensibility. For his investigations on the lymph he had dissected a dog and kept it alive for 3 h. Since, however, one experiment alone offered no sufficient guarantee, he had to do the hard work again, but he noticed: "I must admit that I did not subject them so long to torments without being filled with horror. The Cartesians glorified themselves so much of the certainty of their philosophy, I would wish they should convince me for certain as they are convinced themselves, that the animals have no mind and that it does not make any difference whether one touches, divides or burns the nerves of a living animal or does that to the strings of a self-acting machine. I would then investigate the viscera and vessels of a living animal for hours more often and with greater peace of mind since I realize that there are many things which deserve an investigation and about which one can become certain in no other way."[109] From such experiences in the anatomical field arose an increasingly critical attitude towards Cartesian philosophy and its theory of cognition.

Contrary to the old concept of reality lying between thinking and being, the truth for Descartes consists of his clear and distinct ideas, "which are not the result of reason and experience, or to say it better, all experiences must be

---

[106] <516> OPH I, 120.

[107] <528> OPH I, 136.

[108] It was not possible to trace the source of this quotation. The content accords with N. Fontaine's description of Port-Royal (II, 52–53), cf. Rosenfield, *From beast-machine to man-machine*, 54.

[109] <467> OPH I, 57. See Keele, *Niels Stensen and the neurophysiology of pain*, and Lindeboom, *Dog and frog. Physiological experiments at Leiden during the seventeenth century*.

adapted to them, for example the idea of extension. That has entailed many a sharp judgment against the Cartesian physics which has been compared to a torture which one inflicts on nature, not to discover its mysteries but to force it to tell what one wants to hear."[110]

Stensen seems at first not to have seen any danger for his Christian faith in Descartes and his teaching system although several contemporaries saw in this theory an opening to the rationalism into the faith and to the determinism into the moral. In Holland, the synod of Dordrecht as early as in 1656 published a decree against the Cartesian philosophy, and in 1665 the books of Descartes were put on the Vatican index.[111] Stensen does not seem to have paid attention to all these controversies. Descartes himself remained a believer, if not zealous, in Christ to the end of his life. The young researcher felt his faith threatened rather by Spinoza and his further development of Cartesianism. This happened after they had become acquainted, probably around the turn of the year 1662–1663.

At that time, Spinoza (1632–1677) lived in a small house in Rijnsburg, a village outside Leiden. Excluded from the synagogue in Amsterdam since 1656, after some years in Ouderkerk, he had accepted the invitation of his friends, the Kollegianten, and had come to Rijnsburg. Here, at the presidency of these free minded Mennonites who lived simply and piously without any priest and determined forms of cult, he stayed from beginning 1661 until April 1663, where they provided for his means of subsistence. Some of his most important works were written here.[112] He was the centre of warmly admiring friends, he received many visitors and his treatise "On God, the man and his salvation" possibly stemmed from discussions with students. Stensen may have come to the Spinoza circle through Theodor Kerckring (1630–1693) whom he was to meet later as bishop. The son of a patrician from Hamburg, Kerckring was registered in Leiden on March 12, 1659, but kept his habitation in Amsterdam at the Keizersgracht. He studied medicine and his name survives in the history of medicine because of his discoveries about the formation of bone in the foetus and parts of the digestive apparatus. At the age of 18 he attended Latin teaching by Franz van den Enden, the teacher and friend of Spinoza. Under the influence of the former, whose

---

[110]The quotation could not be verified. Moreover Rossi says (p. 131): "Thus, Cartesian–Spinozian physics in many of its assertions was as far from reality as the medieval, moreover being on a less certain foundation than that of the scholastics—in some respect it was a step backwards."

[111]Thijsen–Schroute, 104, NCNH V, 77 f.

[112]DTHC 14, II, 2489–2506, NNBW I, 1480–1484.

daughter he later married, he converted to the Catholic faith.[113] The diary of Ole Borch gives us some hints about the possibilities of a connection. Already at the turn of the year 1660/1661 on his arrival in Holland, Borch had noted the address of Theodor Kerckring at the Amsterdam Keizersgracht. He then visited him on April 25, 1662, on an excursion from Leiden but seemed interested by some of his more curious ideas. Previously, however, on April 3, his diary remarked about Amsterdam: "There are here atheists, mostly Cartesians, such as van Enden, Glasemaker, etc. They do not recognize themselves as atheists, often speak of God but by God they mean nothing else than the whole universe." Van den Enden was forbidden to dispute because his last disputation seemed to plead for atheism. Borch himself, however, openly bewares of entering in connection with these circles. We know of a visit in Rijnsburg on September 24, 1661, but nowhere is there any question of a discussion with Spinoza.[114] Stensen obviously went his own way as we know from a letter "to the reformator of the new philosophy" in which he, 4 years after his own conversion, wanted to bring the philosopher closer to the Catholic Church and in which he reminds him of their former close relations and expresses the hope that the reciprocal devotion is still alive.[115,116]

At first, Stensen may have been attracted by the noble way of life of Spinoza. His life was in agreement with his philosophy. One has to admire his frugality and self-control towards sinful pleasures as well as his calm and kindness. He contented himself with the small house in Rijnsburg, refused more support and contributed to his means of subsistence by grinding optical glasses. What impression Spinoza made on visitors appears from Henry

---

[113]NNBW 2, 663 f. EP I, 10. He later reminds Bishop Stensen on their joint studies in Holland, EP II, 481. Cf. Francès, 142 f, Thijsen-Schroute, 392. BOD I, 29, II, 109.

[114]BOD II, 92, 228 f. Moreover Spinoza received many students from Leiden and he was easy to find among friends and admirers in Amsterdam.

[115]EP I, 231 f. Details on the friendship are not known. A Calvinist priest, Johannes Sylvius, later reproachfully writes to the convert, Stensen: "You who were worthy to be the friend of Spinoza". OTH I, 95. Spinoza in his library possessed the Observationes anatomicae by Stensen and his De Solido (1669) even in duplicate (Préposiet, *Bibliographie spinoziste*, 342, 340).

[116]*Pertinent to the study of the relation of Stensen and Spinoza is a manuscript by Stensen dated in Rome, September 4, 1677, when he prepared himself to become a bishop. It was found in the Vatican and recorded by Totaro. Stensen dates his contacts with Spinoza to 1661–1662. Spinoza daily attended the anatomical dissections of the brain that the Danish scientist executed "to locate the seat of the principles of motion and the source of the feelings—*per trovare la sede del principio de' moti, ed il termine delle sensazioni.*" Totaro concludes, "Stensen's relation with Spinoza was far from marginal in the years in Holland. More difficult and complex to interpret are the relations in the following years, when Stensen seemed to engage in his most arduous battle against Spinoza." (Totaro, 27–38). See also Israel, Radical enlightenment, 507 ff. Spruit and Totaro, *The Vatican manuscript of Spinoza's Ethica.*

Oldenburg who visited the small house in Rijnsburg in the last days of 1661. This verse can be read there even today:[117]

> Alas! if all men would be wise,
> And would be good as well,
> The Earth would be a Paradise,
> Now it is mostly Hell.

Back in London, Oldenburg, in his first letter to Spinoza dated August 26, 1661, effusively thanks for the new friendship and praises the addressee: "Solid learning, conjoined with courtesy and refinement of manners (wherewith both nature and art have most amply endowed you), carries with it such charms as to command the love of every honourable and liberally-educated man."[118] Even an objector like F.H. Jacobi could exclaim 100 years later: "Be you blessed to me, great, yes holy Benedictus. However you philosophize about the nature of the highest being and may be mistaken in words, his truth was in your mind and his love in your life."[119]

Like the urge for reliable knowledge had led Stensen to the mathematical thinking of Descartes it also led him on to Spinoza who promised even greater reliability. Just at the time, 1663, he published in nearby Amsterdam under his name the *Renati Des Cartes principorum philosophiae Pars I & II, More geometrico demonstratae.*[120] The external aspect of the method of Spinoza is all along mathematical, in which he transferred Euclidian geometry into philosophy. He starts with a series of definitions and axioms, i.e. obvious basic theses. From these the Cartesian system is developed *more geometrico*, proposition after proposition. Corollaries to the propositions are intercalated, and postscripts to complete the demonstrations. When Stensen in the introduction to his myology sings a song of praise for mathematics, he might have used a sentence by Spinoza: "I will, therefore, write on human being as if I had to do with lines, surfaces and bodies." A word which later Albert Burgh happened to hear: "I know that I recognize the true (philosophy)."[121]

The whole order of things leads Spinoza inevitably away from the existence of God as he understood it. The theses about God, the soul, the kingdom of the true and the good are shown with rational sovereignty. He reveals the concept of

---

[117]*The concluding Stanza by Kamphuyzen.

[118]Oldenburg I, 413 ff.

[119]Fischer K ., *Geschichte II*, 92ff.

[120]For a brief overview of Spinoza's writings see, for instance, Ueberweg, 276–294.

[121]Spinoza, *Briefwechsel*, 286.

God as well as the substance, the attributes and modes, the separation of thinking and doing, the determined will. It was rather a *destructio* than a *perfectio* of the Cartesian philosophy. For instance, the difficulty which appears from the unbridgeable large gaps between the world of the body and the world of the mind by Descartes is solved simply by merging both in one substance.

If in later years he emphasized the materialism of the system, Stensen seemed at the time influenced by its spontaneous spiritual character as concerns the self consciousness of the mind and the innate ideas.[122] Only if God and Nature, like mind and body, are one substance, it must be a question of lesser importance, whether the matter is called the mind or the mind the matter. Both form one substance. Meanwhile, not only the contradictions in the system are difficult to overcome, the ambiguity of expressions is disconcerting. Spinoza talks of God, of soul, freedom, virtue, eternity, but these words are used quite differently from the concept of the Christian philosophy of the seventeenth century. Therefore, antagonism towards Christianity and faith in the *Theologico-political treatise* cannot be denied. Spinoza rejects the dogmas but is himself full of dogmatic certitude, not acknowledging the scriptures as divine source of faith.[123] The narrations of the Bible are destined to the common people, the miracles empty inventions, reason presented as kingdom of truth and wisdom, religion only as devoutness and obedience. Religion stands below the State to which all power is given and leaders, enlightened by reason, should overrule it. Already in Rijnsburg, Spinoza nursed the ideas which he anonymously 10 years later expressed in the *Theologico-political treatise*.

Later Stensen would mention several times his relation to Cartesianism; thus in a letter of 1677 to Leibniz. The German philosopher had ironically asked whether Stensen had perhaps found the Catholic truth in the marrow of the bones.[124] Stensen answered: "I can tell you that, in that country of freedom (Holland), I associated with people of a very free understanding, I read all kinds of books and had a very high regard for the philosophy of Descartes and for everybody who was praised for his understanding of Descartes. Then a friend from Sweden brought me the lungs of a deer with the heart attached, to investigate the pulmonary substance. As we had dissected the lungs, we were inspired to boil the heart in order to see whether the

---

[122]Stensen to Spinoza RP I, 231 ff. Dunin-Borkowski, 3, 162 ff., Larsen, *Stensens forhold*, 523 ff., like also Betschart, 135 ff.

[123]Stensen writes, e.g.: *Excuse, quaeso, omnes demonstrationes tuas, et vel unam mihi afferto de modo, quo cogitans et extensum uniuntur, quo principium movens corpori unitur, quod movetur!* "I ask you, put to the test all your testimonies, and give me just one [argument] on the way and manner in which thinking and extension are united, and the principle of motion is united with a body set in motion." EP I, 235.

[124]EP I, 366 ff.

substance was musculous or not. Having boiled it and removing its membrane, the first fibres of the heart which I touched, led me to the lower tip and from the tip upwards again, a truth explaining the whole structure of the heart which up to that moment neither I nor anybody else had ever known. This was in complete contradiction to anything the greatest and altogether most dangerous philosophers held for so obvious a truth that they even had the assertion that those who did not share their point of view on the heart did not understand anything in mechanics. A short time later, an afternoon, I got the desire to compare the structure of the heart to that of the muscles. As far as these were concerned, I considered the system of Mr Descartes as infallible. To this end I chose a leg of a little rabbit which I had dissected a short time before. The first muscle which I tested revealed to me the first step of the structure of the muscle which so far nobody had known and which demolished the whole system of Mr Descartes." From these two discoveries Stensen drew his conclusions. The most important was: "If these gentlemen which the learned world so to speak worships have considered as irrefutable what I in an item prepared in one hour by a 10-year-old boy in a glance can perceive, without words, upsets the whole clever system of these great minds, what certitude do offer me the other subtleties which they glorify themselves with. I mean, if they have deceived themselves in material things which are accessible to the senses, what certitude can they give me against a similar deception if they deal with God and the soul."[125] These reflections, Stensen admits, did not estrange him from the whole of a theory where some scientific findings might be right, but he had freed himself from the overwhelming reverence which he had nursed for them and he had gradually recognized the weaknesses of the human mind and the abysses of the arrogance.

In his *Defensio epistolae de propria conversione* of 1680 he deepens his critique of Cartesianism.[126] First he mentions the positive, the striving for certitude but then he says: "I consider this philosophy as objectionable only where its father, forgetting his own method, presupposes to be true something he has not established on reasonable grounds. Therefore, many people throw themselves in still more serious mistakes and make the whole Christianity, if they do not entirely reject it, at least so powerless that hardly anything remains other than echo and empty shadows. What the so-called completion of the Cartesian philosophy, but in reality its destruction, made materialistic is obvious in Spinoza and his followers. Forgetting their own hypotheses, they extolled the said method as a demonstration and, since they did not want to

---

[125]EP I, 368.
[126]OTH I, 388 ff.

admit, with Descartes, the ignorance of the manner of how mind and body or thinking and expansion unite, they wanted, pulled down into the deepest abyss, to admit that there were different attributes but the same substance, although they have neither produced a proof that the expansion cannot exist without movement nor, for the mode, how the movement exists in the expansion or communicates itself to another expansion. Hence, they have made theology, the practical and the speculative, a mixture of thousand absurdities through these candidly devised hypotheses. Because they indeed admit a substance only, they make God an aggregate of all material things and provide man with freedom, the enjoyment of all delights of sins since the use of praying as well as of punishments and reward after the death are cancelled. But since, among the works of nature, they find nothing that would be preferable to man, they consider those who outdo the others as the noblest part of God and as the first god among gods, briefly the true Antichrist. And that is that danger of Cartesian philosophy from which, as I have said, God has freed me at one time when I dealt even with these reformers or rather deformers of Descartes and, in a really wonderful way, completely by chance, as an actual gift of God, I obtained the knowledge of the true structure of the heart and the muscles which knocked down all the constructions of these finest heads without words, only by the evidence. Hence I came then to two conclusions the first of which is: If men who in the judgment of many are considered divine set up theses so far away from truth in such a clear and easy matter where it is possible to return to the reliable experience, who does guarantee me that they deserve greater credence when they propose their dogmas about God and the soul where one cannot carry out such investigation? ... As for Descartes, I do not reprove his method but the neglect of it, which was the first instrument of the discovery of my mistakes. This, however, would have led me away from the striving for religion, after the example of so many, if God in his great love had not accepted me."

The conceptions of Spinoza about the structure of the muscles and of the heart may correspond to those of the Cartesianism, which in turn derived from Galen. According to Descartes' *De homine* the machine of the human body had its moving principle in the heat which springs up from the fire in the heart and agitates the blood into movement. This fire makes the blood boiling hot so that it expands and overflows into the arteries where it deposits its nutritive substances. From the blood minute particles rush also into the brain and form the *spiritus animales* which flow through the nerves, believed to be hollow, to the muscles, and there they trigger movement.[127] As soon as

---

[127]See a.o. Vugs, 144–147.

1662/1663, Stensen, relying on his research, passed a devastating verdict about all these speculations.

According to the letter of August 26, 1662 to Thomas Bartholin, he speaks of his intention to devote himself to geometry. Being forced, however, by the Blasius controversy, he gave this up again on the advice of his friends. Then after an investigation of the pancreas and the liver in an aquatic raven[128] the movement of the heart particularly enthralled him: "The heart of this bird was indeed so vigorous that its pulsations did not lose the least either of their velocity or of their power until the second hour after the opening of the animal. "[129] This observation, Stensen pointed out to his friends. Besides Sylvius, Piso and the Danzig physician Christoph Gottwald[130] also attended. In the next days, he devoted himself with uncommon zeal to investigations on heart and vena cava, heart valves and auricles, mostly on rabbits and dogs. A series of other records, undated but related to this matter, actually draw conclusions from the observations: that the diminishing movement can become strengthened again by a firm resistance or that the movement comes up not to the whole mass but to the fibres and propagates in waves.[131] It is interesting that Stensen recalling this investigation some 10 years later added: "The following facts furnished to me a very valuable argument for trampling the arrogance of human intellect: 1. After so many centuries of work and speculation the philosophers have so far discovered nothing about the true cause and mode of animal movement. 2. Those who pronounced their opinion on this matter with great authority have all sold us false theories instead of demonstrations. I am far from promising a full explanation of everything. But God must be thanked who at least made obvious to all the falsity of those who are in error. He of old sagaciously said: It is virtue, to fly vice; and the highest wisdom, to have lived free from folly."[132]

The next notes about muscle and heart studies are found in a letter of April 30, 1663 to Thomas Bartholin. Here these matters are made clear and certain, and in a much later published report he calls them "the new structure of the muscles and of the heart".[133] Stensen explains that the fleshy muscle

---

[128]"in corvo aqvatico, *skolfer* vocant Belgæ". According to Maar this bird was a *phalacrocorax carbo* (<513> OPH I, 247). Thus a cormorant, not an eagle as sometimes indicated.

[129]<514> OPH I, 118.

[130]Christoph Gottwald became a famous collector of minerals and natural effects.

[131]<517> OPH I, 123 ff, 157 ff.

[132]<362, 521> OPH I, 127 and note I, 249: Horace, Epistles, I, 1, 41–42.

[133]<543> OPH I, 157 ff.

fibres do not run straight but transversely from one tendon to the other, the different fibres, however, being parallel. He also draws a compound muscle, speaks of the aponeuroses and releases the certain opinion which he will prove in his next work: "As far as the substance of the heart is concerned, I will, so I believe, make it obviously demonstrated that nothing is found in the heart which is not found in a muscle and nothing is absent in the heart which is found in a muscle, if you consider what makes the essence of a muscle."[134]

1.2.6   Journey Through Belgium, 1663

When Thomas Bartholin in the letter of April 7, 1663 mentions Stensen's *itineri succinctum*,[135] he did not think of his former pupil's journey back home but of a holiday journey through Belgium. After all the exhausting work, learned controversies and spiritual fights, this journey provided well deserved relaxation.

To the travelling companions belonged, besides Stensen; Borch and Braem, Hasebard and Walgesten, the sons of Gersdorff, Christian, Caspar Christopher and Heinrich, Matthias Jakobsen, the son of the bishop of Aarhus, and three students of whom we know nothing else: Friis, Meyer and Jochim. The first is possibly Magnus Friis from Scania. The other was called Henning Meyer and was a student from Holstein and Jochim was perhaps the above mentioned Joachim Merian from Frankfurt.[136]

A 3-h walk brought the travellers from Leiden to Delft on May 19, 1663. They went through the town without delay to the landing place of the post boat and came to Rotterdam in 2 h. In the "Schilt van Franckrycke", their lodging, they met the Pensioner of Zeeland, Hubert, who once had been envoy to Denmark and, at the dinner, it came to a discussion of the catholic religion. Hubert expressed his defiance regarding the absolute dependence of the Catholics on the Pope.

The next morning at 8 o'clock the friends took a boat bringing them to Dordrecht by lunch time. They did not stop there but went on over the Sound which hid 70 blossoming villages in its swells, which had lain in lowland. In Gertrudenberg they saw the fortifications that lay in water and

---

[134]<544> OPH I, 158.

[135]EP I, 175. On the travel, see Scherz, *Vom Wege*, 57 f., *Reisen*, 35 ff. BOD III, 9 ff.

[136]Except by Borch the travel is mentioned by Braem in his records (REX Thott 1926 4°), and here the participants are also mentioned. Matisen is identified as Matthias Jakobsen by Gustav Scherz. His contemporaries knew him by the name Matthias Jacobi Mathiades, cf. <470> OPH I, 62 <226>.

swamp and represented a strong bulwark at the frontier where bitter fights so often had raged in the past 100 years. Borch notes the seven bastions of the fortress and the double ditch and wall. A Lieutenant Lauer who in the Swedish war had had his quarter at Thomas Bartholin's was their guide.

Two coaches brought the travellers 10 km south to Breda, which was occupied by people of the General States. Here they lodged in the Belgian hospice in the vicinity of the cathedral and on the same evening—after 7 o'clock—they visited the castle and the place where the peat boat with the charcoal burners had crept into the town to conquer it for the Prince of Oranje. Borch tells that the men of the Oranje who had concealed themselves in large heaps of peat in the middle of the boat kept silent even when the Spanish guards stacked their lances through the peat.

In early morning the company admired the monumental grave of Engelbert II von Nassau in the church, and Braem carefully observes the epitaph which magnifies the feats of the warrior in the Spanish–Dutch war. At 3 o'clock, tired from their long walk, they then rode in two coaches through a magnificent plantation of fir trees of the Prince of Oranje. In Grootsundert they made a midday break and reached Antwerp at about 7 o'clock, where they called on in "den blyvenden Wingaerte" in the Cooperstraat. On the same evening they went for a stroll on the banks of the Schelde and then a "Tour à la mode" in the Antwerp business quarter where Borch among others gazed at the silken head-dress for women which resembled a horn.

The "Queen of the river Schelde" had severely suffered in the Westphalia Peace treaty and the irruption of the Dutch in trade life. Yet, Antwerp still possessed more than enough places worth seeing for the 1-day visit of our friends. In the morning of May 22, they were in the largest and prettiest Gothic cathedral of the Netherlands, in the church of Our Lady. The countless treasures of art such as the paintings of Rubens seemingly enthralled them less than the story of Quinten Matsys, the smith who, for the love of a miller's daughter, left the anvil, which a verse perpetuates: *Connubialis amor demulcire fecit Apellem*. From the old well in front of the cathedral they climbed the 123-foot-high tower with chimes (Braem notices 77 clocks) and enjoyed the beautiful view towards Mechelen, Brussels and the estuary of the Schelde.

In front of the church they then met the abbot of Sibricht, Buch von Patern, whom Borch had met the previous year on his way to Spa near Liège. The Prince was "very affable", Braem says. Later, he ate lunch with them and sent exquisite wine, which he had brought from Cologne, to their quarters. Before that, they went together into the city and firstly to the Jesuits "who met us very courteously and one of the fathers led us around". The Jesuit college already at that time had drawn the attention of the world by the publication of the *Acta sanctorum* of the Bollandists. The famous Daniel Papenbroch, himself an Amsterdamian, had in 1659 joined the work of his order-fellows in this field. Our travellers, however, mention neither him nor the work of the Bollandists, although Borch admired the library of the Jesuits, which comprised four large rooms. After visiting the church and the college from the cellar to the roof, the "chapter house", the paintings of van Dyck, they saw the vestry where the guide showed them a large silver cross with particles of the Christ cross, a part of the spine crown as well as many relics of saints, among those the head of John the Baptist. Borch's report has no critical remark. From this room with many other precious objects, they went down to the burial place of the fathers and then to the church of which the Jesuit François d'Aguilon (1567–1617) had drawn up the planes with Rubens who, as Braem recalled, had contributed to 18 "quite curious pieces" to decorate the church. From the Jesuits they went to the Waterhuis with the large pump works built by Gillis van Schoonbeke, from which the breweries got their water. A machine moved by a horse could provide 860 barrels a day. Before lunch they paid a visit to the small Capuchins' church and saw the library, the dormitory and the infirmary of the cloister. "In the church and library they have vats with flour to spit in." In the afternoon they went to the castle erected by the Duke of Alba: 500 Spanish soldiers, 37 guns and strong bulwarks expressed there the will of the Spaniards to keep the town. The commander permitted a visit under the guidance of a lieutenant. Then the travellers went to the Premonstratensians' St. Michel abbey. The 100-year-old printing shop of J. Moretus, the *Platiniana*, was visited and the twelve presses were admired. Then they saw the stock exchange with its nice colonnades and its four towers. The diary cites the luring inscription: "Learn, o guest, even if the whole earth lies between, here you are nevertheless no foreigner as long as you have money." Together with the prince-abbot, our friends then visited the city hall, which pleased them. They were not less interested in one of the 34 companies of the militia, which were carrying out their exercises in front of the building.

On May 23, the friends, 12 men in a coach, left Antwerp through the Charles V gate and rode to Mechelen where they entered at 12 o'clock. The town little pleased Borch except for the garden of the Capuchins with its crucifixion group and the neat tiny wooden houses in the duck-pond. They rode to Leuven after already 1 h, where they entered at 6 o'clock and they spent the last hours of the day with a visit to the university and the anatomy theatre. In the sixteenth and seventeenth centuries the university, founded in 1425, had a leading position among the high schools of Europe.

On May 24, the Corpus Christi Day was feasted with all magnificence in the Flemish city, 3 years before that other Corpus Christi Day in Livorno that would be so important for the conversion of Stensen. After a conversation with an acquaintance from Antwerp, the painter and wax moulder Froetiers, at 7 o'clock the next morning, the friends went to visit the different colleges. Soon they had to renounce their intention to see all of them since there were more than 80. They visited the Jesuits and the Carmelite nuns, where Braem notes: "We receive a pretty image of Mary from the nuns of the *Monasterio annuntiationis B. Virginis*". They also visited the Carmelites and the Augustinians. A monk offered them a book about the miracle that had occurred in the church where the Holy of Holies was particularly revered. The castle where one listened to the echo from the 130 feet deep well was also visited as well as the cloister of the Dominicans and the city hall.

However, they arrived just in time at St. Peter to attend the 12 o'clock mass and the following Sacrament procession, which lasted until 2 p.m. Borch as well as Braem described it as very solemn. First came the artisans, each corporation with the image and the banner of their protective saint, then followed the members of orders in coloured dresses, each order with the image of its founder in front: in white dresses 32 Carmelites, then 48 Augustinians, 72 Recollects and as many Dominicans. A large crowd joined them. In their hands "burning torches with which they meant to win an indulgence". Then followed the assisting clergy surrounded by 400 torch bearers, and behind them 42 priests of the cathedral in expensive chasubles. After the large brass band accompanying the songs, came the Rector Magnificus under a large canopy, carrying the Holy of Holies, followed by the higher clergy, the prelates, cannons, two abbots, 84 professors and 34 municipal officers. Subsequently a huge statue of St. Christopher, followed by four rifle clubs and finally the champion shot with his councillors. This ended the procession. One hour later the travellers left Leuven and, after a coach journey of 4 h, they reached their inn "Klein Antwerp" in Brussels.

They had reserved an entire day for the capital of Belgium but not more and, already in early morning of May 25, the friends walked between the fierce columns of St. Gudule. Braem paid much less attention to the old Gothic cathedral than to the marble memorial of the Duke Johann II of Brabant or the monument with the gold-plated bronze lion of the Stathouder of the Dutch Archduke Ernst or to the pleasure garden with deer and a 12-year-old ostrich, with a grotto and a water organ. They could not see the residence because it was occupied by the Spanish governor but they made do with the stable with 127 horses and the armoury with equipment and flags from Belgian–Habsburgian history.

One wonders again and again at the visits of churches and cloisters by the company of protestant travellers. Whereas they hardly noticed the house of the general postmaster Taxis or the horse market where the counts Egmont and Hoorn were executed, they visited with interest the cloister of the Carmelites, Notre-Dame des Victoires and spent a long time in the college of the Jesuits where, besides the library, they particularly admired the large cabinet of curiosities. In the afternoon they climbed the 956 steps of the tower of the cathedral and enjoyed the beautiful view. Then they went to the cloister of the Franciscans and Augustinians and finally to the Beguine convent with its 800 nuns, as well as to the pretty city hall of Brussels. On the return to their lodging the friends learned that the Prince of Sibricht had taken residence just vis-à-vis and, before they could visit him, he came in person to meet them and spent 2 h in their company.

On May the 26, at about lunch-time, a coach brought them from Brussels to Alost and at about 6 o'clock to Gent where "de gouden Apple" received them. Still that same evening they visited the castle where the Emperor Charles V was born, to find a glass grinder whose beautiful glasses and crystals aroused their admiration. The next morning they climbed the 396 steps of the belfry in the heart of the city and went to the city hall, "an expensive structure". In the college of Jesuits a bigger thrill than the library waited for them, namely the opportunity to speak with "*Patre a Sancto Vincentio*", an 80-year-old man, "a wonderful mathematician".[137] He was the outstanding erudite Gregory of whom a modern historian says that Huygens as well as Leibniz stand on his shoulders. Leibniz used the extensive work *Opus geometricum quadraturae circuli et sectionum coni* by the Belgian Jesuit and described the author as the new star of the geometry in Belgium. In their conversation with the Father the travellers noticed that he made a bright impression despite his 80 years. He also promised a new script in the course of the year. There was no time for a visit of the big Beguine convent. However, they saw the Corpus Christi procession of the Augustinians in St Bavo, though not the great *processio triumphalis*, which was to take place in the afternoon. At 12 o'clock they were already in the boat on their way to Brugge where they would be accommodated in the "Rood poort" at 8 o'clock.

---

[137]BNBel. 21, 141–171.

The morning again began with climbing of the 117-foot-high belfry and the visit of the cathedral after which they visited a chapel where *Carolus Bonus* stood.[138] Then they went to the city hall and to the Jesuit church, "the most beautiful of all the Jesuit churches in these provinces as far as the architecture is concerned". At 2 o'clock they went to Ostende, where they actually were compelled to present themselves to the fortification commandant before a soldier could guide them on the walls. On the way to Nieuport the next morning the entire coastal area appeared to be under military guard and, on their arrival in Dunkirk, they had to go to the governor, a Marquis de Mompensan who paid them full honour though: "At 5 o'clock in the morning, the French and Swiss drum players came and attended on us." Everywhere they saw traces of war. Before Gravelingen they were searched and obtained a ticket for Calais after which their luggage was entered. At 5 o'clock they were in Calais on French ground and went at once to the stowage room, "where all our goods were deposited except for our travel bags". Then they settled down in the "Gold Dragoon".

Here, from Calais, on the morning of May 30, Stensen was able to see the England he would never enter. It was a bright day in spring. The English coast stood out clearly on the horizon.[139] When they went down to the harbour on their way they saw once more a Corpus Christi procession in which the governor was taking part.

Soon after the company broke up. Borch with the Gersdorff brothers and several friends crossed the Channel to England paying 30 guilders in French money. Stensen, Braem, Friis and Meyer turned back to Leiden. Over Gravelines, Dunkirk and Nieuport they first went back to Brugge where, among others, they visited the Capuchins "where, inside the door, we saw Mary and St. Franciscus made of wax, who were as if they were alive" and they passed through the narrow lane and the place in front of the door of the mayor Lernotti where Christian Ulfeldt had stabbed down the major general Fuchs 6 months earlier. Here the friends missed the boat which, however, they caught up with after a quick march, and they came to Sluys at about 10 o'clock. On the boat they met Heinrich Maximilian von Leuchtenberg, brother of the elector of Bavaria, with retinue. After visiting the castle and the

---

[138]"There was the son of a Danish king honoured as Holy, and his bones are kept there and only once in a year displayed, and that is on the first Sunday of March" (Braem, 71 v), cf. *Bibliotheca Sanctorum*, III, 794 ff.

[139]Braem writes: "In the morning we were in the church and in its tower from where we could see England and in particular Dover" (p. 73).

well, which gave a threefold echo, they went to Kadzand and sailed to Vlissingen where, still the same evening, they saw the harbour and the wall. Next morning, a townsman of the city who had joined them in Calais showed them the city hall and finally the gibbet from which the brother of the duke of Alba had been hanged.

Middelburg was the last place in which the travellers stopped for 1 day because Braem had an uncle who invited all the travellers to his table the next day, Wednesday June 4. Afterwards they saw the many sights of the richest town of Zeeland: the Prinzenhof, the mint, the Giethuys and, particularly, the Indian House. At noon on June 5 they moved on again. After an obviously exhausting walk through the Western coast regions of Holland, over islands and peninsulas, sometimes by boat, sometimes on foot, over Zierickzee, Hellevoetsluys and Delft, they were back in Leiden, where they arrived in good condition on June 7 before the closure of the gate.

Though Stensen actually never mentioned this journey, without any doubt it added a brick to the building of his personality and, as well, it provided the biographer and his readers with a good look at the environment in which the young erudite moved. Apart from the general impulses by the visit of foreign countries and towns, particularly if they belonged to the most interesting in Europe politically, religiously, culturally and artistically, apart also from the learned connections that were certainly also a goal of the journey, the travel reports of Borch and Braem show us three traits in the picture of the young Stensen. They show the high social circle of acquaintances and the attention that they raised. A new world to Stensen was the Flemish–French cultural circle under Spanish sovereignty. Moreover, while studying the colourful popular life and admiring the magnificent architecture and art memorials of Belgium, he met the Church for the first time, to which he would adhere later. Here he might have seen a lively spiritual power in the higher and lower layers of the people, outstandingly represented by venerable churches, blossoming cloisters and efficient men.

1.2.7 Departure from Holland

After the journey to Belgium, Stensen continued his studies in Leiden. We know of his collaboration with two important researchers. Jan Swammerdam, as above mentioned, was registered at the university on October 11, 1661 and prepared himself in 1663 for his examination as candidate in medicine, which he passed on October 11, 1663. Stensen mentions his friend in the letter of March 5, 1663 and says that he has shown him how the air is driven in the midst of lungs and heart.[140] It is the same demonstration of which Sylvius reports that Swammerdam has shown him on January 15, 1663, how the air, during breathing in, comes over the arteries to the heart.[141] In the *Diversity of lymphatic ducts*, which he published years later in the *Acta Medica et Philosophica Hafniensia,* Thomas Bartholin's journal, Stensen brings a drawing which is expressly marked as a contribution of Swammerdam.[142]

Now, Stensen undoubtedly met Reinier de Graaf (1641–1673) who was registered in Leiden on April 5, 1663 and spent the next year there.[143] In the year 1664 he published in Leiden his *Disputatio medica de natura et usu succi pancreatici*, on the secretion of the pancreas, which was much appreciated. In this work, he not only emphasizes the discovery of the *ductus salivales* by Stensen but also speaks of him as *amico meo integerrimo*, which leads us to conclude that they were friends already then. In 1670, as Stensen revisited Holland once more, he also visited De Graaf in Delft where he was general practitioner and both friends could greet each other as faith-fellows since de Graaf was Catholic, which excluded him from any chair at the universities of Amsterdam and Leiden.[144]

News from his home town may have urged Stensen to return home to seek a permanent position. Stensen was dependent on the help of his family and often reminds us of his limited means, in contrast to other compatriots. A royal letter of May 26, 1662 had appointed the son of his fatherly friend, his friend of boyhood and playmate Jakob Henrik Paulli, professor of anatomy. Although Paulli had so far not distinguished himself in this specialty and had only disputed *pro cathedra* in Copenhagen on May 23, 1663, he took the oath of professor on June 17, 1663. Paulli was never obliged to deliver

---

[140]<525> OPH I, 133. *Stensen et al. observes fluctuations in the colour and thus the filling of the jugular vein and the vena cava, changes that follow the animal's respiration and suggest as an explanation that *"thoracis dilatatione distentus pulmo aërem sugit qvasi, & attrahit /* the lung distended by dilation of the thorax almost sucks and attracts air". This was verified in 1856 by Franciscus Donders who measured the negative intrathoracal pressure—Donders' pressure. (TK).

[141]Nordström, *Swammerdamiana*, 22 and notes.

[142]<529> OPH I, 137 ff. *Acta Hafn.* II, 240 f. *Thus Swammerdam illustrates what Thomas Bartholin had described in *Vasa lymphatica*, 33 in words: [Lymphatic vessels with] origin from the limbs above the diaphragm insert into the external jugular vein or at its junction with the axillary vein. They found no vessels to transport chylus from the greater lymphatic vessels to the mammary glands.

[143]DSB V, 484–485. De Graaf, *Opera omnia*, (1677), 508.

[144]EP I, 8 f.

lectures but went abroad and gradually drifted away from the study of anatomy.[145]

To Stensen, the conclusion of the scientific studies in Leiden by an examination was vital. This examination must have taken place in the autumn of 1663. Stensen passed it so outstandingly that he was found extremely deserving in obtaining the highest degree in medicine. It is unclear though why his doctoral promotion was delayed. He was awarded his doctoral degree *in absentia* on December 4, 1664.[146] The obligation to gather a report in print of his scientific activities over the years in Leiden appeared to be more important to the young erudite than the academic degree.

Concerning the immediate release of that publication, Stensen says that he already at the beginning of 1663 had promised his friends in Leiden and Amsterdam for whom privately he had carried out dissections, to publish his results with drawings as soon as possible.[147] The first dissection of the heart had made clear for him that: 1. there is in the heart no parenchyma different from the muscular fibres. 2. no fibre ends at the tip of the heart. The different fibres turn around and rise on the other side. 3. the course of the fibres is neither straight nor circular but bent in their middle only. Afterwards he had striven to a more fundamental knowledge of the muscles, particularly when he had entered on an exact description. He had also made observations of the digestive tube, the lungs, and the *levatores costarum*. He would have reached his goal when an unexpected incident[148] interrupted his studies, yes deprived him of any hope of their continuation. Now he at least wanted to present an outline of his discoveries on muscles and add another one on glands so as to make accessible to his friends his earlier publications on these subjects.

In this *Specimen* from Copenhagen, *On muscles and glands*, he offers a fundamental anatomy of the muscle after a rebuke of the overbearing authoritative belief of the Galenics who ridicule the views of a Hippocrates and Harvey concerning the heart. He built his knowledge on observation in which his manual dexterity in the preparation of the bundles of fibres to the finest detail and his technical method helped him. For instance, he took care to boil the heart in order more easily to disassemble the different parts. He described the mechanism of the muscles between the ribs and gave them the name of *levatores costarum*. Then he entered the theme of the sacrolumbar

[145]Garboe, *Bartholin*, 2, 54 ff.

[146]Molhuysen, *Bronnen*, 3, 194 = EP I, 183.

[147]<549> I, 167 ff, German: *Pionier* 71 ff, cf. Gosch II, 1, 187 ff, Bastholm, *History*, 145 ff.

[148]Maybe the news of the death of his stepfather in Copenhagen. He was buried on October, 29, 1663.

muscles, of the diaphragm[149] and the respiratory muscles. He showed that the tongue is not a gland, as was previously believed, but a muscle consisting of bundles of fibres, and he recalls the two kinds of fibres of the digestive tube.[150]

Stensen threw clear light on the essential and characteristic structure and properties of the muscle in eight conclusions:[151]

1. Each muscle comprises arteries, veins, nerves, fibres and membranes; on the contrary he was not able to ascertain lymph ducts in it.
2. There is no muscle in which the different fibres do not turn into a tendon at each end.
3. The flesh is no parenchyma or kind of stuffing. The same fibres which tightly bound form the tendon, when loosely united, also build the flesh. In this sense it can actually be said that the tendon is a continuous body from the beginning of the muscle to its end.
4. In a muscle one hardly finds a fibre which runs in the same direction in its entirety. The different fibres run at least in three directions and form two angles with each other.[152] Sometimes, however, the fibres form a curve like in the muscles of the back or the tendons which move the fingers.
5. These three courses of the singular fibres in the same muscle have not the same length although all the fibres in most muscles hardly show a significant difference in length. Each tendon consists of as many fibres as there are fleshy threads and they form, in their transverse course from one tendon to the other, the same angle at their exit from one and at their entry into the other.
6. The muscle can be divided into three parts: in the muscle body and the two tendons or in the middle part and the extremities.
7. The membrane not only surrounds the muscle with fibres running transversely but also pushes itself in-between the different fibres.
8. The work of the muscles consists of contraction. It is the flesh which draws itself together[153] and not the tendons. But Stensen does not venture

---

[149]As regards the diaphragm he added important new knowledge, thus that the center of movement is not in the middle of the muscle that on the contrary is tendinous.

[150]<549> OPH I, 168 f. Cf. illustration, *Pionier*, 80.

[151]<558> OPH I, 174 ff.

[152]*i.e. the pennate structure of muscle which is the *structural* basis of Stensen's muscle theory (<558> OPH I, 174–175).

[153]"* ... , sed ut breviores semper fiant inter eadem duo puncta interceptæ singulæ ejusdem musculi fibræ", i.e. the *functional* basis of Stensen's muscle theory (<560> OPH I, 177).

the solution of the question how the principle of the contraction would be explained.

9. The muscle must not necessarily be subjected to the domination of the will since many muscles never move according to the decision of the mind. This can be proved at different muscles of the larynx, the tongue, the back and different members.

Then Stensen turned to the heart and set up the proposition: the heart actually is a muscle; and he proved, based on the points just exposed, that none of the essential properties of the muscles is lacking in the heart and that nothing is found in the heart that cannot be found in the muscle. He particularly described the incurvation of the different muscular fibres at the tip of the heart and how the inner layer in the left ventricle proceeds as an outer layer. So, there is no reason to ascribe special properties to the heart such as the seat of the innate heat or the source of the *spiritus vitales*. Already Hippocrates taught that the heart is a muscle, as Stensen points out, but Galen notoriously denied any credit to those who called the heart a muscle and 17 centuries have subscribed to Galen, including even the teacher of Stensen, Bartholin (Figs. 1.2.9 and 1.2.10).[154]

The second part of the *Specimen* discusses in short summaries the new findings on glands which Stensen had arrived at in his Leiden research.[155] At first he makes a distinction between the *conglobate* glands of the lymph system (the lymph nodes) and the *conglomerate* or proper glands which secrete liquids, together with their discharge ducts. He enumerates the different ones which he has investigated: 1. the parotid gland, 2. the vessels of the cheek, 3. the smaller vessels under the tongue, 4. those of the palate, 5. the anterior duct from the nose into the palate, 6. the vessels of the epiglottis, 7. the vessels of the nose, 8. a vessel in the nose, proper to the sheep, 9. the

---

[154]Gosch II, 1, 187 ff. On this part of the work a newer historian of medicine says (1950): "Steno's work on the structure and function of the heart, which Haller called an *aureus libellus*, was unquestionably his most important publication on muscular anatomy, for two reasons: The result he arrived at (recognition of the muscular structure of the heart), and also the method employed, the stringently empirical, which was so manifestly opposed to contemporary science, on the other hand it was perhaps just as much this sober recognition of what the heart is not that placed him in opposition to his contemporaries. Actually, his studies had led him to a revolutionary opinion—even Harvey had called the heart the source of life. The conclusions at which Steno had arrived concerning the structure of the heart in fact brought the question of the contraction mechanism itself into the foreground. He had shown what muscle was and how contraction took place and now, by the stringent and logical application of his discoveries, he sought to show how contraction might be explained" (Bastholm, *History*, 148 f). Steno returned to this question in his Myology (Florence 1667). See also *Franceschini*, Una misconosciuta priorità di Stenono: La scoperta della torsione a cifra 8 delle fibre musculari del cuore.

[155]<566> OPH I, 182 ff, cf. Gosch, II, 1 180 ff.

If therefore it is certain, as that it is fo beyond all Difpute, that we are convinc'd both by our Senfes and Reafon it felf, viz. that there is nothing wanting in the Heart that is to be met with in a Mufcle; and further, that nothing is found in the Heart but what is contained in a Mufcle, how then can the Heart be any longer a Subftance of its own Kind, and confequently a certain determinate Subftance, as the Fire, the Seat of innate Heat, and the Soul; or the Procreator of fome certain Humour, as of the ¡Blood, any more than the Produftor of certain Spirits called vital? For if the Heart is the fame with the Mufcles in Refpeft of all the contained Parts, the Subftance alfo of the Heart and its Veffels, to wit, the Contained, with the Containing Parts, muft be likewife of the fame Nature with that of the Mufcles; for the fame Thing being furnifhed to either of them, and that by the fame Veffels, muft produce the fame Subftance; to which you may add alfo, that in either, viz. in a Mufcle as well as in the Heart there is to be obferved one and the fame Aftion, that is, the Contraftion of the flefhy Part; the neareft or immediate Caufe of which, whether it can be different from the neareft or immediate Caufe of the Contraftion of the Mufcle, confidering the great Similitude and Agreeablenefs there is

Fig. 1.2.9   Stensen's conclusion on the heart in 1664, p. 476. From the English (slightly abridged) translation of the *Biblioteca anatomica*, etc. (ed. Daniel Clericus and Jacob Mangetus), London 1711, 2, 114. Yale Medical Library. Reprinted in Kardel, *A Specimen of Observations upon the Muscles*, (1986), 115. Translator not indicated

Cela va à renverfer ce qu'il y a de plus conftant dans la Medecine.

Fig. 1.2.10 "This will turn up-side-down what is most firm in medicine" was the comment on Stensen's project on the heart by the reviewer in the *Journal des Sçavans,* Paris, March 23, 1665. Meanwhile the review continues less laudatory: "*il ne resoudra point les objections qu'on fait presentement contre son projet /*he will never resolve the objections presently made against his project."

connections from the eyelids to the nose, 10. the lachrymal ducts, and finally 11. the vessels which make the external surface of the pharynx slippery. It has been rightly said that there is no words here of the fine structure of the glands, not more than in the earlier scripts and letters to Bartholin. All that could be found with the investigation art of the time was that these liquids came from organs which specially prepared them and were not just sweated out from the humid membranes themselves, and he provided the evidence of the discharging canals.

Stensen then enumerates the following new findings which he had arrived at: 1. that all lymph vessels were connected to glands but the proper locations of formation of the lymph are unknown. 2. that all "lymph" vessels which belong to the conglomerate glands, i.e. also their discharge ducts, direct their content into the cavities of the body, like the eyes, ears, nose, mouth, digestive tube, throat, etc. 3. that the lymph vessels which belong to the conglobate glands, thus the actual lymph vessels, discharge their content back into the venous system, directly or through the conglobate glands. 4. that all the glands are organs in which the lymph vessels disperse their roots.

From his investigations of the glands, Stensen now deduces some general propositions. First of all, not only all conglomerate glands discharge their liquid in the cavities of the body but, wherever a liquid which moistens a surface is found in an animal body, it stems from such glands. He explicitly explains this for the following organs: 1. the water in the pericardium, 2. the sweat, 3. the water in the cavities of the brain, 4. the liquid over the viscera of the thoracic and abdominal cavities [not correct], 5. even the liquid with which the foetus is fed, which he believed was separated from the blood of the placenta, particularly because he had succeeded in detaching the foetus from the inside of the uterus without any blood loss. "An important fact in the history of the science and often cited", says Gosch. "He held the villi over the circular placenta and over the wall of the placenta for *eminentiae glandulosae,* a mistake which was excusable as long as one had no achromatic

II. *Nicolai Stenonis de Musculis & Glandulis Observationum speci-men; cum duabus Epistolis Anatomicis.* In the *specimen* it self, the Au-thor, having described in *general*, both the *Structure* and the *Function* of the *Muscles*, applies that description to the *Heart*, to demonstrate that *that* is also a *true Muscle* : Observing *first*, that in the substance of the *Heart* there appears nothing but *Arteries, Veins, Nerves, Fi-bres, Membrans*; and that that, and nothing e ◯ is found in a *Muscle*; affirming withall, that which is commonly taught of the *Muscles*, and particularly of the *Heart's Parenchyma*, as distinct from *Fibres*, is due, not to the *Senses*, but the *Wit* of *Anatomists* : so that he will not have the *Heart* made up of a substance peculiar to it self, nor considered as the principle of *Innate heat*, or of *Sanguification*, or of *vital spirits.* He observes *next*, that the *Heart* performs the like *ope-ration* with the *Muscles*, to wit, to contract the Flesh; which action how it can have a different cause from that of the Contraction made in the *Muscles*, where there is so great a parity and agree-ment in the *Vessels*, he sees not. And as for the *Phænomena*, that occur, of the *Motion* of the Heart, he undertakes to explicate them all, from the *Ductus* or *Position* of the *Fibres*; but refers for the per-formance of this undertaking to another *Treatise*, he intends to publish.

    As to his Observations about *Glanduls*, he affirms, that he has been the First, that has discover'd that Vessel, which by him is call'd

Fig. 1.2.11 From the review in *Phil. Trans.* 1665/66; 1(11), 176–177. REX

microscope".[156] An enigma for him was the apparent lack of discharge for the liquid which for example was secreted in the pericardium or the cavities of the brain. After a general remark about the source of the liquid of the glands, followed special considerations on the tears and the influence on them of the emotions, about the glandular liquid in the stomach, and Stensen designates the milk, contrary to Everaerd, and the catarrhous outflow, contrary to Schneider, as products of glands (Fig. 1.2.11).[157]

    The *Specimen on muscles and glands* thus represents an extremely rich scientific report of Stensen's stay in Holland. His name now had good repute in the learned world. Not only did Danish and Dutch researchers appreciate him, his discoveries were soon mentioned in scientific publications.

---

[156]Gosch II, 1, 183.
[157]*See also Moe, *When Steno brought new esteem to glands.*

1.2.8   Back Home and Disappointments, 1664

At the beginning of 1664 started the journey back home. He arrived in Copenhagen at the latest by mid-March.[158] From Leiden to Hamburg he chose the shortest way of some 10 days like 4 years earlier on his journey to Amsterdam. Had he not visited Utrecht before, at least this time he went through this strongly fortified town with its two large canals. The Utrecht Union was concluded in 1579, which bound together the seven Netherlands provinces and made the town the principal place of business of the Generalstaaten in 1593. The many old Gothic churches like the cathedral of St. Martin with the chimes in the 103 m high tower, the distinguished university and library together with other education and welfare offices made Utrecht a cultural centre.[159] Stensen continued to Bentheim and on to Osnabrück with its magnificent old cathedral and the Charlemagne grammar school—both countries that he would come to know later as a bishop—and further on to Hamburg. From Hamburg he followed the usual post route of some 10 days through Southern Jutland and the isles to Copenhagen. He might even have walked for the first time through different places of Schleswig–Holstein and of his home country: Flensburg, Itzehoe, Rendsburg, Haderslev, and over the Aarøsund to Assens and Odense, as well as Ringsted and Roskilde.

In Copenhagen he was certainly cordially welcomed by his family. However, much had changed during his 4 years of absence. His step-father, Johann Stichman, had been buried. It is likely that he became ill in the first days of the autumn 1663 and died. He found his last place of rest in St. Nikolaj on October 26 (st. v.). Stensen had lived the difficult years during the siege with his step-father but not seen the recognition that he had received after the conclusion of the peace. Stichman had been honoured by the King with a golden chain for his service as captain of the townsmen and was appointed royal mint-master on March 11, 1662. The Royal Mint was then in a court in the Købmagergade near Silkegade. Stichman was also among the 32 "best" men to whom the King, on June 24, 1661, conceded to take part in the administration of the town of Copenhagen.[160]

---

[158]<579> OPH I, 196. He conducted a dissection on March 21/31 that may have cost some time in preparation. On the travel, see *Reisen*, 39 ff.

[159]EP I, 213 f. Close to the gates to the town he excused himself in 1670 to Professor J.G. Greve in a letter because he hastily travelled on, EP I, 215.

[160]EP II, 902.

Stensen's mother, Anne Nielsdatter, still lived in the old court in the Købmager quarter in which our Niels in the year 1646, after the death of his biologic father, had received a credit of 1402 rigsdaler.[161] However, mother and son were only granted a few spring months together. In June 1664 Anne Nielsdatter was buried in the family grave near her four husbands with whom she had shared the goldsmith shop in the Klareboderne.[162]

The heritage must have been considerable. In the last years, Johann Stichman had not only been paid a good mint-master's salary but, since 1652, he also had provided the royal court with gold and silver tableware, which he also maintained in good condition. Orders and payments, though, rarely took place at the same pace and, at the death of the goldsmith, the king owed him 2 500 rigsdaler. The widow might have transferred this debt claim to the account of the Mint and thus saved a part of the money.[163]

According to the laws of this time the male successors received twice as much as females. Niels was the principal heir of his mother, but, probably, he contented himself with a partial payment. Anyway, home and workshop went to his brother-in-law Jakob Kitzerow who had established his own house in the nearby Købmagergade. The goldsmith shop in the Klareboderne saw the growth of a new generation. Sister Anna already had several children of whom the names of Johannes, Anna and Marie are known.[164]

The 24-year-old anatomist of whose discoveries the learned Europe spoke had not returned only to put in order his family business. He might also expect the recognition of his native country by a position matching his abilities. The only possibility lay in an appointment at the university of Copenhagen which, in the seventeenth century might have become one of the leading schools in the field of experimental science based on the work of such illustrious men as Simon Paulli and Ole Borch, as well as Thomas and Rasmus Bartholin, and thanks to the brilliant research of Stensen.

Stensen could hardly personally have undertaken steps in order to secure himself a position or a pension. When later, as a bishop, he looks back and mentions his deeply faithful prayer of praise to the Providence, he reveals that he had made his rule in life to avoid what might stem from human shrewdness: "So, I wanted to do nothing by requests or gifts to obtain

---

[161] See Chap. 1.1.
[162] EP II, 901 f, Hermansen, 77 ff.
[163] Jørgensen, *Nils Stensen*, 72 f.
[164] EP II, 905.

honours, position or even the hope of such. Yes, I urgently prayed my relatives not to approach anybody on my behalf."[165]

On muscles and glands[166] testifies the richness of his first years as a researcher. It appeared in the spring of 1664 both in Amsterdam and in Copenhagen; he certainly achieved the strongest application possible. The dedication addresses King Frederik III, "my most gentle Sire" and the publication presents several discoveries to the Danish and the European public for the first time. That the work made an immediate impression appears directly from the Dutch edition, which Dutch friends had encouraged by their help already the year before.

The introduction certainly suggests a nostalgic chord with respect to human and scientific judgment. As far as the overvalued heart and the undervalued glands are concerned, Stensen writes: "The judgment of men which so often deals disdainfully with those who deserve credit, yes even immerges them in the night of oblivion, whereas it raises less deserving people to the summit of honours, this judgment which disregards the note-worthy and important things, is often not only the fertile mother of mistakes, but also of evil." The heart and the glands would be excellent examples of this. While one attempted to represent the former as the source of heat of the body, the throne of the soul, the origin of the vital spirits, although it is nothing else than a muscle; the glands have been disregarded and undervalued although the good health depends to a high degree on their undisturbed work. Then follows, as has already been mentioned, an extensive presentation of the new discoveries on the muscles and on the system of glands.

Among all the valuable observations on the muscles and a clear definition of the concept "muscle", what made the strongest impression? That was the thesis that *the heart is a muscle* and nothing else than a muscle and not an organ particularly provided with arbitrary properties, "no particular substance, and therefore also *not* the seat of a determined substance like fire or innate heat or of the soul; also, *not* the creator of a determined liquid, like the blood; also, *not* the producer of certain spirits like the vital spirits". A 1000-year-old conception was buried here. As late as 1651, nobody less than the discoverer of the blood circulation, Harvey, had called the heart "the father and king".[167]

Stensen added two letters to this publication. The first, *On the anatomy of the Ray*, is dated April 24th (st. v.) and was addressed to the Amsterdam

---

[165]OTH I, 393.

[166]OPH I, 163 ff.

[167]Maar, *Om opdagelsen af ductus vitello-intestinalis*, 233.

physician Willem Piso. Piso had attended one of the most interesting dissections of Stensen in Leiden and had invited him home.[168] Now he received the report from the dissection of two ray-fishes in the house of Simon Paulli on March 21. Stensen here describes the system of mucous canals in fish for the first time.[169] Further, he describes the situation of the viscera, the glandular layers of the stomach and of the intestines and the inner spiral folds of the intestine and their significance for the digestion, also a quite new discovery. After mentioning the *glandula thyroidea*, which so far had not been detected in fish, he turns to the reproduction organs and recognizes the egg shells of the ray[170] as actual eggs. He also mentions the breathing and the blood circulation; he suspects the right destination of the fish gills (oxygen was not known, thus a proper understanding of breathing was not possible): "who knows whether water does not do for them what air does for us, by yielding to them thinner bodies contained in its embraces, and which are the spirits of something, if they yield something otherwise." Finally Stensen describes the *operculum pupillare*, the light screen which so to speak replaces the missing eyelids, which he had discovered at this occasion, and shows that a comparison between the structure of the mammals and that of the fish demonstrates that Nature, to achieve its goal, is in no way loosely bound to one certain organ but can obtain its goals with the help of different organs. Stensen ends with some modest remarks about his discoveries and excuses this simple presentation by the lack of time and an urgent journey, of which we know nothing.[171] He concludes: "I seem to have presented much about a not very great matter but I should have elaborated even further and added even more, if had happened, as I carried the dissection knife, what the pen following the knife had wished for",[172] a significant remark for his dispassionately critical research. No one less than Gerard Blasius recognized the significance of this presentation of the ray's anatomy, including it unabridged in his animal anatomy.[173]

The second letter of June 12, 1664, *On the Passage of Yolk into intestines of the chick*, deals with the nutrition of the chick in the egg and is addressed to

---

[168]<579> OPH I, 191 ff.

[169]His later Italian student Stefano Lorenzini would explain this more in detail in his *Osservazione intorno alle torpedini* (1678), cf. OPH II, 260.

[170]Earlier authors had described as the uterus what Stensen with a scientific definition rejected being the uterus: *pars, in qua nempe conceptio fiat et conformatio, conformatique nutritio.* OPH I, 200. Cf. Pionier, 150 ff, also Lesky, *Entdeckungen*, esp. 236 ff.

[171]*"urgens iter"* OPH I, 207.

[172]<590> OPH I, 206.

[173]In his *Anatomia animalium*, Amstelodami 1681, cf. Gosch II, 1, 192.

Paul Barbette, the Dutch physician who highly appraised the work of Stensen in Leiden.[174] The passage of the yolk of the egg, which the yolk achieves directly in the viscera of the foetus to the end of nutrition was already known by Aristotle in the birds and chicken even if Aristotle was unclear about its origin. The knowledge was kept during the Middle Ages. Albertus Magnus has probably observed the phenomenon himself and in 1573 Volcker Coiter described it. Three years after Stensen's rediscovery, in 1667, the English anatomist Needham declared that he had known the passage of the yolk of the egg already in 1654, without, however, publishing his finding.[175] Yet in general the passage was oddly enough unknown at the time of Stensen, so that even a researcher in this field like Harvey had overlooked it.

When Stensen discovered this passage, it was to him and his times such an important new finding that his astonishment induced him to some principle comments which show him as a mature researcher and man of science: Anatomy in the seventeenth century has brought forth much unusual, comparable with African monsters: "Now just as fright deceives the Africans and the credulous, so does joy trick anatomists when some observation which at first sight is unusual, favouring their preconceived opinion or offering a point of departure for a new line of thought, impels them to redouble their $Ευρεκα$ and leaves no room for pursuing a more careful investigation. There are sometimes still other impulses of the mind, here to be passed over in silence, that caused many to father monstrosities in their writings. It would be invidious to illustrate my meaning with modern examples, while one who brought forward examples culled from the ancients would seem to be denying them their due respect. Nor is there need to seek elsewhere for a fault which is easy for any one of us to detect in himself.[176] For I do not believe that anyone who is not too slight or too casual a student of anatomy will deny that just as he does occasionally observe some new things and some that are extraordinary, so he not infrequently persuades himself that he has seen what he has not. And when those who are striving to gain an understanding of animals through experimentation are carried away by the rash impulses of their minds as if driven by a sort of whirlwind, what wonder is it that many anatomical monstrosities are hawked abroad which are never to be found by others, no matter how assiduously they may pursue the search? Unless the mind is tranquil, it will by no means be free to apply itself to a close examination of facts which can and ought to be closely examined, and unless every least detail

---

[174]Chapter 2.17.

[175]Maar, see Part III, OPH 17, note 1.

[176]<593> OPH I, 211 f.

is noted in so far as the minuteness of the object or its intricate diversity allows, the pathway to error is downhill and very easy."[177]

Stensen made the discovery of the movement of the yolk of the egg by chance in so far that he had undertaken the dissection of the freshly hatched chick in order to determine the role of the liver in the digestion. According to the firm conviction of his times, the nutriment was carried from the digestive tube to the liver through the *venae mesentericae* and changed into blood in the liver, as Harvey still pretended. Stensen wanted to convince himself that it was so, but found, as soon as he began to dissect the chick, the *ductus vitello-intestinalis* through which the yolk destined to the nutrition of the foetus is led directly into the intestine. Here, Stensen, as always, perceives the implications of his discoveries. How well does matter of different kind penetrate the structures of the body, for instance medicine applied from the outside. Or how can we explain that the urine is not absorbed by the bladder whereas moisture can ooze through an ox bladder, as he has seen. "He does not solve these questions", says Gosch, "but he attacks the problem in a way that one cannot resist the conviction that he might have solved many enigmas which only found their solution much later, if he had remained faithful to the study of nature. Rightly to put a question in science is half the answer, as everybody knows."[178]

The chance for Stensen of an appointment at the university must have been most favourable on the basis of this *Specimen*. First he had asserted himself by his publications in Holland and, as shown by his reception in France and Italy in the years to follow, his name in learned Europe had already good repute. Now, his publication on his research of the glands and the muscles with his sensational discoveries ought to have revealed his fertile genius even in his home country. His teachers were full of admiration when he presented his discoveries, not only in writing but also in his demonstrations of dissections. Thus he undertook an anatomy of the ray-fish on March 21 in the presence of Simon Paulli, and Thomas Bartholin mentions another demonstration which he happened to see when he returned from his country house to Copenhagen. Besides him and Simon Paulli, Willum Worm and Henrik a Møinichen were present. Stensen showed the gland system in a calf head with an amazing dexterity. Further he demonstrated the process of nutrition of a chick in the egg and the structure of the heart. "We left him revigorated", says Thomas Bartholin, "because we had been witnesses of so

---

[177]<594> OPH I, 212.
[178]Gosch II, 1, 195.

many observations which had never been done by others that we all rejoiced in such an outstanding disciple who already surpassed his teachers, and our country for having such an excellent prosector."[179]

Despite outstanding qualifications Stensen obtained no appointment at the university, which was all the more astonishing as the medicine faculty at the time after Thomas Bartholin and Simon Paulli had left, had not had any anatomist or physiologist even approximately of the same value as Stensen. It should not have been difficult to find an opportunity to refill the post.

Already on May 26, 1662 Jakob Henrik Paulli had been appointed *Professor Anatomiae* without having shown more than usual aptitude in his anatomy study. He was, in fact, to quit his professorship some years later. Stensen was admired in Denmark as an anatomist and should have come into consideration. Ole Borch thus wrote from Leiden to Thomas Bartholin on January 9, 1662: "Truly it is a great talent which ripens here for our fatherland."[180] Stensen was far and largely known and appreciated for the *De musculis and glandulis* and for the disputation *De glandulis* of July 1661, which was printed again in 1662, enlarged by three short sections with substantial results in the research on the glands. In the year 1664, Professor of history and geography, Rasmus Vinding who belonged to the Bartholin family circle, intended to retire from practical teaching and to renounce the title of professor by letting the lectures be delivered by an assistant. In the general state of the education of the time and the common practice of the professors of changing from one specialty to a new one and from one faculty to another, it should have been easy for Stensen to obtain financial security, binding him to Copenhagen University.

Nothing like that happened. On the contrary, his fellow student Matthias Jakobsen was appointed university lecturer of geography on August 29, 1664, whereas Jakob Henrik Paulli was to do the lectures of history of Rasmus Vinding. Matthias Jakobsen, neither then nor later outstanding in his scientific performances, was the son of the sister Anna of Thomas Bartholin and the brother-in-law of Rasmus Vinding.[181] It was the general procurator and Professor Peter Scavenius, married with a sister of Matthias Jakobsen, who, on August 26, sent the request to the High Secretary in the Danish chancellery, to attribute the lectures in geography to Jakobsen and those of history

---

[179]Bartholin, *De hepatis* ... 1666, 81 ff., 84.

[180]BEM II, 417 (no. 97).

[181]See the diagram of the Bartholin family in Porter, *The Bartholins*, 77. Cf. Chap. 1.1.

to Jakob Henrik Paulli with cessation of the anatomy lectures of Vinding. These appointments were endorsed 3 days later, fully and completely.[182]

With the powerful position of the Bartholin family, even more the position of Thomas Bartholin at the university, one cannot escape the reproach of short-sighted nepotism. It will continue to be an enigma as long as positive documentation does not exist. It must have been a setback for Stensen and a hard blow to his hopes, though no comments from he himself exist. Stensen saw in such ordeals the hand of the Providence which would lead him into the world to reap rich experiences, new discoveries, great honour and, in the end to what he considered as his great chance, to the Catholic Church and to his conversion. Let us render here his own comprehensive description of his feelings in such event: "No mean is more efficient against the evil shrewdness of the 'educated' than the insight in God's Providence gained from experience, as I obviously gained it in different instances so that, several years before my conversion, I began completely to surrender to it, if actually not always in fact but in my vision in that I always avoided what could remind of human shrewdness and does not agree with the Gospel. From then on I began so to organize my life that every day I did what I held for adequate at the time, the place and my forces. Undisturbed of my future, I wanted to do nothing neither by asking nor by giving to obtain an honour or a position or the perspective of it.

Yes, I urgently prayed my relatives not to approach anybody on my behalf. I wanted to do what I had to do, expecting everything else from God so that I could be assured that what happened to me, came to me, not from hands of men, but by the will of God. In this, I was helped by several incidents, occasions on which my friends had advised me and I thought everything well prepared for a certain zeal. But suddenly unforeseen circumstances intervened which upset the whole plan and at the same time other circumstances turned up which neither I nor anybody else had expected. This happened in such a way that one could not reasonably doubt that it came from God. Often such change of plan or advice was painful for me. But I learned by repeated experiences that the Providence had freed me from great evils and how much good I have received. Finally I have learned to pray this prayer: 'Without a sign from You not a hair falls from the head, not a leaf from the tree, not a bird from the air, no thought escapes from the mind, no word from the tongue, and no movement is made by the hand. You have led me so far on ways I knew not. Lead me now, seeing or blind, on the path of grace. It is

---

[182]Jørgensen, *Nils Stensen*, 77 f. DBL 3rd ed. 12, 663 ff.

easier for You to guide me there, where You will, than it is for me to pull away from where my yearnings draw me'."[183]

From the point of view of Denmark and its university, one must indeed agree with the great Danish historian and early Stensen-biographer, A.D. Jørgensen, when he writes: "We can only deeply regret the stupidity and narrowness of mind with which our high school perceived in this case the true needs of our country, adding an irreparable damage to all our intellectual life."[184]

---

[183]OTH I, 393 f.

[184]Jørgensen, *Nils Stensen*, 69 ff, 78 ff. Garboe, *Thomas Bartholin*, 2, 58 ff., Jul. Petersen, *Bartholinerne*, 86 ff. Also Scherz, *Vom Wege*, 59, cf. with the overview of medical professors of Copenhagen University in the seventeenth century: Slottved, 105–110.

1.3 In Paris and France 1664–1665

Paris, coat of arms

1.3.1 On to Paris

With the appointment of Matthias Jakobsen at the University of Copenhagen on August 29/September 8, any hope for a position suitable for Stensen in his country disappeared. Since on the next day Jakob Kitzerow and Niels Stensen, the heirs of Anna Stichman, were paid 300 rigsdaler,[1] Niels might have decided to take up his study travels again in the first days of that

---

[1] *Anna S. Johan Stichmans hinterlassene Erben Nicolaus Steensen Jacob Kitzerow. Anno 1664 den 30 Augustj betalt her paa—drei Hundrede Rixdr.* EP II, p. IX.

© Springer-Verlag GmbH Germany, part of Springer Nature 2018
T. Kardel and P. Maquet, *Nicolaus Steno*,
https://doi.org/10.1007/978-3-662-55047-2_3

autumn, if not earlier, to ensure the possibility of a life as a researcher abroad. Fr. Sylvius, the Leiden anatomist, convinced him of how much benevolence and consideration, based on his performance so far, he could count on by doing so. He found his reasons not to go back to Leiden compelling. Sylvius advised Stensen to go to Paris and undertook to procure him the degree of medical doctor in Leiden *in absentia*. As he was announced, on December 4, the Senate of Leiden University, on the recommendation of the faculty of medicine and of other professors, agreed and based its decision above all on the disputations already delivered by Stensen and on his published books.[2] The diploma was sent to him in Paris.

Sylvius had certainly recommended Stensen to his friends in Paris. Jan Swammerdam had also been France since 1663. He first lodged in Saumur in the house of Tanaquil Faber, where there was a Protestant academy and, on June 28, 1664, he had sent Stensen a report on his research on insects. However, he soon went to Paris where Thévenot received him in his home and also Stensen soon after.[3] Further Ole Borch had landed in Dieppe on August 26, 1663 after a visit in England with his protégés. He soon came to Paris. Thus Stensen could count on the wide circle of friends of his teacher.[4]

The route that Stensen chose from Copenhagen went through Cologne, which he wanted to visit.[5] To Hamburg he followed the route already mentioned through Ringsted, Korsør, crossing the Great Belt, through Funen and South Jutland. Stensen was able to spend more time to see the small towns of his country than on the likely hurried journey back home in the winter. First there was Roskilde with some 2000 inhabitants and a cathedral with tombs of Danish kings, surrounded by other old churches and a cathedral school tracing back to the Middle Ages commemorating the deeds of Bishop Absalon. There was Ringsted also with royal tombs and bright names like Knud Lavard and Queen Dagmar, which were known by everybody. Slagelse, too, goes back to ancient times of Denmark; a nephew of Stensen, the son of his sister Anna, would soon be dismissed from its Latin school. In Odense the name brought recollections of heathen prehistory and the God Odin. St. Knud's cathedral, named after the Danish martyr king, is located here. Stensen had relatives on his mother's side here; his grandmother was Karen Lethsdatter, the daughter of the burgomaster; and his uncle Niels

---

[2]Molhuysen, *Bronnen*, 117 f = EP I, 183.
[3]Schierbeek, 9 f, cf. Nordström, *Swammerdamiana*, 28 f, 39, 43 ff, 56 (note 42). Swammerdam is mentioned in Borch's diary as being in Paris from September 10, 1664. BOD IV, 121.
[4]BOD III, 75 ff.
[5]OTH I, 190. Also *Reisen*, 41 f.

Bruun was a dean.[6] Further on the route, Schleswig–Holstein also was part of the realm even if its dukes led many a feud against Danish authority.

From Hamburg to Cologne, Lehmann may have been his guide. He went via Celle, the residence town of one of the Braunschweig dukes to which Stensen was to be invited many years later for a discussion on religion. The small town had good relations with Hannover, where just a half year later the convert Johann Friedrich, the brother of the Danish Queen Sophie Amalie, came to power; ten years later he would ask for Stensen to be Apostolic Vicar. The traveller would have seen the residence castle at the Leine River only from the outside. Hameln, the next destination, was then a fortification of the duchy of Hannover. Further, Stensen went over Neuhaus near Paderborn, Soest and Wormkirchen. In Cologne, the German Rome with its many churches, towers and cloisters, our traveller must have seen the cathedral with the chapel of the Three Holy Kings, as well as the town hall, the orphanage and the madhouse. The Jesuit college was the largest of the Lower Rhine province; the fathers were assigned as professors to the university. They ran a big grammar school and were preachers in the cathedral. It is not at all impossible that Stensen met the Jesuit P. Nikolaus Elffen (1626–1706), who was a cathedral preacher for many years.[7] Perhaps it was conversations with this outstanding pastor which so deeply impressed Stensen. Here it came to a discussion of the reproaches of Protestants against the Catholic Church, and the father did not deny that everywhere, even among wise people, ill weeds are also to be found. "Only", Stensen later writes, "evil expands everywhere with the difference that the frequency in the one or the other is often different, according to the place. However, whereas one meets the same vices everywhere, there exists a big difference as compared to the virtuous people. To this testify in our faith the confessors, the martyrs, virgins, bachelors, poor, missionaries and innumerable other examples leading a truly Christian life. Obviously they are among people of all states and stocks. The other confessions cannot point to one single of theirs although it is required from all Christians, at least from the teachers of Christianity to let their light shine for the world. A priest of the Society of Jesus made this objection to me in Cologne in the year 1664. He urged me to show to him among the Lutherans one single follower of Christ from the time that Luther separated himself from the Church."[8]

---

[6]EP II, 901 f.
[7]Koch, L., *Jesuiten-Lexikon*, 1011.
[8]OTH I, 190.

Following the mail routes the journey ahead would usually have led through the Ardennes, first along the Meuse valley and then along the Oise valley. There lay Jülich in the wide, fertile valley of the Ruhr. It was a fortification and residence with a castle and an old church. Further it went to Aachen "with many splendid antiques, particularly the sword of Emperor Charles". The warm baths were also well-known. The author of *De thermis* might well have found interest in them. Via Limburg on the left bank of river Meuse he arrived in Liège with its hundred churches and the bishopric cathedral of St. Lambert. It was the main city of the Walloon country, an important fortification and a significant cultural centre. The fact that the Archbishop of Cologne, Maximilian Heinrich, with two others, also administered this diocese was one of the reasons why Stensen later so energetically opposed him being chosen for the bishopric of Münster. Namur, Charleroi and Mons in Hainaut, having often been fought for, were all very well fortified Belgian towns; but they also stood as the symbol of Belgian industry and coal mining. The traveller would hardly have stopped here before, via the last post-station in Senlis, he reached Paris, the goal of many a student and of all contemporary travelling noblemen.

Until the following autumn, Stensen was now a guest of the realm of the *Roi Soleil* who, three years later, let the neighbours feel the harshness of his power-hungry weaponry. Already then, three years following the death of Mazarin and the take-over of the government by the 18-year-old king, Louis had begun to practise the renowned sentence: *L'Etat c'est moi.*[9]

In the middle of the seventeenth century Paris was a major city of the world with half a million inhabitants. Many of the monumental buildings which are still the goal of the traveller today rejoiced the eye already then, when one, through imposing gates, had entered the town with its nearly thousand streets (in 1667 the police prefect La Reynie alighted them with 2736 lamps). They certainly appeared to contemporaries as a brilliant miracle of progress.[10] Paris was even the town of Lafontaine, Corneille and Molière and of Colbert and Bossuet. Outside its walls, the new Versailles would soon replace the old hunting lodge of Louis XIII.

As has already been mentioned, Stensen lived together with Swammerdam at Thévenot's in the Rue de la Tannerie in the elegant Marais quarter.[11]

---

[9] On France in this period, cf. Gaxotte, *La France sous Louis XIV*.

[10] Poëte I, 105 ff, Lemoine, 146 ff, esp. 157.

[11] Schiller and Théodorides, 156. Thévenot in his *Autobiographie* recounts that Stensen stayed with him. Swammerdam stayed a long time with Stensen in Paris, cf. Nordström, *Swammerdamiana*, 56, note 42. In 1680 Thévenot writes to Leibniz: "... *Ie perderois mon temps quand ie prescherois la meme chose a M.*

When in the summer of 1665 their host moved to his country house in Issy outside Paris, they followed him. Both of them would later show their loyalty and gratitude for what Thévenot had done for them.[12]

Melchisédech Thévenot (1620–1692), this friend and benefactor of Stensen in France, was a universal erudite. His government had sent him to Genoa in 1645 and to Rome in 1652. Two years later he took part in the conclave for the election of Pope Alexander VII. In Paris he devoted himself almost completely to scientific interests. The widely travelled linguist and bibliophile had a library comprising 2000 volumes which were not in the royal library, where in fact he became appointed as custodian on December 4, 1685. Thévenot also possessed valuable collections and he ensured the publication of a series of books on travels. In 1685 he became a member of the Royal Academy of Sciences.[13]

Swammerdam had met Ole Borch in Paris already in September 1664. In October Borch travelled—together with Thomas Walgesten, another friend of Stensen, then in Paris—to the Loire valley and, on October 29, he disputed at the Protestant faculty in Angers. Niels Stensen's return to Paris is mentioned for the first time in Borch's 1664 diary.[14]

As in Holland, the diary of Ole Borch permits a vivid impression of the life and work in the French capital of the seventeenth century, as they appeared to the eyes of a visiting Danish erudite. Borch was in the town since September 1663 and the last entry in his diary stems from June 8, 1665.[15] He thus describes Parisian life not only in the year before the arrival of Stensen but also during the months they spent together.

Borch, in his comprehensive notes, guides us to the Louvre,[16] to the cloister of St. Louis, which joins the professorial house of the Jesuits, to the palace Turenne. He visits the Sorbonne renewed by Richelieu and the church

---

(Footnote 11 continued)

Stenon, vous ne m'en dites rien et l'ay tant d'interet a en sauoir des nouvelles. Que deviendreont toutes ces belles observations qu'il a faites sur les animaux, si vous aués occasion d'entrer sur ces matières le vous prie de luy en parler non seulement pour l'interet du public mais pour le mien propre, car il m'a promis plusieurs fois qu'il me donneroit toutes ces observations et principalement celles qu'il a faites dans le temps ou l'ay eu le bien de l'auoir chez moy." (Leibniz, SSB I, III, 425).

[12]Cf. Stensen's letter to Thévenot, OPH II, 95, EP I, 371 f. Nordström, *Swammerdamiana*, 38 ff.

[13]DSB XIII, 334–337. Schiller and Théodorides, 155 ff.

[14]BOD IV, 121, 148 ff, 163.

[15]BOD III, 79 f—IV, 329. Borch's diary ends abruptly on June 8, 1665 in Paris.

[16]Cf. the list of persons and places, BOD IV, 418 ff.

of which the sepulchre of the cardinal was still missing because of the greediness of his heirs. Notre Dame made an overwhelming impression on Borch, although he found it too dark. The circle of friends made excursions to the royal tombs of St. Denis, to Chartres and to the menagerie in Versailles. They also approached the king and his bright court. Borch saw Louis XIV at the peculiar ceremony of the touching of the ailing people in the Tuileries. A crowd of more or less severely suffering scrofula patients was assembled and four physicians examined the action of the royal blessing. He also attended a military parade in the Bois de Vincennes. On February 12, 1665, the professor and his party, perhaps including Stensen, had obtained an invitation to the *Comoedia regia* at which the high nobility and Louis XIV himself led the dance.[17]

Catholic France must have highly intrigued the foreigners, although they did not miss the divine service on Easter Monday with the Dutch residents at the Lutheran church of Charenton. However, the many Catholic churches and cloisters were interesting to them. Borch took part in a Candlemas procession and in a mass of the archbishop in Notre-Dame. He visited a Carthusian cloister and St. Sulpice, not only the church but also the seminary with its hundred priests.

Borch was often guest of the Jesuits in the college Clermont, which took the name of Collège Louis le Grand in gratitude of many signs of royal favour. Its importance was hardly less than that of the university from which the Jesuits were excluded. Borch took part in the disputations and spectacles and listened with interest when a father told him that the college taught 400 pupils from the highest nobility, or when two Jesuits on an excursion to Orléans told him that they had the care of a total of 2000 pupils. He met Father Philipe Labbé and under his guidance admired the library of the college and conversed with him about the program of the studies of the Society of Jesus. P. Labbé (1607–1667) was the famous bibliographer and historian of the Church to whom we a owe large collection of reports of councils.[18] Such visits led right into the hub of spiritual discussions on philosophy of Life in France of that epoch. Cartesianism was wildly contested. The Church had put the books of Descartes on the index, the Jesuits fought his philosophy, and the government forbade the expansion of his theory. However, among men of letters and in the elegant world Descartes was still often the favourite. Whereas the university paid homage to the

---

[17]BOD IV, 245.
[18]BU 22, 256 ff.

Aristotelians, the connection with the Cartesians, however, was kept open. The *Cogito, ergo sum* was discussed and Picot, the translator of Descartes, relates that Rasmus Bartholin had been entrusted a manuscript of the philosopher.[19] Yet Borch seems to notice with a certain satisfaction that Roberval at a disputation in the college Clermont names Descartes "king of the Chimeras". He reported his vanity and exposed with irony that the philosopher wanted to prove God from nature while he could not explain the movement without having recourse to God.[20]

Borch also encountered Jansenism when he relates that, on October 16, 1663, he visited a nuns' cloister on Montmartre and, on this occasion, engaged in a conversation with a priest about Jansenism.[21] That was perhaps general information but the fight against the "Catholic Calvinism" had then been blazing for a while. Twenty years previously Arnauld had published his rigorous book about the frequent communion and ten years earlier Rome had condemned five theses in the *Augustinus* of Jansenius.[22] Paris was divided into two camps, and when Borch also reported a visit to Port-Royal, he named the stronghold of Jansenism. This held true for Port-Royal-de-Paris to where Angélique Arnauld had led her sisters in 1626 and for Port-Royal-des-Champs, which had been populated again by the nuns in 1648. It also served as a cloister for the "eremites" of Port-Royal. They were erudites from the first families of France who devoted themselves here to study, meditation and teaching. There, seven years earlier, Pascal had written his *Lettres provinciales*, and the archbishop of Paris had just laid the interdict on both cloisters in 1664.[23]

Of course the sciences of nature interested Borch the most. Borch knows the medical faculty of the university with its three professors of medicine, Guy Patin, Jean Moreau and Denis Puglon. He notices that the 250 medical doctors of the faculty are nearly all Parisian-born. He listens to lectures in the Jardin Royal and knows the six personal physicians of the king.[24] He also visits less known academies like those at Del Campos and Du Plessis. Borch takes an interest in the famous lithotomist Charles Collot who carries out four

---

[19]Cf. a.o. BOD III, 302 ff, 304 f, 384 f, Schiller and Théodorides, 161 ff, Ueberweg, III, §26, 242 ff.

[20]BOD IV, 32, 306. Cf. Stensen's letter to Spinoza EP I, 235 with similar arguments.

[21]BOD III, 110 f.

[22]On Jansenism, see *Handbuch der Kirchengeschichte* V, 26 ff, esp. 52 ff. Further DTHC 8^I, 318–529.

[23]Escholier, *Port Royal*.

[24]On doctors in Paris, see Lévy-Valensi, *La médecine et les médecins Français au XVIIe*. Also Fauvelle, *Les étudiants en médecine de Paris sous le Grand Roi*, and Le Maguet, *Le monde médical Parisien sous le Grand Roi*.

operations at the Hôtel-Dieu in his presence and he visits the famous mathematician Pierre Petit at whose home he meets Constantijn Huygens. On this occasion the Huygens' pendulum experiment is carried out. Stensen was now introduced to the learned Paris.

1.3.2   Participant in the Life of the Learned

On November 7, 1664, we find Stensen together with Ole Borch at Morel's.[25] Claude Morel († 1703), a distinguished physician and skilful surgeon, who was not frightened away from the most difficult cases and held public demonstrations, was the right man to introduce Stensen to the Paris of the physicians.[26] As one of the directors of the St.-Cosme brotherhood, the old surgeons' association, he procured access to the Amphithéâtre de Saint-Côme for the Danish anatomist. As chief-surgeon of La Charité he was able to make him acquainted with practical hospital work in Paris. La Charité, in the same place as today, was founded by the sons of St. John of God at the beginning of the century. The big hospital with its high wards for the ill and injured was famous for its lithotomies.[27]

Stensen certainly also worked in the largest hospital of the town, the Hôtel-Dieu at the river Seine near Notre-Dame. It was a very old foundation under the chapter of the cathedral. The hospital was open to people of any confession. More than 2000 proper patients and thousands of out-patients were cared for by Augustinian nuns and supported by the Christian love of the Dames de la Charité de Saint Vincent de Paul.[28] At Thévenot's Stensen met the chief-surgeon of the hospital, Pierre Petit.

Stensen, provided with tough work and stimulating research, also strove for the official place of the learned Paris, as he did everywhere on the changing stages of his travels. "The Danish erudite", writes the *Journal des Sçavans* on March 23, 1665, "is now in Paris where he undertakes dissections every day in the presence of many people anxious to learn, and he has done that also in the *Ecole de Médecine* where he has aroused the admiration of everybody by his new discoveries."[29]

---

[25]BOD IV, 163. One notes that Schepelern identified "Morellus" as the doctor J. Moreau, BOD IV, 391.

[26]Eloy, III, 341.

[27]Lévy-Valensi, 320 ff.

[28]Lévy-Valensi, 302 ff. Fosseyeux, *L'Hôtel Dieu*, Coury, *L'Hôtel Dieu*.

[29]*Le Journal des Sçavans* I, où sont contenues les Années 1665 et 1666, 155 f.

The *Ecole de Médecine* was part of the university of Paris, the foundation of Saint Louis. It also recognized the erudite Dane and granted him access to its jealously protected chairs. This certainly did not yet mean any connection with the actual Sorbonne, the theological faculty which then was intellectually characterized by its opposition against Reformation and Cartesianism as well as by its sympathy for the Gallicans and the adversaries of the Jesuits. The Faculty of Medicine, a corporation proud of its independence from the king, had its buildings on the left bank of the Seine, not far from Notre-Dame. Its dissections were performed in the Amphitheatre Riolan, which, after a reconstruction, from 1747 would carry the name of Winslow. The walls are still partially standing there today. However, anatomy and surgery were also taught and practised in what is today the Collège de France, in the Collège Royal and in the gorgeous Jardin Royal des Plantes.[30]

The French university at the time of Stensen was not involved in the bright development of the country, least in the field of natural sciences. The medical faculty was almost entirely devoted teaching of the ancients and their conceptions. "Medicine is a religion, the faculty is its temple, the gods are Hippocrates, Aristotle and Galen." Riolan and Guy Patin were adversaries of the new, and controversies such as the quarrel about the antimony stood in the foreground of interest.[31]

The young science, the experimental science of nature found its best support in private, learned circles. Stensen often returned to the circle of Pierre Bourdelot (1610–1695). The distinguished free-thinking physician had been called to Stockholm in 1651 where, by his successful treatment, he won the favour of Queen Christine who, after his return to France, secured him the abbey of Macé as a prebend.[32] Soon after, he became the centre of an academy in the sessions of which very significant learned men took part. Men like Pecquet, Auzout, De Graaf, Roberval, Gassendi and Justel regularly went to the house in Rue de Tournon.[33] However, it seems that no lasting connection between Stensen and Bourdelot developed. The medical practice of the latter always comprised some sort of quackery and his dignity of abbot served the free mind to extort from his reluctant monks some money of which he was always in need.

---

[30]Lévy-Valensi, 220ff, Fauvelle, 21 ff.

[31]Cf. Le Maguét, 94 ff.

[32]DBF, 6, 1439 f. Brown, 231 f, Lévy-Valensi, 574 ff. It seems that Stensen later wrote letters to Bourdelot, Oldenburg, IV, 30.

[33]Schiller and Théodorides, 168, note 57. Cf. Stensen, *Lecture on the anatomy of the brain*, 98, note 43.

Among the learned people of Paris in the first half of the seventeenth century there were several groups that finally constituted the foundation of the *Académie Royale des Sciences*,[34] and thus founded a proud scientific tradition. The conferences connected with the names Mersenne, Montmort and Thévenot. The circle around the Minorite Father M. Mersenne (1588–1648), "the secretary of learned Europe", had brought together men like Gassendi, Descartes, Hobbes, Roberval and the two Pascals for uninhibited discussions mainly about mathematical problems and for physical experiments.[35] These sessions became more regular under the old *Maître des requêtes*, Henri-Louis de Montmort, one of the 40 members of the French academy who, himself an enthusiastic supporter of Descartes, drew into his circle such determined Cartesians as Clerselier, Rohault and Cordemoy. However, in this circle the new philosophy seems to have acted like an explosive and only the intervention of Thévenot, who invited the remaining faithful members to his own home, saved the circle from ruin and continued the conferences of Montfort.[36]

The wealthy and busy literate of the group was Jean Chapelain (1595–1674),[37] a born Parisian who called Boileau a *méchant poète, mais excellent homme*. During his life he exerted great influence as literary adviser of Richelieu and Colbert and as a friend of Cardinal de Retz. He was co-founder and member of the *Académie Française* and of the *Académie des Inscriptions et Belles Lettres* and was close to the circle of the *Académie des Sciences*. He showed himself to be very interested in the progress of medicine, physics and anatomy and mainly carried out an extensive exchange of letters with the erudites of his time. He seems to have been very interested in the membership of Stensen in the *Académie des Sciences*, as his own institution followed in the year 1666. Mathematicians like Christiaan Huygens and Adrien Auzout had attended preparatory sessions already in June, and the first plenary meeting in which also the anatomists took part took place on December 22 in the Royal Library. Stensen had been in Italy for a long time when Chapelain wrote to him: "What Mr. Thévenot has written to you of my sincere desire to be useful to you is quite true. Would God that it depended on my desire and my power. Nothing would be more pleasant than to be useful to you and I would not have doubted of the success, if you had been here and one had decided to

---

[34]See Scherz's preface to Stensen, *Lecture*, 62 f. Reference 93, note 2. Also EP I, 12 and Hahn, Chap. I, 1 ff.

[35]DSB IX, 316 ff.

[36]Chapelain, 2, 622.

[37]Collas, *Un poète protecteur des lettres*, DBF 8, 408 f.

create an assembly of true nature researchers and outstanding anatomists. Regardless of what I want to do and say, those present have been preferred to those absent. Mr. Pecquet has been admitted at your expense although he had an original sin which seemed to exclude him. But we lose more than you do, whatever honour could have been profitable to you from it. Then, since you are abler in this branch of science than our most competent ones, the regret is entirely on our side that we have been deprived of your light."[38] Stensen also sent his *Myology* to Chapelain.

The diary of Borch notes the following dates for the assemblies at the home of Thévenot: November 10 and 18, December 2 and 9, further January 20, March 3 and 25, April 13 and May 3, 1665.[39] On March 31, Borch received the whole circle in his lodging and, among those absent, Stensen is expressly named. His name reappears on May 29, 1665, when Stensen and many others were consulted at the death-bed of Caspar Christoffer Gersdorff, the sickly ward of Borch.[40]

We know the themes of discussion of some of these assemblies. On November 18, 1664, Stensen demonstrated the parotid gland on a calf's head and also the *ductus Stenonianus* which he had discovered, further the tear system and finally he made a dissection of the heart. On December 2, Swammerdam says that he injected urine in a dog, as Borch mentions. On December 9, the following points of discussion of this *conventus eruditorum* in the home of Thévenot point at Stensen: 1. The course of the muscle fibres in the left ventricle of the heart. 2. Oval vesicles in the heart of a man at the Hôtel-Dieu. 3. A new-born is dissected.[41]

On January 13 and 20, 1965, physical problems were discussed such as the thermometer and the weight of frozen water. The Malta knight Gourillon spoke of his order and the letter of a Jesuit was read. On March 3, the experiment with the paralysis of the hind legs of a dog by ligation of the aorta was publicly carried out by Stensen.[42] On March 25, 1665, Rohault showed a thermoscope. Then the heart of an asthmatic that had died three days earlier was dissected, perhaps by Stensen, and the vivisection of a bat was undertaken. Of the session of April 13 which Closelier, the publisher of the work of

---

[38]EP I, 190.

[39]BOD IV, 164 f, 173 f, 180 f, 186 f, 216 f, 274 f, 283 f, 297, 317 f.

[40]BOD IV, 163, 287, 319. Borch only mentioned Stensen three times in Paris, probably because he, according to Borch, then belonged there. Cf. BOD I, xxxvii.

[41]BOD IV, 173 f, 180 f, 186 f.

[42]BOD IV, 274 f. Without a doubt, this is the so-called Stensen experiment that is briefly described in the *Canis carchariae caput dissectum*, OPH II, 125. The experiment proves that motor control does not only depend on the brain. See <195, 714>.

Descartes, also attended, the diary of Borch reports an *insertio ductus thoracici in axillarem modo Bilsiano*. Auzout spoke of water salamanders which, thrown in a fire, first revealed a love for this element by their movements and then suddenly fell into ashes.[43]

At Thévenot's and in the other circles in which he moved, Stensen became acquainted with many distinguished researchers, above all physicians and surgeons. Thus, Claude Tardy (1601–1671), professor of anatomy and representative of the progressive trend of the medical school,[44] and Jean-Pierre Martel who is mentioned in the letters of the time as the author of a treatise on heat.[45] Later, Pierre Ganellon claimed to be an old friend of Stensen.[46] Stensen came in closer connection with Pierre Borel (1620–1671), a medical doctor from Montpellier who also had tried his hand as a poet. We owe him a book on Descartes' life and writings. He had come to Paris in 1653 where he was soon promoted to personal physician to the king and later a member of the Academy. Stensen had already known him for a long time because of his book on medical anecdotes.[47] Pierre Petit (1594–1677), the chief-surgeon of the Hôtel-Dieu, mentioned in a letter to Christiaan Huygens that the Thévenot's circle had become deeply involved in anatomical investigations immediately after the arrival of Stensen.[48] Besides these professionals, many interested laymen also took part. Thévenot himself must be called an amateur in the true sense of the word. His friend Chapelain has already been mentioned. One of the more colourful figures of the circle was Balthasar Monconys (1611–April 28, 1665) whose travels had led him far in Europe and in the Middle East.[49] The learned royal secretary Henri Justel (1620–1695) is also often mentioned by Borch.[50] He became a good friend of Stensen and corresponded with him in the following years.[51]

---

[43]BOD IV, 297. This question apparently moved erudites of the time very much. Stensen also mentions similar experiments later (Letter to Croone 25th May, 1666, cf. Rome, *Nicolas Sténon et la Royal Society*, 246 ff).

[44]BU 41, 12.

[45]Cf. Oldenburg, I–II, Register.

[46]Winslow, *L'autobiographie*, 35. In 1680 he published the *Epistolica dissertatio de genuina medicinam instituendi ratione*, in which in different places he refers to Stensen's rules of methodology.

[47]<52> OPH II, 234.

[48]DSB X, 546 f. Huygens, IV, 270 f.

[49]BU 28, 605 f.

[50]Brown, 161 ff. BOD IV, Register.

[51]On this friendship, see Oldenburg, IV, 430. Justel was Oldenburg's regular correspondent in Paris. Sometimes he brought news about Stensen, which must have been from an exchange of letters, cf. Oldenburg, IV, 441, V, 37. One time Justel refers from a letter to Bourdelot, Oldenburg, IV, 30.

One must not overlook the mathematicians with whom the old love of Stensen for this discipline could warm up and who soon thereafter would be incorporated in the newly founded *Académie Royale des Sciences*. The Academy introduced its first session on December 22, 1666 with the following resolution: "If the whole nature consists of innumerable combinations of figures and movements, geometry, which alone can calculate the movements and determine the figures, is absolutely necessary for the physics (natural science)."[52]

Of these first members of the Academy of Sciences, besides the already named, are for example, Adrien Auzout (1622–1691), inventor of the micrometer,[53] and Frénicle de Bessy († 1676),[54] Stensen also knew a friend of Pascal's, the professor of mathematics Gilles Personne de Roberval (1602–1675), an irreconcilable adversary of Descartes.[55] Gérard de Cordemoy (1626–1684), properly a lawyer but philosopher by inclination and one of the most ardent supporters of Descartes was also a member.[56] As a historian, he became teacher of the Dauphin by the recommendation of Bossuet. There was also Jacques Rohault (1620–1675), a gifted mathematician and ardent Cartesian.[57] He opened his house for learned persons every Wednesday. In his *Traité de physique* of 1671 he refers to Stensen and his discourse on the nerves and the brain.

The Parisian sojourn of Stensen is usually marked by his famous *Discours sur l'anatomie du cerveau*—a lecture on the anatomy of the brain—delivered at the Thévenot's in the spring of 1665.[58] This is partly due to the great interest for this work even today and partly to the little that we otherwise know on this sojourn.[59] It is notable that Borch does not say a single word about this lecture and its date can only be approximately estimated from a letter of Chapelain of April 8, which vaguely hints at the *Discours*.[60] One is

---

[52]Fontenelle, T. 6, 19.

[53]DSB I, 341 f, EP I, 216 f. DBF 4, 807 f.

[54]BU 15, 143 ff.

[55]DSB XI, 486 ff.

[56]Mouy, 65, 101 ff.

[57]DSB XI, 506 ff, Mouy, 108 ff.

[58]Chapter 2.18. See OPH II, 3 ff and Scherz (ed.) *Stensen, Lecture on the anatomy of the brain*, with facsimile of the original edition, preface, annotations and English and German translations. A German translation in Scherz, *Pionier*, 98 ff. * French editions, Raphaële Andrault, ed., Paris 2009 and Birger Munk Olesen, ed., Paris 2010.

[59]See, for instance, Faller, *Wertschätzung von Stensens Discours*.

[60]Chapelain, 2, 393. See also below.

thus inclined to set the lecture in February, a month during which Borch apparently did not take part in the sessions at Thévenot's.

The interest in the brain can be found only sporadically in Stensen's earlier work. Undoubtedly he was under the influence of his Leiden teacher François de le Boë Sylvius who engaged himself much in the anatomy of the brain. Stensen claimed that he did not know anyone who had engaged himself more in this subject.[61] At the publication of Descartes' *De homine* in Leiden in 1662, Stensen had written to Thomas Bartholin that, in this book, "there are some not inelegant figures. It is certain that they have proceeded from a clever brain. I doubt however whether such images can be seen in any brain."[62] The following year the problem thrilled him even more, as a letter of March 5, 1663 reveals: "Truly, the more I open the brain as well of other beasts as of different kinds of birds the less it seems to me that the brilliant and quite outstanding construction to explain the actions of the beasts, imagined by the illustrious Cartesius, is correct."[63] Stensen then regretted the lack of leisure and money to undertake more precise observations.

It may seem that Stensen with his *Discours* was able and wanted to present to the world a lot of new observations. Yet, this was hardly the case. The high value of the lecture lies on quite different grounds. One is amazed by the clarity and the frankness with which he sees and admits his own uncertainty and the ignorance of his time. It is an intelligent critique with which he whips abstract reasoning, the commonplace talks in natural science of the time. It is the brilliance with which he gives directives for the investigation of the brain, valid for centuries to come. It is finally the clear refusal of the anatomical and physiological constructions of Descartes.[64]

The deep humility of a true erudite is apparent already in the introduction: "Gentlemen,

Instead of promising to satisfy your curiosity in what concerns the anatomy of the brain, I do confess here sincerely and publicly that I know nothing of this matter. I should wish with all my heart to be the only one forced to speak

---

[61]<604> OPH II, 3.

[62]<516> OPH I, 120.

[63]<516> OPH I, 136.

[64]Cf. Djørup, *Stensen's ideas on brain research.* * See also Clarke, *Brain anatomy before Steno*, and Dewhurst, *Willis and Steno*. The *Discours* was reviewed in *Philosophical Transactions* (1669; 5, 1035): "Next, he entertains the reader with an enumeration of the chief *errors* of anatomists touching the brain. And here he examines particularly the systemes of Dr. Willis and Monsieur Des Cartes." For the full quotation, see Kardel, *Steno—life, science, philosophy*, 33–35. For an analysis of Stensen contra Descartes, see Georges-Berthier, *La mécanisme Cartésien*. Stensen's *Discours* in the neuroanatomy between Descartes and Haller was treated in Mazzolini, *Schemes and models of the thinking machine*. Its role in neurophysiology was treated in Meschini, *Neurofisiologia Cartesiana*. (TK).

so for I could take advantage with time of the knowledge of others and it would be very lucky for the human kind if this part which is the most delicate of all and which is liable to very frequent and very dangerous diseases was known as well as many philosophers and anatomists imagine. Few of them take example of the simplicity of Mr. Sylvius who speaks of this matter only with uncertainty although he worked on it more than anyone I know. The number of those who take pains to do nothing is unquestionably the largest. These people who are so prompt to assert anything will give you the description of the brain and the arrangement of its parts with the same self-confidence as if they had been present when this marvelous machine was built and as if they had penetrated all the plans of its great architect. Although the number of those who pretend to know everything is great and although I cannot be accountable for the feeling of others, I remain convinced that those looking for reliable knowledge will find nothing which could satisfy them in all that has been written on the brain. This is certainly the main organ of our soul and the tool with which the soul carries out admirable tasks. Our soul believes to have so penetrated everything which is outside itself, that there is nothing in the world which could set bounds to its knowledge. However, when it has entered its own house, it cannot describe it and does no longer know itself."[65]

After this declaration he wipes away debris derived from the old terms in anatomy that reminds us of all the superficial ideas of the time and of the products of imagination commonplace in early brain physiology (Fig. 1.3.1). He does not permit any doubt that the way to true knowledge will be long by following proposals that could lead to a better understanding. He speaks of the best art and of the best manner to dissect the brain. He describes how to open the skull and names the best ways to obtain reliable representations of the brain, etc. Of greater importance, however, is his understanding that one must endeavour to determine the course of the brain vessels to which, in all probability, its function is tightly connected. Further, he says that valuable knowledge can be obtained by a comparison of brains of different animals, from the lowest to higher animals, as well as through the comparison of brains at different stages in their development within the same animal from foetus to grownup individuals. He also points out the knowledge that the pathologic anatomy of the brain has provided him with and finally he recommends experiments on live animals, the brains of which should be investigated while medicines and poisons are administered in the usual way or by direct

---

[65] <604> OPH II, 3ff. See Clarke, *Brain anatomy before Steno*.

DISCOVRS
DE
MONSIEVR STENON,
SVR
L'ANATOMIE
DV CERVEAV.

A

MESSIEVRS DE
l'Aſſemblée , qui ſe fait chez
Monſieur Theuenot.

A PARIS,
Chez R O B E R T DE NINVILLE, au bout du Pont
S. Michel , au coin de la ruë de la Huchette,
à l'Eſcu de France & de Nauarre.

M DC. LXIX.
AVEC PRIVILEGE DV ROY.

Fig. 1.3.1 Frontispiece of *Discourse of the anatomy of the brain*, Paris 1665 (printed in 1669), facsimile ed. 1950, 8°

application on the brain. He refutes arbitrary claims of the ancients or by Willis concerning the localization of the intellectual capacity in certain parts of the brain. As the proper method to know the machinery of the brain, he recommends to take apart its different parts. Stensen advocates for good illustrations and for no drawing at all rather than bad ones. "That is why all possible means must be used to have exact pictures to which a good sketcher is as necessary as a good anatomist."[66] He complains that too little zeal is devoted to the investigation of the brain, partly because physicians are busy with practical work, partly because professors satisfy themselves with the transmission of the heritage. "Each part properly to be examined requires time and such devotion of mind that all other work and all other thought must be dropped to accomplish this single task."[67] "Sometimes it takes years to discover that which can then be demonstrated to others in less than an hour."[68]

The strongest impression of the maturity and authority with which Stensen strives to ensure knowledge of the truth and sound methods is provided when he declares war against false currents of the time. The *Discours* is an actual fight in writing which directs its darts first of all at the University of Paris, then still a prisoner of faith in ancient authority. This appears most obviously in a conclusive summary of the contents of the *Discours*: "What we have seen so far, Gentlemen, on the insufficiency of the systems of the brain, on the shortcomings of the method which has been followed to dissect and to know it, on the infinity of researches which should be undertaken on men and on animals and this in the different states in which they should be examined, on how little light we find in the writings of our predecessors and on all the attention necessary when working on such delicate pieces, must undeceive those who keep to what they find in the books of the ancients."[69] Yet Stensen is not an enemy of the ancients. He reproaches Bils for not having made use of their knowledge and says: "It must be admitted that so nice observations are reported in the writings of our predecessors that we should have risked to ignore them if they had not warned us. It even sometimes happened that they told us truths which our contemporaries have not recognized because they did not examine them with sufficient application." Stensen himself will only follow such clear demonstrations that they will force all the others to be of the same opinion on their subjects. "I try to obey the laws of philosophy which teach us to search the truth while questioning one's certitude and not to be

[66] <625> OPH II, 24.
[67] <617> OPH II, 17.
[68] <618> OPH II, 18.
[69] <625> OPH II, 26.

content before having been confirmed by the evidence of the demonstration."
He finds ridiculous these self-confident anatomists who discuss first the
purpose of the part the structure of which they do not know and then present
as the reason for the function which they assign to it that God and nature do
nothing without any goal. They fool themselves with the application which
they put on this general sentence.[70]

The *Discours* first of all settles account with Cartesianism, which made a
great sensation in the Paris of this time, 13 years after the death of the
philosopher, while an ardent fight surged for and against his theories. The
kind and style of this critique thus deserves even more admiration. Stensen
shows himself full of respect for the person of Descartes and interprets his
intention as favourably as possible. Descartes did not want to describe the
actual structure of the brain, but he explains to us a machine which would do
everything that men are able to do.[71] So he had outdone all other philosophers, none of whom had given a mechanical explanation of humans and
particularly of the functions of the brain. Stensen talks primarily of such
followers who have absolute confidence in the words of the master. He says:
"The friends of Mr. Descartes who take his 'Man' for a machine, I have no
doubt, will be good enough to believe that I do not speak here against his
machine, the artifice of which I admire, but against those who undertake to
demonstrate that the 'Man' of Mr. Descartes is made like other men: the
experience of anatomy will let them see that such undertaking cannot
succeed."[72]

We know from the statements of Chapelain how this critique of Descartes
was received in the circle of scientific friends. On April 8, 1665 Chapelain
wrote to the well-known anti-Cartesian and later bishop Pierre Daniel Huët:
"I praise the love you nurse for anatomy. The great Democritus began and
concluded his studies with it. Since you are governed by this passion, you
have missed much by your absence from Paris in the last three months when
Mr. Stensen, a Dane, has given the nicest proofs of his art that man ever saw,
to the point that he forced the Cartesians, those dogmatic ones, those
stubborn ones, to recognize, in the presence of the distinguished people of
this town, the mistakes of their patriarch in what concerns the glands of the
brain and the function on which he nevertheless built up all activities of the

---

[70]<618> OPH II, 18 ff.

[71]<609> OPH II, 8. Rothschuh characterized Descartes as "Systematiker" and Stensen as "Methodiker",
see Rotschuh, *Descartes, Stensen und Discours.* * See also Olden-Jørgensen, *Steno and Descartes.*

[72]<612> OPH II, 12.

reasonable mind. But that is not the only thing which aroused the admiration for this learned Dane and, before his departure, one should try to move him to leave us a treatise on his discoveries with figures (for greater clarity). In this field he leaves behind him the ancients and the moderns and since he is only 30 years old, one can expect from him many certain novelties on the human body and great mysteries, for the best of medicine."[73]

The *Discours*, however, did not appear in print right away. On his departure from Paris, Stensen transmitted the manuscript to his protector Thévenot, who wanted to take care of the printing. The completion was protracted although there was a demand for it, as the request of De Graaf shows.[74] Perhaps Thévenot did not want to complete in the nick of time the final edition of the lecture which seems to have been partially written by his hand. On October 26, 1666, he wrote to Prince Leopold in Florence: "I feel great reluctance to use the liberty which Mr. Stensen has allowed me and to lend a hand to his quite outstanding discourse on the anatomy of the brain but, since your Highness has encouraged me, I will let it be printed as soon as possible with the intention, however, to show it to the author before the publication."[75] The publication with Robert de Ninville as the printer and publisher ensued in 1669 and was followed as early as 1671 by a Latin translation published in Leiden.[76]

Among the many admirers of the *Discours*, J.B. Winslow takes the prize. He states "The *Discours* of the late Mr. Stensen on the anatomy of the brain alone was the source and the general model in all my anatomical works."[77] That is printed in numerous editions in several languages of his famous *Exposition anatomique*. V. Maar further notices: "Much of Steno's *Discours* was, in fact, of so far-sighted a nature that it was not till many years after Winslow's time, partly not till our own age, that scientists were able to follow the instructions given by Stensen."[78,79]

---

[73]Chapelain, 2, 393.

[74]*Huygens*, 5, 488. EP I. 187. Also *Stensen*, Lecture, 84 f.

[75]EP I, 14. The complete letter is printed by McKeon. Also Oldenburg, IV, 84 f.

[76]See Stensen, *Lecture*, 85 ff.

[77]Winslow, *Exposition*, xij.

[78]OPH I, xix.

[79]*Illustrations in the *Discours* display one section cut in the midline and three cross-sections, as such new exposures of the human brain. The anatomist Adolf Faller determined the planes of the cross-sections and identified brain structures, clearly and largely correctly depicted by Stensen. Margrethe Herning compared Stensen's sections and MR-scannings of a live human brain. Except for flattening caused by Stensen's study of an unfixed, soft brain, Herning found agreement. In Maar as well as in Faller and in several editions some cross-sections are displayed upside down or mutually out of scale. See the org. text and illustrations facsimile at http://gallica.bnf.fr/ark:/12148/bpt6k106685b/f1.image.pagination.

During his sojourn in Paris, Stensen worked not exclusively on the brain. Of course he did not give up the problem of the muscles and the organs of reproduction also kept him busy. So Thévenot writes to Christiaan Huygens on April 24, 1965: "We have used the opportunity which the cold offered us in the last months and were busy with the anatomical demonstrations and investigations of the reproduction of the animals."[80] That was a field which was related to Stensen's earlier studies of the *Ductus vitellinus*. Swammerdam was also interested in this problem and in this field a fruitful collaboration may well have existed between the two friends.

In the *Acta Hafniensia* we find two papers that are likely to have originated during this sojourn in Paris but also carry observations from the first time in Italy. One is entitled, *The Uterus of a Hare Dissolving Its Own Foetus*.[81] Stensen, to his great amazement, had found in the *cornu uteri* of a hare, a foetus apparently partly absorbed without any sign of putrefaction. From that he concluded that women also could hope in this way to get rid of a dead unborn foetus. The other paper comprises a short presentation of the development of the hen's egg and certainly required many painstaking observations.[82]

It is certain that Stensen also undertook the investigation of an abnormal stillborn human embryo. With his *Dissection of an Embryo Monster for Parisians*, he published the first description of the heart malformation which is now known by the name the Fallot tetralogy. He gives a patho-anatomical description of the foetal circulation. That he was able to picture the abnormal attests his knowledge of the healthy.[83]

In April a significant guest came to Paris. It was the physician André Graindorge (1616–1676)[84] from Caen whose merits on this occasion are not his scientific works which he had completed, but his correspondence with the natural philosopher and apologist, the aforementioned Pierre Daniel Huët (1630–1721) also from Caen.[85] He gives a lively picture of the learned circles

---

[80]Huygens, 5, 343.

[81]<645> OPH II, 57 ff.

[82]<633> OPH II, 41 ff.

[83]<641> OPH II, 51 ff. Scherz, *Pionier*, 96 ff. * The cardiologist Fredrick A. Willius in 1948 (quoting Erik Warburg's paper in Danish from 1942): Niels Stensen observed and clearly described the tetralogy of Fallot (1888) during the course of his dissection of an embryo monster. See also Andersen and Bruun, *Steno-Fallots tetralogy og Bartholin-Pataus syndrom*, 2015.

[84]BU 17, 314.

[85]BU 20, 101 ff. Please note that Chapelain's letter dated April 8, 1665 is also targeted to Huët.

in Paris. Stensen is mentioned several times. Altogether these letters make an excellent supplement to the diaries of Borch.[86]

Borch even mentions Graindorge in his last entry, that of an assembly at Thévenot's place on May 3.[87] On May 9 Graindorge writes to Huët: "I was at an assembly at Mr. Thévenot's where I had the pleasure to see Mr. Stensen at a dissection. He is outstanding and performed well."[88]

On May 19 Graindorge writes again: "This Stensen is startling. Today afternoon we saw a horse eye. To say it straight, compared with him we are like pupils. I asked him to show me a heart tomorrow morning what he promised me with singular courtesy. He is always active. He possesses an inconceivable patience, and by exercise he has acquired an unusual skill. Neither butterfly nor fly can escape his attention. He counts the legs of a louse if one can speak of legs in this connection."[89] And on May 30: "When I depicted us as pupils in comparison with Stensen, I was right. Never before have I seen such dexterity (Fig. 1.3.2). With the eye, scalpels and a small instrument in the hand which always faces the spectators, he shows us what is worth seeing in the structure of the eye… With him I have no reason to get bored while the usual and particular subjects move at a snail's pace."[90]

In June and July Stensen is not mentioned because he and Swammerdam had withdrawn with Thévenot to the country home of the latter at Issy. Swammerdam mentions their sojourn several times. He also mentions excursions to Meudon and along the Seine where he looked for insects and describes a larva which he found right in front of Stensen's window.[91]

On August 5 Stensen is back in Paris where in dissection he investigates muscular fibres of the heart and lung (the latter requires great application) as well as the lachrymal glands and their canals.[92]

Even when Stensen is not directly mentioned, the letters of Graindorge display an interest in fields which Stensen has already dealt with (vessels of the brain and their course, the crystal liquid in the eye, the eye glands, the lymph vessels, the reproduction organs—or fields which he would take up later in

---

[86]*The Graindorge-letters are published by Tolmer and were used by Schiller and Théorides. See also Scherz, *Da Stensen var i Paris*. The letters are e-accesible in REX, Ny Kgl. Saml. 4660, 4°. www.kb.dk/rex They carry the dates: May 9, 10, 30; June 3, 10, 17, 24; July, 1, 8, 15, 22, 29; August, 5, 13, 19; September, 2, 9, 16.

[87]BOD IV, 317.

[88]Tolmer, 267 f.

[89]Tolmer, 269 f.

[90]Tolmer, 272 f.

[91]Nordström, *Swammerdamiana*, 41, 57 f.

[92]Tolmer, 305.

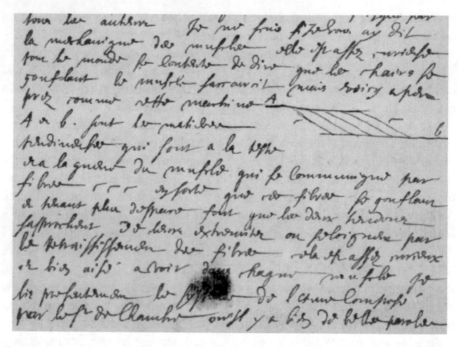

Fig. 1.3.2 In a letter dated Paris July 29, 1665 André Graindorge tells Pierre-Daniel Huet in Caën about Stensen's anatomical presentation of the structure of muscle

Italy—experiments with the blood, a discussion on the role which the Roman opticians Campari and Divini played). The most precious are the references to the discussions on the muscles. They obviously were the starting point of a letter from Florence to Thévenot soon published together with the *Specimen of Elements of Myology* (1667).[93] These demonstrations on the muscles may have been the end of the works of Stensen in that Parisian summer.

1.3.3   Stensen and Catholic Paris

We know very little with certainty of the relations of Stensen to the religious, the Catholic Paris. That he fervently visited the splendid churches and cloisters of the town suggests his pious interest and his travel practice which we are already familiar with and which the diary of Borch describes. Yet what personal connections did Stensen establish? Some 60 years later, Blondel

---

[93]<649> OPH II, 95 ff.

claims that Stensen was strongly influenced by the "Eagle of Meaux"[94] and certainly the Danish researcher knew about Jacques Bénigne Bossuet (1627–1704)[95] who had stayed in Paris almost entirely since 1659 and who was not only a renowned theologian but also a famous preacher to whom the court and all of Paris listened with enthusiasm. Bossuet stood in close contact with the scientific life of his time and his *coetus Cartesianorum* were attended by Stensen's acquaintances, if not by himself as well. In about 1664, Bossuet was particularly interested in winning the Protestants to his side. However, the double *argumentum ex silentio*, which follows the silence of Winslow as well as Arnauld, seems more important and speaks against a friendly relation. The grand nephew of Stensen, who was as enthusiastic about his famous relative as he was about Bossuet who had received him in the Church, spoke of his grand uncle full of admiration. He did not mention in his autobiography that Bossuet exerted any influence on Stensen's conversion. Moreover, Antoine Arnauld (1612–1694),[96] in his *Apologie pour les catholiques* which was published in 1681, points out the influence that a shrewd and pious parson in Amsterdam might have had on Stensen but reveals nothing of any influence from Bossuet, although this book actually wants to win Bossuet's heart and thus to be agreeable to him.

The lack of any testimony of a connection with Port Royal is even more surprising. So much in Jansenism might appear sympathetic to an orthodox Protestant: the omnipotence of grace, the impotence of nature, the harshness against weaknesses, the fear of God, the disregard of the hierarchy, the appeal to the poorly informed pope, the appeal directly to Jesus, the reference to the original Christianity. Yet first of all, Pascal, already then dead but spiritually alive and in many a congenial, ought to lead Stensen to Port Royal which the Frenchman had served during his life with the ardour of his heart and the sharpness of his pen. One would expect a conversation with Mère Angélique and Mère Agnès, the living sisters of A. Arnauld, or with Jacqueline, the sister of Pascal, similar to Maria Flavia, later at the gate of the cloister of Annalena in Florence. Pascal himself (1623–1662),[97] the brilliant mathematician who had rediscovered the elements of the Euclidean geometry, at the age of 16 wrote the most learned treatise on the conic sections since antiquity and became famous by his research on air pressure, had at the age of 30, by his renunciation of a bright learned career for the will of Christ, almost

---

[94]Blondel, 1576.

[95]DTHC II[I], 1049–1089.

[96]Arnauld, *Oeuvres*, 14, 861, cf. Chap. II. On Arnauld see DTHC I[II], 1982 f.

[97]DSB X, 330 ff, DTHC II[II], 2074–2203.

anticipated the course of the life of Stensen. Both men understood the importance of an exact scientific method as well as the idea: "The last step of reason is the insight that there is infinitely much which exceeds our power of comprehension." What Matthias Laros, the German translator of *Pensées*, says of Pascal applies to both of them: "This connection of the scientific with the religious genius has given a unique significance for the world outside the Church and forced the greater minds to a discussion." And: "Philosophically important is his discovery of the function of understanding of the 'heart', the intuition, not in contrary to rational discourse but for the necessary completion as consideration of the appraisal of the first principles of the thinking and above all of the recognition of the infinite."[98] Stensen, as everybody knew, joined the same scientific circle around Montmort and Thévenot to which the Pascals, father and son, had belonged and which familiarly called itself *la compagnie*. It must not be overlooked that Port Royal was then a beacon of Jansenists in their feud with Jesuits. But be it that Stensen was disgusted by the Jansenist solution of the question of the grace: should the grace of God be accessible to all in a sufficient but not compelling manner? Be it because of the rigorism in the question of moral (contrary to the mysticism opened to the world by François de Salle, Olier or Vincent de Paul) as the work of A. Arnauld *La fréquente communion* appeared so clever to him. Be it that Stensen felt disgusted by *Les provinciales* published by Pascal in 1656/1657 against the pretended laxity of the Jesuits, "the typical example of a satirical slaughtering of the adversary", which due to his masterly language, fluent dialectic but also mischievous distortion of the truth has been described as a brilliant mistake, Stensen does not seem to have entered into a connection with either Port Royal or Arnauld. He mentions Pascal only once and that was many years later in the discussion with Leibniz, whose attention he called on the fact that *Les provinciales* by Louis de Montaltes (the pseudonym of the author) had been condemned by Rome in 1657.[99] Later as a theologian, he dissociated himself from the Jansenists, although these regarded his activity as a bishop with sympathy.[100] One will have to think of the differences as well as of the similarities in the spiritual figures of the two men. Perhaps the remark of Pascal, "we come to know the truth, not only with reason, but also with the heart" was too reminiscent of Descartes and Spinoza. Perhaps it opposed the intellectual-historical sense of Stensen who respected the ancients and did not want to know his own work as a total break with the

---

[98]Laros, 992, 993.
[99]OTH I, 337.
[100]EP I, 127 ff.

past. Indeed, Stensen did not like the extreme and suffered from polemics. A squabbler like Arnauld was hardly empathetic to him. The passionate *Les provinciales* have hardly embarrassed his cool critique and did not scare him away from the Jesuits. Our traveller in Paris may not have heard much of the Pascal of the last years, the quiet Pascal who withdrew from the Jansenist partisan mind, chose poverty and self-denial and perhaps reminds us of Stensen's own later years.

The people whose religious connection with Stensen in Paris is testified belonged rather to the opposite side and were in any case not outspoken voices in the intellectual quarrel. Stensen thus visited the sister of the Danish "lensmand" Christian Rantzau, Count of Langeland and Breitenburg, Hedvid Margarete Elisabeth von Rantzau (1625–1706).[101] She had become Catholic with her husband Josias von Rantzau, Marshall of France, on November 10, 1648. After her husband's death she had joined the order of the Annunciation founded by the blessed Maria Vittoria Fornari. The sister introduced him to her confessor Jean Baptiste de la Barre (1609–1680)[102] who counted as an outstanding preacher and as a theologian and an author involved in the controversy. Stensen discussed with him the words of the institution of the Eucharist. P. de la Barre lived in the professorial house of the Jesuits.

Stensen as bishop in Hannover puts us on the trace of another connection when he acknowledged his protector Thévenot on January 25, 1678 for the opportunity of making an acquaintance in Paris and wrote that he himself attributed great importance to his religious development. He writes: "Among the people by whom God gave me particular graces and by whom he prepared me to the position in which he has put me in, you have a large part through the friendship which you have mediated with the maidservant of God, La Perriquet, that God and his eternal glory rejoice. That is the reason why I pray for you every day and I hope that God once grants you that small tour d'esprit which she, the holy soul, wished for you. Well, Sir, all the interesting things of the world are only vanities and the actual knowledge of them is so little compared with what we will see at the first glance on the divine Being."[103] One sees here the worldling Thévenot, much travelled and thirsty for knowledge, interested only in the demonstrations of the Dane, whereas the latter entertains himself with his cousin with religious questions. Marie Perriquet (1624–1668) was a cousin of Thévenot and said to have been a friend of Pascal who confided to her his pangs on *Les provinciales*. She was

---

[101]Krebs, *Maria Elisabeth von Rantzau*.
[102]OTH I, 191, Stensen, *Lecture*, 101.
[103]EP I, 372.

wise and educated. Her personality is depicted as witty, charming and deeply pious. She enjoyed the greatest admiration by the physicist Chr. Huygens who spoke of her as of a saint.[104] Thus one imagines a quite particular scenery. The great traveller and an inquiring mind, Thévenot, at his home full of precious manuscripts, rare books and rich collections, deepened in the conversation with learned friends, and in some embrasure Stensen discussing religious questions with Miss Perriquet, a discussion which was to change the next years of his life.

At the end of the summer of 1665 it came to departures. Ole Borch continued his journey through France and Italy to return to Denmark in November 1666.[105] Swammerdam travelled directly back to Holland.[106] Of Stensen Chapelain writes on July 31 that the Danish anatomist who had aroused the admiration of the French learned men during the winter wanted to pursue his travels through Europe for further education. Before departure, however, his *Discours* was still to be published.[107] As has already been mentioned, this did not happen soon. Stensen gave the manuscript to Thévenot and left Paris.[108] On September 16, Graindorge writes to Huët: "This morning Mr. Stensen has departed, intended for Italy. He travels via Orléans, Bordeaux and follows the German route and he will come back by Lyon. He has promised me to make his way to Caen. He is the openness in person and the most dextrous and the most patient who has ever dealt with anatomy."[109] Stensen took with him written recommendations that Thévenot had given him for his large circle of acquaintances in France and Italy.[110]

1.3.4   Journey in France in the Autumn of 1665

The journey began in the company of Danish friends who had also been with him in Holland.[111] On October 2, Stensen and Jochim arrived in Saumur where they met, among others, Braem and Haisebard. This small town on the

---

[104]EP I, 14, Stensen, *Lecture*, 83 f.

[105]Ole Borch's diary finishes abruptly on June 8, 1665. We do not know when he left Paris. His travel route was: Lyon, Annecy, Genève, Vienne, Tarascon, Nîmes, Montpellier, Arles, Aix, Marseilles, Torino, s. BOD I, xvii f.

[106]Nordström, *Swammerdamiana*, 56, note 41.

[107]Chapelain, II, 406.

[108]Huygens, V, 488.

[109]Tolmer, 321.

[110]<690> OPH II, 95.

[111]Braem who had stayed in France since November 1664 mentions Stensen on October 2, 4 and 6. He and Haisebard left Saumur on October 17. REX, Thott, 1910, 4°. On this see also *Reisen*, 48 ff.

bank of the Loire was then a Protestant stronghold and Swammerdam, for instance, had stayed there in the spring with Tanaquil Faber. Duplessis-Mornay, the "Huguenot pope",[112] resided in this town, which possessed a reformed high school and a college that was founded in the sixteenth century and existed until 1684.

On October 4, Braem and Stensen visited Louis de la Forge,[113] doctor in medicine and professor at the Protestant academy. His philosophical studies had led him to Descartes and he was a friend of the most important Cartesians of his time. In the following year he would publish a *Traité de l'esprit de l'homme suivant les principes de René Descartes*.

On October 6, Stensen with, among others, Hasebrand undertook an excursion to Fontevrault, the famous abbey from the twelfth century, 16 km south of Saumur, which stood so high in the taste of the Plantagenet kings that several of them chose to be buried in its church. The particularity of the abbey was that it comprised a monks' and a nuns' cloister, united under the leadership of an abbess. The abbey had 60 monks and 250 nuns. The members of the order in Fontevrault were recruited from the highest aristocracy of France. The abbesses often stemmed from the royal family or had ranks of princesses.

Stensen and his friends were guided on a day in October by the chaplain of the abbess. They called him Mr. Scherarding and, as always, they visited the cloister exhaustively. The abbess at the time was Jeanne Baptiste de Bourbon. The visitors were shown her audience room, her private chapel and the church where an altar with many relics aroused their attention. The oratory of Madame "where she says her prayers every morning from 5 to 9 o'clock" was also visited. Then they went to the cloister of the monks, its chapter room and garden as well as the library. In the cloister of the nuns, a sister guided the guests to the tombs of the English kings and they then visited the Vendôme hospice where travelling monks could enjoy a free sojourn of 24 h.

From Saumur Stensen continued to Angers. We do not know who accompanied him on this journey. In any case, at his arrival in Pisa in the spring he was presumably alone.

---

[112]BU 29, 347 ff.
[113]Watson, 70 ff.

Angers possessed a Protestant university at which Ole Borch had obtained a doctor's degree after a formal examination in the preceding year. Stensen visited his friend Chapelain and possibly the bishop of Angers, Henri Arnauld, a brother of the above mentioned Antoine Arnauld.[114] Later the journey continued on to Bordeaux, that pretty old city with the important harbour from where Stensen writes to Chapelain and reports the friendly reception which he enjoyed everywhere.[115]

Later he came to Montpellier in Languedoc with the Roman monuments, the splendid churches and a university founded in 1289. The medical faculty vied with any of the capital, yet it was very open and tolerant towards the Protestants. Here lay the oldest botanical garden in France, and the big library with its many manuscripts certainly exerted its force of attraction on the always curious mind of Stensen who may have spent the winter months of 1665–1666 here. He may have found the famous work on sea fauna by the great professor G. Rondelet (1507–1566) of Montpellier and learned about the *Carcharodon rondeletti* which he would mention one year later.[116] The rest we only know about from a letter to Chapelain that of the citizens of the town, Stensen met Laurent Mesme († 1677), a former Cartesian who, after his withdrawal, was called Michel Neuré. He was a universal but satirical mind, particularly active as mathematician and astronomer. Chapelain described him as "an outstanding philosopher who is a right and enlightened judge of scientific experiments".[117]

Notably, here in Montpellier Stensen met a number of Englishmen who were founding members of the *Royal Society of London* or would later become a member.[118] Martin Lister thus writes: "I had the honour, at a lecture of anatomy and at some particular dissections, to assist Mr. Steno the Dane himself in the cabinet of Lord Ailesbury."[119]

---

[114]EP I, 184.

[115]Cf. EP I, 186.

[116]The exact dating for Stensen's sojourn in Montpellier cannot be made. Chapelain's letters dated Dec. 8 and March 8 are addressed to Montpellier. The latter may not have reached Stensen there. On Rondelet, DSB XI, 527 f.

[117]EP I, 186 f.

[118]On Stensen's meeting with the Englishmen, see Rome, *Nicolas Sténon et la Royal Society of London*; Scherz, *Stensens englische Verbindungen*; Eyles, *The influence of Nicolaus Steno on the development of geological science in Britain*; Poynter, *Nicolaus Steno and the Royal Society of London*; * *Steno on muscles*, 24–33.

[119]*Indice*, 292.

Lord Ailesbury, alias Robert Bruce (ca. 1626–1685), the first Lord Ailesbury,[120] was a founding member of the Royal Society, which does not mean any more than that he displayed a certain interest for the natural sciences, which was, however, an interest so great that he invited Stensen to carry out dissections; this time that of an ox skull in which Stensen, among other things, pointed out his duct.

Martin Lister (1639–1712),[121] who describes this dissection, was originally educated in Cambridge but studied medicine in Montpellier from 1663 until 1666. Later he became practitioner in York. Lister is known mostly because of his scientific production on the molluscs of which the *Historia… conchylium* (1685–1692) became very popular. Like Stensen he set a high value on personal observations and so became a significant zoologist. In 1677 he became a member of the *Royal Society*.

He was also attracted by the agreeable behaviour of the Dane. "The demonstrations were neat and clever," he writes, "wherein I much admired the great modesty of the person and this modesty particularly shined through compared to the great impertinence of the French doctor and professor who also assisted the assembly. Later I visited Mr. Stensen whom I found infinitely friendly and agreeable in the conversation. And I saw in him much of the gallant and honest man of whom the Frenchmen speak as of the scholar."[122]

Others attended these dissections. Skippon remarks in his travel notices: "Mr. Stensen, a Dane, stood at the time in Montpellier and he has much success with his anatomical discoveries, like the *Ductus salivaris* from the parotid to the middle of the cheek. We attended his dissection of an ox skull and saw a blade of grass pushed up into this duct. In man the *Ductus* is straight, in the animal it is twisted."[123]

Philip Skippon (1641–1691), after studies in Cambridge, was on an extensive journey through Europe (1663–1666).[124] In 1667 he became a member of the *Royal Society* but otherwise he was himself not remarkable. The interest of his travel descriptions is of particular value for the cover of his travel companions such as the famous John Ray (1627–1703),[125] his former teacher in Cambridge. Ray, a gifted scientist, had to break off a career at the

---

[120]DNB III, 129 f, Hunter, *The Social Basis*, 83 (F56).

[121]DSB VIII, 415 ff.

[122]*Indice*, 292.

[123]Skippon, 123.

[124]Hunter, *The social basis*, 98 (F 227).

[125]Raven, *John Ray*, DSB XI, 313 ff.

university because in 1662 he declined to swear his oath to Charles II, the *Uniformity Act*. In Cambridge he had begun his great work of registration and cataloguing of English plants, a work which in 1670 resulted in the *Catalogus plantarum Angliae*. Later he also became known for his zoological works. Both Englishmen stayed in Montpellier for a long time, and it was there the close friendship of John Ray and Martin Lister began.

The friendship of Stensen and William Croone (1633–1684) probably also started in Montpellier.[126] Croone was a physician in London who displayed a wide interest in natural science. Like the Earl of Ailesbury, he was a founding member of the *Royal Society* and took an active part in its work. The sessions in London had to be interrupted in June 1665 because of the great plague and Croone seized the opportunity to travel to France. There is no written evidence of a meeting between Croone and Stensen but the first letter of Stensen to Croone stems from the spring of 1666. From then on Croone was Stensen's correspondent to the scientific society in England.[127] All these Englishmen must have acquired an extraordinarily favourable impression of Stensen. In the next years the *Royal Society* displayed great interest in his research and eagerly asked for and reviewed his publications.[128]

The sojourn of Stensen in Montpellier is interesting not only because of his encounter with the Englishmen. It may well have had an implication for early geology. For the first and only time some of the most important geologists of the next decades, Niels Stensen, Martin Lister and John Ray were brought together here. In the *Royal Society*, Robert Hooke[129] had been arguing since 1663 that fossils are petrified plants and animals and his view had just been published in 1665 in the work *Micrographia*. Here in Montpellier only William Croone had had the possibility to hear Hooke in person but his book

---

[126]DSB III, 482–483, Payne et al. *William Croone.*

[127]Rome, *Nicolaus Sténon et…*, 245.

[128]On the Royal Society of London, see Hunter, *Science and society in Restoration England,* esp. Chap. 2 and the list of literature, 203 ff.

[129]DSB VI, 481 ff. On his possible influence on Stensen see Eyles, *The influence of Nicolaus Steno,* 173. Hooke's *Micrographia* appeared in January. A few months later in Paris Auzout started his critique on certain points in this work, however not on palaeontology. *Journal des Sçavans* I, 204, 215. Oldenburg, II, 388. Also cf. Tolmer, 273. It is to be noted that Ray did not study *Micrographia* until after his return to England, Raven, 142. * Robert Hooke, *Micrographia …*, Observ. XVII. *Of Petrify'd wood, and other Petrify'd bodies,* London 1665, 111: "… From all which, and several other particulars which I observ'd, I cannot but think, that all these, and most other kinds of stony bodies which are found thus strangely figured, do owe their formation and figuration, not to any kind of Plastick virtue inherent in the earth, but to the Shells of certain Shel-fishes, which, either by some Deluge, Inundation, Earthquake, or some such other means, came to be thrown to that place, and there to be fill'd with some kind of Mudd or Clay, or petrifying Water, or some other substance, which in tract of time has been settled together and hardened in those shelly moulds into those shaped substances we now find them." These were considerations explicitly termed conjectures.

was known in Paris some months after its publication and the question is whether Stensen, Lister and Ray's interest in the fossils stemmed from their mutual conversations, maybe even on the basis of Hooke's conception.[130] Less than one year later, Stensen would be involved in the shark dissection which induced his first written contribution to the debate. Martin Lister and John Ray, despite their friendship, had much different concepts on the question of fossils, and it was Lister who formulated the sharpest and most well-founded attack on their organic origin.[131]

---

[130]Only Croone was in England 1663–1665. Note that at the first meeting after Croone's return in June 1666 he mentioned a letter from Stensen and also showed a stone brought along from Montpellier, Birch, II, 100.

[131]An overview of studies on fossils by Hooke, Lister, and Ray, and their relation with Stensen's research can be found in Rudwick, *The meaning of fossils*, Chap. II. *See also Yamada, T., *Hooke–Steno relations reconsidered*; Brasier, M., *Deep questions about the nature of early-life signals: a commentary on Lister (1673) 'A description of certain stones figured like plants'*.

1.4 The First Year in Italy

1.4.1 In the Circle of the Grand Duke Ferdinand, Early 1666

In the beginning of 1666 Stensen left Montpellier and France to travel over the Alps to Italy, the country which no young scientist should miss. His first halt was in Pisa,[1] the winter residence of the Medici Grand Duke Ferdinand II.

Ferdinand II[2] whose attractive looks in his youth revealed the Habsburg–Lorraine blood transmitted to the Medici lineage through the marriage of his father Cosimo II (1590–1620) and Maria Magdalena of Austria (1587–1631), and through the grandmother, Christina of Lorraine, had enjoyed an outstanding education. Among his teachers and learned friends were Galileo and Torricelli, Michelini and Viviani. After the early death of his father the regency went first to his mother and grandmother, who quarrelled perpetually

---

[1] *Reisen*, 53 ff, Scherz, *Niels Stensen's Geological Papers*, 41, note 53. Italian erudite were regularly in contact with the rest of Europe. Many Danish scientists travelled to Italy, a.o. Thomas and Rasmus Bartholin and also Thomas Walgesten. Meanwhile, there exist fragments of correspondence that do not mention Stensen, cf. Scherz, *Briefe aus der Bartholinerzeit*. On the political history of Italy in the seventeenth century, see *Storia d'Italia* II, 1–2; on the history of culture see Jannaco; on Tuscany, see Diaz, *Il Granducato di Toscana*, I, esp. 422 ff, Barbensi, *Il pensiero scientifico*, esp. 270 ff. The chronology of Stensen's stay in Italy has been revised several times. For an overall view of Scherz's position in 1971 when he died, see Scherz, *Viviani*, 278 which we largely follow. The view in the *Positio* differs several times, cf. *Positio*, Add. I, 1027 ff. More on that to follow. Please note that some of the Italians used the Florentine calender in which—like then in the British calender—the year started on March 25. Thus, when Redi writes "February 23, 1666", it means "February 23, 1667" in our current calender.
[2] EP I, 15 ff, Imbert, *Vita*, 148 ff, Middleton, 17 ff, Diaz, 422.

© Springer-Verlag GmbH Germany, part of Springer Nature 2018
T. Kardel and P. Maquet, *Nicolaus Steno*,
https://doi.org/10.1007/978-3-662-55047-2_4

with each other and their council. They practised a delicate policy in a disturbed time during the fight for power between France and Spain and between the Emperor and the Pope that kept Italian princes continuously in suspense. Soon after his accession in 1628 while a plague[3] began its ravage for 30 years, the young ruler won the hearts of everybody by his clever and courageous behaviour. His intellectual activity, his friendly benevolent and informal natural personality made him popular everywhere.

His Grand Duchy comprised the *Stato vecchio*, the old state with Florence. It counted over 700 000 inhabitants yet it had to register decline in economy as well as in population.[4] However, the scientific and artistic interests of Ferdinand brought culture to a higher bloom.[5] Together with his brothers, as soon as in 1628, he revived the Platonic Academy, the precursor of the Cimento. The Grand Duke was not only a broad-minded patron but he himself an active researcher. For instance, in 1654 he improved on the thermometer of Galileo by closing the glass tube to keep air out and he set up a laboratory in his own palace. The Medici family moreover showed rare concord and harmony. Ferdinand had ungrudgingly associated his three brothers to the government, above all Prince and later Cardinal Leopold, the noblest Medici figure of the century. Ferdinand's behaviour towards his wife, Maria Vittoria della Rovere (1622–1694) was less harmonious. They were married in 1637. The eccentric French daughter-in-law Marguerite-Louise d'Orléans (1645–1721), the cousin of Louis XIV, caused them much worry.

This noble prince received Stensen with pleasure. Still many years later Stensen recalled not without emotion their first encounter: "When I remember", he writes to Cosimo III, "how much I have received from the House of Your Highness from the first day onwards when the Lord did me the favour of seeing Your illustrious father in Pisa until today, I find the greatest reason to be ashamed as well towards God as towards Your Highness because I have enjoyed so much and repay so little."[6]

---

[3]The bubonic plague of 1629–1631 was brought into Northern Italy from Germany by soldiers from the Thirty Years' War. It was responsible for a reduction of the population by an estimated 27%, in Florence by 10–17%. See M. Manfredini et al.

[4]Imbert, *Vita*, 17 f. Moreover, Reumont, *Geschichte Toskanas*, I. Diaz, 363 ff.

[5]In the contemporary letters there exists a vast, sparcely recorded source material that gives an insight in the scientific environment at the court of the Grand Duke. Part of this material has been published but rather sporadically. In the correspondence of Malpighi, used and later published by Adelmann, particularly the letters of Borelli make such an insight, Malpighi, *Correspondence*, Adelmann, I, esp. Chap. 8, 203 ff.

[6]EP II, 652.

Stensen must have arrived in Pisa at the end of March or at the beginning of April 1666.[7] The splendid commercial centre of the Middle Ages now counted about 11 000 inhabitants only.[8] The witnesses of its magnificence still remained: a Roman cathedral with five naves, the wonderful baptistery and campanile attracted travellers, and bright feasts still took place in the palaces of the nobility on the bank of the Arno. The friend of science might devote himself to study at the university and in the old botanical garden with rare and expensive plants. Pisa was the birth place of Galileo. There he had studied mathematics in high school and made his first great discovery of the laws of the pendulum.[9]

The Medicis received Stensen in a very friendly manner and took an eager interest in his anatomical knowledge. Soon after his arrival Stensen made a demonstration at the end of which he showed the structure of the heart. Besides the Grand Duke himself and Prince Leopold,[10] he also met Prince Cosimo (the future Cosimo III) who found the young Dane extremely affable and modest.[11] Among the court officials he formed a friendship with the noble Marquis Luca degli Albizzi (1638–1708).[12] In these first days in Italy he made a friend in the young professor of anatomy, Lorenzo Bellini (1643–1704) who, at the age of 19 had published an important treatise on the kidneys and who had become professor in Pisa at the age of 20.[13]

---

[7]Nordström, *Magliabechi*, 34 f. *Cf., a recently recovered letter by Prince Leopold to unknown recipient dated 1666, April 27, (unchecked transl.): "We have had [illegible] of two persons passing by here, about whom I think I have already written to you, who are Mr. Stenone, Danish anatomist, young of age but remarkable in his business, [illegible] moreover with every sort of erudition, and a good geometrist which much helps him in his business, and the true type of modesty". On recording in REX

[8]Imbert, *Vita*, 17 f.

[9]On Pisa as a centre of science, see Adelmann, I, 137 ff.

[10]His fame had rushed ahead of him, thus Prince Leopold, M.A. Ricci and Viviani knew about him before his arrival, see Scherz, *Niels Stensen's geological work*, 41, Note 53, Sten. Cath. 4, 1958, 71, Scherz, *Viviani*, 274. The *Stenoniana Catholica* was a journal in Danish edited by Gustav Scherz. Ole Borch's diary from his sojourn in Italy is, as has already been mentioned, not preserved. In Borch's autobiography he writes: "As to the other Italian towns, he found none more flourishing and elegant than Florence, the Grand Duke as a patron of learning being a true likeness of Apollo and his brother Prince Leopold, the later Cardinal Medici, receiving Borch like a Mæcenas regenerated. Lorenzo Magalotti, Francesco Redi, and Carlo Dati admitted him to their friendship, giving him access to the acquisite Medicean Library filled with rare manuscripts." BOD, I xviii.

[11]Nordström, *Magliabechi*, 34.

[12]Cf. EP I, 316, 317, II, 622, 741.

[13]Nordström, *Magliabechi*, 35, DBI 7, 713 ff, DSB I, 592 ff.

Pisa, coat of arms

Carlo Fracassati (†1672), the professor of Medicine in Pisa, also came into contact with Stensen. In 1668–1670, he was professor of surgery and anatomy in his birth town and went from there to Messina. He was a friend of Malpighi and worked together with Borelli, with whom he planned a history of anatomy.[14]

The friendship of Stensen and Michel Angelo Ricci (1619–1682), another guest of the city, also began here in Pisa.[15] Born in Rome from poor parents he enjoyed a good humanistic education. He studied physics and mathematics under Torricelli. In 1666 Ricci published his *Exercitatio geometrica de maximis et minimis*, a short but valuable work on the determination of tangents. He was an active councillor of the Accademia del Cimento and was also called by the Pisa university as arbitrator in learned controversies. He further kept a lively exchange of letters with many learned men of Europe, particularly Thévenot and Chr. Huygens, and was actively interested in the foundations of the academies of London and Paris. He also actively helped Fr. Nazzari and G. Ciampini in the publication of the *Giornale dei Litterati,* the scientific journal which appeared in Rome in 1668. Pope Alexander VII made Ricci Consulter of the Holy Office. Pope Clement IX used him to found the academy for studies of antiquity, and Pope Innocent XI in 1681 appointed him cardinal, in spite of his own reluctance, the man being modest and totally devoted to science and piety. In a letter of July 10, 1666 to Viviani, Ricci speaks of his first encounter with Stensen: "The particular respect which I had felt for his person (i.e. by reading Stensen's publications) was confirmed by his affable nature and the friendly traits of his mind, noble gifts which appeared when I had the opportunity to see him in Pisa."[16]

[14]EP I, 212 f, Adelmann, I, esp. 227 f, 375, see also Index, vol. V, 2387.

[15]EP I, 32 f, Sten. Cath. 5, 1959, 11 ff, DSB IX, 404 f.

[16]Sten. Cath. 4, 1958, 71.

The Medici coat of arms

When Cardinal Leopold sent Stensen's *Myology* to Ricci, the Roman erudite thanked him on May 30, 1667 and said he was enthused by the zeal, the gift of observation and the genius of the Danish researcher and praised his endeavours in the spirit of Galileo.[17] In November 1668 Ricci wrote to the cardinal: "Yesterday I saw to my great pleasure Mr. Niccolò Stenone whom I have not greeted since he has become Catholic... I like his modesty and uprightness and his mind which is clear and rich in nature sciences and other nice knowledge."[18]

The court normally spent the winter months from Christmas to Easter in Pisa, thus in 1666 until the end of April (Easter was on April 25). Stensen did not stay that long. On April 11 he journeyed to Florence, the capital of the Medicis.[19]

The town was surrounded by 3.5-yard thick and 30-yard high walls over which towers rose every 200 yards. Together with the church spires they formed a "forest of towers". Today's suburbs did not exist but the four old quarters of San Giovanni, Santa Croce, Santa Maria Novella and Santo Spirito were already there. Four bridges led over the Arno.[20] When he went through the old tortuous narrow streets, Stensen might well have visited them first like Holger Jakobsen, his disciple, to whom he procured the good quarter all'Angelo in Via de'Calceoli, perhaps the same in which he had lived himself during his first sojourn.[21] Jakobsen actively walked around in the town which, with its 85 000 inhabitants, was three times as big as Copenhagen. Jakobsen admired the cathedral made of black and white marble and churches like Santa Croce and S. Annunziata with the image of grace. He assiduously

---

[17]Scherz, *Niels Stensen und Galilei*, 44.

[18]EP I, 33.

[19]Nordström, *Magliabechi*, 34 f.

[20]Imbert, *Vita*, 35 f.

[21]Jakobsen, 106.

visited the libraries, particularly the Laurenziana and the one near St. Maria Novella, the key of which was delivered to him by Magliabechi. Also the picture gallery of the Uffizi, the Palazzo Vecchio and the residence of the Grand Princes in the Palazzo Pitti with the Bobolo gardens aroused his admiration. He knew of the Fonderia of the Grand Duke with its medicines and was impressed by the Ospidale degli Innocenti and the 37 other charitable institutions.

Soon after his arrival, Stensen was certainly visited by Magliabechi. Prince Leopold had urged him to become acquainted with Stensen.[22] Antonio Magliabechi (1633–1714),[23] born in Florence and goldsmith in his youth, soon distinguished himself so much by his knowledge of languages and his other erudition acquired from books that Princes Leopold and Cosimo entrusted him with the supervision of their libraries. From then onwards he spent his life in the libraries and between piles of books in his house in Via della Scala. He was insatiable in the reading of books, the contents of which he stored in his gigantic memory. Hence for learned men of his time he became an oracle never to be ignored to whom anyone from all over Europe addressed questions. Stensen thus immediately obtained the best introduction to the learned circle in Florence.

Now Stensen came into contact with the scientists of the Accademia del Cimento, the founder of which he had already met in Pisa.[24]

Leopold de' Medici (1617–1675)[25] was the greatest of his lineage in the seventeenth century, a harmonious, highly gifted nature, ennobled by high-principled education and self-education. He had a humanistic education —he mastered Latin, Greek and French—accompanied early on by a great love for natural science and mathematics, which he had learned under the guidance of Galileo and his disciples, Soldiani, Michelini and Torricelli. He admired the great physicist and intervened actively for the publication of his work. He also maintained a correspondence with many other learned men of his time and possessed in his palace a rich library for the keeping of which he had discerned Antonio Magliabechi, as mentioned above, and which he gladly made available to learned men. He had a cheerful and humorous nature, even in the later years when he was often ill. Full of interest for theatre, music and every noble art, he himself drew and painted, collected paintings, sculptures, medals and antiques. Many pieces of art known today from all over the world

---

[22]Nordström, *Magliabechi*, 34 f.

[23]Scherz, *Niels Stensen und Magliabechi*, cf. Nordström, *Magliabechi*.

[24]Nordström, *Magliabechi*, 34, Sten. Cath, 5, 1959, 12f.

[25]EP I, 17f, Middleton, 33 f, *Omaggio a Leopoldo de'Medici*, esp. I, 26 ff.

found their way in his time to the Palazzo Pitti. He was also an example for his surroundings from the moral and religious point of view. He showed himself as benevolent as generous without his kindness degenerating into weakness. The Grand Prince already had a successful government of Siena behind him (1636–1641), where he had won the hearts of the people by strong justice coupled with kindness. Once returned to Florence, he actively favoured trade and industry. He was the driving force for the revival of the Platonic academy mentioned above. He helped with the publication of the big dictionary by the Crusca academy of which he had been a member since 1641. He was made cardinal by Pope Clement IX on December 12, 1667. On December 3, 1674, just 1 year before his death, he was ordained priest.

The greatest merit of Leopold concerning natural science was the foundation of the Accademia del Cimento, "the experimental academy", the spiritual father of which was Galileo Galilei (1564–1642).[26] When Stensen arrived in Tuscany, the great physicist had been buried for many years. However, his brilliant thought and knowledge strongly dominated scientific life in the Arno city, and his tragic fate still made the hearts of so many eye and ear-witnesses shiver. Beyond doubt it must have come to a meeting between the spirit of Galileo and the guest from the North whose own research was based so completely on critical observation and experimental method and who had already taken the way towards the Church with the authorities of which the tempestuous Pisano had collided. However, on this spiritual encounter we possess only a very short written testimony, which deserves attention. When Stensen writes about the formation of crystals in liquids, namely in *De solido*, he refers to the *Solidissimae magni Galilei demonstrationes*,[27] with an expression of unconcealed respect. One must realize the background of the spiritual situation to be able to evaluate properly this respectful admiration. Galileo had been registered by his father to study medicine at Pisa university. He had soon turned to mathematics and its applications. He discovered the laws of falling objects and so established the conceptual basis of classical mechanics. Already as professor of mathematics in Padua (1592–1609) he enjoyed European fame but his name flew all over the world only when he, stimulated by the discoveries of others, built his own telescope, directed it towards the sky and made his great discoveries, which he published in the *Nuntius Sidereus* in the year 1610: the mountains of the moon, the moons of Jupiter, the innumerable stars of the Milky Way and of the Pleiades.

---

[26]DSB V, 237–249, DTHC, 6, 1058–1094.
[27] <803> OPH II, 210.

Then Galileo returned to Tuscany. He dedicated the *Nuntius Sidereus* to the Medicis, whose name he also gave to the planets of Jupiter, the "Medicean planets". In 1610 he was appointed *Mathematicus Primarius* at the university of Pisa and mathematician and philosopher of the Grand Duke of Tuscany. Here the battle was waiting for him.

First he continued his march to victory, also when he went to Rome in the first months after his return, to answer his peripatetic opponents. Here even the astronomers of the *Collegio Romano* confirmed his discoveries. Galileo was a guest of the cardinals and he was greeted as the new Columbus of the sky.

Yet, in the time to follow, as he adopted more and more strongly the view that the Copernican theory of the movement of the earth round the sun was not a hypothesis but a doubtless truth, on February 23, 1616, the theologians of the Holy Office declared the view that the sun is in the centre of the world and that the earth moves around it as heretical. Three days later Galileo had to promise not to present this theory either in writing or in speaking in future. Soon afterwards, the books of Copernicus were put on the Index. In the years to follow Galileo provoked his opponents in writings and discussions, particularly in 1632 by the Dialogue on the *due massimi sistemi del mondo, Tolemaico e Copernico*. In 1633 he was forced by the Inquisition, with all harshness and the threat of more severe punishment, to abjure his theories. Galileo spent his last years in Arcetri in a kind of mild imprisonment, partly in the vicinity of his eldest daughter, Sister Maria Celeste (1600–1634).

In the verdict of the Inquisition one can often see an actual religious decision. Yet one must keep in mind that Galileo himself remained a convinced and faithful Christian Catholic to the end of his life.[28] His piety resisted his trial in the conflict with the Christian authorities. Galileo did not consider this conflict as a conflict with the Church. He himself and his friends hardly saw in the decision a proper point of faith or an obligation to believe but a disciplinary prohibition to which they submitted themselves. The problem was little discussed publicly. However, Galileo's friends felt totally justified in conscience to retain the Copernican theory as a hypothesis from a scientific point of view, which actually it still was as long as the laws of Kepler and the theory of gravitation of Newton were not yet known or recognized. When Stensen, after his conversion, names the great Galileo, it is without any indication of a conflict between science and faith.

Galileo required from science the greatest possible certitude with irrefutable experience and logically compelling conclusions. The most mature fruit

---

[28]Scherz, *Niels Stensen und Galilei*, 48 ff.

of Galileo's mind was the Accademia del Cimento, the first famous experimental academy which wanted to serve the blossoming nature research under the device, *Provando e riprovando*.[29] The noble enthusiasm for true research which Stensen found everywhere delighted him. He admired Ferdinand II who already at the age of 32, soon after Galileo's death, founded in his palace a *Conversazione filosofica* to which men like Torricelli, Michelini, Viviani and Redi belonged. In the year 1657 the experimental academy was established under the auspices of his brother. Prince Leopold made rooms available in his palace. He gathered able men informally and without any huge organization. He put means at their disposal. He himself and his brother, the Grand Duke, displayed much understanding for the technical workmanlike side of research and kept artisans for the manufacture of various and outstanding instruments. The subjects with which the academy dealt almost exclusively until the arrival of Stensen derived from the mind of Galileo. There were problems in physics, astronomy, hydrostatics, optics, electricity theory, acoustics and meteorology. The first session of the academy on June 19, 1657 was followed by many others in an irregular succession, taking place more rarely since the beginning of 1666 if one can rely on the diary of the academy which Magalotti kept as the secretary. However, at this time Magalotti travelled to Rome and Naples. In November 1666 two sessions are reported (on the 22nd and 28th), in January 1667 nine, in February two (on the 4th and 9th) and on March 5 the last one. In those months, under the editorship of Magalotti, there also appeared, a famous golden book which recorded the results of the Cimento, the "*Saggi di naturali esperienze*".[30]

As Stensen did not come to Italy until the beginning of 1666, the question arises whether he really became a member of the academy. This question can be answered in the affirmative with great reliability. His name is not in the original members' register which comprises nine learned men, among whom were Borelli, Dati, Magalotti, and Viviani. Redi who undoubtedly belonged to the academy is also missing from this list. Soon after his arrival, Stensen was in active personal dealings with different members. He took part in the sessions and worked for the academy with the agreement of Prince Leopold, as the correspondence of the latter with Guastaferri after the publication of the Saggi shows. In any case, he was more justified as a member of the Cimento than as a member in the language academy of the Crusca in which his admittance on July 16, 1668 is testified. One was happy in Florence to be able to refresh the

---

[29]On the Academy, see Middleton, *The experimenters*. A study of the Accademia del Cimento, 1971. Also EP I, 18 ff.

[30]Middleton, 83–254 is a commented English translation of this work.

lustre of the academy by his fame. Magalotti, the secretary of the academy, therefore, surely reckoned with his membership when he mentions in a letter from Antwerp that the place in the Cimento which had become vacant by the departure of Borelli and others was filled by Stensen. Caverni solved the question by distinguishing four periods of the academy which according to him began with the death of Galileo. The final period would be from 1667 to the death of Cardinal Leopold in 1675. In this period Redi, Viviani and Stensen were the most important representatives of the circle of researchers. Their interests and results embossed the work of the academy.[31]

Stensen developed friendly relations with the already mentioned secretary of the academy. These relations were nourished by common intellectual interests.

Lorenzo Magalotti (1637–1712),[32] the descendant of a Florentine patrician family, was born in Rome and educated by Roman Jesuits. In 1656 he was introduced to anatomy in Pisa by Malpighi, in philosophy by Rinaldini and in mathematics by Borelli. Chamberlain to Prince Leopold since 1659 and to the Grand Duke since 1662, Magalotti had also taken over from Segni the job as secretary of the Accademia del Cimento in 1660 where his manifold education was very useful. Without being the driving force in research, he provided good specimens of scientific investigations on cold, air pressure and sound. He was also trained to be methodical and so appreciated that the Royal Society of London admitted him as a member. Moreover, he had a fine feeling for language and style. For the active member of the Crusca, Latin and Greek were familiar, English and French current, and Turkish and Arabic not unknown. If not exactly a poet, he rightly deserves a place in the history of Italian literature as a prose writer, as proved by the *Saggi* written by him with a Galilean clarity. In the home of the friend beyond the bridge of St. Trinity, Stensen could certainly often admire the erudite taste of Magalotti. Besides the advantage of a richly formed mind, Magalotti also possessed pleasant features, such as a high majestic stature and an elegant affable appearance. He was attracted by the wide world. He wanted to be a *postiglione di Europa* and his travels as envoy of the Grand Duke led him, after the Netherlands and Germany, to Vienna and the far North.

Stensen came to know Magalotti when he arrived at Florence in April 1666 and the count was an interested witness of the demonstrations of the Dane at

---

[31]Caverni, I, 181–205. Middleton lets the academy completely cease already in 1667, and Stensen is mentioned only as a possible member (Middleton, 309 ff, esp. 317, 326 f). Bonelli (p. 260) is of opinion that: "… Stensen was an Academian of the Cimento and all the more so because there never was a real statute that bound the members".

[32]DSB IX, 3 ff, EP I, 23 ff.

the court of the Grand Duke. In May he mentions that Stensen is busy with studying lobster, fishes and other materials from the sea and from rivers,[33] and in August he speaks of a series of demonstrations which Stensen had carried out on the junction of the lymph vessels in the subclavian vein, the thoracic duct of Pecquet and of vessels so far unknown in the eel. Stensen had killed a dog by introducing liquids in the veins. By ligating an artery he had paralyzed the hind legs in a dog. He had made still other observations on the nature of the muscles than those already in print. Indeed, on the whole he offered *una selva di speculazioni*.[34]

Francesco Redi (1626–1697)[35] from Arezzo, who had thoroughly studied medicine and natural science in Pisa and, after further education in Rome, Venice and Padua, committed himself to his native Tuscany, had just then (on November 28, 1666) been appointed by Ferdinand II as physician in ordinary, lecturer at the *Studio Fiorentino* and leader of the Grand Duke's Fonderia.

As a physician he stood out by his simple and genuine methods. He rejected a lot of useless medicines, recommended moderation in food and drink and was very critical of the fancy theories of his time as shown by his manuscript, *De erroribus medicorum*. Behind the parchment face and the dryness of the figure of the young fellow one would not have expected all the intellectual life which stirred in him, of whom E. Menage says that it is doubtful what is greatest in Redi, his "thorough knowledge, harmony of eloquence or amiability of appearance".[36] The genius acquired a name not only as natural researcher. Redi is still today respectfully mentioned for his rejection of *generatio spontanea*, his investigations on the infusorians, on viper poison and on the generation of smaller animals. Italian history knows him also as a poet and linguist.

He must have been a stimulating acquaintance for Stensen. He was a thorough expert of his mother tongue and its dialects. In addition to the classical languages he mastered Hebrew, Arabic, French, Spanish and German as well as the Provençal and Catalan dialects, and, as a member of the Crusca academy, he was highly praised for his dictionary. The hilarious songs of his repeatedly published dithyramb, *Bacco in Toscana,* show in their style of an Anacreon and a Horace that he could introduce his guest from the North into the knowledge of the Italian wines. This poetry, a wanton praise of the wine god, was first published in 1685 but stemmed from one of the meetings of

---

[33] OPH II, 286.
[34] EP II, 922.
[35] DSB XV, 341 ff, EP I, 21 ff, Leikola, 115 ff.
[36] Redi, *Opere* 1, 13.

the Crusca, a so-called *Stravizzo*, held on September 12, 1666, in which, besides many Florentians and foreigners, Prince Leopold and his brother Mathias also took part, probably with Stensen.[37] In the poem Bacchus was presented as a guest in the Medici *Villa Poggio* and celebrated off-hand in verses and rhymes which increasingly remind us of the language of drunkenness. Bacchus curses the bad and weak wines, praises the wines of Tuscany and appeases a violent sea storm with two bottles of the noble juice which are hanging in the yard. In the end, the wine of Montepulciano is declared the king of all wines. These festive evening meals were held in one of the many palaces of the nobility and there, as said, the Crusca academy, the society for the purity of language, which had bran in its coat of arms, was invited. For the feast the humorous and festive frame of a mill was created. The light came from 12 sieves in the shape of stars which hung down from the ceiling. The chairs figured a back-basket and a shovel. The chairman sat on a throne of millstones. The clothes were sacks, the table a dough vat. One drew the manuscripts from a mill funnel, held all sorts of funny talks and each participant had his own academy name.[38]

For the year 1666 and the following time the closest collaboration of Stensen with Redi has been testified. Redi was a competent physician and an active experimenter. He was considered as the founder of the Tuscan school. The cursory recordings reveal many shared investigations. Observations of the so-called *cavalucci* stem from the autumn 1666.[39] On the occasion of a discussion on the skin vessels of fishes, Stensen mentions that, together with Redi, he had observed a passage in the skin of an eel.[40] Redi mentions Stensen in connection with the *Argenteus piscis*, the silver fish (*Argentina sphyraena*) and the live small animals that he found in its stomach, "since this observation was made at the end of 1666 by this very learned prelate in my rooms in Livorno".[41] He also reports experiments on a scorpion, which he let starve from November 1666 onwards until in January 1667 it was stiffened and looked quite motionless. In Pisa, together with a French guest, the surgeon Claude Morel, on February 23, 1667 he then let the animal put to the test the strength of its poison on a young dove, which died after the sting.

---

[37] Massai, *Lo Stravizzo della Crusca del 12. September, 1666 e l'origine del "Bacco in Toscana di Francesco Redi"*.

[38] Imbert, *Vita*, 161.

[39] Redi, *Opere* 3, 121.

[40] <705> OPH II, 117.

[41] Redi, *Opere* 3, 353.

Stensen might have seen this, been stimulated to investigate the blood and viscera, and repeated the experiment with several chickens.[42] Many reports testify the cordial friendship between the two learned men. They were so intimate that they also met at festive "*brindisi*" and Stensen borrowed some money from the physician. Stensen was certainly a frequent guest at Redi's home at the Canto a'Soldani and, from 1672, in the Via dei Bardi.[43]

Warm friendship and noble enthusiasm for mathematics further tied Stensen to the much older Vincenzio Viviani (1622–1703), the last "disciple of Galileo".[44] In the school of the Piarist fathers, Sebastian de Pietra Santa and Clemens Settimi, who had early drawn his attention to the value of geometry, the 17-year-old Vincenzio had already then made such great progress that he attracted the interest of Ferdinand II, who chose him as his mathematician and introduced him to Galileo. Together with Torricelli, Viviani spent the last years of Galileo's life in Arcetri in the vicinity of the master, who had grown blind. He wrote a precious summary of the life of Galileo and considered the publication of the complete manuscripts of the physicist as the great task of his life. The Grand Duke made ample use of the practical genius of the young engineer, among other things for the work of strengthening fortifications, and, following the death of Torricelli, appointed Viviani as a lecturer in mathematics at the *Accademia Fiorentina del disegno*, then in 1649 professor of mathematics of the pages of the court and, on May 15, 1666, as mathematician of the Grand Duke. In the controversy between Rome and Florence about the watershed between the rivers Tiber and Arno, it was Viviani who had to study the hydrographical conditions of Tuscany. By walking up and down the rivers he became an expert of the petrifications, the old Etruscan inscriptions and the landscape of Tuscany.

Viviani and Stensen soon showed much empathy for each other. Ricci expresses much joy about this in a letter of July 10, 1666 to Viviani: "Therefrom, Sir, you may judge how much I liked to see at your home this noble man (i.e. Stensen), worth all the good marks which you dispense him. He says that he is very pleased and thankful for the friendship which you have showed him and he declares that he appreciates your virtue and merit. I enjoy his good opinion of you. It seems to me to be an obligation to the truth to tell you that I hope also to arouse in you the same feeling..."[45]

---

[42]Redi, *Opere* 7, 213 f.

[43]EP I, 22 f.

[44]DSB XIV, 48 ff, Scherz, *Viviani*, 262 ff.

[45]Sten. Cath. 4, 1958, 71.

Already here one meets Stensen as a guest at Viviani's home in the Via S. Antonio where one could see, near the bust and pictures of the life of Galileo, also a statue of Louis XIV. Above all they worked together in the Accademia del Cimento. Viviani was particularly interested in experiments on the balance of the liquids in tubes as well as in the theory of waves and its application to acoustics.

For Stensen, Viviani soon became the always obliging intermediary with the court who cared for lodging, the livelihood and position of his friend, particularly in the absence of Magalotti. Testimonies of the cordial solicitude of Viviani have been kept for us on the first arrival of Stensen in 1666 as well as on his return in 1670.[46] Yet more important than the helpful friendship were the scientific stimuli of the mathematician Viviani, who on his part found in the Danish erudite an expert in zoology and anatomy, the intellectual complement for which, since his feud with Borelli, he had wished so much.

Giovanni Alfonso Borelli (1608–1679),[47] the son of a Spanish soldier, was born in Naples. He studied in Rome under Castelli and as fellow-student of Torricelli. He then worked as a teacher of mathematics in Messina (1635–1656) from where he was called to Pisa and in the following years undertook anatomical studies and became an active member of the Accademia del Cimento. In 1667 he returned to Messina and in 1674 went to Rome where his main work *De motu animalium* was published posthumously at the expense of the Swedish Queen Christina.

The tension between Borelli and Viviani had existed since 1658 when the latter, a great admirer of Greek mathematics, endeavoured to reconstruct the work of Apollonius on conic sections and above all his lost last four books.[48] Borelli hoped to achieve the same task more quickly after he, with the permission of the Grand Duke, had an Arabic manuscript that he assumed to be the whole text of Apollonius and which he let translate in Rome by the Maronite Abraham Ecchellensis. However, whereas Viviani was able to publish his work *De maximis and minimis* in 1659 and in it completed and even went beyond Apollonius, the very difficult translation from Arabic could not be printed until some years later.

---

[46]Scherz, *Viviani*, 275 ff.

[47]DBI 12, 543 ff, DSB II, 306–314. Middleton, 27 ff, Adelmann, I, 144 ff.

[48]Scherz, *Viviani*, 265 f. See also Middleton, 310 ff. He concludes: "Thus I do not believe that Borelli hated Viviani, or that he even envied him more than he did the others. He quarrelled with Viviani, because Viviani was the one with whom he had occasion to quarrel" (p. 341).

At the end of May Stensen went on a planned visit to Rome.[49] He promised the Grand Duke, who had returned to Florence after Easter, to come back to Florence and settle down for some time.

1.4.2   First Visit to Rome in May/June 1666

This was the Rome of the learned and art-loving Chigi Pope, Alexander VII (1655–1667). From the beginning of his pontificate he had shown the energetic will to live and work according to his high spiritual vocation.[50] A coffin in his bedroom and a skull on his desk would remind him of the vanity of earthly matters and staved off, at least in the first years, all nepotistic influence from the Curia. The Pope had shown his most favourable side during the great plague of 1656/1657 during which he bravely remained in Rome and, by his energy and clever measures, prevented further diffusion of the epidemic, which nevertheless carried off 10% of the population of 150 000 inhabitants.

Eleven years before the arrival of Stensen, Christina, the abdicated Queen of Sweden, had made her festive entrance in Rome having adopted the Catholic faith in Innsbruck in face of the whole world.[51] Alexander VII was deeply moved when she, the daughter of Gustav Adolf, knelt in front of him. It was in the name of King Gustav Adolf that the Peace of Westphalia had been concluded so unfavorably for the Catholics, against which the Pope, then Nuntius, had protested. In the first years after her entrance Queen Christina embossed not only the social life in Rome through shining feasts organized in her honour, she was also the bright centre of a circle of artists and learned men. She founded her own academy and assigned rich means as a consequence of her manifold intellectual interest also for buying books and objects d'art. In May 1666 she had the project of undertaking a journey in the north.[52]

The pontificate of Alexander VII in politics resisted the rough will for power of Louis XIV and the threat by the Turks. In the Church field

---

[49]Even before Stensen's departure from Pisa, A. van der Broecke tells that one could expect him in Rome (Nordström, *Magliabechi*, 34). *The precise dates of Stensen's travels in the spring of 1666 are not known. He spent at least one month in Rome. In agreement with Nordström Scherz suggested in 1971 that Stensen only had a short time in Florence before leaving for Rome (*Reisen* 58, 60f, 63). New evidence in *Positio*, 1028 indicates, however, that Stensen stayed in Florence from April 11 to about May 18, after which he left for Rome where he arrived before May 23. (HMH)

[50]DBI 2, 205–15. Pastor, 14^I, 303 ff.

[51]Pastor, 14^I, 328 ff, esp. 342 ff.

[52]Pastor, 14^I, 537. *"Despite an interest in contemporary science, Christina does not refer to Niels Stensen. According to the sources available they never met," Børresen, *Niels Stensen's Theology*, 71.

Alexander fought against Jansenist disorders. He also promoted the arts and science. Rome at the time offered the visitor a picture of active intellectual striving. The Pope himself pursued humanistic studies in the circle of learned friends like the later Prince Bishop Ferdinand von Fürstenberg—who later called Stensen to Münster.[53] The Pope had a taste for the progress of historical sciences and invited Henschen and Papebroch to Rome. The learned Greek Leo Allatius became Custodian of the Vatican, where Stensen must have met him personally.[54] The Cistercian Ughelli, a friend of the Pope, published his famous work on the Italian Episcopal sees. The historian Athanasius Kircher whose work Stensen had read already as a student in Copenhagen, and whom he must have met now, was then highly appreciated in Rome and the Museum Kircherianum attracted many visitors.[55] Above all Alexander VII contributed much to the architectural embellishment of Rome. In addition to the many new churches and palaces which had recently been completed, the pilgrim from the North would have admired the brilliant new creation by Bernini, the colonnade of the Piazza of Saint Peter.[56]

We know very little of Stensen's first visit to the Eternal City. Some certain particulars, however, have been reported such as his meeting with Malpighi, Riva and Fabri.

In his autobiography, Marcello Malpighi mentions that he met Stensen in Rome: "In Rome I had the opportunity to speak with the famous Stensen with whom as well as with Giulielmo Riva I had a meal in the Villa Ludovisi and since then Mr. Stensen has always been for me a dear friend and guest."[57]

This notice shows Stensen as a guest of the leading families of Rome. The Villa Ludovisi is indeed the magnificent property in the gardens of the Pincio that the nephew of Gregory XV, Cardinal Ludovisi, had erected in 1622 and the small but precious art collection is accommodated in the National Museum.[58] The villa at the time comprised three beautiful buildings surrounded by large gardens and parks. Thick laurel bushes, evergreen oaks, dark cypress trees, pines with high aerial tops, long drawn avenues, pools and trickling brooks and springs spread coolness and shade. Antique statues, sarcophagi and other sculptures shone everywhere. The intellectual builder of the villa was then no longer alive. It belonged to the Prince Niccolò Ludovisi

---

[53]Pastor, 14[1], 494 ff.

[54]OTH I, 221.

[55]EP I, 109. DSB VII, 374 ff. *Elsebeth Thomsen, *Steno in the world of museums*, 80.

[56]On Rome see f.i. Petrocchi, *Roma nel Seicento*, and Cesare D'Onofrio, *Roma nel Seicento*.

[57]Malpighi, *Opera posthuma*, 43.

[58]*Enc. Ital.* XXI, 604 f, D'Onofrio, 315 ff.

and his son Gianbattista. A namesake, also Cardinal Ludovico Ludovisi, gathered around him the circle into which Stensen was drawn. He was the cardinal who was the most eager for reform and who in 1683 gave his contribution to the war against the Turks by putting his silverware at disposal.

In Malpighi (1628–1694)[59] Stensen acquired an intellectual peer and a genial friend. Stensen already knew his famous work on the structure of the lungs and the lung vesicles which had appeared in 1661 and which Thomas Bartholin had published in 1663 in Copenhagen.[60]Ferdinand II had called Malpighi in 1656 from his home-city of Bologna to Pisa where he pursued his anatomical research and worked together with Borelli. In the year 1659 Malpighi returned to Bologna but 3 years later, because of dissents, he accepted an invitation to Messina before returning to his hometown in 1666.

When Malpighi met Stensen in Rome he had already carried out the series of works in microscopy which inscribe his name in shining letters in the history of science. He is indeed the discoverer of capillary blood circulation, blood corpuscles, the Malpighi network of the skin and the Malpighi corpuscles of the kidneys. As said, his two letters *De Pulmonibus* dated from 1661. His investigations on the anatomy of plants and animals began in Messina. Then there appeared also the letters on the organs of the senses and on the structure of the intestines. Widely ranging in his interests, highly educated, strictly methodical and indefatigable in experimenting, moderate and mild in behaviour, he was full of enthusiasm for nature. In his different investigations he never forgot the sense of the whole. The behaviour of Malpighi towards Stensen was the most cordial imaginable, based on their relationship of mind and character in general, as specific in their congenial position towards the natural research and its problems.

The above-mentioned Giovanni Guglielmo Riva (1627–1677) was a precious acquaintance for our learned man.[61] The highly respected surgeon at the Ospidale di S. Maria della Consolazione in Rome had made himself appreciated particularly by his research on lymph vessels. He delivered lectures on anatomy not only in his hospital—his most famous disciple was Lancisi—but also in his huge house resembling a museum in the Via della Pedacchia. Here academy sessions took place at which Stensen was obviously a guest. The name Riva is indeed written on the copper engravings that give us an impression of this house.

---

[59]DSB IX, 62 ff, EP I, 30 ff, Adelmann, esp. vol. I.

[60]EP I, 167, cf. above Chap. II.

[61]EP I, 31.

In Rome Stensen also met an acquaintance from Pisa, Michel Angelo Ricci who mentions and praises him in letters to Viviani and to Prince Leopold.[62] Ten years later, at the ordination process of Stensen as bishop, two librarians report visits of Stensen to the large palaces with important collections of books. One was Stefano Grade, the curator of the Vatican library. The other was Thomas de Juliis from the library of Cardinal Flavio Chigi.[63]

Stensen must also have visited the Jesuit colleges. That is how he met the historian and writer Honorarius Fabri (Honoré Lefèvre, 1607–1688).[64] The French Jesuit, a great admirer and expert of natural science was himself famous for many discoveries and in connection with the Accademia del Cimento. After Prince Leopold had sent him Stensen's *Myology* together with another book in 1667, Fabri thanked him: "Both have I not only read but devoured. I like very much the new, quite elaborate, representation of the muscle fibres. But it is not new for me since the author had communicated it to me last summer. Its nature which thus applies simple as well as light principles could hardly have provided a lighter machine for moving the muscles."[65] From co-workers of Fabri, the famous opticians Divini and Campani, Stensen reported optical novelties to William Croone on May 13.[66]

Thorough discussions about faith with Fabri have also been testified and our traveller above all cultivates Catholic connections. This appears from the fact that the later General of the Order, P. Charles de Noyelles, who had been an assistant to Germany since 1661, then became acquainted with Stensen.[67]

We do not know how long Stensen stayed in Rome and its vicinity. In July he was in Florence again.[68] His presence at the Feast of Corpus Christi in Livorno on Thursday, June 24 can be inferred from a letter to Lavinia Arnolfini. Since he furthermore writes that from Livorno he travelled to Florence to learn Italian, he is likely to have been on the way back from Rome in 1666.[69]

---

[62]Sten Cath. 4, 1958, 71, and ibid. 5, 1959, 12.

[63]EP II, 935 ff.

[64]Koch, L., *Jesuiten-Lexicon*, 537 ff, DTHC 5, 2052 ff, EP II, 932. He recommended Stensen to Prince Leopold on July 1, 1666 (Sten. Cath. 4, 1958, 70, 72).

[65]EP I, 109.

[66]Rome, *Nicolas Sténon*, 246 f, Birch II, 102 f, Scherz, *Niels Stensen und Galilei*, 44 ff.

[67]EP II, 540.

[68]Borelli writes on July 10 to Malpighi that Stensen has come to Florence (Malpighi, *Correspondence* I, 318 f); on 24.8. Magalotti writes to Segni that Stensen has been in the city for two months (EP II, 922).

[69]OTH I, 9. Stensen does not specify in which year this event took place. Since it happened before his conversion, it must have been on the feast of Corpus Christi in 1666 (June 24) *or* in 1667 (June 9). Scherz was convinced that it was in 1666. This makes the following chronology: May 1666 in Rome, on the return in Livorno on June, 24, and in Florence some days later. Nordström (*Magliabechi*, 36) and with him the *Positio* (p. 1029) adopt 1667. To wit, on July 1 and 2, 1666 Fabri (Sten. Cath. 4, 1958, 72 f) and

Livorno with its castle and the Marzocco tower at the harbour had always been a trade centre and had developed particularly well in the seventeenth century under the Medicis.[70] The most beautiful Medicean creation, as Livorno was called, had developed into a traffic centre in the Mediterranean Sea already under Cosimo II. The fleet grew. Ferdinand II had added the quarter Little Venice and called in Jesuits and Barnabites to settle there. The open-mindedness and freedom, also in religious aspects, contributed to the fact that many nations were represented in what was the richest source of income of Tuscany. Whereas the Grand Duke and his family had their rooms in the Fortezza Vecchia during their frequent visits to Livorno, under Ferdinand II rooms were also reserved for outstanding visitors in the Palazzo at the Piazza Grande. From his window or from the colonnade the guest could attend the Corpus Christi procession, which attracted many visitors every year.[71]

This experience greatly impressed Stensen: "I was in Livorno for the Corpus Christi feast and when I saw the consecrated wafer carried through the town with such great display of magnificence, this thought arose in me: either this consecrated wafer is a simple piece of bread and those who show so much veneration for it are fools or it is the true body of Christ and why then do I not venerate it myself?"[72] This thought caused him a great deal of concern and stimulated him to the study of the Scripture and the Church fathers.

The stay in Livorno cannot have lasted long. This is suggested in a remark of Maria Flavia with whom Stensen made the acquaintance in 1666 after he had come to Florence "for some days" to spend the "feast of St. John".[73] The *Festa degli Omaggi*, the yearly homage feast to the Grand Duke, is celebrated

---

(Footnote 69 continued)

Ricci (Sten. Cath. 5, 1959, 12) sent letters of recommendation—on his request—to Prince Leopold. If Stensen brought these letters to Florence himself, he cannot have left Rome until in the beginning of July. Based on this, it seems unlikely that the date he was in Livorno was June 24, 1666. On this question Gustav Scherz remarked: "Stensen needed not carry these letters himself from Rome. They were intended to promote and to introduce, and Prince Leopold already then knew and appreciated him well. Rather they are discreet petitions for help and accommodation for the traveller, the precarious situation of whom Ricci and Fabri might have better understood." (Reisen, 63). *Harriet M. Hansen comments: "For my part, I really do not know. Look at his itinerary of 1666, you might prefer 1667. But look at 1667 with strong pains in his teeth and 1666 seems preferable. It all depends on your estimation of his road to the Roman Church: was it a long uncertainty for over one year, or a process of a few months."

[70]Imbert, *Vita*, 20 f.

[71]Berti, *Il Palazzo Granducale di Livorno*, 6 ff.

[72]OTH I, 9.

[73]EP II, 987. *Positio*, 97. Some dates and years in the latter are not correct, thus Stensen's conversion was placed in 1669, the *Positio* expressing doubt at this point.

at the same time. Usually on June 24, this year it may have been postponed for some days because of Corpus Christi.[74]

1.4.3   Studies on Muscles in Tuscany

The next period in Florence was a quiet time after a restless year of travel. Stensen was the regular guest of the court, which cared for him and rented him lodging in town.[75] The hospital *S. Maria Nuova* offered a bright opportunity for anatomical studies and dissections. This "Inn of Our Lady" was founded by Folco Portinari, the father of the Beatrice of Dante. It had been enlarged over the centuries, enriched and adorned with works of art. Since 1650, due to the care of Filippo Ricasoli, the hospital experienced a period of renewal and was an ideal place for nursing and research. It is still in function. From its portico one enters through five portals to the three wings of the hospital, to the men and women cloisters, the care of which was entrusted to the Church, on to the church, to the bathhouse, the funeral chapel and the anatomy theatre. Actually it referred to three hospitals since there were separate departments with wards for injuries and infectious diseases. It could take up to 1500 patients. There was also a college for students. At the gate stood the largest pharmacy of the town.[76] Stensen used this ideal workplace most actively whenever he was in the Arno city and the court very liberally put objects of investigation at his disposal. He often worked together with the Englishman Sir John Finch (1626–1682) who had become doctor of medicine and syndic in the university in Padua, had returned back to England with Charles II in 1661 and in 1663 accepted the invitation to be lecturer of anatomy in Pisa. Moreover, since 1665 he had been a resident of England at the court of Tuscany.[77]

Stensen also went into connections with the Dutchman Tilman Trutwin (†1677), another foreigner at the service of the Grand Duke. He worked as physician and anatomist at the Spedale di S. Matteo. Later Stensen would help him to die as a good Catholic.[78]

---

[74]Imbert, *Vita*, 43 ff.

[75]From a letter by Viviani April 1, 1667. It appears that until then Stensen had received 25 scudi and free accommodation, EP I, 26 f, *Positio*, 53 f. OPH II, 63; cf. Scherz, *Viviani*, 277 and Maria Flavia's account, EP II, 987.

[76]Richa, 8, 175 ff, *Lo Spedale di S. Maria Nuova*.

[77]EP I, 16 f, DNB VII, 18. Malloch, *Finch and Baines*. *Finch was later appointed ambassador of his country to Constantinople in 1671.

[78]EP I, 190.

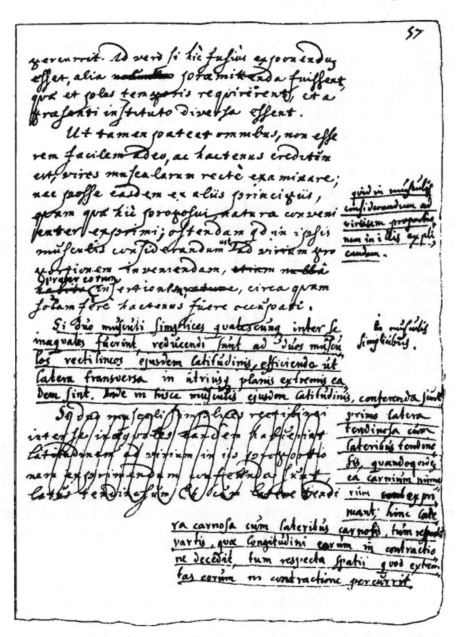

Fig. 1.4.1 Page from *Elementorum myologiae Specimen*. The manuscript was written by a clerk with corrections in Stensen's handwriting. The printer's manuscript, REX

We also find the Lorenzini brothers at the side of Stensen; Lorenzo (†1721), who originally was a mathematician, and his brother Stefano, the physician whose work with the red and white muscular fibres doubtless was inspired by the studies of Stensen for his *Myology*.[79] Stensen was full of admiration for the learned intellectual culture that he encountered in Florence and praised the liberality of the Prince who offered him all means for anatomical research, in addition to different animals, above all corpses, for study at the hospital. He proposed plans to the Grand Duke for the study of the foetus, liquids and blood and also recommended the use of chemistry for therapeutic aims (Fig. 1.4.1).[80]

As literary fruit of his first year in Italy and of his connection with the Accademia del Cimento, in April 1667 Stensen published the *Elementorum myologiae specimen, or geometrical description of the muscle*. The first censorship of October 27 allows us to conclude that the manuscript of this treatise—at the end two shorter works were appended—was prepared in the spring and summer, and written at the latest in the first months of the autumn of 1666, in the restless months of the arrival and during his journey to Florence and Rome, thus before the famous dissection of a shark that deeply influenced his scientific career. The research of the elements of a muscle theory was dedicated to the Grand Duke Ferdinand II as a thankful homage to the "Prince, most renowned the world over among men of letters, for his firm grasp of affairs" who kept at his court very famous men and of whom the guest modestly speaks, who "has chosen to give me, a man from the North who scarcely rises above mediocrity in talent, some share in those hours in which you relax your mind, when wearied by official responsibilities, you take that delight in the mysteries of art and nature which others would seek in games and pleasure."[81]

It was not from the beginning the intention of Stensen to publish it as a "geometrical description" as indicated in the printed title, *seu musculi descriptio geometrica*. This appears from the title of the original manuscript: *NICOLAI STENONIS/de musculis aliisque animalium partibus/observationum anatomicarum specimen alterum/quo/ELEMENTORUM MYOLOGIAE SPECIMEN/et/CANIS CAR-CHARIAE DISSECTUM CAPUT/variis observationibus et figuris illustrata proponentur/ad/SERENISSIMUM/FERDINANDUM SECUNDUM/MAGNUM ETRURIAE DUCEM.*[82]

The manuscript (Fig. 1.4.1) thus had been described as *specimen alterum*, that is as a continuation of the first specimen of 1664. Already as a manuscript it is dedicated to Ferdinand, but it does not yet comprise the third part and above all the designation *Musculi descriptio geometrica* is still missing.

---

[79]EP I, 16 f, 228.

[80]Scherz, *Viviani*, 276.

[81] <649> OPH II, 63 = GP, 69–71 with notes.

[82]On the manuscript with Stensen's reports, see Scherz, *Danmarks Stensen-manuskript*. Also GP, 123, note 1.

What caused Stensen to publish the dissertation as a geometrical description and to affirm this in the title? First and foremost this was an acknowledgement of the Accademia del Cimento and of the mathematical method advocated by it since Galileo. The seventeenth century, in its passionate love for mathematics, often used this purely intellectual discipline, the objects of which are products of human thinking, like points, lines, surfaces, numbers and distances, etc., and of which the properly deductive method is contrary to the induction of nature science, for a formulation as clear and concise as possible of any scientific knowledge.[83] There lay in it a sharp protest against the many ordinary false deductions and against a priori arguments through fantasies and analogies. The tendency to more reliable knowledge determined Santorio to use iatrophysical experiment to make medicine an exact science by introducing mathematical-physical methods to measure, weigh and count, whereas the iatrochemists of the time wanted to explain all biological phenomena by chemical processes. Spinoza had even presented his ethics *ordine geometrico*, and, beyond the form of the axioms, demonstrations, corollaries, had explained the deductive process of mathematical thinking on the whole as the only standard and guiding principle to obtain true knowledge of everything.[84]

Florence had a proud tradition of exact natural science and mathematical principles. Leonardo da Vinci had wanted to see science penetrated by the spirit of mathematics,[85] and Galileo as a pioneer physicist and passionate fighter for the inductive method in natural research outlined mathematically the laws of his science.[86]

When Stensen appeared in the circle of the Cimento he undoubtedly had mathematical discussions on muscles. Stensen had asked for the help of Borelli with geometry, and in the *Myology* he also mentions a problem which had been put forward *in illustri consessu*, thus in a session of the Cimento.[87] The question was whether in muscles, the tendons of which extend beyond the flesh and during contraction form a straight line together with the diagonal, the contraction withdraws from the width of the flesh that which is added to its length. The contribution of Viviani to the *Myology* is explicitly expressed by Stensen. In the last part it says: "So that one actually should not assign this to imagination rather than to experience, I call upon the testimony

[83]Allard, *La science moderne* II, 192 f.

[84]Ueberweg 273, 280 ff. Spinoza published his *Principia philosophiae more geometrico demonstratae* in 1663.

[85]Allard, *La science moderne* II, 24 ff.

[86]Scherz, *Niels Stensen und Galilei*, 53 ff. Meli, *Galileo's role in Steno's Myology*.

[87]Malpighi, *Correspondence* I, 318 f, cf. Adelmann, I, 286. OPH II, 84.

of my dear friend Vincenzio Viviani, mathematician of the most serene Grand Duke, who was present as a keen observer of these facts and of others contained in the present book."[88] In earlier sections he names Redi[89] as well as another member of the Cimento circle, Carlo Dati (1619–1676), whom he calls his very learned friend.[90]

It is all the more surprising that Stensen does not at all name Borelli with whom he ought to have the most in common concerning this issue. The famous work of muscle physiology by Borelli *De motu animalium—On the movements of animals* was to appear in 1679/1680 but the basis was already laid down during Borelli's stay in Tuscany in 1657/1667.[91] This work partly deals with the same problems as Stensen's *Specimen* and actually proved to become a pioneering study by its mathematical applications in biology.[92] Borelli, however, does not seem to see an ally in Stensen. A letter to Malpighi of July 1666 shows that the irascible Napolitano felt the credit of his work threatened: "I am not so naive as to consider him as the embodiment of modesty and good manner for which he is taken at the court. The short epistles which he has published clearly show his willingness to work everything and keep others away. I know that those *Oltramontani* come here to us cautious and deliberate to play with us and in the long run we must see ourselves as beaten."[93] The fact that Stensen was a friend of Viviani and that he was a strong adversary of the theory of *spiritus animales* and inflation of the muscles defended by Borelli indeed did not contribute to a reciprocal comprehension.

---

[88] <739> OPH II, 154.

[89] <731> OPH II, 117.

[90] <700> OPH II, 115. Middleton, 31 f, NBG 13, 156 ff.

[91] According to a letter dated September 11, 1666, Borelli was just then writing the first part of his work on the mechanics of animals' movement. Malpighi, *Correspondence* I, 325. *See Borelli, *On the force of percussion,* transl. P. Maquet, 2015 (org. ed. Bologna 1667), and Borelli, *On the Natural Motions resulting from Gravity,* transl. P. Maquet, 2015 (org. ed. Bologna 1670), particularly in the latter, p. 129, Chap. 6, entitled: "There is neither attraction nor attractive force in Nature". See also a review of the latter work in *Phil. Trans.* 1671 vol. 6 no. 73 2210–2214: "… there is not in nature any proper Attraction or Suction; but things seeming so to be perform'd are done by the Pulsion or Trusion of other Bodies …", doi:10.1098/rstl.1671.0030.

[92] *The basic assumptions of Stensen's model (p. 163 note a) were rejected by Borelli in *De motu animalium,* I, I, prop. V and II, II, prop. XXVIII, (see *Steno on muscles,* 33–37). Borelli's arguments were renewed in 1694 by Johann Bernoulli, the mathematician, in his second doctoral thesis in Basel, *De motu musculorum.* Bernoulli found it ridiculous that a muscle contracts without the arrival of new material. Bernoulli held fibre shortening to be incompatible with the Aristotelian axiom, "Anything which moves is moved by something else" (Bernoulli, § 3, see Kardel, *Prelude to two dissertations by Johann Bernoulli,* esp. 21). Fibre shortening was verified by microscopy but not until the middle of the eighteenth century by Albrecht von Haller (see *Steno on muscles,* 41).

[93] Malpighi, *Correspondence* I, 318 f, cf. Adelmann, I, 286. OPH II, 84. .

Just from the dedication[94] of his treatise Stensen praises mathematics as the guide to exact knowledge in anatomy. He says: "I wished to demonstrate in this dissertation that unless myology becomes part of mathematics, the parts of muscles cannot be distinctly designated nor can their movement be successfully studied." Then follows a praise in which, however, one must notice the wise restraint of the author who sees in mathematics a method only. "And why should we not give to the muscles what astronomers give to the sky, what geographers to the earth, and, to take an example from microcosm, what writers on optics concede to the eyes? These writers treat the natural things mathematically so they may be more clearly understood. Our body is an organism composed of a thousand organs. Whoever thinks that its true understanding can be sought without mathematical assistance must also think that there is matter without extension, and body without figure. Nor is there any other cause of the many errors which have foully defiled the description of the human body than that Anatomy has hitherto disdained the laws of the Mathematics. For, while ignorant of the rule of the legitimate prince, in its, may I not say, blind judgement, Anatomy has governed all things; it has thrust on us the dubious for the certain, the false for the true, the unknown for the known. Anatomy finally has brought the matter to such a point that nothing remains more unknown to man than man himself. How fortunate it would have been for us, how fortunate for all human kind, if our ancestors had decreed that those who spend a life time in the study of anatomy would pass on to posterity only that which is well established. Our knowledge would not be so wide but it would be less hazardous. And if medicine, based on these established principles did not relieve the sick of their pain, it would not add to this pain. Now we have enormous books on anatomy and medicine. Nevertheless, we drag the wretched sufferer among thousand tortures; we even drive him to a tearless death through a thousand torments. Our greatest misfortune is that often when we deem to be helpful, then indeed we are only harmful. And why would it not be permitted to hope for great things, if anatomy was transformed so that experimental knowledge would rely only on well-established facts, and reason accepted only what has been demonstrated; in other words, if anatomy used the language of mathematics."[95]

After all this extensive glorification of mathematics, Stensen will, for all that, leave to others, more qualified, this connection with myology and anatomy, and assigns to himself one aim only: "I expose their true structure

---

[94] <652> OPH II, 63 ff = GP, 68/69 ff. See also Scherz, *Pionier*, 134 ff.
[95] ibid.

by a new method and I demonstrate that the mode of contraction by inflation of spirits such as proposed by the majority hitherto is built on a very shaky foundation", even if he had the intention of publishing a complete theory of muscles and bones by opportunity and on favourable reception of the first attempt. In the first lines of the dissertation Stensen emphasizes that one is here dealing with a methodical mathematical application of his knowledge of the structure of muscles acquired by experience. "Almost three years ago I published a *Specimen* on the muscles of the heart, tongue, oesophagus and some other organs, but without illustrations. I will now illustrate it with figures, partly of already known, partly of unknown muscles. These illustrations are shown here with the aim of making clear that the geometrical structure of the muscles which I am to propose is not just imagination, but derived from experience."[96]

Stensen was brought to this solution of methodical mathematical application of his earlier findings concerning the muscles and to contest mathematically the inflation theory by special motives. From the letter to Thévenot attached at the end of the *Myology* one gathers that he met fierce opposition among adherents of the ancients against his new views on the heart and muscles, particularly in France from the influence of the inflation theory as defended by Descartes.[97] A reproach was raised that the ancients already counted the heart among the muscles, which Stensen did not deny. Since, however, they had not recognized the nature of the muscles and the heart, the controversy between Galen and Hippocrates had remained undecided over the centuries.

He conceded to another reproach which declared his and similar purely scientific works vain and useless that indeed still much remained unclear about the muscles but much had also been clarified. Thus, no one had earlier undertaken the proper dissection and the separation of the muscle and the terminology had been confusing.

Against those people of bold verbosity who name such studies useless in practice, Stensen rebutted: "If they face the fact that their diagnosis is mere words, their cure but guess work, they will admit, even reluctantly, that it serves a purpose to discover what is true and certain in anatomy." The objection that everything has remained the same in the course of so many centuries is right but the answer is, in short: "They all look for remedies and for that part to which the remedies apply, but few attempts to understand."

---

[96] <652> OPH II, 64.
[97] <690> OPH II, 95 ff.

Stensen thus justly rejects the idea of any fluid rushing in from the outside and all speculations and presumptions tied to that: "The often-used expressions like animal spirits, the finest part of the blood, its vapour or nerve fluid are only meaningless words."[98]

Most learned men considered the hardening and increase in volume during the swelling of the muscle as a consequence of an inflow of nervous juice, the so-called animal spirits. Walter Charleton (1658) had already taught that the volume of the muscle does not change. Nevertheless he firmly held to *spiritus animales*.[99] How little a lightly based and contradictory knowledge weighed is shown by the example of the Cartesian Jacques Rohault (1620–1675) who, some years after the *Myology* of Stensen, falsely defended the theory of the *spiritus animales* against which Stensen fought.

Stensen's work[100] first shows in a sketch the current theory of the muscle structure and in another the result of his own observations. He attempts at reducing purely methodically the muscle to a geometrical figure. Relying on many observations and explanations, he defines the muscle structure as fibres forming a parallelepiped. According to him the tendons are parts and parcels of the muscle and consist of fibres that are direct continuations of the muscle fibres. The two tendinous endings into which the middle, softer muscle fibre continues are not linked with it at one point but rather in tendon plates which become thinner the more the fibres become separated from each other and more fleshy. A longitudinal cut through the muscle yields a parallelogram of the proper muscle which is delineated above and by a triangle formed by the tendons. All that is "represented in good Euclidean form with definitions, with accessory and main propositions as well as figures".[101,102] However, it is obvious that Stensen does not think of definitions in the true sense of the word. The explanations of the words are mixed with all sorts of secondary matters. The question for the author was less to give a rigid closed system than to give an adequate explanation for the facts of observation accessible to him. The literal conclusion of Stensen is: "Relying on this basis I represent a

---

[98] <697> OPH II, 106.

[99] Bastholm, *History*, 154.

[100] Mouy, 125.

[101] Gosch, II, 1, 200.

[102] Stensen's geometrical arguments have four steps. 1: In 43 *definitions* a three-dimensional parallelepiped of muscle is outlined from anatomical observations in different muscles. It is illustrated in three plates. 2: Five *suppositions* make the parallelepiped move from relaxation to contraction. This step was hypothetical since microscopy was insufficient to verify fibre shortening. 3: Through six *lemmas* Stensen deduces geometrically that, given these conditions, the model remains a parallelepiped. Step 4 is Stensen's main *proposition* that "every muscle swells when contracting" even if its volume remains the same (Kardel, *Steno's new myology*, 46–47).

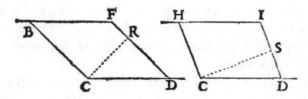

Fig. 1.4.2 CS > CR, which means that the resting muscle B C D F with fibres parallel to B C swells when in action H C D I with fibres parallel to H C even if there is no volume increase. Parallelograms on the same base and in the same parallels are equal to one another, Euclid I. 35

muscle as a collection of motor fibres arranged so that the flesh in the middle forms an oblique parallelepiped and the tendons form two opposite tetragonal prisms. A motor fibre is a definite structure of very tiny fibrils, immediately joining together lengthwise. Its intermediary part differs from its extremities in consistency, thickness and colour, and is separated from the intermediate part of the adjacent motor fibres by transverse fibrils continuing into the aponeurosis."[103] The long series of definitions extends to concepts like *fibra motrix*, *caro*, height, thickness, partly supported by drawings. It is followed by the description of two kinds of muscles: of the simple, straight and of the composed muscle which, however, can be brought back to the above-mentioned parallelepiped of muscle and two tetragonal tendon prisms.

Then Stensen returns to the second and perhaps central problem of his dissertation. He refutes the foundation of the theory of the *inflatio musculorum*, the *spiritus animales* and *succus nervosus* by geometrically demonstrating that muscle swelling is possible without any increase in volume. The demonstration shows that the tendon fibres that form the two ending plates of the parallelepiped in a normal muscle displace themselves parallel to each other when the sides represented by the fleshy fibres shorten and arise so that the acute angles that they form with the lateral tendon plates widen and become almost right angles. However, since the heights as well as the basic surfaces are the same before and after the contraction, the volume must remain the same even if the muscle becomes thicker (Fig. 1.4.2). A swelling occurs when it contracts without the muscle increasing substantially.

Then follows an explanation of three structurally based tables and an explanation of the elements of myology by examples from nature, thus by several muscles in humans, the structure of muscles in fishes, lobster and the almost unknown muscles of the ribs.

---

[103] <654> OPH II, 68.

Stensen's treatise also provided much new and important particular knowledge as the Italian anatomist Franceschini emphasizes. Stensen makes the observation that muscle fibres of animals can have different colours, like red, grey and white. Thus he founded the distinction between red and white muscle fibres, which is usually attributed to W. Krause (1868) and more to L. A. Ranvier (1874) and was claimed by Castaldi (1929) for Lorenzini (*De anatome Torpedinis*. Florentiae 1678).

Stensen defines the phenomenon of the contraction by attributing it to the middle, fleshy part which is quite different from the tendons in structure, thickness and colour. Before Leeuwenhoek, he determines the histological structure of the muscle fibres and divides the muscles into simple and composed ones. Stensen has doubtless made more understandable and explained the proper line of force of every muscle. Franceschini assigns to that the opposition of Borelli who was angry to be surpassed just in his own field of mathematics. A merit of Stensen is also to have shown that the muscle does not change in volume during the contraction. The invariability of the volume was definitely confirmed in 1887.[104]

The *Myology* may be the most controversial work of Stensen. The Galenics and the unconditional Aristotelians opposed it. They wanted to deny mathematics to anatomy.[105]Borelli and his disciples were also hostile to it. However, enthusiastic friends like Viviani who was deeply grieved over this hostility asked Prince Leopold to send a copy to the mathematician P. Stefano Angeli in Venice for an expert opinion. Angeli answered on June 4, 1667. He declares that he does not understand anything in anatomy but he is enthused by the mathematical aptitude of Stensen and deplores the cool reception by those who did understand it, *honeste vestiri, gloriose mentiri*. "The principles of geometry were understood by few physicians only and despised by most of them. Some have scorned my praise for Stensen's book as a novelty and sworn by their respectful beard never to want to see it." Molinetto alone, the anatomist of the university in Padua, was somewhat favourably minded.[106] Michel Angelo Ricci, himself a mathematician, to whom Prince Leopold had also sent a copy, wrote from Rome: "It is many years since someone like Mr. Niels Stensen has come to this country, with whose book Your Highness has honoured me. He unites great knowledge, accuracy in observation with genius and clear vision. No wonder that he makes such beautiful discoveries

---

[104]Franchescini, *L'apparato motore nello studio di Borelli e di Stenone.*
[105]Caverni, 3, 36 f.
[106]Caverni, 337.

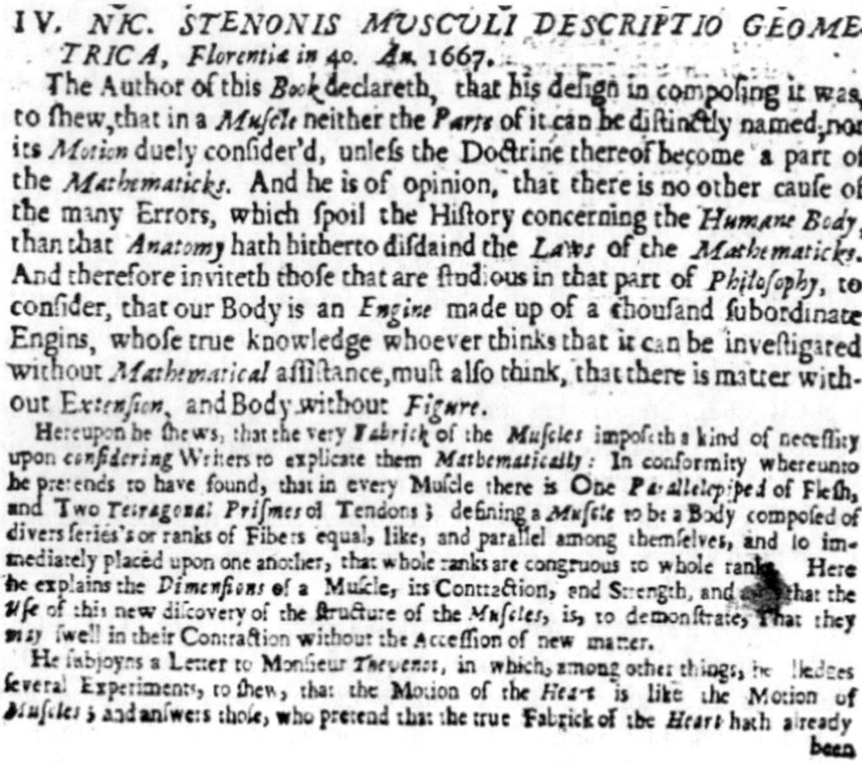

IV. NIC. STENONIS MUSCULI DESCRIPTIO GEOME-
TRICA, Florentiæ in 4o. An. 1667.

The Author of this Book declareth, that his design in compoſing it was, to ſhew, that in a Muſcle neither the Parts of it can be diſtinctly named, nor its Motion duely conſider'd, unleſs the Doctrine thereof become a part of the Mathematicks. And he is of opinion, that there is no other cauſe of the many Errors, which ſpoil the Hiſtory concerning the Humane Body, than that Anatomy hath hitherto diſdaind the Laws of the Mathematicks. And therefore inviteth thoſe that are ſtudious in that part of Philoſophy, to conſider, that our Body is an Engine made up of a thouſand ſubordinate Engins, whoſe true knowledge whoever thinks that it can be inveſtigated without Mathematical aſſiſtance, muſt alſo think, that there is matter without Extenſion, and Body without Figure.

Hereupon he ſhews, that the very Fabrick of the Muſcles impoſeth a kind of neceſſity upon conſidering Writers to explicate them Mathematically : In conformity whereunto he pretends to have found, that in every Muſcle there is One Parallelepiped of Fleſh, and Two Tetragonal Priſmes of Tendons ; defining a Muſcle to be a Body compoſed of divers ſeries's or ranks of Fibers equal, like, and parallel among themſelves, and ſo immediately placed upon one another, that whole ranks are congruous to whole ranks. Here he explains the Dimenſions of a Muſcle, its Contraction, and Strength, and ſo that the Uſe of this new diſcovery of the ſtructure of the Muſcles, is, to demonſtrate, That they may ſwell in their Contraction without the Acceſſion of new matter.

He ſubjoyns a Letter to Monſieur Thevenot, in which, among other things, he alledges ſeveral Experiments, to ſhew, that the Motion of the Heart is like the Motion of Muſcles ; and anſwers thoſe, who pretend that the true Fabrick of the Heart hath already been

Fig. 1.4.3 From the review in *Phil. Trans.* 1667; 2: 627–628. REX

and exposes them with such great clarity and elegance as appears in this new book."[107,108]

In the twentieth century, the myology is often considered to be the weakest work of Stensen. Bastholm nevertheless regards Stensen's *Myology* as a turning point in the research on muscle.[109] Stensen isolated the phenomena and problems from the traditional scientific empty words and simplified them. He further offered, quite apart from his contraction theory, the result of many new observations and experiments and finally his studies are of fundamental

[107]EP I, 33.

[108]Henry Oldenburg received Stensen's *Myology* in February 1668. The book was presented to the Royal Society on February 13 by William Croone and reviewed in the *Philosophical Transactions* (Fig. 1.4.3). *Oldenburg has listed Stensen's *Myology* as the 70th and final item in his *Catalogue of my best books*. Malcolm, *Library of Henry Oldenburg*.

[109]Bastholm, *History*, 158, with a discussion of the myology p. 150 f. Bastholm also held the *Myology* to be the weakest work of Stensen, *like Vilhelm Maar had done in his Introduction to OPH I, p. XVIII.

importance because he motivated them separately and so desisted from pure scientific research without any immediate practical purposes.

J.E. Hoffman, an expert on the history of mathematics in the seventeenth century, writes: "In a reasonable judgement of his muscle theory we should not ask whether it is 'right' in the sense of today or whether it can be brought in agreement with the findings of today (which actually is the case to a large extent), but whether it gives a reasonable explanation within the facts of observation accessible to him and whether what he presented is a comprehensive system to which he has striven. Put in this way, I answer the question of the rightness of the statements of Stensen by a definite yes… The grandeur of his presentation is the simple conception and the outstanding presentation of a model which with a safe hand makes stand out something typical from a lot of so different and apparently contradictory phenomena. And, therefore, beyond the details, I see in the theory of Stensen one of the most significant deeds of the new experimental science which makes the author one of the most important representatives of iatromechanics." (Fig. 1.4.3)[110]

1.4.4   The Huge Shark, Autumn 1666

While the *Myology* was still under censorship, a huge shark was caught outside Livorno, actually a *Carcharodon carcherias*, a kind of white shark that still occasionally appears in the Mediterranean Sea. The dissection would exert great influence on the scientific development of Stensen. He was on the verge of creating a new science. Two dissertations *Canis carchariae dissectum caput* and *Historia dissecti piscis ex canum genere* are appended to the *Myology*, thus beginning a new period of Stensen's research.

Magalotti provides us with a lively description of the catch of the white shark. On October 26, 1666, he writes to the Archbishop of Siena, Ascanio II Piccolomini,[111] that between Gorgona and Meloria, some miles off from Livorno, a French fishing boat had observed over the water the head of a shark followed by thousands of those silly fishes that always accompany it for which reason fishermen call them sea owls. The animal was finally caught in a sling and hauled to land where it was tied to a tree and beaten to death: "The belly was immensely big and was thrown back into the water with the bowels. From the internal organs, the liver was saved. It weighed 300 lb and the whole inside

---

[110]J.E. Hoffman in a letter to Gustav Scherz, September 24, 1953. *See also Meli, *Steno's myology*.

[111]EP II, 922 ff. According to the letter from C. Datis to O. Falconieri dated October 6, 1669 the shark was caught on September 29 (Dati, *Lettere*, 55). On the shark, see Springer (1971), *It began with a shark*.

2400 lb, so that one estimated 3500 lb for the whole fish. Five very big gaps were seen on each side of the head. The head came to Florence under the fine knife of Mr. Niels Stensen, an outstanding anatomist and very amiable Danish learned man who has been staying in Florence for three months, highly respected by the Grand Duke, the Princes and everybody who had the pleasure of testing his rare qualities. He had thus observed three things worthy of consideration. First some vessels which ended in the outer skin and which, through some very large openings, exuded a kind of red slime which serves to keep smooth the otherwise rough and uneven skin of the whole fish. The second are some veins and arteries by which the whole inner substance of the big cartilages, which are instead of bones, is strewn. Their parts spread out in the open air appear strewn with such humidifying vessels with ramifications which appear as a luminous agate as a result of the transparency of the substance. The third realization is that what authors have taken for the optic nerve, an unpierced cartilage the function of which is not yet known and can still only be guessed."

Stensen introduced his report with informal modesty.[112] For a change he wanted to lend variety to readers, tired by his lengthy considerations on muscle and various observations. He then thanks his learned friend Carlo Dati for providing him with an engraving of the head of a shark which he had procured from the manuscript of Michele Mercati from the Vatican collection[113] and which immediately draws the attention to the well-developed slime system of fish, which he had already discovered in 1664 on samples of ray fish, *Raja batis*. His description of these so-called Lorenzini ampullae (named after his pupil Stefano Lorenzini) is correct. However, when he assumes that their function is the production of slime to facilitate the movement of the fish in the water like boats which furrow the sea are coated with grease or smear he could not foresee, in fact, a hydrostatic organ.[114] Then he describes the side channels, which he had seen in the eel with Redi and similar organs in other fishes. He is unclear about their purpose. Today we know that they are also a sensory organ.

In the description he erroneously takes myosepts for tendons, misled by the skin muscles that are often found in animals. The eyes aroused his attention and he was the first to describe, among other things, the noticeable stalk in the orbit cavity of the plagiostoma, which carries the eye, the optical nerve which earlier zoologists had missed and a chiasm which is absent in other fishes. Other particularities of the fish eye and of the eye in general will be published

---

[112] <699> OPH II, 115 ff, cf. GP, 71 ff.

[113]Mercati's work from the end of sixteenth century was published in 1719.

[114]Cf. Springer, 361, and Spärck, in *Stenoniana* I, 50 ff, both to be used in the following.

in the fish anatomy that he had undertaken at Thévenot's, he says. Yet this never happened. Behind the eye he discovered the ear although he was reluctant about its significance. Only with histology has it become possible to determine its function as an organ of hearing and equilibrium.[115] He reflects on the small size of the brain and he sees in this a reason to consider the spinal cord to be an independent centre of nervous activity. In such a big animal all nerves cannot possibly have their endings in such a small brain. For the foundation of these opinions he also discusses the remarkable widening and deepening of the spinal cord in birds, which still carries the name that Stensen gave it: *Cavitas rhomboidalis*. In this context, he mentions an experiment that shows that not everything is directed by the brain. By temporary ligating the descending aorta of a live dog one can temporarily paralyze its hind legs. The dog survives without any after effects when loosening the ligature.[116] The experiment was startling in Stensen's time and from then on carried his name.

Finally Stensen describes the cavity of the mouth with its ranks of teeth and delivers to a closer understanding of this a plate from the already mentioned Mercati's engraving of a shark head with yawning cavity of the mouth and some other illustrations.

This dissertation became famous, not because of its description of the shark, albeit a modern expert called Stensen the first shark specialist[117] It is because in the later part he argues from a comparison of the shark's teeth and the so-called glossopetrae for the organic origin of the latter based on his experience from the examination of layers of the earth in which fossils are found. He thus set the basis for further studies, which has let him be named a founder of geology.

The fossil problem had interested researchers for centuries. The natural explanation of fossils as earlier forms of life already arose in antiquity and further developed in the sixteenth century by men like Girolamo Frascatore (c. 1478–1533) and Bernard Palissy (c. 1510–c. 1590). Fabio Colonna (1567–1640) was the first scientist to demonstrate in 1616 that "glossopetrae" are shark teeth and, with his dissertation *De glossopetris dissertatio*, palaeontology had made the first steps to a position as an independent discipline of natural science.[118]

---

[115]Springer, 316.

[116] <141, 714> OPH II, 125. Cf. Rome, *Nicolas Sténon et la Royal Society*, 254 ff.

[117]Springer, 314.

[118]Scherz, Niels Stensen's geological work, 21 f, further Morello, *De glossopetris dissertatio*. On the interpretation of fossils, see also Rudwick, *The meaning of fossils*, and Morello, *La nascita della paleontologia nel seicento*, 7 ff.

The fossils of sea-animals found far away from the sea, even in high mountains, were an essential problem. Most scientists firmly kept to the conceptions of miraculous intervention of the strong hand of God in such absurdities of nature. In the manuscript from which Stensen took the illustration of the head of a shark, the author, Michele Mercati, warns against explaining glossopetrae as being shark teeth.[119]

With his eyes, open for every natural phenomenon, Stensen had early been interested in geological and related problems without having pursued any deep study. His teacher Thomas Bartholin had written a *De glossopetris Melitensibus dissertatio* already in 1644[120] in which he expressed himself about the tongue stones that he had collected himself in Malta, as Stensen mentions. He might have seen such stones himself in the Wormianum Museum to be included in the Royal Danish Kunstkammer in Copenhagen.[121] A note on fossils is found in his extensive student excerpt from Pierre Borel's *Historiarum et observationum medico-physicarum centuriae*[122]: "Snails, shells, oysters, fish, etc., found petrified on places far remote from the sea. Either they have remained there after an ancient flood or because the bed of the seas has slowly been changed. On the change of the surface of the earth I plan a book, etc."[123] The sentence expressed by young Stensen may sound like a prophecy.

Despite the geological building blocks to which he had already then been exposed, despite observations of the earth's crust, which must have caught his observant eyes on his journeys—one thinks of the mounting or destroying force of water which he experienced during his strolls in North Holland or also of the far-reaching travels through France from the Loire valley to the Alps—, that spark was still missing which in one moment gathers all knowledge into one great synthesis. This did not occur until the experience of the dissection of the shark. In the last months of the year 1666 the geological knowledge of Stensen grew to an amazing clarity, which reveals itself in the *Canis carchariae* despite its unassuming form (he mentions his limited experience, thus that he had never been in Malta). In 11 points in *Historia* he

---

[119]Quoted from Stensen, cf. OPH II, 116 <700>.

[120]Thomas Bartholin's unfinished dissertation on the tonguestones from Malta was lost in a fire, see Bartholin, *On the Burning of his Library*, 28. Thomas Bartholin had studied them at different locations: "In the mountain rocks of Basilicata I found fossil shark's teeth like those of Melita (Malta), although of a different colour, and some shellfish pressed into the stones," Bartholin, *On medical Travel*, 72.

[121]Schepelern, 239, 240, 276 f. *Cf. Elsebeth Thomsen, *Steno in the world of museums*, 77.

[122]*Chaos*, fol. 28 ff. In Paris Stensen consorted with Borel (cf. previous Chap. 1.3). Borel's chemical experiments were mentioned in Canis Carchariae (<726> OPH II, 134 with note).

[123]*Chaos*, col. 16, fol. 31 v, ed. Z, 58.

describes his observations, which are obviously based on studies in Tuscany. On this grounding he proposes six *Conjecturae* or presumptions:

1. Soil from which bodies resembling parts of animals are dug out does not seem to produce these bodies today. 2. The said soil does not seem to have been solid when the mentioned bodies were formed in it. 3. Nor can there be strong opposition to the belief that the said soil was once covered by water. 4. There seems also to be no objection to the belief that the said soil was at some time in the past mixed with water. 5. I cannot see anything to prevent us from regarding the said soil as sediment gradually accumulated from water. 6. There seems no objection to the opinion that bodies dug from the ground which resemble parts of animals should be considered to have been parts of animals.[124]

There can hardly be any doubt that for Stensen these presumptions were his conclusions bordering on certainty.[125] "How well then everything fits together! How unanimously they come together in agreement," he says in Conjecture 5.[126] One must read his final judgement about tongue stones: "That they are teeth of the shark is shown by their shape, since they are closely matched, planes against planes, sides against sides, base against base." That Malta can once have been situated in the sea like several other islands which suddenly have changed their position "by a rapid conflagration of underground exhalations" explains the number of teeth thus found. And he repeats: "Thus since the bodies resembling parts of animals that are dug from the ground can be considered to be parts of animals, since the shape of tongue stones resembles the teeth of a shark as one egg resembles another, since neither their number nor their position in the earth argues against it, it seems to me that those who assert that large tongue stones are the teeth of a shark are not far from the truth."[127]

Yet it is extremely significant for Stensen's strive for security and his cautious method that he, however, still holds back the final judgement and carefully avoids any form of definite certainty. How modest the first sentence sounds: "The controversy to be decided in regard to the larger tongue stones is whether they are the teeth of Canis Carchariae or stones produced by the earth."[128]

---

[124] <700, 732> OPH II, 129–136.

[125] Scherz, "bereits an Gewißheit grenzende Schlüsse". Stensen says next that those who assert tongue stones to be teeth of sharks are not far from the truth, *a vero non multum recedere mihi videntur* (<731> OPH II,139).

[126] <727> OPH II, 135. *Stensen's enthusiasm is caused presumably by a sudden realization of different ways solids can be separated from fluids to form crystals and fossils, thus supporting his hypothesis on accretion.

[127] <731> OPH II, 139.

[128] <718> OPH II, 127.

And he concludes with the explanation: "While I show that my opinion has the resemblance of truth, I do not maintain that holders of contrary views are wrong. The same phenomenon can be explained in different ways, indeed nature in her operations achieves the same in various ways."[129]

When this dissertation was completed, the erudite Manfredo Settala (1600–1680) from Milan came to Florence for a visit. He was known because of his collection of natural curiosities and he could support the theories of Stensen.[130]

Stensen thought he was close to a well-rounded investigation of fossils[131] hardly knowing that he stood at the entrance to a whole new science: geology.

In review, the year 1666 shows a seminal and intellectually rich time under the sun of Italy. A powerful princely lineage took Stensen under its protection and offered him the best working conditions and a carefree existence. He immediately encountered intellectually outstanding friends and the Florentine intellectual culture offered him a powerful fertile soil. The months were filled by bustling work which presented its results in now classical treatises.

1.4.5   Annexe to Chapter 1.4. Troels Kardel: Stensen's Myology After Scherz

Gustav Scherz aptly concluded that the Myology may be the most controversial work by Stensen <191>. At proposal, it received some favourable commentaries but it was considered a mistake when rejected by Borelli in 1680 and flatly ridiculed by Johann Bernoulli in 1694 due to steadfast adherence to Aristotle's physical principle: "Anything which moves is moved by something else" which leaves no place for self-motion such as fibre shortening.[132] In 1711 a new edition of the *Elementorum Myologiae Specimen* was reviewed in the Journal des Sçavans.[133] The reviewer took notice that Stensen explains the mechanics of the function of the muscles per principles of geometry and claims that the contraction of the muscles must not be attributed to the involvement or the rarefaction of the animal spirits. The author's "explanations cannot be

---

[129] <730> OPH II, 138.

[130] <731> OPH II, 139, on Settala see EP I, 33 f and NBG 43, 827.

[131] <766> OPH II, 184.

[132] *Omne quod movetur movetur ab alio,* Aristotle, *Physics,* Ch. 7, quoted against Stensen's myology by Johann Bernoulli (*Dissertations* 1997, 108–109). The Aristotelean influence on muscle theories in the seventeenth century was recorded by Nayler, *Insoluble,* 282, and in SOM, 33–44.

[133] N. Stenonis, *Elementorum Myologiae Specimen,* 12°, Amsterdam Jansson Waesberg, 1711; web-accessible, earlier unrecorded edition, without preface to explain its re-publication. Review in *Journal des Sçavans,* 1711, 523–525.

understood without the help of several figures drawn on paper. Moreover, they require attention, not to say a very particular meditation. We believe it is more convenient to refer the readers to the book of Mr. Steno than to risk boring them with a summary which may only increase the obscurity of the matter already obscure by itself - *une matière déja assez obscure d'elle-même*." The reviewer found little of general interest in the two added shark dissections since, "neither of them contains anything which might excite the curiosity of those who are not versed in the science of anatomy". In concordance, the British surgeon James Douglas in his *Myographia* preferred Bernoulli's account of muscular motion that, "seems to be the most natural and the most agreeable to the Rules of Mechanism of any that has hitherto been advanced."[134]

The esteem went from bad to worse in 1910 when Vilhelm Maar, the editor of Stensen's scientific works, expressed, "the Myology is, perhaps, the weakest in Stensen's writings"[135]; not to forget, Harald Moe, professor of anatomy, in 1988: "Stensen's point of departure, the parallelepiped of muscle, was not correct".[136] Only in the 1980s, groups of biomechanical investigators in several countries showed that actually many skeletal muscles in man and in

---

[134]James Douglas, *Myographiae comparatae specimen, or, a comparative description of all the muscles in a man and in a quadruped*, London: Strachan 1707, p. xix. Several later editions.

[135]Vilhelm Maar in OPH I, p. xviii: "When in spite of much diligence and the most careful proofs [Stensen] still did not arrive at a correct result, this was due to the following two errors. Firstly, his starting-point was a wrong conception, ... that the course of the muscular, as well as that of the tendinous, fibre were each of them rectilinear, forming an angle at the two places, where the muscular fibre became a tendinous one. [...] Secondly, he did not pay attention to the fact that every separate muscular fibre, when shortened by contraction, must need become thicker, and that this in its turn must act on the whole figure of the muscle during the contraction." Maar in *the first argument* clearly rejects the now reinstituted pennate structure of muscle. Maar argued alongside G.A. Borelli and von Haller. "[Stensen's] structure is not true", and his compatriot C.C.A. Gosch, see complete excerpts in SOM, 41–43. In my judgment Maar in the second argument builds on a misreading of Stensen's geometrical model of contraction. In plane geometry, as seen from the side of the parallelepiped, the muscle fibres in Stensen's plane of the order form a parallelogram of flesh in which contractile fibres continue in the tendons at an angle. Stensen describes that having the same base and being in the same parallels, according to Euclid (Prop. I.35) the parallelograms of flesh in relaxation and contraction are equal. Meanwhile C S, the thickness (*crassities*) of the muscle, becomes larger than C R, the thickness in relaxation (Fig. 1.4.2). In this perspective, the muscle grows thicker in contraction due to the angulation of motor fibres towards the tendons, while the area of the contracted parallelogram of flesh remains unchanged. When muscle fibres are then viewed in a plane at a right-angle towards the paper plane, there is no visible angulation of the motor fibres towards the tendons and no change of the width (*latitudo*) in contraction. This is expressed in Lemma II: "When a muscle contracts, its width does not change" <665>. In the latter projection the diminished area from shortening of fibres matches the earlier mentioned increase of muscle thickness. As argued by Stensen the volume of the parallelepiped is equal in relaxation and contraction according to Euclid XI.29: "Parallelepipedal solids which are on the same base and of the same height, and in which the extremities of the sides which stand up are on the same lines, are equal to one another." *Ergo*, without change in volume, "the thickness of the contracted muscle is greater than the thickness of the non-contracted muscle" <670>, cf. SOM, 141. Maar's arguments against Stensen's myology thus appear untenable.

[136]Moe, *Stensen-biography*, 1988 (in Danish), 100; this sentence omitted in the English edition, 1994, 100, with endorsement of Stensen's myology.

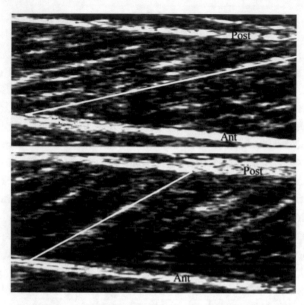

Fig. 1.4.4 Ultrasound scan of the medial head of human gastrocnemius muscle by Chow, Medri, Martin, Leekam, Agur, and McKeeby, 2000, showing shortening of parallel muscle fibres in contraction of the muscle. The pennation angle changes from approximately 15° in relaxation (upper), to 35° when contracted (lower). In Stensen's illustration, the pennation angle is approximately 35°, Fig. 2.22.20, left part. The widening of the pennation angle, the shortening of the motor fibres and the tendons remaining parallel in contraction as recorded by ultrasound are in accordance with Stensen's geometrical model, Fig. 1.4.2. However, an increased distance between the parallel tendons in these recordings is not.Professor Peter A. Huijing explains: The distance between the parallel tendons (the aponeuroses) increases a little due to their elastic properties: As the force exerted on the elastic aponeuroses decreases due to the lower length of the muscle and its fibres, the aponeuroses become shorter and muscle volume would tend to be decreased. One solution to consider is a further increase of the angle between the fibres and the aponeuroses but this would require further fibre shortening, leading to further lowering of the force and decrease in aponeurosis length: The problem would persist; so this has to be solved without additional muscle fibre shortening and the aponeuroses are pushed a little away from each other in that way preserving the constant muscle volume. Stensen's model was not equipped with elastic aponeuroses, and that is why in his model there is no need for this interaponeurosis distance to increase. (Peter A. Huijing, Vrije Universiteit Amsterdam, personal communication 2017. See also Huijing and Woittiez,"The architecture on skeletal muscle performance: a simple planimetric model", Neth)

animal species are pennate "actuators" consisting of uniform parallel muscle fibres at an angle towards parallel aponeuroses or tendon plates.[137]

---

[137]The crucial structural evidence was provided in 1980 by Brand, Beach and Thompson, orthopaedic surgeons who studied dead tissues of forearm and hand muscles in man with the aim of improving methods for tendon transplants. See further references on biomechanical investigations in SOM, 49.

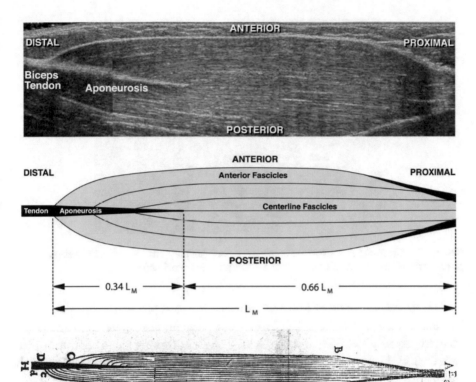

Fig. 1.4.5 Composite ultrasound image (top) and schematic line drawing (middle) of biceps brachii architecture in a well-trained healthy subject as viewed in a sagittal plane by Pappas et al., 2002. The internal aponeurosis spans 34% of the muscle length. Compare with Stensen's sketch of the same muscle (bottom) from a *postmortem* in meager human, <681>. As written, Stensen's sketch is viewed in the plane of the *order* and most likely in the same sagittal plane as shown in the two upper figures. Note that tendons and muscle fibres are drawn alike. The internal aponeurosis spans almost the same proportion of the muscle length. (Fig. 1 in George P. Pappas, Deanna S. Asakawa, Scott L. Delp, Felix E. Zajac, John E. Drace, *Nonuniform shortening in the biceps brachii during elbow flexion*, 2002.)

Lately, ultrasound in vivo examinations in humans incidentally confirmed Stensen's description of the structure of skeletal muscle, thus in the gastrocnemius (Fig. 1.4.4) and the biceps brachii muscles (Fig. 1.4.5).

Numerous reports corroborate the *structural* foundation of Stensen's *mensura* (measure or model) of the contracting muscle which, as held by Stensen, is "obvious to the eye of anyone any time meat is brought on the table" <693> (Fig. 1.4.6). When comparing Stensen's drawing (Fig. 2.22.20 Plate II, Fig. IV) and an ultrasound description of the elbow flexion by de Oliveira et al. 2002, accordance of the inner structure of the human semimembranosus muscle is seen.

Fig. 1.4.6 Bipennate muscle unexpected on display in a *bifteck* served at a post-conference dinner, *Steno and the Philosophers,* Paris 2015

The *functional* basis of Stensen's model is expressed in his Supposition 3, "When a muscle contracts, the individual parts all along the flesh move and shorten equally" <664>. Nowadays, fibre shortening is a commonplace concept, so basic that it escaped recognition until 1990 that well into the eighteenth century this was unacceptable and considered unscientific.

Stensen combined the structure and function by a two-stage geometrical model to show that muscles while shortening can make a swelling during contraction *even without change of volume.* Tradition dictated that muscles swell from inflation of the muscle by animal spirits supplied through the nerve, *or* from swelling by ebullition within the muscle by fermentation. The "inflationist" interpretation was transferred from ancient authors by Persian and Arabic authors and modified by Fabricius ab Acquapendente, Descartes, William Croone, John Mayow, and towards the end of the seventeenth century by Johann Bernoulli. Stensen's early "contractionist" effort was forgotten by researchers when fibre shortening could be observed under the microscope in the late eighteenth century and single fibre electro-physiological examinations took lead. Stensen's myology remained in limbo until late in the twentieth century when, (1) the pennate structure of muscle was instituted, and (2) the machine power of computers became sufficient to match data from observations by data from models resembling that of Stensen in a meaningful way.[138] (Fig. 1.4.7)

---

[138]SOM, 51–55.

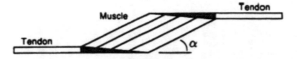

Fig. 1. Schematic representation of a musculotendon actuator. Muscle fibers are parallel and insert on tendon at the
same pennation angle (*α*). Tendon length includes the length
of free tendon and the length of aponeurotic tendon (stippled
area).

Fig. 1.4.7 Sketch of unipennate actuator applied in a computer model by *Hoy, Zajac and Gordon* 1990 showing accordance with Stensen's Plate I, Fig. 4, <679>

Stensen also proposed that the heart contracts through the shortening of its motor fibres: "What has been proposed here concerning the muscles, if it applies to the heart, is sufficient to demonstrate the initial proposal: the heart is actually a muscle".[139] Like on skeletal muscles, the contractionist position on the function of the heart was incompatible with the theory of action by swelling of the walls from inflation favoured by Descartes and Borelli. Adherence to an Aristotelean axioma blindfolded researchers on heart and muscle right until fibre shortening became visualized by microscopy a century later. At that time Stensen's description of muscle shortening from fibre shortening had been forgotten by researchers.

---

[139] <562 > OPH I, 178.

1.5 Researcher and Convert in Italy 1667

1.5.1 Embryology and Geology

The year 1667 was full of tension for Stensen. Major anatomical and geological problems incessantly drove him to investigations and experiments. Often we find him on journeys to Lucca, Livorno and Rome. In the most bristling correspondence with erudite friends, he experienced the final break-up of the Accademia del Cimento. In addition he lived through a difficult conflict between his old and his new faith, an unwavering study of the Catholic religion and finally the last difficult struggle for his conversion.

A letter from the beginning of this year provides a lively picture of Stensen's situation. Viviani writes to his friend, the Grand Ducal Chamberlain Count Molara, who was an outstanding connection with the court.[1] Viviani reports a conversation with Stensen in the evening preceding the departure of the court to Pisa where it usually spent the time between Christmas and Easter. Stensen had apparently remained in Florence to deal with his book, the main part of which, the *Myology*, had been approved by the censorship on December 29, 1666. It is very possible that Stensen remained in Florence to complete two shorter dissertations which appeared together with the *Myology*.[2]

In the course of the conversations between Viviani and Stensen the two men discussed Stensen's projects. Stensen praised the conditions in Florence with the good opportunities for dissecting human bodies and animals. Viviani

---

[1] *Positio*, 49 f, Scherz, *Viviani*, 276 f.
[2] GP, 117.

© Springer-Verlag GmbH Germany, part of Springer Nature 2018
T. Kardel and P. Maquet, *Nicolaus Steno*,
https://doi.org/10.1007/978-3-662-55047-2_5

continues: "I answered him that the use and the keeping of such a nice opportunity depended on him and that this would certainly be offered by His Highness. He answered that he had very well understood that from benevolent expressions and that he felt ashamed by nothing else than that he could not show his gratitude and submission by works. But he thinks of nothing else than of acquiring new knowledge and of drawing benefit from it."[3] Viviani thought that he might conclude from the conversation that Stensen would gladly stay in Tuscany after his return from Naples (did he also want to go to Malta which he had never seen, as he complains in the *Canis carchariae*?) but he also felt obligated to his king and had to think of his return back home to Denmark. Viviani, therefore, proposes in the letter that he and the court should attempt to retain Stensen.

At this point, at the turn of the year 1667, Stensen did not yet feel attached to Tuscany although things were obviously going well for him there. Nothing more is heard of the projected travel. He went to Naples in 1668 but he never went to Malta.

Finally, Viviani states that Stensen, as soon as the work on his book permits, would either remain in Florence or go to Pisa and Livorno where hunting and fishing provide abundant dissection material. The conditions are well illustrated in a letter of February 4, from Redi to Stensen.[4] Redi writes about a new observation, "while I am looking forward to a long discussion about it when you will have come back to Pisa to the court and after lunch and the evening meal we will gather at the fireplace and have nothing else to do." Redi recalls an experiment often repeated by the Grand Duke and Prince Leopold, namely to kill quadrupeds promptly by injecting air in a vein with a syringe. In Florence two dogs, a hare, and in less than 7 min a sheep were brought to death and the same occurred in Pisa with two foxes without any difficulty. After that the princes, as well as other learned men, resolutely, quite generally and without any exception, declared it impossible that a considerable quantity of air can be present in the blood vessels of living beings. Now some days ago fishermen from Porto Ferrajo had presented to the Grand Duke a live sea turtle weighing 90 lb and the Grand Duke had given it to him for observations. He had opened it with the help of Tilman Trutwin and made many observations of which he had given a description, while Filizzio

---

[3] *Positio*, 49.

[4] EP I, 188 f. It appears from the printed editions that the letter is dated February 4, 1667, which is in accordance with Redi's use of February 4, 1668 (Florentine calendar). Cf. Nordström, *Magliabechi*, 35. Gustav Scherz and the *Positio* (p. 50) remain, however, with the year 1667 because, perhaps, the original letter does not exist (possibly a proof may have been made since Prince Leopold in the letter is not yet referred to as being cardinal (he was appointed in December, 1667).

Pizzichi made drawings which he would transmit to Stensen after his return. The most remarkable, however, was that the blood of the turtle had been full of air bubbles: "What do you gentlemen say of that? Oh, how little we know though! And how easily we deceive ourselves when we believe general, quite general principles, to be in the matters of nature and we want to propound them." He had not wanted to be satisfied with one experiment only but he had opened four more turtles, with the same result. He was also to investigate turtles which live in freshwater and on land. Redi ends with greetings from three acquaintances, Count Molara, his friend Alessandro Visconti, and Marquis Giovanni B. Schinchinelle, all at the service of the court.

At the end of February Stensen was back at the side of Redi. On February 23 he was present, together with a French guest, Claude Morel whom Stensen knew from Paris, at investigations carried out by Redi.[5] On February 28 Redi writes from Pisa to Leopold and mentions, among other things, that Stensen has had a fever.[6] Later they are found together in Livorno and when Redi mentions the dissection, undertaken at the end of 1666, of the silver fish with live small animals in its stomach, he means the end of the Florentine year, thus in March 1667.[7]

On March 21 Redi writes to Viviani: "We were just at lunch when the unpleasant news was brought which you, Sir, asked me to transmit to our dear Mr. Niels Stensen... since Mr. Stensen, as you know, makes me the honour of eating at my table in the morning and in the evening, and I have the satisfaction of enjoying his generally virtuous and affable conversation.

We also are not idle but every day we make nice anatomical dissections and nice observations. At the court of Florence I hope to be able to report nice things to you, Sir, and the Grand Duke omits nothing in His unending noblesse to help my investigations. I join the answer of Mr. Stensen to this letter. I do not say anything to Oliva and Borelli."[8]

The names of Borelli and Oliva remind us that the Accademia del Cimento was broken up during those days. According to its diary, the last session took place on March 5. Early 1667 the above-mentioned members, as well as Rinaldi left the learned circle at the Medici court.[9] This was even more regrettable because the *Saggi*, the literary manifesto of the first experimental academy in Europe, was now appearing. The critical review of the first

---

[5]Redi, *Opere* 3, 75 ff.

[6]Redi, *Opere* 6, 326 ff. Redi's February 28, 1666 = February 28, 1667.

[7]Redi, *Opere* 3, 353.

[8]Redi, *Opere* 7, 213 f, EP I, 22.

[9]Middleton, 309 ff, esp. 316 ff.

manuscript by different learned men within and outside the academy had delayed the publication to the displeasure of Magalotti. That the breaking up of all members did not follow is due to the intervention of Redi and Viviani, indeed the most able members of the original group, and above all to the research of Stensen, who then devoted himself with creative energy to the study of the organs of reproduction and now turned to geology.

It is in this context that the third dissertation not yet discussed in Stensen's *Myology* will be called *Historia dissecti piscis ex canum genere*.[10] This third dissertation is not mentioned in the work's original hand-written title page. Thus it must have been written after the two others, probably in February 1667. That also was occasioned by a shark which the Grand Duke had sent to Stensen from Pisa. It was a Scymnus Lichia which Stensen named pork, sea dog or Salvian's pig, referring to an author who had described the animal.[11] "Of this text of Stensen is true", Gosch says, "what is true of almost all his publications, of most of what is written by real geniuses, that it is full of beginnings and almost necessarily leads to further research."[12]

Stensen finds many of his earlier observations confirmed, such as the mucosal system, the purpose of which he says again: "This shows the industry of ingenious nature which oiled the surface of fishes to let them split more easily the opposing waters."[13] One perceives Stensen's joy for the process of nature when he deals with the bladed teeth of the shark's jaw: "One would hardly describe the shape and position of the bladed ones as elegantly as nature shaped and set them."[14] He describes the smell organ very well and emphasizes how nature, always equal to itself, achieves the same in fishes with many skin folds as in the higher vertebrates through spongy and folded bone layers, namely to enlarge the perceiving surface.

Finally, he describes the sexual organs, and by the description of the *oviductus* arrives at a new, pioneering discovery in which he refers to an already published finding in his *On the anatomy of the ray*: "Likewise in the same anatomy of the ray, following the common opinion, I said about the uterus: this does for the viviparous animal all that which is expected from the ovary, oviduct, and egg in oviparous ones. Thus actually, since I have seen

---

[10]<733> OPH II, 149 ff.

[11]Cf. GP, 301. Molara tells in a letter dated February 4, 1667 of the catch of a "pesce porco" provided for Stensen to dissect. From the letter it appears that such a catch was utterly rare, (EP II, 924). Hyppolytes Salviano described the shark in 1554.

[12]Gosch, II, 1, 211.

[13]OPH II, 150.

[14]OPH II, 150.

that the testicles of the viviparous animals contain the eggs in themselves, since I have observed that their uterus opens in the abdomen in like manner as the oviduct, I no longer doubt that the testicles of the females are analogous to an ovary,[15] whatever the way either the eggs themselves or the matter contained in the eggs are transmitted from the testicles into the uterus. I will demonstrate publicly elsewhere, if and when there will be an opportunity, the analogy of the genital parts and correct this error by which people believe that the genitals of females are analogous to the genitals of males."[16] This was quite a new observation for the time (Fig. 1.5.1).

Stensen wanted to test his observations of egg, uterine tube and womb experimentally as thoroughly as possible and endeavoured for clarity in questions of details. He intended to write *ex professo* on these problems, which shackled him extremely. In those first months of the year, we see him actively involved in experiments on the reproductive organs. In this context an undated letter from Redi to Prince Leopold must be mentioned. It must have been written before December 1667 when the cardinalate was bestowed on the Prince. Redi reports that Stensen and he himself were at the time at his home investigating the uterus of a non-pregnant hind and, considering the inner part of the *collum uteri*, they observed four half-moon-shaped parts, each one finger breadth thick, behind the opening. They were thus arranged as if to prevent the semen from entering the uterus. In the next days several hinds would be opened in order to investigate the matter more thoroughly.[17]

The results of these and corresponding investigations were not published until years later and then merely in two short articles.

Stensen clarifies his motives in the *Observationes anatomicae spectantes ova viviparorum*[18] "To confirm and elucidate observations by my friends on the reproduction of animals from an egg, let me add to their works that which divine generosity pointed out to me from the dissections of various animals concerning the eggs of the viviparous. By egg I mean not only the round vesicles full of humour which constitute a large part of the testicles, but also the chorion with all its contents. I use terms which are usual for most people, the testicles of females meaning the ovaries, the uterine tubes and horns, and the uterus meaning the oviducts. The ovaries of the female testicle provide the

---

[15]Stensen's definition, see below at note 17.
[16]OPH II, 152 f. Cf. Scherz, *Pionier*, 116 ff, Lesky, *Ovum uterus*, Lesky, *Entdeckung*, esp. 239 ff.
[17]Redi, 8, 271 f. Cf. Stensen's remarks on the hind, <758> OPH II, 175.
[18]Eggs of viviparous animals I, <741> OPH 25.

principle of the egg. The oviducts, i.e. the horns of the uterus with the uterine tubes, provide what is required for the perfect growth of the foetus."[19]

Stensen then describes the observations from his dissections. One immediately notices the rich material obviously set at his disposal by the Grand Duke's favour. He investigated the organs of five cows, three sheep, one dog, two rabbits, one female bear, one hedgehog, one hare, one salamander and one turtle. He can also present the results of five female corpses, a human illustrative material amazingly rich for that time, which enlightens the free and favourable conditions for research in Florence. Stensen applies different methods, cooks the organs on several occasions and also observes a small rabbit in a vivisection. In the paragraph about females, he seems to distinguish between the vesicles and the eggs. For the fourth cow he notices that the blood vessels of the female animal are not connected with the blood vessels of the foetus.

The *Ova viviparorum spectantes observationes*[20] are obviously a continuation of the previous treatise and contain observations on the fishes blackmouth cat shark, spina, electric ray, silver fish, vipers, and on she-wolf, she-stag, sow, she-donkey, mule and hind, one of each.[21]

The first dissection and its results deserve particular attention. Stensen designates the shark on which he made his observations of the reproductive organs as *Galeus laevis*. He describes the attachment of the egg and embryo to the uterus as in mammals. He depicts how he convinced himself by blowing up the umbilical cord that, beside the blood vessels, it contains a duct which opens up in the intestinal tube of the foetus and comes out of a cavity in the placenta. There is also a drawing of that and of the ovaries from which one concludes that they are united at their upper extreme. He mentions that the placenta consists of two membranes. Stensen further explains the conditions of the foetus in another live-bearing shark, the *Acanthias vulgaris* which has no placenta. The umbilical cord and the deepening which corresponds to that of the placenta contain only the rest of the yolk of the egg, which also arrives in the intestinal tube as he had seen it in the chicken.

The original manuscript of the first part of this publication is available, neat and clear, from his own hand.[22] The text is mostly the same as it is found

---

[19]OPH II, 159. On the problems of terminology, see Lesky, *Entdeckung*.

[20]Eggs of viviparous animals II, <751> OPH 26.

[21]The separation from the previous paper, OPH 25, is accidental, apparently determined by the printing in *Acta Hafniensia*. The dissection must largely have taken place during his first year in Italy and when fish material was easily available. See Molara's letter February 27. EP II, 924.

[22]FBN, Post 21 II, 6 fol. 36 rv, 37 rv.

in Bartholin's *Acta Hafniensia* and in Maar's edition but some discrepancies suggest that the publisher made some changes before the printing and these were not always modified to their advantage. Several of these places are more strikingly and more directly, yes also more correctly expressed in the first manuscript.[23] In some sentences one also sees the researcher quite otherwise at work, for example, where he says in the manuscript: *In such a narrow room they must also turn themselves before the birth. Otherwise the head is the first which comes out at birth.*[24] Such discrepancies raise the question of the reliability of the transmitted printed scripts of Stensen when they depended on the understanding of the publishing contemporaries.

The embryological knowledge of Stensen was extremely important in a time when physiology was in the dark in what concerns the way the foetus is fed in the egg and the uterus. Some thought of a food intake through the mouth, others through the blood vessels and the umbilical cord. The discovery had a particular value by the fact that Aristotle already reported that the egg in the *Galeus laevis* not only develops completely in the uterus, and thus that the fish is viviparous but, like in mammals, it is fixed in the uterus and, therefore, has a placenta as well as an umbilical cord.

It is apparently the merit of the famous nineteenth century comparative physiologist Johannes Müller to have drawn the attention on the perspicacity of two great minds, Aristotle and Stensen, who made the same discovery at an interval of more than one thousand years.[25]

Aristotle mentions in his *Natural history* (book 6, Chap. 10) that, among the viviparous sharks, there are some in which the foetus is attached to the uterus through a placenta like in the mammals. Müller had himself vainly looked for that *Galeus laevis* for a long time. He explains that this discovery had remained unnoticed long after Stensen's observation: "The causes why this object has remained hidden for such a long time lie in its nature itself which suddenly became clear. The *Galeus laevis* of Stensen is one of two species of one and the same family existing in the Mediterranean. They are easy to confound. The other species is completely different as far as the

---

[23]Cf. <751> OPH II, 169, line 15–17: *Unica tantum, eaqve exigua, placenta cuilibet foetui erat, qvae rubicunda adhaerebat oviductui circa inferius orificium, & membrana obducta cavitatem formabat*, and in MS: *Non nisi una cuilibet foetui erat placenta, qvae parvula et rubicanda oviductui annectebatur circa inferius orificium.* OPH II, 169, 18–19: *Vasa umbilicalia per rimam inter binas anteriores pinnas sitam sub diaphragmate in corpus foetus penetrabant.*

[24]*oportet itaqve in tam angusto illo spatio ante partum se intorqveant, si alias caput primum est, qvod in partu exit*, omitted in OPH II, 169, after line 5: *vertebant.*

[25]Müller lectured at the Prussian Academy of Sciences on April 11, 1839 and August 6, 1840, "Über den glatten Hai des Aristoteles und über die Verschiedenheiten unter den Haifischen und Röchen in der Entwicklung des Eis", printed in 1842.

reproduction is concerned being affiliated with the viviparous sharks without connection to the uterus, the *Vivipara acotyledona*."[26]

Quite amazed, Müller says: "The first who after Aristotle once again has seen the one discovered by him, with attachment of the foetus to the uterus, is Nicolaus Stenonis, the gifted Danish anatomist (Fig. 1.5.1). It seems that he had no knowledge of the observation of his great predecessor. It is that more remarkable that his description entirely corresponds to that of Aristotle."[27]

Indeed, Stensen surpasses Aristotle: "Although Stenonis has fallen into the same mistake concerning the egg membranes as Aristotle, he has, however, beside the confirmation of the main fact, the essential merit that he recognized the connection of the hollow placenta with the intestine through a duct contained in the umbilical cord and that he recognized two membranes in the placenta the inner one of which forms the cavity of the placenta and continues in the said duct, the outer one, how he says, forms a coating of the placenta. Two such membranes actually are present. They are the two membranes of the yolk cavity. The dissertation also contains a figure of the two Fallopian tubes which behave like in other sharks. Stenonis undoubtedly has been the first to see that the Fallopian tubes join above and have a common orifice.[28]

Stensen's *Dissection of a dog-fish* and the following dissertations show him as a discoverer of the female reproductive organs, on which he also intended to write a conclusive work. He had dealt with this matter already in Holland. Swammerdam remembers having seen the *tubuli seminales testiculorum muris majoris* in the study room of Stensen already in 1662.[29] In 1664, in Copenhagen, he had written a report on the *Anatomy of the ray* and given a definition of the uterus, deeply unsatisfied by the concepts of his time. Then he discovered the yolk duct in chickens. Yet not until in Italy did he acquire a reliable and clear conception of the ovaries and confirmed it through several dissections of different animals. However, the honour of being the discoverer of the egg follicle was awarded to Stensen's friend from Holland, Regnier de Graaf, with his publication of *De mulierum organis* (1672). This work became the object of a controversy. Swammerdam reproached de Graaf to have appropriated the discoveries of others like those of Swammerdam himself, of

---

[26]Müller, 191.
[27]Müller, 203.
[28]Müller, 204.
[29]Swammerdam, *Miraculum naturae*, 50.

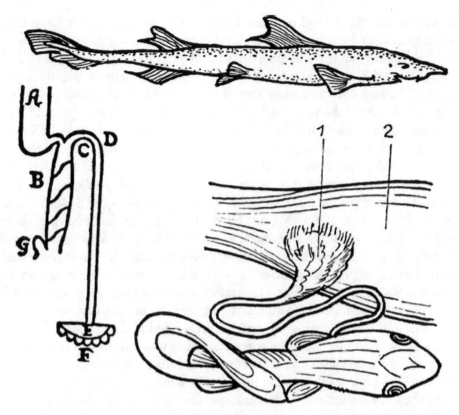

Fig. 1.5.1 Composite figure on reproductive organs from Scherz I, 184. *Top*, a *Mustelus galeus laevis* shark up to 2 m long. *Left*, a sketch by Stensen in which A: belly; B: spiral intestine; C: connection between genital and intestinal tract; E: connection with placenta; F: surface of placenta and oviduct; G: appendix, OPH II, 170. *Right*, after J. Müller: 1. Yolk sack, placenta. 2. Uterus connected with the former

van Horne and above all of Stensen.[30] De Graaf defended himself by saying that, before its publication, he had not known of the studies of Stensen on the shark and the ovaries. He reported that Stensen had been with him in Delft in 1670. He had shown Stensen the preliminary works for his treatise *De mulierum organis* and, at his request, Stensen "had imparted him that he had found eggs of different sizes in different other animals… in marmots, pigs, ravens, badgers, she-stags, hinds, she-wolfs, she-donkeys, also in mules and others." De Graaf contested the priority of van Horne in this question but

---

[30]Swammerdam, *Miraculum naturae*, 50 ff. On the controversy see also EP I, 8 f. The question of priority was also discussed in 1673 in the Royal Society of London that stated Stensen as being the first to describe the *ova viviparorum* (Rome, *Nicolas Sténon et la "Royal Society"*, 226 f.).

said of Stensen that he had not mentioned him, "because I did not know that he had written on this subject", de facto a complete recognition of Stensen.

The honesty of this explanation cannot be doubted if one pays full attention to the word "written" (*scripsisse*) since the results of the investigations of Stensen on the female reproduction organs were known orally in Holland already in the year 1668, at least in the circle of some specialists, as Swammerdam testifies for himself.[31] However, the whole question of priority turns pale when considering the realism and the modesty with which Stensen faces it. He does not move one finger to reclaim the honour of discovery. As Swammerdam announces to him on March 5, 1668 the publication of the preliminary draft of the manuscript of van Horne on the genital organs, it is not loathsome to presume that van Horne precedes him in publishing their common observations, Stensen answers on March 18, 1668 from Livorno: "Dear Friend. With great longing I expect the honoured Mr. van Horne's and your observations on the testicles[32] and it is far away from my intention to take it ill that a teacher or friend forestalls me, so that I solemnly assert that I have not only mentioned him when it had come to my acquaintance but, at the same place as discussing my own made public his observations."[33]

De Graaf even more nicely testifies the noble unselfishness of Stensen. In the year 1670, Stensen never does emphasize himself in his correspondence with de Graaf but loftily sets his investigations at the disposal of the latter.

In the above-mentioned[34] manuscript, *Ova viviparorum spectantes observationes*, there is a series of observations of turtles and a goose, which are omitted in the *Acta Hafniensia*. The paragraph following the description of the *Galeus laevis* is:

"Not much could be observed in the sea turtles. The humid air which is less favourable for anatomical experiments compelled to undertake both dissections very quickly. Thus, the more time was devoted to the investigation of one, the less time remained for the investigation of the other."

The structure of the oesophagus was elaborate. It was rugged by innumerable prongs turned towards the stomach and I have no doubt that what has been said about them is right: they prevent what has been swallowed from being spat out again through the same way. It could be assumed that, on the weak base of these prongs, certain moisture is prepared which is pushed out by the passage of hard bodies when swallowing.

---

[31]Cf. EP I, 205 f.

[32][= ovaries].

[33]EP I, 205 f.

[34]See note 5.19.

In the stomach and in the gut one found some wood gnawed by decay and a small piece of coral, otherwise nothing firm. The part of the gut the closest to the stomach, one Florentine yard long, was provided outside with a strong muscle. Inside it had a crust with thousand holes so that it resembled a honeycomb.

In the heart there were five openings although there was one chamber only. Two took-in the blood which was driven out from the auricle of the double chamber. The three others led the blood driven out from the heart further into three different arteries. There were ten valves, two for each opening, all of them shaped somewhat like a half moon; although in those valves opposite the auricle the half moon was in reversed position. A tendinous body which divided the entry to the auricle in two equal parts replaced the function of two valves.

The cleft of the throat was extended in length like in birds. There was no epiglottis. The rings of the trachea were all cartilaginous. The lungs were separated from the belly.

Blood taken from a scarcely fed goose had little serum and this was quite watery whereas the serum of a goose well fed for days was abundant and quite milky without anything watery. Herewith one easily explains the white colour of the big liver although plunging the still warm liver of the animal in water to draw out all the blood contributes much to this colour. With the same argument one can very likely approve the argument of those who think that the milk of women is diluted from the arterial blood. Anyway as I drew the red substance from the artery of the goose there was no sign of milky serum so that everything was intensively red. But as the warmth progressively went away, the colour also changed until the blood coagulated on the bottom of the basin was clear milk. Its serum parts were separated from the cheesy parts by pouring vinegar".[35]

There can be no doubt that, at the beginning of 1667, Stensen was also dealing most actively with the geological questions that had caught his eye and alerted his mind during the dissection of the giant shark, the *Canis carchariae*. It is not without reason that we see him on hikes that inspired the introductory words in *De solido*: "While travellers hasten over rough mountain tracks towards a city on a mountain top, it often happens that they judge the city, at first sight, to be close to them; constantly, numerous twists and turnings along the route delay their hope of arrival to the point of weariness, for they see only the nearest peaks; in fact, those things hidden by the said

---

[35]For the excerpt (Latin), see the German edition of Scherz I, 210–211, note 30.

peaks, the height of hills, the depth of valleys, or the level of plains, whatever they may be, far exceed their conjectures, and they, deceiving themselves, estimate the intervening distances from their own desires."[36]

We thus see Stensen on the way to Lucca with two letters of recommendation, which Redi had delivered to him on Easter Monday, April 11, after the months at the court in Pisa and Livorno. He wanted to observe the Liberty feast which the town had celebrated every Whitsunday since 1570 to remember the liberation from the domination of the Pisano. Yet above all he came to meet the friend of Redi, Francesco Maria Florentini.[37]

Born in Lucca, Francesco was a very distinguished researcher and physician. Popes Urban VIII and Alexander VII offered him positions as general physician and professor at the La Sapienza in Rome, which he declined. He wrote several works, among them *De genuino puerorum lacte, mamillarum usu... disquisitio* (Lucca 1653). He was also a competent botanist. We have an interesting testimony of the scientific interests which both learned men shared. In an underlined yet unprinted part of the summary in the manuscript of the *Canis carchariae caput dissectum*, Stensen remarks that Fiorentini had first shown him briefly in Lucca the vertebral spine of a fish from the land of Malta which had confirmed his conjectures about the origin of the Malta fossils.[38]

The second letter of recommendation was to Lavinia Arnolfini, the wife of the envoy of Lucca in Florence. This letter shows that Stensen, only 1 year after his arrival in Italy, had become acquainted with Lavinia Arnolfini who would intervene so deeply in his life. Redi wrote: "A man born a subject of Denmark–Norway will soon visit the nice town of Lucca and bring with him the most gallant and friendliest of men that are to be found in that far land. If, indeed, he had not the Luther dogmas engraved in his mind, I would be sure that Your Grace would consider him as a man of extraordinary perfection. I recommend him to your protection, Madam, and to his Excellence the Envoy and I am convinced that you will bestow on him every help so that he can content there that innate thirst for science which has made him a pilgrim of the world. The man about whom I write to you is the famous Mr. Niels Stensen to whom my master, the Serene Grand Duke, shows more than usual esteem, and of whom, Madam, I once spoke to you in Pisa."[39]

---

[36] <765> OPH II, 183, GP, 137.

[37] EP I, 37. The letter is dated in Pisa. BU XIV, 142.

[38] Transcript in GP, 131, note 139; copy of MS, 312.

[39] EP I, 37. The letter is dated in Pisa.

Lavinia Cenami (1631–1710),[40] born in Camajore, a village in the region of Lucca, was awarded a quite unusual life. She spent her youth in Paris and actually wanted to join a nunnery when, after the death of her sister, out of consideration for her family, she married Silvestro Arnolfini (1604–1685), who was more than twice her age. He was a nobleman from Lucca, who had fought on several battle grounds at the service of Mazarin. She bore him four children. Silvestro Arnolfini was Lucca's resident in Tuscany from 1665 until 1674. His home was a high-class palace in the Via de' Bardi on the left bank of the Arno in Florence. Stensen's first visit to her seems to have been nothing more than a formal visit of which we know nothing more.

That visit in Lucca cannot have lasted long,[41] and on his return to Florence Stensen moved into a new and better house. Viviani had recommended to the Grand Duke to improve the conditions of lodgement of Stensen, not least to rid him of a none too pleasant hostess.[42] On April 20, he obtained not only a room in the Palazzo Vecchio but also the necessary furniture like a canopy-bed with three mattresses, stools and tables as well as bedding, a dinner set and pictures—the inventory as a whole shows that Stensen was quite pleasantly accommodated.[43]

The *Specimen of elements of myology* was published at the end of April or in the beginning of May 1667. Stensen still succeeded in adding the notice above on Fiorentini's fossils from Malta. On May 21, Redi sent two copies of this work to Lucca and at the same time gave his thanks for all which Fiorentini had done for Stensen.[44] Viviani sent a copy to Father Stefano de Angelis who, as has been mentioned, expresses his enthusiasm about the geometrical method in his thanking letter of June 4.[45] For the rest it appears that none less than Prince Leopold sent the work to the most different learned men in the name of the Medici house as well as of the Accademia del Cimento. Cardinal M.A. Ricci, enthused by the reception of the book, wrote his thanks on May 30. It is very interesting that the book, which arrived in the hands of the Pope elected on June 21 1667, was sent to him by Prince Leopold, himself to be elected Cardinal on December 21, 1667. G.B. Copponi (1620–1675),

---

[40]EP I, 36 f.

[41]Scherz (*Reisen*, 65) mentions that Stensen stayed in the city about 1 month; this is however improbable because already on April 17 Fiorentini wrote to Redi: *Il Sign. Niccolò Stenone era così degno dell' ossequio di tutti* (Coturri, 74), cf. also Positio, 1031.

[42]EP I, 26 f.

[43]Sten. Cath. 3, 1957, 36.

[44]Manni, 58. Bellini could already before May 16 refer to the contents of the book to Borelli (GP, 42, note 56).

[45]Scherz, *Viviani*, 283 f.

professor of anatomy and philosophy in Bologna, received the work as a gift
from Prince Leopold on June 24, 1667.[46]

At the end of May the work arrived in Paris, where Chapelain thanked
Stensen for having sent him the copy and promised to transmit letter and
book to Thévenot as soon as the latter returned to Paris.[47] Other copies
procured by Prince Leopold were probably not sent out until the end of the
year. In any case, the work was difficult to procure: Auzout sent his copy
before having read it properly to the impatiently awaiting secretary of the
Royal Society of London, Henry Oldenburg.[48] In July Settala who is himself
mentioned in the book had announced the publication to him but the work
did not arrive in England until February 1668. It was immediately reviewed
in the *Philosophical Transactions*.[49]

The publication of this work, the first of Stensen in Italy, did not make
him rest on the laurels. At the beginning of June we find him in Rome at
Ricci's to whom he brings a parcel from Viviani. In his letter of June 7, 1667,
Ricci as usual has only praise to say of the Dane.[50] This journey of which we
know little apparently had a geological aim.[51] Molara thus writes to Stensen
on June 14 from Florence: "I am delighted by your progresses in the investigations of interesting matters, particularly of the mussels and I am even
more pleased that you have made satisfactory observations which confirm
your view."[52] However, his journey was not devoid of problems: the Count

---

[46]Fabroni, 2, 156 f. Sten. Cath. 6, 1960, 60 ff.

[47]EP I, 192.

[48]Oldenburg, III, 578, IV, 23, 62 f, 78, 84 f, 87, 123.

[49]Oldenburg, III, 455, VI, 123, 145.

[50]FBN, Cod. Gal. 255, cc. 43a. Notice that Gustav Scherz differs from several scholars that hold Stensen
to be in Livorno at the Corpus Christi festival in 1667 on June 9. See, for instance, *Positio*, 1032 f. Scherz
knew all the arguments but steadfastly withheld that the Corpus Christi that made such a powerful
impression on Stensen was that in 1666. If Stensen was to have been present in Livorno by June 9, he
must have left Rome not later than June 5. In Scherz's opinion there was not enough time for Stensen to
experience the heated time during the sedivacance and the conclave, starting on June 2. HH notes that
Stensen's itinerary in June 1666 and 1667 cannot be settled without doubt. Both positions are, therefore,
perfectly possible. What matters more, however, is the implication on the timing of Stensen's conversion.
Some prefer the relatively short time from June to November 1667 for his doubts and worries. Scherz saw
the process as a long time of reflections and studies until the culmination in November 1667. Two minor
facts might support Scherz's position: On March 30, 1667 Redi predicted his conversion (EP I, 22), and
on June 11, 1667, in a letter to Count Molara Stensen mentions a cold and the misfortune from the
extraction of a tooth (EP I, 192). Would such personal discomforts leave him enough energy to
experience a tremendous upheaval during the festival? HMH.

[51]In the *Prodromus de solido* Stensen refers to deposits of ashes "that I have seen outside Rome" (GP, 162
and 226, note 64), possibly on Monte Mario, an observation that stems from this travel to Rome. He was
hardly engaged in geology then.

[52]EP I, 192 f.

expressed his concern about Stensen's cold and an accident by a tooth extraction. He writes that, after his return, he wants to offer him the muscles of a lamb, a monster with two mouth openings that has been given to the Grand Duke.

At that time, the whole of Rome was preparing for a papal election. Pope Alexander VII had passed away on May 22, 1667, and the conclave and the Curie preceding the choice of Clement IX (1667–1669) from June 2–20 gave rise to many attacks and rumours in Protestant and places hostile to the Church, above all in Holland.[53] In his publication *Occasio sermonum*, Stensen recalls this sojourn and the rumours, misinterpretations and nasty calumnies which made quite an opposite impression on him. Even Queen Christina later turned to De Witt to counter them. Stensen writes: "As far as the Pope election is concerned, I have been myself, then not yet Catholic, in Rome during the vacant see after the death of Alexander VII of blessed memory, and I have learned by personal experience that all the rumours of those days partly were political, partly are inventions of men who wish ill to what is holy, as even the work of Christ was laid out as sorcery, violation of the laws and seduction of people. On one and the same day the most contradictory things were heard and those who stood far away from the matters largely knew more than those who had to do with the matters."[54] Stensen himself found it normal that in the choice of a man the grounds and interests of human prudence were put forward (knowingly, once again Spanish, French, German and Italian candidates were then opposed to each other). As always his view was directed to the positive and there was much to be pleased of in a college of cardinals counting men like Barbarigo, Altieri and Palloto.

For the next 6 months in the life of Stensen we have no sources of information at all. However, based on the work on the *Prodromus* written in the summer of 1668, he must have been continually busy with geological problems and have undertaken travels in Tuscany. We can also easily suppose that he carried out other dissections for the court. Yet, apart from this, during the autumn he struggled through to a conversion, which should so decidedly determine his future.

---

[53]Pastor, 14,[1], 527.
[54]OTH I, 189.

1.5.2 Stensen's Conversion

(a) Initial Doubts

Stensen himself has sketched for us his way to the Catholic Church in his *Epistola de propria conversione.*[55] It was surely written 4 years after the decisive events but offers most concisely a clear and reliable picture of the spiritual evolution, together with the commentary of the *Defensio* of the letter and the *Defensio* of the *Scrutinium reformatorum* and with what the friends said.

There can be no doubt that young Stensen did not only come from an orthodox Lutheran country and stemmed from a family of very believing ministers, but that he also was personally deeply faithful in his childhood and youth. This means that he consciously and convincingly adhered to the Lutheranism in theory and in life; much inner and warm piety still flourished in a century full of strict orthodoxy and theological controversies.

Stensen was well-taught in his faith. The *metae*, the religious lesson of the school of Our Lady, was not small and the life of the schoolboys was marked by the faith from morning until evening. One only needs to read the notes of the student to be convinced of his piety.[56] We have no reason to question Stensen's words when he says that he held Luther's translation of the Bible and his Catechisms, the small and the large, for the true and unaltered word of God, that he always faithfully adhered to the faith of his fathers,[57] and that he had always vehemently opposed the Catholic doctrine, particularly in discussions.[58]

He kept his Protestant Lutheran faith intact until the end of his studies in Copenhagen and until his first stay in Holland. The student's reading and excerpting of a Catholic work like the *Joseph Aegypti prorex* of the Jesuit J. Drexel influenced anew his moral aspiration so that one would be tempted to find parallels between the Egyptian Joseph and Stensen's Christian life as we know it. Yet it was an ascetic treatise that was read for its literary quality, its pithy logic, its lively examples, its classical quotations and its paternalism, and was anyway much less dangerous for the orthodox Lutheran faith than, for instance, the controversial theology writings of Bellarmin which then often circulated in the Protestant world.

---

[55]OTH I, 126 ff. Also editions of Stensen's account of his conversion in Danish, German and Italian. On the conversion, see *Positio*, 60ff. *Defensio* of the letter, OTH I, 371–437; Defensio of the scrutinium, OTH I, 225–289.

[56]Cf. Chap. 1.1.

[57]*Patris … institutis semper adhaesi*, OTH I, 126.

[58]*acriter … opposui*, OTH I, 126.

In Holland that orthodox Lutheran faith, as Stensen knew it, was for the first time challenged and even shaken, and this as a result of three experiences. First, the relations with Cartesian philosophy and its representatives were not without effect. In Cartesianism Stensen praised and liked primarily its method to discover prejudices and the rule of doubting everything. Not because he wanted to surrender to a universal doubt but because, as a man of science, he strove for certainty and clearly recognized the human capability of errors, he eagerly seized any means that promised certain knowledge.[59] Stensen's scientific method was strongly influenced by Descartes who taught him critically to test even the foundations of his faith and thus took him away from strict orthodoxy quite so thoroughly that it almost cost him the Christian faith. Stensen, as can be seen above, actually reproached Descartes, and even more so Spinoza, to have reached, against their own method, an idealistic rationalism from which he himself only escaped as if it were by a miracle. "There is this danger of the Cartesian philosophy from which God, as I have said, has liberated me at that time in which I was associated with the reformers, or rather the deformers, of Descartes. He gave me, against all expectation, the knowledge of the true structure of the heart and of the muscle which alone, at a glance and without a word, overthrew all the constructions of the subtlest minds."[60] This experience aroused in Stensen a certain distrust of the theories of philosophy concerning God and the soul when philosophers went so much astray in the field of science which is accessible to experiment. He felt strengthened in his faith and overwhelmed by the knowledge of the particular love of God that had led and protected him and invited him to reciprocal love. He was more ready than ever for a critical test of his childhood faith.

Then he came across the second experience, which is the multiplicity and the disagreement of those who appealed to the Holy Scripture in Holland. This disunion must have worked overwhelmingly on the guest from the North. In Denmark, the leading theologian of the seventeenth century, Jesper Rasmussen Brochmand, whom Stensen had know already as a schoolboy, wrote in the dedication of his big *Dogmatik* to Christian IV, in 1633: "From Your arrival at the government You have worked so that all Your subjects should think and speak unanimously of God and the divine matters and that You have done with such a result that those who deviate in their religious

---

[59] *Quin illam* (Cartesian philosophy) *repono primo loco inter alia, quae effecerunt, ne amplius primo judicio inhaererem*, OTH I, 388.
[60] OTH I, 399.

thinking go away, now exiles, far from Your kingdom and realm."[61] This praise depicts, as C.S. Petersen emphasizes, more an ideal state than reality. To this must be compared what, for example, A. Otto writes about the Jesuits and the quiet reformation in the time of King Christian IV.[62] However, the royal authority had undoubtedly for the time being repressed the disintegrative effect of the Protestant scriptural principle and ensured the unity of the Lutheran doctrine.

It was quite different in Holland where Calvinist orthodoxy had attempted to achieve exclusive dominance. The Arminian controversy, i.e. the theological battle between the Leiden professor Jakob Arminius, the most significant opponent of the Calvinist doctrine of predestination, and the strict Calvinist Franciscus Gomarus ended with the victory of Calvinism at the Dordrecht synod of 1619, politically and socially with the execution of the noble council pensioner Oldenbarneveldt. Yet there was nevertheless no question of an exclusive dominance. The Dutch republic tolerated the Armenians, Lutherans, Mennonites, Socinians, Jews, Labadists and other sects. When Stensen arrived in Amsterdam there was a multiplicity of religious trends like nowhere else on the continent. Greater pressure persisted only on the Catholics. Although the Peace of Westphalia ensured them a certain tolerance, the about 300 000 Catholics were not allowed to organize any processions or public ceremonies. Still in 1708 it came to the expulsion of Jesuits and the closing of their churches.[63] However, otherwise the whole subjectivism of Protestantism assailed the student. Sects who quarrelled among themselves and finally called forth indifference of faith prepared the unbelief of the next century. It is not by chance that Stensen, in his later works of controversial theology, deals almost exclusively with the Gospel principle of Protestantism as the main source of the error and the persisting splitting. "They have remained separated from the Church which has raised them and from the simultaneously established societies which like the Church itself promised a reformation on the same principle so that the catechesis of one contradicts that of the other and the one who joins the altar of one does not dare joining the altar of the other."[64]

The *modus multorum vivendi politicus*, as he says, finally contributed to the weakening of Stensen's prejudice.[65] He thus names those "Christians" that

---

[61]Petersen, C.S., I, 470.

[62]Otto, 7 ff.

[63]AGN 8, 322 ff. Pastor, 14, 1014, TRE 4, 63 ff, 66 ff.

[64]OTH I, 116.

[65]OTH I, 126.

might be called disciples of Spinoza or followers of Hobbes but in whom Christianity declined the more they were considered clever by the judgement of the mass. Most of them recognize one religion only which, against all reason, allows unbridled passions. Others let themselves be guided by reason exclusively, follow their inspirations, live friendly with all religions, even adopt from time to time the external forms of all. They have thus no other concept of the religion than that of a bond of society that the simple-minded hold for something divine. Such people generally contribute to disturb the divine faith in the heart of the people. "If for instance a young man who has no faith in God sees that learned persons, clever in the judgement of the world, high-ranking, sometimes even standing in the holy service, think nothing holy, and that all those of them who actively confess their religion and do not strive in the spirit of the wrongheaded human cleverness for carnal desire, wealth and honour, are considered as simple-minded, ignorant and stupid, it is difficult without particular grace of God to sail round these rocks without shipwreck."[66] Thus it seems that Stensen also felt threatened for a time by those lukewarm Christians and the spirit world was a temptation for him.

The consequence of all those experiences in Holland admittedly was a deepening life of faith in general, above all of his faith in Providence. Stensen writes that there are no more efficient means against that shrewdness of the world than to come to know the Providence of God, "as she became so obvious to me in some instances already some years before my conversion when I began to throw myself in her arms, indeed not always in fact but at least in desire, and avoided everything which stemmed from human shrewdness that did not correspond to the Gospel. Therefore, I organized my life so that every day I did what I thought was in accordance with the time and place and my forces. I cared so little for my future that I did nothing by requests or gifts to obtain honours, position or even the hope of obtaining such. Yes, I urgently asked my relatives not to approach anybody on my behalf. I wanted to do what depends on my forces but to expect the rest from God in order to be able to be sure to have obtained everything, not by human favour, but by divine vocation whatever fell to my share at its time."

He had been led to that by different instances in which unforeseen hindrances had suddenly upset all his preparations achieved by friendship, whereas at the same time other prospects had as unexpectedly occurred. "Often I found difficult to change my decisions and projects but, taught by

---

[66]OTH I, 391.

many experiences of how many evils God had spared me and of how much good he did me, I finally learned this prayer: 'Without a sign from You not a hair falls from the head, not a leaf from the tree, not a bird from the air, no thought escapes from the mind, no word from the tongue, and no movement is made by the hand. You have led me so far on ways I knew not. Lead me now, seeing or blind, on the path of grace. It is easier for You to guide me there, where You will, than it is for me to pull away from where my yearnings draw me'."[67]

From what has been said one realizes that the natural and Christian faith of Stensen became firmer and more conscious but that the strict Lutheran orthodoxy of his youth underwent a weakening.

On the other hand Stensen admits that by meeting many noble Catholics with whom life brought him together he had acquired a certain affection for their faith. "God had moreover bestowed on me certain knowledge of the phenomena in nature which everywhere attracted not only learned and experienced men but also many pious men who had heard of the signs of favour of God which he had granted me by the study of nature. Full of zeal for my salvation, they looked for an opportunity to discuss with me intimately and to explain the supernatural to me, who made clear to them the natural matters so that it seemed God had given the natural gifts to me so that I had an opportunity also to acquire the supernatural ones. In this way I could often discuss intimately with many people during my journeys everywhere and see the big difference between the striving for science and that for piety, and I attached myself with a certain respect to those whom I had put to the test on this matter. Since I found among the Catholics, several who would not have met me otherwise, I often felt a clear inclination to Catholics."[68]

Such declarations are easy to understand against the background of the confessional situation of his time, i.e. the successful advance of the Catholic reform of the Baroque period, whereas the Protestantism often decayed and fell apart. Stensen came to Rome through France and Italy. In France the spirit of François de Sales and Vincent de Paul, of Bourdaloue and Bossuet, and of Pascal was alive. A religious flowering was unmistakeable in the life of the orders, in the missions, in the charity and in science. France also presented other phenomena: Gallicanism, Jansenism and Quietism, and even atheist and libertine currents. Yet the clear glance of Stensen was always directed to the beautiful, the true and the essential. In Italy there was also still

---

[67]OTH I, 393.
[68]OTH I, 394.

a brilliant Catholic culture. Many complaints are known of the last period of Stensen's life when his field of activity was North-West Germany then suffering religious disunion as a consequence of the spiritual damage of the Thirty Years' War. In his adult period, the time of his scientific discoveries, the time of his conversion stands the more brilliant before his eyes.

The problem of Catholicism versus Protestantism had hardly disquietingly come to the foreground until Stensen arrived in Italy. Without being strictly orthodox, the traveller felt himself firmly anchored in his Lutheran faith. This does not exclude that he by and by had come to know Catholic life closely. Discussions on faith certainly occurred often. It is known that in Cologne he had a conversation with a Jesuit already in 1664.[69] During his journey in Belgium he visited Catholic churches and cloisters and attended a Corpus Christi procession. In France the celebrated erudite experienced a brilliant display of Catholicism. Conversations on faith are expressively testified with Mrs. von Rantzau in the cloister of the Annunciates, with her confessor, the Jesuit La Barre, and with Miss Perriquet at Thévenot's. Stensen later thanked the latter for the mediation of the friendship with *la servante de Dieu*, who had prepared him for the status of a Catholic clergyman.[70]

(b) Experiencing Italian Catholicism

It was not until some months after his arrival in Italy that the confessional problem stood before him, demanding an answer with all determination. This was when he attended the festive Corpus Christi procession in Livorno on June 24, 1666. When he saw the consecrated wafer and the respect of the people he asked himself whether they were or he was himself reasonable. "By this thought which bestirred my soul, on one hand I could not accept that a great deal of the Christian world, as the Roman Catholics are with that many bright and learned men, was blinded; on the other hand I did not want to condemn the faith in which I had been born and was brought up. It was clear for me that there was here a dilemma and that there was no possibility to reconcile the two theses opposed to each other. Neither was it possible to consider a religion for the true one, which went astray in a so essential point of the Christian faith and led its followers into error."[71]

At that moment Stensen's deeply searching mind began to work. He began by looking backwards. "It was not only the Church community of Livorno which passed before his eyes", says A.D. Jørgensen, "but also an immense

---

[69]OTH I, 190.

[70]EP I, 372.

[71]OTH I, 9.

amount of people, century after century, generations from the rise of the sun until its last setting. He stood before the silent testimony of history and time. Did this testimony rest on a deception? Or were the apostasy and the rejection of it in the Protestantism an aberration?"[72] He deepened the truth of the belief in the Eucharist and derived from it the whole problem of the Church. Looking back he later said: "After I had convinced myself and made sure about the truth of the matter of which I speak, I no longer doubted my duty to have to leave the Lutheran confession. When a religion is mistaken in an essential point of faith, it certainly cannot be of God since, by the force of his infinite wisdom, He cannot be mistaken, so by the force of His highest truthfulness He is unable to lie in His revelation or to deceive us with His utterances. Therefore, each sect which deviates from what stands firm for us as a revelation of God to His Church must be a pure invention of men."[73]

With this doubt in his heart, Stensen went from Livorno to Florence to spend there some time "because of the Italian language", as he says, "which is renowned for being spoken purely there, and then to continue my journey and see the other important towns of Italy."[74] He was thus then still a mere traveller. The stay in Florence changed his plans. More than anything new scientific problems disturbed them. However, in the abundance of new works his restless mind sought a certitude concerning the secret of the Eucharist. "I endeavoured very actively to seek the truth, trusting on that God would guide me by his light to the discovery of the truth which I sought with a sincere heart although the Lutheran education of my childhood retained me, yes, urged me to the opposite and to a hardening of my old opinions."[75] Not satisfied with the learned discussions on this subject, Stensen wanted also to seek clarity in the questions that beset him by a deepened study of the original texts, of the Gospel and of the Church fathers. He pursued this study particularly in the Biblioteca Laurenziana with its famous old Greek and Hebraic manuscripts. He also went on his own way of observation and experience.

Jørgensen says: "It is in complete agreement with the whole personality which we have come to know in Stensen as a man of science that, after that strong impression, he not only discusses the theological controversies with his acquaintances, but also he keeps on observing exclusively. Exactly as in his studies, after having become aware of a fact, he directs all his observations and all his sagacity to its proper understanding, without bothering one instant

---

[72]Jørgensen, *Nils Stensen*, 118.
[73]OTH I, 8.
[74]OTH I, 9.
[75]OTH I, 9.

about what others have thought and meant. So does he direct his attention to the Catholic faith itself, as he tells his acquaintances… He hides his unrest and his doubt for those men whom he frequents who avoid mentioning religious questions in his presence, according to the usual rules of good education. But he is himself in constant unrest and seeks the truth, relying on God who had enlightened his soul to distinguish the right which he was looking for with a sincere heart."[76]

We find him in this disposition soon after his arrival in Florence in the pharmacy of the Annalena cloister with Sister Maria Flavia. How he made this acquaintance, whether at the occasion of a visit to the Palazzo Pitti, whether as a friend of Redi, the physician of the cloister, whether quite by chance he wanted to buy ointments and essences, from now that cloister pharmacy became the frequent aim of his strolling. The Annalena cloister lay in the Via di S. Maria and Via Romana, below the Palazzo Pitti. It got its name from the noblewoman Annalena di Malatesti who transformed her house into a cloister after the death of her husband who died there in 1453 and was buried in the choir of the beautiful church. Beside a large cloister from where one came into a memorial oratory with a much honoured miraculous Mary statue (Maria SS. della Palla), the building complex comprised also *la casa di Baldaccio* in which Annalena had lived originally. Right there, close to well organized wards for patients, was a "spezeria" with manifold remedies and a "fonderia" including a cellar in which the essences were prepared.[77] Maria Flavia's maiden name was Settimia del Nero and she was the seventh daughter of the Florentine Senator Alessandro del Nero (1586–1649), a distinguished diplomat. Jørgensen tells of her: "That woman who stemmed from a rich and prominent house, whose lineage owned a castle in Florence and was patron of the Annalena cloister, had taken the veil 34 years ago. She now sat there and sold herbs and ointments or gave a picture of the crucified to the indigents, strong in her simple faith, eloquent in her zeal for the Church and its means of grace."[78] Near her Stensen undoubtedly got a strong impression of the renunciation and peace of soul in the cloister which was strongly contrary to his own unrest.

When the nun heard that Stensen was a heretic it caused her much pain,[79] "because, from the little that he spoke to me, I found in him many good

---

[76]Jørgensen, *Nils Stensen*, 118 f.

[77]Richa, 10, 119 ff.

[78]EP I, 270 ff. Jørgensen, *Nils Stensen*, 122.

[79]Her report was written 2 years after Stensen's death at the request of her spiritual guide. In this she delivered two letters from Stensen. EP II, 987 ff.

qualities, particularly a great modesty". She felt herself driven to tell him without further ado, "that he does not know the good Catholic faith and is on the way to hell". He was not infuriated and quietly listened to her although she repeated the same several times. "To me he seemed willingly to listen and admitted that it was totally his taste to speak but not to dispute about matters of faith. When I heard that, I summoned up courage to say something to him. Thus I told him that he should pray God every day to let him know the truth, what he promised to do. He did it indeed every evening (as his servant reported to me)". Since Maria Flavia did not feel able to discuss faith with a learned man, she wanted to put him in connection with others. Stensen appeased her and confessed his own great aversion to speak about that with others. He felt full of shame and embarrassment regarding conversations about religion. He urgently implored the sister not to let anything be disclosed about that. She contented herself further to speak to him in all simplicity and confidence.

The following declaration is also indicative: "As he was at the grid one morning, while it just rang bells for the Ave Maria, I requested him to pray with me. He thus prayed with me to the middle, to 'fruit of your womb'. As I asked him to complete the prayer,[80] he made some difficulty because he contested the intercession of the most holy Virgin and of the saints. It was enough for him, he said, to pray until the above said words. I asked him to visit the Santissima Annunziata.[81] He also went there, ensuring that he did everything for me. When I saw that, I gave him certain prayers to the holy Virgin. He said the prayers every day. Similarly I advised him to be abstinent and eat fish on Fridays and Saturdays. That too he did, as the above mentioned servant reported to me. I put him forward that Luther was bad and had rebelled against the Church. But he refused to speak badly of him and said that he had been a good man."[82]

From this simple report one gains a quite clear picture of the spiritual state of Stensen when, disquieted by the Corpus Christi procession in Livorno, he came to Florence and devoted himself to the study of the Gospel and the Church fathers. He obviously wanted to learn about Catholic life, readily spoke of religious matters but avoided proper discussions and he was afraid that his spiritual turmoil would be known. Very indicative of this is his

---

[80]Second part of Ave Maria: Holy Mary, Mother of God, pray for us sinners, now, and at the hour of our death. Amen.

[81]The famous Maria Grace image from c.1341 in Capella della Santissima Annunziata close to the entrance of the Basilica (p. xxxii). Richa 8, 1 ff.

[82]EP II, 988.

request that nobody should speak of his conversations about the Catholic religion. He wanted to keep his freedom. Perhaps he did so out of respect for his family and country. He did not want to bind himself but to investigate theory and life. He readily participated in general Christian exercises like prayer for enlightenment, fasting and visit of churches, but he kept his Lutheran point of view and Maria Flavia, despite repeated exhortations to conversion, noticed so little progress that, partly driven by her superiors, she directed him to a priest. To her repeated exhortations, he answered: "I come, I come, pray for me." However, to her he did not seem to come closer.[83]

One of the priests was P. Leopold Leonelli, a Barnabit, born in Florence in 1620 and consecrated priest in about 1642.[84] Stensen spoke with him several times without it coming to a closer exchange. The Father, like others, had the impression that Stensen, tempted, "like by the devil", suddenly balked at treating matters concerning the Catholic religion and there was little hope of Stensen's conversion. Through the intermediary of Maria Flavia, Stensen also came in contact with P. Savignani. The nun managed it so that Stensen came to the house of the Lucca envoy and requested Lavinia Arnolfini to receive him at her home. Mrs. Arnolfini answered that she knew the good qualities of Stensen but she did not find herself to be appropriate to discuss such matters.[85] However, she had already recommended him to her confessor. When Maria Flavia told this to Stensen, he immediately went to P. Emilio Savignani (1605–1678), a Bolognese by birth. He was the Rector of the Jesuits' college in Florence but already on July 27, 1668 he was appointed rector and magister at the Borgo Pinti in Florence where he died.[86] He was known and loved in Florence, particularly as confessor. Stensen was charmed by his love and goodness and opened his heart to him. It came to long conversations, which dragged out but led nowhere.

Most people considered the case as hopeless. Even the good Maria Flavia lost patience. This state of doubt and indecision seemed so long to her that she later estimated it quite excessive at more than 2 years. As a warning she told Stensen the example of a heretic prince who delayed his conversion but to be sure always kept two Jesuits at his side, eventually to let himself be converted at the hour of his death. However, at the decisive moment, as by higher guidance, they were nowhere to be found. The instance made a deep

---

[83]EP II, 988.

[84]EP II, 988.

[85]Bambacari writes: *nè le fu difficile l'obbligarsi il di lui genio, già ammiratore della sua virtù, non meno che dell'ingegno.* (p. 37, also *Positio*, 102).

[86]EP I, 327, note 9.

impression on Stensen, as he recognized later. In this situation the personality of the envoy's wife Lavinia Arnolfini intervened so deeply and fruitfully in Stensen's evolution that later he always called her his spiritual mother.[87] She excused him because his study of nature took all his time. "But the study of natural science then hindered my serious examination of the religion. I was completely involved in this study until the divine mercy pulled me, in the completely unexpected following way, from the other studies to that of the salvation."[88]

Despite the difference in age, the Arnolfinis formed such a harmonious couple that, at the death of Silvestro in 1685, Stensen wrote of Mrs. Arnolfini that, during his long illness, she had served her husband "like Jesus Christ in person and according to the holy laws of a Christian marriage".[89] Girolimo, the son of Silvestro, joined the Jesuits in 1677. Two daughters, Teresa and Caterina, took the veil in San Domenico in Lucca. On his death bed, Silvestro recommended to his eldest son, Attilio Francesco: "See always your mother with respect and follow her advice; she is a saint."[90] As wife of the envoy she was the witty and affable centre of the refined circle who met in the Palazzo in the Via de' Bardi on the left bank of the Arno. Her biographer depicts her not so much as the *Dama del mondo* but as the *Serva del crocifisso*. She quietly lived a strictly ascetic life and her healthy constitution permitted her work and care into her old age. Her affable behaviour was also appreciated in the daily nuisances of which she used to say that one ought to acquire an unshakeable peace of mind in the fight against them. Her devotion to the poor and needy was unlimited. Her home had a hidden gate through which all the indigents could turn to her. Her heroism in the hospitals reminds that of saints of the Middle Age; even leprosy could not deter her. With piety and warm human love, she had a sound, natural behaviour that was totally adapted to the practical virtuous life, and her dispassionate engagement even towards the eccentric showed itself among other times when she immediately saw through and disapproved as fallacious the Sister Francisca Fabroni who, in a cloister in Pisa, enjoyed the reputation of saintliness and displayed ecstasies and illuminations revered by pious and eminent people.[91]

Of their lengthy discussions on faith, we have details of two of them—the one that provoked Stensen's crisis and the final one that concluded his

---

[87]OTH I, 127.

[88]EP I.

[89]EP II, 836.

[90]Bambacari, see *Positio*, 102.

[91]EP I, 112.

doubts. Stensen writes of the first: "A noble dame renown for her holy way of life had had with me several conversations about the religion and, as she asked me whether I did not feel a weak longing for the Catholic religion, I answered that I had noticed in the Catholics much which had not displeased me and in the Lutherans some things which did not please me, but that I had so far seen nothing which would compel me to leave the religion to which birth and country had engaged me. After these words, this servant of God, driven by a great and true Christian zeal exclaimed: "Oh, I wish that my blood would be enough to prove this necessity to you. Be God my witness, still at this instant I would give my life for your salvation." Moved by that unexpected proof of Christian love, I answered that so far I had not yet found in anybody such a love of God and fellow-men. Yes, I admitted that until this time I had spent more trouble and sweat for other studies than for the salvation of my own soul. But I promised, from that instant, to submit the religion to a serious scrutiny."[92]

Stensen's reaction reveals a particular hindrance for his conversion. With all his sympathy for the Catholic truth and holiness, he felt the negative incentive from the practical consciousness and the personal experience of Protestant failure. It is not by chance that he points out that association of birth and country. His faith in Christ and that of his family were undoubtedly deep and serious. Already before his conversion he lived an exemplary Christian life, as testified several times. He had actually come to know the division of Protestantism in Holland, but in his own country he had not lived the dynamite of the Protestant subjectivism.

That the ardent confession of Lavinia made such an impression on him is coherent with the fact that one credits her straight person with no sentimental flower of speech or sentence. She must have proved the seriousness of her mind also by acts. Stensen thanks her in the last letter he wrote to her a couple of weeks before his death on All Saints Day 1686 and says: "God has inspired you to make personally and with joy a lot of exercises of devotion as well as exercises of love and mortification to lead me out from the power of darkness to the kingdom of the Son of His love."[93]

The most valuable immediate consequence of the conversation was the active study of the ecclesiastical questions to which Stensen applied himself as he says. He actually spent the early hours of every day for the study of the controversial questions and of the holy fathers. He also declared himself ready

---

[92]EP I, 258, OTH I, 127, *Positio* 71 f.
[93]EP II, 891.

to accept them as soon as he knew them. He also immediately asked for books from his friends, not only Catholic books but also the Magdeburger Zenturien, the main Protestant historical work of the sixteenth and seventeenth centuries, which contained almost all that a passionate defender of Lutheranism and adversary of the Catholic Church might collect of attacks and blames against the latter from archives and libraries, and, on the Catholic side, led to the big annals of Baronius.[94]

Here one should mention the motives, which Stensen later explained as logically decisive for his conversion, which then occupied his mind and which also later flashed up again and again in his writings and by which the genius of his mind appears by concentrating on few essential truths. Stensen starts from the fact beyond contradiction that the true Church of Christ once certainly was the Roman Catholic Church. That was when our pagan forefathers accepted the faith from it and swore obedience and fidelity to the Roman bishop. Then, Luther and the other reformers founded their confessions and sects that now exist and, placed in front of their multiplicity, it is to be asked in which of them the true Church of Christ is to be found. Nobody could accept the equivalence of all reformatory directions; then the spirit of truth could not be the originator of Churches opposed to each other. However, nobody knows which of them is to be preferred since there are no characteristics which could indicate one as divine rather than the others. All boasted of the purity of their doctrine, all appealed to the Gospel and called their Church apostolic. Yet they could not appeal to anything other than their own authority, if it was the question to bring proofs of the validity of the Gospel and its translations, or for the practice of the sacraments or the catechism doctrine and its origin. Then their common answer is: read the Gospel and you will recognize us as the genuine ones. Everything thus would finally depend on the only question: which Gospel's explanation of the points of issue is of divine origin, that of Luther, that of Calvin or another reformer, my own one or that which is stated by the testimony of so many holy fathers and the blood of martyrs, of that Church thus from which our fathers have received their faith? As a particular argument, Stensen still emphasizes the holiness and apostolic character which the other Churches could not reach.

Moreover, what he observed with his own eyes in his immediate surroundings, especially among his friends, was of decisive importance.

---

[94]Bambacari, 37 (quoted in *Positio*, 102), OTH I, 9, 127. Stensen borrowed books through Magliabechi, the book collector and librarian of Prince Leopold and later Cosimo III; EP I, 193 f. Cf. Nordström, *Magliabechi*, 27, note 4.

Again and again we see Stensen returning to the argument of holiness. It is so right what A.D. Jørgensen says about his conversion: "Stensen's theological polemics show us fully what attached himself so strongly to the Church: he points out again and again its historical right, its inner harmony and consequence, its largescale coherence in the course of time, its wealth in great intellectual men, saints, missionaries, martyrs and Church fathers, the strict seriousness of its requirements, the gentleness and mercifulness in its care for the individuals, the soothing and appeasement in its means of grace, its intercessions and its charity work. That was the picture which formed in his soul and demanded subjugation or rejection. What would Protestantism set against that?"[95] His friends were obviously very interested in the conversion. Redi, in a letter of March 30, 1667 to V. Inghirami, even too optimistically hopes for Stensen's conversion: "I have him so in my hand that I can predict it with certainty."[96] Viviani felt a deep joy about the completed conversion.[97] Yet nowhere does one have the impression of any strong influence by words. Stensen was very thankful for this discretion and it has attracted him not so little. When later, as bishop, he had to defend his Church against Protestant reproach of proselytes' intrigues and inquisition coercion, he pointed out his favourable experience: "I lived in places where the inquisition exists, in Rome and Florence as well as in Pisa and Livorno, and I saw everywhere that the non-Catholics enjoyed the greatest freedom of life if they did not do anything objectionable. Yes, even at the Pope court I heard people freely dispute with prelates against the faith, the ecclesiastical hierarchy or the monarchical regime."[98]

Of the practice at conversion, Stensen said: "When it is question of winning a soul, one begins 1. With prayers as well one's prayers as those of others, if one knows souls which burn with zeal. 2. With frequent communion to which one also invites friends to make prayers more efficient. 3. With fasting and alms. They attempt with the greatest zeal to bring them closer to the Catholic truths through men distinguished by their righteous life and erudition. And that not for days or weeks but for months and even years in the course of which one uses neither violence nor cunning but rather applies great zeal so that, when somebody leads conversation on faith from less right intentions, they are led back to right intention to seek first the empire of God and his righteousness, so that, at the time of the trial, they do not find a less

---

[95]Jørgensen, *Nils Stensen*, 135 f.
[96]EP I, 22.
[97]EP I, 27.
[98]OTH I, 286.

merciful God because they had come in less pure mind towards God. Not only religious persons practise this zeal to win souls but I have known people who live at the court and because of their job were totally involved in worldly businesses and, however, were so preoccupied by the improving of wicked Catholics and by the return of non-Catholics to the Church that this seemed to be their first duty. People also come together to holy places to apply themselves to pious exercises and encourage others to prayers, fasting, penitence and other exercises of love and mortification. One actually writes to persons very zealous for the souls ardently to recommend the matter. I knew such people in Italy and elsewhere also who obtained from God an improvement of life and of the doctrine and they were so far away from ambition and greed that they refused to have anything to do with honours and wealth, and they were so courageous that they would rather have taken pangs of death for themselves than behave against their conscience."[99]

The less Stensen's friends tried to influence him by persuasion and coercion, the more they exerted influence on his final decision by the example of their life: "The life of some Catholic friends exerted a great influence on me as neither the philosophers promise anything like it nor had I observed anything like it in acquaintances of other religion. Already then I had also recognized that, as far as the truth of the doctrine is concerned, one must not give attention to the ignorance and wickedness of the people who confess the doctrine."[100]

The Catholic friends whom Stensen means here certainly were first of all his colleagues of the Accademia del Cimento. They united in themselves an unusual high intellectual culture with a true aristocracy of mind and deep religiosity. Redi, the son of pious parents, the friend of P. Segneri and P. Serra, was besides the fruitful, trustful researcher, a modest, affable, always righteous character. Extremely liked by the court, he readily served others and established the peace in many discords. His prayer in the disease was: *Sit nomen Domini benedictum.* His religious poems show a soul turned to eternity.

Magalotti, another friend of Stensen of those days, was well a sensuous child of the world who liked a tasteful home, distant travels, art and enjoyment. However, one also noticed in him the deep influence of the faith. He read the breviary every day, readily gave alms when he was not himself in money embarrassment, which often occurred. He wrote letters against

---

[99]OTH I, 286 ff.
[100]OTH I, 126 f.

atheism, translated the Imitation of Christ and the Castle of the Soul of Saint Theresa and finally attempted at joining an order. Viviani also, a brilliant mathematician, was a deeply religious nature animated by the striving for moral perfection. His notes reveal the spirit of self-observation and self-discipline but also a kind heart, a hand ready to help. And Stensen later recalled thankfully the Marquis Luca degli Albizzi and emphasized his piety.[101]

These experiences made a deeper impression on Stensen than all the arguments of the textbooks. He became an experienced observer and experimenter concerning the faith. When a Protestant opponent later wanted to dismiss the reference to the characteristic of the holiness with one word of the falsehood and the hypocrisy of the Papists, Stensen answered: "Hereby I must think of an experience which I have often made earlier in anatomy. How often didn't I hear people, who themselves looked at the prepared specimens or at the preparation itself, hold long and in their own eyes learned discourses on these pieces and the method to prepare them, of the groundlessness of which some more patience and observation would have convinced them. It is so easy to fall into the biggest mistakes if we set ourselves up to judge in matters of which we have no idea. This happened to Alexander the Great when his remark about the paintings and the colours in the workshop of Apelles called forth the laughing of the apprentices."[102]

"After I had weighed all that", Stensen writes, "I concluded that I could not follow any of the reformers since none of them could claim a divine authority for his reform; none of them could prove to possess more truth than the others. Therefore, I decided to return to that Church to which our forefathers owed their christianization, from which all the other Churches stemmed, which alone proves that it is apostolic and distinguishes itself by the other characteristics which indicate the supernatural presence of God. In it I have already seen then those proofs of holiness that are promised by none of the other churches, far less to be found in them."[103]

Stensen says that he could still present several other motives: conversation on the faith, exhortations from different sides, which are still more efficient but the reasons thus given are sufficient. To be sure, he adds, they offer a moral security only but the divine security is totally a gift of the grace.

---

[101]EP II, 741.

[102]OTH I, 396. * Stensen makes reference to the parable of Alexander and Apelles three times, <427, 429, 590>.

[103]OTH I, 128 f.

That in any case the reasonable motives alone provoked no decision appeared clearly in the time before the decisive step. Despite all reasons and conclusions, Stensen, uncertain, balanced to and fro. He felt himself torn apart in his mind. He says: "Although the above-mentioned motives immediately urged me, so many different cares assailed me and I was so entangled in the strings of obscurity that, not master of myself, I could not find the end of my misfortune until, at the feast of all piously dead souls towards the evening, suddenly so many proofs and circumstances met so that I was finally clearly conscious that I had to admit that God himself was taking me by the hand and leading me to his Church: 'You, oh Lord, you have torn my strings'."[104]

Stensen actually felt torn inside that autumn. This is testified by an event that Leibniz recounts and which shows the twisted soul in a disconcerted state in which one pays attention to signs and coincidences.[105] It was well in that mood that Newman felt himself in the last time before his conversion and compared this to a death bed on which the membership in the old confession wastes away. "A death bed doesn't know any history... It is a time when the doors are closed and the curtains are drawn and in which the dying will, nor can, distinguish the different steps of his pain."[106]

Which "many different troubles" Stensen means, is difficult to tell. According to the report of the Nuncio Trotti, Stensen had "different very subtle difficulties concerning the Sacrament, the purgatory and the papal authority".[107] His professional career was not threatened, his scientific career rather became ensured by his conversion as the matters were in Tuscany. An external constraint is not to be considered. The heaviest pressure came from the consciousness of being separated from his family and country by the change of faith, all the more that an order of Frederik III to return home and the offer of an honourable position were not in his hands but were on the way, which he must have been aware of. The frustrated attempt at travelling home, delayed by the lack of a boat in these days, was certainly due to such preliminary information.[108]

Stensen was clearly conscious of the hostility to which he would be exposed from his earlier friends and this was not indifferent to his sensible mind. He writes: "Above all it was in the last days of my hesitation when I pondered

---

[104]OTH I, 127.

[105]EP II, 938 f.

[106]Newman, *Apologia*, 147.

[107]EP II, 925.

[108]Cf. Maria Flavia's account, EP II, 989.

how much evil I had heard about other converts and foresaw, that I did not dare hoping any better treatment, that my mind, strengthened by many experiences in the judgement of God, only found rest to which the following truths prepared often consolation: the One who has given me the ability to think sees all my thoughts. Or also: You may say anything You want about me, You may do with me what pleases You, God sees everything."[109]

(c) Recognition, November 1667

On All Souls' Day in 1667, which brought the decision and which in the future Stensen called "the day of the dead, the day of my life in the spirit", Mrs. Arnolfini had apparently invited Stensen to her table and, after the meal, the conversion was discussed.[110] Again she presented many motives for it which should have convinced Stensen and addressed him the most pressing exhortations which love and zeal for souls inspired her. The result was meagre. As always Stensen seemed to be undecided. "Then she exclaimed, as under the inspiration of God, to give him the last incentive, with a certain holy annoyance: 'Sir, the visits and talks which I granted you against my habit have no other motive than the zeal for your eternal salvage and are only an outflow of the love which could gain you for the faith. But, since you do not want to give up to the recognized truth I will no longer lose my time. Do not come any longer to me if you are not decided to become Catholic'. She accompanied those energetic words with an inner regard to God to whom she turned her heart so that he bless this unexpected outcome of the matter with his omnipotent grace. The peremptory dismissal appeared to Stensen as a flash of lightning of divine anger. He went. Yet in his soul the divine grace was at work. Having decided to go to P. Savignani, he met him passing by chance. Both, talking of indifferent matters, went to the cloister of the Father. While the Father, who also wanted to convince him, was going to his room to fetch the suitable authors, Stensen remained alone behind. There he felt his heart at once and so strongly converted by the grace of God that, when the Father came back with the books, he declared that hereafter he did not need witnesses nor reasons to convince himself of the truth, which had been clearly unveiled to him by inner enlightenment of God."[111]

The Father embraced him with great joy and Stensen left him, as Maria Flavia tells, to hurry to Mrs. Arnolfini who, as soon as she had become aware of the desired decision, went to the chapel to pray the *Te Deum laudamus* in

---

[109]OTH I, 392 f.
[110]EP II, 825, 891 and 730. Bambacari 38, quoted in *Positio*, 102.
[111]Bambacari, quoted in *Positio*, 102 f.

which Stensen, still quite stunned by his own inner change, joined.[112] On November 3 he asked from Maria Flavia certain relics and a picture of St Annunziata that she had promised him if he converted. He also sent her 50 scudi to have a pair of silver candlesticks made for the altar of the figure of Our Lady in the Annalena cloister, in front of which he had often prayed. It was that above-mentioned miraculous image in the Annalena cloister that Manni saw one century later with the inscription: *Di Niccolò Stenone. Di Annalena per il Coro vecchio.*[113] Deep peace progressively entered after the victorious fight, particularly when he presented himself to the Father Inquisitor P.M. Hieronymus Baroni da Lugo, OSF on November 4 and had completed his abjurationon November 7.[114] "Hardly had he completed the act that a change immediately appeared in him, as the wife of the envoy and I noticed. We saw him more serene and filled with the greatest desire of perfection."[115]

At the feast of the Immaculate Conception 1667 Stensen confirmed his abjuration in front of the nuncio and received the sacrament of confirmation. Nuncio Trotti was very pleased and reported the conversion to the secretary of State Decio Azzolino and to the Pope. He hints at the faith crisis in Holland when he writes in his report: "that Stensen has precisely studied all the sects during 15 years and has found in all of them vanity and poison and thinks that the conversion would deserve a proper statement in which the kind work of the Holy Spirit in the search of the truth would be told, which he has researched so laboriously with so great moral excellence for so many years and finally wanted to receive the sacrament of confirmation from the nuncio himself to point out his faith in the infallibility of the Pope."[116] Among those who enjoyed the conversion of Stensen the most was Prince Leopold to whose letter Magalotti answered from Antwerp on January 6: "Of all the news… this one alone is capable of filling the heart with infinite joy, and not only because of the joy which his true well-being gives me but also when I think of the satisfaction which the incomparable zeal of Your Highness will draw from that, and of the profit which that period of his life at the court will have from it. After the current dissolution of our academy by

---

[112]EP II, 989, cf. OTH II, 436.

[113]Manni, 84 f.

[114]EP I, 363. * Stensen's three descriptions of his religious conversion are analyzed by Olden-Jørgensen, "Jesuits, Women, Money or Natural Theology? Nicolas Steno's Conversion to Catholicism in 1667", *in* Andrault and Laerke, *Steno and the Philosophers*, 45–62.

[115]EP II, 989.

[116]EP II, 925.

the departure of Borelli, Oliva and Rinaldi, in my opinion, nothing more desirable could happen and, if the two other places can be filled up relatively as well, we could have good reason to console ourselves of the suffered damage though, it must also be admitted that it is a considerable loss. Then, if Rinaldi and Oliva are given all the recognition and honour which they rightfully deserve, Borelli was in fact a learned man who awarded lustre to a court because he had perspicacity and discernment, even if he was an unpleasant, I might almost have said unbearable, man."[117]

Viviani reported that, on the very day of the confirmation, Stensen had received from his king the order to return back home.[118] Without knowing the actual circumstances, it seemed that King Frederik III had recalled Stensen back to Denmark on October 19/29 and offered him a provisory annual pension of 400 rigstalers.[119] Viviani says: "He will not make up his mind before having heard from His Majesty whether he agrees with this change of religion and, since it is not believed that this will be the case, one hopes that we can enjoy the continuation of his presence."[120]

The disappointment of the Protestant acquaintances of the scientific researcher then on the top of his fame was not less genuine than the joy of the Catholic Italian friends was great. This is revealed by the discussions with Johannes Sylvius during the visit of Stensen in Holland in 1670.

In the following decades, even more in the following centuries, judgements on the conversion are found at all nuances, from the highest enthusiasm to the deepest bitterness and accusation of the lowest motives. However, there were also judgments from opposite sides, which would pay unconditional attention to the character of the convert also if they discarded his Church. The literature historian N.M. Petersen thus said: "It is clear that Stensen left the faith of his fathers because Protestantism did not satisfy his deep mind, because he did not observe any improvement in the life and the original Church moved him through its martyrdom. Without any doubt this apostasy was an aberration. Its motive however, the martyrs' life which called it forth must be considered with respect and not pulled down in the sphere of the lowest life."[121] With even greater insight in Stensen's psyche, his biographer

---

[117]Fabroni, *Lettere* I, 295 f, *Positio* 67.

[118]EP I, 27, *Positio,* 66.

[119]EP I, 194.

[120]Cf. also Justel's letter to Oldenburg: *On offre à M. Sténon des appointements considérables en Dannemarck ou il ne veut pas retourner qu'il n'en soit bien asseuré et qu'on ne luy accorde la liberté de conscience,* Oldenburg, IV 441).

[121]Petersen, N.M., 282.

A.D. Jørgensen judges the conversion: "The explanation has been sought in different ways, then as well as later, but always at the expense of his character or his perspicacity, totally in contradiction with what one otherwise hears of him.—According to all what is at hand, the answer must come from the opposite direction. It happened precisely because he was such a clear mind, such an energetic thinker, precisely because he was such a pure character, his will so strong and his longing so high. Though he was mistaken, he, however, reached certainly his goal—and who does question the length of the way, its thorns and pains, if the goal is reached?"[122]

The convert himself often looked back in his mind at that so deeply indenting step in his life, always with the feeling of warm gratitude, as two witnesses of his last year of life testify. On March 24, 1686, on the eve of St. Mary's Annunciation, he writes in his letter from Schwerin to Cosimo III: "In remembrance of the graces which were obtained from so many friends during the visit of the SS. Annunziata whom the Lord glorifies so much in Florence, in front of whose image they obtained for me the grace of the conversion and that I now live as a son of the holy Church. I admire the divine mercy which moved and moves the hearts of so many people to collaborate at my true well-being. I am deeply ashamed when I consider my ingratitude towards God and towards the men. Yes, I am afraid that God to punish my sins of ingratitude would let me linger deprived of all spiritual help and of any advice, in these regions, whereas I had hoped to obtain some new zeal in the divine service and love in those holy places. When I think of the God-fearing and God-respecting persons whom I have seen and met in that, my spiritual country, as guardian of the precious treasure mentioned above, I remember some episodes of the history of the apostles and I find a relation between the local Catholics and those whom I knew then, that is the relation which, when I still lived outside the Church, seemed to exist between the Christians of today and those of whom the history of the apostles speaks. Truly, if God had not shown me such examples and such information and if, in His mercy, He had not moved so many hearts to intercession, who knows in what spiritual misery I should have fallen again. I dare not even pretend that I should have remained at that place of His mercy from which I have once gone out."[123]

Stensen wrote to his spiritual mother Lavinia on the last All Souls day which was allowed him on earth: "God has inspired you and, through yourself and your friends, has let many works of devotion, of love and

---

[122]Jørgensen, *Nils Stensen*, 121.
[123]EP II, 862 f.

mortification be exerted to move me from the reign of obscurity to the reign of His beloved Son."[124] Accordingly, we must trust that Stensen's own subsequent attitude to his conversion is certainly sincerely expressed in his letter to the reformed preacher Johannes Sylvius in which he exclaims his full gratitude towards God: "Be loved in eternity the name of the Lord who has called me from the obscurity to the light, from death to life. May He confer to you and all our common friends, yes all men, the true peace of mind within the bosom of the true Church."[125]

---

[124]EP II, 891.
[125]OTH I, 129.

1.6 In Italy 1668. The Geological Dissertation, the *Prodromus on a Solid Within a Solid*

1.6.1 Preliminary Geological Studies

On December 8, 1667, Stensen, after renewing his taking of the Catholic confession, received the sacrament of confirmation from Nuncio Trotti. Immediately afterwards he received the order from King Frederik III to return to Denmark and soon after, the chamberlain of the Grand Duke, Bruno della Molara, communicated to Stensen the court's intention to depart on Sunday after lunch for its usual winter move to Pisa, adding that this could be discussed in more detail in the palace at midnight. Stensen himself, if he wanted, could leave on December 9 to embark on his "mussels journey".[1]

What was meant by "mussels journey" is not specified. On January 4, 1668, we find Stensen at the court in Pisa, where he apparently remained for the next months, with a visit to Livorno in March.[2] Molara's mention of the journey, however, shows that Stensen was very busy with his geological investigations, which had been put somewhat aside during the time before his conversion.

It would appear that the air in Pisa did not do Stensen any good. At least we hear of a fever during his winter sojourn in 1667. January 1668, he also complains of a headache, which deters him from his study. In the beginning of February Stensen writes: "I have spent more time in distractions than in orderly study in order not to force my mind during my ill health. The greatest evil of this ill health is that it is not accompanied by any other evil which

---

[1]EP I, 194 f.
[2]EP I, 195–205.

© Springer-Verlag GmbH Germany, part of Springer Nature 2018
T. Kardel and P. Maquet, *Nicolaus Steno*,
https://doi.org/10.1007/978-3-662-55047-2_6

would allow finding out its true cause."[3] Perhaps it was the result of a state of exhaustion, which would be understandable after his strenuous scientific work in anatomy and geology and the strong spiritual tension of his conversion.

Prince Leopold had been made a cardinal on December 12, 1667, and was preparing then for the spiritual promotion. From March 5 to July 17, he stayed in Rome where he was presented with the cardinal hood. Stensen remained in lively communication with him and with the Accademia del Cimento, and was apparently in great general repute. When the physician Don Girolimo Bardi (1603–1670) wanted to ask a favour from the new cardinal, he was advised to use the intercession of Stensen, which he did.[4]

A letter from Stensen to Cardinal Leopold of January 14 shows that the Medici patron offered Stensen a copy of the *Saggi*.[5] Lorenzo Magalotti, the secretary of the academy, had already completed this, the *Saggi di naturali esperienze*, the famous report on the experiments of the Accademia del Cimento, at the beginning of 1665, but it did not appear until the end of 1667.[6] The first edition was widely distributed as a gift from Leopold in these winter months. Stensen was able to appreciate this famous golden book of experimental science and the value of its experiments, which only very few were able to at that time. He wrote to the cardinal: "Excellence, among all the books which I have seen so far about such experiments, I did not find any which would be of equal quality. Anybody who enjoys it, Enlightened and Reverend Sir, must feel most deeply obliged that, in this fruit of your academy, not only you brought to the knowledge of the erudite so many truths so far unknown but also, from that you offered to the world, a rule which all those who attempt at discovering the hidden truths of nature must apply. The small importance which befits my weak assessment does not allow me to continue the praise that such a perfect work deserves, above all at a time when the greatest minds of all the experimental academies vie with each other to celebrate the noble undertaking of the Accademia del Cimento, whereas they do not see without deep pain, hindered by unexpected misadventures, a study in the reconnaissance of nature the great accomplishments of which were carried out in the shortest time, and which could be a clear sign for everybody of what its continuation could hold for the future."[7]

---

[3]EP I, 196, cf. EP I, 22.

[4]EP I, 195.

[5]EP I, 196.

[6]On the lengthy process of its writing and printing the *Saggi*, see Middleton, 65 ff, esp. 74 ff. Some samples carry the year 1666, others 1667.

[7]EP I, 196.

On the instruction of Leopold, Stensen also sent a copy of the *Saggi* to the Danish king and writes: "I am convinced that I will favourably dispose Your Majesty toward me by sending the book which His Eminence has destined to this end". This must be seen as relating to his recent conversion. We do not know why Stensen would have sent a copy of the *Saggi* to the District judge of Jutland, Villum Lange (1624–1682), nor do we know when he sent the *Saggi* to Leiden. Stensen also became the advocate of the *Saggi* against attacks and undertook its scientific defence. The Roman physicist and mathematician Fabrizio Guastaferri,[8] an acquaintance from his first journey to Rome, had through a common acquaintance, transmitted to Stensen some doubt about four of the experiments described in the *Saggi*. Stensen answered, after first presenting his explanations to Cardinal Leopold, who transmitted them to Viviani to be checked. The questions concerned three experiments on air pressure which the *Saggi* had emphasized, and a fourth one about the increase in volume in the freezing of water.[9]

Of great interest are his remarks about vacuum, which he holds as possible, contrary to Aristotle's position. Stensen previously wrote in the *Chaos* manuscript: "Experiments prove that vacuum exists. Today the view that there is no such vacuum has been almost overcome."[10] In the home country of Torricelli, Stensen of course held even more firmly to the possibility of the vacuum. Yet Stensen was in the dark concerning its nature, speaking in his cautious manner of a substance that is not yet sufficiently known. The state of chemistry at that time precluded a clear understanding.

In a letter from Stensen to Cardinal Leopold of February 4, 1668, we meet the considered geologists whose fanciful propositions of the time he confronts with smiling irony. The cardinal had sent Stensen seaweed and reported the amazement of Mutolino (possibly a cover name of Borelli) who thought he recognized in it an animal shape with head, back, feet and tail. Such shapes then led to the profoundest speculations. Stensen saw in it nothing more than a strange joke of nature and makes fun of it: "It would be a nice painter's joke if mandrake men with seaweed dogs and horses and other plants organized a hunt of animals of similar shape in a forest of men who are about to change into trees. Your Highness has done well to send me the whole as a joke without asking me for an argument which would have embarrassed me very much. I can only see that the apparent hair is nothing else than the remnants of leaves from previous years, as is seen at the stem of the palm. One

---

[8]EP I, 205.

[9]EP I, 198 ff, 200 ff, 204 f.

[10]Chaos, col. 156, fol. 68 v, ed. Z, 369. Also Scherz, *Niels Stensen und Galilei*, 16 ff.

furthermore notices in the leaves that are still attached a line between the part of the leaf which is already beginning to fall and the other which remains, as if nature had already taken the measure of the hair with which she wanted to leave the stem coated to present through it a hairy animal."[11]

At the end of March Stensen undertook a journey that would lead him to the island of Elba.[12] This time not only "mussels" interested him, but also mines. In his later work he would deal with topics apparently unrelated to fossils but yet had to be dealt with for the discussion of fossils. His interest in petrifactions led him to examine the earth layers, which contained fossils. He also had to consider the inorganic processes behind the origin of stones, such as the growth of crystals. On Elba he found a great variety of material: fossils, volcanic stone, and different kinds of crystals. Preller had called the island the Pearl of the Mediterranean (Fig. 1.6.1).[13]

We know of this excursion through Magalotti, to whom Stensen wrote on April 20: "What I have seen everywhere comforts me in the opinion without any fear of contradiction, or better in the opinion of the ancients which I have defended in my last dissertation (namely about the origin of the mussel shells, molluscs, and of the fossil teeth which were found in mountains)."[14] On Elba Stensen was nearly arrested as a French spy by the Spanish commander of the Spanish harbour; this indicates that he must have visited the eastern part of the island, including the iron mine near Rio and Monte Calamita.[15]

On the mainland he also stayed in the little town of Volterra, south of the Arno river, where he was apparently the guest of the fortification's commander, Raffaelo Maffai (1605–1673), who was also responsible for the salt mines in the region. Raffaelo Maffai had himself written some dissertations on metals and mining, as well as literary works. His son Ludovico was Stensen's guide to the area's curiosities; the fossil localities in the deposits underlying the town, the salt mines in the Cecina valley and the lagoons boiling with boric acid. A greeting to Signor Cavalcanti indicates that he also visited Libbiano, the domain of that family.[16]

One month later, we hear again from Magalotti that Stensen is preparing for a new journey, this time to the north.[17] He will travel to Seravezza and

---

[11]EP II, 204.

[12]On the travel, see *Reisen*, 68.

[13]Preller, 190.

[14]Oldenburg, IV, 345 f, Scherz, *Neue Stensenbriefe*, 169.

[15]On Elba see f.i. Serpell, esp. 11 f, 90 f.

[16]Oldenburg, IV, 345 f, *Indice*, 294.

[17]Oldenburg, IV, 431, Scherz, *Neue Stensenbriefe*, 169.

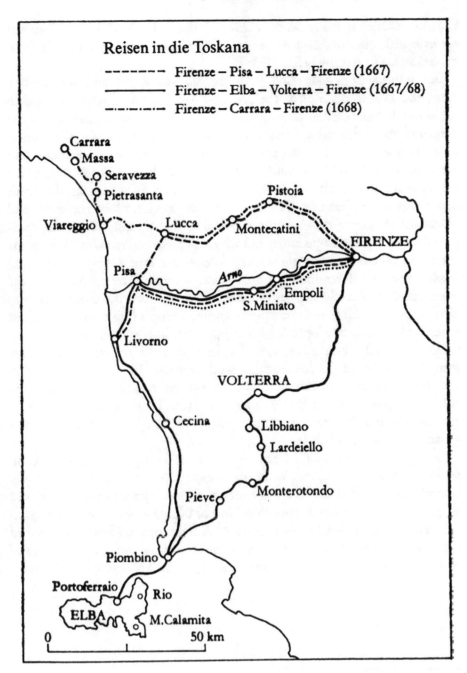

Fig. 1.6.1 Travels in Tuscany, Scherz, p. 217

Carrara to visit the many *marini e minerali* there. The region is well-known for its marble quarries, silver mines (near Pietrasanta), and vitriol plants (near Strazzione). He returns before June 8.[18]

For investigations extending from Tuscany to Rome and surrounding areas, the Tuscan plain offers the geologist a rich field of research. Stensen could hardly have desired better ground.[19] The whole plain, irrigated by the Arno and its neighbouring streams, is formed from alluvium. On the borders there are many hills consisting partly of schist and tuff, with horizontally and vertically folded strata permitting good insight into their structure. An immense abundance of fossil shells of all kinds and of fossil remains of animals and plants are at hand everywhere. The hills continue into the main and secondary chains of the Apennines, which are the result of greater compression and higher temperature, and which lack organic remains. Within the caves and crevices a great variety of minerals are found. Hot sources and sulphur springs still remind us of their volcanic origin. The main mass of the strata consists of sandstone, gneiss, and limestone from the Cretaceous and Tertiary periods. Here and there the limestone changes into marble of various kinds. Mines were also operated for copper ore, iron glance, silver ore, cinnabar and metallic quicksilver, as well as huge quantities of quartz. There are many indications of the deepreaching work of water. In the caves, stalactites and stalagmites can be be studied. There are several sources of abundant lime and iron deposits; others contain much sulphur. There are also hot springs but no volcano is active; there are only extinct ones, but Stensen was apparently unaware of these.

During Stensen's geological expeditions he collected all kinds of material: fossils, crystals, minerals, etc., which were important for his work. In this way he gathered a collection that was unusual for his time because it did not simply consist of fortuitous curiosities, but of unique objects which served as an essential foundation for his research.[20] A transcript has been preserved of the catalogue that he established some years later of his own collection and of that of the Medicis. In this catalogue, called the *Indice*, one obtains a survey

---

[18]FBN, Cod Gal. 252, c, 141 a., cf. *Reisen*, 69.

[19]Cf. Rodolico, *L'Esplorazione naturalistica dell' Apenino*. See also Rodolico, *Die Florentinische Landschaft* and v. Borsig, *Die Toscana*.

[20]Cf. Spärck, quoted in *Indice*, 199.

of what he sought.[21] A review of the contents gives the following approximate groups, which are not strictly delineated from each other:

Nr 1–29 Crystals of quartz

Nr 30–92 Emeralds, diamonds, other crystals as well as stones and ores

Nr 93–114 Marcasites and some other minerals

Nr 115–216 Objects found in the sea, mussels, snails, fossils, etc.

Nr 217–255 Corals

Nr 256–304 Samples of soil and stones, ores, mussels, crystals, fossils, mainly from the long journeys 1669–1670

The catalogue also contains pieces of information about the places of origin of the different stones, complementing other information we have of his journeys in Tuscany. He collected silver in Seravezza, quartz in Canali delle Viti in Pietrasanta, lead in Bottino and quartz with superficial veins of tin in Levigliani—all of these north of the River Arno, a region he visited during a journey to Lucca in the spring of 1667 and in another to Carrara one year later. Moreover, the catalogue also includes material collected south of the Arno: antimony from Siena, petrified wood from Querceto, a stone formation from Monterappoli, black crystallized sulphur from Libbiano, cast iron from the Tuscan Maremma where he may have arrived during his journey to Rome or during a trip to Volterra and Elba. Finally, a series of names can be traced to the region of Livorno, such as the alumina from Monterotondo and lead ore and vitriol from Campiglia. The greatest part of the first 255 numbered items must have been collected in 1667–1668, since they are used in the *Prodromus*.

On May 18 Stensen writes to Magalotti about his projects as follows: "I am quite certain that I will not leave Florence this year since I have promised his Eminence to complete my treatise on the earth and the bodies found in it, and also certain experiments on the blood."[22] This promise to the Grand Duke was kept, although in a somewhat modified form, because Stensen wrote his *Prodromus*, the *De solido intra solido naturaliter contento dissertationis prodromus,* in the summer 1668, intended as a preliminary to a no longer existing dissertation on solid bodies naturally contained in other solid bodies.[23]

---

[21] *Indice di cose naturali, forse dettato di Niccolò Stenone,* printed in Scherz, *Vom Wege,* 141 ff. (with German translation), and *Indice,* 201 ff. (with English translation), comparecf. Scherz, *The Indice of Nicolaus Steno* and also Chap. 1.8. * See also Cioppi, Elisabetta and Stefano Dominici, *Origin and development of the geological and paleontological collections,* Firenze 2010, esp. 21–24.

[22] *Oldenburg,* IV, 431.

[23] What we know about the preparation of the *Prodromus* comes entirely from its Preface and Dedication to the Grand Duke, <763> OPH II, 183 ff. = GP, 137 ff.

1.6.2   On a Solid Within a Solid

The immediate reason for drafting *Prodromus* was a new call from Denmark, probably during the summer of 1668. In any event, Stensen justifies the haste by his duty of obedience to his sovereign. It is likely that, after the first call reached him on December 8, 1667, he had informed his home of his conversion and now he had evidently received a favourable answer from Copenhagen. On July 4, 1668, he left his rooms in the Palazzo Vecchio[24] and, on August 30, Viviani had read and approved the work. Consequently, it must have been written in a very short time.[25] This quick drafting gives the dissertation its somewhat sketchy character.[26]

If Stensen preferred to deliver a provisional work, rather than waiting to return to Denmark and preparing it there completely, this was due to his fear of new works hindering the completion of previous studies. "While I was intent on classifying all the glands of the body, the death of my relatives interrupted my already begun research on the marvellous structure of the heart. To take me away from a detailed account of the muscles, a shark of prodigious size was thrown up by the sea and now that I am wholly dedicated to my present experiments, I am summoned to deal with other matters by the one to whose intimation the law of nature commands to obey and to whom the great favours bestowed on me and my family urge to obey."[27]

In the light of those facts, one understands better why Stensen in *De solido* writes of the *praecipitata scriptionis brevitas* of his dissertation and concludes it with the affirmation: "This is a succinct, not to say disordered, account of the chief things",[28] namely the proper dissertation, to which a list of the places of discovery would be added. Knowing the short time during which it was drafted is very important for understanding many features of the *Prodromus*.

From the beginning, Stensen had considered writing the work in Italian, not only to please his patron but also to show to the famous academy, whose members had accepted him in their circle, his zeal for learning the Tuscan language. He thus mentions being admitted in the Crusca Academy on July 16, 1668.[29]

---

[24]FAS, Medicei, Guarderobo 754, fol. 11 ff. Cf. Sten Cath., 1957, 36. Stensen stayed in Florence until November, but we do not know where he lived.

[25]OPH II, 333 ff.

[26]Cf. OPH II, 183 ff and GP, 135 ff (with references and annotations to text that differs from the manuscript). The *Prodromus* has been translated into English, German, Danish, Italian, Russian and Japanese.

[27]<767> OPH II, 184 f.

[28]<822> OPH II, 224.

[29]EP I, 19.

The work is dedicated to Ferdinand II. In the Introduction Stensen asks the Grand Duke to excuse him for having protracted his investigations over more than one year, contrary to the initial claim that he needed only a few days to solve the problem presented by the shark teeth. He experienced, as he writes in the Introduction, what the traveller experienced who thought himself to be quite close to a town high on a mountain whereas in his march he passed through deep valleys and distant heights which slowed his progress. His doubts had been like a Lerna snake: he had felt as though he were in the windings of a labyrinth. Now he must fulfil his promise to the Grand Duke, and if it cannot occur with the complete treatise, he must give him what he has.

The contents of the *De solido* are so rich in such a short space that an exhaustive review would reproduce the larger part of the dissertation. Stensen himself, however, divides the prospective proper work into four parts, which form the framework of the *Prodromus*.[30]

The *first section* deals with the scientific method.[31] One should not simply deal with the solution of a partial problem, but the problem must be seen and solved in its context. He did not want to treat only the difficulty of the ancients, how bodies originating in the sea could be found very far from the sea, but to investigate all solid bodies which are included in other solid bodies, their nature, the place of their discovery, and how they came into being. One should answer all those questions in order to solve the problem. That basic rule in methodology guided him above all precedents, and it also fortuitously raised special considerations from a geological standpoint. This has often not been sufficiently appreciated.

Stensen then addresses those who do not clearly distinguish between that which is certain and that which is uncertain in natural science. "Indeed, the advocates of experiments have rarely had the restraint either to avoid rejecting entirely even the most certain principles of nature or to avoid considering their own self contrived principles as proved." Therefore, like Seneca in the moral field, he will, in the natural science field, select as the best those theories that are generally accepted by everybody, whatever school one belongs to, be it progressive or conservative.[32]

Stensen then presents some general considerations on physical bodies, such as the difference between solid and fluid, and the origin of motion and alludes to the fundamental elements of matter as "imperceptible small parts".

---

[30]For an overview in German, see Stensen, *Das Feste*, 21 ff, and in English GP, 25 ff.

[31]<768> OPH II, 185 ff. "I had divided the dissertation into four parts."

[32]<771–722> OPH II, 188. On Stensen's and the contemporary opinion on the structure of matter, see also GP, 222–223 (notes 27, 29, 31–33) and Schneer, *Steno: On crystals*, 293 ff.

He refers to the original and unknown cause of motion, saying: "Certainly to deny this cause the power of producing actions contrary to the usual course of nature is the same as denying man the power to change course of rivers, of struggling with sails against the winds, of kindling fire in places where without him fire would never be kindled, of extinguishing a light which would not otherwise vanish unless its fuel supply ceased, of grafting the shoot of one plant on to the branch of another, of serving up summer fruits in mid-winter, of producing ice in the very heat of summer, and a thousand other things of this kind opposed to the usual laws of nature. For if we ourselves, who are ignorant of the structure of both our own bodies and the bodies of others, alter the determination of natural motions each day, why should not He be able to alter their determination who not only knows the whole of our structure and that of all things, but also brought them into being. Moreover, to be ready to admire the genius of man as a free agent in things made by human skill, and to deny a free mover, to things produced by nature would indeed seem to me to show great simplicity in pretended subtlety, since when man has performed most skilful things, he cannot, except through a fog, make out what he has done, which instrument he used, or though which causes the said instrument moves."[33]

In the *second section* Stensen first discusses the origin of different bodies and proposes three theses:[34] (1) "If a solid body is enclosed from all sides in another solid body, the first of the two to harden was that one which, when both touch, transferred its own surface characteristics to the other." The relative age of fossils and crystals and of the surrounding earth can be determined by this proposal. That is, Stensen determines the relative age of two adjacent bodies, observing that crystals with smooth surfaces had already hardened when the rock containing them was still fluid, but in other cases the veins of mineral crystals formed after the containing material had already hardened. (2) "If a solid body resembles another solid body, not only as far as the state of the surface is concerned but also in the finer arrangement of their parts and their smallest parts, it is also the same as far as its origin and the place of its production are concerned, except for insignificant particularities due to the site, particularities which one often finds in any other locality." From that, Stensen draws corresponding conclusions about the origin of crystals and of fossils resembling plants and animals. (3) "A solid body produced according to the laws of nature is produced from a liquid." This, Mietleitner describes as a

---

[33]OPH II, 188 f.
[34]<776> OPH II, 191 ff..

sagacious proposition for that time. In connexion with that, Stensen divides the fluid in an *externum fluidum* from which the body is built and an *internum fluidum* in plants and animals. The fluids in animals are then subdivided into three groups, the outer one on the surface of the mucous membranes, the inner common ones, such as the blood, and an inner fluid specific to particular organs such as the muscle fibres. Interestingly enough, Stensen mentions the fluid in *capsula ovis.*[35] Stensen here for the first time and before de Graaf, hints at knowledge of the egg and the egg follicle (Fig. 1.6.2).[36]

The *third section* is devoted to the special study of those different solid bodies such as incrustations, sediments, crystals, and objects with the shape of sea-fishes, conches, and plants.[37] The incrustations consist of lamellae which are parallel but not in a single plane, producing convex or concave shapes, as in an agate, depending on whether the inner or the outer part was formed first.

Then follows Stensen's theory of strata, accompanied by a great number of observations and incisive conclusions.[38] He demonstrates in three points that the earth's strata are sediments deposited from a liquid, considering in turn the sedimentary structure of strata, the distribution of material according to its weight from the finer material that surrounded the larger whole body.

In eight points he then draws conclusions from the material composition of strata and from the bodies contained within them, such as the remains of animals and plants, sea salt, or spruce cones. Here Stensen also recognizes volcanic forces, as when he infers the occurrence of fire from the presence of charcoal, ashes, pumice, bitumen and dross. The evidence is more certain, when a whole stratum consists only of ashes or charcoal: "Such a stratum I have seen outside the city of Rome, where the material for burnt bricks is dug."[39] He attributes stone-like strata to sources with petrifying water or to "eruptions of subterranean vapours (*halituum subterraneorum eruptiones*)".[40]

---

[35] *Capsula ovi* is presumably the egg follicle. <780> OPH II, 194.

[36] GP, 153–157. For the section of Oldenburg's English translation transcribed in 'Steno on Compartmentalization of the Water Space in the Living Organism', see Kardel, *Steno—life, science, philosophy*, 81–83. The aim of Stensen's lengthy account on liquid spaces in animals placed among geological considerations may be found in the author's point that worms and stones (gall- and bladder-stones) are generated in the so-called external fluid outside the body proper. He may thus have researched the formation of crystals from an interest in stone formation in diseases. Another valid feature is that Stensen suggests that particles are separated from the external fluid being sieved into the internal fluid, and the exchange between the inner common and specific fluids are at the capillaries.

[37] <783> OPH II, 196 f.

[38] OPH II, 197 ff.

[39] <786> OPH II, 198.

[40] OPH II, 199 = GP, 165.

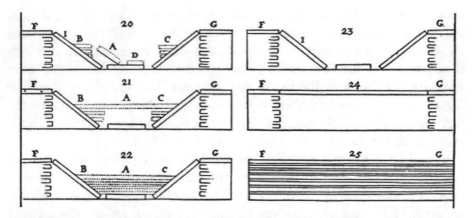

Fig. 1.6.2 Superposition of strata at origin (25) and "back-stripping" a landscape in Tuscany to present days (20), depicting the principle of original horizontality and the principle of lateral continuity in six cross sections. From the *Prodromus on solids*, OPH II, 226

Concerning the conditions under which sedimentary strata are formed, Stensen presents four separate conclusions. For example, he states that ground of firmer consistency must have been present underneath at the time these strata were formed, and also that the strata must have been limited laterally. In an important realization of the time frame revealed by the stratification of the sediments, Stensen concludes that the strata when originally deposited must have been parallel and horizontal. Where strata today stand perpendicular or oblique to the horizon, their original position cannot originally have been such. Precipices, channels, valleys, caves and underground passages are the consequences of the explosion of underground gases or also of water erosion of its supporting foundation.

There follows a section on the formation origin of mountains, the essential cause of which is the modification of the position of strata resulting from erosion by flowing waters and breakdowns within the earth. This is demonstrated in seven points, along with traces of underground fire in or near rocky mountains.[41] Mountains can also arise through such fire outbreaks. Two kinds of mountains can be distinguished, namely those consisting of strata, i.e. earth or rock strata, and others consisting of remnants of strata. Since Stensen made his observations mostly in the mountains of Tuscany, which are built from sediments, it is quite understandable that volcanic forces fall back in his conception. He did not think, at least then, of a formation of

[41]<788> OPH II, 200 f.

the mountain as a consequence of the fall of temperature and contraction by wrinkling from below. We do not know his views 1 year later after his journeys in the Alps and the Carpathian Mountains.

In five points, Stensen makes statements about mountains that in relation to more or less fantastic time conceptions must have appeared sensational, for instance, that no mountain has existed ever since the creation of the world, that mountains do not grow like plants, that the stones of the mountains, except for a certain hardness, have nothing in common with animal bones, "neither in material nor in manner of production nor in construction nor function; if one may be allowed to make any pronouncement on a subject so little known as the function of things".[42]

When Stensen asserts that crevices, passages, etc., in the rock have come into being only through later changes, and consequently that also the multiplicity of minerals has appeared only after its formation and that, furthermore, the material of those minerals stems from the adjacent rock, Mieleitner notices: "Herewith Stensen becomes a direct precursor of the lateral secretion theory of the nineteenth century." (Fig. 1.6.3)[43]

Stensen then[44] talks in three points of the way of realization of the strata changes when water, air or vapours resulting from conflagrations flow out from the inside of the earth, and of the formation of minerals in the empty spaces and interstices between the strata and refutes the superstitious search for minerals with the help of roots and branches, being "as dubious as the opinion of certain Chinese is ridiculous with regard to the head and tail of the dragon that they use in discovering a favourable position for burial places in mountains".[45] Most minerals pursued by men would not have existed from the creation of the world and all minerals likely stem from vapours. A lot of phenomena herewith find their explanation.

Next Stensen addresses the various solid bodies that are included in other solid bodies, as referred to in the title of the book, which has inscribed the name of Stensen into the annals of crystallography.

His discoveries, therefore, must be seen against the right historical background of science of which Johnsen says: "On the whole, all the Middle Ages and also the antiquity are said totally sterile as far as crystallography is concerned."[46] Stensen's chapter on rock crystal is quite different. He indeed does

---

[42]<789> OPH II, 201.

[43]Stensen, *Das Feste*, 26. * The theory of lateral ore genesis was put forward in 1847 (Britannica.com).

[44]OPH II, 201 ff.

[45]<791> OPH II, 202.

[46]Johnsen, 407. See also Tertsch, *Niels Stensen*. Schafranovski and Burke in GP.

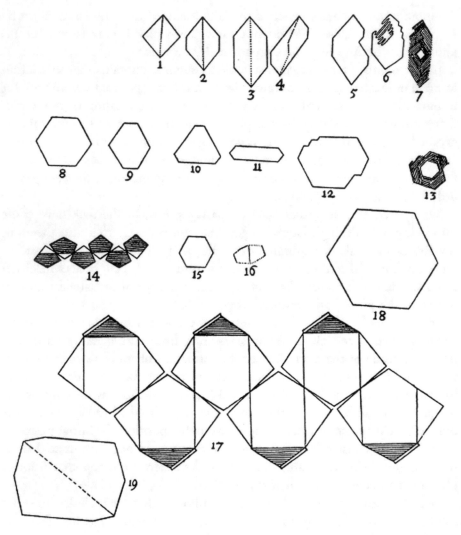

Fig. 1.6.3 Illustration from *De solido*. Printed together with Fig. 1.6.2 these diagrams show rock crystals described in the text

not venture to express his opinion about the first laying-out of the crystal but he refutes several contemporary, purely speculative interpretations and begins with a description and definition of his own expressions. In seven points he ascertains a series of facts concerning the growth of the crystal, namely that the mountain crystal grows through deposit of material on the sides of the crystal, thus not like plants. "This difference between the growth of the inorganic bodies by accretion from the outside in sediments, incrustations and crystals and the growth in organic materials from inside is one of the most fundamental crystallographic findings which we owe to Stensen."[47] The new crystalline material is not deposited on all sides but often only on their peaks or on the external surfaces. The special shape of a crystal is attributed to the action of a liquid flowing out of the crystal. Mieleitner says: "This 'liquid' which Stensen accepts on theoretical grounds is identical to atomic forces coming from the crystal surface which regulate deposit of the material and condition its passage into the state of crystal, i.e. condition the regular arrangement of the atoms. Here also an extremely sharp observation which makes honour to the erudite of the seventeenth century."[48]

Stensen is also convinced that the crystal, as it is separated from the lye, can also be changed back into it, "for it is certain that just as a crystal has formed from a fluid, so that same crystal can be dissolved in a fluid, provided one knows how to imitate nature's true solvent".[49] Stensen thus explains the properties of the crystal conditioned by its origin.

He then describes the crystals of haematite, of diamond, of iron pyrites which, depending on the time of their origin, the site, the kind of precipitation and the perfection of its shape differ from rock crystal.

Stensen is struck by the importance of the crystallographic studies: "If an accurate investigation of angular bodies were begun, not only of their composition but also of their dissolution, we should soon acquire a sure knowledge of the variety of motion with which the particles of both the tenuous and surrounding fluids are agitated."[50] Several other observations are reserved for the proper dissertation. One is amazed again and again by the abundance of correct observations among which one single conception conditioned by the time is quite striking: that glass similar to crystal could perhaps be produced in a cold way.

The same is true for the following palaeontology section on shells of mollusc, the original and most difficult problem.[51] Stensen first investigates shells from the sea and then compares them to those taken from the mountains. In three points he describes the shells and the threads from which the shells are formed as an animal product. In four further points the way that animals produce fibres and scales through a kind of secretion is explained.

"According to our current knowledge, the shells consist of three layers, namely, above, a very thin skin (cuticle), then the prisms layer which is

---

[47]Johnsen, 407.

[48]Stensen, *Das Feste*, 27.

[49]<798> OPH II, 207 and 298 f.

[50]<804> OPH II, 211.

[51]<805> OPH II, 211 ff.

formed by parallel small rods of calcium carbonate which all stand at a right angle to the surface, finally of pearl-mother layer, which is the thickest and consists of transparent lamellae which are also formed from calcium carbonate and are generally parallel to the surface. Stensen observed the prisms layer and the pearl-mother layer as well as their composition and existence on the upper surface of shellfish. He even knew the presence of the *cuticle* even if he did not find it everywhere."[52]

Oysters and other shellfishes stem from eggs and not through decay, Stensen boldly says in defiance of an old error. He then explains the multiplicity of shell shapes and the formation of pearls by the way they come into being. "The difference between the integuments of pearls and the *testulae* in shells of pearl-bearing molluscs is merely that the filaments of the shells are located as it were in the same plane while the integuments of pearls have their filaments distributed over the same spherical surface. An elegant example of this was provided by one pearl among those that I broke at your command; this pearl, although externally a glittering white, enclosed within itself a black body that resembled a grain of pepper both in colour and size, in which the placing of the filaments, with one tending towards the centre, was very clear, and the rows of spheres of the same filaments could be distinguished".[53]

After explaining different irregularities, such as an uneven upper surface, yellow colour, etc., Stensen considers shells and their problem with great acumen and through references to examples. He then considers animal remains in the earth, like shark teeth, ray teeth, fish spine as well as bones of terrestrial animals.[54] Either they wholly resemble actual parts of animals or they are different from them in weight and colour only.

The problem of why so many shark teeth are to be found in Malta is debated, also why some bones are so large. Stensen refutes the objection that the fossils ought to stem from the Deluge, in which case the shells would certainly already have been disturbed. The durability of fossils depends totally on their composition. The enormous bones discovered near Arezzo are attributed back to the raid of Hannibal.

The explanation of fossil plants follows the principles valid for animals, "for either they resemble true plants and parts of plants in every respect, which is a rare occurrence, or they differ from them only in colour and weight, which is a common occurrence, sometimes burnt to charcoal, sometimes impregnated

---

[52]Stensen, *Das Feste*, 28.

[53] <807> OPH II, 213.

[54] <811> OPH II, 216 ff.

with petrifying fluid or they correspond only in their shape, there being a great abundance of such cases in various places."[55]

Stensen concludes this third section: "I have, previously, reviewed the chief bodies for which the place of discovery has caused many to be indecisive about their place and production; and I have, on the same occasion, indicated how from that which is perceived a definite conclusion may be drawn about what is imperceptible ... How the present state of anything discloses the past state of the same thing is made abundantly clear by the example of Tuscany, above all others."[56]

Supported by drawings Stensen asserts: "Thus, we recognize six distinct aspects of Tuscany, two when it was fluid, two when flat and dry, two when it was uneven; what I demonstrate about Tuscany by induction from many places examined by me, so I confirm for the whole earth from the descriptions of many places set down by various writers."[57] Among those writers the Bible is introduced with particular motivation:" But lest anyone be afraid of the danger of novelty, I set down briefly the agreement between nature and Scripture, reviewing the main difficulties that can be raised about individual aspects of the earth."[58] Following this reasonable method Stensen explains for instance the possibility that the earth could have been covered by water, deals with chronological questions and even indicates how the rising of the water can be explained by the laws of nature. Like so often in his writing where great knowledge is fortuitously mentioned, so also here: "Change is indeed continual in the things of nature, but nothing in nature is totally destroyed."[59]

Stensen reserves for the full dissertation detailed proofs of the modification of the terrain in Tuscany, the land thus between the Arno and the Tiber, and concludes the *De solido* with the following words: "And this is a succinct, not to say disordered, account of the chief things that I have resolved to set down in the Dissertation itself, not only more distinctly, but also at greater length, with in addition, a description of the places where I have observed each item."[60]

The figures illustrating the *Prodromus* and the captions are of great importance. Stensen felt that clarity of the text might suffer from its shortness

---

[55] <814> OPH II, 218.

[56] <816> OPH II, 219. * The quotation demonstrates that Stensen was conscious of what was perceived and what was imperceptible in his considerations in earth science. Cf. Kardel, *Prompters of Steno's geological principles*.

[57] <817> OPH II, 220. On Stensen's view on the geology of Tuscany, see Rodolico, *Niels Stensen*.

[58] OPH II, 220 f.

[59] <818–819> OPH II, 222.

[60] <822> OPH II, 224. This final remark must always be kept in mind, when judging the *Prodromus* (Scherz).

and, therefore, decided, "among a great amount of other figures" to add the following ones.[61] The remark: "I have explained the letters of the diagrams in the dissertation itself, in the order in which the diagrams follow each other",[62] allows us to assume that the proper dissertation was already partially at hand. However, the most important sentence is this: "Diagrams 5 and 6 are of the class, countless numbers of which I could present to prove that in the plane of the axis, both the number and length of the sides are changed in various ways without the angles being changed."[63] Herewith Stensen demonstrated that in any case he knew the law for quartz, saying that each kind of crystal had a well-determined angle between the sides (the law of constancy of the angles).[64]

1.6.3   The Scientific Value of the Treatise

It is not easy to evaluate the genius of the *De solido*.[65] This work did not have its like at its time and in the history of sciences there is hardly any comparable dissertation that offers such an abundance of new and epoch-breaking scientific findings delivered in such a short space.

It not only deals with particular discoveries among a lot of errors but, apart from a very small number of errors which, partly, were methodically unchallengeable and thus unavoidable, there are many correct concepts so

---

[61]OPH II, 225.

[62]OPH II, 226.

[63]OPH II, 225.

[64]The law is described by Romé de l'Isle in 1783. Stensen is often mentioned as the real discoverer of this rule, thus by Hobbs in *Stensen, The Prodromus*, 171, Johnsen, 413, Tertsch, *Niels Stensen*, 127. Meanwhile cf., Hoykaas, 182 ff. For a discussion see also Burke, *The work and significance of Nicolaus Steno in crystallography*, 169 ff, in which it is shown that the rule was known by several people in the seventeenth century, but remained unappreciated until it became of practical importance.

[65]Gustav Scherz openly declared himself a non-specialist in geology who built on the interpretations of others, in recognition that a thorough and embracing scientific evaluation of Stensen's geological works, of what he owed to his predecessors and contemporaries, remains to be written. In this connection it is remarkable that Stensen is so often described as if his writings arose within an empty space. What is known in general about Stensen makes it unlikely that he did not have a broad knowledge of the existing literature. Admittedly no one has substantiated a direct linkage with the English writer Robert Hooke; in 1670 the Sicilian writer Agostina Scilla (1629–1700) published his now classical work *La vana speculazione disinganata dal senso*, without knowing the writings of Stensen. A seminal exchange of experiences and thoughts took place through travels, through books and, not to forget, through correspondence in spite of geographical separation. In Amsterdam and Leiden, in Paris and Montpellier, in Rome and in Florence Stensen communicated with the leading scientists. The elucidation of a common interest in geology, in particular at the court of Ferdinand II, would indeed contribute to improve the understanding of the prerequisites of Stensen.

that frequently the baselines and the outlines of a new science rightly face us. It is certain that Stensen here alternates as a pioneer geologist, palaeontologist and mineralogist. It must also be emphasized that, although he came to speak of domains that are very different today, Stensen presents the contents of the *Prodomus* as a unit, a coherent argumentation in which all important matters for the presentation of the problem are included. Burke calls it "an example of clarity which contains a wealth of precise observations and correct scientific conclusions."[66] One understands the word of the German translator, the geologist Mieleitner: "It occurred to Stensen as to Midas, king of Phrygia, what he touched became gold under his fingers."[67]

In order to understand what is unique in the *De solido* well, one must think of the historical background. The pre-Socratic natural philosophers had dealt with geological problems and had considered the maritime fossils in the mountains of Asia Minor or in Malta as an indication of them having been covered by an earlier sea. Beckmann says: "But, if we consider all the witnesses of a geological kind from the following 1000 years of the antiquity and Middle Ages, then only the statement remains that nowhere an essential thorough knowledge of the history of the earth is to be found, that never a serious step has been made forwards since the pre-Socratics, even not during the Renaissance. On the contrary, steps have been made backwards. Even Leonardo da Vinci did not go further than some occasional ingenious statements. One can rightly point at single documents of the earth history but nobody has succeeded to draw therefrom a consistent picture of the historical development of the earth, generally to address nature with questions concerning the history of the earth. And nobody could succeed based on the particular attitude of this epoch towards nature and because of the lack of a true historical attitude."[68]

Here the method of Stensen was of decisive importance. As Professor Hobbs says: "One of his propositions could well be printed in big letters on the walls of our laboratories and lecture-halls as a warning to those who pursue scientific research. 'The source of doubt seems to me to be the fact that, when asking nature, the points which cannot be determined for sure are

[66]Burke, 163.

[67]Preface to Stensen, *Das Feste*, 24. * Research by Dominici and Cioppi clarifies several questions about the *Indice*-catalogue and on the fate of the Grand-Ducal collection of fossils and minerals. Dominici, S., and Cioppi, E., All is not lost: History from fossils and catalogues, *in* Rosenberg, G.D., and Clary, R.M., eds., *Museums at the Forefront of the History and Philosophy of Geology*, Geological Society of America Special Paper 535, https://doi.org/10.1130/2018.2535(05) 2018, (in press).

[68]Becksmann, 329.

not distinguished from those which one can solve with certainty.' How much annoyance could have been spared if the erudite of our time had kept this point in mind!"[69]

For Becksmann, the greatness of Stensen resides in "the clarity and methodical logical consistency with which he considers the documents on the history of the earth and in the loneliness of his research of the history of the earth, based on that and carried out in a true historical setting. History of the earth is not for him something of the past, second-hand, but the ground on which the present rests, from which he seeks to understand the picture of today in a true historical context. As founder, not of a new science, but of a new way of thinking and of seeing the history of the earth, he stands before us as a 'powerful autochthon', to apply to him an expression of Dilthey."[70]

The Russian translator, the geologist Schafranovski asserts: "(Stensen) can be considered until today as an example of a true mineral-crystallographer. Ideal acuity and precision of the observation; careful investigation of the crystalline material, beginning with its occurrence in nature, heightened attention to the smallest deviations of an actual crystal from the idealized model; active interest for the questions of the crystal genesis; competent generalization of the observed facts, which leads him to the discovery of basic laws—these are the traits which characterize the success of N. Stensen."[71]

Today Stensen is considered as the first to have recognized the principle of actualism in geology, *i.e.* that the past events can be explained by forces that are still acting now, unlike the catastrophe theory attributed to Cuvier, which relies on geological forces that are quite unknown to us. Actualism was not widely recognized until the nineteenth century with Lyell and von Hoff. "In Stensen, the actualistic method of seeing was obviously justified also by the fact that, in his comprehensive work, he wanted to bring evidences for geological changes in historical time not only from Tuscany but from the whole Italy", Bülow says.[72]

According to the Danish geologist Hilmar Ødum, "the most amazing in Stensen is his perfectly modern concept of the geological forces: his disposition for that is what we now call the principle of actuality. He did not need to grasp fantastic speculations to explain upheavals in the past. He understood

---

[69]Preface to Stensen, *De solido*, 170.

[70]Becksmann, 336.

[71]Schafranovski, 258.

[72]Bülow, *Der Weg*, 161.

that the forces of the past are the same as those of the present time and that the forces which have produced the mountains and valleys, are still always working to change the aspect of the earth."[73] Stensen herewith introduced the historical perspective into the consideration of the history of the earth. As the Italian geologist Francesco Rodolico says, "for the first time in history he completely grasped the whole idea of a comprehensive cycle of evolution of the landscapes, in that he inductively reconstructed them from their past apparent shapes and determined their successive phases for all time."[74] The German Kurd von Bülow says: "To recognize the temporal in the spatial— nobody had done that before Stensen—, from the whole rock to read a dynamic course of time, has since then become and remained the main object of scientific geology."[75]

The harmony between faith and science which appears in the *De solido* is also noticeable and amazing. It must be kept in mind that Stensen's fundamental geological research occurred in that year when he converted to the Catholic faith. He wholly attained what A.D. Jørgensen says of the conversion of Stensen precisely at that time: "It will always belong to the greatest triumphs of the Catholic Church that Niels Stensen, at the same time, passed over to its doctrine through a rich inner evolution and after a strong spiritual struggle, and founded a new science with rare discernment and noticeable genius ... In the strictest sense the convert Stensen has created the geology."[76] Did the convert not feel any tension between his religious conviction and his exact research? Not at all. The deep piety which animates the sharp-witted researcher can be observed in many places. It is quite natural for him to call God the Mover, the original and unknown cause of movement. Even the pagans believed that in such things there is something divine. It is not difficult for him to believe in the miracle. "Moreover, to be ready to admire the genius of man as a free agent in things made by human skill, and to deny a free mover to things produced by nature would indeed seem to me to show great simplicity in pretended subtlety, since when man has performed most skilful things, he cannot, except through a fog, make out what he has done, which instrument he used, or through which cause the said instrument moves."[77]

---

[73]Ødum, *Niels Stensens geologiske syn og videnskabelige tankesæt*, 53.

[74]Rodolico, *Niels Stensen*, 242.

[75]Bülow, *Steno*, 157.

[76]Jørgensen, *Nils Stensen*, 143.

[77] <773> OPH II, 189.

Ødum notices that the Creator and his omnipotence are always found in the unimpeachable logical context. God never appears as *deus ex machina* when Stensen comes up against something which he cannot explain. Never does he use God as a helper in need.[78] For Stensen, there is no contradiction between faith and science, as was mentioned above. This appears from the fact that precisely in the *De solido* he speaks of the "great Galileo".[79] The attitude that appears from this respectful qualification written in the circle of the Accademia del Cimento in Florence reveals a mind that clearly distinguishes between the knowledge through faith and that through science.

The same is true for Stensen's position about the Bible. Passages in the *De solido* in which Stensen points out the concordance of his representation of the geological condition in Tuscany and the mosaic account of Genesis and the great Deluge have frequently been deplored. This is often considered a retreat from the high position of science in the other parts of the book.[80] Nothing is more absurd than that. No interpretation of the Scripture forbids the faithful and no scientific method the researcher to lean on historical sources as long as his own observations do not contradict them.[81] People then disposed of none or of too limited observations and had no idea of the length of the geological periods. When Stensen related the elephant bones in the Arno valley to the war elephants of Hannibal's fight with the Romans at the Trasimene lake near Arezzo in about 200 years B.C., he felt so little impressed by the chronology of the some 4000 years with which the then biblical and profane history counted, that long before that he endeavoured to demonstrate to his contemporaries that the fossil elephant bones of Arezzo could have resisted the destructive force of 1900 years.[82] What Stensen turned on in geological observations could be interpreted in whole accordance with the Bible and, therefore, did not give him any reason to reject or to question the Bible.[83]

---

[78]Ødum, 58.

[79] <803> OPH II, 210.

[80]On the relation to the Bible, see Stensen, *Das Feste*, 29 ff, also Rudwick, 68 ff.

[81]See Krogh and Maar in *Stensen, foreløbig meddelelse*, ix f.

[82] <812> OPH II, 217.

[83]As emphasized by Rudwick (p. 69 f), the biblical account of generation was no hindrance for Stensen and like-minded geologists. It rather supported an understanding of the fact that the earth indeed has a history of generation. Remacle Rome said aptly: "Sténon n'était pas plus naïf que les savants du siècle dernier qui comptaient en millions d'années des périodes géologiques que l'on évalue actuellement en milliards d'années." (*Nicolas Sténon paléontologiste*, p. 97).

1.6.4   Remembrance in Posterity

The publication of the *Prodromus* took place in the spring of 1669. Stensen had then begun his great travel of November 1668 to July 1670 and it was Viviani who cared for the practical affairs as written in his hand in the manuscript to be printed on the first page[84]: "This was printed under my supervision." (Fig. 1.6.4)

The first review by the censors is dated August 30, as mentioned above, and here Viviani explains that the work was worth being dedicated to the name of his Eminence. The official permission of printing followed on December 13 only, according to the report of Redi.[85]

The book must have appeared by the end of April 1669. Malpighi then possessed his own copy, and there were two more for Riccioli and Montanari already on May 7. The Grand Duke put a great number of copies at the disposal of Stensen. In Stensen's absence, Viviani took it on himself to distribute these according to the wishes of Stensen. Thus the surgeon Guglielmo Riva thanks Viviani on May 25 from Rome for ten copies of the *De solido*, which he would obviously distribute in the Tiber city. Athanasius Kircher apparently received one of these.[86] Orazio Ricasoli Rucellai, the prior of San Casciano, obtained one copy for himself and one for a friend at the end of June 1669. Carlo Dati, who had on his part provided Stensen with illustrations for *Canis carchariae*, was in possession of the book on July 23. The Frenchman Adrien Auzout received his copy in Rome on August 23. A Mr. Vabel expressed thanks for the book from Venice, and J. Hertz acknowledges receipt of 12 copies, destined for the bookshop.[87] Malpighi ordered four more copies from a Florentine bookshop on October 26, 1669.[88]

Interest in the work was shown also outside Italy. For example, Spinoza had not less than two copies in his library.[89] In England, it was translated at the beginning of 1671 by Henry Oldenburg, the secretary of the Royal Society. In the Introduction Oldenburg points out that this brilliant script, which amounts to a report on the beginning and formation of earth mass and

---

[84]On the following see Scherz, *Viviani*, 285 f. * The printer's manuscript is all orderly written in Viviani's hand (Fig. 1.6.4) without any change or note by Stensen. Supposedly Stensen had left the manuscript in a mess when departing for his grand tour.

[85] OPH II, 333 f.

[86]EP I, 208.

[87]Scherz, *Vom Wege*, 87.

[88]Adelmann, I, 350. [aut]Adelmann, Howard B.

[89]Préposiet, 340.

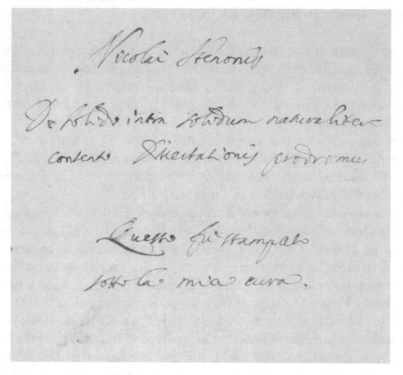

Fig. 1.6.4 The title page of the *Prodromus* in the hand-writing of Viviani, the printer's manuscript, BNF: Gal. 291

the products in this, should be made known to as many people as possible.[90] In the following year Robert Boyle included the translation as an appendix in his work *Essays of effluvium*.[91]

Doubt has been often expressed, with some good reason, whether the *Prodromus* of Stensen also exerted an influence on his contemporaries and on the development of geology.[92] Here the editions that turn up again and again through 300 years testify for the enduring interest in the work of Stensen on the part of specialists.[93]

An English translation appeared as soon as in 1671, as has been mentioned. A duodecimo edition appeared in 1679 in Leiden by Jakob Moukee. The eighteenth century, which is often thought to have ignored the

[90]Included in the American edition, Stensen, *Prodromus*, 199 ff, also Eyles, *Influence*, 170 ff.
[91]Cf. Stensen, *Prodromus*, 198 f.
[92]Among several others Zittel, 35 f, Garboe, *Nicolaus Steno and scientific geology*, 118, Frängsmyr, 204 f.
[93]Cf. Stensen, *Das Feste*, 128 f, *Prodromus*, 194 ff.

fundamental geological findings of Stensen, still presents some editions of the *De solido*. It appeared in French translated by Gueneau in the meritorious *Collection Académique* in 1757, however in a rough translation and without figures or captions. There was a new edition of the whole Latin original text in Pistoia in 1763. Here it is indicated on the title page that the book is also available at Vincento Lendi's in Florence.

In the nineteenth and twentieth centuries attention was particularly drawn to the work since L. Elie de Beaumont printed the dissertation in the *Annales des Sciences Nouvelles* (Paris 1832) and the most important excerpts in the *Fragments géologiques tirés de Sténon* (p. 337–377). Ten years later, an excerpt provided by Leopoldo Pilla appeared in the *Tipographia Galilaeana* in Florence. Friedrich Hoffmann in his *Geschichte der Geognose*, Berlin 1836, 40–45, reproduces the most important passages in a literal translation. Carl Vogt does the same in the *Lehrbuch der Geologie und Petrofactenkunde*, vol. II, 384–390, Braunschweig 1847. The growing interest of the twentieth century is testified by an often quoted Danish translation by the Professors A. Krogh and V. Maar in 1910, and in 1904 a facsimile edition of Ed. W. Junk appeared in Berlin. With the Maar edition of all the natural science works of Stensen in their original languages in the year 1910, the *Prodromus* became easily accessible and subsequent translations can be considered as a fruit of that edition. A modern English translation appeared in 1916: *The Prodromus of Nicolaus Steno's dissertation concerning a solid body enclosed by process of nature within a solid* by John Garrett Winter with a foreword by William H. Hobbs. New York 1916 in: University of Michigan Studies, Humanistic Series, vol. II, 163–283. In addition to a genuine Introduction, this edition comprises a precise bibliography. In the series, *Ostwald's Klassiker der exakten Wissenschaften Nr. 209*, Karl Mieleitner provided a German translation with the title *"Vorläufer einer Dissertation über feste Körper"* (Leipzig 1923). G. Montalenti published the Italian translation *Niccolò Stenone, Prodromo di una dissertazione sui corpi solidi naturalmente inclusi in altri corpi solidi. Tractato dal Latino con Prefazione e note a cura di Gius. Montalenti* (Rome 1928). (In: *Universitas scriptorum, classici della scienza*. Nr. 7–8) (Fig. 1.6.5).[94]

A Russian edition appeared in 1957 published by the Soviet Academy of the Sciences. The 1960s brought new editions of Winter's English[95] and

---

[94]Italian translations of *De thermis, Canis carchariae*, and the *Prodromus* can be found in *Niccolò Stenone —opere scientifiche I–II*, edited by Luciano Casella and Enrico Coturri in Florence, 1986.

[95]This facsimile edition from the University of Michigan Humanistic Studies, vol. XI, 1916, came in the series, *Contribution to the History of Geology*, vol. 4, 1968.

Mieleitner's German[96] translation as well as a comprehensive Latin–English parallel edition with translations by Alex J. Pollock in Gustav Scherz, the *Steno, Geological papers*, published by Odense University Press.

The specific influence of the geologists of Stensen's own time and of later generations also appears to some extent.[97] As Krogh and Maar in the Danish edition said: "The two countries from which the most important contributions to geology came in the next century were England and above all Italy and it can be shown that Stensen's work exerted a considerable influence on the evolution in the said countries, as far as Italy is concerned with certitude and as far as England is concerned with great likeliness. The ways of development can often be determined quite concretely.[98]

Since then any doubt about the influence of Stensen on England has been done away with.[99] As was mentioned above, a very fruitful discussion about fossils took place stimulated by lectures of Robert Hooke at the Royal Society.[100] *Prodromus* was translated into English in 1671 (Fig. 1.6.5) and provoked an exhaustive response from Martin Lister, an acquaintance of Stensen from Montpellier. Lister possessed a comprehensive knowledge of zoology and fossils and, based on his own observations, knew that the resemblance between still living animals and fossils was often superficial only. The idea of the extinction of species was remote at that time and Lister firmly maintained that the English fossils were not of organic origin.[101]

---

[96]Mieleitner's edition is edited with preface by Gustav Scherz in the series *Ostwalds Klassiker der exakten Wissenschaften*, N.F. vol. 3, 1967.

[97]For background information the reader is referred to Schneer (ed.), *Toward a history of geology*. Also Hölder, *Geologie und Paläontologie in Texten und ihre Geschichte*; and older reviews like Zittel, *Geschichte der Geologie und Paläontologie bis Ende des 19. Jahrhunderts*. See also Hölder, *Stensens Bedeutung für die Begründung der Geologie und Paläontogie*. On Stensen's *Prodromus* in posterity also cf. Stensen, *Das Feste*, 32 ff.

[98]Stensen, *Foreløbig meddelelse*, x.

[99]On the following see Eyles, *Influence*, also Haber, 23 ff, Rudwick, esp. 61 ff.

[100]Together, Hooke and Stensen are considered the greatest geologists of the seventeenth century. Hooke's main work in geology was not published until 1705, thus after his death. However, his concepts were known already in the 1660s from his readings in the Royal Society. Possibly Stensen may have become acquainted with these concepts in France (see Chap. 1.3), although it has not been possible to trace this influence (Eyles, *Influence*, 173 ff). Drake and Komar, *A comparison of the geological contributions of Nicolaus Steno and Robert Hooke* makes a comparison and concludes: "In a comparison of Steno's geological contributions with those of Hooke … Hooke emerges as having made a more extensive and profound contribution … Robert Hooke, like Steno … , deserves recognition by geologists as a founder of their science" (p. 127). See also Drake, *Robert Hooke and the foundation of geology;* * and Yamada, *Hooke-Steno relations reconsidered.*

[101]Letter to Oldenburg dated August 25, 1671, published in Philosophical Transactions, October, 1671, VI, 645 (excerpt).* Full text of letter in Anna Marie Roos, ed., *The Correspondence of Dr. Martin Lister (1639–1712)*, Volume One: 1662–1677, Leiden 2015, 381, (Oldenburg VIII, 212 ff), see also DSB VIII, 415 ff, Rudwick, 61 f.

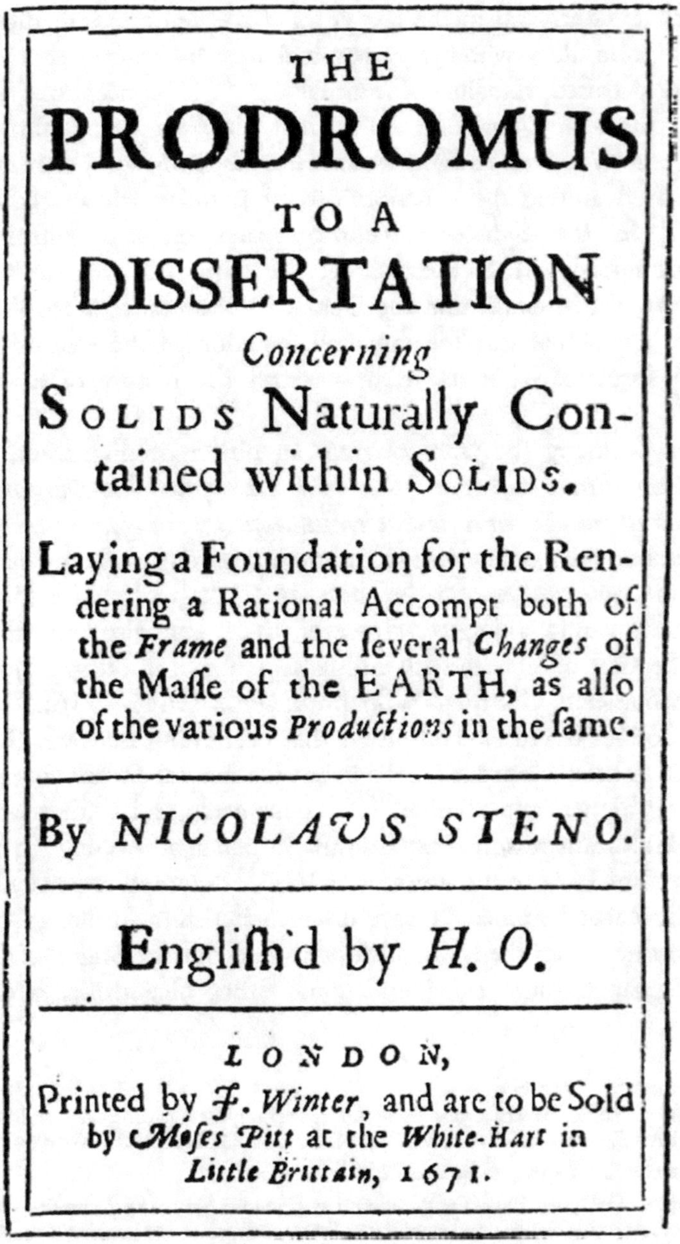

THE
PRODROMUS
TO A
DISSERTATION
Concerning
SOLIDS Naturally Contained within SOLIDS.

Laying a Foundation for the Rendering a Rational Accompt both of the *Frame* and the feveral *Changes* of the Maffe of the EARTH, as alfo of the various *Productions* in the fame.

By *NICOLAVS STENO.*

Englifh'd by *H. O.*

LONDON,
Printed by *J. Winter*, and are to be Sold by *Mofes Pitt* at the *White-Hart* in *Little Brittain,* 1671.

Fig. 1.6.5   Frontispiece of the English translation of *De solido*, London 1671, by Henry Oldenburg (H.O.), 8°, REX

The debate continued in the last years of the century with different contributions. John Ray, whom Stensen had met in France, considered that fossils are petrified remains of animals.[102] The conceptions of Hooke resembled those of Stensen so much that Hooke accused him of plagiarism.[103] John Woodward's *Essay toward a natural history of the earth* (1695) must be cited among the contributions of popular science. It was much propagated in the eighteenth century, even on the continent. Here Woodward agreed with Stensen on certain points but without mentioning Stensen, like, for instance, that the Deluge was the cause of fossils found far away from any actual sea. Yet above all, he adopted the idea of Stensen of using the sequence of strata to understand the history of a determined region.[104]

Also in Germany the work became an object of discussion, as in the *Commentatio physica aequae ac historiae de Glossopetris Luneburgensibus Auct. M. Jo. Reiskio, qui Maltenses simul ceterasque quotquot cognitae sunt glossopetras paucis exposuit et omnes lapidum fossiliumque generi secus quam Fab. Columna et Nic. Steno voluerunt asseruit* (Norimb. 1687),[105] with which most learned men certainly would still agree. However, there were also learned men who received the ideas of Stensen with enthusiasm, above all the great philosopher and mathematician Gottfried Wilhelm Leibniz (1646–1716).[106] He thus writes in 1681: "It must be admitted that everything that Mr. Stensen has achieved in natural science is excellent but his treatise *De solido intra solidum* deserves the highest appreciation."[107] Those geological studies were one of the most important reasons that Leibniz, in personal meetings in Hannover from the end of 1677 to the summer of 1680, attempted to persuade Stensen to return to natural science. "I have often incited him further to carry them out and to draw from them conclusions to find out the origin of the human kind, the general water flood and some other nice truths which would

---

[102]"Observations", 1673, "Miscellaneous discourses concerning the Chaos, Deluge and Dissolution of the World", 1692 ff, comparecf. Raven, *John Ray*, Chap. 16: "Of fossils and geology", 419 ff.

[103]Eyles, *Influence*, 174, Drake and Komar, 129 f.

[104]On Woodward, DSB XIV, 500 ff. On Woodward and Stensen, Eyles, 182 ff, also Rudwick, esp. 82 ff. Woodward did not admit that he owed anything to Stensen, cf. Jahn, *Some notes on Dr. Scheuchzer*, 193 f. *Woodward's scientific credibility was rebuked by John Arbuthnot in *An Examination of Dr. Woodward's Account of the Deluge, & c., with a comparison between Steno's Philosophy and the Doctor's, in the Case of Marine Bodies dug out of the Earth*, London 1697.

[105]It considers Johannes Reiskius, *Die Glossopetris Lüneburgensibus*, Lipsiae 1684. On Reiske (1641–1701), see ADB 28, 128 f.

[106]On Leibniz's geology, see Chap. 1.6.

[107]Leibniz, SSB II. I 518.

confirm what the Holy Scripture tells of that."[108] The first indications of Leibniz's interest in geology can be dated to January 1678.[109] In his main work on natural science *Protogaea*, Leibniz relies much on *De solido*.[110] Despite several gradual deviations in detail one may say that at least 21 of the 28 chapters of the *Protogaea* could stem almost word for word from the pen of Stensen, as well as a lot of particularities in the other chapters.

The traces of Stensen's influence on the development of geology in the eighteenth century lead from Italy to England via Switzerland and also to France.

The prolific Paduan professor of medicine Antonio Vallisnieri (1661–1730)[111] has so sound views on the organic origin of the petrifaction and the action of water in the formation of the earth crust that he could be easily called a disciple of Stensen. This is also true for Domenico Guglielmini (1655–1710) with his studies of crystal owed to Stensen.[112] The influence of Stensen on the "Prince of the Tuscan nature researchers of the eighteenth century", Giovanni Targioni-Tozzetti (1712–1783) is quite beyond doubt.[113] This physician, book-keeper and traveller knew like no one else the history and geography of his homeland and his *Viaggi in Toscana* reads like a lively repetition of the journeys of Stensen in the country. When he refutes the imaginative theories of his time concerning the origin of the mountains, he warmly emphasizes: "I will do justice to the immortal Niels Stensen who, after sharp-witted observations of the ground of Tuscany, arrived at a clearer knowledge of the earth sphere than anybody else."[114] He attributes to Stensen, among other things, his ideas on the original mountains and their kinds of rocks to differentiate them from the later heights and their strata of clay and sand.

In Switzerland, close to Italy not only in space but also in mind, Johann Jakob Scheuchzer (1672–1723), the city physician of Zürich and a great expert and lover of the Alps, knew and cited Stensen in his *Ouresiphoites Helveticus*,[115] as did J.H. Hottinger (1680–1756) in his *Cristallologia*.[116]

---

[108]Leibniz, SSB I, III, 456 f.

[109]Sticher, *Leibniz' Beitrag zur Theorie der Erde*, 249.

[110]Bülow, *Protogaea und Prodromus*, esp. 205. See also Hölder, *Leibniz' erdgeschichtliche Konzeptionen*.

[111]DSB XIII, 562 f. Also Rodolico, *La Toscana descritta dai naturalisti del settecento*, arich sourcebook for the Prodromus-tradition in Italy, likewise Rodolico, *L'Esplorazione naturalistica dell' Appenino*.

[112]BU 18, 76 f, Tertsch, *Das Geheimnis der Kristallwelt*, 122 ff.

[113]DSB XIII, 257 f, Rodolico, *La Toscana*, 7.

[114]Quoted in Stensen, *Das Feste*, 34.

[115]DSB XII, 159, in which Scheuchzer is mentioned as the "Founder of European paleontology". He translated Woodward's essay into Latin. See also Jahn, *Some Notes*.

[116]Hottinger, esp. 29 f.

They testify the knowledge of the *Prodromus* in Zürich. In 1763 the erudite Bernese preacher E. Bertrand published his *Dictionnaire Universel des Fossiles propres et des Fossiles accidentels* (La Haye, 1763). Therein he defends the view of Stensen on petrified teeth against men like Reiskius and says: "Almost all the learned men confirm today that such fossils are petrified teeth of sea-fishes."[117]

Geological research in France in the eighteenth century was marked by a speculative school with G.L. Leclerc de Buffon (1707–1788) as a leading figure.[118] This group did not understand Stensen's ideas in that field at all and characteristically reproached him for never having produced a theory. There were also other, perhaps less illustrious, but therefore deeper-minded naturalists who stood more firmly based on facts and whom half the translation which, as mentioned, appeared in 1757 in the *Collection académique, partie étrangère*,[119] a series, provided with the documentation for the bright development of natural science. The genius of Stensen was highly appreciated and the publisher Gueneau says: "In that work it is demonstrated that Stensen was the man of his century who knew the best the right method for the study of nature and who knew how to re-unite theory and observation in an outstanding way."[120]

This edition was used, among others, by the great French mineralogist Jean Baptiste Louis Romé de l'Isle (1736–1797)[121] and, in his *Crystallographie* of 1771, he says: "I introduce here that more gladly these remarks of Mr. Stensen since they are made with much acumen, show a similarity between bodies the shape of which otherwise seems to be so different, can provide the reason of the varieties which one so often perceives in the crystals of one and the same kind and finally since the iron-rusty crystals which it is question of are the most developed among all those thus known."[122] Among other

---

[117]Bertrand, I, 247 f.

[118]DSB II, 576 ff.

[119]*Collection académique* IV, 377–414.

[120]*Collection académique* IV, xxvi.

[121]DSB XI, 520 ff, Burke, 173, Romé de l'Isle refers in some places directly to the Latin edition, cf. Burke, 174.

[122]Thus in the German translation by Christian Ehrenfried Weigel, Greifswald 1777, p. 358 f. In the second edition Romé de l'Isle says: "Quant à la forme cristalline du quartz ou cristal de roche, observée par les Anciens mêmes, les Modernes ont eu recours pour l'expliquer, à diverses hypothèses qu'on peut lire … surtout dans Sténon, mais je crois devoir placer ici quelques observations trèsjudicieuses de ce dernier sur la manière dont se forme le cristal de roche. Ces observations m'ont paru propres à jeter du jour sur la théorie de la cristallization en général, & sur-tout à faire connoître certaines particularités qui distinguent le cristal de roche des autres sels-pierres, également formés par l'intermède du fluide aqueux." Then follows a long quotation from Prodromus (Romé de l'Isle, *Cristallographie*, Paris 1783, I, 63 f).

achievements, Romé de l'Isle, based on the crystal chapter in the *Prodromus*, formulated the law of the angular constant.

The period around the turn of the eighteenth to the nineteenth century is designated as the heroic age of geology. This marked a break with the tradition, which attempted to explain the origin and evolution of the earth through speculation, unrestrained by natural laws and empirical experience. The "greatest of all teachers of mineralogy and geology", the Freiburg professor A.G. Werner, stimulated his disciples for the exact observation method. The knowledge of fossils gained among researchers like Cuvier a leading position in the theory of the formation of geologic strata. The fossils were used as "commemorative medals" only, to determine the age of the sedimentary stones. A bustling activity developed in all domains, rich in results and in well-known names of erudite men. In the first half of the nineteenth century these positions of scientific geology became more and more established. This spread to the universities; laboratories and collections were set up and, with the growing quantity of work and number of workers, specialization also made greater progress. A complete understanding of Stensen's geology, his methods and results, occurred in that effervescent uprising time.

It is no wonder that A.G. Werner (1749–1817)[123] himself never wrote about Stensen. Werner worked more through the spoken word and indeed had an abhorrence of writing and writers. However, Werner knew *De solido* very well—he actually even possessed two copies. The treatise belonged to the geological works that he recommended to his students. His theories, in part, much resembled those of Stensen.

Another significant geologist of his time was the Scottish James Hutton (1726–1797) who published his main work *Theory of the Earth* (1795) near the end of his life.[124] He often referred to the concepts of Stensen of the geological phenomena and of the history of the evolution of the earth and, through further work with the same fundamental principles, he formulated the first modern theory of the unity of the earth.[125]

The Frenchman Georges Cuvier (1769–1832),[126] who had amazing knowledge and talent as writer, made himself particularly appreciated in the palaeontology of vertebrates and in the classification of living animals. Cuvier knew Targioni-Tozzetti and cites *De solido* once. Cuvier claimed to have

---

[123]DSB XIV, 256 ff, Zittel, 85 ff.

[124]DSB VI, 577 ff. Bailey, *James Hutton—the founder of modern geology*.

[125]Bailey, 122 f.

[126]DSB III, 52 ff.

discovered a new method of research in palaeontology, but it has been often pointed out that his method ("when he does not speculate but works", as Huxley expressed it) was nothing else than that of Stensen.[127]

Only the next generations would up to a point draw attention to the importance of Stensen for geology. The erudite and witty Italian G. Bocchi (1772–1826), with his epoch-making work *Conchiologia Fossile Subappennina*, demonstrated the engagement of Stensen in palaeontology.[128] The origin of the mollusc shell from a substance that the body of the animal exudes is called a particular discovery of Stensen. Further, it is said that Stensen drew the attention to the different stages of petrifaction in the earth.

Alexander von Humboldt (1769–1859), the greatest disciple and later opponent of Werner, was well the last true universal erudite in the field of the physical sciences.[129] In his *Essai Géognostique* (Paris 1823) von Humboldt emphasizes that Stensen distinguished between original rocks not containing any animal remnants and those full of animal and plant remnants, and *De solido* is directly cited.[130] In his *Kosmos, Entwurf einer physischen Weltbeschreibung* (1844), Humboldt displays his high regard for Stensen by speaking of him as a "man of the all-embracing knowledge",[131] which was by all means of the greatest value if one thinks that this work was a milestone in the history of the physical geography, a mirror of the world which gave an overall picture of the nature scientific knowledge of all civilized races in the middle of the nineteenth century.

L. Elie de Beaumont (1798–1874),[132] the French precursor of volcanology, is often named as being the one who drew Stensen into the limelight. It seems that it was Humboldt who drew his attention to Stensen, whose assessment he extensively cites in his *Fragments géologiques* in the *Annales des sciences naturelles* XXV (Paris, 1832). Here Elie de Beaumont puts Stensen among the epoch-making authors who rush far ahead of their century and have a foresight of later great discoveries.[133] He then gives a comprehensive overall view of Stensen's ideas with long quotations, in which he sees, for instance, the first germ of the discoveries of Romé de l'Isle.

---

[127] Huxley, 453.
[128] DSB II, 480 ff, see Brocchi, 60 f.
[129] DSB VI, 549 f, Baumgärtel, esp. 20.
[130] Humboldt, *Essai géognostique*, 38 f.
[131] Humboldt, *Kosmos*, II, 267.
[132] DSB IV, 347 ff.
[133] Beaumont, 337.

After the publication of the *Principles of geology* in three volumes in London (1830–1833), Charles Lyell (1797–1875)[134] took a front position among the geologists of his time and his work presented the exemplary success of 12 editions during his lifetime. In the short historical introduction Lyell counts Stensen among those nature researchers who recognized the nature of the fossils and says that the significant work of that time of Stensen should be published and herewith the priority of the Italian school be recognized.[135]

The early-deceased Friedrich Hoffmann (1797–1836)[136] calls Stensen the "creator of modern geology" in his lectures in Berlin, which were published after his death by H. von Dechen as *Geschichte der Geognosie*. Hoffmann thus explains in his skilful presentation that Stensen, first among all, made the exact composition of the earth's crust in his surroundings the object of his study and from that he deduced important general claims for the way it was formed. "He exposed his researches in a book remarkable for all time but rare now."[137] Hoffmann exhaustively divides Stensen's discoveries and concludes with the confession: "I have been more comprehensive in the explanation of these views than in that of others because the fact is that, of the successors of Stensen, no mountain researcher, until Werner and his contemporaries, grasped the science from such a fruitful point of view. No one has initiated so many propositions in that science, which were confirmed later. Most of his successors have remained far behind that creative genius."[138]

In summary, the merit of having appreciated for the first time the discoveries of Stensen in crystallography—which had now become an independent science—is due to Franz von Kobell's *Geschichte der Mineralogie 1650–1860* (Munich, 1864). Kobell shows how Stensen studies the laws of growth, draws the attention on the striation of the surfaces of the pyramids, knows the constancy of the angle, the parallelism of the surfaces, and lets the crystal be formed from the liquid. "For crystallography the dissertation of Stensen is more important than the observations of his predecessors going in that direction. It gives the reason why surfaces otherwise of the same kind can appear so differently shaped, and shows the regularity in the increase of a crystal through the invariability of the angle by which it abides. The

---

[134]DSB VIII, 563 ff.

[135]Lyell I, 25. Like many contemporaries, Lyell meant that Stensen was influenced by a certain fear of a conflict with the Church by publishing the scientific results.

[136]Hoffmann, 40.

[137]Hoffmann, 41.

[138]Hoffmann, 45.

importance of the stripes is rightly recognized, at least in the mountain crystal", he says.[139]

1881 was the year in which T.H. Huxley delivered a famous lecture on the "Origin and progress of palaeontology". Huxley asserts that "the principles of research which Stensen exposed and illustrated since then have been, consciously or unconsciously, the guiding rules for palaeontologists".[140] In the same year, an international congress of geology took place in Bologna. Homage was paid to Stensen's memory by his colleague specialists, who carried out a pilgrimage to the site of his tomb in the Basilica of San Lorenzo in Florence.[141] On its cloister wall they placed a memorial plaque with the following inscription below a bronze relief:

"Traveller, you see here the bust of Niels Stensen which was placed on October 4, 1881 by more than 1000 men of science to his memory. Geologists of the world, after their congress in Bologna under the direction of Cavaliere Giovanni Capellini, undertook a pilgrimage to his grave and, in the presence of representatives of the city and erudite professors of the university, have paid tribute to the memory of this man who is extremely famous among geologists and anatomists."

1.6.5   Annexe to Chapter 1.6. Troels Kardel[142]: Stensen's Geology after Scherz

Gustav Scherz edited a unique collection of essays by multiple authors on "Steno as Geologist"[143] that was published soon after his death. It was a pathbreaking work in Stensen research, along with his "Steno – Geological Papers", the richly annotated and illustrated edition with English translations published two years earlier. Since then, scholarly works have added new aspects on Stensen's geology.

In an influential essay, Stephen Jay Gould in 1981 reassessed the significance of the *Prodromus De Solido intra Solidum* … : "The *Prodromus* is as its title states, about a solid in a solid[144] and their proper classification by mode

---

[139]Kobell, 19.

[140]Huxley, 453.

[141]Stensen, *Prodromus*, 185 ff, Congrès géologique international. Compte rendu de la 2me session, 249 ff.

[142]The author kindly acknowledges suggestions from Jens Morten Hansen, Alan M. Cutler, Gary D. Rosenberg and Arne F. Petersen.

[143]Scherz, *Dissertations on Steno as Geologist*, 1971.

[144]*Solid* is corrected from *solids*, being singular case in 1669 ed. and MS.

of origin. It is founded upon two great taxonomic insights: first, the basic recognition of solids within solids as a coherent category for study and, second, the establishment of subdivisions to arrange solid(s) within solid(s) according to the causes that fashioned them. Steno uses two criteria for his subdivisions. First, in what might be called the *principle of moulding*, Stensen argues that when one solid lies within another, we can tell which hardened first by noting the impress of one object upon the other. Thus, fossil shells were solid before the strata that entomb them because shells press their form into surrounding sediments just as we make footprints in wet sand ... The principle of moulding allows us to establish the temporal order of formation for two objects in contact. This criterion of history eventually forced a transposition in thought."[145]

In addition to the principle of moulding, Gould recognized another of Stensen's insights: "Past processes cannot be observed in principle, only their results remain. If we wish to infer the processes that formed any geological object, we must find clues in the object itself. The surest clue is *sufficient similarity*—part by internal part—with modern objects formed by processes we can observe directly. Similarity can be misleading but our confidence in common origin mounts as we catalog more and more detailed similarities involving internal structure and chemical compositions as well as external form."

Gould concludes: "With these two principles – moulding and sufficient similarity – Steno established both prerequisites for geological or any historical reconstruction: he could determine how and where objects formed, and he could order events in time." This principle shows how to determine the relative ages of two adjacent structures—either geological or of any solid structure examined. As indicated by Gould there is even a clue to this principle hidden in the book title, *On a Solid Within a Solid.*

More recently, the Danish geologist Jens Morten Hansen reemphasized that certain knowledge of past processes is based on solid evidence classified according to what he calls Stensen's chronology and recognition criteria. They are basically the same as the principles derived from Stensen by Gould. Traces impressed or contained in a solid material can convey information from the unobservable past. Solid material was the category studied by Stensen as indicated by Gould and implied by the book title, *De solido intra solidum* etc. Hansen adds a *preservation criterion* of solids as revealed from geological structures such as unconformities and, most evident perhaps, when crucial

---

[145]Gould, *The Titular Bishop of Titiopolis*. Gould's essay was also published in a book-collection, *Hen's teeth and horse's toes*, marking the centenary of the death of Charles Darwin.

items are missing links to look for among recordable structures.[146] These criteria were prerequisites for Stensen's proposals on the dynamic development of the earth's crust illustrated stepwise from geologic cross-sections of Tuscany (Fig. 1.6.2). Even so, Stensen's principles are basic for determining the earth's chronology and for the study of the evolution of life, which was the context of Gould's assessment of Stensen's geology.

Furthermore, Hansen and Kardel both emphasize that by expressing the biological origin of fossils as conjectures and by concluding that there seems to be no objections to this opinion being *verosimiliter*–Popper: *verisimilitude*; or Stensen: *a vero non multum recedere,* not far from the truth[147]—Stensen practices a scientific method resembling the hypothetical-deductive method advocated by Karl R. Popper.[148] In Stensen's answer to the third objection on muscle concerning the oblique angles, Stensen proposes a model, *mensura,* of his conception of its structure.[149] In a similar vein Popper speaks of models as means of realizing ideal measures, for constructing instruments of measurement, based on implicit definitions: "Models are attempts at realizing the ideal concepts."[150]

Among other recent commentators, François Ellenberger, in his *History of Geology*, vol. 1, devoted the bulk of his chapter on seventeenth century geology to a detailed discussion of Stensen's *Prodromus*, calling it a "decisive breakthrough".[151]

In 2003 the American geologist and writer Alan Cutler published a scientific biography of Stensen, *The seashell on the mountaintop—a story of science, sainthood, and the humble genius who discovered a new history of the earth*, focusing on Stensen's geological accomplishments. Cutler's book has been translated into eight languages and is probably the book on Stensen's research that has reached the widest audience. The author described the conflicting reactions over decades on Stensen's two publications in earth science.[152]

---

[146]Hansen, *Stregen i sandet, bølgen på vandet.* See also Hansen, *Steno's modern, but forgotten theory of science.* Kardel, *Steno—life, science, philosophy*, 73–74 and 96–97.

[147]OPH II, 139 <731> .

[148]K. R. Popper, *Conjectures and Refutations*, 1989, 221–222.

[149]"However, I do not wish to impose on the reader the impression that I have examined all muscles in all animals and that I believe for certain that the position of the surfaces in relation to each other is everywhere such as I have described. I claim only with certitude to have found this position of the surfaces in many cases. The demonstration of this simple and regular structure of the flesh is based on these observations. It is thus not without reason that I propose it as a model for all others" <656> .

[150]"Modelle sind Versuche, die idealen Begriffe zu realisieren", K. R. Popper, *Die beiden Grundprobleme der Erkenntnistheorie*, 1979, 181. Citation and translation by courtesy of Arne Friemuth Petersen.

[151]F. Ellenberger, *History of Geology*, vol. 1: "From ancient times to the first half of the XVII century".

[152]Alan Cutler, *The Seashell on the Mountaintop*, Chap. 12, "Shell Game", 123ff., and Chap. 16,. "World Makers", p. 174ff.

Alan Cutler now comments on a recent article[153] by the late Martin Brasier who discussed Stensen vis-a-vis Martin Lister, who was a vocal critic of the biological origin of fossil shells. Brasier points out that in a 1673 publication.[154] Lister argues in favour of the biological origin of some echinoderm fossils based on taphonomic criteria. Brasier cites this as "the earliest known example of taphonomic reasoning in a scientific paper." In fact, Steno had previously published taphonomic observations in both *Canis* (1667) and *De Solido* (1669).[155] In Conjectures 1 and 2 in *Canis*, Stensen addresses the question of whether shells and tongue stones are preserved animal remains, or if they grew in situ due to plastic forces in the earth. He uses the preservational qualities of shells and their lack of distortion in hard versus soft ground matrix to argue against their in situ origin. In Conjecture 6, he briefly discusses fragmentation, burial, and diagenesis of fossil remains. In *De Solido*, Stensen adds to these ideas, describing different modes of preservation of shells (original material, molds and casts, permineralization) <802–804> . For one specimen he uses his observations to deduce its taphonomic history,"it is possible to conclude with certainty that the shell had been left upon the land by the sea, covered up again by a new deposit and abandoned by the sea". Though Stensen's taphonomic ideas remained undeveloped, they were clearly an important element of his reasoning.

In 2004 Toshihiro Yamada defended a thesis on Stensen's geology and published a translation in Japanese of *De solido*. In 2017 Yamada continued with The Transformation of Geocosmos: Perception of Earth from Descartes to Leibniz with chapters on Kircher, Varenius, Hooke, Spinoza and Steno.[156]

In two publications by the Geological Society of America a number of essays have brought Stensen's geology into new perspectives.

1. In 2006 Gian Battista Vai and W. Glen E. Caldwell edited *The origin of geology in Italy*, with 13 contributions that follow Italian geology from the time of Leonardo and the coining of the term geology in Bologna in 1603 by Ulisse Aldrovandi up to the nineteenth century.[157] Stensen's position in Italian geology is emphasized by several authors.

---

[153]Martin Brasier, *Deep questions about the nature of early-life signals: a commentary on Lister (1673) 'A description of certain stones figured like plants.'* Philos Trans. A Math Phys Eng Sci. 2015.

[154]Alan Cutler, personal communication. Martin Lister, *A description of certain stones figured like plants, and by some observing men esteemed to be plants petrified*. Phil. Trans. 1673; 8, 6181–619.

[155]Cf. Hooke's contribution, <160, note 121> .

[156]Toshihiro Yamada, [*The emergence and development of the theories of the earth in seventeenth century Western Europe: a special reference to Nicolaus Steno's works*] (in Japanese), University of Tokyo. *De solido* in Japanese, ISBN 4-486-01,668-8. The Transformation of Geocosmos ... (in Japanese), Tokyo, Keiso 2017, ISBN 978-4326148295.

[157]Vai and Caldwell (ed.), *The origin of geology in Italy*, 45, Table 1.

2. Gary D. Rosenberg edited *The revolution in geology from the Renaissance to the Enlightenment* featuring essays dealing with the geology of Stensen [box].

Gary D. Rosenberg, ed., *The Revolution in Geology from the Renaissance to the Enlightenment.* Geological Society of America. Memoirs, 2009, vol. 203.

Gary D. Rosenberg. The measure of man and landscape in the Renaissance and Scientific Revolution, p. 13–40.

Elsebeth Thomsen. Niels Stensen in the world of collections and museums, *p. 75–91,*

Kuang-Tai Hsu, The Path to Steno's synthesis on the animal origin of glossopetrae. *p. 93–106*

Toshihiro Yamada, Hooke-Steno relations reconsidered: Reassessing the roles of Ole Borch and Robert Boyle, *p. 107–126.*

Troels Kardel, Prompters of Steno's geological principles: Generation of stones in living beings, glossopetrae and molding, *p. 127–134.*

August Ziggelaar, The age of Earth in Niels Stensen's geology, *p. 135–142.*

Alan H Cutler. Nicolaus Steno and the problem of deep time, *p. 143–148.*

Sebastian Olden-Jørgensen, Nicholas Steno and Rene Descartes: A Cartesian perspective on Steno's scientific development, *p. 149–157.*

Jens Morten Hansen, On the origin of natural history: Steno's modern, but forgotten philosophy of science, *p. 159–178.*

Frank Sobiech, Nicholas Steno's way from experience to faith: Geological evolution and the original sin of mankind, *p. 179–186.*

Gian Battista Vai, The Scientific Revolution and Nicholas Steno's twofold conversion, *p. 187–208.*

Rosenberg emphasizes that "conceptualization is a prerequisite for science, and that anatomy and geology are no exceptions: Art history records that the Western concept of landscape preceded the science of landscapes, and that the study of geometry and the structure of the human anatomy were essentials in renaissance painting as precursors to research", a position Stensen dwelt upon: "There are almost as many structures as there are pairs of muscles. I am not surprised that they have not been observed by those who do not look beneath the muscle's surface. But it surprises me that [artists] who have drawn the muscles often have been more exact than [anatomists] who described them in words, and that the industry with which painters have approached Nature's skill had not the power to urge them to admire such a work of art, which is the first step to investigation." Rosenberg does not find it surprising that Stensen, the anatomist, was to publish the founding principles of stratigraphy in geology which are statements about the structure of landscape changing in time like the body changes in motion and time.

Kardel emphasized the similarity of Stensen's method used in biological and geological research.[158]

Ziggelaar summarizes on an often-raised question, "Stensen shared the belief of his days that the Deluge in Holy Scripture had to be understood literally. Not only did it fit well in his system of geology, but he also found it confirmed in his research. Leibniz, his contemporary, was of the same opinion. The teaching of the Church was no constraint on ideas of the length of time involved in the history of the earth ... Only after 1840 did diluvianism die out with the acceptance of Hutton's plutonic theories and independent evidence from Agassiz's discovery of glaciations."[159]

Steno's importance is also highlighted in a paper on the role of the Mediterranean region in the development of sedimentary geology by Alfred G. Fischer and Robert E. Garrison.[160]

[158]Kardel, On deducing the unobservable, 2015.

[159]Rosenberg, G.D., *The Measure of Man and Landscape in the Renaissance and Scientific Revolution*, ibid. 13 ff. Ziggelaar, A.Z., *The age of Earth in Stensen's Geology*. Ibid. 135 ff.

[160]Fischer and Garrison, *Mediterranean region in the development of sedimentary geology*.

1.7 The Great Travel 1668–1670

The great travel of Stensen through Europe from November 1668 to July 1670 is enveloped in enigmatic obscurity.[1] Judged from the Foreword of the *Prodromus*, Stensen looks forward to travelling back to Denmark soon after finishing the manuscript (the first censorship stems from August 30, 1668).[2] He remained in Florence until November, and thereafter he travelled south. Not until in January 1669 did he begin slowly to move north. He took his time in Northern Italy and Central Europe to arrive in Holland at the end of the year. From there, this journey took him back to Florence in only about 1 month. He saw his home again not until 1672. The traveller seems to have carried out, at the service and at the expenses of the Grand Duke, a geological journey in an active attempt to expand and deepen the findings of the *Prodromus*. The scarce documents from those years fail to explain that long journey and deny that Stensen, actually or erroneously, received a call from his fatherland and that, thereafter, he wanted to hurry home after a short visit in southern Italy. In the south contradictory information met him, but nevertheless he decided on a greater geological journey to Northern Italy and Austria serving his research, always awaiting a renewed call and keeping open his way towards home. The letter to Viviani of April 20, 1669 throws some light on his situation: "I do not know whether I told you that somebody wrote to me in the last letter from home that I should receive new letters from His Majesty. In another, that the one who promised me a letter from the king

---

[1] *Reisen*, 71ff.
[2] Justel writes likewise in September, 1668: "M. Sténon a écrit qu'il quitteroit bien tost l'Italie", Oldenburg V, 37.

© Springer-Verlag GmbH Germany, part of Springer Nature 2018
T. Kardel and P. Maquet, *Nicolaus Steno*,
https://doi.org/10.1007/978-3-662-55047-2_7

does not understand the style of the court, etc. I do not know what God has in view with all that…".[3] The Catholic faith of Stensen was indeed the great hindrance and free religious practice in Denmark was a necessary condition for his return.[4]

The scientific interests of the Danish King Fredrik III had led him to the foundation of the great Royal Library. In Peder Schumacher, the later chancellor Griffenfeld, the king had a free-minded advisor, who was certainly friendly inclined to a return of his famous subject. However, with the new absolutist constitution, an intolerant wind blew. When Stensen's protector, Hans Svane, died in 1668, Professor Hans Wandal was appointed bishop of Zealand. Wandal had expressively pointed out in a book that it was the duty of a king to protect in his kingdom the purely orthodox Lutheran religion, to forbid all sects, to banish their adepts and to refuse free practice of religion.[5]

In that autumn of 1668 we have no trace of Stensen before Viviani, on October 27, writes to the astronomer Gian Domenico Cassini in Bologna: "To satisfy the thirst of knowledge of His Serene Highness my Lord, observations of Jupiter were carried out on the 23rd of this month in the night by the highly honoured Sir Cavalieri Molara, by Mr. Auzout and Mr. Stensen and by myself with the 10 and 20 yard long telescope of Campani and the said Mr. Auzout to check the ephemerides of Your Highness."[6] On this occasion we also see the Frenchman Adrien Auzout, whom Stensen knew from Paris, in Florence again. Auzout had left France in 1668 to settle in Italy.[7]

1.7.1   Through Italy

In the middle of November we meet Stensen in Rome, which was then still under the sign of the short pontificate of Clement IX. This highly educated Pope proved to be as favourable to artists and the erudite as he was an

---

[3]EP I, 207.

[4]Cf. Viviani's letter, December 13, 1667 (EP I, 27), and Justel's dated June 13, 1668 (Old-enburg VI, 441). Blondel correspondingly knows that "il l'assura cependant que si on lui accordoit la liberté de professer en son pays la religion catholique, il se rendroit encassement à ses ordres" (p. 1580).

[5]Helweg I, 458.

[6]Scherz, *Viviani*, 204 f.

[7]EP I, 216. In September he passed Bologna (Adelmann I, 346). By the end of October he was in Florence and later at the recommendation of Cardinal Leopold he was to travel to Rome where Ricci mentions him in a letter in February 1669. Maybe he was in Rome at the same time as Stensen.

amicable sovereign for the Roman hospitals. On November 22 he received the Swedish Queen Christina back in Rome after 2 years.[8]

Stensen did not see that. On November 17, Michel Angelo Ricci writes to Cardinal Leopold: "Yesterday I enjoyed seeing Mr. Niels Stensen whom I had not greeted since he had become Catholic. After his return from Naples I hope to be able to keep him in Rome many days. His modesty pleases me, and his honesty and his mind which is clear and rich in natural science and other nice knowledge."[9] Ricci then writes of the bad roads to Naples and of the dangerous influence of the air over the swamps. On the whole, this journey seems to have been full of risks. When Ole Borch had followed the same track some years earlier, people gathered in groups of 32 on horseback for reasons of security (Fig. 1.7.1).[10]

The main goal of the journey to Naples was certainly the investigation of volcanic phenomena in the region of the famous Vesuvius. Again in Rome, Stensen writes to Cardinal Leopold on December 22: "I will see how much can be taken from the Tuscan observations to clear up the said search."[11] He otherwise mentions the stay in Naples only in a short notice 15 years later for his "conscience search"—he recalls then having dismissed a beggar in Naples.[12] It must have been a fairly rough scene that had imprinted itself so deeply in his memory.

It is very likely that Stensen knew the Neapolitan erudite physician Lucantonio Porzio (1639–1723).[13] Porzio, later professor of anatomy at the Sapienza in Rome was then physician in Naples and the central figure of an academy for nature researchers and mathematicians, the Accademia degli Investiganti, a principal stronghold of the Italian Cartesianism,[14] to which Tommasio Cornelio, a friend of Ricci, like Porzio himself belonged. In the mind of the Accademia del Cimento, they particularly undertook experiments on the nature of liquids. In 1667 Porzio had just published on this subject. His method was based on practical clinical observations and appeared when he went to Austria during the Turkish war of 1683 to exert his art in the Austrian army. Thereafter he published a book in 1685 in Vienna, on the

---

[8]Pastor 14[1], 536 ff.

[9]Fabroni, *Lettere* 2, 163.

[10]Kock, *Oluf Borch*, 76.

[11]EP I, 206.

[12]OTH II, 437.

[13]In a sample of the Prodromus that was in the stock of Munksgaard Antiquarian bookstore in Copenhagen in 1940 was written: *Exlibris Licii Antonii Portii ex gratia auctoris, quem Romae cognovi*. On Porzio, see Mosca, *Vita di Lucantonio Porzio*.

[14]Besaucèle, 1 ff.

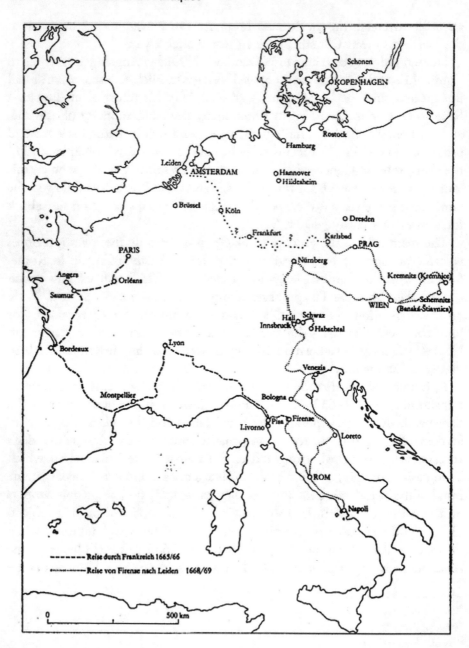

Fig. 1.7.1 Through central Europe 1669, Scherz I, 246

care of the injured and on the hygiene in war—*De militis in castris sanitate tuenda*—which was often translated. In 1736, after his death, his *Opera omnia medica, philosophica et mathematica in unum collecta* appeared in two volumes.

Stensen was in Rome again before Christmas. On December 22, from the eternal city, he thanks Cardinal Leopold for his recommendation to Felice Rospigliosi (†1688), who was then the archpriest of S. Maria Maggiore and was made cardinal some years later. He secured Stensen the permission to read "forbidden books of different kinds". Stensen also thanks Cardinal Leopold most warmly for the occasion of gaining the favour of his king by sending the *Saggi*. King Frederik III has much enjoyed the gift of the *Saggi delle esperienze* and, what is more, he has kept for himself the second copy destined to the professor and librarian Peder Lauridsen Scavenius, as the latter has reported.[15] There is apparently nothing new in the above-mentioned letter from Denmark concerning his summoning back home.

At the end of 1668, Viviani, then busy with the printing of the *Prodromus* on the instruction of the Grand Duke, writes to the secretary: "Mr. Niels writes to me that he thinks of travelling from Rome to Venice at the New Year. I would like that he finds the passport at his arrival, as well for himself as also for Mr. Pecorini, yes, and for everyone of his company, for all the princes and lords through whose states he passes, furthermore also every recommendation letter which he needs to visit the mines of Innsbruck and Hungary, while I hope in any case that he does not go further of his own choice."[16]

One thus sees that Stensen projects a longer journey with geological aims and this entirely at the service of his Grand Ducal patron. He travels with a Tuscan passport, all connections are prepared to receive him and the expenses taken care of by his princely friends, the Medicis. Francesco Pecorini,[17] a physician and disciple of Redi, accompanies him.

In Rome Stensen was also together with Rudolf Christian Bodenhausen (†1698), the renown German mathematician and later the teacher of the son of Cosimo III. Bodenhausen writes in a letter to Viviani: "You should know that Mr. Niels has left on January 19 in the direction of Loreto."[18] Our traveller probably chose the route over the Umbria highland and the Middle Apennines, even if there is a possibility that, combining his geological

---

[15]EP I, 206. Scavenius was the brother-in-law of Matthias and Holger Jakobsen and belonged to the Bartholin circle, DBL, 3rd ed. 12, 663 ff.

[16]EP I, 213. End of 1668 = December 1668 or March 1669?

[17]EP I, 227. He considered himself as Stensen's student, EP I, 224.

[18]EP I, 216, cf. Sten. Cath. 5. 1959, 13, *Positio*, 337.

interests with a pilgrimage, he went to the eastern coast over the Abruzzi. The route over the Middle Apennines offers many natural and artistic impressions. It leads over the highly situated Civita Castellana with its beautiful cathedral, over Terni and Spoleto with rich historical curiosities, over Foligno and the heights of the Apennines to Tolentino and the magnificent Macerata with its university. For the first time he set foot in Loreto, famous for its basilica of artistic skill, the greatest treasure of which is the holy house of Nazareth.

Stensen then continued his journey on the Adriatic coast and already on February 9 we meet him in Bologna, which then belonged to the Church state, a large town that was worth seeing. However, more than the kilometre-long Porticus, the towers and churches, the Archiginnasio and anatomy theatre, it was certainly his learned friends who attracted Stensen into the university town, above all Malpighi whom he had not seen since 1666.

Malpighi, who taught anatomy and practical medicine in Bologna, had carried out studies on the silk worm. For this, the Royal Society of London made him a member in those days and took care of the printing of his work. Réaumur said of that work: "It is nothing but a web of discoveries, a treatise in which one may obtain a greater knowledge of the admirable inner structure of insects than in all the works which have preceded it."[19] The priority of the discovery and of the drawings became the object of a fierce argument with Swammerdam, which erupted in 1671. Malpighi was certainly also interested in the geological investigations of Stensen but the main scientific problem that preoccupied them was certainly of an embryological kind.[20] Malpighi had already earlier, like Stensen in 1667, made many observations on the follicle, the oviduct and its liquids and he said: "I reported this and similar matters to the very learned Stensen, my guest, who communicated them to Reinier de Graaf when he returned to Belgium. This one even promised to the same Stensen that he would send to me before publication a comprehensive dissertation on the same subject which he was working on."[21] A diary entry of Malpighi reports on investigations of the two friends on February 9, 1669.[22] Together they investigated the muscles moving the eyelid of a chicken and Stensen made several remarks which Malpighi wrote down,

---

[19]From Cole, 183 f. On Malpighi and the Royal Society, see Adelmann, I, 669 ff.

[20]Later in the spring Malpighi was to receive a copy of the *Prodromus*. He was also interested in liquids in organic and inorganic bodies, see Belloni, 144 ff. He had not any real insight in Stensen's geology, see Adelmann, I, 502 ff.

[21]Malpighi, *Opera*, 47, EP I, 32.

[22]EP I, 32. Adelmann, I, 348 reads various sections differently and reaches a different interpretation.

among others the following: "He reported that the muscles have two kinds of fibres, the first of which form the bulk, and the others exist with and continue into the tendon of the surrounding part of the animal. —In the stomach there are glands, particularly in the birds. In eels the foetus is likely formed in the belly, i.e. in the viscera. To bleach the bones, one should wash them every day and expose them to the sun, after having removed the marrow.[23] —That the oviduct does not correspond to the ovary[24] and in the fishes there is a certain condition. He reported often to have seen a foetus in the uterus of a fish. The foetus was hanging from the uterus itself without chorion or amnion…"

Among the erudite friends whom Stensen visited in Bologna, was Carlo Fracassati whom Stensen knew from Pisa, who had accepted a teaching chair of surgery and anatomy in Bologna in 1668, as well as Silvestro Bonfigliolo (1637–1696), also a Bolognese and friend of the natural sciences in cordial association with Malpighi.[25] He wrote two years later to Carlo Dati and in his letter cites Magalotti, Redi and Stensen, *tutti Accademici Gelati*, members of the Bolognese academy in which Stensen thus must have been accepted then.[26]

Stensen only remained some months in Northern Italy. We do not know of his sayings and doings there.[27] On April 6 we find him in Murano, a small island outside Venice with 24 glassworks.

The lagoon city then counted about 300 000 inhabitants and was admired more than ever.[28] The so-called arsenal possessed the best shipyard of Europe in which 4000 workers toiled, as well as large ordnance depots. The many churches, 26 cloisters, 17 hospitals and the romantic Venice with its hundreds of bridges and thousands of gondolas then, like today, exerted a great attraction on the visitor.

---

[23]Adelmann has here: … *tendines musculorum qui ossibus applicantur inseri alte in ossis substantia opposita via.*

[24]Adelmann reads: *Quod tubae uteri respondeant ovario*, cf. his commentary I, 348 f.

[25]EP I, 213.

[26]*Reisen*, 77.

[27]On March 1, Grand Duke Ferdinand writes a recommendation letter for Stensen to his sister Anna in Innsbruck, cf. EP I, 17. Stensen delivered it to her. He can have received it himself in Florence or he could have received it by carrier. It seems that Stensen made an excursion in Frioli, north-east of Venice in early April, 1669 (FAS. Post 19 III 4. Letter from Druyvesteyn to Viviani, April 13, 1669).

[28]Holger Jakobsen in his travel book: *Urbs potentissima* and the verse of a poet that Rome was built by men, but Venice by gods, *Jakobsen*, 140.

In Murano Stensen would have seen Montanari's experiments on glass beads together with Count Carlo Rinaldini Geminiano Montanari. The glass beads were brittle like those imported from Holland.[29]

Montanari (1633–1687), professor of mathematics in Bologna since 1664, was an extraordinary, much travelled erudite, the inventor of a micrometer and a telescope as well as a correspondent of the Accademia del Cimento.[30] He carried out studies on glass and made experiments on blood transfusion. A letter of Montanari to Cardinal Leopold of September 1, 1671 testifies of his interest in Stensen's geological research. In this letter he reports from Mount Cimone in the Modena region: "A botanist also comes there at his own expenses and Mr. Stensen could have made observations there like anybody else since one sees the strata of those rocks sloping down from the summit to one side arranged one above the other, just as he taught it." Montanari could also prepare the friend for his journey to Austria. He had indeed been bestowed the hood of doctor of law in Salzburg by the Archbishop Guidobal von Thun and had explored the mines of Steiermark, Kärnten and Hungary in the company of Paolo del Buono.

Count Carlo Rinaldini (1615–1698), whom we find likewise at the side of Stensen, stemmed from a lineage of counts. Given to the study of theology, philosophy and mathematics, he had at first brandished the sword at the service of the Pope, but in 1649 Ferdinand II had called him to Pisa as first lecturer in philosophy and later appointed him teacher of Crown Prince Cosimo. As a member of the Cimento his activity was not outstanding. Stensen may have known him in Florence and met him again as a lecturer in philosophy at the Studio in Padua.[31]

In a letter from Venice on Holy Saturday, April 20, Stensen writes to Viviani not only that he wanted to continue his journey east but also reveals, as can already be noticed at the beginning, that he is still waiting for a letter from his king which would call him back home.[32] He speaks there of "somebody" who promised him that letter whereas "somebody else" obviously believes in a misunderstanding, provoked by the ignorance of the kind of writing at the court.

It is difficult to say who those correspondents in Copenhagen were. It is hardly question of just names, which were also known to Viviani, like Ole

---

[29]Fabroni, *Lettere*, I, 159. * One may wonder that Stensen took time at Murano while barely visited nearby Padua, and also that he never returned to France on his long travels.

[30]EP I, 35, DSB IX, 484 ff. * See also Fabroni I, 165.

[31]Middleton, 34 f.

[32]EP I, 207.

Borch or one of the Bartholins. One thinks rather of Stensen's brother-in-law Jakob Kitzerow, who was then an esteemed townsman and was in relation with Copenhagen's castle as goldsmith and purveyor to the court, or of the ageing registrar and wholesale merchant Jørgen Carstensen, Stensen's former guardian. Even as a Catholic Stensen remained in constant connection with the Carstensen family. As things were at home, it must have been painful and bitter for our traveller to be deprived of even a faint wink from the home-country, while Denmark did not seem to have any interest in him.

Stensen saw himself attended to with quite another love and care by his Florentine patron. When he left Venice at the end of April and travelled through Padua, Verona, Trident and Bolzano to the capital of Tirol, he found such a cordial welcome that, on May 12, moved by such kindness in Innsbruck, he wrote to the Grand Duke Ferdinand: "I can only say that the generosity of Your Highness shows itself that so more noble, whatever little the person deserves it, to whom it is exerted, and that I will implore the divine Majesty during all my life to pay back this with all desired luck as well for your Person as for the illustrious house of the Medicis."[33]

1.7.2   The Travelling Geologist in Central Europe

Innsbruck, coat of arms

Soon after his arrival in Innsbruck, Stensen had asked P. Athanasius Kircher for recommendations for Vienna and surroundings, which he had been promised at his departure from Rome. Through a friend, probably Riva,

---

[33]EP I, 208.

he had his *Prodromus* sent to P. Kircher and modestly added: "When you receive this book, I ask you to excuse its imperfection."[34] This letter is the last testimony that Stensen was also acquainted with the polyhistor, perhaps the greatest of his time. Probably he had visited him on all his journeys to Rome, thus in 1666, 1667 and 1668. In the year 1665 Kircher had published his *Mundus subterraneus*, the large-scale history of the earth, often quite imaginary and based on pure speculation, which is so wholly in contradiction with Stensen's script with proofs based on observations.[35] Now the founder of modern geology sent his *De solido* to the father with the request for patience for its shortages. Stensen has certainly received the asked-for recommendations and these undoubtedly helped him to establish useful connections in Germany and Austria. Kircher enjoyed high respect in Vienna and had many connections to the Imperial Leopoldian Academy for nature research, with which Stensen now came into contact.

It was the sister of Ferdinand II, Anna de' Medici (1616–1676) who received Stensen in Innsbruck, at the warmest recommendation from her brother. In 1646, the affable and deeply religious lady had married the Archduke Ferdinand Karl of Austria, the Governor of Tirol, who let her take care of many government matters. She became a widow in 1662 and in 1669 she no longer occupied any official position. Her letter of June 16 to her brother gives a hint as to Stensen's sayings and doings in Tirol: "He also gave me your letter of March 1 and I enjoy his visit and his knowledge which he delivered in a sound lecture after his visit of the pits of Hall and the saltworks of Schwaz. I also expect to hear on the emerald mines whereto he went to see what can be done. He has not yet sent to Your Grace, as he has promised me, the precise description of the dissection of an animal, which he undertook on a monstrous calf, as Your Grace hears from himself and temporarily gathers from the attached drawings. In the head, instead of brain, four litres of water were found wherefore it could not lift it."[36]

Hall is an old picturesque town situated at the river Inn 10 km East of Innsbruck. The Romans formerly got salt from here, which came dissolved from the 10 km distant Salzberg and its mines to the saltworks, was purified and went into the trade from Salzspeicher on the Inn river bank. Although hit by a recession in the seventeenth, Hall still had a great importance and silver, lead and tin were also mined. The town of Schwaz was situated still further to the east, in the lower Inn valley and was once famous for its silver mines.

---

[34]EP I, 208.

[35]On Athanasius Kircher, see <178>, EP I, 109.

[36]EP I, 17.

The Fugger house reminds us that the Augsburg patrician lineage gave work to 30 000 miners.[37]

After those journeys Stensen went back to Innsbruck where he delivered his extremely interesting lecture. We do not know its contents. In the middle of June he must have been on an emerald journey. It is very noticeable that otherwise no emerald mine is known in Europe in the seventeenth century. The only possible location seems to be the little known Habach valley in the region of Salzburg, some 60 km from Innsbruck.[38] Stensen had to achieve a likely mountain tour through the Ziller valley and the Salzach valley in the high Pinz district where the Habach valley, from the village of the same name, branches off towards the high Tauern.

Stensen did not leave any report on his experiences in those mountains. Only in his *Indice* is there an account of different minerals, crystals, etc., from Tirol.[39] On the contrary, there is much information pertaining to the said description of the abnormal calf, the dissection of which from the beginning may have satisfied him less than it excited the onlookers hungering for thrill. It was published by Th. Bartholin in the *Acta Hafniensia*.[40] The calf head was likely offered to him for dissection by the circle of the Archduchess. Stensen's genius brilliantly acquitted himself of this task and provided a drawing, which shows his high steadiness.[41]

Refuting the conception of his time that congenital deformities may be the consequence of some fantastic imagination of the pregnant mother, from the dissection of the monstrous calf head, Stensen states that the true cause is an illness of the foetus. He then describes the dissection of the calf head in which he finds two abnormalities: namely a large quantity of water and a tumour at the root of the nose blocking for the flow of liquor from the ventricles of the brain. As conclusions and basis for further investigation he establishes:

1. One must put into doubt all accounts that a part of the brain has been partially or totally dissolved into water.

---

[37] *Reisen*, 81.

[38] Scherz, *Niels Stensens Smaragdreise*, esp. 52.

[39] Cf. Stensen, *Indice*, Nr. 261/12–13, 284, 287, 298, 300 with notes.

[40] OPH II, 229 ff. Stensen's letter to Ferdinand II was written in Italian and translated into Latin for publication in *Acta Hafniensia*. The Italian version has not been found. * See Stensen, *Steno on Hydrocephalus*. Stensen's clear description antedates those of liquor cerebrospinalis flow through passages named after successive investigators, see P.J. Koehler, G.W. Bruyn and J.M.S. Pearce, *Neurological Eponyms*, Oxford Univ. Press, 2000, 24 ff.

[41] The drawing is made in red chalk. No other such drawing exists from Stensen's hand. It may well have been provided by a court artist since it is found attached to the letter by Anna to Ferdinand II. (HMH + TK).

2. Also such stories that part of the brain has come out through the nose.
3. The union of the lateral parts of the brain by way of the corpus callosum, the septum lucidum and the fornix is not absolutely necessary for sensations and the movements of living beings.
4. Also not the connection of the cavities through the substance of the brain since this animal has lived some time while these were open.
5. The brain substance could endure a great pressure on its lateral sides and on its base without complete loss of sensation and movement.
6. It is very likely that, with time, one could find, in the location of the tumour, a vessel which serves to the evacuation of the water.[42]

Stensen finally concludes this discourse, having shown "how the brain resists the greatest accidents although being the noblest and most delicate organ and how a disorder in the foetus itself can sometimes be the cause of these deformities which would be said to be produced by the imagination of the mother."[43]

Stensen left Tirol probably at the end of June and walked over the Bavarian mountains, perhaps via Munich or Augsburg, to Nuremberg where we meet him in the circle of erudite friends.

The free imperial city had suffered severely from the Thirty Years' War but was already, like before, one of the most beautiful and richest towns of the empire, as the place of residence of old patricians and tradesmen and as a city of art with venerable churches and buildings. The social circle of Stensen was here Johann Georg Volckamer the Elder (1616–1693) as well as the physicians and erudite of the Nuremberg Collegium Medicum and of the Leopoldian-Carolingian academy of nature researchers. Volckamer was himself a Nuremberg patrician. After a careful education he had travelled through half of Europe. The able physician, who later became imperial physician in ordinary, also dealt with physics, built telescopes and studied the declination of the compass needle. Stensen was received here with proper respect: "The marks of kindness that your Eminence and your friends have shown me in Nuremberg were so great and accompanied by such kindness that I cannot thank you enough therefore."[44] He sent particular greetings to the executive mayor Joachim Nützel (1629–1671), the patron of the painters academy. Volckamer also acknowledged Stensen charmingly for his visit:

---

[42]The interventricular foramina (or foramina of Monro) are channels that connect the paired lateral ventricles with the third ventricle in the midline of the brain described by Alexander Monro in 1783.
[43]OPH I, 237. On teratogenesis, see also OPH 2, 39.
[44]EP I, 210.

"I have often thought of you and I felt wonderfully moved by the great benevolence which you have displayed to me and my family... I still always enjoy the association of the blood circulation, the *ductus chyliferus* and lymphatics which you have shown us in alamb."[45]

From Nuremberg the journey probably went over Regensburg and at the end of July on the Danube to Vienna, where his presence is testified on August 3 for the first time. Immediately after his arrival he had to thank the Grand Duke in Florence for a quite unexpected gift of 400 florins, which was already awaiting him in Vienna. "Since I do not know how I should otherwise show my gratitude, I will endeavour to do everything I can somehow to accelerate the begun works and, if I do not succeed in bringing them to perfection, at least to do them one way or another so that it becomes clear to everybody with what generosity Your Eminence has contributed to their completion."[46]

The Vienna of Emperor Leopold in which Stensen made his entry was not much bigger than the Copenhagen of that time, but the Danube city was the biggest fortification of the German empire against the Turks. Powerful bulwarks and towers greeted travellers already from a distance. Stensen had recommendations to the artist from Nuremberg, Benedikt Winkler the younger, and to the physician Georg Fabricius who had been imperial notary since 1653 and resided at the court as agent.[47] He went on a journey to Hungary, more precisely to Schemnitz (Banská Štiavnica), already on August 14. Countess Katharina Zriny, born Frangipani who was married to the Panus of Croatia, Peter Zriny, was travelling with her 12-year-old daughter Aurora Veronika to her estate Murany in Upper Hungary. She had offered him a place in her coach. Stensen could hardly foresee in which tense political atmosphere his companion undertook that journey. Hungarian noblemen had then formed a big conspiracy to liberate their country from the bond with the Habsburg house. Leaders were, among others, the court judge Count Nadasdy and the Margrave Frangipani whose high-treason plans were already known by Emperor Leopold. Husband and brother of Countess Zriny would be imprisoned in the next year and executed 1 year later. The Countess herself died in 1673 in Graz, and her daughter later joined the order of the Ursulines.[48]

---

[45]EP I, 209.

[46]EP I, 210.

[47]EP I, 211.

[48]EP I, 211. On the conspiracy, see Redlich, esp. 196 ff. On the travel to Hungary, see Vida, *Niels Stensens Ungarnreise im Jahre 1669*. The Hungary of Stensen is in present days Slovakia.

The immediate aim of Stensen was the town Schemnitz in the south-western part of the Slovakian Erz mountains, which together with Kremnitz (Kremnica) had the largest mines in Hungary, rich in gold and silver. Deep shafts and tunnels led to the inside of the earth and many melting furnaces sent columns of fumes to the sky. Stensen stayed about 2 months in Hungary and must have spent much of the time at the foot of the high Tatra and the Beskiden mountains. Otherwise we know hardly anything of that time and it is only from the *Indice* that some names are known at all, telling just a little about his route and scientific aims. Here Schemnitz and Kremnitz are cited several times. Also a silver mine of a mountain covered with marcasite and with stone full of fluorite is mentioned (261/4–5). A spongy stone coated with chalcedony stems from the gold and silver pits of Kremnitz and its Emperor Leopold's shaft (261/1). From the vicinity of that main locality, Neusohl (Banská Bystrica) is cited, where one gets a stone which is transformed into melting fluorite which is then used in the furnaces to obtain silver (261/11). Herrengrund, north of Neusohl is also on the list, thus the mining town that had produced copper ore since the thirteenth century. Hungary is also generally described as the place of discovery of a series of minerals.[49] It remains undecided whether from the mention of Sieben-bürgen in connection with lead ore and from that of Poland in connection with rock salt, a visit can be concluded, for instance in Wieliczka near Krakow. The only report on that journey in Hungary is in a letter of October 27 to Malpighi, immediately after the return to Vienna.[50] Stensen writes: "My journey to visit the mines gave me much satisfaction, not so much because of new observations which were but few, than because I saw with my own eyes those things which one not easily understands by reading the specialists in metallography. I have, however, also seen something to confirm my conceptions concerning the change in the earth, that in the same places one finds strata of the greyish blue hard sand-stone inclined on the horizon, which cannot have been naturally created in that position. It also seems to me that there is no longer any doubt that the gold and silver veins and their ramifications are only a filling out of the space between sandstone and sandstone and of clefts in this sandstone." It appears that our traveller adds confirmation to the observations of the *Prodromus* and especially to that section in which he speaks of "the origin of the different coloured stones and of the spaces in which minerals originate." Here it is said in point 3: "Many things may be

---

[49]Stensen, *Indice*, Nr. 268, 269, 273, 274, 289, 298.
[50]EP I, 212.

detected in the examination of rocks that are attempted in vain in the examination of the minerals themselves; since it is more than probable that all the minerals filling either the clefts or wide spaces in the rocks obtained their material from vapour forced out from the rocks themselves, whether this occurred before the strata changed their position ... or after this event..."[51] The first remark belongs to the reflection in the *Prodromus* on the modification of earth strata, partly by violent rising of strata as a consequence of underground explosion, partly by collapse after removal of the supporting strata.[52] Malpighi sent Stensen a report, probably based on the description that Borelli provided Cardinal Leopold with, on the eruption of the Etna, the most violent that occurred to the biggest volcano of Europe. This had just happened at the beginning of the year 1669. Stensen also showed a lively interest in that event and asked for a closer report on a book *De fluido*, also a subject which the *Prodromus* thoroughly deals with.

No decisive information had yet arrived from Denmark and Stensen was close to giving up completely the idea of returning home. He writes to Malpighi on October 17: "I hope to complete my journey this autumn but I am now as far with my request as I was when I was with you. That means that little is missing before I give up my idea of returning back home."[53] Thereafter he offers Malpighi service for Holland and he asks to send further letters to the Netherlands via the Grand Ducal agent in Venice.

The journey from Vienna to Holland went through Bohemia. Stensen later speaks in a short remark of a visit to Prague[54] and his collection of minerals shows that he also there made observations. He had obviously been in the mines near Kuttenberg (Kutná Hora) and in Joachimstal (Jachymov).[55] He probably also visited the famous spa Karlsbad (Karlovy Vary) with its warm springs.[56] In the collection recorded in *Indice* there is also silver-, zinc- and iron-ore from Bohemia.[57]

---

[51] <792> OPH II, 203.

[52] <791> OPH II, 200.

[53] EP I, 212.

[54] OTH II, 437.

[55] Targioni-Tozzetti, *Catalogo*, III; 299, N 723 and II, 83 N 77, cf. notes.

[56] Stensen, *Indice*, Nr. 298.

[57] ibid. Nr. 268, 271, 272.

1.7.3   Defender of Catholicism in Holland in 1670

One can really only guess why Stensen then went to Holland.[58] Perhaps was it in the service of the Medici house. Grand Prince Cosimo had travelled in the Netherlands for the second time in the summer of 1669 and was interested in the rarities cabinet of the Swammerdams, father and son, which he had looked over together with Thévenot during his first visit. He had offered to the young Swammerdam alone 12 000 florins for the collection of natural objects, but on the condition that the erudite himself should bring it to Florence and take residence there.[59] Perhaps Stensen was to exert his influence in this direction.[60] Or perhaps he wanted to meet Thévenot who was just then staying in Holland. Perhaps he had an urging invitation from his many Dutch friends.

We do not know when Stensen arrived in Holland, but his presence is testified in Leiden at the latest on December 30,[61] by a visit to Johann Friedrich Gonovius (1611–1671), the important historian and philologist who was professor of history and Greek language in Leiden.[62] However, he seems to have stayed more in Amsterdam than in Leiden. It is likely that he lived at the Tuscan Resident Ferroni in Amsterdam. Francesco Ferroni was a Florentine tradesman and banker who, as agent of the Medicis since 1667, used his dwelling at the Keizersgracht and went back to Tuscany in 1672.[63] In any case, Stensen speaks of a man in whose house he lived and who could send his letters to Florence.[64]

It is very likely that Stensen then saw his scientific friends in Holland again, but no testimony of such meeting has been found. In 1667 Jakob Golius had died. Johannes van Horne died on January 5, 1670. Stensen certainly came in contact with François de le Boë Sylvius. Also he was to pass away by Stensen's next visit. He mentions the two Swammerdams and recounts that he spent some months in Amsterdam with Thévenot.[65] Reinier de Graaf reports that

---

[58]On the journey from Bohemia to Holland, see *Reisen*, 96 ff. Stensen counted on Holland as being his destination with a return from there to Florence, thus with no visit to Denmark, already in the summer of 1669, Jarro, 17.

[59]Hoogewerff, xlix f.

[60]Little is known about Thévenot's sojourn in Holland. He arrived before June 28, 1669 (Hoogewerff, 319) and was together with Stensen for some months in the spring of 1670 (EP I, 219).

[61]GP, 33.

[62]NNBW I, 989–992.

[63]Hoogewerff, xxv.

[64]OTH I, 195.

[65]EP I, 214, 219.

Stensen visited him in Delft in 1670 and talked with him about the reproductive organs,[66] a subject that was debated passionately. De Graaf was now a
practical physician who was in keen demand and, as a Catholic, he was twice
bound together with the convert.

Stensen himself tells us more thoroughly of the daily debates concerning
faith during his stay in Amsterdam. They would deeply intervene in his life.
The problems of pastoral activity that would progressively absorb his life met
him here with demanding power.

Six years had passed between his first and second visit in Holland. In these
6 years he had not only become a famous scientist whose *Discours* and
Prodromus had made his name renowned in the scientific world, but he had
also joined the Catholic Church out of deep conviction. For the first time he
would personally explain that step to his previous fellows-in-faith. They
immediately rushed to him with all possible reproaches about dogmatic errors
and moral scandal. One reproached the idolatry and audacious confidence in
good deeds of the Catholic Church. Church immunity covered the worst of
crimes. One spoke of the venality of the Italian public servants and knew of
the moral dangers of the public houses to which their sons were exposed in
Italy. Also the rumours that were spread about the conclave for the election of
Clement X gave rise to mistrust and suspicion.

Stensen was an affable and penetrating apologist. He particularly saw two
reasons for their aversion against the Catholic Church: "One is common to all
non-Catholics and goes back to the evil tricks of those who first left the
Church and taught their followers to attribute to the faith itself all the human
errors stemming from evil and ignorance."[67] Another reason was the local
historical circumstances, namely the hate against everything Catholic from
the time of the Dutch liberation war. Stensen remembered that he himself
had overcome this last difficulty only with pains and with the help of God's
grace.

It was not difficult for him to dismiss the different reproaches. He soon
pointed out that one should not attribute to the doctrine human nuisances,
which it very deeply abhors and emphasized that the Church immunity
granted to the fugitives in the holy places is a mildness of the highest wisdom
and equity, regardless of human weakness and capacity of error. He desired
also for the Church that indulgence which the Lord grants to the weeds
among the wise ones, and he refuted, based on his own experience, the

---

[66]EP I, 9.
[67]OTH I, 188.

rumours about the conclave. About the question of immorality he explained: "For the rest, I have observed the same vices everywhere in the large towns but not the same virtues. They are not, as it is held against me, vices of the Italian towns since one meets the same in the Dutch towns. It is true that the civil laws are stronger in one place than in another, which does not hinder that the same opportunity of committing delinquencies is found everywhere as well as the same frequency of a certain scare of them. I have known many of my friends whom the moral licentiousness of Amsterdam has corrupted although it is forbidden by the laws." Stensen comes back to his experience again and again: "But if one also comes everywhere across equality of vice, the confessors, the martyrs, the maids and spinsters, the poor, the missionaries and innumerable other examples of true Christian life of our religion testify that there is a very great difference in virtue among people of the same rank and sex, examples of which they could point out no one similar among their clergymen although it is required from Christians, at least from the teachers of the Christians, to let shine their light."[68]

Concerning the reproach against the doctrine of the Church that it is a human invention, Stensen points out that from the Catholic side there is not one article of faith of which one could prove that it was introduced by a man after the apostles, whereas contrariwise, as far as the Protestant doctrines of faith are concerned, that could be said of all, and that, for instance, before Calvin there had been no Church, yes, not even one man from whom he could have taken over all of his controversy doctrines.

When the discussion arrived at this point, Stensen's friends proposed to discuss the questions with one of their friends, Sylvius, to which he agreed very willingly. This Johannes Sylvius (†1699) had been preacher at the German Reformed Church in Amsterdam since 1667 and had otherwise become known for some controversy scripts against two Catholic priests.[69] Stensen saw the Calvinist preacher for the first time at a lunch in a garden in the circle of friends. Sylvius seemed surprised that a Dane could be Catholic and this ensued in a discussion about faith during which Stensen explained that all the problems could be reduced to one, namely: "What interpretation of a passage of the Holy Scripture among the different interpretations of the different associations which call themselves Christian, has to count as the most certain?"[70] And he quoted the same example that he had discussed with Father Jean Baptiste de la Barre, S.J. in Paris in the time, on the stimulation

---

[68]OTH I, 190.
[69]OTH I, 21 f.
[70]OTH I, 191.

of Mrs. Hedwig M.E. von Rantzau, namely how the word of Christ: "This is my body" was to be understood in the sense of the Catholics who believe in the real presence, while the appearance of the bread remains, whereas the Lutherans teach the presence of the body and bread, and the Calvinists assume the real absence of the body but the real presence of the bread.

Sylvius defended only his Calvinist conception and it came to a long dispute. The debate must have been unusually lively, until the night broke in and the company had to move off. Some days later Stensen visited the preacher at his home where he also met some other acquaintances. One finally agreed on the question: What should a doubter of the true religion do? Whether the single comparison of passages of the Holy Scripture is enough or the subjective opinion, both of which Stensen denied, whereas he answered in the affirmative the question of whether one could arrive at a greater certitude from the examination of the practice and of history. Thus, it should be clear for everyone what one must believe if one can lead back his practice only over one or the other century whereas, for the other, one can indicate no creator of his practice since the days of the apostles. The question led to a true argument during which particularly an old man named Oseas, an ex-Jesuit, passionately but so irrelevantly distinguished himself so that Stensen finally refused to continue the discussion with him. It became midday and one moved off but Stensen still pointed out the word of Augustine[71]: "Indeed, I would not believe the Gospels myself if the authority of the Catholic Church did not move me to do so," and he advised his host to distinguish the word of God itself from the word of God in the mouth of various interpreting people. At the same moment as they were still disputing at the door of the house and Sylvius was testifying his interest in the method of the brothers Walenburch, two persons passed in the street, of whom the whole of then Amsterdam spoke. They were Anna Maria von Schürmann (1607–1678),[72] a highly gifted, universally educated Colognese and Calvinist who had joined the Labadists and written an apology for them, and the founder of the sect himself, Jean de Labadie (1610–1674), a previous Jesuit from Southern France who had left the order in 1639, indulged in mystical-quietistic doctrine, was not morally unobjectionable, passed to the Lutheran doctrine in 1650 but was deprived of his charge as preacher by the synod of Utrecht since 1650. He had then founded an evangelical community in Amsterdam in 1670. Labadie partly followed the reformed doctrine, partly the Catholic

---

[71]Augustine against the letter of Mani, 5, 6.
[72]OTH I, 196.

cloister ideal and above all he followed his own inner inspiration, which he set above the Bible and Church.[73]

Stensen was deeply shaken by thinking of the spiritual situation of these four persons: "That one hour showed me four persons who deserve true compassion. Two of them, after having become old in the true faith and in the devotion to a holy order had ultimately fallen off from the Church and one of them had found disciples at that time already, and the other was still looking for some. The last two, grownup in the Calvinist dogma, already then had obtained the grace of God to recognize the uncertainty of their own dogma, even if one remained with his first conception, not because he considered it as certain but because he considered everything else as uncertain as well. The other was a virtuous maid educated by the study of languages and sciences, who had left the liberty of Calvin and embraced that discipline which the fallen-off old man had kept, even if not quite within the spirit of the Church dogma."[74]

In the first week of the year 1670, Spinoza's *Tractatus theologico-politicus* had appeared in Amsterdam anonymously. It aroused the public opinion extremely by its denial of biblical inspiration and of the revealed religion. It triggered off a storm of aversion, which caused the court of Holland to demand punishment of its author, of its publisher and of its propagators. De Witt was still able to protect Spinoza temporarily, whereas on June 30, 1670 the Church council of Amsterdam announced its *gravamina*.[75]

There is no doubt that Stensen at that time became attentive to the book, also that he had certainly not visited his previous friend who, until the beginning of 1670, lived in Voorburg and then in Den Haag. However, some months later, in the tranquillity of his work in Florence, he wrote a letter to the "reformer of the new philosophy",[76] a vain appeal to the proud intellectual Spinoza for whom Christianity was a web of errors that may be useful to the people whereas the wise "keeps to the clear and distinct concepts, the religion which is nothing else than mathematical method adapted to metaphysics".[77]

His other friend Swammerdam was also a child of sorrow. The great entomologist and one of the greatest zoologists of all time was of a deeply religious nature but so emotional that he would have needed clear guidance.

---

[73]DTHC 8[II], 2383 f. SHKG 4, 189.

[74]OTH I, 196. SHKG 4, 186 ff.

[75]Meinsma, 401 f.

[76]OTH I, 95–102.

[77]The quotation is not verified.

However, with a certain stubbornness he resisted all attempts of Stensen to influence him. He was on the way to the visionary Antoinette Bourignon of whom he would blindly become a companion later. When he left Amsterdam, Stensen, deeply stirred by all that spiritual need, promised his friends, who wanted to continue the discussions on faith, a written continuation and also put on paper a series of theses. He would have taken to the pen anyway. The discussions, namely, had not satisfied him. "I had observed that the truth is darkened by dispute rather than clarified because the theses are often changed and intermixed with digressions and this not only interrupts the series of conclusions but rather breaks them off in the middle and that hardly ever clarity is attained about one discussed thesis alone."[78] Stensen, therefore, laid his own method in order, which consisted of determining first the object of controversy clearly with regard to the different wordings applied, then stating the first argument likewise according to its wording and without taking any other argument before the former had been agreed upon.

Days and weeks thus passed quickly in Holland and it was impossible for Stensen to accept all invitations. On April 20, therefore, from Amsterdam, he asked Greve in Utrecht to excuse him.[79] Johann Georg Greve (Graevius, 1632–1703) was one of the most important classic scholars of the seventeenth century. Born in Naumburg in Saxony, he studied in the Netherlands where he finally became professor in Deventer and, since 1661, he taught politics, history and eloquence in Utrecht. Greve's enormous correspondence shows him to be a universally interested and kind man who wanted to associate Stensen, and perhaps also Swammerdam, to Utrecht, which was far behind Leiden. He offered Stensen his house to live in. Stensen declined because of the care of a severely ill friend. Stensen writes: "Nothing otherwise would be more desirable for us than such an erudite college where the education struggles for the palm with the kindness" and he complains that, because of his long and burdensome travel, he could not send to Greve his "discursus", meaning the *Prodromus*, and also that he had only second-hand reports about the eruption of the Etna and the destruction of Ragusa (Dubrovnik). Stensen also addressed greetings from Thévenot as well as from the two Swammerdams, but asked Greve, on the other hand, to address his greetings to de Bruin and other friends. Johannes de Bruin (1620–1675)[80] from Gorinchem was professor of natural sciences and mathematics in Utrecht

---

[78]EP I, 213 f.
[79]EP I, 213 f.
[80]BU 6, 75.

since 1650. In addition to scripts about the gravity of bodies and the origin of light, he also published a defence of the Cartesian philosophy.

In May 1670 Stensen received from Florence the announcement of the severe illness of his great patron, the Grand Duke. Blisterings in the thorax and lungs had violently affected him. At the end of April, Ferdinand received the sacraments of the dying and died on May 24, attended to the end by his physician in ordinary, Fr. Redi. Of course Stensen did not know that when he wrote to Viviani on May 25 and imparted his intention to depart that week still and to return to Florence.[81] A letter written right in front of Utrecht's gates shows how Stensen hurried away from Holland: some hours later in front of Utrecht again he asks Greve to excuse him for the hasty continuation of his journey.[82] One month later he was in Florence again.

It was a long and hard journey that the much travelled researcher had behind him. It went through half of Europe, from Italy over the Alps to Austria and Hungary, further to Bohemia and through the whole of Germany and then from the Netherlands, Belgium and Germany back to his beloved Florence.

However, much greater were the spiritual realms through which Stensen had passed. He had walked over the Alps, the Carpathian Mountains and other mountains as a passionate researcher who wanted to confirm and extend his *Prodromus*. His notes from this trip are lost to us, but we know from his own pen that the journey did not offer him much new, but had, however, confirmed his first discoveries. This shows so significantly that the theses of the *Prodromus*, surely originating on Tuscan soil, offered a certain general knowledge already in the first draft, which was only confirmed by his journey in Europe's highest mountain chains.

Stensen came back to Florence as a theologian. The encounter with Protestant friends in Amsterdam had opened new worlds for him. Stensen felt the spiritual need of his time very deeply. He would devote months to his apologetic scripts in which he undoubtedly established his great aptitude for that kind of work. The clarity of his thoughts, the heat of his convictions, the calmness and mildness of his being most certainly worked benevolently on his opponents and, in any case, in that second visit to Holland one can see a turning point away from science into that of pastoral activity.

---

[81]EP I, 27.
[82]EP I, 215.

1.8 Under Cosimo III in Florence

When Stensen informed Viviani that he was preparing for the journey back to Florence, he also expressed the desire of obtaining a room at his disposal in which he could carry out experiments on the liquids of plants and their relation to the liquids of animals. Grand Duke Cosimo was immediately ready to satisfy Stensen's request and charged Viviani to find a house with a small garden on the other bank of the river Arno. Since it was not ready for occupancy before November 1, Cardinal Leopold temporarily accommodated Stensen in the house of a chaplain.[1] On July 1, Viviani expects Stensen to arrive within the next eight to ten days and on July 11 he informs the court administration: "On the order of His Highness one shall give to Mr. Niels Stensen from Denmark who resides in the Casino of S. Marco by favour of Mr. the Cardinal's dean...", followed by an impressive list of objects that apparently was established for a large company: 24 French napkins, 16 large and 16 small plates, copper tableware, bed curtains, cupboard, table and tables, pictures, etc.[2] The first reception thus already showed the benevolence with which the new Grand Duke met Stensen and which he would maintain during his entire life.

Cosimo III (1642–1723)[3] may not have met Stensen very often before his accession to the throne because from October 1667 to May 1668 and from September 1668 to October 1669 he was travelling on distant journeys to Holland and Germany, Spain, France and England. The motive of these

---

[1]EP I, 27.
[2]Sten. Cath. 3, 1957, 37. By Casino di S. Marco is understood the Medici Palace in Via Cavour built in 1574 by B. Buontalenti.
[3]EP I, 37 ff, *Positio* 168.

© Springer-Verlag GmbH Germany, part of Springer Nature 2018
T. Kardel and P. Maquet, *Nicolaus Steno*,
https://doi.org/10.1007/978-3-662-55047-2_8

journeys was not the most delightful. The Prince had benefited from a careful education which provided his passion for hunting as well as the knowledge of languages and a deep piety. When 18 years of age he met the misfortune of his life in the person of the 16-year-old Princess Marguerite Louise d'Orléans (1645–1721). They were married in Florence in 1661 at a brilliant wedding. The pretty and cheerful young girl easily won all hearts but lacked the intellectual education and strength of character to confront the disappointments of life. Brought up as the future wife of the Dauphin and full of swarthy daydreaming for the French court standards, she was little satisfied with the title of Italian Grand Duchess. However, the most painful sacrifice that the political marriage required from her was to renounce the man of her passionate love, the General Charles de Lorraine. She did not have the courage to resist the dictate of Louis XIV, who ordered one marriage and forbade the other. In Tuscany she showed no restraint and caused the court of the Medicis and more so her husband all the torments that a whimsically passionate and hysterically unrestrained female nature can cause to her surroundings. It was attempted to improve the atmosphere between the couple by a temporary separation and Cosimo's journeys mentioned above served this goal. After the accession to the throne the behaviour of the new Grand Duchess soon became even more unbearable and Cosimo finally broke off all companionship and let her return to France in 1675. Stensen never mentions this tragedy.

Back home, Cosimo did not enjoy ruling the state. His government occurred in a period of decline in Italy and was not able to stop the regression of Tuscany. The will of peace of the Grand Duke, laudable as it was, made the country despicable to other Italian princes by the lack of military readiness and put it at the mercy of foreign rivalling powers, France, Spain and Germany. Cosimo's ambitious projects of marriage for his children—the three children who sprouted as witnesses of short periods of improvement of the couple's relations (August 9, 1663, August 11, 1667 and May 24, 1671) were Ferdinand, Anna Maria Luisa and Gian Gastone—ended in unhappy connections. Within the country, agriculture and industry burdened by heavy taxation regressed. The autocratic government inevitably failed, despite Cosimo's personal activity and sense of duty. Finally, over his last decades, there lay a dark cloud, the conscience of the downfall of the Medici dynasty without any prospect for Tuscany of recovering its past freedom.

The years that Stensen spent in Florence under Cosimo, or those during which he later corresponded from Germany with the Grand Duke, did not give, by far, the impression of the later decline and fall. Only respect for the personality of the ruler is felt. Indeed, Redi wrote to Cardinal Corsini in

1677: "Oh, Monsignore, how great are the goodness, the piety and the righteous purpose of His Serene Highness the Grand Duke. I have no word properly to describe them… But do not let off praying God for the salvation of a saint, righteous and so fair and pious Prince."[4] Outstanding men like Magalotti and Viviani, Noris and Segneri served the Medici court, which basked in the old magnificence.

Above all a shining merit of Cosimo will remain forever: he rightly appreciated Stensen's nobility of mind and soul and unflinchingly stood at his side as a true friend, adviser and helper. It is not least honourable for the prince and his adviser that they properly appreciated the geological discoveries of Stensen and did everything to help him complete his great work. Stensen, moved, thanked Cardinal Leopold on July 22: "Back to Florence, I hear from Mr. Vincenzo Viviani of the liking which Your Eminence bestows on me, as well by remembering me, your unworthy servant, as also by providing me with the house which I now enjoy and for which I say my humble gratitude to you, most Honourable and Serene Lord. The Serene Grand Duke has allotted me the pension which was granted to me previously by his Serene Father, to whom the Lord gives his glory, and has expressed his desire that I shall continue the work, *De solido intra solidum*."[5]

Since Stensen, as he himself mentions in the above letter to Viviani, had returned to Florence with concrete projects of scientific work, all conditions seemed to be given for a period of research. The immediate goal, of course, was the geological main work that Stensen had promised in the *Prodromus* and for which he had collected rich new materials during his long travels.

1.8.1 Theological Writer

On his departure from Amsterdam Stensen had promised a distinguished, old but otherwise unknown lady, who was very well-meaning to him and was sorry for the controversy with Sylvius, to complete the faith discussion in writing. He would have liked to carry out his purpose already during the journey. That is what he attempted but "disturbances on the roads," he writes, "and in the inns prevented me to put into a determined order the many and different matters which I had to consider every day."[6] He was hardly accommodated in Florence and had satisfied his first duties towards

---

[4]EP I, 38.

[5]EP I, 219.

[6]EP I, 217. On the controversy with Johannes Sylvius, see also OTH I, 21 ff and *Positio* 123 ff.

the court and old friends when he quickly wrote down his most important thoughts on the subject and sent them in a letter to Sylvius on July 15, the *Examen objectionis* as "a kind of prologue". The whole title of the letter, which we possess only in a later printing of an enlarged version, immediately reveals his purpose and its main contents: "An investigation of the argument concerning different passages of Holy Scripture and their interpretations presented as divine by different Churches,... all of which consider themselves as reformed and pretend, to us who never needed a reform of the faith, that they alone are certain and believe in God whereas our faith leans not on a divine but on a human authority."[7]

In this letter Stensen uses the same method that he always used after fruitless disputes in Amsterdam, i.e. first to determine the assertions of the opponent in clear definitions and then to draw from them the suitable conclusions.

In the determination of the concept of the opponent view of the faith principle, first the attribute "Roman" attracts attention, to which Stensen adds that he speaks of Roman Catholics and Roman non-Catholics, in this case reformed. That should remind his opponent of the common mother's lap from which they had come at the time of unity of faith, of the sacraments and of the love, a time in which there were no Lutherans, Zwinglists or Calvinists.

Both parties pretended to possess the true word of God, the unaltered revelation. Roman Catholics recognize as sign and proof of the true word the Holy Bible and the tradition that he receives from the Church, which is of divine origin and, in faith matters, he relies on his own reason as far as this coincides with the doctrine of the Church. The Roman non-Catholic simply considers the Holy Scripture as the word of God, which he receives in translation or in original text from learned men, in which he finds the right doctrine and feels the power of God without relying on man or society.

After the statements of both parties about themselves, Stensen ascertains purely rationally and based on experience: each word of God, each revelation of God, in whatever society, needs for announcement certain signs of mediation, just like any human thought must be communicated by oral or written signs. One could, therefore, rightly differentiate between the word of God itself and the human signs which propose it. In fact any society

---

[7] *Examen objectionis circa diversas Scripturas sacra et earum interpretationes tanquam Divinas a diversis Eclesiis propositas ... ubi omnes, qui reformatos et credunt, nobis nulla unquam fidei reformatione indigis objiciunt: Se solos certos esse, quod Deo credant, nostram autem Fidem non Divina sed humana autoritate niti,* OTH I, 39–48.

distinguishes three signs by which the word of God is transmitted, namely the written word of the Holy Scripture, the different interpretations of the Holy Scripture and those principles that are due to tradition. The one who considers his interpretation as a principle of faith is forced to consider it as a gift of the Holy Spirit. In fact, all human beings received sign of God's word from their Church, orally or written.

Furthermore, everyone must be convinced of believing in God and of possessing the unaltered truth in the signs delivered to him by his Church. Whereas the Catholics attribute this integrity to the Church and prove this by the acts of the Church, the non-Catholics do not accept such infallibility and also do not prove the integrity of the sign. They, therefore, confound the concepts and attribute to the sign what belongs to the word of God and they recognize to their own judgement an authority which their Church has not. The truth is that the Catholics lean on divine authority whereas the non-Catholics lean on human authority, and, what is worse, on prejudice.

Stensen sent this *Examen objectionis* to his friends in Holland on July 15, 1670, first of all to J. Sylvius asking for his opinion. The letter also makes some remarks on concupiscence and the Walenburgh method and concludes by warning: "But human reasons are fruitless, if we do not obtain the grace of God to recognize the error of handed over opinions, to recognize the truth, and to understand what has been recognized, a grace which I implore for everybody with all my heart."[8]

The letter offered an appropriate basis for discussion. It reminded of common good but also clearly raised the contradictions and reduced the controversy to a single question of principle. Yet if Stensen expected a significant progress in the debate from his method, he was profoundly disappointed. Instead of an answer, the letter in return of September 25 from Sylvius brought insinuations and added plenty of new questions.[9] The parson of the "German Church" revealed that Stensen's letter had been transmitted to him "by the most famous and erudite councillor of this city". That may very well mean Pieter Blaeu, the son of the known book printer Johannes Blaeu. Pieter had been in Florence, had stayed at the side of Cosimo III during his visit to Amsterdam and was appointed secretary of the city in 1668.[10] From the very beginning Sylvius termed his opponent as "unworthy an exchange of letters" and reproached him to be blindly obedient. He claimed that the Protestants followed their own judgement and not the

---

[8]EP I, 218. OTH I, 31.

[9]EP I, 220 ff.

[10]Hoogewerff, xxv f.

authority of the Church and then, with a series of passages of Scripture, attacked the Catholic faith doctrine of the mass, the good actions, Mary, the mediation of Christ and of the Church to come back, repeating himself, to the principle that the Scripture alone is the only source of faith.

If Sylvius alone had been involved, Stensen would hardly have continued the letter exchange. However, behind the reformed preacher there was a series of dear friends and, therefore, Stensen writes: "When instead of an answer I received another question which on the one hand pointed out your prejudices about our doctrine, I could not miss with good conscience such a convenient opportunity to bear witness to the truth. But I held also for improper to interrupt without permission the study to put in order certain observations promised to His Highness the Grand Duke. Therefore, I told him of the shortly begun conversation about the truth of the religion and he allowed me in his great wisdom crowned by piety not only to interrupt my other studies for a while but decided that this is the only highest necessary problem to solve. To this end he offered me his library and, seizing this favourable opportunity for rest and books, I wrote several scripts in which I concentrated more on the truth than on the style…"[11]

The first three great apologetic treatises by Stensen thus stem from this time. Stensen did not undertake the study of the relevant theological litera-ture lightly, as is shown by a brief exchange of letters with Magliabechi, the great bibliophile and eccentric of Florence. Although most of the letters of Stensen to him are undated, their contents point out to these months. Stensen made use of Magliabechi and his books so much that, at the end of that period (on May 12, 1672), he could thankfully write: "I send you, Sir, the books which you have lent me. They make up half a library and for most part they were not in the hands of their proprietor for almost 2 years. They were in the hands of someone who has little endeavoured to gain from them the fruits which would certainly have been richer in the hands of others."[12]

The first product of this thorough study was the paper: "Niels Stensen's letter to an erudite with whom he would like to conclude eternal friendship in the unity of the Holy Roman Church, or a method to convince a non-Catholic according to Saint Chrysostom, Hom. 33 in Acts of the Apostles"[13] This quotation from Chrysostom which Sylvius had appended to his *considerationes conclusae*, to the letter of September 25, as testimony of the Protestant Scripture principle reads as follows:

---

[11]OTH I, 197 f.
[12]EP I, 262.
[13]OTH I, 54–70.

"There comes a heathen and says, "I wish to become a Christian, but I know not whom to join: there is much fighting and faction among you, much confusion: which doctrine am I to choose?" How shall we answer him? "Each of you", he says, "asserts that 'I speak the truth.'" Indeed: this is highly in our favor. For if we told you to be persuaded by arguments, you might well be perplexed: but if we bid you believe the Scriptures, and these are simple and true, the decision is easy for you. If any agree with the Scriptures, he is the Christian; if any fight against them, he is far from this rule."[14]

Stensen immediately seizes this quotation because here is the core of the controversy as he points out in the determination of the *status questionis*: "In our case the question is: in matters of faith, whom must one believe, is it the word of God as one reads it in the Holy Scripture without taking account of whatever transmitted or declared authority, or the word of God as far as it is presented and explained by a faithful society."[15]

The quotation, however, was also of particular importance because it stemmed from John Chrysostom, that is "golden mouth", who enjoyed the highest consideration in east and west. His works were more widely spread than those of any other. He was an expert on and preacher of the Scripture like hardly any other.

The words of this doctor of the Church, Stensen says, actually bear witness to the Catholic principle and he produces as proof of that first of all the words of the quotation themselves.[16] Then in a short meaningful summary he renders the dialogue between the Greek and Chrysostom[17] and explains his own method[18] and finally applies it to the actual case.[19]

From the wording it is already clearly apparent to what end the lecture of the Scripture is proposed to the doubting one, "namely not for the investigation of the faith but of the Christian, not for the truth of the word but of the agreement of the word preacher and word interpreter with the word",[20] which the Catholic still does today since he does not fathom the truth by a comparative study of the passages of the Scripture but by a comparison of his Church and Church doctrine with the Scripture. The summary of the

---

[14]OTH I, 55 ff.

[15]OTH I, 55.

[16]OTH I, 56.

[17]OTH I, 56.

[18]OTH I, 56–66.

[19]OTH I, 66–70.

[20]*Scilicet non fidei, sed Christiani investigatio, non verbi veritas, sed verbum proponentis et exponentis cum verbo consensus*, OTH I, 56.

dialogue between Chrysostom and the Greek underlines this conception: The Church father urges him to find the marks of true Christianity in the Scripture and to apply them to the disputing parties.

Finally Stensen comprehensively explains the method of Chrysostom with regard to the time and the context of the whole homily. With the Greek is meant a pagan who wanted to excuse his absence of faith and his scepticism by the schism of the Christians who surround him, thus of Arians and Catholics. He seems to have good reasons at hand. Both parties came with the same claims, for instance that one would have the true interpretation of the Scripture and the other a distorted one. However, Chrysostom calls that pure pretexts (*dubia dubitationem amantis, non quaerentis veritatem*). Because of a mess of meanings one does not hesitate in practical life to form for oneself a certain meaning. No customer hesitates to buy wares because of the entangling claims of the merchants. In questions of faith one may also form for oneself a judgement either by an indication of the contradictions of one or the other doctrine of faith or by the contradictions between doctrine of faith and person. The latter would be easier, for instance the solution of the question whether the Arians or the Catholics have been separated from the original Church. However: "Both ways are in contradiction to your principle according to which it is prescribed the doubting person to find the laws of dispute and faith, without any human authority, only by a comparison of the authors, and thus to give preference to the reasons of the authority, whereas the Church teaches that the authority has priority over the reason, i.e. that one must first establish the authority of the teacher and then ask him the reasons for the matters to learn."[21]

If Chrysostom had acknowledged the Protestant principle of faith, he would simply have presented to the doubting ones the Scripture in which everybody believed and of which he himself said that it was true and simple. Yet he never produced one word of the Scripture to prove it. It was obviously not his intention to recommend the lecture of the Scripture to those who doubt, and even less was there for him a lecture of the Scripture free from any authority. When Chrysostom terms the Scripture clear and simple, he understands by that the Scripture on the basis of the Church. Understanding unquestionably depends on the mental capacity of the reader and the Church father explicitly says in the following homily that reading the Scripture is not less dangerous for the ignorant than when somebody prescribes for himself a

---

[21]OTH I, 63.

purgation of which he does not know anything or put on an armour that does not fit him.

After this presentation of the method of the Church father, Stensen only draws his conclusion for his own controversy. Calvin teaches to disregard any other human authority. Everybody should seek by himself the truth from the Holy Scripture by comparing different passages, thus seek the doctrine from the doctrine. Chrysostom, on the contrary seeks, like the Catholics, the marks of the true Christian. The teacher first, then the doctrine.

After having briefly drawn the attention to the language of the marks, which clearly show that Calvin has separated himself from them and that before him nobody has followed his doctrine and practice, Stensen points out the contradiction that on the one hand one does not owe the faith to any human being but owes it to the Scripture only, but practically like the Catholics draws it from the catechisms, the sermons and theological compendia and interprets the Scripture according to the rules of these teachers. Learned men also may seek their faith according to the Protestant principle but common people, usually ignorant of foreign languages, cannot do that any more than the Catholics. The only difference is that the Catholic openly recognizes this contrary to the Calvinist and explains it as reasonable and wanted by God. The Catholic thus believes in his Church, the Calvinist in men, who are more or less subject to errors.

Stensen then exhorts Sylvius to draw the consequences for his own salvation and to apply by himself the same method that has been the method of Christ, the apostles, the Church fathers and the whole Church, namely "through the works of God, the clearest and most believable proofs to possess the word of God. To embrace these works of God, not only the miracles in the proper sense of the word and the other characteristics of the true Church, but also the greatest of all miracles, namely the examples of an apostolic life in true imitation of Christ which are found in the Catholic Church in every century until today, despite the furious hands and minds which fight against the truth or despite the pleasures of a long peace which enervate any other virtue".[22]

Stensen finally mentions the actual example of a conversion of the reformed preacher Daniel from Béarn, which in its time aroused great sensation. His son, indeed, had first become a Barnabit to finally bring back the father and the whole family into the bosom of the Church.[23]

---

[22]OTH I, 68.
[23]OTH I, 69.

In its shortness and precision the treatise reveals characteristic qualities of Stensen. He wants here to attack the *omnium errorum fundamentum*. He proposes the proof based on one quotation only of one Church father only who deserved great importance so much that Chrysostom in the relevant homilies of that 15th chapter of the Acts of the Apostles describes the apostles' council and says that it settled a conflict which had been brought for ecclesiastical authority.

We do not know the reaction of Sylvius' circle in which the treatise had been communicated. On the other hand, the critique to which Leibniz submitted the epistle and which appeared in print in 1675 cannot be forgotten. The *Lettre à un amy* in which Leibniz gives his opinion well stems from 1677.[24] The ducal librarian who served the same court in Hannover as Stensen writes, after having testified his high appreciation for the anatomist, that, in the opinion of common Parisian friends, he could still make many more important discoveries: "His zeal and his piety deserve praise but I must say that one of his letters has almost provoked my irritation. In order namely to weaken the force of a passage of S. Chrysostom, he uses such a bewildered subterfuge and presents so noticeably false pretexts that I nearly doubt of his sincerity…"[25] Of what the bewilderment of Stensen consists does not, however, appear from the words of the philosopher, who does not take pains to refute the arguments of Stensen thoroughly. Leibniz thinks that a calm and dispassionate reader may have said that the passage of Chrysostom could be used to the advantage of the Catholics as well as to that of the Protestants. Those who could judge the doctrine of an exegete from the Scripture may well, if even perhaps more uneasily, take the doctrine directly from the Scripture. Furthermore the Protestants are no self-taught persons, they have also their doctrine of faith. The whole matter is far from the reasoning of S. Chrysostom and from the controversy with the Protestants. It is here question of a pagan who does not know whether he should accept the doctrine of a Catholic or that of an Arian. Chrysostom of course does not advise the pagan to take the faith from the Bible where there is a much easier way, that of the confrontation of his doctrine with the bible. Leibniz concludes: "I know that Mr. Stensen has nice enlightenment and the sentiments of a high virtue. But I think that the zeal for a party which he holds for the best lets him say things which he would have himself strongly blamed in other circumstances."[26]

---

[24]Scherz, *Gespräche zwischen Leibniz und Stensen*, 84. EP I, 66 ff.

[25]*ibid.*

[26]EP I, 67.

The occasion for Stensen's second irenical writing which we shall in brief call, *De interprete S. Scripturae,*[27] was the book by a reformed preacher, Johannes van der Waeyen (1639–1701)[28] who died as professor in Franeker and who also dealt with the interpretation of the Holy Scripture. Stensen is likely to have read it during his visit in Holland. Van der Waeyen had opposed a book by the reformed preacher Ludwig van Wolzogen, who in his time had attacked the disciple of Spinoza, Lodewijk Meijer, because, in his book *De philosophia S. Scripturae interprete exercitatio paradoxa,* 1666, he wanted to make philosophy the guiding principle of life. However, van Wolzogen himself had stirred so much hostility among his colleagues that more than 20 Protestant theologians, among them also van der Waeyen, opposed him and the controversy showed the whole weakness of the reformed basic principle. Against that background of Protestant disunion in Holland, Stensen wanted to present the core question of all controversies, the Scripture principle as opposed to the Catholic Church principle and from the Church fathers show that one must keep first to the Church and its authority before being able of certifying the truth of the Scripture. The Calvinist principle is that every believer must take the true interpretation of the Scripture, not from reason, not from the Church, but from the Scripture itself. The doctrine of the fathers is that one should indeed explain the Scripture, if possible, by the Scripture but not every passage of the Scripture can be explained by passages of the Scripture and it is not the concern of every one. However, every believer must obtain the true interpretation of the Scripture from the Church of God. This is proved by passages of four Greek and two Latin Church fathers, whereas the opposite assertions, above all that of van der Waeyen, are disproved. If one misunderstands quite clear passages of the Church fathers, what is then with much more difficult passages of the Scripture? If preachers cannot acquire clarity, what with their disciples? And he concludes: "May God open the eyes of the mind for him and all who pretend to find in themselves the sense of the Holy Scripture so that they recognize their own misery and seek the true interpretation of Holy Scripture in the true Church."[29]

---

[27] *Nicolai Stenonis ad virum eruditum, cum quo in unitate S.R.E. desiderat aeternam amicitiam inire, epistola, detegens illorum artes, qui suum de interprete S. Scripturae errorem sanctorum patrum testimonio confirmare nituntur,* OTH I, 76–90.

[28] On the background of this work, see OTH I, 73 ff.

[29] OTH I, 90.

The *Scrutinium reformatorum*[30] belongs to these apologetic dissertations, but we do not know the background for this treatise aimed at practical life. Here again Stensen starts from the original historical community between Catholics and Protestants, which the latter had left because of certain lacks. Depending on whether this lack concerns the doctrine or the practice, one distinguishes two kinds of reformers. The holy ones reform life according to the doctrine of the Church, the others, like Luther and Calvin, reform the doctrines according to the Scripture. The former begin by reforming their own life and pass from a freer to a stricter life. The latter do not reform their life but pass from a stricter to a freer life and preach to others the same freedom so that their life does not surpass the achievements of the philosophers.

Stensen now compares the reform of life to the reform of doctrine. The life reformers would have changed much in relation not only to the commandments but also to the counsels of Christ and to an apostolic life as lived in the orders. The doctrine reformers despised the counsel, even spoke of the impossibility of keeping to the commandments and called themselves apostolic without showing any sign of an apostolic life. Conversions of life reform lead from a sinful to a holy life, whereas conversions from a strict to a freer life have nothing marvellous, nothing going beyond the philosophers. Life reform keeps the value of priesthood so high that even off-spring of royal houses consider it an honour to belong to it and to renounce all the other honours to announce the Gospel to the poor. The doctrine reform, on the contrary, has diminished the value of priesthood so that high-ranking persons deem it below their dignity to belong to it. In life reform one keeps the unity of faith, of love and of sacraments all over the world, whereas the doctrine reformers were erstwhile and are now in discord between each other. After a comparison of the deeds of the life and doctrine reformers with the words of the Gospel, the author draws the conclusion that life reformation stems from God, but doctrine reformation stems from men.

All these treatises by Stensen are characterized by their pleasant concision, which was quite different from the habit of the time and from the large books of his antagonists. Stensen immediately goes to the essential. Sylvius handles a lot of questions of peripheral significance. Stensen deals with the fundamental problems. He always has a pastoral goal in his view. He wants to convince, to teach. In this, mildness is the mark of his controversy. He readily leads the opposite conception *ad absurdum* but no word which hurts is expressed.

---

[30] *Nicolai Stenonis Scrutinium Reformatorum ad demonstrandum Reformatores morum in Ecclesia fuisse a Deo, Reformatores fidei non fuisse a Deo*, OTH I, 105–120.

In the year 1671 it also came to the writing of a letter to Spinoza and the publication of the *Nicolai Stenonis ad novae philosophiae reformatorem de vera philosophia epistola*.[31]

The time and circumstances throw light on the questions of the motive of the letter. After his return from Holland Stensen dealt with apologetic writings. In Amsterdam he had been able to determine the aversion that arose against Spinoza's *Tractatus theologico-politicus*, anonymously published in the first weeks of 1670.[32] With this treatise written from 1665 to 1670, Spinoza had descended from the calm room of the erudite into the arena of passionate fighters. Here the critic of religion, the politician, the teacher of law and state spoke out. (The main work of Spinoza, the *Ethics*, found its final edition between 1670 and 1675 and appeared after his death).[33] The Church council of Amsterdam had let its *gravamina* be known on June 30, 1670 and the court of Holland required the punishment of the author and of the publisher on April 16, 1671. Much later the treatise was called the book of the free-thinkers but immediately after its publication it one saw in it a declaration of war against Christianity and religion. In this situation there was reason enough for Stensen personally to take position in regard to the book and creed of his previous friend, the more so because he saw his Dutch friends influenced or threatened by that book.

Spinoza himself refuted with indignation the accusation that he seemed to have rejected all religion: "What he (the critic) understands by religion, I do not know, what he understands by superstition, I do not know. Has he rejected all religion the one who puts up the principle that God must be considered as the highest good and as such must be loved in free will? And that our highest luck and our whole freedom consist of that? And furthermore that the reward of virtue is virtue itself, the punishment of foolishness and of incapacity is foolishness itself? And finally that everyone must love his fellow-creature and obey the orders of the highest power? And that I have not only said expressively but moreover proved with the strongest reasons."[34]

---

[31]OTH I, 95–103 = EP I, 231–238, cf. *Positio*, 155 ff. In Danish: *Stenoniana* I, 114–125; in German: *Pionier*, 279–287. The date is approximately given by: nondum qvartum in ecclesia annum absolve (OTH I, 98 = EP I, 233) – thus being before November 2, 1671 (Stensen considered Nov. 2, 1667 as the day of his conversion). Since he travelled in July and August 1671 in Northern Italy, the letter must be dated soon before or after that.

[32]Meinsma, 401 f.

[33]Spruit and Totaro, *The Vatican Manuscript of Spinoza's Ethica*, Leiden 2011.

[34]Spinoza, *Briefwechsel*, 194. * The present translation is based on Scherz's German Biographie with an eye to Letter 67b from Nicholas Steno to the Reformer of the New Philosophy Concerning the True Philosophy", Edwin Curley, *The collected works of Spinoza*, Princeton and Oxford, 2016, 451–458.

Stensen in his letter immediately attacks a main thought of the *Tractatus theologico-politicus*, namely the concern of Spinoza about the public and personal security: "In the book, the author of which you are, according to what others said and to what I myself for different reasons assume, I see that for you everything turns about the public security or, much more, what according to you is the goal of public security, your own security, although you use the means contrary to the desired security and you neglect completely for you that side of security of which one should only strive to." Stensen sees that spiritual security in danger above all by the fact that takes into account the worldly authority: "While you allow everybody to think and talk of God at leisure, if it only does not happen so that it abolishes the obedience which, according to you, is to exhibit not so much God as man."[35]

Stensen immediately accompanies his unequivocal statement with a declaration of warm friendship and sincere benevolence: "As I see lingering in that obscurity a man to whom I have been a close friend and who, as I hope, is not hostile to me even now (I am namely convinced that the remembrance of the former devotion still retains our mutual liking), and as I think that I have also once found myself, even if not quite the same, in very severe errors, so, moved by the greater compassion on you, the greater the danger from which I was freed which made clear the mercy of God towards me, I wish you the same Heavenly Grace which I have myself obtained, not by my merit but only by the goodness of Christ. To join acts and prayers, I offer myself to you most readily to investigate with you all the proofs which you may subject for examination in order to discover and follow the true way and the true confidence. And even if your books show that you are far away from the truth, so the love for peace and truth which I once noticed in you, and which, despite this obscurity, is not completely extinct, gives me hope that you will lend a willing ear to our Church, if what she promises, what she offers to those who want to approach to her, is sufficiently explained to you."[36]

To the philosopher Stensen presents the Church as the school of perfection, the true eternal certainty and the peace bound to the truth. It also gives the means to acquire such great goodness by pardoning misdeeds, the most perfect rule of life and the true practical perfection through this rule. Moreover, the Church addresses not only the erudite, the subtle minds and the people far away from the world, but everybody: "The aim of this rule is that the man directs all his deeds, not only the external ones but also the most

---

[35]OTH I, 95 = EP I, 231.
[36]OTH I, 95 f = EP I, 231 f.

secret thoughts according to the order constituted by of the creator of the universe, or, which is the same, that the soul in every action considers God, its creator and judge."[37]

Stensen thus distinguishes, quite untraditionally, four steps of striving for perfection. The first, the state of blindness, in which man handles in everything as if his thoughts were not subjected to any judge, the state of the non-baptized and others hardened in sins, should surely turn Spinoza's look into his own soul. Stensen says: "This state does not exclude that one discusses much and often also true arguments about God and the soul while one, however, deals with that as if it was with matters far away and external. This results in doubt, many contradictions and frequent faults."[38] The second step of perfection resembles infancy. Man learns to recognize his sins against the true and the good and turns to God who bestows on him invisible grace through the sacraments. The third step requires constant exercises in virtue, which in turn prepare for the fourth step, which is the acknowledgement of the mystery of the Holy Scriptures which are accessible only to a pure heart. One begins by looking at God and lives in complete union with him. The goal of Christianity is nothing less than awakening the soul from death to life and to direct to God the spiritual eyes which were directed to the error: "By all-sided right reflection you will find the true philosophy in Christianity alone, which teaches about God what corresponds to the dignity of God and about man what corresponds to man, and leads its adherents to the true perfection of all deeds."[39]

Stensen then talks with great enthusiasm of the Catholic holiness which he sees shining in all the centuries and which he has himself observed several examples:

"I have not yet lived in the Church for 2 years and have already seen so great examples of holiness that I must truly exclaim with David: Your testimonies are quite believable."[40] He particularly points out two classes of men, namely conversions of great sinners from the depth of vice to the height of holiness and the holy life of quite simple people, workers, domestics, "according to their way of speaking so-called simple heads, who without any study have obtained elevated knowledge of God at the foot of the crucified."[41] Many other questions are only skimmed. It is thus said of the

---

[37]OTH I, 96 = EP I, 232.
[38]OTH I, 97 = EP I, 233.
[39]OTH I, 98 = EP I, 233.
[40]OTH I, 98 = EP I, 233 f.
[41]OTH I, 98 = EP I, 234.

authority of the Church that "it permits one head only, the authority of which does not indulge in the arbitrary change of whatever matter, which is a defamation of the opponents, but is destined to obtain necessary conditions always unchanged for those who rely on the divine right, but to change the matters of human right or indifferent matters, as appears expedient for the Church for righteous reasons."[42]

After this positive presentation of the ecclesiastical striving for perfection Stensen attacks the system of Spinoza. He exhorts the philosopher to seek deep into his own soul. With frank conscientious research he will find it dead. He lingers on the material moved and pays little attention to the mover. He teaches a religion of the bodies, not of the souls, looks after the behaviour of the person and the society but pays little attention to the knowledge and love of the Creator. He denies Grace, recognizes only rational certainty, ignorant of the certainty of faith, which surpasses all proofs. And what certainty does he offer himself? With his insight Stensen points out the central difficulties of Cartesianism in the Spinoza version: "Please consider all your proofs and furnish one only according to which the rational and the extensive unite, according to which the moving principle unites with the moved body. But what do I request from your proofs for these matters, of which you even cannot explain to me the likely *modi*, which results in that you cannot explain without imputation the feeling of lust and of pain and not the movement of love and hate and, therefore, the whole philosophy of Descartes, however carefully discussed and reformed by you, even cannot by demonstration explain only this simple phenomenon, namely how the impulse of matter on matter can be conceived by the soul united with the matter. But I ask you: which other knowledge of the matter do you give us apart from mathematical investigation of the quantity as related to the figures which are only hypo-thetically proved for any kind of particle? But what is more unreasonable than to deny the divine words of the one the divine work of whom is obvious to the senses, because they are in contradiction to human proofs which rely on hypotheses? And whereas you do not even understand the material condition through the intermediary of which the mind perceives the material objects, how can you make a judgement over that condition of the body which, glorified by transformation of the corruptible into the incorruptible, should be re-united with the soul."[43]

---

[42]OTH I, 100 = EP I, 235.

[43]OTH I, 101 = EP I, 235 f.

Stensen concludes, highly interestingly, with an acknowledgement of the *philosophia perennis*: "Truly, I am quite convinced that it is one and the same as to find artificial principles to find new principles to explain God and the nature of the soul and of the body, since sheer reason tells us that it is in contradiction to the divine Providence, that the true principles about that would have remained hidden through so many thousands of years to the holiest people, whereas they would have been first discovered in this century by men who have not at all attained the perfection in moral virtues. Therefore, I consider as true only the principles about God, the soul and the body which are kept since the beginning of the created things until this day in one and the same company, i.e. the city of God."[44] Stensen then mentions a passage in Justin, which likewise attributes great importance to the uninterrupted and common tradition of certain basic principles and in which the holiness of life is also appreciated as witness of the truth of the doctrine. Stensen finally proposes that Spinoza closely show him the contradictions and uncertainty of his doctrine and exhorts him to quit and, after the conversion like another Augustine, to bring back to God the thousands and thousands whom he has led astray from the knowledge of God.

We do not know whether Spinoza ever received this letter and if so we know even less of how it worked on him. A conversion or even a strong influence on the philosopher was as good as excluded. One has only to consider his proud word to young Albert Burgh: "I do not pretend to have found the best philosophy but I know that I recognize the true one."[45] Moreover, Spinoza's system did not at all justify such a dangerous self-conscience and belief in his infallibility.[46]

From Stensen's side this was in any case a serious attempt spiritually to help the erstwhile friend, to unite Christian benevolence and sober expertise.

Many passages first aim at gaining the trust of the opponent. Stensen reminds him of their former friendship. He admits to have been ensnared in similar errors himself. He points out Spinoza's love of peace and truth. The prominence of the Church as the guide to God corresponds well to Spinoza's love for the God and ruler of the Old Testament, and when Stensen wanted to make him "another Augustine", indeed no wish could have been more honourable.

If Stensen offered himself for further lessons, he had good reasons, as Spinoza knew as little as nothing of the Catholic Church. As

---

[44] OTH I, 101 f. = EP I, 236. * Justin, *Dialogue with Trypho*.

[45] Spinoza, *Briefwechsel*, 286.

[46] Meyer, 173. Gustav Scherz's reference. The source was not found. See German ed. I, 298, note 44.

Dunin-Borkowski understands it: "Spinoza had never heard a strictly scientific word about Catholicism. Despite the relatively high number of Dutch Catholics, as far as we know, he had never had an opportunity to hear more about Catholicism than about any unworldly sect."[47]

It would have been quite useless in such circumstances to engage in a discussion of controversial religious questions. It reveals his fine understanding of the subject that Stensen counted on the Spinoza striving to perfection and, therefore, presented the Church as a guide to holiness, and thus brought forward the respectable side of Christianity. Spinoza's personality with his love of truth and distinguished human attitude had in his time worked more on Stensen than his system. Spinoza's philosophy, "strove to the knowledge, not for itself, but with the aim of practical interpretation, with the aim of transformation into action, and also not to serve a will of power and domination but to serve their own and others' way of life. At the same time he strives to agreement between knowledge and life, between life and doctrine. His philosophy is a philosophy of life. The ethical and religious motive is the first and essential which he sets in the foreground much more than Descartes and Leibniz."[48]

The polemic part of the epistle is relevant, sharp and unerring. It unveils the profound weaknesses, the arbitrariness and contradictions of the system. This particularly applies to such assertions that Spinoza is above all considerate of public security, that Spinozism is a *systema ex suppositis formatum*, that he addresses almost only those acquainted with philosophy and leaves ordinary people to themselves, that he does not know the light of the grace. There were then concentric attacks when Stensen speaks of the weak basis of the *certitudo demonstrativa* of Spinoza or also when he demands one single proof that *cogitans et extensum* could go up into a higher unity, or an explanation of how material pleasure and pain-feeling or opposite impulses of the mind bound to the matter could be perceived, and above all when he asks: "But which other knowledge of the matters do you (and Descartes) give us, please, except for a mathematical investigation of the quantity of figures which are very hypothetically demonstrated as far as the kind of the particles."[49] Leibniz equally says: "The author seems to be very cranky. Seldom does he proceed on clear and natural ways. He arrives at his goal only by leaps and bounds and

---

[47]Dunin-Borkowski, 119.
[48]Meyer, 145. See above note.
[49]OTH I, 101 = EP I, 236.

through roundabouts. Most of his demonstrations outwit the mind rather than they enlighten it."[50]

Stensen did not limit himself to considerations on nature history and experiment but, in striving for truth, he kept to the historical overall view.[51] The fundamental truths concerning the nature of God and the human body–soul nature actually belong to the foundation of human cognition like the natural law and its main requirements of moral common goodness of all times and in all areas. Franz Werfel says: "The time is the sharpest acidity of the world, an aqua regia, in which only the purest and most solid gold has an existence. Any light metal, even if it also keeps its intrinsic value, becomes corroded and finally dissolved. Most of what has irritated people during one day is already a dream the next morning… In a large space of time there is more knowledge than in the shrewdest intelligence."[52]

It is the particular Catholic attitude: "The Church … has the overview over the centuries, distrusts those who believe now that they have the key to all enigmas in their small hands."[53]

It is not new that the recognition of moral and religious truths is influenced by the ethical goodness of man, as the Holy Scripture says: "Fear of the Lord is the beginning of wisdom" (Proverbs 9, 10), or: "A soul that plots evil wisdom enters not, nor dwells she in a body under debt of sin." (Wisdom 1, 4).

Leibniz read the epistle of Stensen about 6 years later, immediately after the death of Spinoza and appreciates it in a *Lettre à un amy*. After a general affirmation that one finds in Spinoza "many excellent ideas among a great number of unacceptable and very false assertions", he says, referring to the condescending pastoral tone of Stensen's letter, which is obviously distasteful to him: "Because of the great difference between their opinions, Spinoza was not particularly affected by the exhortation of Mr. Stensen. Actually it seems to me that Mr. Stensen assumes much too much to convince a man who believed so little."[54] Then Leibniz summarizes the contents of the letter in ten points and thus reveals that he must have spoken of the letter with Stensen. He indeed knows not only that the miracle of holiness means the zeal of Lavinia Arnolfini but also mentions the example of the conversion from evil to better life, which Stensen had before his eyes. It relates to the Maltese knight Buonacorsi, who in a brothel, probably in Florence, was so moved by a

---

[50]Meyer, 147 f. See above note.

[51]Cf. Pionier, 272 ff.

[52]* Franz Werfel, *Das Lied der Bernadette*.

[53]Pflieger, 177.

[54]EP II, 929.

report of the martyrdom of Father Bressani that he gave up his light-minded life and became apriest.[55] Francesco Giuseppe Bressani, S.J. (1612–1672),[56] came from Rome. He became a zealous mathematician and went to Canada where, on the way to the Huron Mission, he was captured by the Iroquois and cruelly tortured. He was at length ransomed by the Dutch and was sent to France from where he returned to the Huron Mission which was again ravaged by the Iroquois. Bressani died in Florence where Stensen certainly had heard him in St. Giovanino where the Father, despite disease and weakness, preached and presided exercise for good death. In the absence of an answer from Spinoza, Leibniz himself took position for the ideas of Stensen but very superficially. When, for instance, at the very first point, he disposes of the testimony of great zeal for the souls with the sentence: "Those who quietly speak are usually moved by passions and consequently little enlightened."[57] Moreover, Leibniz explains the passage in which Stensen turns down new fundamental principles about God and the soul as if he wants to refute all new ideas about these subjects. The final judgement is characteristic of the self-confident philosopher who, however, is touched by Stensen: "Far from reproaching Stensen, I can say that I appreciate him, yes, if I may say so, I like him. Then I believe to recognize in him a zeal which is inspired by true love. I do not wonder about his aversion against the philosophy because he has not yet experienced the strength of the metaphysical proofs."[58]

In relation with the Spinoza letter, a short letter of November 24, 1671 to Malpighi is particularly valuable.[59] Stensen writes: "My greatest difficulty concerns how man's soul, which is spiritual, can perceive the change of movement in a corporeal object, thus in man, and how the animals which one imagines without any soul can have a perception of that change of movement which occurs in the nerves. What you say of the vibration and of the completion of the organization is clever and probable but, as you rightly notice, it is a difficult matter to imagine. Yes, it seems to me downright impossible to arrive at the way of nature to this end, not because I personally cannot arrive at that—I will admit my weakness—, but because I see that no philosopher has determined it so far. All in all, if it was possible, I would readily do one of both: either recognize whether it can be explained with reliable evidence, or

---

[55]Stensen first met Leibniz in November 1677, thus after this letter. The information on Buonacorsi was written in the margin. Cf. EP I, 67.

[56]DBI, 14, 194 f.

[57]EP II, 930.

[58]EP I, 231.

[59]EP I, 248 ff.

recognize why our intellect is unable to do so. Nothing is more certain than that we and the animals perceive the movement of two material things bouncing on each other but how that movement can be perceived, either by a material principle like in animals, or by a spiritual principle like in man, seems to me to be such a wonderful action of the omnipotent Creator that I cannot wonder enough about that. Just think of what astonishment that propriety of the soul must call forth, by way of which it comes from a perceptible matter to one imperceptible and from the matter with compelling necessity to the immaterial. And if the natural abilities of the soul have in themselves such wonderful and so inconceivable proprieties, those of the grace are that more worth admiration that they make fit to see God and his mysteries."

Stensen continues to Malpighi: "I have certain friends in Holland who have devoted themselves so much to the Cartesian philosophy that they want philosophy to be the judge of the recognition of the grace. To find in philosophy itself the means to the recognition that the natural light has no more relation to the objects of the grace than the sense of touch has to the objects of the sense of vision, I seek all the weaknesses in philosophy to find, in recognizing them, what can be determined and what cannot, and the reason why it cannot be determined. And since you enjoy investigations of nature and have retired from occupations of practice, I ask you on your side to show always what the light of nature can and what it cannot so that both of us with God's help can be for those souls the cause which leads them back to the true Christian humility and a complete devotion to the word of God. And they then become thankful for the blessing which God shows to them through our intermediary. For us it will be a new motive of eternal joy in that triumphal Church."

The exchange of letters with the circle of Johannes Sylvius in Amsterdam, was, however, not yet finished. Stensen's letter of January 6, 1672, *De infelici Silvii responso ad duos ipsi propositos syllogismos, unum contra credentem Sacrae Scripturae, alterum contra negantem eam, epistola*, has been preserved. It concerns the answer of Sylvius to two syllogisms proposed to him, one for a believer in the Scripture, the other for a denier of the Scripture.[60] It immediately begins with the question: "What is it that matter which needs such a defence?" This direct beginning, the title under which this and the conversion letter were published 5 years later (*ex pluribus scriptis*) supposes the existence of the other writings. They were addressed to the entire circle of Stensen's

---

[60]OTH I, 234–236.

friends in Holland with all its diversity of faith and with some indications that lead towards the high circle.[61] *De infelici responso* first shows to the believer in the Holy Scripture, from the Bible, the characteristics of the true Church, such as for example its unity and holiness, and refers them to the existing Church and to the whole Church history. Stensen draws the attention of the one who denies the Scripture quite generally to the magnificent appearance and reality of the Church, in which one must recognize the hand of God. Stensen once again turns against the reproach that his faith is rationalism and shows the union of the natural and supernatural in the act of faith. The Catholic does not understand the truths of the faith without divine light. He does not believe the understood without divine impulse and, in the life based on faith, he counts on God's assistance. God speaks from the outside through the members of the Church whereas he enlightens us inside through the Holy Spirit, moves our will and anticipates our deeds.

Some days later, on January 12, 1672, Stensen then writes his *De propria conversione epistola* to the Amsterdam preacher and the Dutch circle of friends.[62] This letter was caused by a remark of Johannes Sylvius who had denied the Catholic faith principles any strength to convince. It could neither lead the doubting ones to the Church nor make the born Catholics certain in their faith. No "turn-coat" could rightly name the motives for his conversion to that Church. Stensen took the pen in his hand but in regard of the importance of the conversion so deeply encroaching on his life it was an amazingly terse presentation. Stensen himself draws attention to his motive: "Since the mail leaves immediately and does not permit a thorough presentation, I accepted here in a short summary what God gracefully let come to my mind." This was followed by the motives that we have already come know at his conversion; first the three points that detached him from the orthodoxy and disposed him for the conversion: the philosophy of Descartes which taught him to doubt, the entangling multiplicity of the sects in Holland and the *modus vivendi politicus* of many people. Then the practically decisive motives are named: the pious life of Catholic friends, the spiritual zeal of Mrs. Arnolfini, the fundamental study of the religion and the coincidence of many impressions on All Souls' Day. Finally, the logically decisive motives follow: whereas the reformers could not indicate the divine origin of their novelties, the Catholic Church is apostolic, all stem from it and it has all the characteristics of the Church of Christ, particularly that of holiness. Stensen ends

---

[61]Cf. OTH I, 130 ff.

[62]OTH I, 126–129 = EP I, 257–260.

with an exclamation: "Be the name of the Lord praised in the eternity, who called me back from the obscurity to the light, from the death to the life. I offer you and all our friends and even to all men the true rest of the soul in the lap of the true Church."[63]

Stensen helped converts not only in writing but also in practical life. This is shown by a short letter of thanks on December 22, 1671 to Cardinal Leopold, who had given a gift to an indigent converted woman. Here one also finds the interesting sentence: "And together with me Mr. Adrian will be obliged to do the same, whose conversion shows steps to a true Christian life."[64]

Mr. Adrian seems to be the Dutch copperplate engraver Adrian Haelwegh (about 1637–about 1696), who had activity in Amsterdam but worked in Florence in 1670. He may well have been converted to Catholicism under Stensen's religious influence.[65]

1.8.2 Geologist of the Grand Duke

Concerning the scientific work of Stensen in those years we know that he was busy making a system for and cataloguing the Grand Ducal collection of minerals to which he contributed his own material. Still 100 years later Manni could see in the royal natural science collection in the Palazzo Pitti "different minerals and fossils with the relevant inscription, written in the hand of our nature researcher with such precision and elegance that they present as the fruit of a genius."[66] Targioni-Tozzetti also points out that Ferdinand II acquired the quite nice collection of minerals, crystals and different petrified items which the famous Stensen had set up during his study.[67]

In the beginning of the year 1671 we meet Stensen studying in the museum of the academy in Pisa to which Ferdinand I already in 1595 had brought a sample of his collections. Felice Marchetti, the active assistant of Stensen, writes to Cardinal Leopold: "I have enabled Mr. Niels Stensen to examine all the objects in the Galleria de'Semplici and to write down these observations on paper, as he wanted. He told me that he had found a series of extremely interesting objects, particularly in the minerals of which he will

---

[63]OTH I, 129 = EP I, 260.

[64]EP I, 250 f.

[65]Cf. Nordström, *Magliabechi*, 13 f, 29, 40.

[66]Manni, 132 f.

[67]Scherz, *Vom Wege*, 130. Cf. for the following ibid, 128ff and Scherz, *The Indice of Nicolaus Steno*.

write a detailed report".[68] In October of the same year Stensen, answering a scientific question, writes to Paolo Boccone: "In the cabinet of Pisa I have noticed three kinds of corals of the same nature. I have called them *bie-cheianti*-corals in the catalogue which I have prepared by order of His Eminence." [69]

After the death of the nature researchers Th. Bellucci and P. Nati, Stensen was sent to Pisa in April 1672 by Cardinal Leopold in search of objects of exposition for the collections in Florence and at the same time to prepare a new catalogue of the thus reduced Pisa collection. Probably because of the foreseen journey home, he completed this work already on May 7.[70] At that time Stensen seems to have had the supervision over the grand ducal natural science collections. Magalotti complained about that in June 1671. Not only had he lost a friend but also he had to take over work that did not say much to him.[71]

This catalogue of the year 1672 has been preserved in the catalogue series (*Inventario*) of the Galleria de'Semplici in Pisa. It is a register with precise and comprehensive descriptions of the different specimens including the minerals as well as a short list of the stones which were to go to Florence, but for this part of the collection, Stensen writes that he has completed a particular catalogue.[72]

One of Stensen's catalogues has been kept thanks to Targioni-Targetti, who introduced it in 1763 in his register of the museum in Florence. This catalogue, *Indice delle cose naturali, forse dettato de Niccolò Stenone che feci copiare dell' originale, existente nella R. Galleria*, is mentioned several times above in our Chaps. 2.6 and 2.7.[73] The particularity of this *Indice* as compared, for example, to the Pisa inventory is that Stensen seems to have been responsible not only for the registration but also for the systematizing and even so, which is particularly interesting, that he had collected a great many of the objects himself. Among these are samples from England, Spain and Egypt, where he never travelled. The abundant grand ducal collections were also at his disposal.

---

[68]FAS, *Arch. Med. Filza* 3318, fol.. 652, Indice, 192.

[69]EP I, 247.

[70]Livorno e Pisa, 595 f, moreover 515, 551 f.

[71]EP I, 23.

[72]Excerpt from *Inventario fatto dal Signore Stenone delle cose naturali*, see Livorno e Pisa, 595 f, dated 1673. The date may only refer to the fair copy. Bellucci, who died in 1672, also had an index from 1675. * The Pisa version is transcribed and introduced by Stefano de Rosa, *Un inventario*. It is also found in the printed catalogue for an exhibition on Steno in Florence, see de Rosa 1986, pp. 63–80, continued in de Rosa, *Niccolò Stenone ... I suoi rapporti con lo studio Pisano*. Cf. Cioppi and Dominici, *The Museum of Natural History of the University of Florence. The Geological and Paleontological Collections*, 2010, 21–24.

[73]English translation, *Indice*, 202 ff.

The *Indice* is a catalogue of a natural science collection with the description of single objects in notices, such as for Nr. 93–94: "Two spheres of dark marcasite" but also quite comprehensive ones such as for Nr. 29: "A beautiful piece of manifold colours which divide into several parallel veins. From them there are several columns united together in their length, which, at their ending, are united with another body, and have each their pyramid at the other ending. And the order of the colours is this, if one begins at the surface of the pyramids: the first opaque white; the second transparent and colourless like water; the third, that of the thickest vein, is strongly amethystine; the fourth somewhat transparent; the fifth white with black peaks; the sixth transparent colourless; the seventh opaque yellowish."

The contents is divided into two parts: the first and largest (Nr. 1 to 255) is systematically subdivided into quartz, other crystals, minerals, sea findings including fossils, and corals. This part was without any doubt collected and put in order in preparation of *De solido*, and thus stems especially from the years 1667/1668. In the second part with the heading "Catalogue of the other interesting things which are to take along", are objects which are written down table for table; thus table Nr. 1 with earth strata (Nr. 262–266); Nr. 2 sea bodies (267) etc. It ends with the Nr. 304: "Below the table, samples of earth layers of different substances."

Where the first part normally registers simple pieces, only groups are mentioned in the second part, like for instance Nr. 298: "Accretions of different colours, formed in water in the vicinity of Tivoli, in Karlsbad in Bohemia, in the silver mines of Bohemia, Tirol, Hungary and in other pla-ces." As shown by examples, in this part of the catalogue the sites of the findings are often indicated, obviously because many of them stem from the long travel of Stensen.

It is hardly possible to determine precisely which collection *Indice* records, although its principal matter is clear: it deals with the collection that Stensen actively used in his research. It seems that the first part comprises the col-lection in Florence and the second part is the above-mentioned special cat-alogue that Stensen prepared for those objects that were to go to Florence.[74]

---

[74]There are thus references to several catalogues by Stensen, but we know fairly little about them. In his letter to Boccone, Stensen mentions a catalogue of the Pisa collection with an entry of *coralloides biecheianti* (EP I, 247), presumably the same catalogue mentioned by Marchetti in May 1671 and part of the Indice in an early version. Years later, Boccone refers to an inventario of the Pisa collection by Stensen, called "molto erudite". * Boccone was in Florence in 1669–1670. If he indeed saw the catalogue himself, it must have been made before Stensen left Tuscany in 1669. In the second half of the Indice, Stensen refers to another registration with number (see nos. 263, 266, 267, 278). When Targioni-Tozzetti worked with the Florence collection in the eighteenth century, he refers to some "small and unmethodical notes on stones and crystals made by Niels Stensen" (Indice, 194). (HMH).

One thing only is certain: when Targioni-Tozzetti, as mentioned above, registered the imperial collection in the following century, he saw many of the objects described in the first and second part of the *Indice* and in his inventory he brings a series of indications on Stensen's notices from the *Indice* and from other sources which he still found, mostly loose labels.[75]

In the summer of 1671, Grand Duke Cosimo III and the Accademia del Cimento dispatched Stensen to definitely refute a theory originating from Aristotle that had been rejected by the academy. It was the so-called *Antiperistasis*, which referred to the concentration and strengthening of one's own forces to repulse an attack of opposed forces. With this principle one sought, in the idea of a superficial analogy, to explain the most divergent phenomena, from the fever in lung inflammations to the observation that cellars are warm in winter and cool in summer. Reports of Alpine caves that produced ice only in the summer were brought forth again and again as proof of that *Antiperistasis*.[76] The court supported Stensen's journeys with money and recommendations, particularly to the Count of Castelbarco and friends in Milan like Buondichi, Count Alessandro Visconti and Manfredo Settala.

In June 1671 Stensen took off headed north. He first visited the caves above Gresta, not far from the Lago di Garda. If one travels from Riva towards Torbole, the road gradually begins to rise and one arrives at a beautiful mountain lake near the village of Loppio. The Gresta valley is a side-valley to the Loppio valley and extends over some 18 km into the mountain. In the seventeenth century the entire region with the localities of Ala and Mori was under the rule of the Counts of Castelbarco. Near Pannone there are still ruins of the castle in which Count Francesco di Castelbarco (1626–1695) and his wife Claudia Dorotea, born Countess of Lodron, received the guest. Thirty years later the valley would become a theatre of war. In his march the Duc de Vendôme, Field Marshall of Louis XIV, destroyed the small manor. The cave itself was situated higher in the valley near the twin villages Ronzo–Chienis, between the two mountains Monte Stivo (2059 m) to the north and Monte Biaveno (1617 m) to the

---

[75]Cf., Scherz, *Vom Wege*, 218 ff, *Indice* 278 ff.

[76]On Cimento Academy and Antiperistasis, see Middleton, 207, 246, 340. The two journeys are only known from the two reports to Cosimo (OPH II, 241 f, 245 ff. Cf. also *Reisen*, 101 ff. Gustav Scherz gave a lively description from a visit to the caves nearly 300 years later in his 'To berømte grotter i Alperne'. English review in Axel Garboe, *Niels Stensens grotto letters 1671*. <839–848>.

south-east.[77] On the slope of the latter the entire mountainside seems pierced by gaps and caves leading into the inside. Glacial air flows out from many crevices and cavities so that the mountain seems to breathe with ice lungs. One of the greatest and best known caves from which the inhabitants often take ice in summer fits the description given by Stensen, for instance, concerning the draught from a cavity situated far inside the mountain, a draught so powerful that the flame of a match was strongly blown outwards. The temperature was close to freezing point.

Stensen thus found his way to that Gresta cave. Bad weather rendered the investigations rather difficult. First he drew a ground plan and a cross-section of the cave and, so doing, he observed a draught from the background. As he had found the holes from which the draught was blowing it appeared clear, "that the cold which one notices outside the cave does not stem from ice within it and that the ice in it is not provoked by a cold which forms inside as a result of the strong warmth which one notices outside but that so cold air escapes from remote cavities of the mountain obviously through a cleft that the small instrument suddenly fell by 13°, i.e. below freezing-point ... [78] In order to discover the source of this cold wind it suffices to consider the time when it is strongest, i.e. when the sun is warmest, and to bear in mind the conditions in the deepest excavations when a mine is being dug. The only cold is the cold which enters from outside in summer and winter alike through the deepest clefts observed there. It would require a cold substance on the ground, such as water or ice, to cause a cold wind which would correspond to the coldness of its source. It is therefore highly probable that the rocks, which are continually warmed on both sides of the mountain by the sun, melt the masses of snow and ice deposited the previous winter in the cavities, which are connected to the cave by clefts. This melting process has two results: cold air is sent out, and the rocks on the floor of the cave are covered with ice. It can be said, then, that the water in the cave is frozen partly by cold air which blows over it, and partly by the coldness of the rocks beneath it."[79]

To determine the time of freezing, whether it occurs at the beginning of summer or at the time of the greatest heat, Stensen still wanted to wait for

---

[77] On the geology of the region, see Corra and Ferrari, *Itinerari di Stenone nelle Prealpi Tridentine ...*, esp.176ff.

[78] On Stensen's thermometer cf. notes EP I, 240, GP 247.

[79] OP II, 241 f = EP I, 239 = GP, 237.

observations that Count Castelbarco had promised him. To conclude: "This interest shown by Your Highness will enable us to come to a final decision about the century-long discussion on Antiperistasis."[80]

After examining the cave of Gresta, Stensen went to Milan where all help was bestowed on the friend of the Grand Duke and co-member of the Accademia del Cimento. He praises Francesco Brondichi, the Florentine resident in Milan, Count Alessandro Visconti and Manfredo Sottala, who helped him in the preparation of the journey to the cave above Moncodeno. The Moncodeno cave lies high over the eastern branch of the Lago di Como[81] which is dominated by the 2400 m high massif of Grigna. If Stensen, as seems probable, started his walking tour in the heights from here, he climbed up from Varenna along vineyards, olive groves and chestnut woods to the 913 m high mass of rocks Esino. From here it is about a 3–4 h walk to the Alps of Moncodeno. On a footpath for mules, which winds around magnificent chestnut woods and larch trees, one arrives at a wider Alpine pasture before the 1296 m high Cainallo pass. Alpine roses and edelweiss greet the hiker. The more romantic part of the walk begins in the pass, of which Stensen writes: "When I reached the cave, exhausted by the walk past frightening deep drops above and below the cave, as well as by strenuous climbing, and overwhelmed by all the new impressions, I did not think of making many observations which might occur to me now. I would perhaps have made them there if the spot had been nearer some inhabited place, and not, as was the case, inhabited rather by goats and chamois than by people."[82] To the left the gorge goes steeply down into the valley. To the right high rocks rise up. The Grigna peak rises over the whole. High above, during almost the whole climb, one faces the green Alpine pastures where cattle browsed on the grass already in the time of Stensen; herdsmen told him of their experiences over a glass of milk and named the ice of the Moncodeno cave an eternal ice. As they put it, "the ice has been there since the beginning of the world."[83] From the pasture it takes only 15 min to the entry of the cave, a cleft in the rock, about 8 m high in the middle of the thin wood. The cave is some 10 m high, 15 m wide and 5 m deep. The ground of the cave is covered with ice on which stalagmites stand like columns with a width varying between some centimetres and 1 m. The ice is clear and hard. Most

---

[80]OP II, 242 = EP I, 239 = GP, 237.

[81]On this cave, see Corra and Ferrari, 183 ff. Only in EP I, 245 and GP, 249 are printed Stensen's drawings appended with the letter.

[82]OP II, 246.

[83]OP II, 246.

of the columns are high. They are formed by drops falling down from the ceiling. To the right of the steps there is an abrupt abyss of rocks of which the herdsmen told Stensen that stones that are thrown down provoke a long rolling noise. Stensen offers a precise description of the cave illustrated by a longitudinal- and a cross-section. He describes the columns of ice in the middle, the ice formations on the sides the clearness of which he points out, and the shape of the ice figures. No draught was observed but an intense coldness which penetrated the feet of the visitor while with sparse light he attempted to seize the picture of the cave in rough drawings.

Stensen was very pleased with his visit and admits that the cave of Moncodeno has by far surpassed his expectations: "I was able to confirm by observation the view I had begun to form by a way of reason in the cave of Gresta."[84] The formation of ice reminded him here in some pieces of the formation of crystal so much that he understood that several people have taken the crystal for hardened ice because they found similarities in transparency and shape. Two experiences certainly preserved Stensen from this erroneous conclusion. One was negative: namely he had never heard that a crystal has been found in ice in any of the caves. The other was positive: crystals are found also in places where there is no ice to be found.

Further conclusions drawn by Stensen from his observations are the following: 1. That it is not warm in a cave if it is cold outside. Otherwise snow could not accumulate in it. It also contradicts the assertions of the herdsmen who call the snow in the cave eternal. 2. That ice forms also in the summer. This is also testified by the herdsmen who led their cattle in these mountains in the summer heat and took ice from the caves. Moreover, one could gather that from the ice formations on the walls. 3. That the water that freezes into ice does not come in abundance but unnoticeably from the air. 4. Finally, as decisive finding against the *Antiperistasis*: that the cold in the cave is not like a kind of cooling concentrated inside fighting against the heat concentrated outside, but stems from the coolness of the snow at the entry of the cave.

Stensen had no more time for a written report about the Plinius source north of Como, the interruptions of which were already known at the time of the Romans, and about the *Fiume latte*, the milky stream, also known since ancient times, which flows foaming in the Lago di Como from a 300 m high rocky declivity from March to October and is completely dry during the winter. In a letter of August 14 from Milan to the Grand Duke, he promised an early oral report, pointed out the merits of Fr. Buondichi for the

---

[84]OP II, 245.

completion of the tour in the Alps and seized the next opportunity to return home to Florence via Bologna.

At the end of September 1671 Stensen received through Cardinal Leopold a letter from the erudite Sicilian Boccone. Paolo Boccone (1633–1704)[85] from Palermo was a competent botanist who had made a name by great works on the flora of Europe. On his last journey at the end of 1669, he had come to Florence, where Ferdinand II had appointed him botanist of the court, which can certainly not have been a well-paid position since the far travelling Boccone wrote from Paris, complaining: "If I had had protectors in Italy I should be no orphan child in France." On February 20, 1671 he wrote to Stensen from Paris that he had heard from Pecorini that one of the greatest difficulties in the exploration of nature is to find the principles and origin of the regular figures in minerals. Obviously Pecorini thought of the crystals about which Stensen had just written in *De solido*. Boccone will only provide a contribution to the resolution of that problem and relates the regularities *de la pierre étoilée de Sicile*. He is obviously thinking of the family of star-like stones of Sicily (*Astraeidae*) with masses of polyps and laminated walls or individual cup corals, which are found in the Red Sea and in the Mediterranean. He has apparently overlooked their organic origin. At least he wants with their help to find out "the causes of the other regular figures in some other stones" and seeks the advice of Stensen.[86]

In his answer, Stensen understandably shows little desire for further discussion.[87] Seeking the causes of the regular figures of the stone, Pecorini would certainly have thought of the figures of a crystal, which often has 18 angles, often even more but sometimes less, and Stensen says: "I also notice that, among the authors who have spoken about that, I have found none who, in so different kinds of crystals, has determined the cause of the shape of the individual crystal and the cause of the figure of the kind and, however, the assertions are based not only on speculation but on investigations of more than 1000 years."[88] On this subject Stensen would think of the passages of old writers mentioned by Boccone. The observations of star corals were nice but were also valid for the coralloids. "Even their experiments on petrified mussels do not indicate the cause. They inform us from the matter of the living to trace their fundamental causes but the origin of these organisms remains hidden. I admit that the natural objects are often recognizable during

---

[85]DBI II, 98 f.
[86]EP I, 224 ff.
[87]EP I, 246 f.
[88]EP I, 246.

their growth but we hardly know anything about their origin or about that initial structure which they have received from nature..."[89] Then Stensen once again names the coralloids that he saw in Innsbruck at Pandolfini's and put in order in the grand ducal collection in Pisa. He sends greetings to Bourdelot and the friends. Boccone answers on January 8, 1673 that he has felt extremely honoured by the writing of Stensen, but otherwise still remains in the dark about the essential difference between corals and crystal.[90]

Likewise could Stensen in given circumstances obtain other tasks. Thus, among the Viviani manuscripts there is a statement from which it appears that Stensen had received the work of Nicolaes Witsen from Amsterdam, *Scheepsbouw en Bestier* (Amsterdam 1671). During a visit from his friend, Viviani was eagerly interested in a section of the book, "How the sails are unfurled at best against the wind" (p. 141). Stensen then translated the five propositions about the clash of forces.[91]

Stensen and Malpighi kept a professional and personal correspondence. Stensen felt such a cordial friendship with Malpighi that he would never want to see clouded by the shadow of any doubt. On September 30, 1670 he writes: "As far as you are concerned, I should wish that you never come to me with excuses. You must believe me that I am so ensured of the degree of your friendship that nothing can happen of which I should not be convinced that it has occurred with good reason without that you ever tell me. After God nothing in the world seems to me to be more sacred than a true friendship in which any shadow of distrust never can be found." At this opportunity we come to know of an intermittent fever of Stensen which, however, was then already overcome.[92] On December 6, 1670 Stensen transmits some books which had been underway for six months to his friend in Bologna. They comprise inter alia Malpighi's *Dissertatio epistolica de bombyce* which had been published in London in 1669 as well as a book by Kerckring, probably his *Spicilegium anatomicum* (Amsterdam 1670) and *Osteogenia foetuum* (Amsterdam 1670) in which precious observations on the *vena cava* in the horse and the development of the foetal skeletons are found.[93]

In the letter already mentioned to Malpighi of November 24, 1671 we see the only medical advice that we know of from Stensen and which he only hesitantly gives. Malpighi, the good and friendly man completely devoted to

---

[89]EP I, 247.

[90]EP I, 281 f.

[91]EP I, 29.

[92]EP I, 223.

[93]EP I, 224, see p. 10.

his work, had much to suffer from intrigues from colleagues at the university and their adherents. Twenty years later he still complains: "At my age, which will soon attain 62 years, I have learnt at my expense that a so-called friend can do more harm than many declared enemies. The persecutions which I bear more and more since 20 years are the sequences of a friend whom I highly appreciate and whom I respect to the death … I am on the brink of the grave. A persistent disease of the kidneys which always caused me turbid and often bloody urine has made me practically a dead man since I cannot carry myself further than the next church."[94]

Malpighi's face was pale and meagre and, whereas in his younger years he suffered from stomach pain, fever and inflammation of the eyes, he had in his old age frequent kidney and stone pain and rheumatic ache.[95]

Stensen's advice refers to the importance of food: "I should, therefore, not absolutely doubt of your complete recovery if you would simply make up your mind to think of your diet."[96] He knows three persons whom the physicians had given up. In at least two of them a change in diet restored them to perfect health. He himself had seen a man become so strong after a change to frugal food that he could walk without a cane after having passed his 90th year. He was to abstain from food that may excite gastric juice. And he advises against milk, cheese and whey cheese. In the letter of December 22 Stensen further explains this dietetic advice:[97] "As far as wine is concerned, I do not know whether it is beneficial to your constitution, also not apples. But I know somebody in Florence who instead of wine took meat broth and beer and, instead of other fruits, raisin or dried grape and he felt himself much better with salted juices which tormented him before, as also freed from difficult evacuation of the bowels." Stensen also pointed out the solace of faith: "And then the uppermost physician often helps not the little when we work together with Him also in the domain of nature, through His extraordinary assistance and through such a help the well-being of the soul is also served."

In the activity of Stensen in those years one looks vainly for any sign of work at his own *De solido*. The Grand Duke was, as mentioned, in a very positive mood for this work and he provided Stensen with good working

---

[94] Atti, 309.

[95] Atti, 409, cf. the reference in Adelmann V, 2418 (on the health of), and I, 659 (on the postmortem of Malpighi).

[96] EP I, 248, cf. Adelmann I, 375.

[97] EP I, 251.

conditions.[98] March 8, 1671 one hears from Paris that Stensen is busy with a
Traité des coquilles (Vernon to Oldenburg, VII, 498) but we do not know
how much came out of that. Years later Leibniz indicates that Stensen had
given his geological records to his disciple Holger Jakobsen who spent a half
year in Florence with Stensen in 1676 and then visited him in Pisa in
February 1677.[99] It is not unlikely that Stensen, who then faced the episcopal
consecration, transmitted his records to his disciple. We do not know their
nature, their quantity or their final destiny.[100]

1.8.3   Departure, Beginning 1672

At Christmas 1671 Stensen faced a difficult decision because on December 23
he finally received the call to go home in the expectation of which 2 years
previously he had started out on his long journey through Europe. He
immediately wrote to the Grand Duke Cosimo III: "This morning I received
letters from Copenhagen which asserted a communication of His Majesty to
my family. They wanted to call me back to the fatherland by a letter as soon
as I was inclined to accept the call. They would have sent the letter by the
same mail if the difficulties which I had raised because of the call by the father
of the present king 4 years ago had not let apprehend something similar this
time again. The cause of these difficulties, however, should be removed as, on
the 16th of the last month, liberty of conscience was published in
Copenhagen for everybody who considers living in the city. By this move
they hope that, from political motive, the city will enlarge and religious
activities grow for the benefit of the Church."[101]

We do not know from other sources the ordinance to secure the liberty of
conscience of November 16, 1671. Yet, there can be no doubt that the situation
in Denmark was more favourable than ever for Stensen's return home.[102]

Christian V was not personally unfriendly towards the Catholics and all
three members of the chamber of commerce, Griffenfeld, Ahlefeldt and
Gyldenløve, for reasons of trade policy, were for moderating the strict regulations against believers of other faiths. A rescript of December 14, 1670 had

---

[98]EP I, 27.

[99]Scherz, *Indice*, 190, note 1.

[100]Cf. Garboe, *Niels Stensens geologiske Arbejders Skæbne*, and Garboe, *Niels Stensen's lost geological manuscript.*

[101]EP I, 252.

[102]On the following see EP I, 76 ff and in the following Chap 1.9.

already given the Portuguese Jews the right of sojourn and free acquisition. Now on September 26, 1671, the French ambassador Terlon de Bau was allowed an embassy church in Copenhagen as well as the erection of a Catholic cemetery. Three to four priests were allowed to work at the embassy. These and similar free trends certainly provoked a sharp and partially successful protest by the Protestant bishops and parsons but at the turn of the year 1671/1672 the situation was such that it is likely that there was a promise to the relatives of Stensen.

The optimism of the apostolic vicar V. Maccioni, who was at that time staying at the castle of the Queen Mother in Nykøbing, in the retinue of Johann Friedrichs from Hannover, foresaw a rosy future, all the more as the Catholic community had received a zealous missionary, Father Johannes Sterck, who is first testified to be in Copenhagen on April 18 at a baptism ceremony. By "those of mine" of whom Stensen speaks, we must understand Anna Kitzerow and her husband Jakob who then went into connection with the court of Christian V as a distinguished goldsmith, and even more Jørgen Carstensen, the secretary of the royal register and Stensen's former guardian.

Stensen was happily surprised by the call. In the two short years after his return from Holland he had settled down well in Florence and Cosimo III had immediately entrusted him with the charge of education of the crown prince. However, the letter that he immediately wrote to his patron left the decision only formally in the hands of the Grand Duke: "If this proposition had come to me from any other side, I should have rejected it but since it comes from the Prince whose subject God has made me by birth, and since I am thus set between two Princes, I am obliged to one of them as a result of nature, to the other because of his good deeds. I do not venture to take a decision by myself but I expect, from the verdict of the one whose good deeds I enjoy at the moment, an instruction on how I must behave towards the one whose subject I have been made by nature..."[103]

Cosimo III understood the situation and deferred the decision to Stensen in this matter "in which his comprehension and his intelligence will know the appropriate measures to take up better than any other", and comforted him thereabout: "There is one thing which I can assure you, that I will gladly see you anywhere and that I will keep for you the same good will and liking which moved me to keep you at my court with so much joy..."[104] After quiet reflection Stensen made his decision on December 28.[105] He was convinced

---

[103]EP I, 252 f.
[104]EP I, 253 f.
[105]EP I, 254 f.

that Christian V would eventually renounce his return home if he requested it, in view of the proofs of extremely great favour of the Medici. There would also be no personal interest to induce him to depart, "because the proofs of the generosity of Your Highness towards me are so great that I could neither expect nor want more for me. It only remains the motive of the faith which Your Highness himself pointed out", which must be understood as the service of the Catholic cause in the community of Copenhagen.

Stensen requests to be able to use two to three months after the letter to his king, which he sends on December 29, for work still to be specified to him in the service of the Grand Duke, after having spent his spare time in the study of religion.

He also does not give up the hope to be able to return after some years with the permission of the king and to assume again his service to the Grand Duke, "which, in your great goodness, you have offered me from your own impulse". As reasons for his gratitude to the Medici house he asserts: "because God let me quite unexpectedly find in this court an extremely honourable living; because I was there almost unconsciously drawn to the true faith and I learned to know so many servants of God that, among the Catholics of Tuscany, I mean to see those who present us the history of the first centuries of Christianity as an example of true piety."[106] Cosimo praised Stensen's decision and assured him to be willing to reserve him the education of his son. He adds: "With this prospect your separation becomes less unpleasant and your return that more agreeable."[107] As far as the studies are concerned the Grand Duke granted him the greatest freedom and cordially thanked him for a drawing by Swammerdam, which Stensen had sent him on January 2.

The next day Stensen immediately wrote his consent to Copenhagen. A royal rescript followed on February 13, 1672, which announced his yearly pension of 400 rigsdaler from the day of his return to Denmark: "Then it is Our gracious will and order that you immediately start the journey to our realm Denmark."[108] This rescript was sent via Holland and only arrived in the hands of Stensen on April 25. The next day he thanked the assessor in the State college, Griffenfeld, for his benevolence from which it can be concluded that the assessor was the main driving force for the call to return.[109] Stensen promised to hasten his departure as much as possible but more than 3 weeks still elapsed before he could leave Florence. Among other things, he had to

---

[106]EP I, 255.

[107]EP I, 256.

[108]EP I, 261. Almost verbatim the home call dated October 29, 1667 st.v. (EP I, 194).

[109]EP I, 261 f.

complete his catalogue of the collection in Pisa.[110] This was signed on May 7 and on May 12 he sent his books from the grand ducal library back to Magliabecchi: "they form a half library."[111]

Magliabechi depicts the mood, remarkably seized, at the court in Florence which he left, in a letter of May 17, 1672 to Angelico Aprosio in Ventimiglia.[112] Stensen had just visited him and moved him to tears by his extraordinary humility and his devotion to divine matters. Would they make him a martyr in Denmark, would he achieve much success or soon come back? He might well have been called back home, Magliabechi states, but only for the "simple teaching of anatomy" and for a very modest salary. Magliabechi was surprised that he would want to go. Magalotti also deplored the departure—he lost not only his friend but also his spiritual father. There only remained for him the difficult task "the supervision of the museum of natural objects which Cardinal Leopold has made peremptory on order of the Grand Duke", as he wrote to Falconieri.[113]

In this letter, as well as in one that Redi sent shortly after the departure, Stensen is presented first of all as the servant of God and not as a scientist. Redi's letter is an answer to the repayment of 18 doublons, which Stensen had lent. The letter deals above all with money. Redi apparently means that he is too petty and wants to improve. "Still before tonight I will send a certain little alms and I will so go on every day", he writes. "I thank you for the opportunity which you offer me (by his return) and I thank you for that with all my heart." The letter indicates that Stensen and Redi had had long conversations about the Christian way of life and that Redi saw in Stensen a spiritual father. He asks: "Help me with your prayers and ensure me further the honour of your letters which will always be for me an incentive and a help." Finally he says: "Addio, Niccolò, do well. May God keep you for the good of all. Addio. Addio."[114]

On May 19 Stensen left Florence without waiting for his passport. He asked the Secretary of State of the Grand Duke, Apollonio Bassetti, to send it to him in Innsbruck, which also occurred.[115] The letter of thanks by Stensen

---

[110]Cf. above.

[111]EP I, 262.

[112]*Positio* 174 f, Nordström, *Magliabechi*, 37 f. From the letter it appears that Stensen left a manuscript on *Origines* († 234), the Greek Father of the Church, that he already then thought of publishing.

[113]EP I, 23.

[114]*Positio*, 175 f, cf. Sten. Cath. 4, 1958, 5 ff.

[115]EP I, 263. See also *Reisen*, 104 ff.

from the capital city of Tirol reveals the haste of the journey. After a hurried audience by the Grand Duchess the coach carried him further on towards Augsburg.[116]

From Dresden he wrote an interesting letter to the Grand Duke about his travel. In Bologna he had visited Malpighi "who gives up his hermit life in the countryside. He makes every day extremely interesting observations on the liquid from which plants form and grow, as well as on the formation of the chick in the egg. He has already made different drawings and, if he continues like he has started, he will provide a precious work."[117] Stensen expressed great compassion for the sickly state of Count Marescotti, a friend of Malpighi. Full of admiration, he then speaks of a not further named servant of God whom he met in Venice, a man without education but full of love for God and an expert of the heart. "If I enlarge so much upon this subject, the reason therefore is the comfort which I feel only at the idea of the mentioned conversation."[118] Stensen also travelled for the Grand Duke and his collection of natural objects. He was thus promised interesting objects by the Grand Duchess Anna in Innsbruck and, in Dresden, Stensen obtained, though the intermediary of the fellow-traveller Gabriele Angelo Battestini, an audience with the Elector of Saxony, Johann Georg II (1615–1680), who had ruled since 1656 and embellished Dresden with magnificent buildings and was also an admirer of the Italian opera. Saxony's considerable coal mines as well as the clay and kaolin works near Meissen could certainly increase the grand ducal collections with many interesting pieces. Characteristic of the time were the miraculous signs and extraordinary events of which Stensen tells in the letter. He reports them with a certain reserve and adds: *relata refero*. From Dresden the journey possibly went through Hannover and Hamburg to Denmark.

[116]EP I, 263 f.
[117]EP I, 264.
[118]EP I, 264 f.

1.9   The Two Years in Copenhagen
1672–1674

Copenhagen, coat of arms

1.9.1   Copenhagen 1672

On July 3 (July 13 st.n.), 1672, after having been absent for 8 years, Stensen saw his hometown again.[1] During those years much had changed in the town, in the country, and in his family and circle of friends.

On February 9, 1670 Christian V (1646–1699), then hardly 24 years old, had succeeded his father on the throne to be crowned and anointed on June 7 the next year. The young king was a passionate hunter and rider without deeper intellectual interest. He possessed, however, a sound understanding of

---

[1]KRA, Hoffuit-Besoldings Baag udi Schat-Cammerit 1670–1677, fol. 107. st. n. means calendar in new style (Gregorian); st. v. is in the old style (Julian).

© Springer-Verlag GmbH Germany, part of Springer Nature 2018
T. Kardel and P. Maquet, *Nicolaus Steno*,
https://doi.org/10.1007/978-3-662-55047-2_9

people, was duty-conscious, gallant, good-natured and affable, and he enjoyed the favour of his people.

Christian V was above all anxious of maintaining and increasing absolute power, the divine character of which Hans Wandal (1624–1675), the bishop of Zealand, had magnified in his stately *Jus Regium* (I–VI, 1663–1672), which was to prove its vitality.[2] In the fight for influence on the young king, first the triumvirate of Frederik Ahlefeld, Ulrik Fr. Gyldenløve and Peder Schumacher won. They actively undertook reforms of all kinds[3] and Stensen alludes to these in his first letter to Cosimo III.[4] The ordinance of May 15, 1671 about a new rank of counts and barons as well as the introduction of the Order of Dannebrog in October 1671 show the care for the external lustre of the new regime whereas the striving to increase trade and industry led to those tolerant laws that had made the call for Stensen possible.[5]

This call was due first of all to Peder Schumacher Griffenfeld (1655–1699)[6] whose rapid ascension to power Stensen witnessed in those years, as the Secret Council promoted him to Count of Griffenfeld in November 1673. He was appointed Chancellor of the Kingdom, the function of which he took over in July 1674 after the death of Peder Reetz. One can hardly imagine a closer and warmer relation between Griffenfeld and Stensen than that revealed by the two preserved letters to the powerful minister.[7] Griffenfeld wished to improve scientific life at the university and he was determined not to let go of Stensen as a bright star on the Danish firmament.

The city of Copenhagen had also changed during the past years. Everywhere courts and new houses had arisen. Not only single buildings like St. Nikolaj, the church where Stensen had been baptized, which had been provided with a new spire, and the beautiful new Nørreport, the northern gate, completed in 1672. The bellicose mood of the young king aimed at reclaiming lost Danish provinces ensured that the fortification of Copenhagen was built out, particularly between Nørreport and Vesterport where many private houses were expropriated and destroyed. A royal order of March 30,

---

[2]DBL, 3rd ed. 3, 311 f, Jørgensen, *Peder Schumacher* I, 261 ff.

[3]Cf. KU 1479–1979, V, 165 ff.

[4]EP I, 269.

[5]Cf. Jørgensen, *Peder Schumacher*, Chaps. 9, 10, 12. On the new nobility, esp. 388 ff, 400 ff, commerce and industry, 424 ff.

[6]DBL, 3rd ed. V, 290 ff.

[7]EP I, 261, 293.

1672 (st. v.) also required the removal of all houses outside the gates near the Lakes and the embanking of St. Jørgen's lake.[8]

Findings during these excavations immediately absorbed the geologically-minded Stensen. He wrote about this to Cosimo III on August 6 (August 16 st. n.): "During the set-up of the moats around the city, they found in the earth two strata, separated by a stratum of white sand. These two strata are full of pieces of wood and coal below which there is a great abundance of amber or amber mass in different colours. I have seen among those samples mosquitoes and other insects included, inter alia a big piece with a drop of mobile air in a liquid, the same as is seen in crystal. Many conclusions can be drawn from that, such as, among others, that the whole island of Copenhagen consists of sea sediments, and this in a region where there was previously no sea but forests and that the amber is a product which was often present in these forests, be it a juice of plants like turpentine, be it something else which has then changed its position in the earth as a result of the liquid of the earth, be it that the amber which is found below Pomerania has not sprung up there but comes to the light when the alluvial earth flushes to the surface that which it held buried in its clasp."[9]

The home in Købmagergade at Klareboderne had developed with Stensen's sister as a house-wife and Jakob Kitzerow in charge of the goldsmith's workshop. Here six fellows and apprentices as well as two maids worked. Kitzerow had become the master in 1655 and had married Anne the following year. After the death of Stichman he had successfully managed the goldsmith manufacture. His accounts in the bank, mostly written in German which may indicate his origin, mention large deliveries to the court, especially in the years after 1670. He possessed two more houses in Copenhagen. In one of them in Studiestræde with a garden, experiments of Stensen are testified. The couple had several children, Johannes who became student in Slagelse, and some daughters.[10] Stensen's older step-brother had then been vicar in Scania for a long time. He was a widower, had little fortune and some of his sons seem to have befriended Stensen.[11] The former guardian of Stensen, secretary of the register Jørgen Carstensen, was still alive and had, beside his other proprieties, a court in Boldhusgade in which many children of his third marriage bustled about. Stensen was accommodated in the house of his brother-in-law very agreeably and at no cost. He writes about that to

---

[8]Cf. Bruun, *Kjøbenhavn*, II, Chaps. 3–4, esp. 288, 240.
[9]EP I, 269.
[10]EP I, 905.
[11]Cf. EP I, 295.

Maria Flavia on August 20 (August 30 st. n.): "I live at my sister's quite peacefully. No one contradicts me since many are of the opinion that everybody becomes blessed in his faith if he only lives well. And, as they do not blame me, they also will not hear anything else although some people also speak well of us."[12] His care for the souls of his relatives would not be entirely unsuccessful.

The situation at the university was not favourable, especially with regard to the actual goal of his call, namely the wish and will of all sensible men and the rulers to raise the level of medical studies. The state of the Anatomical House was a witness of the decay of the medical studies. This *Domus Anatomica* at the edge of Nørregade and Our Lady's churchyard, in the middle of the university quarter in between the residences of the professors, had been erected in 1644 by the friend of Stensen's father, Simon Paulli. There had been a time of great progress of anatomical studies. However, except in Paulli's time, it had only been used in the years 1649–1654 when Thomas Bartholin with the prosector Michael Lyser at his side had carried out anatomical work. Otherwise the Anatomy House was mostly closed; yes, it rapidly decayed and Bartholin's program for the public lecture of Stensen states that "*per plusculos annos pulvere conspersum et obmutescens*", for years it was covered with dust and mute.[13]

The medical faculty possessed properly speaking good heads. Nevertheless some professors at the university in 1664 had by all means quite failed. Jakob Henrik Paulli, who had been appointed *Professor anatomiae* in 1662, had first travelled outside the country. He soon turned away from anatomy and he never came to teach it. Mathias Jakobsen, who had obtained his provisory entrance in the university as vicar in 1664, on that day, August 29 st. v., when Stensen had left Denmark, was then and remained insignificant for the development of medicine.[14] However, the official professors were active men, such as Thomas Bartholin, his brother Rasmus, as well as Vilhelm Worm[15] who supported Ole Borch in the philosophical faculty. However, Thomas Bartholin, who had completed his third rectorship on June 6, 1672 st. v., and his brother were relieved from giving lectures for other functions. The medical activity of Vilhelm Worm was insignificant and Ole Borch was by far a more zealous teacher and above all he taught Latin, chemistry and botany.

---

[12]EP I, 271, cf. 275.

[13]On *Domus Anatomica*, see also Thomas Bartholin, *Cista medica*, 185 ff, V. Maar, *Et blad af Domus anatomicas historie.* * Thomas Bartholin, *The Anatomy House in Copenhagen*, ed. Niels W. Bruun (2016).

[14]Panum, 1 ff, Garboe, *Thomas Bartholin*, 2, 53 ff.

[15]DBL, 2nd ed. XXVI, 292 ff.

In these circumstances lay a wide working field for Stensen, and he was determined to fill it out. As always, he concentrated himself so much that he had little time to do anything else. Five years later, he excused himself for an imperfect report to the propaganda congregation about the situation of the Catholic community in Copenhagen because the accomplishment was made "in a time which kept me busy with other studies."[16] But he missed friends, as can be seen from a later letter on the care of a commission of Cardinal Leopold.

Despite the unfavourable weather in the autumn there were a series of dissections, of which Th. Bartholin speaks. Almost half of all the important demonstrations were carried out in the first months. Stensen himself wrote already at that time, on November 9 (November 19 st. v.) 1672 to Cardinal Leopold about his time schedule: "The opportunity to obtain objects and the interest of our people have constantly given me much to do during many weeks being enough to keep me busy with anatomical works in the weeks to come."[17]

This was all the more admirable since Denmark provided the most unfavourable beginning by keeping Stensen completely in the dark concerning his situation. His return order had been silent about that. Only a salary of 400 rigsdaler was spoken of, and this moreover, was not paid until 1 year after his return, on July 29, 1673.[18] Nowhere is there a mention of the question of his title and rank, of his rights and duties, or of a clearly delineated mission for his teaching. Stensen was here and there named as the *Anatomicus Regius*,[19] for instance, in the *Acta Hafniensia*. That seems to have been just a manner of speaking without any official character. It can be imagined how uncertain the modest Stensen must have felt, how hindered at each step he took on own initiative.

The painfulness of his situation becomes clear through many letters that Stensen wrote to his Italian friends; thus to Cosimo (who had paid for his travel expenses and was ready to support him further) on August 6 (August 16 st. n.) 1672: "so far nothing has been decided about my position ('accomodamento') since the person who has taken on himself to find a way out to my satisfaction is very busy."[20] On October 5 (October 15 st. n.) the situation had not yet improved. "My relatives and friends show to me so far

---

[16]EP I, 338.

[17]EP I, 278.

[18]EP I, 261. KRA, "Hoffuit-Besoldings Baag udi Schat-Cammerit" 1670–1677, fol. 107.

[19]Thus in *Acta Hafn*. I, 249, II, 81.

[20]EP I, 269.

the greatest kindness and nobody attacks me for my religion. However, it is right that they find no way to appoint me but I do not pay attention to that and do not bother them with requests. I leave everything to their discretion, quite convinced that, when God's hour comes, the position will be found which is the best for me."[21] The fortuitousness of the whole situation is apparent in the manner how it came to the dissection of a human corpse: "I had been invited to an improvised dissection of a human cadaver in the anatomy theatre. I believe that God has decreed that so to accustom me to speak publicly and to teach me through the experience that I am not good for the anatomy theatre since they will discover in me many and big faults, partly because, by using some modern languages, I have forgotten the use of Latin, partly because I feel in myself ardent impulses to please people even in matters in which I am truly despicable, *erubesco videri talis in oculis hominum, nec erubresco esse talis in oculis Dei.* (I am ashamed to appear so to the eyes of men, but I am not ashamed to be so in the eyes of God.)"[22] The letter of November 9 (November 19 st. n.) shows that people were conscious of the untenable situation: "Many pity me again and again because they still see me in the same situation as I was at my first arrival although truly I do feel nothing of the pain which they assume. I feel more pity when thinking of their souls for which they care that less that they are consumed by the desire to increase their cares for the body."[23] Had no one really thought of that difficulty before and found a practical solution that would make the convert able to reach the goal of calling him home? It seems that it was not the case or, to say it better, one feels behind all this that a fight went on about that call. The reasons for that noticeable handling were certainly of a religious nature and are not difficult to understand by the then dominant intolerance in general and by tradition at the university in particular.[24]

As the university resurrected in the year 1537 after the implementation of the Reformation, it was a teaching institution for educating priests under the guidance of the State where Lutheran orthodoxy ruled. The limits for scientific studies, says a modern historian, were "narrower than in any Middle-Age university, not only as far as the form was concerned, but also in actual life."[25] The State had taken the place of the Catholic Church. The government also took over the religious matters of the people. From 1625

---

[21]EP I, 274.

[22]EP I, 274.

[23]EP I, 279.

[24]Cf. KU 1479–1979, V: "Det teologiske fakultet", 130 ff., 156 ff.

[25]Norvin, I, 30.

and formally even until 1872, the professors of the university had to acknowledge in writing the Evangelical Lutheran religion. Moreover, according to the statutes the academic grades could not be attributed to any suspect of false doctrine.[26]

Stensen himself was submitted to such disgraceful handling, as one can see, with patience and humility. Yet the scientific work of the anatomist must have been very difficult in a situation like this, which obviously gave him as little as possible contact with the students.

However, several remarks show that joy for his return was also at hand. Thus Rasmus Bartholin writes to Oldenburg on August 14: "You should know that our Mr. Stensen has returned from Italy, and that his intelligence and his knowledge now have contributed with us to the benefit of the students."[27] Jørgen a Møinichen, the parson of Hyllinge, congratulated Stensen for his return on July 17 (July 27 st. n.), 1672 in a long Latin poem of well wishes. He requested: "Explain to the dear fatherland the structure of the heart and the true structure of the earth ground. So will your fame equal that of Alexander and your name, o great ornament, will rise higher than the Alps."[28]

On October 21 (st. v.) Georg Huber, the publisher of the newspaper *Mercurius*, wrote a praising poem on the first dissections in the Anatomical Theatre: "But what do you say, Goddess of music about our Stensen? Do not praise him since this wise man flees praising discourses. Say only this that everybody who has pleasure in seeing remarkable things should see Stensen, ask and read him."[29]

1.9.2   Royal Anatomist

The disciple of Stensen, Holger Jakobsen, has given us a valuable survey of the anatomical activity of his master in his manuscript, *Lectiones anatomicae publicae et privatae prosectoris incomparabilis Nicolai Stenonis*.[30] Jakobsen (Latinized Jacobaeus) (1650–1701), the younger brother of Mathias Jakobsen and a nephew of Thomas Bartholin,[31] had joined the university in 1666 and

---

[26]Matzen, I, 45, Norvin, I, 184.

[27]Oldenburg, XV, 201.

[28]Garboe, *Præsten Jørgen a Møinichen*, II. Garboe, *Niels Stensen (Steno) set fra en sjællandsk landsbypræstegaard*.

[29]Quoted from Jørgensen, *Nils Stensen*, 240 f, note 180, Cf. Huber in DBL, 3rd ed. VI, 594.

[30]REX, Ny kgl. Saml. 309aa 4°. Printed in part in OPH II, 297 ff, EP II, 927 f.

[31]DBL, 3rd ed. 7, 209 f.

studied under the guidance of Thomas Bartholin before continuing his studies in Leiden. Because of the invasion of Holland by the French, he returned to Copenhagen in August 1672, just in time to be present when Stensen was Royal Anatomist. On May 30 (June 9 st. n.), 1674, almost at the same time as Stensen, Jakobsen left Denmark again to spend three more years abroad, during which he would meet his respected master again. Maar says: "Jacobaeus was the only real disciple of Stensen in those years and he was strongly influenced by the whole personality of his teacher, an influence which persisted later under quite different circumstances (Fig. 1.9.1)."[32]

Stensen's former teacher and private preceptor, Thomas Bartholin, also acquired great merit because of his disciple in that period in Copenhagen. His outstanding talent as an author was then praised with the publication of the *Acta Medica et Philosophica Hafniensia*, the first Danish scientific journal, a counterpart of the *Philosophical Transactions* of London and *Journal des Sçavans* of Paris. The first volume appeared in 1673 and the most precious contents of its five volumes published in the years 1673–1680 consists of works inspired by Stensen or his genius.

Ten works by Stensen published in Thomas Bartholin's *Acta Hafniensia*, Fig. 1.9.1[a]	
The city and year of study indicated by from Maar. Titles from the translations by Paul Maquet. Work numbers in Part II	
Human embryo with malformations (Paris 1665)	2.20
Uterus of a hare (Paris 1665)	2.21
On a calf with hydrocephalus (Innsbruck 1669)	2.28
On egg and chick (Paris 1665)	2.19
Observations on movement of the heart (Leiden 1662)	2.10
Eggs of viviporous animals I (Florence 1667)	2.25
Eggs of viviporous animals II (Florence 1667)	2.26
Diversity of lymphatic ducts (Leiden 1662/62)	2.12
Muscles of an eagle (Copenhagen 1673)	2.34
Preface to anatomical demonstrations in the Copenhagen Theatre in 1673—*Prooemium*	2.32

[a]*Open access* www.books.google.com

With Holger Jakobsen and the *Acta Hafniensia* as guides, one obtains a reliable first-hand picture of the anatomical work of Stensen in those 2 years, of the objects of his dissections, the donors and the audience, the times and sites of action and partly of the results of this research.

The first dissection, that of a hedgehog, took place on September 3 (September 13 st. n.) in Kitzerow's house in Studiestræde. Present were

---

[32]Maar in Jakobsen, iii.

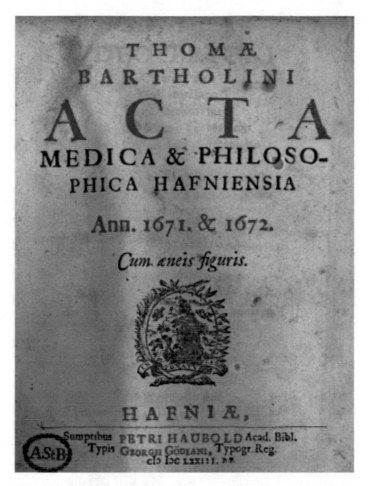

Fig. 1.9.1   Title page of *Acta Hafniensia*, TUB (For a colour version of this figure, please see the colour plates at the beginning of this book.)

Mathias Jakobsen, Daniel Protten, a physician from Göttingen who later practised in Elsinore, and the two nephews of the mayor of Slangerup, Hans and Jørgen Atke.[33]

On September 20 (September 30 st. n.), the dissection of a bear, which the French ambassador had presented as a gift, took place in Kitzerow's house, actually in his garden. In addition to the brothers Jakobsen and Ole Borch, the following were also present: Professor Bertel Bartholin, a brother of Thomas, the city physician Kaspar Kölichen from Silesia, Mathias Moth, the

---

[33]OPH II, 297, EP II, 927, Acta Hafn. I, 175 f.

Fig. 1.9.2 "Among other animals which the very famous anatomist Mr. *Niels Stensen*, having come back to us, publicly and privately dissected with his undefatigable hands and displayed to onlookers, a reindeer deserves the first place, such an animal having never been subjected to the knife of an anatomist so far," are the words of introduction by Thomas Bartholin in his report on Stensen's dissection in *Acta Hafniensia* 1671 & 1672, 274–278, which included three illustrations. REX. The style of drawing is similar to that of Jacobæus in his *Exercitia academica*. An English translation by Paul Maquet is published in the e-journal *Rangifer*, vol 32, No 1 (2012), see http://septentrio.uit.no/index.php/rangifer

son of the late court physician Paul M., Hans Atke, the Royal Botanist Peder Kylling and two surgeons. Stensen injected *spiritus vitrioli* (sulphuric acid) to coagulate the blood.[34]

The first public dissection of a male corpse took place from September 23 to 30 (October 3–9 st. n.) in the Anatomical Theatre. In between there was a dissection of a deer and of a carp. The corpse was that of a soldier hanged for robbery.[35] This resulted in a ghost story. One evening an old townsman was shocked by a supposed apparition, and many took the ghost to be the dissected soldier.[36]

Then on October 27 and 29 there followed a series of private dissections of a dormouse, a cat and a little squirrel.

On October 21 (October 31 st. n.) a greater demonstration again took place in the *Theatrum Anatomicum*. For the first time, a reindeer was cut open by the knife of an anatomist. It was the gift of no less than Griffenfeld. Finally, a she-bear, a gift of Terlon, was dissected. Georg Huber used this dissection as homage in his newspaper.[37]

After the anatomy of a hare on November 4 (14 st. n.) and that of a long-tailed monkey on November 16 (November 26 st. n.), the gift of the organist of Our Lady's church, David Bernhard Meder, who was present himself, in addition to Dr. Mathias Moth, Christian Bartholin and others in the house of Holger Jakobsen, the last dissection took place in 1672 on November 26 (December 6 st. n.) on a reindeer, also a gift from Griffenfeld.[38] Thomas Bartholin was present at the dissection of the reindeer. He reports that one was a male and the other a female. The remark witnesses the abundance of Stensen's work: "When his works permit it, Stensen will himself communicate other observations on the two reindeer, and also on other animals, which our fellows eager for knowledge dealt with under his supervision, as well as on a shortly dissected human corpse."[39]

Activity in the new year began on January 7 (January 17 st. n.) in the house of Ole Borch with the dissection of an eagle, a dog, a calf, an Icelandic she-fox and a marten (a gift provided by the city mayor Peder Resen). Jakobsen

---

[34] OPH II, 297 f, EP II, 927, Acta Hafn. III, 32 ff.

[35] OPH II, 298, EP II, 927, Acta Hafn. I, 277 f.

[36] Acta Hafn. I, 168 ff.

[37] OPH II, 299, EP II, 927, *Acta Hafn*. I, 274 ff. In Scherz, *Vom Wege*, 116, it appears as if two reindeer were dissected on October 31 and a third reindeer on December 6, altogether only two reindeer were dissected. * One dissection was reported by Thomas Bartholin and illustrated probably by students. See Kardel and Maquet, *Anatomy of a reindeer*, Fig. 1.9.2.

[38] OPH II, 299, EP II, 928, *Acta Hafn*. I, 278 ff, 314 f.

[39] *Acta Hafn*. I, 277 f. * English translation by Paul Maquet, see e-journal *Rangifer* 2012, 32(1).

points out in the dog that Stensen prepared the thoracic duct and demonstrated the milky vessels and, in the calf, "the greatest among prosectors, Nicolaus Stenonius, showed the entrance of the lymphatic vessels into the thoracic duct."[40]

On January 29 (February 8 st. n.) the great public anatomical demonstration finally began. It continued until February 8 (18), except on Sundays.[41] It was considered the official beginning of the teaching activity of Stensen in his hometown and represents in its formal perfection and intellectual maturity a summit of development as "one of the most beautiful testimonies which his noble and elevated mind has left. With simple, grasping eloquence he touched the hearts of his listeners by exposing the final aim of research, the mystery of all beauty, the basic foundation of all thoughts and impulses of life."[42] It should be the public endpoint of the life of a brilliant researcher taking place in that house in which 14 years earlier he had seen and carried out his first anatomical experiments as an unknown student. That *Theatrum Anatomicum* in the *Domus Anatomica* in the university quarter was a square two-story building with two large attics and two gables built in Dutch renaissance style. On the ground floor four doors led partly to preparation rooms where the prosector and anatomy servants subjected the corpses to examination and where the anatomist could warm up at an oven during cold winter times. The second door led to a boiling room with a big copper cauldron and the fourth door to the royal lodge. Through the third door one entered into the *Theatrum Anatomicum* proper, a large and high square hall with four ranks of wooden banks and tables arranged as an amphitheatre. In the middle there was a table that could be turned around in all directions. In addition to skeletons of animals, there were two human skeletons, named Adam and Eve, with a tree and the snake of paradise between them, the hallmark of character.[43]

On January 28 (February 7 st. n.), as Dean of the medical faculty, in a programme Thomas Bartholin invited to an act for which a female human corpse had been obtained. He described it as an "immense blessing of the eternal God and the sublime King that the Anatomical Theatre was again awakened to life". He passed over the dead period after his own anatomical activity with a short ambiguous sentence. "Different hindrances, whether due

---

[40]OPH II, 299 f, EP II, 928. The dissection of an eagle is found in *Acta Hafn.* I, 6 ff, in which it is ascribed to Ole Borch.

[41] OPH II, 300 ff, EP II, 928.

[42]Jørgensen, *Peder Schumacher*, 182.

[43]Bartholin, Thomas, *Cista medica*, esp. 200 f.

to the violent circumstances of the war or to hardship in peace time, have had as a consequence that those who should have been most suitable to work have kept their hands away." He excused his own staying far away with his other work, weakness of health or, to speak more openly, with the aversion of the people for the despised and neglected art, an aversion which Stensen also had met. He explained that together with Simon Paulli he had always nursed the ardent wish that anatomy would again revive. May this wish now be satisfied. Through the grace of the King, Dr. Nicolaus Stenonius, "the new Democritus of the century", is back "to convince his homeland that he considers the fame which he gained in the erudite world through wonderful discoveries and treatises which his zeal breathed, not as being to his own but as being to public benefice." So minded, immediately after his arrival, he has unveiled 'without envy' to the academic youth the mysteries of the body and, during the former autumn, he has zealously tackled dissections the results of which were published in the *Acta*. Now he will show the anatomy in a female human corpse and then demonstrate anatomically a reindeer, a gift of Griffenfeld. The century is rich in discoveries: "Among them our Stensen, who is not surpassed by any prosector in dexterity in dissecting and in the facility in discovering, and has worked with such fruitful mind at home and abroad that he has thrilled the erudite world to admiration."[44]

In this opening discourse,[45] Stensen immediately and with noble humility turns the attention away from himself and attributes all honour to God, the king and his listeners: "It has pleased God to disclose to me, although to one not seeking but rather reluctant, many things long denied before me to other anatomists far more worthy than myself. It has pleased the King to open today the anatomical theatre of his fathers, closed for many years, to show publicly the observations of others as well as mine. May it please you to pay attention not to the words and hands of the demonstrator, but to the reve-lation of the wonders of God in his works … The anatomist is a pointer or a rod in the hand of God, pointing out the curiosities of the body like the guide of an exquisite museum. The anatomist himself sometimes deserves to be noticed because of the elegance of his diction and dissection. Such praise belongs to my teachers, my most famous predecessors in this place.

---

[44]Thomas Bartholin, *Universitetsprogram*, 28. Jan. 1673. See Sobiech, Ethos, …, 2016, 92.

[45] <763> OPH II, 249 ff, Acta Hafn. II, 359 ff. German: *Pionier*, 288 ff. Danish: 1. Knud Larsen, *Stenoniana* I, 102 ff. 2. * A. Kragelund, *Den humanistiske renæssance*, 226 ff. Italian: *Stensen, Opere*, II, 257 ff. English: *Steno, Life, Science*, 112 ff. with facsimile from *Acta*.

Sometimes, however, the slip of his tongue and the clumsiness of his hands —this I acknowledge for myself—would rather offend than delight, if by itself the skillful structure did not rivet all the attention of the spectators by itself."

Then Stensen turns to the object of the dissection, "the dreadful picture of the death", and wants to raise the listeners above the bad impressions of the senses in the dissection room through a glance at the inner beauty in the structure of the organs. Nature does not disappoint: "The world only promises more and greater things than it comes up with. Nature comes up with more and greater things than it promises." He reminds us of the value of uncut diamond and unpolished pearls which must be laid bare. The outer beauty often is only a fraction of the inner. A pasture pleases the eye even from far away, but when one bends over the pasture itself to look more closely at the flowers and leaves of the single plants, then the differences and the loveliness of the shapes appear so clearly that one unwillingly exclaims: "from a distance they appear beautiful, but nearby they are far more beautiful!"[46] What joy if one precisely examines one single plant, the single parts, their interplay, the flow of the sap, its growth from seed to fruit (Fig. 1.9.3).

Here Stensen is only thinking of human beauty: all those who remember once having looked at an exciting figure with a mind that was insufficiently equipped against temptation will admit how strongly human appearance acts on the senses of people. This is only comparable to a glance at a distant pasture. Even a microscope will show us only the surface of the human skin, somewhat as if we saw the spikes of grass swaying in a distant grain field. "But actually if this hand, the pretty and well proportioned external aspect of which so often draws the entire attention of the observer, at the same time, transparent like a crystal, revealed also the colour of the tendons hidden there, rivalling pearls, and their artefact which surpasses every ingenuity, who would thence not promise the minds of the spectators a far greater pleasure? Yet if one is allowed to proceed into these same parts, the skin and the tendons, and to look closely at the most skilfully wrought textures of the fibres, the complexity of their course and the intricacy of their labyrinths of which we grasp very little and this only by conjecture since they escape every sense, who would cling any longer to the sensible perception of the external appearances alone and, from the attractiveness or the unpleasantness of that perception, judge the rest? Yes indeed, after having rejected all the errors of the senses, who would not repeat: beautiful is what appears to the senses without dissection; more beautiful what dissection draws forth from the hidden inside;

---

[46]<854> OPH II, 252.

Fig. 1.9.3 Copenhagen's Anatomical Theatre where Stensen delivered his *Prooemium*, from Th. Bartholin's *Cista medica*, REX

yet by far the most beautiful is what, escaping the senses, is revealed by reasoning helped by what the senses perceive."[47] The pagans fight against

---

[47] <856> OPH II, 253.

forbidden loves because they turn their attention to what debases. Asked for a mean against such loves, the anatomist may, instead of going down to faults, raise the loving sense to nobler motives. "We have reason as a judge of the perceptible, thanks to which a reliable approach to the imperceptible is given us. Far be it for us no longer to be men and put ourselves below the beasts. Pursuing the most certain truth by pondering in frequent meditation, let us rather rise from ignorance to knowledge, from imperfection to perfection, and raise in ourselves thoughts worthy of man about his own true dignity. If the smallest part of the human outer aspect is so beautiful and so greatly affects the observer, what beauties would we see, what pleasure would we experience, if we contemplated the entire structure of the body, if we gazed upon the soul which so many and so ingeniously constructed instruments obey, if we considered the dependence of all this on the cause who knows all that we do not know. Beautiful is what we see, more beautiful what we know, but by far the most beautiful is what we do not know."[48]

Finally Stensen speaks of the goal of anatomical science. "Through the stunning work of art of the body to lift the spectators up to the dignity of the soul and, consequently, through the miracles of both to the knowledge and love of the Creator." Here Stensen inserts a defence of the purely scientific anatomy which serves the immediate knowledge, contrary to a purely practical-pathological method of his time which dealt with healing, often without any preliminary knowledge. "Those who make it merely the servant for the prevention and healing of disease are in error, therefore, and treat Anatomy beneath its dignity. It does have a certain use there, but not as much as we think, since the recognition of an extraordinary state cannot extend beyond the knowledge of its natural state. Since the latter is yet quite limited, the former cannot advance its limits much further. True anatomy, however, which fits all observers, is a method by which God leads us first to the knowledge of the animal body, and thence of God Himself by way of the hand of the anatomist."[49]

And then Stensen again points out the relation of all anatomical work to the Creator and the final goal: "For the anatomist should not praise to himself either in his discoveries or his demonstrations. He is just a creature of God, engaged in the work of God, who not only watches on but also operates His own work. He cannot rightly attribute to himself apart from God anything except his own deficiencies and errors. Therefore, I want to ask all of you to

---

[48]OPH II, 254.
[49]OPH II, 254.

praise with me the divine Goodness if you see something worth your expectation, but to ascribe all errors of my tongue and hands to my impatience or a pride hidden for myself, which perhaps wishes more or greater or at least other things beyond the will of God and to which then is deservedly denied even that which I should have obtained easily otherwise."

He then passes on to dissection and, contrary to most professors of his time, Stensen is his own prosector. He summarizes his method in a classical final paragraph: "And so, with God as my guide I shall proceed to the anatomical demonstration of the present body. I will do everything to display to your eyes and to your minds what we know hitherto about the body through reliable researches or reasoning." Then he speaks of the sources of error in the recognition of the truth. With all his strength he rejects any one-sidedness in the understanding. "The belief that it is sufficient for the anatomist to explain the parts prepared for the eyes and that the rest can be completed by the spectators through reading or thinking at home, is silly. I would gladly admit that assumption if there was nothing written on Anatomy by our predecessors as being true which we in our time have recognized as erroneous, or if the mind engaged in the search for truth proceeded free from all prejudices. Now this is far from being so and, since nothing is more difficult to give up than prejudice, even what is carefully published today is not so deprived of errors that preconceived opinions do not leave their traces, and if I exempted myself from such prejudices, I should deserve the stamp of the most insolent arrogance."

Stensen concludes this scientific swan song by presenting his method of scientific cognition which strives to higher certainty, takes into dispassionate consideration the whole material and intellectual nature of man. Full of joy for the achievements of his time, but also with respect for the great men of the past, he mentions this method, which he has himself so successfully applied and will transmit as the experience of his life to the young generation of his fatherland as a representative of that philosophia perennis which he had also paid homage to already in his most brilliant script De solido and which he points out at the moment. "Indeed to defend this pursuit of the truth to the best of my ability, conscious of the easy risk of error, and to avoid the mistakes committed by others, I will not keep to experiments alone nor bring forward arguments alone, but I will seek such a mixture of both that, by the reckoning of everyone, if not most, at least much will possess a demonstrative certainty. To this end from the general science of bodies I shall put forward only that which is common to all philosophers, even to the most opposed among them as I have explained elsewhere in one of my writings."

After this introduction Stensen passes on to the dissection with a short sentence, fundamental for system and order: "I shall present the parts of the body not according to their different positions, but according to their substance and their functions, so as to be mindful of both brevity and clarity. I will be rather moderate in the refutation of the errors of others, remembering the words of a man as pious as he was wise: The knowledge of the truth, he says, is sufficient to discern and overturn all errors, even those formerly unheard of, if they are only brought to light."[50]

In this foreword Stensen appears as quite imbued with the profession of anatomist and nature researcher, considered as highly valuable and intimately bound to his whole philosophy of life. Observe, for instance, how enthusiastically the speaker has presented the ideal of pure science without wanting with that science to break away from life or from faith. As Gosch says, Stensen is a "nature researcher *ex professo* who studied the different branches of natural science for their own sake. In this regard he stands as the precursor of a new time and as one of those who emancipated zoology from medicine."[51]

Attention must be paid to the highly philosophical point of view about the theory of scientific knowledge at which Stensen has arrived in the service of truth and certainty, which he already presents in *De solido* and which he will transmit to his disciples as the most important gift.

Attention should also be paid to the wise, calm and elegant simplicity that appears in the lines in which Stensen speaks of sexual morality and chastity in connection with the public dissection of the female organs of reproduction and birth. It was self-evident in his time that so-doing he did not cowardly drift away from the ethical and pedagogical sides of sexual questions. One can feel how elegantly matter-of-fact his conception based on nature and faith is if one compares to the embarrassed but fussily provoking language which Th. Bartholin sometimes uses for similar admonitions. Stensen recognizes the problem and says: "All who recall having looked closely at the loveliness of a figure, with a mind not sufficiently prepared against enticements, will acknowledge how great the power and the influence of the human figure have on the minds of men." To subdue the covetousness, the pagans would have pointed out the short-comings in the object of their love. The *Prooemium* does not wish any distorted sexual life. It wants to ennoble the lowest by the highest: "The anatomist, however, asked for remedies for such desires, should

---

[50] <859> OPH II, 256. * According to Niels W. Bruun, 2008, 129, note 39, a quotation from Augustine, *Epist.* 118.

[51] OPH II, 256.

not stoop to reproaches, but would raise the loving soul to nobler reasons of love, provided his soul is not completely unable to raise itself a little above the senses."[52] Jakobsen has noticed for us the wisdom which he himself, the pupil, brought with him lifelong from that dissection on February 6 (February 16 st. n.):[53] "The foetus is fed by and grows from a serous humour transmitted through the placenta from the mother's blood into the blood of the foetus and excreted in the outer surface of the foetus which transmits its parts within the amnion in a continuous circulation from the outer surface to the inner and vice versa, until finally the more bitter returns render the foetus restless by their stimulation, from which movement the uterus is stimulated to expel (Fig. 1.9.4).

This very deed, considered in itself most worthy of all deeds that of preservation, is also accompanied by the very great pleasure put there by the Creator in reward of those who carry it out properly. The harshest symptoms are set forth as a punishment for those who abuse it so that they finally take refuge in *mercurialia*. To them, nature proclaims that if pleasure is not enough for arousing love, pain is able to arouse fear." That is a language which is still understood today.

A.D. Jørgensen rightly says that this first great lecture was the swan song of the great anatomist: "His inaugural discourse which should open a new era of the anatomical study actually was a parting discourse to science which had offered him so many lucky moments." But strong doubt can be nursed about the statements according to which Stensen's activity in Copenhagen shows that he had essentially ceased to be a nature researcher.[54]

Whether Stensen as pioneering researcher arrived then at new scientific knowledge we simply do not know due to the lack of sources. The only paper of that year stemming from his hand—the *Prooemium*—shows him above all as a teacher. His appearance was, according to Jacob Bircherod's words: "small, delicate and always weak",[55] but also intellectually towering most university teachers.

However, Stensen had not dropped his old projects of natural science even if, quite naturally, they also tended more and more to a philosophical and theological synopsis, still being full of deep reluctance against proud apparent science and imaginary speculation disregarding facts. This appears, for

---

[52]Gosch II, 1, 245.

[53] <856> OPH II, 253 f. Fig. 1.9.4* Translation by Paul Maquet in Kardel, *Steno—life, science, philosophy*, 139.

[54]Jørgensen, *Nils Stensen*, 186.

[55]EP II, 902, note 1.

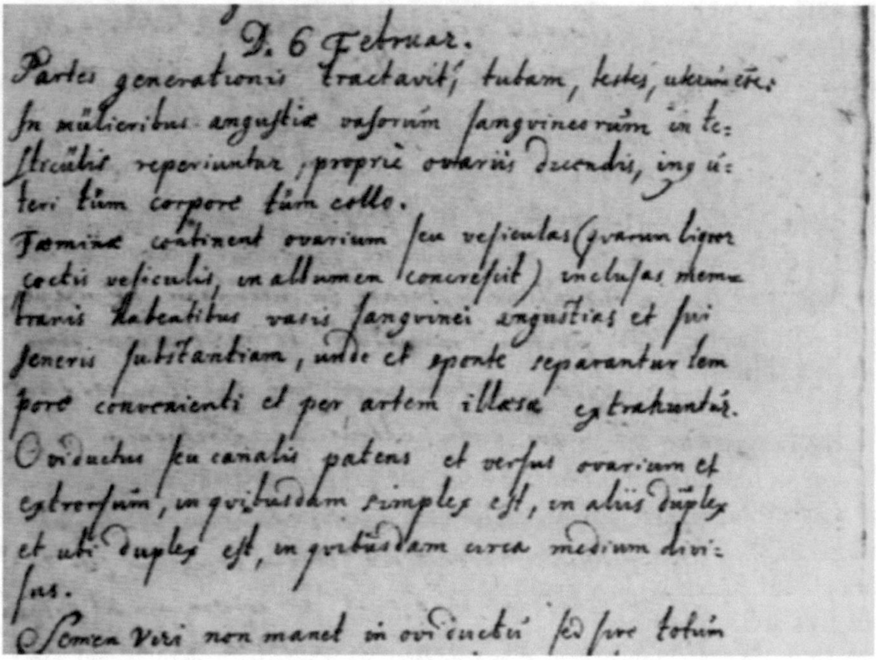

d. 6 Februar.

Partes generationis tractavit; tubam, testes, uterum: In mulieribus angustiæ vasorum sanguineorum in testiculis reperiuntur, proprie ovariis dicendis, inq̃ ut feri tum corpore tum collo.

Fœminæ continent ovarium seu vesiculas (quarum liquor coctis vesiculis, in albumen concrescit) inclusas membranis habentibus varis sanguinei angustias et sui generis substantiam, unde et sponte separantur tempore convenienti et per artem illæsa extrahuntur.

Oviductus seu canalis patens et versus ovarium et extrorsum, in quibusdam simplex est, in aliis duplex et ubi duplex est, in quibusdam circa medium divisus.

Semen viri non manet in oviductu sed fere totum

Fig. 1.9.4 Jakobsen's MS in REX, see Kardel, *Steno—life, science, philosophy*, 145

instance, from the conclusive words of a dissertation on the movement of the heart printed in the *Acta Hafniensia*: "Other things could also be concluded from this. They will be presented in more detail if the Lord let me complete what I undertook on the heart in the past. Meanwhile others will be able to experiment as much as it will be given to the researchers to progress in the way the animal moves and on the cause of the movement. The following facts furnished to me a very valuable argument for trampling the arrogance of human intellect: 1. After so many centuries of work and speculation the philosophers have so far discovered nothing about the true cause and mode of animal movement; 2. Those who pronounced their opinion on this matter with great authority have all sold us false theories instead of demonstrations. I am far from promising a full explanation of everything. But God must be thanked who at least made obvious to all the falsity of those who are in error. He [Horace] of old sagaciously [said]: 'To flee vice is the beginning of virtue, and to have got rid of folly is the beginning of wisdom.'"[56] Here the scientific plans and projects disappear for the solution of which Stensen had found no

---

[56]See p. 438. *Acta Hafn.* II, 147.

time in the course of his astonishing quick journey as discoverer, practically 10 years altogether. He was still in the Leiden discussion of the heart and the muscles and of the muscle problem as debated in Florence with Borelli and Viviani.[57]

But his great new interest was for the anatomical investigation of a number of animals, which he undertook soon after the public demonstration. On March 5 and 7 (March 15 and 17 st. n.) followed the dissection of a parrot and a jackdaw and on April 4 (April 14 st. n.) that of an eagle which is described in the *Acta Hafniensia*. It is a freely sketchy dissertation without illustrations, the last from Stensen's hand. It concludes with these words: "This imperfect description of the muscles, perhaps not without its own errors, is not less boring to readers than their preparation has been delightful to observers. The most delicate, skilfully crafted structures very often met in them are indeed only obscurely described in words. But the flesh revealed to the eyes by the course of its fibres, the colour of the tendons, the proportion of the insertions and the distribution of its pulleys surpasses all admiration. If it be God's will that I complete the Myology of many animals already begun, what now appears to be a sterile study will be abundantly fruitful, and by means of this the true causes of discrepancy in the outer shape of different animals are to be understood. Then by this the mechanism is to be brought to light, indeed, especially to defend the freedom to act of the universal cause against necessity claimed by modern authors, who, while in the trickery of their arguments, seem to take away all freedom. Themselves, in the choice of the most exquisite charming words, too often use the greatest liberty, and thereby destroy in fact what they try so laboriously to ascertain by words."[58]

It seems that there were not many dissections in the following time. On May 4 and 6 (May 14 and 16 st. n.) dissections were performed of a tailed monkey and a hound, on June 1 a dormouse and on June 21 (July 1 st. n.) snakes from the egg. The summer 1673 was spent very quietly. On September 9 (September 19 st. n.) a hound was dissected, partly by Holger Jakobsen who also dissected the dog of Jens Ølby on October 22 (November the 1 st. n.) and a heron was dissected on December 4 (December 14 st. n.).[59]

On September 30 (October 10 st. n.), 1673 Th. Bartholin, applying a new ordinance for physicians and midwifes in furtherance of Griffenfeld, held an examination for 15 midwives in the Anatomical Theatre in the presence of

---

[57]Cf. Stensen's remark on the examination of muscle from November 19 (November 29. st. n.) 1672, EP I, 278.

[58] <871> OPH II, 259–277, *Acta Hafn.* II, 320–345.

[59]EP I, 928, OPH II, 308 f.

Ole Borch, Henrik a Møinichen, Kaspar Kölichen and Niels Stensen. Almost every one of the midwives had themselves given birth to more than 12 children and thus shown their practical skill. However, the anatomical knowledge was so poor that they were ordered to attend Stensen's anatomical exercises in the coming winter and to immediately to announce to the anatomist if a death occurred in childbed among the poor population, so that he could grant them occasional teaching free of charge.[60] Since there were almost no anatomical exercises in the following winter, these decisions can hardly have been carried out.

Of the dissections of the year 1674, Holger Jakobsen cites on December 29, 1973 (January 8, 1974 st. n.) that of a parrot in his room, as well as that of a small pig, on January 21 (January 31 st. n.) that of a sheep's head and on February 12 (February 22 st. n.) that of a crow, on February 18 (February 28 st. n.) that of a duck. Most were carried out by the disciples, above all by Holger Jakobsen; on February 27 (March 9 st. n.) a magpie, an owl and a fox were carried out by Caspar Bartholin. On March 26 (April 5 st. n.) followed the last of the dissections known to us, that of a peacock dissected by Caspar Bartholin and a parrot by Holger Jakobsen.[61] The intention of Stensen to write a book about myology seems to have been widely known. This may coincide with a rumour that had forced its way to J.G. Volckamer in Nuremberg. Th. Bartholin intended an up-dated edition of the *Zootomy* of Marco Aurelio Severino, which the latter, however, contested. It could not come into his mind, even in dreams, how the discoveries in the field of zootomy had now grown unendlessly. He then continues: "What we can expect from our Stensen, who has already returned to Florence, is quite uncertain. One thing is quite certain, that he can contribute much to that contention."[62]

Even if it did not come to the formulation of that myology or of any animal anatomy, the more we might call the *Acta Hafniensia* a testimony of the uninterrupted researcher mind of Stensen. What is found as scientifically valuable in those Danish *Philosophical Transactions* stems more or less from Stensen, above all the dissertations by Holger Jakobsen, where already the chronological succession reveals the common work with the master in Copenhagen or Italy. Jakobsen was a disciple who, despite all enthusiasm, could not always easily follow the flow of ideas of his master. Still in 1677, as he dissected a *Pesce porco* shark in Italy, he wrote: "It is doubtful whether the tongue stones (*glossopetrae*) mentioned by the authors are teeth of that fish or

---

[60]*Acta Hafn.* II, 53 ff.

[61]OPH II, 309, EP I, 294.

[62]*Acta Hafn.* II, 35 f. Bartholin's letter is dated January 15, 1675 (st. v.).

products of the earth. Many people think with Stensen that those dug out from the earth and parts of animals of similar bodies are the remnants of animals which lingered in these regions, others that they originated in the earth without any animal influence." He himself keeps to the famous Bartholin and his dissertation about the Maltese tongue stones.[63] A modern history of comparative anatomy, like that of F.J. Cole, pays all due respect to his and other reports of the *Acta Hafniensia* but adds that the scientific value of the *Acta Hafniensia* did not essentially differ from such works which were then published in Germany, Holland and Denmark, "except for the fact that the German society in the first days could not sing the praise of a Stensen or a Swammerdam." Cole justly points out the interest for biology and zootomy and recognizes Stensen's intelligence in practical service of the Danes: "The genius of Bartholin was not profound, and displayed little searching of the mind, and it is clear that he relied for inspiration on his famous colleague Stensen, to whom he frequently refers in terms of unusual praise as a 'great prosector, subtle of hand and modest in speech'. Indeed in some papers under his name he is only reporting on dissections and drawings made by Stensen, who is occasionally mentioned in *Acta* as the source of much work to which his own name is not attached."[64,65]

Cole expresses the highest admiration for the investigation of the eagle and its muscles. He writes: "Stensen's work is one of the most remarkable essays in zootomy published up to this time, and it is perhaps more detailed and reliable than almost any other, but the unfortunate absence of illustrations handicaps the labours of his interpreters. Among the muscles of the head he describes those of the nictitating membrane with considerable precision, and shows how the tendon of the pyramidalis muscle is threaded through that of the quadratus, in order that the optic nerve may not be stretched or abraded when the nictitans is drawn across the eye, in all of which he was anticipating the Paris

---

[63] *Acta Hafn.* V, 251.

[64] Cole, 369 f.

[65] Stensen's eye for the study and drawing a well-documented conclusion in comparative anatomy is demonstrated when he writes: "Nature to so great an extent is not tied to certain organs to act but is able to carry out a function with diverse instruments. This is demonstrated here by the uvea, the lungs, and the intestines." <590> OPH 1, 206. Combining structure and function, sometimes in a framework of time and relating results with those from studies in other animals or places, Stensen remained aware of own limits and most often stayed within them, "[In sharks] the cavity which, concealed behind the eyes in the cartilaginous skull, undoubtedly serves to hearing. It was very beautiful.... I will deserve indulgence if I present nothing here on the function and mode of action of this canal by which the fishes hear since I do not even know how we hear." OPH 2, 170 <751>. In OPH 2, 151 <735> Stensen also describes the organ in a shark which he considers to be that of smell, and he reflects, "the larger the surface which is touched by the ambient fluid, the more corpuscles can act together on this part. I do not dispute here whether odours are perceived through a humid medium. Something must be assigned to this action."

Academicians. The muscles of the tongue, hyoid, crop trachea, neck, thorax, abdomen, uropygium, wing and leg are all explored so efficiently that this work must rank as the first comprehensive monograph on the muscles of a bird."[66]

It must be kept in mind that in those years Stensen was above all a teacher. He was called back to this end. To that he applied all his knowledge and ability. Records of Jakobsen point to a versatile, stimulating and thorough teacher who organized exercises to familiarize his listeners with the experimental method. "To explore the essence of coldness and heat with the help of experiments…" So begins a section of the *Addictata Stenonii* for the observation of the different properties of ice and vapour. "As far as snow is concerned one can distinguish the certain from the uncertain as follows…", is found in another place.[67]

There can be no doubt that in the course of few years Stensen brought the whole study of anatomy at home to a quite different and higher level. Take only the *Riolani Enchiridion Anatomicum*. Still in 1673 Th. Bartholin recommends that as a textbook for the study of anatomy. Stensen reviews it so critically in Holger Jakobsen's *Addictata* that hypotheses and false conceptions are refuted one after the other. Copenhagen at that time had not yet heard such bold and clear words against what is false and distorted: "Blunders never occur more easily than when one tries to explain the function of an organ. Thus many mistakes have been committed by the ancients and are committed almost dayly by men of our time. Very many have pronounced their judgement without having previously well understood the structure of the organ since they cannot make sure the function of an organ if they had not beforehand grasped its whole structure, the correlation and the concurrence with the other organs."[68] The student must have gained quite an excellent overall view of the state of the contemporary human anatomy, of the organs of the chest and the abdomen, the brain, the eyes, the ears and glands and finally of the reproductive organs and the heart, which Stensen systematically demonstrated and explained. One recognizes his methodical caution, his strict distinction between what is certain and uncertain. One dwells on some single interesting observations and remarks such as about fat, blood, drunkenness, etc. (thus about drunkenness: "People become more easily drunk by small gulps than by full mugs, if namely a large quantity is taken in at a time"). Again and again the doubt arises whether such a formulation actually corresponds to that of Stensen.

---

[66]Cole, 383 f.

[67]Cf. OPH II, 287 f.

[68]This can be found in Holger Jakobsen's recordings, cf. OPH II, 293–297, *Acta Hafn.* II, 122.

Where does the foetus acquire its matter from? What goes into the arteries when a person blushes or turns pale? Where does the thirst of the dropsical patient arise from? The question is asked whether a blood transfusion can be undertaken safely. The answer is typical of Stensen's cautious manner: "We do not know the nature of the blood and even less the signs which distinguish one kind of blood from another and therefore one cannot safely undertake a blood transfusion since, if the nature of bloods to be joined is not known, the sick is exposed to thousand dangers. But every direct injection into the blood also is dangerous and alien to the arrangement of nature which has so well enclosed the blood."[69]

The circle of students and co-workers who Stensen gathered for his anatomical demonstrations was very interested but not very large even if the notes of Holger Jakobsen or the *Acta Hafniensia* do not always name all onlookers. Beside the introduction, Th. Bartholin's own presence is testified only once for a great demonstration, and that of his brother Rasmus not at all. The most zealous and more successful pupils were Bartholin's nephews Holger and Mathias Jakobsen and his son Caspar Bartholin, Jr. A short, overall view of the study of medicine given to a friend in September 1673 shows that Thomas Bartholin had himself a good understanding of the new possibilities of the study of medicine. He typically presents there first of all a long series of textbooks, almost the same titles that had once appeared in Stensen's notes during his study time. However, he says that work will progress only slowly if one does not exert oneself as it is done at the university of Copenhagen. He then cites the public and private dissections, the botanical excursions, the chemical exercises and so many other helpful means, which all remind us of the word of Seneca: Long is the way through prescriptions, short and efficient that through examples.[70] Bartholin immediately engaged his disciple Stensen to work with him at the *Acta Hafniensia*. Already in its first volume, which appeared in 1673, there are a series of Stensen's dissertations, which the latter had obviously brought with him from Italy. It is that more noticeable that the Foreword to this volume, which was written at the end of 1672, does not comprise a single word of welcome or introduction. The second volume, in its spirited introductory homage to Griffenfeld, names Stensen who searches the inside of man and animals "with skilled hand and modest discourse" but Stensen was then already in Florence again and Bartholin mostly wanted to thank the King's Chancellor for the nomination of his son Caspar as a professor.[71]

---

[69]From Jacobæus *Exercitia Academica*, OPH II, 291 ff.

[70]*Acta Hafn.* II, 123 f.

[71]Cf. Garboe, *Thomas Bartholin*, II, 154.

Stensen's old friendship with Ole Borch with all their differences of mind certainly also continued in those years and his presence is testified twice, whereas nothing is heard of Jørgen Ejlersen, who was rector of the university in 1672/1673.[72] Physicians of Copenhagen showed interest, such as Daniel Protten[73] of Göttingen, who became city physician of Elsinore in 1674 but died already the year after. The Silesian born Kaspar Kölichen († 1699)[74] was a distinguished physician who was promoted to doctor in Copenhagen in 1665. He was city physician from 1668 and later became professor. Henrik a Møinichen (1631–1709)[75] from Malmö, the physician in ordinary of Christian V, was a respected erudite, outstanding by his noble character and his affability. There was also Jens Foss (1629–1687)[76] from Lund in Scania, doctor of medicine, then assessor of the supreme court of justice. One also finds Peder Kylling (about 1640–1696),[77] who was among the greatest Danish plant experts of his time, as well as the organist at the church of Our Lady, David Bernhard Meder.[78]

The names of some younger attenders are also known, such as Hans Atke (1651–1692) and his brother Jørgen († 1682),[79] Mathias Moth (1649–1719), the future upper-secretary and zealous lexigrapher,[80] the two sons of Holger Vind (1623–1683), the noble vice-chancellor, who was known as the protector of the sciences, and Poul and Jakob Vinding, the sons of Professor Rasmus Vinding (1615–1684), who exerted great influence in the government and as a teacher and lawyer.[81] The later bishop Frands Thestrup mentions his studies with Stensen.[82] The young German physician Johann Valentin Wille from Colmar (1615–1684) was also a pupil of Stensen.[83] The university of

---

[72]Slottved, 224.

[73]Ingerslev, II, 58.

[74]Ingerslev, I, 518 f.

[75]Ingerslev, I, 508.

[76]DBL, 3rd ed., IV, 505.

[77]DBL, 3rd ed., VIII, 396, cf. KU 1479–1979, XIII, 174.

[78]EP II, 928.

[79]Cf. EP II, 927. Hans and Jørgen Atke were the sons of Karsten Atke, the brother of whom, Hans Atke, was Mayor in Slangerup, cf. Sundbo, 162 ff. Hans Didrik Karstensen Atke was apparently converted to Catholicism by Stensen before he left for Leiden for further studies. He ended his life in 1692 at Peterswardein in Serbia as a Jesuit and a field chaplain of the Imperial army, see Holzapfel, *Unter nordischen Fahnen*, 195 f.

[80]DBL, 3rd ed., X, 79 f.

[81]EP II, 928.

[82]DBL, 3rd ed., XIV, 439 f.

[83]Høeg, *Licent. Med. Johann Valentin Wille*, esp. 20 ff.

Copenhagen then counted about 150 newly registered, and as a whole about 500 students, so apparently there was no great rush for Stensen's exercises.

It did not depend on the teacher, on his ability or will, that the Anatomical Theatre was very rarely used and finally fell dormant again. If ever the fierce words *Etiam hic dii fuere* written above a picture of the *Domus Anatomica* were justified, it was then. One of the greatest Danish nature researchers stood here in the maturity of his intellectual strength, ready to transmit his knowledge to the younger generation. The youth was ready, many of the best in the country wanted it. All the more, nobody could replace him. Griffenfeld and many far-seeing men in the kingdom of science recognized the value of Stensen and Stensen himself, in many passages of his letters, mentions the circle of "mine" who had managed his return (Fig. 1.9.5).

What was then the cause that Stensen this time again could not take roots in his own country? Most directly the failure of those persons in the administration and at the university who had in their power and duty to produce to him a satisfactory ordinance. In his unassuming way Stensen excuses the person who had taken on himself to find a satisfactory solution by much work.[84] Who was this intermediary? Perhaps Griffenfeld himself, perhaps it was Rasmus Vinding, perhaps another professor? We only know with what little zeal the whole matter was treated and we know with how much more energetically the reigning erudite family behaved when it counted to push forward their own offspring. A.D. Jørgensen indicates that in May 1674, as soon as it was known that Stensen would go back to Italy, Rasmus Vinding asked Griffenfeld to bestow the professorship on his step-son and nephew Holger Jakobsen. Griffenfeld did not answer. Jørgensen writes angrily: "Griffenfeld wavered here rightfully. From Stensen's side there was not yet made any request of relief and thus it must appear unseemly to forestall such a request. Rasmus Vinding and his influential friends from the Bartholin family are responsible for the way the brilliant nature researcher has been dealt with and there was not the least reason at hand to speed up the decision, the definitive loss."[85] Griffenfeld let 3 months elapse before he

---

[84]EP I, 269.

[85]Jørgensen, *Peder Schumacher*, II, 155. * A newly recovered unpublished autograph letter in Italian by Stensen dated in Copenhagen on April 28, 1673 to Vincenzo Viviani shows that Stensen already then had thoughts of returning to Tuscany (unchecked transl.): "... The reason of delaying writing to you even further was the hope of hearing perhaps tomorrow the outcome regarding my position, that even now is still in doubt. Yesterday a grand minister of the King told me that tomorrow he wished to speak to me at length and see what could be done. But God knows what will happen. And God may be blessed however it turns out. Meanwhile, with his will being done, everything will be for the good of whoever fears him. Since I cannot gain a stable position, for now, I cannot think either of Signore Lorenzo or Signore Giovan Battista. God arranges everything with them, and with me, according to his Holy will.

dispatched the nominations by which Holger Jakobsen and Caspar Bartholin, Jr. became professors. It took a trimester and yet, how quick it was compared to the waiting time of Stensen.

The principal fault must without any doubt be attributed to the dominant religious intolerance of the orthodoxy for which the Catholic and convert was certainly a splinter in the eye. His work and stay had to be impeded in any way. One only has to think of the official representative of that fanatical opposition, the bishop of Zealand, Hans Wandal (1624–1675).[86] The erudite and dialectically clever theologian, the author and leader of the Church, was "a characteristic representative of an epoch of strict orthodoxy for his incompatibility towards all other theological directions, and a stiff defender of the state church." Already in 1665, in his book *Nubes sine aqua* Wandal had polemized against Valerio Macchioni whose influence on Frederik III he feared. He also wrote sharply against the Jesuit missionary Heinrich Kircher's *Nord-Stern*. Even the small Catholic community in Copenhagen seemed to be a danger to the alert enemy of all free movements and the spiritual father of all ordinances against believers of other faiths; and the more so since this community enjoyed slightly more freedom in those years. This fear was perhaps justified if one thinks of the 200 Danish converts whom the Dominican Hermann Krattmann († 1704) alone had won in Paris.[87]

1.9.3 Amongst Catholics in Denmark

Stensen's arrival in Copenhagen coincided with a period of relative freedom for Catholicism in the capital, on which the apparition of the famous erudite throws particular light.[88] The brave founder and protector of the first Catholic community after the Reformation, the energetic Spaniard, Count Bernardino de Rebolledo (1597–1676) whom Stensen had seen as a student before leaving Denmark in 1659, had found a believing, powerful and wealthy successor in the ambassador of Louis XIV, Hugues de Terlon (about

---

(Footnote 85 continued)

Signore Bartholino is a Professor, Signore Scavenio is a Procurator General, Sig. Langio the Provincial Judge. If I live to the Holy year, I hope to go to see them, to ask permission to come to the service of the Lord Prince, as His Serene Majesty has very kindly told me that he would like me to serve him. ..." To be recorded in REX.

[86]DBL, 3rd ed., XV, 270 f.

[87]DBL, 3rd ed., VIII, 267.

[88]On the following, see EP I, 75 ff and Holzapfel, *Niels Stensen und die katholische Gemeinde in Kopenhagen*.

Fig. 1.9.5 Title page of *Nord-Stern*, the German edition of Heinrich Kircher's *L'Etoile du Nord* Source: *Europäische Bildung in der Residenzstadt Schwerin. Buchausstellung zum Vorseminar der Schweriner Jesuiten 1739 bis 1788 in der Historischen Bibliothek St. Anna*. Schwerin, Heinrich-Theissing-Institut, 2003, 68. About the "Nord-Stern Case", see Clausen, *The written word*, 38–41

1620–1690).[89] De Terlon not only played a great role politically but became, above all, the protector of Catholicism in the North. Catholics could attend the mass with French and German sermon in the chapel of the Embassy. After the death of Frederik III, Terlon was assigned an extraordinary mission by the successor Christian V, namely to obtain free exercise of religion for the Danish Catholics. Terlon was then the French resident ambassador in Copenhagen from 1670 to 1676.

The situation thus was relatively favourable since the new king was not hostile towards the Catholics. The most powerful men of his government, Griffenfeld and Gyldenløve, for political reasons of trade, tended to ease the strict regulations against the sojourn of believers of other faiths. A rescript of December 24 1670 gave Portuguese Jews permission to stay and free acquisition. This tolerance arose the indignation of the orthodox clergymen. When the newly founded college put to the king the request of 20-years' privileges and free exercise of their religion for Christians of all confessions, this led to angry protests from bishop Wandal against the pure "atheists". This was followed on October 24, 1671 by an interdiction of the spreading of German religious books which did not conform with the state religion.[90] However, on October 6, Terlon obtained the privilege of building an Embassy church in Copenhagen. Three to four priests were allowed to work at the church. Even the storm that the script *L'Etoile du Nord* of the Jesuit missionary Heinrich Kircher (1608–1676)[91] set loose by asserting that the Protestant preachers lacked the apostolic vocation and consecration (this was stamped as an attack against the royal sovereignty) provoked little reaction. Bishop Wandal actually published a controversy in which he defended the valid vocation and ordination and Kircher had to leave the country. However, his colleague, Johann Sterck (1630–1692)[92] from Aachen, who had just been expelled from Sweden where his zeal had almost cost him his life, worked all the more actively in the Danish capital city from 1671 to 1679. Every Sunday there were two masses, in the afternoon vesper and every week German catechizing took place. The yearly mission report on ecclesiastical matters for 1672 mentions 25 baptisms, 340 Easter communions and 19 conversions. The optimistic letter of July 7, 1672 which the then apostolic vicar V. Maccioni—who in the retinue of the Catholic Duke Johann Friedrich of Hannover had sojourned for 1 year at the court in Nykøbing of the Queen

---

[89]EP I, 75 f.
[90]Cf. EP I, 280 f.
[91]EP I, 76 f, Holzapfel, *Niels Stensen und die katholische Gemeinde in Kopenhagen*, 31.
[92]EP I, 77 f.

Mother Sophie Amalie, the Duke's sister—lets us conclude with great confidence that there was optimism among the Catholics of Copenhagen, which however the actual situation did not match. Maccioni recounts, for instance, that Sophie Amalie had successfully provoked a discussion on *De notis verae ecclesiae* in front of the king to whom, on that occasion, he had made urging propositions as if it were to his father. Maccioni also gave a catechism and different scripts of controversy to the Queen Mother.

Stensen immediately took an active part in the community life of the Catholics in Copenhagen. The first report to Pallavicini, who soon after assumed the post of Nuncio in Cologne and kept it until 1680, was very optimistic.[93] "The Catholics in this kingdom are not accounted as evil, as there are noble people who believe that salvation could also be in the Catholic Church," which perhaps hints at Frederik Ahlefeld. He recalls the conversion of noble people that occurred again and again abroad.

Stensen highly appreciated that the Catholics could now be baptized and married in the Embassy chapel in Østergade and that the new privilege opened the possibility of a Catholic burial. One could quite freely attend the divine service. The chapel was very small and one had hopes for a bigger church from the generosity of the Roman and French Catholics. It was not even rare that Lutherans came to the services. However, most of the few Catholics were foreigners. The Hamburg community was blossoming. He asks in a letter Sister Maria Flavia in Florence for her prayers, "so that God take pity of those northern countries and offer the life to so many souls".[94] He recommends for her prayers the little successor to the throne, the later Frederik IV. He also asks his princely friend Cosimo III for prayer for the "poor souls of this country which, outside the grace of God and the true Church, live in a so deep and certain sleep as if they were already at the gates of paradise. They object to me the evil which they have seen in Italy but give God that they, like they have followed the evil, also exert the good that they could have learned there if their wish and search of the true joys had been as great as their wish for transitory joys."[95]

On October 5 (October 15 st. n.), 1672 Stensen is dispassionate and tells Pallavicini that the new war (Louis XIV's invasion of Holland which engulfed much money) will hinder the building of the new church "without which I feel great trouble. If we have first the herdsman and the sheep, then the stable will well be found. It is quite true that a public and spacious church would

---

[93]EP I, 274 ff.

[94]EP I, 271.

[95]EP I, 273.

often offer many people of different faith an opportunity to get in and together with the outer word to become aware of the inner. But I see difficulties which remove all hope in regard to that." For the rest he would rather see that the church would be built by a great private benefactor and thus be free from the political constellation. "May the dear God send us many good Catholics and apostolic herdsmen so that they think of nothing else than the salvation of the souls, *omnibus omnia fiunt, ut omnes Christo lucrentur*. I have sometimes heard that one speaks of missionaries and money for their subsistence, but *salvo aliorum judicio*, I hope for myself no other actual achievement than from such missionaries who follow the example of Christ, the apostles and apostolic men like S. Ignacio, S. Xavier and others."[96] In order to arouse zeal, Stensen proposes two indulgences in relation to the Holy Communion, which is too rarely received. The indulgences would be bound to the communion and a prayer for the conversion of the poor souls.[97] Stensen was overcome by much sympathy by thinking of the souls of his fellow-townsmen and he writes to Cardinal Leopold on November 29: "When I see so many souls from the whole North receiving at each moment the gifts of God without reflecting on the proper way and manner of recognizing Him, I cannot keep off feeling pity, the more so that there is behind that no actual malignity as appears from their honest manner to handle everything else except religion. When I speak of that, my language seems quite barbarous to them, *barbarus hic ego sum, quia non intelligor*."[98]

The parish register of the 1670s[99] is preserved and informs us of the conditions of the Catholic community and Stensen's personal acquaintances. Thus at Christmas 1672 (Julian calender.) he stands as godfather of Georg, a son of the surgeon Georg Urstet and his wife Katherina. The parish register says: *Patrinus clarissimus D. Doctor Stenonius et Christina, uxor Domini Philippi Hacquart*. On April 14 (April 24 st. n.) 1673 P. Sterck baptized twins of the Spaniard Antonio Madra from the family de la Colonna. This time Stensen and Mrs. Hacquart were godfather and godmother of the first child Nikolaus, and Jakob des Arcis and Mrs. Hacquart were godfather and godmother of the other boy. The baptism of the son Nikolaus of the Gothenburgan Gerard Kanter whose wife was French took place on June 20 (June 30 st. n.); and Stensen and Dorothea Mars were godfather and godmother. Then for the fourth time the famous anatomist was godfather of the

---

[96]EP I, 274 f.

[97]EP I, 277.

[98]EP I, 279.

[99]Cf. EP I, 79 f, Holzapfel, *Niels Stensen und die katholische Gemeinde in Kopenhagen*.

daughter, Jeanne, of a French couple. The father was René Bureau and his wife belonged, like the godmother Maria Nicola la Rose, to the royal household of the Queen Mother.

The best known of the cited names is undoubtedly that of the Royal Surgeon Philipp Hacquart (about 1616–1698)[100] who, deprived of his property by the Thirty Years' War, had come to Denmark and served under three kings. He enjoyed high consideration and was called inter alia to the sickbed of Queen Christina of Sweden in 1661. Since his salary was paid as belatedly as those of other functionaries of the court in those years, in addition to his medical practice he also ran a trade in greater style. He even equipped ships, inter alia for whaling in Greenland. The wealthy man, who possessed several houses in Copenhagen, was a true Catholic and his second wife since 1658, Christine Vogt (about 1629–1684), a lady's maid of Princess Wilhelmine, the daughter of Frederik III, was also Catholic. Of the sons of Hacquart, one was city physician and another one was also an active physician.

Stensen often went to the French Embassy where he was a very distinguished guest. Particularly the nephew of Terlon was attached to him.[101] Stensen, later in retrospect thought very sternly about that society: "When I went back to my home country, how lukewarm, how faint-hearted, and how little sincere I was though! How often have I been missing the measure and modesty at the table of such great men!"[102]

The motives for going were less the pleasures of the table than the conversations at the table of Terlon, which were as passionate as they were fruitless. In his report of 1677 to the propaganda congregation about the situation of Catholicism in Denmark Stensen writes: "They do not like the scholastic way and manner of debate because they have no clear concepts. Therefore, one passes over unexpectedly into objections against the use of philosophy in matters of faith. In debating with them, one does not come one step further, as I saw it in those years at the table of the ambassador Terlon who had two missionaries very skilled in controversies and every day held open house for twelve persons. There it was endlessly debated, but I do not remember that in the whole time it resulted in one convert, although many others were converted by the services, the Christ doctrine and trustful conversations with the Fathers. Yes, all those debates rarely ended without that a word would escape that embittered. Therefore, I decided no longer to let me be drawn into a dispute without simply giving an explanation of our doctrine,

---

[100]DBL, 3rd ed. V, 458 f.

[101]EP I, 277, note 4.

[102]OTH II, 437.

i.e., instead of debating, to take a piece of paper, to fold it over its length, to write the theme in the middle and on one side the pros, and on the other side the contras."[103]

A survey of the Catholic community with which Stensen concludes this report is, moreover, very instructive. He mentions that his knowledge of the situation is limited, because, as he has said before, other studies have kept him busy, and it is valid only for Copenhagen, partly because hardly any Catholic community in the rest of the country can be spoken about, partly because he has spent the 2 years exclusively in Copenhagen and has not been in any provincial city. The freedom of conscience in the capital city is then greater than ever and quite dependent on the necessity of the king to have Catholic allies, on the good services of the ambassadors to the court and government and on the zeal of these ambassadors themselves. Each conversion provoked complaints by the Lutherans and any political turn-around may be a calamity for the community. To remedy this uncertainty one has thought of different means. One of these "was proposed by the First Minister of the kingdom (that is Griffenfeld) when I was there. He promised to give the Catholics in Copenhagen the same privilege which they enjoy in the free ports if sixteen well-to-do Catholic trade families either from Amsterdam or from elsewhere were willing to settle down in the town."[104] It seems that this proposition was dropped because of the outbreak of the French–Dutch war. Most Catholics were foreigners, among them some wealthy ones who possessed civic rights. However, most of them were courtiers or soldiers, among them some colonels. "Among the Lutherans several ones were very well disposed and, if one wins their sympathy, they may often be very zealous. That cannot be said of all the Catholics who were there in my time. Little obedient to the commandments of the Church, they gave scandal to everybody and caused great sorrow to the missionaries." Stensen then describes some particular instances. There should at best be two missionaries in Copenhagen permanently and an anonymous priest could solve different problems in Copenhagen and in the country. Thus, to serve high-ranking persons who wanted to convert but who feared the consequences, or also to read the mass and administer the sacraments to faint-hearted Catholics who otherwise crept to the chapel before daybreak or in bad weather. Stensen advises to send missionaries of strong personality only and gives inter alia the following pastoral and moral argument: "Above all the meals of the Lutherans are full of traps from which the

---

[103]EP I, 343 f.
[104]EP I, 340.

one who has not made good progresses in the reunion with God will not escape. The men consider as a duty of courtesy to invite people to drink and the women show themselves offended if one does not eat from all which is brought on the table. And who does not resist these sirens will even less resist the third one, namely the very great intimacy in the commerce with women, at the table as well as otherwise. This has contributed to corrupt more than one and, even if nothing bad happens herewith, the opportunity will be given to nasty comments. That is how a Lutheran preacher who debated with me viewed a missionary who had, however, always lived a holy life."[105]

In the autumn of 1673 there occurred a controversy with Brunsmand, a typical orthodox Lutheran. After his theological studies during which he appeared as a zealous debater, Johann Brunsmand (1637–1707),[106] the son of a preacher from Trondheim, in 1668 became rector of Herlufsholm, an old private school in Southern Zealand. The 9 years of his rectorship conveyed a muddy impression because of economical and personal religious contests, like his whole life had something of gloominess and inadequacy. As author of a popular and scientific kind, he showed a talent for style and erudition but was known above all for violent polemics against Catholics and reformers and above all by his several times translated ferocious history of devil and witches, "A dreadful house cross" of 1674. Brunsmand knew the Kitzerow family. In a letter he calls Stensen's sister "highly respected, distinguished, and a good friend", and her son Johannes was instructed privately by him in the elements of logic.[107]

In the year 1673 Brunsmand published in Copenhagen *Francisci Spirae Fortvifflelsis Historiae*, which was reprinted several times. The goal of the book was obviously strongly intended to warn against the passage to the Catholic Church because, as the professor of theology Christoph Schletter says in the Introduction to the first edition, the "Babylonian whore" sends out so many spies and defrauders to mislead ignorant people. Also according to Brunsmand's own words, the fate of Spira was a warning first of all for those who have been born, educated and taught in the pure evangelical doctrine and, however, abjure this quite simply and deny it and pass to horrible papistry or other similar harmful heresy and sects, from greediness for temporal honours, power, for advantage and pleasure or from fickleness and mania for innovation.

The story of Francesco Spira, in the presentation of Pietro Paolo Vergerio and in different translations, had served as a warning example of conversion

[105]EP I, 343.

[106]DBL, 3rd ed. II, 590 f.

[107]OTH I, 141.

for the reformed circles in the whole of Europe already for a whole century. The Italian lawyer of Citadella near Padua had, indeed, taken a lively interest in the Calvinist doctrine in the middle of the sixteenth century and his home soon became a centre of Protestantism in the town. Cited alone before the inquisition court of Venice he felt bound to abjure his false doctrine, to pay a fine and publicly to reject his conceptions. In the follow-up he developed a deep depression, attempted suicide several times and died soon after.[108]

Now Brunsmand presented the story to Danish readers and, in the middle of the year 1673, he sent it to Stensen, who answered on December 7.[109] This occurred privately, at least the ensuing exchange of letters was first published 6 years later in which Stensen simply pointed out that the sentence that Spira allegedly had to abjure was Catholic, namely: "We are justified and saved in God's son Jesus Christ only. We cannot rely on any human work. Nevertheless we dare not for this reason neglect piety and holiness of life which on the contrary we should strive to with the highest zeal so that the will of God which has ordered that fully occurs."[110]

Stensen proves this by using his method. He indeed compares Brunsmand's words and those of the Tridentate council. He concludes: "According to our doctrine Christ only merited for us the justification and the eternal life but the merits of Christ are for the elected ones without their own merit by way of the double gift of the Holy Spirit, namely the gift of the faith or of the divine recognition of the obvious truths, and the gift of the deeds, as well the inner ones like acts of faith, of hope, of charity, repentance, spiritual meditation, as also the outer ones which stem from the inner life and soul, like the oral prayer, alms, mortifications and deeds of repentance..."[111] The disunion would be a consequence of the false conceptions of *meritum, peccatum* and *concupiscentia*, and these in turn a consequence of the one-sided search of the Scripture. The orthodox Protestant interpretation of this concept would kill human responsibility.

Stensen retorts: "I do not pay attention in the same way to all passages of the Scripture where it is dealt with grace and sins, but rather to those which handle responsibility and irresponsibility and human weakness, every word of God and very true, than to those passages which do not explain the whole matter because they need other passages to be fully understood, which handle

---

[108]OTH I, 139 ff.

[109]OTH I, 147, EP I, 283 ff. Thomas Bartholin thanks him for the receipt of his sample on September 18 (September 28 st. n.) 1673, cf. Sten. Cath. 7, 1961, 70.

[110]Brunsmand, 2.

[111]OTH I, 149 = EP I, 285 f.

the transformation of man, the strength of Christ in ourselves, the collaboration of man, the dignity of the laws, the good deeds and their reward."[112] This doctrine prevents many good deeds and at least any striving for perfection.

Brunsmand sent his answer some months later; there he endeavours to say that even the Catholics do not assign their salvation to Christ alone. He also names a series of other Catholic doctrines which appear objectionable to him. One should reproach him to have renounced a woman because of inborn religion. He enumerates a series of Protestant theologians whom he recommends to read.[113]

Some time later Brunsmand himself came to Copenhagen and had a conversation with Stensen, after which he wrote to the sister of the latter, Anne: "From the conversation which I had with your brother I saw well that for him there is no longer any hope to come back to the right religion. He is so firmly ensnared in the papistry that nothing else pleases him than what serves to embellish it one way or another and therefore it also happens that, in that intention, he often says and tells something which is obviously untrue."[114]

A.D. Jørgensen rightly writes about Brunsmand's script: "Stensen speaks above all from his own experience and presents a personal, clear and convincing conception of Christianity. On the contrary Brunsmand gives us a whole book seven to eight times as long as the letter he received. It is filled with theological erudition and plain, in part rude, polemic. He lets his adversary understand that one must be erudite to be able to discuss religion, exactly as one must be erudite to write anatomy books... The philosopher Leibniz also said of Stensen that, from being an outstanding anatomist he had become an average theologian. But in this regard neither he nor Brunsmand have hardly taken the time to distinguish between different things. In fact Stensen was as genial a theologian as he had been brilliant as an anatomist and geologist. His method was the same: to see with his own eyes and then to himself look for the strictest possible study among the sources. Therefore, as anatomist, he never relied on the investigations and descriptions of others. He investigated himself and any theory which was against what he saw lost for him its value, whoever had set it up. He behaved in the same way as theologian with the only difference that the object of his research was the personal life, his own and other people's attitude to God called forth and nursed

---

[112]OTH I, 154 = EP I, 290.
[113]Brunsmand, 24 ff, cf. OTH I, 143.
[114]OTH I, 144.

by the Church. His utterances, therefore, bear almost always a personal imprint; that is the way he lived it, that is the way it took importance for him, that is the way he has learnt it in the history of the Church. That this was called mediocre theology 200 years ago corresponds to everything we otherwise know of the time and its obscure spiritual products, but as a result his utterances have still today the freshness and fullness of life, exactly like his scientific reports, …[115] Alongside the importance of the individual life as a condition for the salvation, Stensen points out the sanctification, the work of the personality with oneself inside and outside, like it had been explained from the beginning of Christianity as a condition … It is just to say that he gave his life to provide a factual proof for the faith which he had won in his personal experience."[116]

1.9.4   Departure from Denmark

At the beginning of 1674 the decision to return to Italy matured in Stensen's mind. The proper aim of his call, the revival of the anatomy study, proved to be an illusion. The mission in itself was not easy and imposed great demands on his knowledge and will. The older Danish generation of erudite had culminated, the circle of younger physicians was neither very fruitful nor large, thus the *Acta Hafniensia* would not survive the publisher, Thomas Bartholin. Stensen certainly was a teacher and researcher who could and would create new blood, but he did not receive the means to do so. Even a modest honorarium of 400 rigsdaler[117] was paid not until 1 year after his arrival and the second annual salary was to follow after his departure.[118] Thus the "Royal Anatomist" in Copenhagen lived, in fact, from the benevolence of his sister and a pension which Cosimo III liberally continued to pay him. Stensen had hardly complained of that but rather of the lack of a secure position in a function with a certain power. This deprived him of any regular possibility to take an influence on the anatomical study. This meant a loss of contact with the academic youth from which, well, he should be kept apart. Stensen obviously never disposed freely of the *Domus Anatomica* and

---

[115]…, although here and there this and that is mistaken and misleading. [the rest of the quoted sentence].

[116]Jørgensen, *Nils Stensen*, 192 ff.

[117]The court physician J.H. Brechtfeld received 600 rigsdaler (Carøe, 41), Professor Rasmus Vinding 500 (Vaupell, 39). Magliabechi mentions this in May 1672 to be a 'provvisione scarsissima' (Nordström, *Magliabechi*, 37). A direct comparison of salaries of the professors is difficult since tarifs did not exist, cf. Matzen, II, 33 ff, Slottved, 22 f.

[118]KRA, Hoffuit-Besoldings Baag udi Schat-Cammerit 1670–1677, fol. 106 b to 107 a.

the means for a demonstration of which only few were carried out in the second year. The disappointment over teaching activity in Copenhagen undoubtedly contributed to Stensen's decision to become a priest, which inevitably led him to give up nature study. Yet nothing reveals that he already nursed such a project in 1672 and that the call at first hand provoked an unmistakable flaring up of an old love for research. In the years of his conversion it came to a remarkable stimulation and deepening of his religious philosophy of life, which stepped forth more and more strongly. Pay attention to the strongly methodical search of certainty, which is unmistakably present in Jakobsen's notes about Stensen's teaching activity, and also the apostolic impulse he received in Holland. Both worked further and were well-behind an unconscious evolution to priesthood. And, however, his friends in Italy at his return still expected that he would complete his geological works for which he had had no time earlier.[119] He himself in his parting letter to Griffenfeld still spoke of the *recherches naturelles*".[120]

Like the first travel abroad from Denmark in 1660 and the second in 1664, this one was also the beginning of a whole new chapter of his life, which led him from natural science to the holiness of the Church. This does not exclude other motives. Stensen longed for the rich Catholic life, which he had come to learn in Florence and Rome. The small Catholic community in Copenhagen had little to offer, and he himself could be of little use to it. His respect for the Grand Duke also required that he accept the function of educator of the Grand Prince.[121] Florence could issue renewed invitations more easily, having received information that Stensen's situation at home was quite unsatisfactory. The invitation certainly occurred shortly before Magalotti's journey through Denmark in 1674.[122]

A letter of invitation has not been preserved but, on May 29, Stensen thanks the Grand Duke for his great goodness and "the offer by letter which goes far beyond what I dared promise myself,"[123] which certainly indicates the position as educator for the Prince under very favourable conditions. Cosimo had already on May 22 transferred a travel sum of money of 300

---

[119]Cf. 21.

[120]EP I, 294.

[121]EP I, 293.

[122]Magalotti was travelling in North and Western Europe in 1674 and passed Denmark on May 20 on his way to Sweden. Apparently he did not meet Stensen during his short stay in Copenhagen. On June 2 he wrote to Cosimo: "I had myself on the way to the Court in Sweden the orders to prepare his return from Copenhagen," Magalotti, 222, EP I, 300, note 13.

[123]EP I, 293.

talers to be paid in Hamburg[124] and on June 5 Stensen presented his parting request to Griffenfeld in a letter which throws clear light on the mood of Stensen in those days: "When Your Excellency sent to Florence the king's order for me to come back here, His Most Serene Highness of Tuscany had already proposed me his idea to let me instruct natural philosophy to the Prince, his son. Therefore, when departing from that court, I promised to come back once as long as the King, my master, would grant me that grace. Actually, being ensured by repeated letters that His Highness would like me to go back now, I humbly pray Your Excellency to intercede for me by His Majesty so that I get permission to go and give expression of my gratitude to that prince for the graces which I received from him. That work could make me more able for the service of His Majesty when he will order me to come back. I also hope not to live there idle in regard to the King, my master, as well through what I may achieve in nature research as for what opportunity may arise of facilitating the first entry for the ones of ours who would consider trading in those seas like those from Holland and Hamburg who each year touch at the harbour of Livorno. This second year has passed without any service to the public for lack of subjects for the anatomical theatre. I hope nevertheless that Your Excellency will be kind enough to obtain for me the grace of the pension with a passport necessary for travelling in time of war. I will pray God every day that he gives Your Excellency all happiness, and for the time being and for the eternity."[125]

The passport and permission to travel abroad which Stensen received on July 14 suggest that he wanted to visit foreign countries again. They do not speak of a return.[126] The nomination of the 19-year-old Caspar Bartholin as *Professor anatomiae ordinarius* on November 7, and also that he was granted a 3-year vacation, shows that Stensen's return was neither planned nor desired. On that same November 7 Holger Jakobsen was nominated as professor of history and geography.[127]

Stensen probably left Copenhagen in mid-July, 1674. Without awaiting the payment of his pension on September 16, he went to Hamburg where we meet him on August 11. In his company there was, in addition to Terlon's nephew, his own nephew, the son of his half-brother Johan who had lived in poverty in, the now Swedish, Scania. Stensen wanted to provide Johan with the possibility of further surgical education at St. Maria Nuova in Florence.

---

[124]EP I, 292.

[125]EP I, 293 f.

[126]EP I, 294 f.

[127]Garboe, *Thomas Bartholin* II, 139.

There were also two converted young girls who would go to a cloister school in Cologne or Hildesheim.[128]

The next goal of the journey was Hannover through an invitation by Duke Johan Friedrich who, during a visit to his favourite sister, the Queen Mother Sophie Amalie, had charged the royal surgeon Hacquart with sending Stensen to Hannover at the first given opportunity. The Duke received Stensen extremely benevolently and urged him to stay for 15 days. Stensen reported to Cosimo III: "In that time, he invited me three times to his table to discuss religious matters with some of his courtiers, the first time with the superintendant of all Lutheran preachers of the state and with Mr. Grote, the second time with the Court Master Mr. Moltke, the third time with a Calvinist. Then he wanted me to carry out an anatomical demonstration, which I did, namely on the blood circulation and the structure of the heart."[129]

The superintendant was no one less than the famous syncretist and Union theologian G.W. Molanus (1633–1722)[130] who had just assumed his function and was also Abbot of Loccum. Otto Grote (1636–1693),[131] Baron von Schauen, was Silesian born and related to the Danish Great Chancellor Frederik Ahlefeld. He had first served Queen Sophie Amalie, then come to the service of Johann Friedrich and had proved to be a very able functionary and diplomat who, as far as religion is concerned, was a convinced Protestant. Gustav Bernhard von Moltke,[132] the Court Master of the Duchess, was married to a Catholic and converted himself the following year.

Further, Stensen met the above-mentioned Apostolic Vicar Valerio Maccioni, whose function he would take over 3 years later. He also made acquaintance with the Capuchin who served at the court church.

Through a Capuchin to whom he had confessed, on the morning of his departure, Stensen exchanged a gold gift received from the Duke into money for the indigents. He writes about this to Cosimo III: "His Highness gave me a gold medal of a value of 25 ducats with his picture and 50 ducats to buy a chain for it… I ask my Capuchin confessor to give the medal back to His Highness after my departure with the prayer of exchanging it for alms, partly because it is above my situation, partly not to risk that, during the journey it

---

[128]EP I, 295 f.

[129]EP I, 297 f.

[130]DTHC 10[II], 2082 ff.

[131]NDB VII, 163.

[132]EP I, 299 f.

falls in unworthy hands. This His Highness conceded and gave me the equivalent to be spent for poor converts."[133]

First, Stensen had to take care of the two girls from Copenhagen, whom he led to Hildesheim to entrust to the custody of Mrs. Rantzau,[134] whom he knew from Paris. In 1666 this lady had founded the cloister "Little Bethlehem" in Hildesheim, the superior of which she remained until her death. In the hospice she achieved the conversion of a young man of whom Stensen writes: "a student of law and son of a mayor from Bremen, directly sent to the hospice by God, who told me his thoughts and, after 3 days of retreat with the Jesuits, made confession of the Catholic faith and showed signs of a vocation to a spiritual life." He followed Stensen to Cologne as reported on September 25.[135]

In Cologne a letter from Swammerdam changed the whole travel plan. By letter, already sent from Hamburg and Hildesheim, Stensen had invited his old friend to travel with him to Italy after Magalotti had informed him of a wish of the Grand Duke to that end. Swammerdam, however, asked Stensen to visit him in Amsterdam where a letter from Cosimo was waiting for him. He thought himself of leaving his relatives and native country and to offer his whole life to God. Stensen would visit him come to please him. The latter writes to Cosimo: "In any case I will go there in the hope, if it pleases God, also to lead this soul to the Church and that Your Highness considers the time and the expenses for this journey as well used. He already writes me that he would apply his skills and efforts only for the honour of God and for the reconciliation of his sins. There is no longer anything, like previously, to impede him to converse with me and with Catholics. Only he may blame me for willing to serve at a court. But I hope that he will make no difficulty to come with me after he will have heard from me what Prince has to be served and that at the court one can live without using the tricks of intrigues of the courtiers which only fit those who set their luck in wealth and honours. May it please God that I achieve all that to the end of which he has called me to his holy Church in his infinite goodness and again let me leave the homeland."[136]

Stensen was in Amsterdam before October 12.[137] Swammerdam ardently desired the arrival of his friend. This is comprehensible in the great spiritual need in which he had come. The groundbreaking research on bees,

---

[133]EP I, 298, cf. EP I, 296 f.

[134]Krebs, *Maria Elisabeth von Rantzau*, esp. 104 ff.

[135]EP I, 298.

[136]EP I, 298 f.

[137]Sten. Cath. 4, 1958, 72.

undertaken between 1668–1673, had exhausted and weakened him physically, all the more as he was subject to malaria attacks. His gloomily pondering mind had then, in 1673, come under the influence of Antoinette Bourignon de la Porte (1616–1680), a mystical visionary who, under the impression of supposed apparitions, as a "new Eve", gathered followers around her. Swammerdam had expressed the course of his life to her in a letter, which she judged as follows: "From the record of your life which you have sent me, I see that everything was nothing else than the pleasures of Satan." That had so deeply disturbed him that he had decided to give up his research as being sinful and to submit himself to her judgement. In 1675 he was still writing about the life of the dayfly, ingeniously as ever, but with innumerable more or less appropriate religious views and biblical citations. He would readily retire into seclusion and, to ensure his living, intended to sell his collection of insects, estimated at 12 000 florins to Florence. However, neither that exchange nor the journey occurred. Antoinette Bourignon dissuaded him with all her forces.[138]

It would likely have been a benediction for Swammerdam if he had surrendered to the influence of his much clearer and exceptional friend and if under the spiritual guidance of the Catholic Church he had severed the bonds to A. Bourignon. The two friends might certainly understand each other in the wish as expressed in 1675 by Swammerdam: "I am now resolved to commit my thoughts more to love the Creator of these things, than to admire him in his creatures,"[139] even without that Stensen would have declared nature research vain or totally sinful because man dares not compete with the All-mighty in the perfection of his works. After all it is not excluded that Swammerdam was then influenced by his friend. In any case, one can see that he remained in connection with him and Swammerdam was accommodated in the home of Stensen's sister when he came to Copenhagen in March 1676. He also saw Antoinette Bourignon more soberly and left his writings for Thévenot, partly striving for their preservation. He particularly wished that the dissertation on bees be published, "because the wise and all-mighty God is so clearly represented therein". Certainly he then fell into some sort of indifference.[140]

The only preserved letter written by Stensen in the last months of the year 1674 informs us of Stensen's own position towards natural science, which he was to abandon. It is addressed to the auditor at the Cologne nunciature,

---

[138]Schierbeek, 45 f.
[139]Cole, 273.
[140]Cf. EP I, 897, note.

Giovanni Battista Pacichelli (1636–1695).[141] In the summer 1674 Pacichelli had ended his work in Cologne and, before going back to Italy, he wanted to visit Scandinavia and to get some information from Stensen. The letter was sent to Stensen and Pacichelli included Stensen's answer of October 28, 1674 in his description of the journey, *Memorie di viaggi*.

In his letter, Pacichelli claimed that Stensen was "born in Florence and brought up in Rome" and this called forth a sensible comparison between life in the service of God and in natural science. Stensen writes back: "Certainly God has given me in Italy that life without which the other one would have been for me the cause of eternal unhappiness, and truly Rome, Florence and other towns have procured heavenly food through the hand of several persons loved by God. Also the divine goodness has shown me different of his wonderful works in the natural processes so that, after having seen such great works of nature and grace, I should keep busier to nurse the life which was given to me in Italy so that the other life could become as if nothing had ever existed of that life which I earlier sought in the discourse of people. A life which deceives the more, the less one discovers the danger therein, coated as it is by many proud titles and by so many examples of persons who are lucky according to the world. Fantasy fills and entertains the soul against its will with such titles and examples, and it impedes itself from directing its energy to itself and to its God to become careful while it is still time for the life it has been created for and for which it is continually called by its God. But even how much I admit the happiness which was earlier mine, so the same things which should serve me for food in the new life renew the old one, and, instead of reminding me of the presence of God that speaks so clearly in the wonders of nature as in our ability to explore them, and in the satisfaction which follows the perception of any singularity, I keep completely to the matter whilst I may forget God because of that which is created only as a sign to remind us of God."[142] Half a year later such thoughts were to lead Stensen definitively from science into priesthood.

---

[141]Scherz, *Una lettera di Niccolò Stenone particularmente attuale*, cf. *Positio*, 214 f.

[142]*Positio*, 215 f. That this letter was written in Copenhagen must be considered to be an error of the editor of the *Positio*. Obviously it was written in Amsterdam. From the rather effusive Pacichelli it seems that Stensen had not only instruction to bring Jan Swammerdam to the Medici court but even the famous mathematician Canon of Liège, René François de Sluse (1622–1685) to become a teacher to the Prince (cf. DSB XII, 459 ff, BNBel 21/22, 716–731). However, the "Oracle of the Netherlands" refused.

1.10   Educator and Priest in Florence
1675–1677

Florence, coat of arms

1.10.1   Teacher to the Prince

We do not know by which way Stensen travelled from Holland to Florence. He may have arrived at the town on the Arno at the turn of the year 1674/1675[1] and, as educator of the Crown Prince, may have been attributed well-furnished lodging in the Palazzo Pitti.[2] Teaching soon began, since Stensen excuses himself to Viviani for not having visited him yet and he adds: "I have the instruction to begin tomorrow at 18 o'clock my service to his Serene Highness, the Prince. Wish me all the best and pray for me so that I accomplish what God has called me for in this position."[3]

---

[1]EP I, 300, Nordström, *Antonio Magliabechi*, 10 and note 7.
[2]EP I, 305, Nordström, *Antonio Magliabechi*, 10, 11.
[3]EP I, 300.

© Springer-Verlag GmbH Germany, part of Springer Nature 2018
T. Kardel and P. Maquet, *Nicolaus Steno*,
https://doi.org/10.1007/978-3-662-55047-2_10

Prince Ferdinand (1665–1713)[4] was 12 years old, a well-built, quite mature boy. He took an interest in physical exercises as well as in his studies. His father took care of providing him with the finest teachers and educators. Stensen shared this task with the Court Master, Marquis Luca degli Albizi, whom Stensen much appreciated, and with the later Archbishop G.A. Morigia, with Redi, Viviani, the brothers Lorenzini and the later Cardinal Noris. The result did not match the efforts of the teachers. Ferdinand should later turn away from the serious and strict behaviour of his father to acquire enthusiasm for many of life's lighter ways much like his mother who was just then spending her last months in Italy. On June 14, 1675 she left for France for ever. Loved by the people and an enthusiast for art and science, the Prince Ferdinand lacked a firm character and soon surrendered to dissipation from which even a sound and intelligent wife, Violante Beatrix from Bavaria, whom he married in 1698, could not divert him. He died childless and sunk into a sickly melancholia before having acceded the government.

Stensen's task, as he wrote to Griffenfeld, consisted of teaching *philosophie naturelle*,[5] which might mean all natural sciences and, moreover, to give his pupil a solid intellectual framework on a Christian base. Two years later, as he was leaving, Stensen wrote to him: "When the Serene Highness, the Grand Duke, the father of your Highness, spoke to me for the first time of my service to Your Highness, this occurred with the precise words that I should teach You the sense of the Christian philosophy, and when I set myself to execute this order, he told me a second time that I should make well understood that there is still another and higher Prince to the authority of whom all Princes are subjected."[6] It was teaching aimed at the knowledge and the behaviour of the Prince and which Stensen, at the end, summarized in a *Trattato di morale per un principe*, which he dedicated to the Crown Prince.

Unfortunately, this *Trattato* has not been preserved but there exists a short description of precious parts, which was written for the young prince.[7] This description shows how Stensen in practice understood to follow the indications of the father by setting his geological knowledge into a greater context.

---

[4]EP I, 39 f, cf. Pieraccini 2, 717 ff.
[5]EP I, 293.
[6]EP I, 322, Nordström, *Antonio Magliabechi*, 14.
[7]Cf. Scherz, *Niels Stensen's Geological Work*, 36.

This presentation has already for a long time been drawing the attention of many people because of its contents related to mineralogy.[8]

In a short annexed notice Stensen writes: "It is useless to be taught anything about places of finding, substance and growth of different stones if we do not know their main and accessory goals in order not to misinterpret them. If we consider the way of life of the great majority of men, we see how those stones are put at the service of arrogance. People wrongly imagine to be the higher considered the more of those stones they possess, the more they ornate themselves or their houses and altars."[9] At the very end of the manuscript he writes: "I have put forward all these points so that, fascinated by the artistic workmanship of beautiful objects, we do not let the results of the curse be the objects of our highest desires."

Stensen begins as follows: "Every ornament, ecclesiastical or secular, is a reminder of the curse which rests on mankind."[10] It is always exhausting and often dangerous to procure them. The experienced far-travelled geologist knows that mines can collapse on the men, that poisonous vapours flow out and can kill in some instants and that other conditions permanently threaten the health of man. In all that one can see the sentence about Adam and his successors: 'In the sweat of thy face shalt thou eat bread, till thou returnest unto the ground'. "In order to earn their living, the hard-working poor must sacrifice their energy and life for what the idle rich use to adorn themselves."

Further he says: "Many of the situations mentioned offer still another proof of the curse in so far as they usually originate in places not created by God, but after the malediction of earth, take shape in crevices, cracks, landslides or produced in some way in subterranean caves. This holds for diamonds and all precious stones whose matter certainly was created at the beginning of time with the other material of the universe, and which was mixed with the other particles of solid and fluid bodies until, after the destruction of the earth it was deposited in old subterranean caves and was fitted by human activity to man's purpose". Stensen finished the chapter with this assertion: "They are all mined from the earth with infinite pain and by risking lives, where the belly is the teacher, and they in reality are monuments reminding on the human curse."[11]

Stensen explains: "A proper use of ecclesiastical and secular ornament is to point to the inner ornament of the soul, the striving to the perfection of the

---

[8]The treatise, the *Ornamenta* <899>, is printed in GP, 250–262 with English translation from a MS with Stensen's own corrections, cf. GP, 264, note 1. It was earlier printed as Sermo XL in OTH II, 342–349.

[9]Scherz, *Niels Stensen's Geological Work*, 47, note 159.

[10]OTH II, 342.

[11]OTH II, 343.

virtues of the true likeness of God." He thus points out to the future Duke that: "Only princes possess the carbuncle, just as this gemstone colours everything else with its light, so should the prince be aware that the morality of his subjects are determined by his own virtues and vices." He warns: "From what is mentioned above we see how unhappy those princes and great men are who blindfold the eyes of their subjects with outer splendour but herewith damage their own mind in the eyes of God by vanity and sins." Further he asserts: "Only those princes who strive before God and their subjects to attain the spiritual qualities of which the external objects given to them by God are a reminder, are worthy the name of princes. For this reason, the prince requires among his necessary knowledge that of the symbolic character of visible objects so that he can look at himself from time to time and at once remember what some particular ornament means and why it is carried in particular occasions." Stensen then gives examples of such meanings and concludes with the remark that the precious stones "must move us to repent our sins, to practice virtue and to yearn for *visio beatifica*."

The preceding considerations leave open the question why God has created a wealth of objects which at the instant of creation apparently had no purpose. Here Stensen mentions our own ignorance. "Whoever considers the problem seriously must admit that the greatest part of earthly objects is, to our knowledge, superfluous. However, no one can say whether they are really superfluous or useless because of this. It is one thing to consider such objects within the bounds of our own intelligence, and another thing to examine them in relation to the total sequence of created objects. It would be equally stupid to call useless something the purpose of which we are ignorant of, and deny the existence of objects we have never seen. This would be the same as an apprentice who denied that his master had other instruments in his workshop than those which he saw there in the first days, or that all instruments whose purpose was unknown to him were superfluous. If we turn to the Creator Himself, we can see the same wisdom on His part with regard to objects with which we are familiar, and we realize that we cannot accuse him of stupidity in the face of the great number and variety of objects, but must believe that there is not a grain of sand or dust missing in the whole universe, or one that serves no purpose or is superfluous."

We are not aware of a later contact between the teacher and his pupil. More than 10 years later Stensen heard of an illness of Ferdinand and wishes him well in a letter of December 9 (December 19 st. n.), 1685 to Cosimo III to whom his first born son already then had caused much concern. He says: "May the Lord give him full health of the body and perfect health of the soul for the consolation of Your Highness and for the actual as well as the eternal prosperity of the Eminent House, and of the whole state, yes, of the whole holy Church because the welfare of one part of the body overflows into the needs of the whole body."[12]

1.10.2   Priest and Missionary

Soon after he began his teaching activity Stensen completed the step to priesthood, which affected his life so deeply. It is very likely that the ordination took place on Holy Saturday, April 13, whereas the first mass was certainly read on Easter Sunday, April 14, 1675.[13]

The ordination just 3 months after the return to the Medici court is quite astonishing. It was not the usual practice of the Church to be satisfied with a preparation of at most a few months. Yet no less than the Archbishop of Florence himself indicates to us the spiritual line of development which quite naturally led to this imposition of hands. It was Cardinal Francesco Nerli (1636–1708),[14] a Roman born whose family palace, however, was in Florence. He had worked as papal legate in Poland, Vienna and Paris, finally became State Secretary, in addition to being at the head of the Archdiocese of Milan from December 20, 1670 until January 11, 1683. This prelate is an example of a priest. He was conscientious and of angel-like manners. As a contemporary says: he was feared for his strictness. He was more competent than many others to form an opinion of Stensen's behaviour in Florence. He writes: "Having become Catholic, he who had also lived innocent in the false sect and had acquired many moral virtues, proposed for himself a strict way of life and observed it so faithfully that in a short time he arrived at a high degree of Christian perfection and very soon became known as a man of prayer, many tears and constant union with God. He was himself quite mortified, full of love for his fellow men, particularly for those in hospitals and prisons whom he assisted as well for their spiritual as for their material needs. His zeal for the honour of God and the salvation of the souls was so great that he sought every opportunity in which he could come into a friendly relation with Jews and heretics that came to this town for their trade. By his pleasant manners and his efficient, truly admirable, art of convincing, he arrived at converting some Jews and many heretics. Some of these, outstanding men, remained in Florence not to subject themselves to the danger of abjuring if

---

[12]EP II, 840.
[13]EP II, 768.
[14]EP I, 46, Pastor, 14[I], 643.

they returned to their own country. They were lovingly received and richly taken care of thanks to the generosity of the Prince. This way of life has acquired for him the love and respect of everybody without, however, lessening the bad opinion which he had of himself. With his great humility he knew how to do so that, even after long personal contact, nobody would have thought that the man who spoke so humbly of himself knew anything, although, according to the specialists, he was the first anatomist of Italy, one of the outstanding philosophers, a great lingvist and, therefore, had been appointed teacher of the prince. If he talked with religious people or wrote to friends, he always described himself as the most miserable sinner who needed the prayers of all. And, however, those who dealt with him for a long time or who lived with him are ready to swear never to have observed the least passion in him."[15]

This beautiful picture, drawn in the hue of the time and place, and stemming from the hand of an important person procures us the strong impression which the life of Stensen made on his surroundings in Florence and which naturally shows that he was quite directly striving to holiness, the more so that Stensen's theological studies were obviously considered as quite fundamental. Like Gregory of Nazianz said of his friend Basilius: "He was a priest before becoming a priest."

It seems that Stensen made himself the decision to be ordained priest, after which he opened his heart to his confessor P.E. Savignani and obtained his approval.[16] Since the ordination certainly occurred in the cathedral, it fell to the cathedral parish priest Ippolito Tonnelli (†1696)[17] to prepare Stensen for the ordination.[18] Tonnelli was well-taught in liturgy. He had, among other things, published a *Sacrum Enchiridion* about the ecclesiastical ceremonies. Since he had some doubt about the validity of the baptism of Stensen, Stensen was baptized again in all discretion and privately *sub conditione*. After due preparation, the permission for ordination as a priest as well as the right to hear confessions was given based on his generally recognized erudition.[19] After the spiritual exercises of S. Ignatius[20] and other pious exercises, Stensen,

---

[15] EP II, 923 f.

[16] Thus may Nerli's account be understood: *ricevuta l'ubbidienza dal Padre Emilio Savignani suo confessore di farsi sacerdote.* EP II, 933.

[17] EP II, 586.

[18] Cf. EP II, 584.

[19] *per il concetto universale di sua dottrina*, EP II, 933.

[20] The expression EP II, 933 *lunga preparazione con gli esercizii di S. Ignazio* may refer to the great 30-days exercises.

in virtue of an apostolic breve was ordained sub-deacon, deacon and priest on three feast days.

He celebrated his first mass, as said, on Easter Sunday 1675 at the altar of grace in the SS Annunziata, the richly decorated church which Michelozzo had erected in the middle of the fifteenth century to replace the original oratory of the seven founders of the Servite order of 1250. On the left just after the entry there is a small marble sanctuary with the grace picture of the Annunciation, which Bartolomeo might have painted in a wonderful manner and which is surrounded by exquisite pieces of art and dedicated gifts.[21] Pious people of ancient times had come in pilgrimage, Saints like Carlo Borromeo, popes and princes and many others. Here Stensen's friends prayed for his conversion. Here he himself rushed after the decision to convert had been taken, to give thanks, and here also he wanted to celebrate his first mass.

Stensen himself has explained to us the deepest motive for the choice of the priestly ministry, in a letter of May 25, 1675 to P. Athanasius Kircher, in which he thanks the Father for his congratulations in his new position: "Truly, after I had constantly experienced the good deeds of God towards me (and I shall not be able ever to weigh them), I found them so great that I could not help feeling driven with all enthusiasm to offer him the best in the best way as much as my frailty allows. Since I now recognized the dignity of priesthood and that, through the priest as well thanksgiving for benefits as pardon for sins, as other things agreeable to God are offered at the altar, I begged for that and obtained also that I would be allowed to offer the eternal Father the immaculate sacrificial lamb."[22]

A moving word in all simplicity at an important turn of life. A happy soul parts from the past, a soul which, by thinking at the many wonders, natural and supernatural in life, exclaims: "What return can I make to the Lord for his generosity to me?" and which then also finds the answer of the psalmist: "I shall take up the cup of salvation and call on the name of he Lord." (Vulgata Ps 115, 12–13, 17). Stensen had devoted his youth to the striving for truth and certainty and, as researcher attained high pinnacles of science. At his conversion, the holiness of human life, not so much the religion as science, but the lived religion, "the profusion of the redemption" (Eph 3,19) had played a decisive role and made him richer. When he looks back on his rich life before the renunciation of his research career, there is neither resignation nor disappointment in him but, full of deep happiness, he will, thanking

---

[21]<xxxvi> EP II, 770, Richa 8, 1 ff.
[22]EP I, 301.

God, give the best and the highest, "the best in the best way". However, feeling powerless, his glance falls on the Saviour, on Christ who came to deliver and to sanctify the people and finally summarized his life in the sacrifice of the cross and renews this sacrifice bloodlessly in the Eucharist. He thinks of the high dignity of that sacrifice which the last of the Old Testament prophets has already foreseen: "From the rising of the sun to its setting, my name is great among the nations; And everywhere they bring sacrifice to my name, and a pure offering" (Mal 1,11). And through him, with him and in him (*per ipsum, cum ipso and in ipso*) as it is prayed at the mass, he will thank God, fight the devil, gain grace. He became priest to "offer the immaculate God's lamb for himself and others."

The idea of sacrifice became the leading thought in the shaping of the personal life of Stensen the priest. He felt himself drawn to the very ancient Catholic expression of this life, the vows, as is apparent in the further text of the report of Nerli: "Since, in his new dignity, he perceived a motive to greater virtue, to the vow of chastity which he had already previously kept unhurt, he wanted to add the vow of voluntary poverty which he always strictly observed. Because, from the 40 scudi which he received from His Highness as a salary every month, he kept only six scudi for his scanty living and for clothes. He devoted the rest to works of charity with the necessary permission in regard to his vows. He would readily have added even another vow and he urgently went to his confessor, aiming in everything to do the most perfect and to promote the greatest glory of God. Since the confessor did not allow him, he was satisfied by obliging himself not to do anything which would not be to the honour of the divine majesty or the good of the neighbour."[23]

For Stensen it is not only matter of renouncing the material goods as a means to perfection, not only of perfect chastity and celibacy and liberation from all worldly care and distraction somewhat in the sense of the words of Christ: "If you wish to be perfect, go, sell what you have and give to the poor, and you will have treasure in heaven. Then come, follow me." (Mt 19,21), but Stensen would, through a pledge, bring a particular sacrifice, do those acts of which S. Thomas, contrary to the reformers and quietists, teaches that a work with vows is better than one without vows because it leaves the liberty to God, strengthens the will, raises the honour of God and deepens the devotion of the men. The new priest thus wanted to oblige himself by a vow, always to do the most perfect of two things. That was a promise which

---

[23]EP II, 933.

S. Theresa had made and which brought her into so many qualms of conscience and embarrassment that she was dispensed of it by her confessor.[24]
P. Savignani also dissuaded his zealous penitent. Then Stensen pondered
entering into an order. Upon examining his conscience, he asks, reproachful
at the thought of that time: "For what reason did I not enter the congregation
of S. Paul and of the holy Re-deemer?" [25] Stensen's attention had been drawn
by the Barnabite P. Leonelli and the future Archbishop and Cardinal J.A.
Morigia, then in Florence, whom he knew well, to the here-mentioned
regular clerics of S. Paul, also called Barnabites, an order of the Italian reform
movement of the sixteenth century, which was devoted to the education of
youth, the popular mission and the conversion of believers of other faiths.
The congregation of the holy Redeemer also was widely spread in Italy.

Stensen's position at the court excluded the usual, regular care of souls even
if at times he took part in it. He was asked to become the confessor to some
Polish ladies who could not speak Italian and he obtained, therefore, a
confessional in the church of the Theatins: "It is not possible to describe with
what preparation and love he administered this function and what spiritual
profit he offered to those who entrusted the care of their souls to his
direction."[26]

One readily had recourse to Stensen for difficult cases. He thus assisted a
former colleague and co-worker as a priest on January 30, 1677. He helped
the above-mentioned Dutch anatomist Tilman Trutwin. About the peaceful
death of a colleague in the lap of the Catholic Church, Dr. Jacopo del Lapo
writes: "A Franciscan Father whom the Grand Duke sent to him and the
good, reverend Mr. Stensen who is today one of the best priests of our own
helped him to die as a good Catholic."[27]

It is striking how successfully the priest worked at the conversion of
believers of other faiths. As he said once, it was his particular vocation of
missionary "through the procuring of conversion of others to thank God for
having converted me to his holy faith".[28] Much understanding and a warm
heart for the material and spiritual needs of people who were in the same
struggle as himself was a good prerequisite for that spiritual care of the
converts. Where later he asks for money and alms from the congregation of
propaganda, he does it, not for himself, but: "I hope some help for the welfare

---

[24]Teresa de Jesus, 350.

[25]OTH II, 437 and note 8.

[26]EP II, 933 f.

[27]Manni, 159 f.

[28]EP II, 667.

of those souls which God has given me so that others who are of good will be not afraid if they abandon that position and see themselves forced to another position which is far below that which was secure for them in the Lutheranism or if, forced by the need, they must live among their family with the danger of abjuring, an easy matter for tender plants who miss the opportunity to priests and sacraments."[29]

Stensen's apostolic drive to lead believers of other confessions to the Catholic Church and his special aptitude to do so were so obvious that one readily turned to him for difficult cases. During a sojourn in Livorno, Stensen wrote to the Superior of the Barnabite college S. Sebastiano, P. Amadei Antonietti: "I can, however, tell you that Mr. Antonio has received the wine and has sent a letter to Your Eminence through a friend. He shows himself more and more inclined to accept our holy faith and I hope that not much time passes until he abjures."[30]

Another person who converted in Florence in 1675 was Philipp Senger whom Magalotti had taken with him from Hamburg while his mother, his sister and his brother-in-law would be taken into the Church some 4 years later by Stensen as apostolic vicar in Hannover.[31] Stensen obviously proposed the Exercises to Senger, and one can see how conscientiously he counselled his soul from a series of rules about the behaviour in important matters according to the will of God and according to the laws of Christian equanimity, obviously an indication for the convert in his doubts and fights before deciding.[32]

Jakob von Rautenfels (†1681)[33] from Mitau, the son of a Courland princely upper-secretary and uncle of the Superintendant of Courland, seems to have been drawn already in his young years to seek his luck in the wide world. He knew Eastern and Northern Europe had seen Riga, Uppsala, Copenhagen and Hamburg and had spent 2 years at the home of the Tsar's physician in Moscow. His motto *Spero in spem contra spem* (I hope in hope against hope) lets us suppose that luck was not very favourable to him. In any case he suddenly appeared in Italy in 1673 without any means and in Rome in 1674. In view of founding an existence, he presented himself in the capital of Christianity as a Catholic and went to the famous Jesuit P. Athanasius Kircher with a plan for the recovery of orthodox Christianity. P. Kircher was

---

[29]EP I, 401.

[30]EP I, 318.

[31]EP I, 304, cf. 413.

[32]EP I, 303 f.

[33]EP I, 41 f, cf. Nordström, 18 f and note, esp 36 and 38, p. 31.

immediately enthusiastic. In the return of the Oriental Church to the Mother Church he saw the most important aim of his zealous research of hieroglyphs and he put Rautenfels in connection with Cardinals Rasponi and Altieri and with the influential Mgr. Ravizza. A Russian embassy, seeking western help against the Turks, had been friendlily received in Rome the previous year. The young man from Courland, who reported travels through the kingdom of Russia and boasted of being a trusted friend of the Tsar, managed his business so cleverly that he was asked to write a plan and instructions for the missionaries who were to be sent for the recovery of the Russian Christians. However, he does not seem to have achieved proper co-operation. Rautenfels travelled to Naples and his friends, among them Count Valerio Zani and P. Kircher, seem to have sent him to Stensen in Florence. Stensen not only looked through but also shook Rautenfels and converted him in the autumn 1675.[34] P. Kircher comments on the conversion in a letter of November 7: "Until he was finally touched by the sublime rays of the divine light and after the recognition of the infallible truth of the faith, through you, my Stensen, as interpreter of the divine will and tool of his providence, he was received in the lap of the holy Mother Church."[35]

P. Kircher had known what happened through a letter, which Rautenfels himself, full of remorse, had written and sent to Rome and he put into clinging Latin a bright testimony of the zeal and the aptitude of the priest to gain converts: "It may above all be your mission, my Stensen, and you must consider that yourself you have been in such doubt and condition of the error from which you have escaped by the grace of God, to show to others, who are in the shadow of the death, the way of the eternal salvation by a constant example of behaviour and life. That you do that very positively, I have heard not only from the mouth of our Fathers but it also appears entirely in your small work which you have published, on the true and proper meaning of the Holy Scripture against the adversaries, and which I have read with great joy."[36]

After his conversion, Rautenfels closely attached himself to Stensen, who obviously took good care of him assisted by the Grand Duke, and took him to Hannover and Münster. Rautenfels died in 1681, probably of an epidemic disease during a journey to Florence as a companion to a noble protégé, sent by Stensen.

---

[34] Nordström, *Antonio Magliabechin,* 18.

[35] EP I, 307. It may have been for Rautenfels and his problems that Stensen borrowed a book on religious controversies with *De unitate et Schismate* from Magliabechi in June 1675, EP I, 302.

[36] EP I, 308.

The conversion of a young Dutchman from a distinguished family did not proceed less dramatically. This led Stensen, at least indirectly, to new contact with Spinoza. The person in question was Albert Burgh (1650–1708),[37] son of the general treasurer of the United Provinces, Conrad Burgh van Kortenhoef, and through his mother the grandson of the famous poet and historian Pieter Cornelisz Hooft. Albert enjoyed a very careful education, studied for 5 years in Leiden (1668–1673) and then undertook the usual grand tour to Italy. Stemming from a strictly believing and active Calvinist family, he seems to have then become estranged from the faith under the influence of Spinoza. At least he mocked Catholic practice. This young sarcastic person, however, became pensive after a conversation with a cleric of a regular order in Venice and a visit in Padua to the famous basilica of San Antonius moved him so much that he made the decision to convert, or in any case to seriously consider conversion. In this mood he came to Florence and in contact with Stensen, who reports the following in a letter to Cosimo III: "As I was leaving the palace this morning, I met Mr. Burgh who, quite contrite, was about to write a letter in which he unveiled to me his inner agitation and explained that he would readily have gone today with me to the P. Inquisitor. He still feels some reluctance but notices comfort if he surrenders to the idea of becoming Catholic, and fears when he again turns to doubt. It would be nice if nobody else hears of the matter before it is all completed but he is determined to give up everything to arrive at the truth. I hope that God gives him the grace to that end and I may say that I have never met a similar example of conversion. Beside the clear and strong reasons which he has already for our side, he feels such a strong impulse that it does not let him think of anything else since he felt it when he saw the picture of S. Antonius in Padua. He does not seize to wonder enough about it because, for several days later he actually felt constantly impelled to defend the Catholic Church and to want the truth. It is the more amazing if one remembers that several times he has called the named Saint a Saint in the vogue. I have promised to come back this evening at about eight o'clock and I will see in what mood he feels himself. If he continues like this morning, we can make the first visit to the P. Inquisitor today."[38]

Unfortunately we do not know the precise date of that "today",[39] but there is no doubt that Burgh converted and remained in Florence for a long time

---

[37]EPI, 43 f, Emmen, *P. Franciskus Burgh*. See also Geurts, *Niels Stensen en Albert Burgh*.

[38]EP I, 305.

[39]It was apparently in the month of June that Stensen let Burgh read the famous and very controversial work of A. Arnauld and P. Nicole, *Perpétuité de la foi de l'Eglise touchant l'Eucharistie*, vol. I–III, Paris,

under the guidance of Stensen.[40] From December 5–20, 1676 he was in Rome, where he is found in the register of the converts' hospice, which Clemens X had then generously settled in the Palazzo del Ripetta.[41]

In Rome he came to know his countryman, the Dominican P. Martin Harney, under whose spiritual guidance he put himself and he felt himself powerfully enticed to the ideal of the order of Saint Francis of Assisi which he met in the so-called Reformella, a more severe trend of the Franciscan order, which Bonaventura from Barcelona had founded in Rome. Uncompromising, he immediately chose a life according to the recommendations of the Gospel, dismissed his servant, dressed poorly and lived so that he did not need to request anything from anybody.

At home, in Holland, the announcement of his conversion had set his family and friends in agitation. Instead of support, threatening letters ensued. The situation did not improve when, at the beginning of 1677, he went back to Holland as a poor pilgrim in Franciscan attire. His parents were at first furious and horrified. They hardly wanted to see him again. Later they attempted to retain him in the country. However, Albert declined such proposals and returned to Rome to join the Franciscan order on December 30, 1677. After completion of philosophical and theological studies he was ordained a priest in 1682. He was called Francis in the order and worked as preacher and author of anti-Lutheran controversy until his death in 1708 in the cloister St. Francesco a Ripa.

The conversations between Stensen and Burgh up to the conversion must also have impressed Spinoza, whose friendship and doctrine had exerted such a great influence on both men. Stensen certainly presented to the young Dutchman the fundamental ideas and dangers of the Spinozean system as he had lived them himself. Stensen let him read a letter that he had addressed to the philosopher many years earlier. Burgh on his part decided to write to Spinoza and carried out his project on September 3, 1675.[42] After the

---

(Footnote 39 continued)

1669–1674, in which the Catholic teaching by the presence of Christ in the Eucharist and its agreement with the teaching of the Orthodox church in the Orient, the Holy Scripture and the Church Fathers is defended, but also was reproached for its excessive strict view of practices in early Christianity. EP I, 316.

[40] On September 3, 1675 Burgh wrote as a Catholic to Spinoza from Florence, cf. Spinoza, *Opera* 4, 280 ff.

[41] *Pastor*, 14[1], 638. At the name Albert Burgh is written: *Convertito gia in Fiorenza da Monsig. Stenon, insieme con Michaele nostro coco, di cui era padrone*—thus a further conversion (EP I, 44).

[42] Spinoza, *Opera* 4, 280 ff.

arbitrariness of Spinoza's assertion had finally come up, his sanguine and irascible nature made him attack the former master as violently as he had obviously earlier been enthusiastic. With the ardency of the young he rejected various opinions in Spinoza's doctrine and invited him to convert. Different sentences in which he criticizes some points of the system of Spinoza or emphasizes the holiness of life, are reminiscent of thoughts of Stensen. In the letter in which Spinoza repulses the attack, one finds the self-conscious sentence: "I do not claim to have discovered the best philosophy, but I know how to recognize the true one", and Burgh is to remember that he himself in his time condemned Stensen whose traces he now follows.[43] It is very likely that it was Burgh's conversion and the exchange of letters with Spinoza that incited Stensen to publish his own letter to the former friend. It appeared at the Grand Duke printer's N. Navesius at the end of 1675 together with other papers.[44] These minor papers had been written already long before they were published in the first year of his priesthood. Both the *De interprete S. Scripturae* and the *Epistola exponens methodum convincendi acatholicum juxta D. Chrysostomum* appeared in October 1675 at the expense of Cosimo III. It may be assumed that the greater part of the edition went to Holland.[45]

From the reaction of one of the up-coming leading reformed theologians in Leiden it is evident what sensation Burgh's conversion and that of Stensen, as well as Stensen's scripts, made in Holland. The ardent young convert had also sent to his philosophy professor in Leiden, Dr. Kraenen, a thorough report on the reasons and the course of his conversion,[46] which is more or less identical with the *Méthode très facile pour convaincre les ministres*,[47] which Friedrich Spanheim (1632–1701),[48] the historian of the Church and passionate polemist of Leiden, took as motive for a dissertation: *Lettre à un amy, sur les motifs, qui ont porté un réformé à se rendre à la Communion de Rome. Où l'on répond aux illusions d'une nouvelle méthode.*[49] Spanheim otherwise says some

---

[43]Spinoza, *Opera* 4, 316 ff, esp. 320. * See also Israel, *Radical enlightenment*, 228.

[44]The letter to Spinoza was published on December 10, 1675.

[45]The date of the censor on the letter *De interprete* … is September 3, and it appeared before October 14, 1675; for the letter on the method of Chrysostom the date is September 14 and that letter appeared in the week of October 29, cf. Nordström, *Antonio Magliabechi*, 281.

[46]Spinoza, *Opera* 4, 281.

[47]On this work, see Räss, 12, 275.

[48]LTHK 9, 938.

[49]Spanheim, *Opera theologica* III, Pars II, Col. 1147 ff. The letter is dated November 8, 1675. Spanheim, however, speaks generally on a *Methode facile pour trouver la vraye Eglise par la lumière de la raison*, Sp. 1152, but since the script was occasioned by the conversion of a young Dutchman to the Catholic church and since Stensen and Florence are mentioned as mediating factors, there can scarcely be any doubt that here Burgh's conversion is being dealt with.

years later that he would have quite particularly aimed at Stensen to whom he reproached the assertion that the true Church could be recognized with the light of reason only and namely because he explained that the Roman Church has so many bright saints, knew so many shining miracles, was proved by so many texts. He, Spanheim opposed it: "But we have removed the masks of the men and we have undone the paint through our treatises which some years ago we have written in French against those new philosophers."[50]

This erudite and eloquent publication gives a good picture of the bitterness that Burgh's conversion and Stensen's activity in Florence as well as this controversial letter released, and it unveils the mentality that Stensen also points out in his writings. That, from the beginning Burgh was characterized as a man who was going from Jerusalem to Jericho and fell in the hands of robbers, as a man going astray who suddenly finds himself at a place full of vipers and precipices, as a lost son and a Joseph in Egypt, corresponds to the opinion about the Catholic Church that existed then: "That artificial Church has the means to blindfold all sorts of minds and to satisfy all sorts of passions… It draws the simple-minded by an appearance of piety, the superstitious by the ceremonies, the credulous by the report of miracles, the chaste by a pretended celibacy, the melancholic by distractions and concerts and finally it draws the ignorant by illusions and imagination."[51]

Bitter and sarcastic remarks were poured out about the novices who, in Italy, without deeper studies, were immediately turned into fanatical missionaries, about the boasting of holiness, about the stories of Saints Francis and Antony, about false miracles, etc.: "I cannot wonder enough that in the heart of that Italy in a country of pleasure and dissipation, in the magnificence and in the pomp of the world, under the noticeable horrifying superstition, that it is precisely there that your man saw himself on the way to God, that he has found the way to the faith and believed to have escaped the world."[52] If so little suffices to win our youth, this youth should certainly not see Italy. There are right there certain diseases which are very infectious. The air there has always been considered as very dangerous. It is not necessary that it approach a man, a philosopher, a runaway in a town where the Poggis and Aretinos, the Vanini and Macchiavelli are so well-known "to pass over from the reformed religion to indifference and from indifference to papistry."[53]

---

[50]In *Specimen Stricturarum* from 1681, Spanheim, 1145.

[51]Spanheim, 1147.

[52]Spanheim, 1152.

[53]Spanheim, 1155.

Stensen's ardent care for the non-Catholics was, by the way, not approved of everywhere in the Catholic world. Magliabechi, for whom the piety spread around from the court of the Medicis appears particularly hypocritic, expresses himself in a letter to his trusted friend, the Dutchman Jacob Gronovius, in extremely harsh words against the efforts of Stensen.[54] He asserts that a Protestant can stay only a short time in Florence. Very soon he will be visited by Stensen and, if he does not convert, he will very soon be forced to leave town for some reason.[55] In several of his letters Magliabechi mentions Protestants whom Stensen attempted to convert without any success, such as the Dutchman Bruno van der Dussen, the German Gerhard Meier from Bremen and the Pole, Bogdan Lubieniecke.[56]

1.10.3   The Former Scientist

Stensen's scientific researches move almost completely into the background at that time.[57] They are mentioned only in the wishes and presumptions of others. Thus when Jacopo del Lapo wrote to Fiorentini on September 29, 1674: "It is believed that he will write here his natural history of the muscles and other similar matters by order of the local Highnesses"[58] or when the Venice envoy speaks with Magalotti and says that it would be the wish of the Padua university to transmit the chair of anatomy to Stensen.[59] In the relations with his former friends of research, the religious element stood out still more strongly. Thus when Stensen mentions his Dutch friend in a letter of July 18, 1675 to Malpighi: "Mr. Swammerdam has sent me the figures herewith to be transmitted to you if you want to accept them since he is in the mood of completely abandoning the study of nature. He has written a dissertation on the same subject but he has then torn it apart and has kept the figures only. He is seeking God but not yet in the Church of God. Pray God for him and let your friends, whom you know as true servants of God, pray for him."[60] We do not know which dissertation is meant here. Swammerdam

---

[54]See further Nordström, *Antonio Magliabechi och Nicolaus Steno.*

[55]Nordström, *Antonio Magliabechi*, 23.

[56]Nordström, *Antonio Magliabechi*, 16, 20, 22 f.

[57]On Stensen's transition from science to religion and his conversion, see the dissertation *Herz, Gott, Kreuz* by Frank Sobiech and Sobiech, *Radius in manu Dei. Das medizinische Ethos in Leben, Werk und Rezeption des Anatomen Niels Stensen*, 2012.

[58]Manni, 315.

[59]Magalotti, 222.

[60]EP I, 306 = Malpighi, *Correspondence* II, 715.

was then, like Stensen, in a development from practitioner of science to deeper faith.[61]

A letter of Ole Borch written in the autumn of 1675, probably in November, gives an insight into Stensen's position towards one of his friends and to his country, towards those whose faith he then shared and towards his own conversion. Borch had taken over the rectorship at Copenhagen University in May 1675 and certainly missed the teaching capacity of Stensen. The greatest academic event of the year was a hearing on June 2 (June 12 st. n.) on a publication on matters of controversy by Bishop Wandal who had died on May 1 (May 11 st. n.) to counter Heinrich Kircher's earlier mentioned book *Nord-Stern*. It was followed by the solemn promotion of 15 doctors on June 5 (June 15 st. n.).[62] Jesuits had apparently also been invited to the debate but had declined the invitation. Whether stimulated by the confessional controversy or angered by the loss for Denmark of the friend and researcher, the old friend and formerly his teacher seems to have reproached Stensen for his conversion and its consequences for his person, his family, science, and fatherland, and asked him to retract after the example of S. Augustine.

Stensen's answer was unusually warm and cordial and showed, in a mixture of Latin and Danish, which is unusual for Stensen, his mindfulness when thinking of home and the past, and the profound values of his personality.

First he thanks God who has led him to the oldest Church which alone possesses the characters of the true Church of Christ, "in which also our ancestors have worked for their salvation for seven centuries", then he confesses his love for family and friends: "I recommend to God almost every day *in memento pro mortuis* (part of the Holy Service) my deceased kinship and my friends without judging anybody, not as if I believed that anybody may become saved outside the Church, but, however, in the hope that, perhaps because of the lack of opportunity to learn *explicite* to know the Catholic truth, they believed with a *fide implicita* that which the holy Catholic Church believes and, at the hour of death, they could obtain from God the grace of an act of repentance …"[63]

He could perhaps have retracted one or the other of the Lutheran prejudices and philosophical mistakes, which he perhaps still entertained. Yet as Augustine never become Manichaean again, he hopes that God will preserve him from a desertion from the Church and, as always, he shows himself

---

[61]Cf. Swammerdam's letter to Thévenot from the year 1678, EP II, 897.

[62]EP I, 81.

[63]EP I, 308 f.

deeply impressed by the holiness of that Church, "in which also today the same holiness of life, wonderful actions, preaching of the gospel in the whole world are blossoming, and where one ascribes to the merits of Christ, not only the pardon of sins but also the works directed to the eternal salvation, the words and feelings of the holy and righteous. Oh veritably holy Catholic Church. Oh veritably blessed sons to whom the eternal mercy offers You as mother. Oh do recognize, my friend Borch, the gift of God and what it would mean to serve God in the lap of the Catholic Church."[64]

Stensen feels himself deeply obliged to his country and recognizes his debt to it, but these are the self-accusations of a saint and they are on a very different level than what Borch thought. "I have sinned, my country, my country, I have sinned because, before my conversion, I have offended God by so many sins and made me unworthy and inadequate to receive the grace of God. I have sinned, my country, because I resisted so long God calling me to the true Church and I have come so late to that state in which I can generously pray for my country. I have sinned, my country, because, after receiving the grace of conversion, I have progressed so halfheartedly and slowly on the way to God and I have performed so little with the heart, the tongue and the hands to obtain for my home country the true goods of the generosity of the divine mercy. May I, however, from this moment, begin to consecrate with true zeal all the gifts of God to me for the salvation of my country. My dear friend let us strive to love God with one heart, from one mouth, in one faith, after having rejected the vanities and errors of the world. What joy will be ours in eternity if all of us serve Him here and induce sinners to renounce their sinns and alone to do God's works. You have broken my chains, Oh Lord, I will bring you the sacrifice of praise and I will implore the name of the Lord."[65]

Stensen's educational comprehension and pastoral sense came into play in a blessed manner in those last 2 years in Florence, since he became employed as adviser of the government and secretary of the Grand Duke. Tuscany had then suffered a great loss by the death of Cardinal Leopold, of whom Stensen says in the letter to Borch: "We have recently witnessed here the truly Christian death of Cardinal Leopold de' Medici."[66] Cardinal Leopold had been consecrated as priest a year before his death. Being the founder of the *Cimento* he now lived his spiritual vocation exclusively. In the first days of November 1675 he became severely ill, so that his chamberlain A. Chigi,

---

[64]EP I, 309.

[65]EP I, 309 f.

[66]EP I, 310. Cf. note 9.

from whom he had asked this service many years earlier, drew attention to the proximity of his death. He was quite calm, embraced the one who administered to him, received the sacraments in full conscience and died on November 10 after an overall blessed and noble life. For the country the death was a severe loss, the more so because the Cardinal had remained at the side of his ruling nephew as an adviser, and the Grand Duchy badly needed reforms. Cosimo was not unaware of this. A report by Stensen of January 1676 to the Grand Duke about a proposition of reform of the legal procedure and taxation shows him involved in reform work. Stensen had discussed the conditions in the Tuscan administration and the heavy taxation with a foreign friend and referred to the views of his informant. He writes: "Concerning the heavy burden of taxes, which would be the [worst] evil to cure, sometimes it is impossible for the poor to carry on in this manner. In such a holy work Your Highness would have no harder thing to do than to choose adequate workers, i.e. of a few but such who possess the true love."[67] Concerning the law students he proposed to send these young fellows into practice immediately after their studies to let them gradually advance and occasionally to take them to Florence. Stensen even had a proposition ready against foreign policy and against the hangman.

As was mentioned earlier, we hear almost nothing on work of natural science in those years. When P. Athanasius Kircher interrogated him about a book *Magnes luminaris*, the answer was quite repellent: "I have not seen the *Magnes luminaris* and I do not know whether a book with so much pledge deserves the time spent for its reading. The more it deprives us of our time, the more of eternity is stolen from us."[68]

When his former students from Copenhagen visited him, he promoted their studies again. The 26-year-old Holger Jakobsen and his cousins Caspar and Christoffer, the sons of Thomas Bartholin, had left Copenhagen at about the same time as Stensen.[69] They had studied in Holland and talked with Swammerdam, had traversed Belgium and then spent 1 year in Paris. From there they went to Italy and arrived in Florence on May 1, 1676 where they stayed for half a year. Stensen furnished them at first a good hotel, an audience with the Grand Duke, and they received a house for accommodation with their own cook. Magliabechi provided them with a key to the

---

[67]EP I, 311.

[68]EP I, 314.

[69]Garboe, *Thomas Bartholin*, II, 139. About this travel see Holger Jakobsen, *Rejsebog*, 73 ff, cf. Garboe, II, 146 ff. On Thomas Bartholin's advice to the young travellers, see Bartholin, *On the burning of his library and on medical travel*, 2nd part.

Grand Duke's library. They surely saw Florence through the eyes of Stensen. In his travel diary Holger Jakobsen says of SS Annunziata: "Annuntiata, a very beautiful church where there is a Mary Picture which has created many miracles. At this picture there is the greatest devotion in Florence. Dozens of silver lamps are hanging all around and they are always burning. It is the richest place in jewels after Loreto."[70] On October 18 they left Florence and arrived in Rome on October 22. On February 11, 1677 they went north again over Sienna to Pisa, where they stayed with Stensen. About those days Holger Jakobsen writes: "As long as I was in Pisa we stood every day in the palace and dissected several rare fishes which the Grand Duke let bring from Livorno. What was prepared during the day I demonstrated to the Grand Prince in the evening. Usually Francesco Redi, Gornia and Niels Stensen were present."[71] On March 6 Jakobsen travelled at the expense of the Grand Duke from Pisa to Volterra in order to see different "wonderful things *in naturalibus*".[72] This must mean the fossils and geological curiosities that Stensen had observed. On March 12 they were with Stensen and Redi in Livorno and Jakobsen writes: "With Stensen and Fr. Redi I then made the ichthyotomy."[73] It was particularly about a torpedo fish. Its electrical discharges interested the Grand Duke. On March 16 they were again in Florence and then travelled north and back home via Bologna and Venice.

From a letter of the later Cardinal Enrico Noris to Magliabechi of March 8, 1677, we see that the friends were also influenced on religion. "Sir, you cannot expect Mr. Holger and Mr. Christoffer (Bartholin) now because they follow the court to Livorno. Mr. Stensen instructs the former in such a manner that he attends his mass with devotion, goes to the sermon in the cathedral. Both are accommodated in the house of the catechizer."[74] One day later he makes the same remark and finally: "Mr. Caspar goes to Livorno in a hurry to incite the other two to the journey back. I wonder however, whether Mr. Stensen, hoping to win Mr. Jakobsen, will not oppose,"[75] thus exposed as an escape from the religious atmosphere. Unmistakable, as Jakobsen testifies it, is also Stensen's religious approach to nature. For example, he begins a description of the "wonderful structure of the tongue of a black wood-pecker" with these words: "Like the piece of art of the highest Creator and of

---

[70]Jakobsen, *Rejsebog*, 107.

[71]Jakobsen, *Rejsebog*, 137.

[72]Jakobsen, *Rejsebog*, 137.

[73]Jakobsen, *Rejsebog*, 139.

[74]Manni, 163.

[75]Manni, 164.

nature lights up other creatures, thus the spectacle which this small bird offers to the one eager to know raises the observer's mind and eyes to admiration and shows in the smallest the essence of the whole and the sagacity of its mindful architect."[76]

Leibniz mentions great projects that Stensen had with his disciple: "I have heard that the late Mr. Stensen, as he got rid of the profane literature, entrusted Mr. Jakobsen, a Danish physician, to carry on his thoughts about natural geography (*géographie naturelle*) and the physical changes of the surface of the earth. I should like to know whether that learned man is effectively working on that."[77] It is not excluded that Stensen actually gave his geological records to his enthused student who was also to teach geography in Copenhagen. Stensen was already then aware of his nomination for bishop, which would definitely deprive him of all possibilities of work in science. However, Jakobsen certainly did not show himself able to usefully to pursue the work of his master. After the time of collaboration with Stensen, his scientific research came to an end.[78]

Christoffer Bartholin (1657–1714) also seems to have been interested in geology.[79] In a letter of September 12, 1676 he tells his father that he is busy with a dissertation on crystals and their formation "following the famous method of Stensen in *De solido*."[80] However, no outcome of this interest is known. Christoffer soon left science to become a magistrate in Jutland.

It appeared later that his brother Caspar (1655–1738)[81] also benefited from the geological teaching of Stensen since, in Copenhagen, he was not only professor of anatomy but also of physics. In his teaching of geology he followed exactly the *De solido*.[82] But he was mainly anatomist, and his name is recorded as eponym for the *glandulae Bartholinianae*.[83] The sojourn in Italy was very fruitful. He was provided with several manuscripts and gained the

---

[76] *Acta Hafn.* V, 249.

[77] Scherz, *Indice*, 190, note 1.

[78] Cf. Garboe, *Niels Stensen's (Steno's) geologiske arbejdes skæbne*, and *Nicolaus Steno and scientific geology*, 117 f.

[79] DBL, 3rd ed. I, 474.

[80] *Acta Hafn.* IV, 1677, 64 f.

[81] DBL, 3rd ed. I, 472.

[82] Garboe, *Nicolaus Steno and scientific geology*, 118, and Garboe, *Geologiens historie i Danmark*, I, 90.

[83] Caspar Bartholin is eponymist for two structures and a disease: *ductus sublingualis major*, Bartholin's duct, and the *glandula vestibularis major*, Bartholin's gland, the latter being the seat of an infection, *Bartholonitis*.

friendship of Malpighi. Part of their correspondence has been preserved.[84] Caspar Bartholin admired Stensen as a scientist. He called him his teacher in anatomy. Yet he could not understand the convert Stensen.

Only sparse records have been preserved from the 2 years of Stensen's priesthood in Florence. We are not quite sure whether he carried out the intention of a pilgrimage to Rome in the jubilee year, this time as a priest. It is very likely that he was among the hundreds of thousands of pilgrims who despite the disarray of wartime marched to the eternal city to devote themselves to the exercises of piety as prescribed by the zeal of the old Clement X, who himself visited the seven main churches of Rome five times.[85]

The main event during those 2 years was undoubtedly the priestly ordination, which outshone all other happenings and revealed the life and striving to perfection of Stensen in the spirit of the gospel advice, with a strong urge to the contemplativeness of cloistral life. It shows, moreover, the pastoral care, which he chose all by himself and to which he would have devoted himself further if he had had the choice. It was care of the individual soul that particularly appealed to him, for the salvation of those who confessed other faiths. He obviously felt called to this mission and in this work he achieved an amazing number of great results. The obedience and later life put Stensen in other positions, which he managed outstandingly. Yet it is fair and right, when judging his last years, to think of the field in which he mainly intended to work.

His spiritual writings, the so-called *Opere spirituali*, were found in Florence in a transcript by Hippolito Noferi. For the most part they originated at the time when Stensen was in Florence. They witness the deepened spiritual life and striving of Stensen in that period; thus a dissertation, *Spir. 10*,[86] "About the examination of the conscience", for the monthly conscience account to the spiritual leader. It shows the spiritual problems with which Stensen wrestled at that time. Now the priest at the Medici court and educator of the crown prince, Stensen was striving to the Christian perfection even at the cost of the most strenuous efforts. Here there are questions of breviary prayer and mass saying, of study of morals and of the Holy Scripture. The courtier anxiously wonders whether he in conversation fled into dissipation or whether he offended modesty by being too loud in speaking and laughing, whether he has also visited Christ in the poor and the diseased on days in

---

[84]Scherz, *Briefe aus der Bartholinerzeit*, 175 ff. Here the letters are first published (p. 192), and later also found in Malpighi, *Correspondence*, cf. register.

[85]Pastor 14[1], 638 ff.

[86]OTH II, 55 ff.

which he visited high-ranking persons. Also whether, in the conversations, he avoided to come back over and over to points of religious controversy, to diseases and scientific questions. One obtains a picture of his liveliness, influenced indeed by the Italian surroundings, when he enquires whether, in speaking, he has not gesticulated too much. He critically asks himself whether, in his relations to God and with people he has been as respectful as he was with the Grand Duke and with the Prince, and whether he has been humble enough to avow his ignorance in front of the disciples. Briefly, it is an intimate representation of the work of the whole day, seen from the spiritual and ascetic side.[87]

Also the other spiritual treatises in the collection help us to draw the spiritual picture of the priest. *Spir. 1*, "About the daily spiritual exercises", reveals his schedule of the day and the spiritual exercises that he links with it. *Spir. 2* is an overall view on the way and manner of his meditation. *Spir. 3*, "About the regulation of our actions" and *Spir. 4*, "The behaviour in enterprises of great importance", shows the behaviour that he requires from himself and from the others when actions have to be carried out and in the approach of matters of great importance. *Spir. 5*, "Thoughts about the exercise of the presence of God in our actions", and *Spir. 6*, "About the union with God", display his behaviour in front of God, the exercise of the presence of God and the union with God. The priest has certainly arranged the section on confession and communion for his penitents. One understands very well that Noferi was thrilled by these indications: "I have drawn this quite small work from his manuscripts from the arrangement of which it can be concluded: that he has applied it to himself and used it as a rule for his actions as one reads about Carlo Borromeo, the Bishop of Milan; that he carried with him that small but true golden booklet of the moral philosopher Epictet, as a kind of viaticum. If, therefore, you read it you will recognize how much his mind was always thoughtful of the service of God, and learn from his example to control your actions in order to carry them out with that perfection as the state of a good Christian requires."[88]

Altogether, the energetic striving of Stensen to Christian perfection is evident. What he does or does not do is under the sign of piety. From early morning until late evening prayer, his life is a search of God. Quite in agreement with the recent ascesis, a strong trend appears to be systematizing. What already appeared towards the end of the Middle Ages reached its

---

[87]OTH II, 61 ff.
[88]OTH II, 56.

summit in the spiritual exercises of S. Ignatius of Loyola, which Stensen often used, and what in mysticism brings forth the great systematic of contemplation, John of the Cross, applies also to Stensen. It is not the monk of a contemplative order who speaks out through him but an active man staying in the middle of the world who, nevertheless, will serve God entirely and, therefore, precisely divides his time and methodically builds his religious life.

The unmistakable influence that the Italian ascetic school exerted on Stensen is also a characteristic of these manuscripts. Italian asceticism developed in the sixteenth and seventeenth centuries first of all to fight against the pagan thirst for pleasure of the Renaissance and against the influence of the Protestant reformation. The idea of the spiritual struggle found an enduring expression in the *Combattimento spirituale* of the Theatin, Lorenzo Scupoli, whose book appeared for the first time in Venice in 1589. It is inspired by the idea of God's greatness and the nullity of man, by a deeper love of God and by hate of the lower I, by the struggle between the higher and the lower forces. It is not difficult to find in Stensen reminiscences of those ideas or of passages from the famous Cistercian Cardinal Bona who has exerted such a great influence on modern ascesis.[89]

---

[89]OTH II, 68 f.

Part II

Steno—Scientific Papers—OPH Numbers

2.0 Scientific Disputation on the Subject of Hot Springs

To be defended publicly by

Niels Stensen,

a Dane from Copenhagen,

on July 8, at the usual time and usual place,

under the presidency of the most distinguished and learned

Arnold Senguerd,[1]

Master of Liberal Arts, Professor Primarius of Philosophy

at the renowned University of Amsterdam.[2]

Gustav Scherz rediscovered Niels Stensen's first dissertation in 1960, It has therefore no OPH number. It appeared in the *Journal of the History of Medicine and Allied Sciences* 1960, Vol. XV, No. 3, 247–264. The present text and annotations are transcribed from Gustav Scherz, ed., *Steno—Geological Papers*, Odense University Press, 1969, pp. 49–63, by permission from the publisher.

---

[1]Arnold Senguerd or Senckward (1610–1667), born in Amsterdam, from 1639 professor of metaphysics (from 1641 professor primarius) in Utrecht, and from 1648 at the Athenæum of Amsterdam, where he became gradually more interested in subjects of natural philosophy and science. The funeral orator said of Se. that he was a follower of Aristotle, yet also interested in the new currents, and, in general, a man of piety, mildness and sincerity. (A. J. van der Aa, *Biographisch Woordenboek der Nederlanden*, vol. 17. Haarlem 1874, pp. 613 ff. —D. J. van Lennep, *Illustris Amstelodamensium Athenaei memorabilia*. Amsterdam 1832. pp. 119 ff.).

[2]The Athenæum, the first university of Amsterdam, from 1632 began to develop in the old chapel of the St. Agnes Convent. The rather few lectures in the very first years became better organised in Steno's time, yet without examination and promotion (I. H. van Eeghen, *De eerste Jaren van het Athenæum*. In MDCXXXII. *Illustere Begin van het Athenæum* MCMLVII. Amsterdam 1957, 8 f.).

Printed in Amsterdam by *Jan van Ravensteyn*,[3]

Typographer to the City and the illustrious School,

MDCLX

To a very distinguished and learned man, M. ARNOLD SENGUERD, Doctor of Philosophy, Professor Primarius of the same in the renowned Lyceum of Amsterdam, Prefect of the Library, and most worthy Curator of Public Schools, AND ALSO To a very celebrated and highly experienced man, M. GERARD LEONARD BLAES, Doctor of Medicine, a very successful practitioner in the most flourishing Republic of Amsterdam, where he professes medicine publicly,[4] To the former, his president, to the latter, his host, and both, his instructors, NIELS STENSEN presents himself.

Proposition

I.

The word "thermae" is derived from the Greek verb θέρω, "thero", meaning *I make warm*, whence we obtain the adjective θερμος, "thermos" meaning *warm*[5]; hence, the waters that we now consider are called thermal because all, or at least the majority of them, have the power to warm. Those which are actually cold on the surface of the earth are believed by Fallopio[6] to

---

[3]Jan van Ravensteyn (1618–1675), belonging to a family of booksellers, was a private printer and bookseller who had a monopoly of printing for the town and the Athenaeum (M.M. Kleerkooper en W. P. van Stockum jr., *De Boekhandel te Amsterdam voornamelijk in the 17ᵉ eeuw* I—II,'s Gravenhage 1914–16, p. 578 ss.).

[4]Gerard Blaës (Blasius. c. 1625–1682), son of Leonard Bl., Denmark's King Christian the Fourth's Dutch architect. Bl. was himself a physician in Amsterdam, from 1660 extraordinary and from 1666 ordinary professor of medicine. Author of bibliographical and original works on comparative anatomy (*Nieuw Nederlandsch Biografisch Woordenboek* 7, 138 s., and Francis J. Cole, *A history of comparative anatomy: From Aristotle to the 18. Century*, London MacMillan, 1944, 150 ff.).

[5][θε]ρω (thero), the usual form of (thermo) or [θε]ρμαίω (thermaio). Cf. H. G. Liddell-Scott, *A Greek-English Lexicon* 1 (Oxford 1940). - *Thesaurus graecae linguae* IV (Parisiis 1841). *[θε] missing in Scherz, 1969, 62.

[6]Gabriele Falloppio (1523–1563), the great Italian anatomist of Padua, and one of the greatest surgeons of the century, also wrote a significant book *De thermalibus aquis libri septem. De metallis et fossilibus libri duo*, Venezia, 1564 (G. Fallopii, *Opera Omnia*. Francoforti, 1660 pp. 193–268). The last 32 chapters describe the Italian hot springs. The passage Steno refers to is from chapt. I (p. 196): *Nec sine mehercule ratione, ab eo qvod calefaciant, et thermae, et thermales aquae sunt vocatae, qvoniam omnes fere aquae istae calidae sunt, aliae tamen magis, aliae minus, ut aliae sunt subtepidae, aliae ferventissimae; dixi autem fere omnes, quoniam reperiuntur nonnullae (sed rarissimae) quae non sunt calidae, imo frigidissimae sunt. Verum licet quae frigidae actu sunt, in terrae superficie sint tales, tamen credo qvod in suo principio, hoc est in locis subterraneis a terrae superficie, id est in canalibus ipsis, calidae et ipsae sint, et quae me movet ratio est, qvia video Aquarianam aquam quae frigida actu, ut nostis, est, servari incorruptam per spatium centum et plurium annorum ... Nec hujus rei alia, ut arbitror, est causa qvam optima mixtio. Sunt enim in aqva elementa adeo commixta, ut nec ex*

be warm at their source or in lower subterranean regions; he considers that it would not be possible for these to remain unchanged for a long time, that is for many years, away from their original source, unless the waters were very well mixed, and since intermingling is a result of the digestive and divisive action of heat, he thinks that heat may also be attributed to those waters.

II.

The places where such waters are contained are called *baths*, and the waters themselves are known as bath waters.[7] However, some distinguish between "thermae" and baths because "thermae" are properly waters that we go into to induce sweat, and are hence called "sweat-baths"; but those that we go into for the sake of washing are properly called "baths". We can entrust this controversy to the philologists.

III.

They are also known as *mineral waters*,[8] because many of them, if not all, contain some minerals. Because of their efficacy and use in medical treatment, they are also called *medicinal waters*.

IV.

It is most uncertain what is the source of the heat in hot springs.[9] Water is cold by nature; this is evident from the fact that when it is left by itself and the cause of its warmth eliminated it becomes cold again even though it may have been extremely hot. It is certain that this cold is not imparted to it by the air since water becomes colder than air; the air is, in fact, unable to impart a more intense cold than it has itself. It is necessary, therefore, to seek an external cause of this heat.

V.

Perhaps the cause of this heat is friction[10] of the water and its frequent incision into rocky places? This suggestion is not completely devoid of probability, because bodies do grow warm through motion, and indeed if

---

(Footnote 6 continued)

magna commotione agitationeqve nec temporis diuturnitate possint dissolvi. Sed cuinam vestrum, juvenes praestantissimi, clarum non est, optimam mixtionem ex maxima caliditate dirigente et concoqvente provenire? See also P. Capparoni, Profili bio-bibliografici di medici e naturalisti celebri Italiani *(Roma, 1928) 2, 46 ff.*

[7]Today these waters (supposed to have temperatures above 20°C) have also different names: *thermal waters, hot baths, hot springs*; the last one is used here.

[8]The word *metallum* may not only mean substances such as gold, silver etc., but also any substance dug from the earth (Aeg. Forcellini, *Lexicon totius latinitatis* III, 233).

[9]Thermal waters are, of course, much better known to day, partly as to the origin of their warmth, and especially as to their chemical composition. Cf. e.g. M. Messini. G. G. Di Lollo. *Acque Minerali del Mondo.* Roma 1957. Yet there are still many doubts.

[10]The Latin word *attritio* may be translated by attrition or abrasion.

things turn out otherwise, this is accidental. Moreover, it is usually in rocky places that the springs of such waters exist.

VI.

Against this proposition there seems to be an objection that if such be the true cause of the heat of hot springs, then springs of water that flow cold at present ought also to be hot, for they too are beaten and agitated and abraded as they pass through rock fragments and mountain crags. There is no great difficulty here; since various mountains consist of rocks of different nature, shape, and juxtaposition, the variety of interstices that result and the varied sharpness of the stones could be the reason why hot water flows from some mountains of rock and stone, but from others, on account of the large size of the spaces between the stones and a paucity of sharp edges, the water is not divided in the same manner and comes out cold.

VII.

There is, alternatively, the question whether this heat arises from sulphur. Against this is the fact that sulphur is extinguished by water, and, as long as it is not alight, does not produce heat; hence, it seems that sulphur might, indeed, have the aspect of a material cause but not of an efficient cause.[11] The same may be said of bitumen.

VIII.

Who can say whether subterranean fires dispersed through certain areas in the heart of the earth produce this heat or transfer it to the waters? That there are subterranean fires is sufficiently clear from mountains that spew out fire. An objection to this opinion seems to be that every fire needs fuel, and there scarcely seems to be any cause that could provide a limitless supply to these subterranean fires, since bitumen, sulphur, and other similar materials are consumed by burning.

IX.

Or again, do lime and ashes heat those waters in the bosom of the earth? We observe fire, heat, and the boiling up of water when water is poured over quicklime. And here lies a difficulty—where could such a great quantity of lime arise? Besides, from this it would be hard to derive a reason for lasting heat.

X.

It does not seem contrary to reason to assert that these various causes come together in the heating of hot springs. It is indeed certain that heat arises from the movement of bodies. It is trifling to deny that heat is generated from the abrasion of sulphur and from the agitation of the abraded pieces, since heat that is potentially in the sulphur is brought into actuality. That there is

---

[11] *Causa materialis* and *causa efficiens*, the scholastic expressions for substance and cause of a thing.

sulphur in hot springs is most clearly shown, over and above what is observed by smell and taste, by the abundance of sulphur that is found in the cisterns and channels of the springs. From the testimony of experience it cannot be doubted that there are subterranean fires. It is likewise certain that water stirred and mixed with lime becomes hot. But what, and how much, each of these causes contributes to the warmth of hot springs, and how each does it, we shall explain more fully in the course of the disputation.

XI.

Besides heat, usually, various mineral qualities and influences are observed in hot springs. The question is, where do hot springs obtain these properties? I answer, generally from bodies in which the hot springs are contained and through which they flow.

XII.

If the question is asked, how do those bodies share their powers with the hot springs, I say, just as there are in animals parts that contain, parts that are contained, and parts that produce force, so a distinction should be made between parts that are gaseous, liquid and solid. It is probable that parts of each type are shared with the waters, but in such a way that blue vitriol,[12] sulphur, alum, and bitumen share their whole substance with water, since they readily become liquid. But in the case of metals and stones, because these do not liquefy so readily, some particles are eroded.

XIII.

The odour of the waters, a result of exhalations, is sufficient indication that gases are united with them.

XIV.

It is evident from the taste of those waters, especially those found in abundantly watered regions, that there is a liquid or some denser mineral substance in them.

XV.

It is clear that solid material more dense than watery fluid is present in these springs from the ochre that is found in their channels.[13] The same thing is shown by the quite large pieces of sulphur that are found in the cisterns. The same moreover is evident from the rock-hard substance into which the canals through which these waters pass harden.

XVI.

As for the gaseous portions, it is certain or probable that the minerals themselves emit gases which mix with the waters as they go along. I pass over

---

[12] *Chalcanthon* is used by Dioscorides to designate the sulphate of copper (blue vitriol), or even the sulphate of iron itself (green vitriol). J. M. Stillman, *The Story of Alchemy and Early Chemistry*. New York 1960. 42 f.

[13] *Canalibus illorum* in the printed treatise is probably a misprint for *canalibus illarum*.

the fact that gases and exhalations surrounding the minerals penetrate into the pores of the minerals and thus become intermingled and saturated with the properties of the minerals.

XVII.

An objection may be put: if a solid, dense material were joined to these waters, then the waters would become dense and turbulent, but they are clear. My answer is that those parts which are joined with the waters are agitated and separated into tiny particles; because of this there is not such a great thickness developed as can be discerned by the eye, just as the salt water of the ocean is clear though it is mixed with salt material. But I would add to this that it is true that those waters in which lesser amounts of solid material are mixed are of a finer and more tenuous nature.

XVIII.

Here is another difficulty that can be put forward as a contrary opinion: metals are too hard to allow particles to be worn from them. I answer that this objection is easily countered by declaring it to be false, since the removal of scrapings from gold teaches the opposite. Secondly, it is probable that some particles can be abraded easily from metals since those in the heart of the earth are not as hard and solid as they are after being worked skilfully. Thirdly, since falling drops can hollow out stones, it is certain for metals, just as for stones, that something can be abraded by frequent motion, to become mingled with waters in motion. Fourth, the same is evident from the erosion of the hardest rocks covering the canals[14] of hot springs owing to the abrasive action of exhalations from those waters.

XIX.

If, on this occasion, it is asked whether it is possible for bodies to become imbued with the properties not only of imperfect metals but also of perfect metals by contact, I reply that the waters, by contact, take on the properties and powers of perfect metals also, and that this is demonstrated by aeruginous water which flows through veins of copper, washing away the adhering verdigris. The same is true of water containing extracts of gold, silver, and steel, as well as water that has been kept too long in vessels of bronze and lead.

Corollaries

I. Are the parts of gold heterogeneous? Yes.

II. Are iron and steel different in species? No.

---

[14] *Illorum canales* in the printed treatise is probably a misprint for *illarum canales*.

2.1 On the First Discovery of an External Salivary Duct and on the Experiments of Bils

To Thomas Bartholin in Copenhagen

When I consider how well you, very famous Sir, are disposed towards me, and since I believe not to deserve such consideration, I wish I could honour your extreme kindness with an equal respect. The evidences of your love indeed appear so numerous that it makes clear to me that the ancients rightfully

wanted teachers to take the place of a venerable father.[1]

I think that the day when I was adopted as one of your disciples must deservedly be regarded as equal to my birthday. What else indeed does your very kind letter demand of which I should answer with much gratefulness to all the parts, if not each of them were so large that it would itself require a particular answer? But because my want of eloquence does not allow me this, I find it better gratefully to remember this rather than by thanking with common words to proffer something which is lighter than the matter is worth of.

OPH 1 vol. I, 3–7: "De prima ductus salivalis exterioris inventione & Bilsianis experimentis" is found in *Th. Bartholini Epist. Medicin. Cent.* III. Hafniæ 1667; 86–95, numbered *Epist* XXIV. Further annotations by Vilhelm Maar are found in the transcripts, www.extras.springer.com.

---

[1]Juvenal, *Sat.* vii.

© Springer-Verlag GmbH Germany, part of Springer Nature 2018
T. Kardel and P. Maquet, *Nicolaus Steno*,
https://doi.org/10.1007/978-3-662-55047-2_12

Since, however, by the same letter you advise me to publish a picture of the external salivary duct, I cannot but expose to you briefly also the envy which this small discovery caused against me and, along with it, what profit I drew from this envy, not to seek fame in trifles but to reject the hateful crime of plagiary. I indeed regret that I am forced either to speak much of a matter of not so great importance or to be the victim of an infamous stamp of ignominy. Rightly considered the thing itself does not at all deserve that, since above all a duct totally similar has been discovered previously and even what is dealt with now, was observed by *Casserio* although being called a muscle (which I learnt from your *Institutiones* in the description of the buccinator muscle). Since, however, the accusation which is charged against me because of that should not suffer in silence, I will tell you the whole matter like a disciple to his teacher, and I will leave to your judgement what must be decided from that. This duct had already quite often been shown to people by the most illustrious *Sylvius* and by the most famous Mr. *van Horne* when, some weeks later, the very illustrious *Gerard Blaes* who had become hostile to me for a reason which I do not know yielded so to his feelings that, although he had never applied his hand to the investigation of this duct, although he had not succeeded in indicating what it was when I showed to him this duct divided in two, although in a letter to *Eyssonius* his own brother attributed the discovery of this duct to me some days before these reproaches were notified to me, and finally although he himself, in the last writing, the *Medicina Generalis*, assigned neither the true origin nor the true orifice to the thing which he says he has discovered: nevertheless he notifies to the very famous gentlemen *Sylvius* and *van Horne* through his brother and thereafter to everybody in the mentioned book, that this discovery is due to him. To make matters more clear, give me the permission to expose in a few words the occasion of the discovery and what happened thence.

A year ago, as I had been received by *Blaes* as my host, seeing a good opportunity of having anatomical subjects in that series of lectures which he finished in the third week of my arrival, I asked the very famous gentleman to allow me to dissect with my own hand what I would provide for myself. Having obtained that, I was so lucky that, at the first opportunity, on April 7, when I was dissecting alone in the small study, I found in a sheep's head a duct which nobody had described as far as I know. Having removed all the common wrappings, I considered dissecting the brain when casually I decided to examine first the vessels passing through the mouth. Thus, exploring to this end with a probe pushed in the paths of the arteries and veins, I observed that its tip no longer was retained inside the straits of the tunics but roamed in a large cavity. While pushing the metal, I soon heard the teeth resound. Amazed by the novelty of the thing, I called my host to hear his opinion. He at first accused me of having forced my way, and then he resorted to the all too frequent plays of Nature and finally checked in *Wharton*. But since nothing turned up there and the rather carelessly treated vessels did not permit further investigation, I decided to examine them another time with more attention. Some days later I succeeded in doing so in a dog head, although less clearly. Since thereby a great relationship with the lower salivary duct suggested the function of this vessel, I told in the same month my very close friend Mr. *Jacob Henrik Paulli* that I had discovered a small salivary duct and added a description of it. But since something similar had already been discovered and I could not guess whether this had been observed by other people, I kept silent until I had an opportunity to consult the very famous *Sylvius* on the matter. After hearing me, he thought that this duct had to be looked for in man and, having found it, he showed it several times to onlookers.

This is, very illustrious Sir, the story of the discovery. But to become more informed on the matter, see here the arguments given by the very persons who accuse me. The younger *Blaes*, Doctor in Law and barrister who lived in Amsterdam until May of that year, came to our home every day and attended all the dissections carried out by his brother and which I watched. Having heard mention of this duct done publicly by Mr. *van Horne*, he sent to *Eyssonius*, the very famous professor in Groningen, a letter on the subject attributing the invention to me. Who will thus believe that a brother, while he lived in Amsterdam with his brother and spoke with him every day, as keen on news as one ever was and competent in anatomical matters, will not attribute to his brother what the latter had discovered? There you see one weapon with which they themselves have supplied me. It would be possible to present several others if it were worth the effort. *Blaes* in private letters did not defend himself with arguments but with insults and called me a liar being unfair, malevolent, swollen by the ferment of envy, and I do not know what not, so that he has deserved that I should bring forth not only all what pleads for me but also that I should turn to himself all that which he charged me with; though I know that he will never prove what he accused me of, that what must be rebutted is so well known by most people that they do not need me as a public crier. But nevertheless, if my name had not been associated to this duct by the very famous Mr. *van Horne* in such a noble place in such an assembly of all kind of learned people, and this even again and again, I should have willingly withdrawn my claims. Yet actually so that modesty should not be considered as a sign of guilt, I am forced to complain, although reluctantly, of the insult thrown upon me. Finally, to get on to other matters, let me add one so far which I judge very strong. *Blaes* himself in the *Medicina Generalis*, that treatise which he wrote, reveals with a clear sign that he never investigated this duct. Indeed he assigns to it neither its true origin nor its true orifice. Even to the gland itself from which the duct has its origin he ascribes such an obscure function that, if I had not been sure of having shown it to him myself, I should assert that he had never seen it. This will appear more clearly in that disputation which I have at hand. Since indeed affections of the parotids and diseases of other adjacent parts, on which these and some others of my observations seem to throw some light, require a more careful examination, I did not think it would be irrelevant if meanwhile, for the sake of exercise, I prepared a disputation on some glands of the head and their ducts as found in a calf. I indeed owe to his unjust accusation that, as I examined the head of a calf in which he said to have discovered this small vessel according to his description, not only I did not observe a shade of that which appeared clear to him but I also noticed some new lymphatic ducts. Indeed the examination of the parotids and a comparison with the lower of the maxillary glands led to an investigation of three glands. They differed in nature from the first but communicated with those through vessels. Salivary glands united with two of them, an internal with one, and an external with the other. The third one situated in the neck by its upper part which is a hump receives vessels from the first two and sends downwards a new fairly conspicuous canal from its lower and concave part. The head being separated from the trunk precluded to trace it further. But more about these in the said disputation, if only it will be possible also more clearly to notice the same things in another subject, moreover with another duct which I saw in the thorax to be different from the thoracic duct as far as position and colour of the contents are concerned, of which I observed nothing except a fairly long duct in the left side of the thorax situated beyond the internal branch of the ninth pair of nerves and rising on to the jugular glands. If it is given to follow it more accurately from its source up to its orifice I will present this duct and other things, to you who were the first to discover these ducts and to whom all that is due which others added as rivulets, as the sign of a grateful mind, since there is no other way of showing myself grateful.

But, since I have recalled that vessel, let me bring forwards an experiment which, when I looked in vain in another dog for what was mentioned above, at a first glance appeared to be evidence for the opinion of *Bils* more than it was in reality. While indeed I opened a dog immediately after a big meal, at once it displayed at a first glance to the people present the extraordinary spectacle of the intestines being coloured everywhere by white loops of milky vessels so that I think that the milk running back from there can hardly be less than the blood brought in there. But since the receptacle is not yet coloured by milk and the thoracic duct is very small, after having put a tie on this, I lay aside the dog until the next day. The next day having sought in vain for other things, returning to the cistern I find everything changed. This cistern, indeed narrow the day before and showing nothing less than milk, was now swollen, distended by white humour. The thoracic duct, then hardly visible, was now also turgid and showed several small branches returning, after a short parting, to the trunk from which they originate. But what is most important in this matter: a vessel from the receptacle to the inguinal gland was seen completely full of milk, although some of the rivulets which from another source go to the same vessel disclosed lymph of aqueous colour. At the beginning I wondered why the milk deviated in the gland was not retained by the barrier of the valvulas but soon I saw that valvulas sometimes wasting away in a dead are not suitable to resist, above all if they had been compressed at one side of the vessel. This I heard from the very illustrious *Sylvius*. I have no doubt that these things mislead *Bils* who compresses the vessels to express the juice. Hence the valvulas brought to the sides make easy passage for the liquid moving to and fro. As far as the other secrets of this man are concerned, they seem to become little by little of less importance as the hand embalmed by the very famous Mr. *van Horne* makes the palm of the victory for the mummies of that man questionable and a dissection without blood would not be so difficult to carry out. Since indeed he told us that he uses ligatures, it easily appears how blood can be deviated from one part of the body to another one in the live animal. If indeed the descending trunk of the aorta is ligated in the vicinity of the heart with the ascending vena cava, I do not doubt that the heart emptying the blood from the lower region of the body will fill up the upper region. In this way it will be easy to dissect at leisure the parts emptied of their blood and, on the other hand present for eyes' sight the vessels which otherwise escape being seen swollen of humour. I should have tested these things if I had been more skilled in vivisection and if the smallness of the house had not diverted me from these experiments which also require money and time. Nevertheless, your last letter inspired great hope to me. In this letter you wanted to incite me to study these matters and to ascertain that you will be mindful of me, given an opportunity. I know indeed that Maecenas will not default the one to whom such a gentleman promises to give his attention in this matter. But I retain you far too much. Do well, very famous Sir, and continue to find me worth your favour.

Given in Leiden, April 22, 1661.

Yours respectfully,
Niels Stensen

2.2 On the Glands of the Mouth and the New Salivary Ducts Proceeding from Them

To the very illustrious and generous hero,
Mr. OTTO KRAG,
Lord of Wolberrig,[1] Great Royal Senator,
Excellent Former Ambassador in the Dutch Republic,
Governor of the Fortress and Territory of Nyborg in Funen

As well as to the venerable and most vigilant Father,
the Reverend Lord HANS HANSEN SVANE,
Most eminent Archbishop of Denmark and Norway,
Assessor in the State College and Chairman of the Consistorial College,

---

OPH 2 vol. I, 11–51: *Anatomica de glandulis oris, &, nuper observatis inde prodeuntibus vasis* was published in Leiden in 1661. It was reprinted in 1662 as the first of the four treatises found in *Nicolai Stenonis observationes anatomicæ, quibus varia oris, oculorum, & Narium vasa describuntur, novique salivæ, lacrymarum & muci fontes deteguntur, & Novum nobilissimi Bilsii De lymphæ motu & usu commentum examinatur & rejicitur.* Lugd. Batav. [= Leiden] 1662. The text of this second edition, which was corrected and somewhat expanded by *Steno* himself, was the basis of Maar's edition. Further 90 annotations by Vilhelm Maar in the Transcripts, www.extras.springer.com.

---

[1]Voldbjerg.

To the very famous and most experienced gentleman,
Mr. Dr. THOMAS BARTHOLIN,
Very illustrious Honorary Professor of the illustrious
Majesty King of Denmark and
Norway at the University of Copenhagen,
and Dean of the Faculty of Medicine,
And to the Maecenas, Patrons and Teachers [I am] strongly attached

With respect for ever,
Humbly and obligingly
NIELS STENSEN
D.D.D.[2]

Among other matters which we reckon marvelous without understanding them,, the one which deserves the highest admiration is the capacity conceded by God to the human mind by which this imagines, whenever it likes, the figures of things perceived through the senses, and absent things as if they were present, and it gazes in imagination at all the parts observed in them previously, with the same shape, size, colour, position in a much better way than if they had been drawn from a living model by a Protogenes whose even incomplete works vie in veracity with nature. Indeed, we will wonder that the one who, eager to examine thoroughly the reasons of natural matters, spent his life in experiments, when old but enjoying vigorous old age goes in his mind through the huge universe extending over distances almost infinite and through the parts of this universe, inside a sphere of such small capacity, inside his skull. This one, flying up to the heavenly bodies will explain to us the constant order of the fixed stars, the undeceiving wanderings of the planets, the excursions of the comets deprived of any law. Hence, sliding down he will range in a moment through the air and will depict nice varieties of colours, amazing forms of volcanic eruptions showing themselves in these regions through intervals. Hence descending to the earth he will describe the different works of nature most carefully elaborated which present themselves there and images of art hardly yielding to these works of nature. Finally, he will penetrate in the inside of the earth and will reveal the concealed mysteries of minerals. Another one has all these ideas submitted to his will as if the included macrocosm were hidden from the microcosm. Yet, though it is amazing how easily we can view all the images received, it is, however, fraught with much labour and difficulty, and almost impossible, to appropriate them one by one, if the object is to be made properly intelligible. For the mind, which seeks pleasure in variety, is so impotent in the midst of all its power that, when it is occupied with investigating the objects, it

---

[2]*dat, dicat, dedicat* means gives, devotes, dedicates.

cannot order to itself to "do that" nor make itself free from other thoughts so as to devote itself to an idea alone.

The exactness combined with the multitude of the parts of which natural bodies are composed is such that it eludes and frustrates even the most attentive. Thus, even if from all eternity many people have acted with a maximum of work and indefatigable application to represent the absolute image of the anatomy of the animals in all its details, why do we wonder that nevertheless even now it is apprehended partly and imperfectly only? If indeed the ones who exactly represent the external features of a whole animal with a painter's brush, never observe everything all around without that another, more skilled in the same art, when arriving, finds some things to be deleted, others to be completed, others to be changed, why do we expect more perfection from those who, besides the proportion of the external parts in relation to each other, attempt at inscribing in the brain the pictures of all the internal parts? The skilful texture of the individual parts, the cunning connection of the attached parts are so much enveloped to the onlookers, they show such an abundant crop of things to be investigated that, even if the work of many combines into one, even after a long series of years, one can, however, hardly expect a trustworthy knowledge of them. There are some people, I must say, who are convinced that an easier way is open for achieving the image about which we speak so that there is no need that everything be subjected to the external senses. They claim that reasoning alone can supply all the rest which is missing. However, according to this reasoning, I should expect that, to attribute the heroic importance of Alexander the Great to a mute picture of Apelles, a hurried look at the breathing face would have been enough, he would have been able to supply from his natural capacity what he could not obtain at the same velocity. I am afraid, however, that I should not deserve the censorship of those who would say that heroic greatness is expressed in this way, but not the heroic greatness of Alexander. Thus, in the present matter also, however great the power of perspicacity to create new images by combining and separating those previously observed, and although nothing is easier for it than to conceive different causes for the same thing, where the matter itself remains silent, however, perspicacity even when speaking with the strongest arguments only shows that such may well be the case but it does not claim that it is so. I should explain this more broadly if, Gentlemen, in a matter which concerns the forces of the mind, your knowledge as perfect as human condition permits did not make me blush, and impose silence upon me. You indeed, informed of the works of the ancients as well as experienced by a long utilization of the matters, you have penetrated into the sanctuaries of this goddess and, moreover, you see more thoroughly what is required to form the purest concepts possible. It will thus be enough to mention some examples taken in the history of the lymphatic vessels. Who during so many centuries, even most ready by natural inclination or through dreams, thought of the lymphatics before, you, most famous

Bartholin, you presented to the eyes of everybody what you observed with your eyes, not those of the mind but those of the body? Since then, who succeeded in revealing through his intelligence alone the matters which are still hidden?

Who will demonstrate with trustworthy arguments from where the last lymphatic vessels originate, in the liver or in the rest of the body? The origin of the saliva confirms the same need of observations: reasoning deprived of the work of the senses did not find the paths carrying the saliva into the mouth. Neither will the paths which carry its substance to the glands ever be certain for us, as far as I can see, if they are to be expected from thinking alone. Observation must lead to reasoning and the object itself must be examined, as far as possible from all its sides so that its figure be imprinted in the mind corresponding truly to the object itself. While everybody does that in this world to the extent of his power and presents his observations to be examined to this end, I could not refrain myself either from presenting publicly what I took care to bring together to clarify the function of the lymphatic vessels, not only that of the afferent ones sufficiently confirmed previously, but also the number of the efferent ones which has to be increased, and thus from adding a little to the abundant crop which others have heaped up. So that this manuscript should not be without protectors I have dedicated it to you, bright lights of the fatherland, and to your fame since the natural light of your divine mind would be convincing, since your sagacity acquired in many functions would be urging, and since both would command great blessings partly for myself, partly for my family. Thus accept with a serene face these presents of the fortune of my works, kindly striving for principles. I wish you a happy and long life and the immortal glory of your name finally take care of the breath of your favour.

Leiden, *1661*

§1. Whoever would admit with *Aristodemos* convinced by *Socrates* that which all those who have common sense must admit and will not deny, that "the structure of animals is the work of a wise and life-giving craftsman", since all their limbs and the parts contained in the limbs almost say themselves that they were made foreseeingly, since nothing seems to be so small as not to be destined to its function and nothing to be so mean that it does not teach the wisdom of the Creator demonstrates it, just as in times past the most simple line of art, ridicule for the ignorant, [showed] the dexterity of *Apelles*. When indeed people imagine images destitute of mind and movement, they take great care that nothing is lacking, nothing is too much, so that all things, even the smallest, are placed under the eyes as highly wrought as possible; when an architect considers it his fault if in an edifice which he has built up, he leaves the very smallest space unoccupied and not serving the scope in view, who will then believe that he, whom all the most accomplished masters propose themselves to imitate, although with unequal success, placed an useless work in any place and that he produced something in vain? Who will not rather judge that in the smallest details the greatest, admirable ingenuity is concealed?

§2. Therefore, it is deplorable that there have been among men of great reputation some who kept saying that not so few things were made in animals all but for the sake of distinction, since nothing more than that is against divine mind and thinking. Even more deplorable are those who feel so contemptibly about some works of the highest architect of the earth that, because things appear small and at a first glance do not display a great apparatus, they judge them unworthy of spending time for their investigation and even reject them in almost abusive words. It belongs to human weakness that those who do not use but misuse their senses eagerly pursue only what is delightful and gorgeous.

Revering a monkey clad in purple while disregarding the satyrs of *Alcibiades* they would have been most diligent examiners of Nature, if (as *Cicero* liked to speak) our Providence had presented itself as Epicurean. However, when they not only neglect that but also think that they have to criticize what is not understood when it is observed, this is no longer a sign of weakness but of malice. More sensible was what once that great wise man said, when asked his opinion on a book: what I understood, he said, was good and perhaps also what I did not understand.

§3. Among the different parts of animals, which suffered an injustice of fate, the *glands* come in the first place. They could often not escape the contempt of laymen nor avoid the censorship of those initiated in the mysteries of science, for an unsaid cause and because a searching examination was neglected. Though, they are often at the origin of such diseases that, those whom their simple and rude appearance had not won over to their examination, those whom the greatest skill of the Creator obvious in other things had not drawn along into a similar opinion about them, necessity could, or should, have forced them to this opinion. This we learn from tonsillitis, scrofula, parotitis, buboes, and other evils which intervene in our life often with the utmost danger and not rarely provoke amazing turnings away of humours from one part to another so that their investigation is most excellent for the knowledge of the noblest body of the whole world and necessary to preserve and to restore health.

§4. What the ancients knew of them, if the matter itself is considered, seems to be very little. Since indeed some points would seem to be common to some glands and to sponges, they stopped at this similarity, and assigning to them the task of sponges, they believed to have finished their task and discovered the whole matter. It is of course familiar to the human mind that, if something affecting our senses in that object which we project to examine occurs in a way similar to something perceived previously, that which was known at first comes immediately back to the mind even unwillingly. Hence, by attributing to the new object the properties of the thing more studied, we desist from further examination. Thus, having observed the force of attraction of a magnet, some people have attempted at explaining everything by magnetism. Thus others, by imagining for themselves a hearth and a jar in man, others not to recede from similarity, find also spoons, twigs, lids, cooks. Thus, if people compare the vessels carrying lymph to pumps, they imagine that they pull lymph from a well (thus they speak of the receptacle of chyle) like people who are pumping. For no other reason, when they see a rare texture in the glands, above all the tonsils, they say that they are sponges like the other ones. If you want to hear about the thing itself, apart from these similitudes, you will find that the ancients knew very little about these glands. However, this is not to say that I will detract from the authority of the ancients, which I always respect, always honour, but I am without any doubt convinced that nothing has been so far discovered altogether and brought to perfection. They kindled the light. Our task is to keep it shining and to make it gradually more glowing.

§5. This veil of similitude which hid knowledge of the glands with its cover began little by little to be removed after some peculiar paths were found in them. Hence it appeared that indeed they do not drink-in superfluous moistures like a sponge nor are they common dregs of the viscera and all the body, but they are devoted to a by far nobler function. Since then working on these mysteries, I could notice in these glands some vessels (not yet described by anybody as far as I know) which, I think, not only support their new significance but also throw some light on the art of medicine and discover to us paths which previously were known by nature alone, I thought that it would be useful to publish these few observations so that those who have more time and are better with their hand investigate these matters more accurately and more profitably for medicine.

§6. To comply to the opinion of the famous teachers *Bartholin* and *Sylvius*, I had lately decided to provide a description of the external salivary duct, of vessels observed in the innominate gland of the eye and of others, when, noticing that Mr. *Gerard Leonard Blaes*, at one time my host and beloved teacher, in his treatise, the *Medicina Generalis*, mentions this duct which I had shown to him previously, I thought that this labour had partly been taken from me. I had believed indeed that he was to pursue the matter with an accurate description in this book after having described, as he says, a new and accurate method of which he claims in the preface to be the inventor, apart from which he declares himself that the book contains hardly anything new, the more so since not only a convenient opportunity presented itself to him, while the undertaken description of the parts of the body seemed necessarily to require that.

§7. However, as I pondered more accurately his words and I saw that he had not been able to ascribe a true origin nor a true orifice to this duct, so short, so obvious (as will appear in § 15), that he even had destined the gland itself from which humour proceeds in such a delicate confluent of very abundant rivulets, to such a low function (as is obvious in § 17) that, if I was not sure to have shown these matters to him again and again, I should never believe that he has seen them, I wondered greatly that he was so little consistent in such a small thing.

§8. I owe much to the famous gentleman because he gave me an opportunity not only of claiming my discovery but also of finding something else. While indeed in the Easter holidays this year, I look in a calf for the said parotid glands where he said they are, and follow the vessels where they lead me: I observe, not without admiration, a delicate communication of various glands in the neck, through peculiar vessels. Having seen these, to know more about them, I dissect a second calf head, then I open dogs and finally also the head of an ox, where several salivary ducts unexpectedly present themselves. Thus, to gratify the teachers, so that others, who have better opportunity, examine these matters more accurately, I decided at this occasion to give somehow a description of my observations, assuring for myself forgiveness by more skilful people beforehand if ever it occurred perhaps that I made a mistake.

§9. The glands which supply the mouth with continuous moisture are not one nor are they situated at one place. Indeed, besides the maxillary and tonsillar glands described by the famous *Wharton*, I have observed others below the ears, in the region of the cheeks, beneath the tongue, in the palate: all these have in common that they are made of several quasi-fragments of glandular flesh, even of balls connected to each other by way of fibres, nerves, vessels and membranes, and thus of the kind of those which the famous gentleman *Franciscus Sylvius* calls *conglomerate*. They are all indeed *glands* in the proper sense, although their parenchyma is of its own kind, not bloody but white, fatty, provided with veins, nerves, arteries and lymphatic vessels, present with differences depending upon their various basic constituents: none of these differences comes nearer to the matter than what the famous gentleman mentioned above proposed. He indeed described[3] not the vessels inasmuch as they are not yet found in all, nor the function, which is little known so far, but the conformation itself as far as there are two primary kinds of them: "*There are indeed others which rise up from various conglomerate parts and smaller glands, joined together with some inequality of their surface; others have actually an equal surface and are observed to be made of one substance almost continuing itself, and to be conglobate*". It is also peculiar to the latter kind that its convex side in most cases receives lymphatic vessels, whereas at its concave side it emits other lymphatic vessels originating inside from several thin rivulets. However, like for the glands, there are also two kinds of lymphatic vessels so called by the famous *Bartholin* because of the colour of their contents. Those which are seen in the conglobate glands carry their liquid back into veins whereas those which proceed from the conglomerate glands lay down their contents in noticeable cavities such as the eyes, the nose, the mouth and the small intestine.

However, although the *glands of the throat* (thus all those indeed which function close to the mouth can be designated by this common name) are of the same kind and have similar vessels, they are, however, different by the number of vessels. Some indeed are provided with one excretory duct only, such as those which are found below the ears and beneath the jaw. Some are provided with several excretory vessels, such as the other four, those which are observed in the cheeks, beneath the tongue and in the palate, such as also the tonsils. As, however, *Wharton* has described the tonsils together with their vessels and the inferior maxillary vessels, satisfied of having presented those that we know have not yet been presented by others, we do not want to waste our effort.

§10.   *The glands situated below the ears*, as they well [enough] found them did not display to the ancients anything peculiar and distinct from other [glands]; one thus finds in the ancients no proper name for them. It seems that *Hippocrates* in *On glands* meant them in this description: "glands below the ears, here and there, on both sides, in the throat of the neck". If however the listing made here of these glands is compared with their short enumeration presented later in the same treatise, it

---

[3]Franciscus Sylvius, *Disp. Med. 5 de Lienis et Glandul. Usu th. 26 & 27.*

easily appears that this description is general and common to all the glands of the throat known at that time which, a little further he calls παρισθμια just as *Zwinger*[aut]Zwinger, Theodor also presents in tables, and as the author himself, the best interpreter of his words, manifestly declares when he says that he will deal with the integrity of the glands of the neck. However, those which the ancients called parotid were not the glands themselves but tumours observed below the ears or behind them, which they called otherwise swellings and tumours. Thus those near the ears of which there is so often mention in *Hippocrates*, considered properly and for themselves, indicate what is normally found there. If, however, the divine old gentleman is consulted himself, he wishes this word to mean something abnormal. Hence, in *Epid.* 1, Sect. 3: "in the sheep they are near the ears", where it is obvious enough, not to mention other things which could be told, that this name indicates something ill, i.e. the same which *Celsus* and *Pline* called parotida, adopting a Greek word in Latin language. Although thereby this was the meaning of the word for them, since, however, it is logically suited to the subject of its disease not less than to the disease, it could be conveniently admitted for the glands themselves as it is commonly, if several glands and glands of diverse kind were not found at the same place. Last year indeed, as the very famous *Franciscus Sylvius* in the hospital taught medical practice daily, he showed both to his disciples and to others who might be interested, among other cadavers which, given an opportunity, he opened, that in one all the conglobate glands were affected by scrofula, not in the remainder of the body, but about the region of the ears, whereas both salivary glands and the pancreas were intact. So to distinguish them, the ones which we speak about could be called *parotid conglomerate*, leaving to the others the name of *parotid conglobate*.

§11. The *parotid conglomerate* gland (as it was observed in a calf) visible in the pit below the ear, between the posterior side of the mandible and the process corresponding to the mastoid, is situated above the horn of the hyoid bone. It has an irregular and flat shape inasmuch as the asperities and eminences of the surrounding parts permit. It must be noticed, however, that its lower part is long and thin whereas its middle part is thicker: hence the anterior part expanded above into the edge conceals almost all another and conglobate gland in its concavity. In front of this concavity various paths are noticed for the passage of vessels, but mainly nerves, which are large and abundant, and are distributed through this gland into other parts.

§12. Mr. *Wharton* gave the weights. Comparing this with the others, he observed that the parotid gland weighed 17.6 g, the internal maxillary gland 9.8 g. However, he discovered that in the foetus of a cow the weight of the latter was 11.7 g, that of the former 7.8 g. He reports that in a horse he dissected one which weighed 340 g, and yet was not abnormal or bigger than usual. The famous gentleman would seem to have squeezed out an abundance of matter from these glands in non-determined quantity, unless perhaps he would have held that it already was established that the matter appears to be arranged in the same way in both, which he also hints at in ascribing a similar substance to both. However, besides bigger and more numerous nerves carried through the upper gland, the smaller gland enclosed in the bigger, which, as far as it is not distinct from the other, I think has increased the weight of the parotid gland, makes me believe that the proportion of the internal maxillary to the parotid gland has not been exactly observed. I indeed found in a calf that the said parotid gland, free from vessels and from the conglobate gland lying beside it, weighed 141 g, whereas the inferior maxillary gland weighed only 125 g.[4]

§13. It is attached to the adjacent parts by way of fibres and vessels, and mainly of nerves. A very hard branch of the nerve of the fifth pair, perforating the gland at various places, sends sprouts upwards as well as downwards, which, intertwining with each other, at various places and mainly between the gland and the remainder of the head to which the gland is adjacent, form a network. Besides these, others, reflected from the branch which passes forwards between the temporal muscle and the jaw, are carried backwards into the gland itself to the side of the excretory duct. The *vessels* which are found there are of three kinds. Besides the veins and the arteries supplied by the external branches of the external jugular and carotid vessels, there is indeed also a *peculiar little vessel* which I found in the head of a sheep last year in Amsterdam. When received by Mr. *Ger. Leon. Blaes*, the famous professor at the college of Amsterdam being my host, about the feast of Easter in a series of lectures which he finished with the sixth lecture after my arrival, I noticed then a good opportunity of having animals suitable for anatomical training. I obtained from him permission to dissect privately subjects bought by myself so as to repeat mentally and to imitate manually what I had observed so far done by diverse gentlemen very

---

[4]Weights as transformed by Maar in OPH 1, 227–228.

skilful in anatomy. It is for this reason that on April the 7th, I prepared alone in a small cabinet the head of a sheep which I had bought for myself, to dissect the brain, when in its examination, veins and arteries, running through the mouth drew my attention. When investigating them with a probe in different directions, I felt that, being carried through some vessel into the wide cavity of the mouth, it hit the teeth. Amazed by the novelty of the thing, I called my host to hear his opinion. He at first accused me of having forced a passage and then resorted to a play of Nature. Finally he judged that the book of *Wharton* was to be examined more accurately.

§14. Now, this *salivary duct* originates inside the often-called parotid con-glomerate gland from several rivulets flowing together in one cavity. From there, in the calf as well as in the sheep, this cavity extending downwards passes from the angle of the mandible in a depression carved in the side of the mandible. From there, rising up obliquely forwards, it finally opens into the external part of the mouth by a fairly large orifice dug in the upper and posterior of the papillae situated near the second molar tooth. Besides a proper tunic, various nervous fibres are noticeable in it. Displaying several filaments they run on either side of the duct and, hence, they surround the middle of the duct, inter-twined with each other. It is thus no wonder that a probe is so uneasily pushed in the cavity itself through the divided tunics since, sticking among the nervous fibres this is prevented from proceeding further. If we consider its straight course in man by which it passes between the gland and the middle of the buccinator muscle, it seems to be like a strong cord which, originating laterally from the centre of the bucci-nator muscle, crawls through the bone of the cheeks and ends in a small and thin muscle directly opposite the cheek, as *Casserio* writes. Here I will not add more on this, as it is observed in man, although a little different from what was said, since I know that the famous *Chairman* will deal with it in his comments of *Vesalius*.

§15. Let us see how Mr. *Blaes* describes what has been said above. In *Medicina Generalis* on page 63, one reads this: *Saliva is a humour, etc., made mainly of serum, separated from the mass of the blood in the max-illary glands by the action of heat, and from there carried through a salivary duct noticeable in the upper and equally in the lower jaw, to the anterior glands of the mouth, as from there it is expressed by the movement of the tongue and serves to the taste,* etc. Thus it comes from the maxillary glands and it goes to the anterior glands of the mouth. I will not mention the movement of the tongue which does little for the expression. However, in the same book, on page 23, he thus describes the maxillary glands: *The maxillary glands, as well the external ones, situated at about mid-length of the mandible, as the internal ones showing themselves from the parotids to the chin with various size and colour, have a singular duct called salivary from its function* etc. As *Wharton* and the famous Chairman *Johannes van Horne* after him, in the first disputation on the salivary ducts, have shown that an internal duct proceeds from the internal glands, according to him an origin of an external duct from the external glands must be deduced. Certainly, if he had examined more accurately the words of *Wharton* from which he took this description, he would easily have seen that they do not correspond to so big a duct by any means. Thus the often mentioned very learned Gentleman in Chap. 21 indeed outlines them: *The external maxillary glands are very small and of less importance; they are situated externally at about the middle of the mandible.* Yet is there a need for words when experience itself speaks? He should put his hand to the investigation and he will easily find how true these words were; and there is no reason to believe that he counts the external salivary ducts among the external glands. So that indeed he could not be excused for this reason, he himself excellently bewares in the next chapter, in which he distinguishes them from the first ones by the name of parotids. It is with the same credibility that he proposes the orifice: he indeed saw how the lower glands end; hence he thought that he would not be mistaken if he ascribed a similar insertion to the external ones. Yet the famous gentleman would have written very differently and he would not have imagined glands from which saliva is expressed by movement of the tongue, if he had ever introduced a probe in the duct through its conspicuous exit either in animals or in man. However this is enough. Let us return to the matter.

§16. I should not doubt that there are other vessels besides those already mentioned, since not only has *Wharton* also observed that a duct runs from a similar conglobate gland situated near the internal maxillary gland to the maxillary gland itself, and I have seen it in the head of a calf, but I have also seen vessels running from the conglobate gland itself, which the external salivary gland encloses, into the salivary duct. Since, however, I did not observe them distinctly because they were too short and too small, I shall leave that undecided. I will, however, add this on the conglobate parotid gland; in a lamb I saw some lymphatic vessels running in it; they originated from the extremities of the nostrils and other anterior parts and they passed over the muscles in a straight course. I could not find their primary origin. However, all those which I lay open seemed to draw their origin from the anterior glands of the mouth.

§17. *After the description of these parts has been presented, it remains to consider their function.* Now, the very learned *Wharton* ascribes three functions to this gland: *1. to withdraw some superabundant liquids of the very hard branch of the nerve of the fifth pair and to carry them back to veins; 2. to heat the internal and external ear by its own heat; 3. to fill up and flatten that pit in the vicinity of the ear.* However, it is no wonder that these functions were proposed by the famous gentleman since, except for vessels common to the other parts, he did not observe anything in this gland. This appears from the preceding paragraph of the same chapter. *Actually,* he says, *if authors claim that these glands moisten parts by pouring proper peculiar humour through some vessel, they must show us the kind of humour* which is to originate from it *and how the part receiving the humour gets rid of it.* Since this has now been discovered, it appears easy that a nobler function of this gland is to be found. I do not see why Mr. *Blaes* can be excused, who feeds on the fruits of the discovery of the gland, who, although he boasts to be the discoverer of the duct, none the less, also makes his the functions attributed by *Wharton,* even though incompletely: he proposes the first, although truncated, and he omits the third. He indeed says in *Medicina Generalis* on page 24: the glands with the same substance as the maxillary glands are mainly about the root of the external ear; from there they surround all the external ear *to receive overabundant liquids* (not saying wherefrom and whereto), *and warm up the ear by their heat.* Among those who saw the salivary duct continued up to this gland, who observed the considerable roots of the same duct distributed in great number very delicately in this gland, and who will believe that Mr. *Blaes* would ascribe such low functions to the parotid glands if he had ever put his hand to investigating the salivary duct? To the parotid glands, I say, which, although he had seen that their substance is the same as that of the maxillary glands, he could, however, not have surmised that their function was the same. If the matter was difficult to investigate, if the duct was long and had to be followed through various bends, if it had been soft and prone to rupture, it would have been possible to bring forwards various excuses. However, since it is short, since it is in the way of the external glands, since it is thicker and stronger than any other, since *Blaes* himself in his letter to me recognizes that it is an easy and always obvious matter, he should himself consider how consistent he is. However, I will leave this matter. The true function of the parotid conglomerate glands is to prepare that saliva which is excreted through the external salivary duct in the external cavity of the mouth. We shall deal more with this matter below, where we shall describe the other salivary glands.

§18. Among the glands which are provided with several excretory ducts, those which are *in the cheeks* take the first place. As compared to the others, they are the most conspicuous by their size and by that of their vessels. Now, these occupy the whole region of the cheeks, surround everywhere the external cavity of the mouth and, reaching to the gums above and below, they are between the muscles and the tunic of the mouth. They are thicker in their inferior part. From there on they become thinner and yield free passage through their external surface to veins, arteries and nerves sent either into this gland or in the anterior glands. In the same way as the vessels to be described soon, it could have been divided into an inferior part which is thicker and stronger than the others, a middle part which is very thin and a superior part which is in between. In a cow it was of the same colour as the parotid gland.

§19. Besides these peculiar vessels, on May 21, I observed *ducts* in those parts, as I cut open the transverse and thicker part of the gland a little obliquely. Seeing indeed that this gland was conglomerate, I thought it was also provided with its own duct like the conglomerate glands. I was not deceived. Hardly indeed was it divided when a probe sent through the small vessel divided by the same cut entered freely into the cavity of the mouth. By opening from there all the body of the cheeks from the mandible near the gums and by stretching out a little the membrane, I saw on the same straight line several small gaps and when the probe was sent through them, it found easily a passage into the gland itself. But these ducts do not originate and proceed in one way only. Originating inside the thicker part of the gland, which is inferior, from several roots joining into singular trunks, as they are more capacious where they ascend to the internal membrane, they perforate this membrane by four orifices at about the level of the lower gums. The orifices are narrower than the duct itself and arranged on the same line over the whole length of the mandible. At some distance beneath them, small semi-globular protuberances appear, none of them acute. They are sometimes in great number one fingerbreadth above the orifices mentioned previously. In the middle part there are also many excretory vessels, but very short, and hardly visible unless by expressed humour, whereas those descending from the superior part are a little more visible.

Last year the famous *Sylvius* observed glands in humans intended for saliva although the vessels were not yet discovered. Since indeed he had a diseased woman suffering from ulcers of the cheeks, and he saw that limpid water resembling saliva trickled from there, given an opportunity of examining cheeks in the hospital, he showed to us their glandular flesh. Also what is read in *Aquapendente* on the aqueous humour dripping day after day from a fistula after the healing of a wound of the cheeks seems to prove the same.

§20. In the third place, the *sublingual glands* present themselves. Situated on both sides of the tongue, they are of the same kind as the previous ones except that they are provided with excretory vessels, not shorter, but narrower. They rise up from small rivulets inside the gland itself and hence, parallel to each other, they recede from the tongue to the gums where they gape in orifices through the tunic at about one finger-breadth from the teeth, hardly visible unless they are compressed. There are no protuberances except at some distance from the orifices but, as said for the vessels of the cheeks, wherever the said small vessels penetrate the tunic it is even and flat. I could not observe whether there are small depressions in the papillae. I compressed some of them but did not express anything. Hence, by examining cutoff papillæ, I saw that there were small bits of gland which, emerging above the others, raise there the tunic sharply.

§21. It has seemed right to me to call the last ones palatine since actually all the flesh of the palate, in animals as well as in man, is nothing else than a conglomerate gland continuing in the tonsils. Countless small ducts proceed from it. Perforating the membrane, they make it a kind of sieve. I noticed them from the first time on May 27 as I had already dissected almost all the head of a cow. Then indeed, as I cut out the tonsils, I saw that, by compressing the vicinity, sticky humour was sifted out. Hence all what constituted the ceiling of the mouth, cut away and compressed by the fingers, displayed countless droplets of glutinous humour breaking out through very narrow holes. This was plainly different from the slime of the palate of which *Wharton* in Chap. 22 of *Adenographia* makes the third kind of spittle. This appears obvious from his explanation since he establishes that this slime of the palate is drawn out from the nose above the palate and he calls it slime of the brain, which does not fit the humour which we obtained by compressing the palate.

As I was ready to commit the present text to the printer, I received the third book, *De Catarrhis*, of the famous *Schneider*. In Sect. 2 of

Chap. 3 he mentions a *pituitary membrane* observed at the end of the palate, from which he saw slime oozing. However, besides the membrane itself perforated by many small holes, it is *glandular flesh* which is actually found beneath all this membrane.

§22. So far we made a *description of the glands of the throat,* about the function of which we need not add much. Since indeed peculiar vessels have been discovered, nobody can question that their task is *to prepare the humour which is always found in the mouth.* Now, in order that the upper parts moisten as well as the lower ones, the internal as well as the external ones, there are several small vessels in the mouth to transmit humour equally to all parts. Through them abundant saliva can also be excreted when needed. Thus the famous *Bartholin* reports, among others, the rare story of a noble man who is affected by abundant salivation whenever he eats, but he does not spit at all outside the meals nor when he speaks, even if he drinks very liberally when he entertains friends. There is thus no need to resort to hidden paths in order that the humour which is excreted in *salivations,* whether arising spontaneously in some diseases or whether provoked by the use of sciatalogues, should come from there.

§23. As far as *the chewed matter* is concerned, whether it is spread over the palate, or whether it is stirred with the teeth when the mouth moves, it must enter the subtlest parts by way of diffusing heat through the holes open to glands provided with shorter excretory ducts, and enhance secretion of viscous humour. Thus also *errhina* attracted in the nose and penetrating the glandular parts which are found there, contribute much to the drawing out of serous humour. One should not be surprised that through the same vessel I make some medicine be brought into the glands and excretion drawn out from them. Since indeed the excretory vessels are gaping and short, since the medicines doing that are bitter, consist of subtle parts and are easily mobile in the heat, neither the arrangement of those nor their strength can be questioned.

§24. In what concerns *salivations* with which diseases sometimes are resolved, consideration of the said small vessels seems also to provide a clearer explanation of them. I should indeed think that the water which the illustrious gentleman *Carolo Piso,* in *De Morbis a Seros. Coll.* Section 1, Theor. 4, thinks to grow imperceptibly and to rain from the head through membranes and nerves, must come from the glands mentioned above. Thus, in Sect. 2 of Part. 1, c. 2, he reports in a splenic diseased always spitting patient that the cheeks sometimes considerably but painlessly have swollen, and that this almost daily swelling is resolved either imperceptibly or with much salivation. A little further down he says: tumours of the cheeks as well as those of the glands about the ears and the painless tumours of the glands of the neck, aroused from exposure to the sun or from the autumnal cold, and growing into a monstrous mass in some hours are sometimes returned to normal in one day, or little by little, or with much salivation. Because when these parts are exposed partly to the ambient air outside, partly to the air which passes inside the mouth while air is attracted by respiration, they are easily affected by its harmful properties in a susceptible subject. If thus some matter is retained in these parts affected by heat or cold, there are obvious paths through which it can be excreted from there.

§25. However, great doubt occurs concerning the origin of the humour which moistens the mouth. Some people derive it from the *brain* through concealed paths, others from a juice carried through *nerves*, some from arterial *blood*, the noble *Bils* and, supporting *Bils*, the famous *Anton Deusing* in *Exercit. Phys. Anat. de Alimenti in Corpore Depur.* § 83 from a dew-carrying juice brought through *vessels* commonly called *lymphatics*. We will for the present only show that *arterial blood* is suited for this task; that nerves, however, destined to other functions, are not provided in vain to glands; that this function, however, can only be attributed to lymph or dew-carrying juice is against all reasoning and experience. Before tackling this, let me just say a few words on this humour since its more extensive consideration requires chemical anatomy.

§26. It seems convenient to call *saliva* all the humour which is normally found in the mouth since no other more appropriate word presents itself. Those, indeed, who call it sputum do not mean by this word the humour which is normally present but that which is excreted in unhealthy conditions. The one which is dealt with here is of better quality than the one which must be excreted otherwise if one has kept oneself inside the limits of a diet. This appears from the example of those who have never in their life sifted out anything through their nose or their mouth. Among these were first and foremost the Persians: indeed it was forbidden to them to spit or to blow their noses. This law was of course easy to observe for frugal people who ate only once a day. According to the authority of the divine *Hippocrates* indeed: slimes and saliva are the result of repletion. For them there was no danger of repletion (Fig. 2.2.1).

pag: 21

Fig. 2.2.1 Dissection of a calf

§27. It would seem questionable whether all *saliva* is of the same nature since all is not prepared at the same place. If you consider the kind of glands and that of vessels either afferent or efferent you will make out no difference. All are indeed conglomerate and are found in all vessels of the same kind. There are three points, however, which make some difference: 1. The *colour* which is not always the same in all glands; the parotid glands are sometimes red as well as the inferior part of those

Explanation of the Figures

Fig. I.

a.   Parotid conglomerate gland.
b.   Parotid conglobate gland.
c.   Lymphatic vessel proceeding downwards from the conglobate gland.
d.   Roots of the external salivary duct.
e.   Trunk of the salivary duct.
f.   External branches of the jugular vein.
g.   Nerves which within the gland and also outside this place as in h.
h.   intertwine with each other.
i.   Nervous fibres accompanying the salivary duct.

Fig. II.

a.   Orifices of the vessels proceeding from the inferior gland of the cheeks, into some of which stiff hair has been introduced.
b.   Aperture of the external salivary duct occurring at the top and back of the papillae. The other points denote holes through which viscous humour gets out under compression.

Fig. III.

a.   Sublingual gland.
b.   Its vessels.
c.   Small orifices of excretory vessels.
d.   Depression observed near the side of the tongue.

Fig. IV.

a.   Holes in the palate, through which sticky humour is expressed.
b.   Tonsils.

Fig. V.

One small vessel among those which proceed from the inferior part of the gland of the cheeks.

Fig. 2.2.1   (continued)

situated in the cheeks; in their superior part they tend to yellow with the tonsils, the palatine and sublingual glands, whereas the inferior maxillary gland is mostly pale. 2. The *conformation of the vessels* in so far as there is only one in some glands such as the inferior maxillary glands and those which are found below the ears, there are several in others and these vessels are either small as in the palatine glands, tonsils and the superior glands of the cheeks, or somewhat longer as in the sublingual glands and the inferior glands of the cheeks. 3. The *consistence of the humour* which is found less viscous in longer vessels, more viscous in shorter vessels. Since this diversity in colour does not always occur—I indeed saw in a calf and then also in a lamb that all were characterized by the same pallor—, and since the difference in consistence is not the same in all—more than once I indeed expressed sticky and viscous matter from a large duct as well superior as inferior—, a difference of vessels remains consisting of their size alone, from which it is not permitted to infer a diversity of the content.

§28.   If saliva is examined when it proceeds from the mouth, it resembles water somewhat by its colour and its clearness. Rays of light in it undergo little more change than in water. When spit is full of foam, this is not due to [the fluid] itself but to the movement of the tongue and mouth. In the same way milk, beer and even water itself should be said foamy. If you explore it with your fingers, you will feel it sticky and viscous, and with a power of lubrication, and therefore less mobile than water, more difficult to divide. In the saliva of healthy people you will find no savour as well as no odour. So whereas tasting and smelling find it deprived of quality, seeing and feeling judge it less simple than water and actually if one likes to consult other effects, it will be possible to consider a singular composition in it. Simple liquids such as occur in the examination of natural things do not easily admit others to be mixed with themselves unless they are of their own kind or of a middle kind; and if, by chance, some heterogeneous substances have been mingled with them by some agitation, without the intervention of a third substance, as soon as they are left to themselves, either, if, having been joined in the cold, they are subjected to a light fire, or if, having been mixed in intense heat, they are cooled, they lay down the inconvenient burden. Thus, besides salt and spirit, or substances of similar composition, water allows nothing to be mixed with itself. Water admits oil but one has not yet found anybody who has united oil with water without an intermediary. Spirit could be joined to water and oil but not with salt except by a singular artifice which the chemists keep for themselves among the secrets of their art. If, by distillation you pour rectified spirit of wine into oil of tartar, however you have indeed agitated, the spirit at once rises to the top, and the tartar is precipitated to the bottom. Acids are easily mixed with salts but, even if this is done carefully, not without effervescence. However, saliva is able to tolerate everything. Whether indeed you examine it inside the body or outside, you will find nothing which it rejects, nothing with which it will not associate without struggle. Hence anything taken in the mouth is at once united to saliva, as a universal solvent, as soon as it enters. Hence also saliva which is continuously swallowed without that we give any attention is joined to what is contained in the stomach to enhance dissolution of the food. From what has been said previously, saliva appears not to be a simple liquid but a mixed one and this in well-determined proportion. There are also other arguments at hand which demonstrate that it is not a simple liquid: indeed evaporation produced salt for me (as once in the morning I subjected to the test of fire saliva taken out in a glass, which was limpid and free of all association with mucus); and mobility of mercury being impeded by mixing with saliva (which people call mortification) is evidence of the presence of an oily [component] moderated by acid. Not to mention the various forces song off in medicine and which do not act without the virtue of active principles. Thus, I cannot but admit that what the famous *Sylvius*, most skilled in the chemical analysis of the humours of our body no less than in the anatomical exploration of the body, suspects is more than likely. He indeed thinks that *in the saliva there is much water, a little volatile spirit and very little lye salt mixed with, and moderated by, a trace of oil and spirit of acid.* This could be explained more profusely with various examples if our purpose did not summon us elsewhere.

§29. The saliva having thus been examined with the glands, it remains to investigate the paths which bring its material. To do that more easily, we shall present the single parts with which the glands have to do and we shall enquire by what means either they receive something from there or they send something there, so as to assign its function to each of these parts.

§30. Firstly then, since they have veins and arteries, nobody who knows the circulation of blood questions that they receive something from the heart and send something back to it. Now, because some people assert that the glands do not owe anything to the blood except its heat, yet in general people summon from there altogether the nutrient and the humour which is separated in them, we embrace for the time being the opinion commonly accepted until another theory is proved by reliable experiments. Not to seem either to suppose something or to rely on authority alone we shall present the reasons which induced us to judge that the position of the ancients should not too easily be given up.

§31. If we consider the blood and the parts which either nature or artifice showed in it, it seems reasonable that this be called the cellarer and steward of the whole body. Since all which is found in the parts of the body, whatever elements you wish, either of the ancients or of the moderns, can also be demonstrated in the blood; since this blood containing in itself principles of every kind is carried to every part; and since the chyle which originates from the food continuously repairs that mixture of seeds to be carried to every part, I see nothing against that all things which the single parts require for their nutrition as well as for elaborating peculiar humours, could be taken from the blood.

§32. Some people object here that blood is not carried down into all parts and prove it by the testimony of the senses. However, if it had not been obvious otherwise, the research alone of the anatomists would have taught how much weight negative argumentation has, relying on the senses, and even not those of all people but of some. Indeed, this does not hold: I or somebody else has not yet seen it, thus it does not exist. If indeed people have not seen what is big and what presents itself by its colour such as milky vessels and other things, unless it had been observed at first by others by accident, is it a wonder that small vessels are not always perceived? Others, however, have seen and we ourselves have also seen obvious blood vessels in the middle body of the brain, in the spinal cord, in the membranes themselves which envelope arteries, in the membranes of the inferior salivary duct and even between the nervous filaments of the fifth pair wrapped in a common membrane as well as in the filaments of the third pair, in the presence of the famous gentleman, Mr. *Ole Borch*, Royal Professor in Copenhagen, at one time my teacher, to be honoured forever, who, together with other friends, attended most of the experiments which I mention here. Why shall we deny blood vessels to other parts if not even the very white tunic of the eye is not deprived of them, as it is obvious in inflammation of the eye?

§33. Nor should we attribute to colour so much that when blood appears different from the brain, the tunics, the humours of the eye, the tendons, etc., we should look also for peculiar humours to be carried with it through peculiar vessels, as if all the things which are hidden in this red juice, should also be red, and white things could not be concealed under a purple cloth. Who, I ask, among those who admit the combination of things and their separation, who attach some value to chemistry, ignores how unreliable it is to believe colours? Who did not see white change into black, black change into white in a short time, not to say instantaneously, at very little admixture? How often is it possible to see fabrics shining with the most delicate red, as people say, become almost white in a moment? Not to mention other things, the mineral bezoardicum will provide a unique, but, I think, very evident example: who would believe that, beneath the brightest and manifest red which appears if you pour nitric acid onto trichloride of antimony, such white matter is concealed? What wonder then that, although blood itself does not appear similar to the seminal parts to be fed, other things of various colours appropriate to these parts can be mixed with the blood, an opaque body abounding in thick and fibrous parts. I will not say anything on the experiment which *Bennett* did on blood and will not report what *Pecquet* observed in serum. Therefore their argument that blood cannot feed all parts because it is different in colour from many parts, has little value.

§34. Even less convincing is the argument of those who will say that this white which is secreted by the blood is *chyle*, and infer from this that a useless work is ascribed to nature. Nature at first would mix chyle with blood, and then separate it again from blood. Nothing is more against the sagacity of nature which does everything straightforwardly, as they say. Yet they fight against their own shadow. Not everything that is separated from the blood, even if that matches chyle in colour, deserves to be called chyle therefore. How many milky liquids are there indeed which nobody, except perhaps some *Democritus*, would distinguish from each other by their colour alone? One can give as an example virgin milk [tincture of Benzoes], as the chemists call it, milk of euphorbiacea, milk of compositae, various emulsions, and to remain in the animal sphere, although the milk which comes from animals of various kinds, is little, if at all, different in colour, it is well known that they, however, differ in abilities. Therefore, the chyle which mixes with blood, and the nutritious juice which is retrieved from blood, although they have the same colour, can be very different in other properties. Let us consider once more the mineral bezoardicum: is what remains not white when nitric acid has been taken off, although nobody would have observed it in the very red and very transparent liquid? Therefore, should I not say that it is not different from trichloride of antimony? I will not mention that it has not yet been observed that what goes to feed solid parts, completely matches milk in colour as there are various and almost infinite degrees of whites. Then, if we attribute to heat the first role in alterations, by what means will chyle mixed with blood be able to be carried with it through the heart and the remainder of the body and still keep the same nature, i.e. not being more cooked, attenuated and rendered more suited to nutrition? Even if it is changed in other accidental circumstances, although the same colour comes back, this mixing and separation must not be taken for a useless labour.

§35. When people say that blood is more volatile than the parts to be fed so that more solid food is required, this would appear a very strong argument if all the blood was so volatile. Since themselves they admit parts of various kinds in blood, and they do not exclude chyle, it appears easily that more solid parts also would be present there if they were necessary for nutrition. However, one could question, not without reason, that there is a great difference in the solid parts and, moreover, that various humours and even spirits point out to something by far different: not to roam too far away, since there are various parts to be nourished, it is adequate that various parts are also found in the blood. Expressions such as benevolent juice and closer to chyle are partly metaphoric, partly not consistent with necessity. Neither, indeed, is it sufficiently apparent what they mean by its benevolence unless perhaps something has to be conceived opposite to a predatory force of the blood (about which, in the next thesis). Moreover, if it is supposed that a benevolent juice is required, it does not, however, follow that it is chyle or something close to chyle. All benevolent juice indeed is not chyle.

§36. Those who claim that blood, like a glutton, exhausts the nutritive substance of the parts by its heat, attempt at sustaining their opinion by the example of hunger and fever. However, these famous gentlemen will forgive me if I bring in the discussion a few things which seem to oppose this. Firstly, since in hunger and in fever various signs occur which demonstrate pungency of the humours, and since the blood is always made subtler by continuous effervescence in the heart, and nothing is provided to it except for lymph flowing back from the parts, it cannot be concluded that blood, which is normally temperate, to which new things, which soften pungency, if there was any, are always supplied, therefore plunder so much the nourishment of the parts, because blood made more pungent does so. Then we do not see in these instances that the substance of solid parts is consumed: it is true that the muscles subside, fat diminishes, but the other parts do not suffer noticeable loss. What if, in such instance, even these two parts are not always diminished. We saw among other cadavers which the famous *Sylvius* opened less than 6 months ago in the hospital two opposite examples. In one all the fat had wasted away but all the muscular flesh was present. In the other which had been lying for a long time and which, for a whole month before dying, had taken almost no food, the muscles were considerably wasted away but abundant fat showed itself everywhere. However, we concede that worn-off particles (i.e. which, laid beside others since some time already, are more or less weakened by continuous afflux of heat) are expelled, heat being supplied through the blood. Therefore it would not be absurd that we attribute to the same heat the force of substituting other parts not yet weakened to this point, or torn apart into simpler ones, instead of the first ones, if only it keeps itself between limits appropriate to whatever subject. If these limits are transgressed it is no wonder that some other things are also expelled. Thus an artifice which imitates nature makes with moderate heat summer flowers blossom in the winter and eggs to be hatched in an oven. If you increase this heat, the plant dries up and the white of the egg hardens with the yolk.

§37. Finally, the vessels seem to present some difficulty. The arteries brought into this gland are indeed much fewer in number than the salivary branches and, besides this, there are veins enough to retrieve the blood brought in through the arteries. Neither urges very much. Since indeed saliva does not flow into the mouth with the same velocity at which blood arrives, the delay of the saliva in its flowing could compensate the paucity of blood arriving more quickly. It is reasonable that the veins do not retrieve all which is brought through the arteries. The arteries have something which they do not transmit to veins, the veins have something which they did not receive from arteries. This has been proved by the famous *Bartholin* in *Spicilegium de vasis lymphaticis* I. c.7 where he shows that *"there is normally more copious serum in arterial blood"*. This appears from the lymphatics and milky vessels which lay down their liquids in veins, and also from these refuses of old flesh which *Pecquet* has shown in *De circulatione sanguinis et chyli motu*, c.5. to be mixed with the blood flowing back so that as much is returned from elsewhere as what is taken off in those glands from the arterial blood.

§38. From what has been said, it is clear that the arteries supply to the glands, besides heat, also nutriment and together with it the matter of saliva. This is also confirmed by the great salivations, either arising spontaneously or provoked artificially which cannot derive from the nerves nor from elsewhere.

§39. Another part with which glands communicate, is the brain. One questions why the glands receive nerves from there since neither sensation nor motion is observed in them. Yet, although nobody has noticed movement with his eyes we have effects from which it can be clearly concluded that in the salivary and in other glands there is movement which is not opposed to their structure. How indeed does more abundant saliva flow in the mouth when pleasant and delicious things are presented to us, if the glands did not move? More or less blood cannot be expected to be propelled by the heart. Otherwise more abundant blood would be propelled not only to the salivary glands but also to the lacrimal glands and to other parts and then by far more signs would be observed. Since thus there is a particular effect, it requires the particular determination of a more universal cause to be sought for from the part itself. Since this outflow of saliva follows a movement of the mind, it easily appears that animal spirits flowing in through nerves dispose the vessels of the glands in such a way that more than usual is excreted through the lymphatic excretory vessels. Before this could happen, other paths through which its matter can flow out, i.e. veins, must be made tighter. This means that, since whatever is brought to the salivary glands through the arteries is partly turned back in the veins, partly pass in the excretory vessels, it is necessarily required that there are paths opening from the arteries into the vessels of either kind. Since the humour which is excreted through the excretory vessels is homogeneous and equally mixed, and does not spontaneously separate into heterogeneous parts, but that which is actually turned back through the veins is heterogeneous, and left to itself, immobile in the open air is transformed into heterogeneous parts which are observed also among serous parts, it appears clearly that the channels which extend into the salivary ducts are such that they admit parts of one kind only (made of several, however) but the other paths continuing in the veins are larger and are not destined to receive well determined humours. Thus as long as the blood passes freely in the veins without delay, little salivary matter is excreted. For this to be excreted more abundantly, the other paths must at first be made narrower. It is likely that this occurs by the action of nerves, as said above, since we observe that this is produced by a movement of the mind. And since thus it appears unquestionable that there is movement, although sensation has not yet been observed, it must be admitted that it is not in vain that nerves are supplied to the glands, unless another liquid besides animal spirits would be moved through them. Even if movement as well as sensation were denied to the glands, the spirits themselves, however, carried in the glands and joined to the saliva according to the opinion of the famous *Sylvius*, while they enhance fermentation of food, sufficiently prove that the function of the nerves must not be disdained.

§40. Glands are the third with which it seems these have to do. The vessels indeed which lead from the parotid conglomerate gland into the adjacent conglobate gland, are so short that, to decide something reliable about these glands, I should seem to remember too badly the one[5] who says in 1. *De natura deorum* that it is *inconsiderate to defend without any doubt that which has not been perceived and acknowledged certainly enough.* However, as I consider that vessels run from a similar gland set near the inferior maxillary glands into the maxillary glands themselves, as I see that vessels run from these conglobate glands into another conglobate gland, as I see a duct originating from this common gland which by its lower part extends to the vicinity of the jugular and axillary vessels, and movement of humour in it, I should think that it is not deprived of all probability if I should suspect that something is transmitted from the salivary glands through lymphatics by the intermediary of other glands, to the venous blood. To make this clearer, I will shortly outline some special vessels which I observed to run between conglobate glands in the calf.

§41. As some days ago in a calf another oblong conglobate gland inside a conglomerate gland offered itself to me while examining the external salivary duct, I investigated it and saw that it was convex at its anterior side, concave at its back side. In the concavity countless fibrils joining finally in a membrane gathered by way of some intermediary small branches into a *duct* which extended down in the convex surface of another *conglobate gland more round than the first* where, ramified in three or four branches, it entered the gland itself. While I examine with great pleasure the limpidity and transparency of the lymph visible in these branches more accurately, I see in the same convexity of the gland two other ramifications similar to this one, no less diaphanous and delicate. While I follow where the *trunks* of these ramifications lead, I see *one* proceeding forwards enter the venter of the *conglobate gland situated near the inferior maxillary gland*, and another different from the first be carried from this conglobate gland into the inferior maxillary gland, which the famous *Wharton* also seems to have observed. I see *the other* trunk extend upwards to the inside. As my hand had been less careful in dissecting this subject, I observed in another that the trunk entered also the *conglobate gland situated above the throat close to the end of the nose* and not far from the tonsils. From the inferior concave part of the common gland in the hump of which I said that the three vessels mentioned above scattered, each with its own roots, a *fourth vessel* extended downwards. As it could not be followed further because the head had been separated from the trunk, I opened a dog where this

---

[5][Cicero].

common gland, not so round but oblong, sends a *small vessel* proceeding in a similar way from the inside downwards to the bifurcation of the vena cava in a jugular and an axillary branch where, joining with other lymphatics arriving from elsewhere, it pours its content in the vena cava. Lymphatics proceed to this place from the head and neck, from the arms as well as from the external and internal parts of the thorax. They join together so that a path opens now and then from one to another. This happened to me already three or four times: the origin of the jugular vein very often showed to me various small branches glittering with lymph, and soon becoming red, the bloody vessels having been agitated a little. However, I could not follow them to their insertion itself. Recently, however, as I followed the course of the thoracic duct and introduced a probe into the thoracic duct from vessels destined to carry lymph, liberated from the surrounding of adjacent parts, trying to find a path into the vena cava, I noticed that the probe getting out of the way rose in the lymphatic which descended from the head, and this up to the middle of the neck where a valvula prevented me from pursuing my attempts. This unusual passage from the inferior to a superior duct offered an opportunity for various suspicions. To free myself of these, I liberated the vena cava itself at this place from the fat and adhering membranes, and at the same time I separated as carefully as possible from each other the aqueous ducts. Soon it could be clearly seen that not only the ascending thoracic duct joins the descending duct of the neck but that many vessels also converge to the same place from the back as well as from the forelegs, together with some from the thorax (in fact, I dare not affirm with certainty whether they all equally verge towards the posterior part since, except for one or two, I had not observed that at the time, but I had noticed previously a fairly considerable number in the thorax itself), and almost conspire with each other. From the mutual meeting, however, a short small canal arises which, bent around and a finger breadth long, inserts into the vena cava itself. But since, although I tried to do the same several times thereafter, I succeeded two or three times only, it can rightly be questioned whether a valvula situated at the meeting of these ducts, forced through, did not concede free passage into the superior vessels to the probe arriving from the opposite side. From where these three *glands* mentioned above which discharge in a *fourth* one which we called *common*, receive their contents is not so clear. It is certain that they do not get it directly from the blood. I indeed saw in the presence of the famous *Borch* several lymphatics proceeding from the anterior glands enter both the parotid conglobate gland and the other round gland situated above the throat, but it is uncertain wherefrom these lymphatics have their origin. Howsoever it is, *experience shows that the lymph returns from the exterior parts to the glands mentioned above, from these to a common trunk, from there to the vicinity of the axillary and jugular veins, and from there, with the blood flowing back from the upper parts, to the heart* (Fig. 2.2.2).

§42. I see that the noble *Bils* not once only protests against this and disagrees. Dissection showed the quickness of his hands, his words and writings truly his character. This gentleman, more perspicuous than all the anatomists, as he claims to be considered, not only says that the lymph or dew-carrying juice is carried off for the nutrition of the seminal parts but he deduces from there all the aqueous humours in the body and will thus that saliva also is to be drawn from a chyliferous or, as he says, dew-carrying duct, but not from the blood. The famous *Anton Deusing* approves this opinion on saliva in *Exercitationibus physico-anatomicis de alim. depurat.* p. 191. To prove this paradox, *Bils* first rejects all the anatomists without exception and then attempts at proving by an experiment that the dew-carrying juice is carried to the liver. To prove the first point, he is bombastic tragically, he boasts about his experience, he exalts up to the sky his way of dissecting, he despises now the anatomists, then the practitioners, all the ancients and the moderns together without exception, as ignorant of their profession, as compared to him, without bringing forwards reasons to the contrary but by guarding himself with weapons taken from the public street (which bolder persons would call buffoon-like in somebody else). To prove the second point, he opens one dog in one blow of the knife quickly, I must say. Then with a flying hand, he compresses a considerable lymphatic branch proceeding from the liver, wiping it clean two or three times upwards and downwards with a linen cloth, the sides of the vessel being pressed against each other so as to express the lymph. He soon tells those present to look with attention at the rising of the humour towards the liver. If you ask why he does not use ligatures, you hear at once that those who proceed in this way are mistaken; that the ligatures are adequate in those places where the humours are carried in a circularly movement from one vessel into another but, where there is no such circular movement, like in dew-carrying vessels, ligatures must not be used; that the humour contented in these is moved by the attractive force of the external parts, hence, when a vessel is ligated, the commerce of the pulling part with the well or spring is interrupted and

Explanation of the Figure

Explanation of the Figure showing the conglobate [lymphatic] glands

a. Parotid conglobate gland.
b. Conglobate gland near the inferior maxillary gland.
c. Other conglobate gland situated above the throat.
d. Common gland from which a lymphatic vessel:
e. proceeds to the vicinity of the jugular and axillary vessels.
f. Three lymphatic vessels proceeding from three different glands a, b, c into the common gland d.

thus, traction stopping in the whole part which is between the ligature and the spring; what remains falls back towards the ligature and raises the tunics in swellings. If you question that, he answers that he does not want to contend with reasons, it is sufficient that he can demonstrate this to the eye. These are the evidences of the noble man, by which he holds for unquestionable that he has demonstrated the true movement of the lymph so that, on this foundation, he declares that the universal medicine so far full of errors must be rejected and a new one be introduced. This is the movement of the chyle which Mr. *Bartholin* wishes to be demonstrated to himself in *Spicilegio secundo* as well as in *Reponsio de experimentis anatomicis Bilsiani*; which *Bils* adds in *Epistolica dissertatione*; which he wanted to demonstrate in a living dog to Mr. *Borch* in my presence, so that I wonder that Mr. *Deusing* occupied at explaining the name of chyle from the meaning of *Bils*, accuses

Bartholin of injurious request. He indeed speaks of this humour which is contained in the mesenteric milky vessels and which *Bils* undertook to demonstrate that they are carried partly from there through lymphatics to the liver.

§43. As far as the first point is concerned, I have no doubt that rapidity in dissecting acquired in a long experience can be outstanding and I readily admit that he has deserved great praise from that point of view. I believe that, if his experience, whereas he covered with votive tablets the knees of the goddess hostile to all good things, had propitiated the Divine Wisdom, and, whereas he collected wagons of abuses—I do not know wherefrom—had armed itself with explanations, this experience would have enhanced much the progress of medicine. Since he despises all the others with such haughtiness and appropriates for himself the control over all, I fear that this story will finally be lead to a tragic-comic catastrophe.

§44. In his other argument, the experiment itself deserves to be considered. The reason indeed why he rejects the experiments of others, which are carried out with ligatures and which show opposite results, is not so important, as he himself recognizes, by meeting the opposite explanations brought forward, as many as they are, with one argument. He says they are words of mistaken people and opposes his experiment to those who contradict him. I was present once when he wanted to show the movement of lymph into the liver of a dissected dog, but I must say that this was done at such a velocity that neither my eyes nor those of the attendants could discern this movement. Although I did not see by myself, I will, however, not deny the phenomenon because more than one explanation of it can be given. We know how little the eyes must be believed in things which are carried out quickly, although we look closely at them. How much less can they be believed here where we do not see the movement of the humour but that of the tunic only. He sees that the sides of a vessel, after being compressed and driven to each other, in turn recede from each other. Hence, depending on whether this movement is observed first from the upper or from the lower part, he concludes that the humour is carried upwards or downwards. Anybody slightly skilled in optics easily sees on what a strong support that relies. The vessels could indeed be handled with the fingers alone in such a way that, despite the humour being moved downwards, the tunics would recede from each other and be raised into swellings starting from the lower part. The attendants ignorant of the artifice would assert with certainty that the humour rises. Moreover, the tunics of the vessels, accustomed for such a long time to their cylindrical rotundity, if they are flattened, although there is no humour, resile spontaneously and this starting from that part which is remote from the last compression. Although it is clear from this that it is in the power of the one who compresses to display whatever movement to the onlookers, I will not push this argument in the present instance. I will concede to him that the humour actually rises after compression and I will show that, despite the valvulas, the lymph can be propelled in the lymphatic vessels in a violent movement towards the liver, the chyle towards the lumbar glands, and the blood towards the gland of the neck. So that these experiments be better perceived, I will show that the structure of the vessels is such that it can open to any thing, that opposite movement in the same vessel is not opposed to reason. So that one of them is in agreement with Nature, the other actually is to be expected only from that which hinders Nature.

§45. If we supposed that there are no valvulas in the lymphatics, as *Bils* does who, except for one visible in the thoracic duct, does not acknowledge any, and who told the famous *Henrik a Móinichen* in my presence that he would recognize that he had been wrong and that *Móinichen* alone had understood nature, if, except for the one mentioned above there were valvulas in the lymphatics, then it would be easy to explain why lymph seems to rise towards the liver in that experiment. Since lymph indeed proceeds slowly from the liver, that which is in the vessel is propelled towards the inferior parts if there is quick agitation. The tunics themselves resiling to their normal rotundity, since they cannot accept so much from the liver at this velocity, make that what is pushed downwards is impeded to go further by the weight of the incumbent intestines. It rises little by little to fill the capacity of the vessel. It is thus no wonder that the lymph can be pushed back to its origin in a violent movement if we suppose there are no valvulas. Even if there are valvulas, as I will soon demonstrate there are, the humour can be moved back, despite the valvulas, if a force had been applied.

§46. Valvulas are nothing else than very thin semilunar small membranes, fixed to opposite sides of the veins. It is clear therefore that, as long as they are intact and expanded, they oppose backflow of the content. If, however, while the vessel is compressed and the content expressed, they are laid down over the side to which they are attached, they cannot carry out their function. The vessel resiling either by its own effort or by the force of the compressed humour, the humour, because it is more mobile, rises upwards more easily than the valvula adhering to the side of the vein recedes from this side. But even if the vessel is not compressed, the humour, its normal displacement being impeded if only it moves slowly without great impetus, in that part where the valvula touches only a vein, could win its way little by little in dead subjects since there is there no connection but merely a slight contact which, the vessel being distended by abundant humour, is easily taken away. I will say nothing of the rupture of the valvulas, of the alternate extension and relaxation of the vessel by which ways, and perhaps several ways, contrary movement of the humour could be effected. It is enough for me to have shown that the forced ascension of the lymph towards the liver, which *Bils* showed, does not prove that which *Bils* concluded from it and which is of such value for his followers that they judge that a new medicine is built on this foundation. If indeed, even in the presence of valvulas, the humour could be propelled towards its source, if only some force is added, how could he from this experiment which he does not perform without an obvious force, conclude at a normal movement of humour in a vessel where he supposes that there are no valvulas?

§47. So as not to seem to have shown only that the thing can occur, I will present two examples by which it will appear clearly that it does occur. Once I opened a pregnant dog immediately after a meal, I found the intestines down from the pylorus interwoven almost everywhere with very delicate rivulets of milky vessels. There was a very small receptacle. The thoracic duct was little conspicuous and not yet tainted by any colour of milk. I thus ligated the thoracic duct at dusk and everything was left in place. I left the cadaver until the next day. The next morning, the chyle which had clung about the intestines, propelled into the receptacle, not only had distended the thoracic duct so that all its ramifications could be seen, but also, since further progress had been denied to it on this side, tending to the lower parts and distending little by little the tunics between the valvulas and the sides opposite to these, which are joined by mere contact, it had opened for itself a path all the way to the inguinal glands in the left side. Various small branches which coming from elsewhere enter the lymphatic duct before it reaches the receptacle were swollen even yet with the most limpid lymph, an obvious sign that the lymph was abnormally tainted by milky colour. If indeed chyle ought to be carried there what is then the cause that smaller branches were so deprived of chyle in such an abundance of this chyle that distended the receptacle beyond normal?

§48. Not so long ago, as I investigated in a dog, a lymphatic extending from the common gland of the neck downwards, blood in the jugular [vein] flew back more than I intended as a result of violent separation of the thorax and the parts attached to it. I saw the lymphatic where it reaches the vein take a red colour. Having thus observed the accordance of this lymphatic with the jugular vein, to know more about it, I derived little by little some blood in this lymphatic now by compressing the jugular vein and now by stretching the lymphatic. I pushed the blood with my fingers to propel it towards the gland. But as knots arising here and there delayed the rising blood, I followed another procedure by compressing first the lymphatic and then stretching it again, and by repeating this manoeuvre several times, I progressively stained the entire vessel with a red taint diluted, however, as compared with the veins. I ligated the small vessel thus tainted and soon the jugular [vein] was liberated of the obstacles opposing the backflow of the blood. What was beyond the ligature almost instantaneously vanished whereas the other part of the vessel which was between the ligature and the gland swelled more and more. As thus milk had been pushed downwards to the inguinal gland and blood had been pushed upwards to the common [gland] of the neck, why not would it be that lymph also in a violent movement can be pushed to the liver? Anybody will easily judge the value of this experiment which he always boasts about, by which, if it pleases the gods, he overthrows the explanations of all others.

§49. Various people are favourable to *Bils*, among whom even the famous Mr. *Anton Everaerts* who in his recent writing *De hominis brutique exortu*, p. 17, reports that, in rabbits fed with milk, he has clearly observed an abundant distribution of milky vessels upwards, downwards, to the sides, into all the parts mainly glandular of the body. The milky vessels themselves showed up whiter than snow so much that they drew to themselves the eyes of everybody and almost clouded over the appearance of the blood vessels. Therefore he rejected arteries and nerves and substituted milky veins through which he thinks that chyle is carried for the nutrition of the seminal parts, and the matter of semen to the testicles, that of milk to the udders.

§50. Although the famous gentleman has seen milky juice in the lymphatics, he has not yet demonstrated a movement of white humour which is the most controversial, unless he will that any white humour in the body is chyle and that such white humour cannot be separated from blood which, as I have proved above, does not agree with reason. If chyle were carried from the milky vessels through glands and lymphatics to the parts, what is then the cause that, while it is fed with milk, it would be seen in all the lymphatics but when it receives other nutrition, although it appears milky in the milky vessels of the mesentery and in the thoracic duct, in those vessels however which are in continuity with the thoracic duct without any intermediate gland, it would have been changed into so limpid water that with the interposition of the thinnest fence from one side it would appear as a milky humour, from the other side as clear and transparent water without any tainting of milk? This is clear from the experiment presented in thesis 35, and I saw another fairly clear sign of this fact. Once I had opened a living pregnant dog in the presence of the famous *Borch* and other friends; after I had moved the intestines to and fro, looking for something else, I saw a lymphatic as thick as a small goose feather swollen with milk near the receptacle while the remainder towards the inferior parts was full of lymph. It was a pleasure to see milk and lymph almost touching each other without being confused. One indeed was separated from the other only by a sigma-shaped line beyond which the milk could not drive on whereas, however, it could be pushed back very easily. The evidence of the presence of a valvula was the sigma-shaped separation between the humours of different colour. I cannot, indeed, imagine anything that would change chyle into limpid water or that could have held the displacement of chyle towards the inferior parts back in such an instance. Suppose this movement of nutritive juice which *Everaerts* will. The intestines which prepare in the first place nutrition for all parts would be deprived of food if matter to feed the seminal parts was to be expected from the milky vessels only, unless perhaps, besides milky vessels carrying juice from the intestines to the receptacle, other vessels stretching from the receptacle towards the intestines were to be conceived. At least so many experiments establishing a movement of lymph from the circumference inwards, which I have seen carried out by different gentlemen the most skilled in anatomy and which I have also somehow attempted after them, seem to be more convincing than those which must be disregarded [being based] on such slight evidence sought from colour alone. I should rather think that lymph flowing back from the parts in very small animals fed with milk only carries with it a small portion of very thin chyle cooked a little in the heart. Mostly because rabbits are so voracious that I should wonder that all the blood does not whiten in them since they fill themselves up with milk so much that the abdomen of some of them burst as I saw it.

§51. As far as the chyliferous vessels which the same illustrious gentleman leads to the udders are concerned (from which it would result that not all which is in the receptacle of the chyle rises into the vena cava nor that all the useful humours which are found in the body must be derived from the blood), since various very learned gentlemen had also previously embraced this opinion and the famous *Everaerts* himself confirms it by its own experience, to wish to attack this opinion with arguments based on reasoning alone would border temerity. On page 282, indeed, he claims to have observed *in a rabbit, a female of course, which was pregnant and altogether gave suck, chyliferous or milky vessels running above the muscles of the abdomen over the fat, creeping into the glandular substance of the udders, which afterwards joining together formed some milky tubules, and finally a common canal, and thus provided milk to be drawn through the papilla.* Although, according to his own words, the famous gentleman observed this once only, and it would be possible to make mention here of various things which are lacking in the experiments, I will nevertheless dismiss these and present in a few words what has occurred to me in a similar case. I also, indeed, in the presence of the famous *Borch* and other friends, experimented on a dog some days from its delivery as fleshy rings functioning as a placenta still adhered to the uterus and there were many signs of progeny. Separated from the puppies for 24 h she had collected such a copious amount of milk that the udders amazingly increased in volume could hardly support the slightest contact without giving milk profusely. Therefore, conceiving the hope of learning more about the milky vessels in the udder, I began the dissection slowly and carefully. Soon, hardly had the skin been separated that I observe countless rivulets completely turgid with milk. In some places also they were so dilated that you would have believed that peculiar receptacles were formed. They originated from countless roots disseminated through the glandular crust. A probe introduced through the gaping orifices of the papillae where all converged was sent out at once. Since it appeared obvious that at least these canals of milk draw their origin from the glands of the udders, proceeding further, I undertook to investigate vessels carrying to the udders a snow-white matter. To achieve better results, I separated the udders progressively from the underlying muscles, moving from the sides towards the middle, expecting that, if some milky matters proceeded from the inside it would not finally escape our seeing, since there was so much milk fodder in the udder and the animal had eaten at the convenient time. Finally my hope was frustrated. Besides large and countless blood vessels, a few small lymph ducts situated in the region of the groins not carrying at all the whiteness of the milk found either in the udders or in the receptacle, I did not find anything. Although various things could thus be gathered from this experiment, since however it had been done once only, more were required for a reliable confirmation of any opinion, I will keep silent about these things.

§52. Since thus neither the experiment of *Bils* nor that of *Everaerts* is sufficient to demonstrate a contrary movement of the lymph, it remains me to show that the opinion commonly accepted also holds for the lymphatics of the neck as the distribution of the vessels in the glands proves and as the site of the valvulas in the vessels demonstrates.

§53. If it pleases you to examine a common gland, you will see that the vessels descending from the upper glands do not converge into one trunk but remain separated from each other. Whereas they are undivided on the way, where they arrive to the convexity of the gland, subdivided into several small branches, they finally are hidden to the eyes. If they carried out something from this gland upwards, why would they not join at first into one trunk, which we see occurring in other parts? If you like to consider the valvulas, you will observe that the humour pushed upwards is prevented from proceeding further by the tunic of the vessels raised into knots at different places as I have shown in thesis 47.

§54. found a more obvious sign of the valvulas in a cow in this way. I opened at two places the vessel which runs from the round gland adjacent to the internal maxillary gland to the gland which we call common, the first close to the common gland in which a probe introduced and pushed upwards at once found a small obstacle; from there in the vicinity of the superior gland, whence it passed through pushed freely downwards all the way to the inferior orifice. Then, having retrieved the probe, trying a path through a second hole above the first, I struck the same obstacle as previously and thus, by moving probes three or four times alternately upwards and downwards, I observed that I always failed to break through downwards, the other passing through the same obstacle. If thus *Bils* has not observed other dew-carrying vessels, neither will he demonstrate a movement of the lymph outwards nor that the matter of the saliva must be derived from there.

§55. *I thus conclude* that the humour secreted from the arterial blood in the glands of the throat and expelled in the mouth through excretory lymphatics by the action of animal spirits flowing into the glands and adjacent muscles, constitutes the saliva. The round or conglobate glands which are found in the vicinity of the former ones pour back into veins the lymph received from the external parts, to mix it with the blood flowing back to the heart.

Corollaries I

1. We suspect that the various movement of blood in feelings of the mind and the changed proportion of humours in the blood can be explained through the movement of the vena cava.
2. For indeed it appears, partly to the sense, partly to reason, that the vena cava moves at certain placesFor indeed it appears, partly to the sense, partly to reason, that the vena cava moves at certain places.
3. It is obvious to the sense that its fleshy portion near the heart pulsates.
4. It is in agreement with reason that the movement begins from the extremities of the veins, by the action of which the blood is impelled towards the heart.
5. Thus, depending on whether the extreme [veins] are more or less contracted at various places, humours to be secreted from there and into those parts are either excreted in greater abundance or poured back to the heart through the veins.
6. Thus also the fleshy portion of the vena cava not only supplies the heart with more or less blood but can also change the ratio of the ascending blood (which almost all acknowledge to be the least bilious in irate people) to the descending (mixed with lymph, sometimes also with chyle).
7. Hence it will be possible to conveniently explain the signs which are observed in anger, sadness, joy and other disturbances of the mind.

II

1. We think it is not absurd to consider the glands as the springs of catarrhs.
2. The paths indeed which lead from the brain have not yet been sufficiently discovered.
3. Neither does what the famous Schneider proposes about the arteries or pituitary membranes explain the whole thing.
4. Indeed, besides vessels bringing the matter to be separated, besides a place in which the secreted matter will be poured out, another place must be presented in which it is separated, for which we have glands. It is possible to present these as being among the other springs, also in those which the famous Schneider lists.

III

1. *It is against reason that there would be in the body such a kind of filtering that, the subtler part being retained, the thicker one would percolate.*
2. *They do not succeed those who attempt at defending the contrary opinion with the example of mucus.*
3. *For although mucus already excreted appears thicker than blood, we deny that while being excreted it is thicker than blood.*
4. *They achieve much less those who state that the soul of man which they place in blood inasmuch as it seems convenient for it, sift out this, retain that.*
5. *Indeed as I concede a great force to the soul, I am not convinced that humours undergo something immediately from it.*

IV

For the present we shall deny that salty water becomes sweet by filtering although many people assert that.

Thomas Bartholin's letter upon reception of these dissertations: EP6.

2.3 Various New Observations in the Eyes and Nose, Etc

To Thomas Bartholin in Copenhagen

That I am so slow to return to a task interrupted for a long time, is due partly to a journey, partly to the very poor health of friends. Wandering indeed in the noblest cities in adjacent provinces required much time and from our journey we brought *Hasebard* ill with us back to Leiden; in Amsterdam we found *Walgenstein* [or Walgesten] seriously ill and in bed. Thus visiting alternately either this one in Amsterdam, or that other one in Leiden, we could not devote as much time to our business as we wanted. However, both are recovering, *Hasebard* with the help of *Borch, Walgenstein* with that of *Borri*. During our journey, instead of *Deusing* himself whom I looked for in vain, I found his *Vindications* in Groningen, in which, although he attacks my words, he assails, however, even Mr. *van Horne* and vomits all his bile against him. He must be of very heatedly disposition being sensitive even to the slightest touch. I said that he favours *Bils*. I said that he agrees with *Bils* in his opinion on the origin of the saliva secreted in the glands. I said that I wonder that, occupied at explaining the meaning of the word chyle according to *Bils*, he accuses *Bartholin* of an unfair request, and nothing more on him. These words tormented him as if I had said the harshest things. Among other things,

---

OPH 3 vol. I, 55–58: *Variae in oculis & naso observationes novae &C.* is found in *Thomæ Bartholini Epist. Medicin. Cent. III.* Hafniæ1667. Epist. LVII p. 224. Further annotations by Vilhelm Maar are found in the transcripts, www.extras.springer.com.

in the controversy which he started against you, he turns himself about amazingly and, by taking out of its context that which you said concerning other matters, he defends that the state of the controversy between you and *Bils* is on the movement of chyle through the red mesaraic vessels.[1] He seems to ignore, or to keep silent about, the occasion of *Exequies of the Liver* and, moreover, to have tumbled through your *Spicilegium* with an absent-minded eye. Indeed, the true movement of the lymph which you were the first to observe throws the liver from the throne, also in *Spicilegium secundum* where you deal with the movement of the lymph according to *Bils*, you say yourself that the soundness of your observations is hinged on that and you ask that he keeps his promise soon. Yet *Deusing*, without considering these facts, sticking to the interpretation of the chyle by *Bils*, says that you made an unfair request, that *Bils* should demonstrate the movement of chyle from the milky vessels to the liver, as if *Bils* had not promised that, since in *Epistolica Dissertatio* he says that the juice contained in the milky vessels is carried partly into the liver, partly into the thoracic duct and promises to demonstrate it for the eyes. You will see that and other matters more profusely in the *Reponse to vindications* which, God willing, you will have soon together with the *Disputations* and the treatise *On the glands of the eyes* added to them. You would have had them for a long time already if various hindrances, besides those mentioned above, had not delayed me. Yet while, busy here, I appeal sometimes to dissections, some things not yet seen present themselves to me. Besides various salivary ducts and those lacrimal vessels observed previously in the external gland of the eye, I indeed observed also two vessels of the same kind in the lacrimal caruncle, one on each side of the cartilage. They admit a probe without difficulty. On the very obvious continuation of the lacrimal point I will add nothing since I think it has been described for you by the famous *Borch* and you are to see more of this shortly in the treatise *On the Glands of the Eyes*. However, on the other excretory vessel which I observed not so long ago in the tunic of the nose, some doubts arise on which I should wish to have your opinion. As indeed, to examine more accurately the continuation of the lacrimal point in a sheep, I had broken the bones of the nose, after having removed these, I saw in the tunic of the nose a small glittering canal like lymphatics run forwards and, therefore, I began by taking it for a retrieving lymphatic. As, however, the nose being opened by the dissection, above the obvious orifice of the canal continued in the lacrimal point, I noticed a narrow and black small pit or furrow through which a hair found an

---

[1] [mesenterial blood vessels].

open path and I examined more accurately the place beyond which the hair was not able to penetrate, it appeared to be an excretory lymphatic from the obvious roots which are there. It is amazing that such a delicate duct proceeds in the thin tunic from these minute portions of glands. Yet I wonder more to what end the humour secreted there is carried to the extremity of the nose as its orifice is at a distance of hardly a finger-breadth from there. I should say it is excremental but in other parts of our body we see humour excreted through vessels of the same kind destined to a well-determined function and they are certainly not excremental. However, here it is hardly outside its canal when it is also outside the body. So I do not see to what function it could be [destined] and, therefore, I do not find a name for it. If thus you would be kind enough to let me know your opinion on this matter, you would greatly increase my debt of gratefulness.

Besides that, as I was in Amsterdam, I have repeated the experiment of *Bils* on the movement of chyle. I did not find this difference in the blood although I kept alive to the third hour the dog which had spent the whole day in this torment. Since it is not enough to have tried once to conclude something reliable, at the first opportunity I will turn the same stone, although I must say that I do not torture these animals in so long torments without horror. Cartesians boast much about the certitude of their philosophy: I wish they would convince me as certainly as they are convinced themselves that in animals there is no soul and that it makes no difference whether you touch, you cut, and you burn the nerves of a live animal or the cords of a running machine. I should indeed search more often and more willingly during several hours the viscera and vessels of live animals, as I see much to be investigated which cannot be expected in another way. I saw some other time in the same subject that the thoracic duct was continued with a lymphatic descending from the glands of the neck (a probe indeed introduced into it rose to about its middle, a valvula preventing further progress) and that it did not discharge itself directly in the axillary vein but that another small canal was formed at the meeting of the two ducts (where several others also flow together). This small canal bent in a circle about the vicinity of the axillary and jugular veins discharges itself in the blood. To know more about it, I finally forced blood to flow back through this canal from the veins now into the thoracic duct, then into the other lymphatics by moving these vessels variously with a certain force. Previously, in dogs, I very often tried in vain to find with a probe a path into the axillary vein through the thoracic duct and I could never continue the progress of the probe as clearly as this time. If it pleases God, at the next opportunity I will see whether it is possible to observe the same things more often in some certain way.

Bils yet keeps silent. I do not know whether the dog days have brought a vacation to his cursings or whether he is to abstain from that business for ever.

In what concerns public matters, there is nothing to write about by now except that here fevers have befallen very many people; many among them succumb to them not so much because of the illness as by their own fault. No malignity and no suspect spots indeed have yet been observed. However, in Amsterdam they suffer more than here and the famous *Golius* told me some days ago that a physician of Amsterdam had told to him in a letter that in one and the same night more than hundred people had been affected by fever, however without any more serious symptoms. Yet perhaps I take too much of your time, famous Sir. That is why I stop here and I wish you and your family the best of health, begging most kindly that you continue to favour me by your love and protection and that you refrain from being angered by the faults which a hasty hand could not avoid.

Leiden in Holland, September 12, 1661

The famous *Borch* greets you and your family most courteously.

Your very obedient disciple,
Niels Stensen[2]

---

[2]For Thomas Bartholin's reply, see EP 4.

2.4 Response to Vindications for Rehabilitation of the Liver

To the famous and experienced,
Mr Anton Deusing,
Doctor in medicine and illustrious Professor,
with greetings from Niels Stensen

It was very agreeable to me, famous Sir, that it pleased you to examine my two disputations. It would be even more agreeable if the occupations in which you spend your days had let you a little more time so that you would have more thoroughly understood the mind of those who are involved and what I think. I learned from others that your practice takes much of your time. Actually, from the writings which are published every day, it is very clear that you are very busy reading and writing on different matters. But it is obvious to everybody that a mind tired of other cares and works is seen to be less sharp-sighted in some matters and its very acute sharpness is blunted by uninterrupted speculation on different matters. Whereas the eyes distinctly perceive simply one thing from another at a certain time, they are blinded

---

OPH 4 vol. I, 61–73: "Responsio ad vindicias hepatis redivivi" is the second of the four treatises in *Nicolai Stenonis observationes anatomicæ* ... Lugd. Batav. 1662. It is a reply to Deusing's "Vindiciæ hepatis redivivi", directed against van Horne, whom Deusing took to be the real author of *Steno's disputatio anatomica de glandulis oris* ... Lugd. Batav. 1661. Further annotations by Vilhelm Maar are found in the transcripts, www.extras.springer.com.

© Springer-Verlag GmbH Germany, part of Springer Nature 2018
T. Kardel and P. Maquet, *Nicolaus Steno*,
https://doi.org/10.1007/978-3-662-55047-2_15

even on a bright day when they carefully examine different things together without interruption. I thus have no doubt that, if you had had the leisure of looking more carefully at my matters, [your] *Vindications* which were published recently would never have been published and that other experiments or arguments which would be more serviceable to pursue the truth would have been presented. Indeed, in the *Vindications* you do not attack the matters themselves and you do not defend the movement of the juice contained in the milky vessels through the lymphatics to the glands of the mouth. But, by passing over these matters in silence, you go down to the words and the meaning of the words and you take much pain to show that we did not carry out the task of a good interpreter. Although this examination of the words, if it is looked upon simply, is of such little weight that the time spent on it can be considered as lost without any gain, since, however, it pleased you to make a tempest in a teapot and to attack furiously, not so much me as the famous Johannes van Horne, who is absolutely guiltless, the sympathy due to a teacher on one hand, the love for a respectable name which everybody owes to himself on the other hand compel me to present the whole matter more clearly and to show that neither the famous *Chairman* nor myself deserve such censorship.

As I have declared to be the author of the theses, not only in the dedication but in the theses themselves, such as 4, 13, 14, 15, 32 and others, I gave clear indications that nobody can attribute them to the famous *Chairman*. I do not see what induced you to use the title: *Vindications of Rehabilitation of the Liver, referring with a slight correction to the wrong interpretation of the famous Mr* Johannes van Horne. Why, without any consideration for what has been said, do you attribute all to Mr *Chairman*, as you suggest in the title and confirm on all of the pages that follow? Do not believe, famous Sir, that I am so avid of praises that I should claim as mine, the works of others, and that I should claim to be not only the defender in a disputation but also the author of what was written by somebody else. I learned from the example of others to despise titles begged for and, if there was nothing else, the little crow of Horace taught me what it is to shine with stolen colours. But it is possible to prove that the experiments are mine, and that the observations which were presented are mine, by the testimony of those who were almost always present when I dissected, such as the famous Ole Borch, Royal Professor in Copenhagen, Christian Rudnick, doctor in medicine from Bütow in Further Pomerania, Matthias Jacobsen Matthiassen, medical student. I can also produce as witnesses professors of this place, the illustrious Mr Sylvius and Mr van Horne to whom both it did not displease to watch occasionally my first endeavours. The famous Golius, professor of mathematics and of oriental languages, an honourable teacher, did not disdain either to watch when I prepared the salivary and lacrimal ducts in a calf. If you do not trust the words of a man whom you do not know, believe at least these gentlemen, and object to me, if you have objections, rather than blaming the *Chairman* who, because of his other public commitments, had scarcely time to solve the problems which are to be ascribed to me. Or if you keep to the rule set forth on p. 31 of the *Vindications*: I should have gone astray, the *Chairman* would have commended the contents by his presidency, the errors, however, should not be ascribed completely to him as author, his name must not be exposed to the eyes of everybody, in front of the work. I am really sorry that a famous gentleman, for doing a favour to me, is attributed such things, that he is reproached a *wrong interpretation*, and that that which can contribute *to engage learned people in discussion with each other was reproached to him as inconsistency, sluggishness, imprudence, a not sufficiently free mind, a lack of sentiments,* etc. I am sorry, I say, that such things are said about him, since he himself did not commit anything [of this] and what stirs so much the famous gentleman was not presented in the theses. Indeed, the main controversy about the mesaraic veins because of which you assailed so much Mr van Horne, has nothing to do with my discussions so that, without violation of the truth of the theses, I could friendly agree with the *Chairman*, with you and with Bils, if the truth itself agreed. Indeed I deal with the lymphatics only and I show that Bils' opinion about these vessels is less in agreement with the truth whence this only results: *Nothing is carried from the intestines to the liver through the milky and lymphatic vessels.* But your opinion is: *The chyle is carried to the liver through the mesaraic blood vessels.* What is the discordance here? If I deny that the chyle is carried from the intestines to the liver through the milky vessels, does it mean that I deny that it is carried through the blood vessels? If thus it is not carried this way, why can it not be carried that other way? I do not find here any contradiction and I do not see for what reason, nor moved by what arguments taken from my disputation you accuse the famous *Chairman* of *inconsistency.* You say that he *seems to limp on both sides,* you urge *to declare publicly that you wonder that he made new efforts in a matter which is clear concerning the liver itself,* etc. Since the discovery of the lymphatic vessels, the famous van Horne defends the movement of the lymph which I establish. He considers as likely that of the chyle through the mesaraic blood vessels. I do not touch upon this by a single word. What is thus this inconsistency? Is there any occasion to take up that discussion concerning the mesaraic veins? What is the reason for saying such things by which you *may appear,* as you say on p. 5, *to wish to engage learned men in discussion* or *spread the seeds of discord between those who you do not ignore are friends?* But if I had said the gravest things against you, the ties of friendship between them would be too sacred to you to bring forwards such things. As far as the movement of the chyle to the liver through the mesaraic blood vessels is concerned, even if I say nothing about it in the thesis, since it pleased you to stir this controversy here, I will say in a few words that I have not seen presented either an argument or an experiment sufficiently strong to sustain it. But I do not understand why, on p. 26 of the *Vindications* you say: *I indeed cannot rely here on reason alone otherwise than in the movement of the chyle to the liver through the mesaraic red vessels.* If you attribute so much to reasoning that you rely on it alone for the movement of the chyle to the liver, why do you not concede to it the same power for the movement of the lymph? If you expect here experiments of others, who did assure you that experiments contrary to this opinion could not be shown, given time? You grant a too absolute authority to reasoning in a matter which, after the discovery of the milky vessels, finds more opponents than advocates. The famous Pecquet already showed an experiment contrary to your reasoning in *Nova de thoracicis lacteis dissert.*, Experiment 1. If you had imitated his experiment, you would not have bestowed so boldly the dictatorship on reasoning alone. Hippocrates behaved better. At the beginning of *Precepts* where he dealt with reasoning, he finally concludes: *But let this be enough about this matter. It is not possible to deal with it as settled only by reason but as proven by act.* But I wonder even more that you undertake to defend against the famous Bartholin that which you attribute only to the experiment of Bils on the function of the mesaraic vessels, which you never tried yourself nor saw attempted by others. You discuss the colour of a matter which you affirm not to have seen, a colour which others call "vry swart" (rather black). You interpret that as only a little dark. But I am sure that, after having seen it by experience, you will change your mind completely. I carried out an experiment in the presence of friends. I observed the method prescribed by Bils. I ligated the arteries. I maintained a first dog in life during 3 h, another during 4 h, and it could have lived a whole day in this condition. I opened again the abdomen which had been sutured. I exposed to the air blood removed from the portal vein, from the vena cava and from the aorta. But they all equally coagulated soon. They had the same colour. They were equally dark. What do you conclude from this? But the argument of Bils by which he proves that the liquid contained in the branches of the portal vein has an affinity with the nature of the blood, is brought forwards gratuitously, since not only we concede this but we even admit that it is blood. He thus does not achieve anything with this reasoning. Neither does what he infers about the colour, dense, almost dark, obscure, ashy, prove anything. Please, examine the blood let from the division of a vein and see how variable colours it is possible to observe in different subjects. But I dismiss that.

So it is evident that that by which you wanted to attain the *Chairman* in a gentle language is not his, but you touched him harshly and what you report about the movement of the chyle through the red mesaraic vessels does not at all belong here. Therefore, I must take for myself all by which you shattered him and say that, after changing the name, everything was said for me. But I will soon proceed to show that neither did I deserve what you reproach me. After I shall have presented beforehand all what is due to your sentiments, perhaps not to yourself, the equitable reader will judge what is the sluggishness, the imprudence, the distorted eyes, the oblique looks, the iniquitous dispute, and all the rest said either absolutely or under condition. I will pass over all that and honestly present what is really important in the matter. What indeed is accomplished by words of this kind other than stirring the minds more and more, and so the matter itself is not investigated with the care which is appropriate? I should not bear it if in the theses I could have pronounced one single word about you in an angry mood. But I did not say about Bils anything else than what you would not deny if you spoke with him. On that account I said that his character must be determined from his sayings and writings. But what is this compared with the words with which he harshly attacked my teacher Bartholin without any reason? You take for yourself what I intended for Bils. I do not see why you do that. I did not say that you are favourable to all what he says but to some points of it. Who would blame me if, in a matter which is true, I agreed with a man, even the most wicked of all men? And should I thus be considered as being of the same kind? Do not think, famous Sir, that any just arbitrator of the matters (I do not mention others since even the best sayings can be interpreted very badly), would extend this favour which you bestow on Bils so far that he would assign even the faults of Bils to you. But spare me to be angry who did not say anything about you except what is clear to anybody from your writings. I said in thesis 25: the *noble* Bils and the *famous* Anton Deusing *favourable to* Bils; in thesis 42: *to whom the famous* Anton Deusing *gives his acquiescence for his opinion about the saliva*. What in these words does deserve such a censure? What is the cause here of conceiving such suspicion? What if you yourself say that you are favourable to Bils, if you undertake to defend his opinions against his opponents, as far as they appear true to you? What is there to irritate you against me when I say what you have said yourself? You set up for us too restrictive writing rules if you do not permit to cite the authors as witnesses in what concerns the matter. Indeed on page 30 you say: *Why am I involved personally where it is to be dealt with the matters?* What if your acts were examined according to this rule? But, as far as I see, I made a mistake in that it should have been said that you did not agree with Bils but Bils agrees with you, as appears on page 28: *Though here where we actually agree, rather he has the same opinion as myself, and he seems to confirm by his experiments our opinions presented publicly even before the name of* Bils *was known by the learned people.* As if it also was to be contended about this: who is to be said to agree with the other, and could it not be that you agree with each other. Let it be that you were the first to have seen these things in imagination and Bils was the first to have observed the same with his own eyes. Is it then not to be said that you approve his experiments whereas he approves your thoughts? Everybody knows that Bils has never read your works and you claim that you never saw his. Does it then matter who is to be said to agree with the other, above all since in this instance your reasoning and his experience are equally important?

But let me dismiss that and proceed to the place where I am said to have distorted the words and the meaning. Let me show that I have understood the mind of Bartholin, yours and that of Bils and that I have presented it as it is. To this end, it is necessary that I present first my meaning and the occasion of what was said.

A good interpreter considers not only the words as such but also confers them with what precedes and what follows. When dealing with a matter, it is better to explain all what the author means than to present his words as such unless one adds, if needed, at what occasion the words were pronounced. On page 31 you say: *An honest gentleman mentions the words themselves of others whom he wishes to praise or to scorn (that is what I do towards others while I am seeking the truth) or at least does not distort the words and their meaning, nor does he create prejudices by questionable discourses.* Let us thus see which one behaves more fairly, you who present the words themselves of others, or myself who give their meaning. On page 19, to prove that Bartholin unfairly requests the demonstration by Bils of the movement of the chyle from the milky vessels to the liver, you cite the words themselves of Bartholin in *Spicileg.2* where he determines the state of the question. From the movement of the chyle through the mesaraic vessels to the liver, which he denies, you infer what follows: *But if this is the state of the controversy, as presented and recognized by* Bartholin *himself, why does he request from Bils to demonstrate to him the entry of the chyle in the liver through the milky veins, where an access to a rehabilitation of the liver appears and if so whether an input for making blood is brought to the liver? Is it not an unfair condition, whereas Bils promises to demonstrate the immediate access of chyle from the stomach or the intestines into the liver through the mesaraic red vessels (which Bartholin denies), to request from the same Bils that he demonstrates the access of the chyle into the liver through the milky vessels. Since, according to Bils, nature itself denies such movement to the chyle or nutritive juice, of which it is only question (and not of the lymph or moisture)? Must this condition, under which the acquiescence to the rehabilitation of the liver is presented as not difficult, not be considered as striking a target?* Who, among the readers, who has not studied the *Spicilegia* of Bartholin, will not give credence to your words and believe that Bartholin is in contradiction with himself or has forgotten this state of the controversy in his *Spicileg.2*, whereas he claims something by far different in his *Responsio*? The one who will have looked into the *Spicilegium* itself will have seen that Bartholin undertook to refute two things. Busy in examining the first one, after having seen that what he was not looking for appeared from the first experiment, he said: *Although we should concede with others that the chyle is mixed with blood, this, however, is not looked for at this place. The question is: does this nutritive juice, called chyle, from the stomach go to the heart first,* etc. This refuted, he tackles the other point and presents different things about the movement of the lymph which we will present in due place. The reader will easily see that the state of the controversy dealt with in the first part is very different from what is dealt with in the second part and must not be reduced to these questions concerning the lymph. If thus your words were to be supplied according to the rules of a true interpreter, they would be such. Where Bartholin debates with Bils about the function of the mesaraic red vessels, this is the state of the controversy, as exposed and recognized by Bartholin. When the function of the milky and lymphatic vessels is dealt with, why does *Bartholin* request from Bils to show him the entry of the chyle into the liver through the milky vessels whereby an access to the rehabilitation of the liver appears, and the burden of blood making is assigned to the liver? And anybody who reads the *Dissertation of Bils in the shape of a letter*, who knows the publication of the *Exequies of the liver*, will also see whether this condition which Bartholin proposes to Bils and of which more will be said below is fair or unfair. But at the same time he will also see with which honesty the words of Bartholin were presented and whether it is always reliable to believe those who prove with the words of other people what they say about these other people. Many things are indeed true in context with other things but, if presented isolated, are very false. Thus concerning the first part refuted by Bartholin in *Spicileg.2*, it is true that this is the state of the controversy. If we take it as such, it is not true at all. Since indeed Bils asserts two uncommon opinions, there are also two different states of the controversy. Since he says that this is the state of the controversy, he does not deny that it is different at another place. Reluctantly I stir this controversy on the role of the interpreter, a controversy which I should have passed under silence if your *Vindications* had not, almost forcefully, extorted it from me. Indeed, in order to show that I wrongly interpreted your words, that I distorted your words into a different meaning, etc., you cite the very words of Bartholin, those of Bils, yours and mine so that you find credit with the readers who ignore at what particular occasion they were said. And so as to exculpate myself, the first way which you used to accuse me was to be examined. Let us tackle what I proposed in the theses and see whether I did distort your words in a different meaning or rather whether you did not at all understand mine.

Since I had described everything like an historian in *Disputatio de Glandulis oris & nuper observatis inde prodeuntibus vasis prima*, it remained for me to add something about the function. Hence I added a *second* discussion where straight away, at the beginning, I presented what resulted spontaneously from the observations that *the task of the glands is to prepare this humour which is always found in the mouth.* I also added examples to show that one has not to resort to other paths to let down saliva directly in the mouth. It remained to inquire whence the saliva would have to be led indirectly, which is very questionable. It is possible to demonstrate to the eye that saliva proceeds from the glands in the mouth, but not from where the humour prepared in the glands and moistening the mouth originates. It could indeed be led off from the brain through nerves or over other hidden paths. It could be diverted from the heart through arteries. It could also be carried from the glands through lymphatic vessels. I listed these possibilities in thesis 25. And since the opinion about the function of the lymphatic vessels was very recent, I mentioned its advocates, first its author, Bils, and Deusing who agrees with him on the present matter. But to make it clearer, I said that I dealt in thesis 25 with the paths into the glands, not into the mouth, and I will quote also thesis 29 where I tackle what I promised in the former to do: *The saliva having thus been examined with the glands, it remains to investigate the paths which bring its material. To do that more easily, we shall present the single parts with which the glands have to do and we shall enquire by what means either they receive something from there or they send something there, so as to assign its function to each of these parts.* Thus I examine separately the heart, the nerves, the conglobate glands, as I had promised in thesis 25. In thesis 42, when I had come to Bils, I said that he derives the aqueous humours in the body and the saliva itself from a moisture-carrying duct, but not from the blood and that you give him your agreement on this opinion about the saliva. As Bils also establishes that the saliva originates directly from the glands, but indirectly from a moisture-carrying duct, and as you say that it comes from the mandibular glands and that aqueous material seems to be carried normally to the glands through moisture-carrying ducts (but we speak here of the normal inflow only), but you do not mention blood, is it not clear that you approve his opinion about the indirect origin of the saliva? But without considering these points, in the first of the *Vindications*, on p. 23, you question whether I did not substitute the meaning of Zassius to that of Bils. Hence you show, by using your own words, that you establish not one single but a threefold substance of saliva. And thus you finally complain that, despite clear words, we say that you are favourable to Bils and that you approve his opinion about saliva which derives only from a moisture-carrying juice brought in through vessels commonly called lymphatic. As far as the first point is concerned, by questioning what is not questionable, you show that you have not yet understood what Bils means. Is it not clear that he derives the saliva and the other aqueous humours from there? Let us see his *Dissertation in the shape of a letter* where he speaks of the distribution of the moisture-carrying duct: *And I shall not distinguish all; finally it flows in or ends everywhere in the salivary, lacrimal, pituitary and all other glands.* And towards the end, about his moisture: *This will be easy to understand if it is shown that the tears, the phlegm, the saliva, the sweat, the hair, the nails come from it, and all the body is made able to nourish itself by its circulation, and all ferments of nature derive from it: what does happen,* etc. Is there any doubt here where the words are so clear? In what concerns the second point, you approve what I do not deny just as if I had done so. I speak of the saliva prepared in the glands, which your own words show you derive from the lymphatics, and not from the arteries. But as if I had dealt with the paths carrying the saliva directly into the mouth, you confirm that you mentioned several ones, which I did not deny you did. Thus if you want to obtain something, you have to prove it: *you and Bils, you do not derive the saliva secreted in the salivary glands normally from moisture-carrying branches solely, but from the blood.* When you believe that by deriving part of the saliva from the blood you approve me, you are far from understanding what I mean. You indeed propose two direct paths for the saliva, the salivary ducts and arterioles, and you want that a small more aqueous portion exudates through arterioles to the mouth and palate, distinct from that which proceeds from the gland, whereas I do not lead arteries directly to the mouth and I do not oppose salivary ducts but I subordinate the vessels inasmuch as they supply the ducts with material. You cannot be more in disagreement with me who have not yet seen approved by anybody the exudation of serum from the arteries, which you allude to here, which the famous Schneider expressly described in Vol. 3, Sect. 2, c. 3, page 509, and which I think must be considered as a pure product of imagination. At all places indeed which the praised famous Gentleman holds as sources of the catarrhs, on page 554, I have observed and showed to friends fleshy glands even before I saw this third tome. But also in the epiglottis there are not only glands but even excretory vessels penetrating through its cartilage at the opposite side, obvious for the one who examines them carefully, and, in the membrane of the nose at the side of this duct which runs from the lacrimal point to the anterior aspects of the nose, I have recently observed another excretory vessel originating from these small glands so that I concluded that the conglomerate glands provide us with the true springs of the catarrhs. I will speak more of this matter elsewhere if it pleases God. But while I acknowledge the diverse nature of saliva and recognize that is a mixed humour, I do not find that you are more favourable to me than to Bils. From this it does not indeed result that I refer to anything else than arterial blood. It does not result from the fact that the saliva is a mixed humour that it does not originate from blood alone. About the third substance of saliva, which you assert flows from the brain to the glands, I said in thesis 22 that we vainly have recourse to hidden paths when there are obvious ones. The famous van Valckenburg in his *Letter to Jan Neander on smoking tobacco* enumerates many open paths from the brain into the mouth and vice versa from the mouth to the brain. But it seems that he has considered the bones of the head bare rather than the head provided with all its parts. I will not deny that such paths can be found but I do not believe there are some before having seen them. If thus you show, by any undisputable reasoning or by a reliable experiment, a passage of this kind from the brain to the mouth, you will acquire great gratitude among all lovers of the truth. Since I establish that nothing is carried directly either from the brain to the glands or from the arterioles to the mouth, but well from the arteries into the glands, it is clear that we totally disagree on that matter. But since you say that what is secreted in the salivary glands seems to be carried normally through small moisture-carrying branches without mentioning blood, it is clear to everybody that you approve Bils.

The opinion of Bils on the movement of the lymph has been presented in the same thesis 42 and an experiment has been described to which he always appeals. I added that *this is the movement of the chyle which Mr Bartholin wishes to be demonstrated to him in Spicilegium 2, which Bils adds in his Dissertation in the shape of a letter, which he wanted to demonstrate to Mr Borch in a dog alive in my presence. I am thus amazed that Mr Deusing, busy explaining the meaning of chyle according to Bils, accused Bartholin of unfair request. He indeed speaks of this humour contained in the milky vessels of the mesentery and which Bils undertook to demonstrate that from there it is carried to the liver partly through the lymphatic vessels.* I will not report here that they are the same as those for which you attack me in the examination of these matters: I indeed said that I will ascribe them to your sentiments, not to yourself. I will not deal with an improper expression that you blame profusely, since it is known to all who are familiar with the letters of the Ancients, that friends in letters blamed each other without breaking their friendship and that the expression of accusation was not restricted to narrow limits. And if to some extent you soften your expression on the unfair conditions in which Bartholin, according to you, makes demands that Bils did not promise, and if you explain them as friendly as possible, you will never find me opposing. I indeed never proposed that in order to make you hateful to Bartholin since I know that this kind of go-betweens is hateful to gods and men. But, this dismissed, I will show that Bartholin did not claim anything either in *Spicilegium secundum* or in *Responsio de experimentis anatomicis Bilsianis* (mention of the place of this was omitted in the theses) which Bils did not promise so that it can be clear to everybody that you were never busy in a clear matter to be obfuscated by words. Let us see the words of Bartholin. After having presented the opinion of Bils on the chyle and lymph in the last chapter of *Spicileg.2*, on page 100 and following, and having examined the details, he adds finally at the end of the treatise: *He promised to demonstrate to the eyes some other time that all the body is made suitable to nutrition with this* (i.e. this moisture), *its circulation and fermentation. Although I do not object against the moisture alone, if it is different from the chyle, since however he has another opinion, we shall expect these demonstrations with a great desire to examine better, after removal of the curtain, what Nature will have revealed to the most noble gentleman. We ask with as much kindness as possible that he redeems soon his promise. We wish him to have success in the undertaking,* etc. *In Responsio,* however, on page 10, he says: *I will happily persist in applauding if, besides other points, you demonstrate to these Professors of Anatomy the entry of the chyle into the liver, but directly from the milky veins, not from the red mesenteric vessels through long roundabout ways, in which, mixed with blood, it lost its proper appellation of chyle.* Which equitable arbiter of the matters does not see that Bartholin asks here that the movement of the contents of the milky vessels be demonstrated to him, above all because he expressed himself so explicitly? Actually in the first place he speaks of the total distribution of the moisture in the whole body, in the second place of its partial distribution in the liver. But what else is moisture here for Bils than the juice contained in the milky vessels of the mesentery? And this juice, whenever it looks like milk, has acquired the name of chyle since the first discovery of the milky vessels almost 40 years ago and is known as such by everybody. I will not repeat here what I said above, why you brought forwards against me these words of Bartholin about the state of the controversy, as if it were question in *Spicileg.2* of the red mesaraic vessels only, since it is clear to anybody who looks through the *Spicileg*. that it is question also of the milky and lymphatic vessels and, if one is considered, the other is not excluded. It will be enough for me to have shown that Bartholin, not only in *Responsio* but also in *Spicilegium*, deals with the juice contained in the milky vessels and asks that its movement be demonstrated to him. In the *Dissertation in the shape of a letter*, Bils promises to demonstrate that this chyle according to the meaning of others, this moisture according to his own opinion, is carried from the milky vessels and into the liver and in the rest of the body. *I thus proceed*, he says, *this liquid which, because of its function, we call moisture invades the milky veins through their openings in the intestines and divides itself into two paths in canals separately associated. One path leads to the gland which is attached to the portal vein and from there extends forwards lymphatic vessels to the liver. The other comes into a glandular receptacle of the mesentery from where it repairs the whole body by obvious branches above and below*, etc. But, you will say, it is question of the rehabilitation of the liver and, therefore, of the movement of the true chyle and properly so-called, but Bils denies that the contents of the milky vessels is chyle. Hence, you add on page 17: *Meanwhile, let not the famous gentleman believe that Bartholin is mad, who is easily ready to admit the rehabilitation of the liver, if Bils proves to him that this humour contained in the milky vessels of the mesentery which Bils does not distinguish from the chyle, is carried into the liver through these milky or lymphatic vessels since not even Bils himself does recognize this humour as appropriate to generate blood*. But, as far as I see, either you ignore or you pretend to ignore the explanation of the *Exequies of the liver*. If you had considered this, you would never have pronounced these words. Please, look at Chap. 8 in the *Vasa lymphatica* of Bartholin in *Animantibus inventorum* where, among other things, he says: *Consequently, if I always observed some small vessels near the liver and heretofore I took them for milky vessels, justifiably, until a favour of Nature would convince me otherwise, I concluded that the chyle is distributed partly to the liver, partly to the heart. Since, however, very recently we were the first to see in Copenhagen, as we were searching in animals carefully eviscerated, what these small vessels are, from where they come, what their function is, we did not want to cling too obstinately to an obsolete opinion nor to support any longer the tottering cause of the abandoned liver. We saw that these vessels near the liver are properly of their own kind, which we call lymphatic vessels because of the contained liquid, carry water from the liver to a receptacle. If they are ligated, they swell near the liver and they are empty beyond the ligature. They are similar in substance, colour, contents and function to those which we were the first to discover in the limbs and elsewhere in the lower abdomen. With this new discovery, the liver praised with applause for so many centuries lost all hope of having the function of making blood, and no longer deserve this attribution,* etc. Since the movement of the lymph provided Bartholin with the opportunity to write *Exequies of the liver,* if this movement was denied, he would have written these funerals in vain. Since Bils denied the said movement, not without reason, Bartholin, not to be considered as a fool, requests from Bils that he makes the demonstration of the absence of this movement if he will show himself ready to admit the rehabilitation of the liver. Let 73 Bils prove to him that this humour contained in the milky vessels of the mesentery is carried by these to the liver. Is there anything clearer? *As Bartholin observed that the contents of the milky vessels are not carried into the liver, he discovered for us the lymphatic vessels and wrote Exequies of the liver. Bils asserts that the contents of the milky vessels are carried into the liver and promises to demonstrate this.* Thus, if this is true, the *discovery of the lymphatic vessels* tumbles down and so do the *Exequies of the liver.* The famous Bartholin himself saw this. In the second part of *Spicileg.,* on page 104, which deals with the moisture, he adds: *On this cardinal point, the value of our observations hinges.* Thus since the matter is so clear, why do you cling to the words and why do you urge the interpretation of the chyle by Bils as if true designation of the things were to be expected from Bils? Movement of the milky humour is the matter, not its name. If Bils shows this movement according to his promises, he will be approved and applauded.

It is sufficiently clear that what I presented on Bartholin, on You, on Bils conforms to your meaning and to your words, although the shortness appropriate to a discussion does not allow to add the words themselves. I will pass over the rest and will entrust to the equitable reader the judgement on the suspicions which you develop for yourself. Meanwhile, if it pleases you to examine my works further, give yourself enough time, I beg you, to consider the context of the words together with the words. However, I should prefer that you attack the matters themselves and if you see that anything among either the experiments or the reasoning brought forwards by myself is wanting, tell it. You would indeed find me as prompt to yield to your arguments if they rely on reasoning and on experience, as I am to oppose them if these are lacking. Greetings.

Leiden

November 28 ≈ 18, 1661

2.5 Anatomical Observations on the Glands of the Eyes

To the illustrious and excellent gentlemen

Mr. SIMON PAULLI,

Royal Physician and Prelate at Aarhus,

Mr. JØRGEN EILERSEN,
Mathematician and Scholar,

Mr. OLE BORCH,
Historian and Natural Scientist,

The first one past professor, and
the other [gentlemen] present
professors at the Royal
University in Copenhagen,

Mr. FRANCOIS DE LE BOË
SYLVIUS,
for Practical Medicine,

Mr. JOHANNES VAN HORNE,
for Anatomy and Surgery,

Mr. JACOB GOLIUS,
for Mathematics and Arabic,

The illustrious professors
of the fine University
at Leiden in Holland:

All my highly deserving teachers, to whom honour for all time.

OPH 5 vol. I, 77–90: "De glandulis oculorum novisqve earundem vasis observationes anatomicae"
appeared as the third of the four treatises in *Nicolai Stenonis observationes anatomicæ. ..* Lugd. Batav. 1662.
Further annotations by Vilhelm Maar are found in the transcripts, www.extras.springer.com.

Some new matters which are here presented to you, my teachers, *are not commended as new as if they were brought forward for the first time. They are old, even older than those Arcadians who boasted to have existed before the moon and thus older than the human kind since they were born with the beasts before man was created as the Scriptures testify. Nor do I proclaim them as new because I think they have never been observed before. Although there are as many and as different opinions on the origin and the paths of tears as there are rivers which run in the sea, they do not seem to demonstrate the subject. The most talented Stevin believes and others suppose that there has been a golden age for sciences. And I have no doubt that, if all the texts of the Ancients were at our disposal, many things which we admire today as new and which are investigated with much labour could be learned from them without sweat or blood. But they are new to me since neither the voice nor the hand nor the writings of anybody disclosed them to me, and because the famous Anatomical Theatres of the most noble University of Leiden list them among the new. It would be heavy evidence of light-mindedness if I did not express with words the matter as it is. Nor indeed do I care for the arrogance of those who rush on the works of others but never put their hand to work. If they have been bold, I counter them with your approval and your authority,* very illustrious *gentlemen. Protected by this shield, I will present in a few words what I could have done in many, either if my mind had been set to wasting time by developing profusely that which appears directly from the spoken words or by compiling the written works of others, or if I intended to bore the reader to nausea and to earn for myself the reputation of a Crispinus. Therefore, accept this tiny work carried out for you and continue to surround me with the affection which you have shown hitherto,* (Fig. 2.5.1) Very illustrious gentlemen,

Leiden, December 6, 1661.

Your devoted disciple,

Niels Stensen

What practice and perhaps also observation of animals has taught the mechanics, that, *to facilitate movement the things to be moved should be smeared by some oily humour*, the most ingenious mechanic of all provided more than perfectly in the first structure of animals. The mechanics have seen that, if a third body intervenes between the body to be moved and that fixed over which the movement must occur, the task proceeds far more easily. Hence, as they push a boat over a flat surface by means of rollers laid underneath, so they facilitate also the rotation of a wheel by smearing with an unctuous fluid the axle about which the wheel rotates. This is how they carry out with less

85

OBSERVATIONES
ANATOMICÆ
DE

Glandulis Oculorum, novísque earundem vafis.

Uod Mechanicos ufus docuit, forfan & in animalibus facta obfervatio, *ut ad motum faciliorem reddendum res movendas humore unctuoso oblinerent;* hoc in prima animalium fabrica Mechanicus omnium ingeniofiffimus quàm perfectiffimè obfervavit. Viderunt illi, fi movendum inter & fixum, fuper quod motus fieri debet, tertium motu facilius intercedat, opus longè commodius procedere, hinc, ut fuppofitis cylindris in æquora navem propellunt, fic &, fuper quem rota volvitur, polum liquore pinguiori inungentes gyrationem facilius expediunt. Sic & alios motus, quorum exemplis officinæ hinc inde abundant, quiefcentem à mobili fuperficiem intercedente fubpingui liquore dividentes, minori cum moleftia perficiunt.

Fig. 2.5.1 Facsimile edition 1951

trouble other movements the examples of which are numerous here and there in workshops: they separate a surface at rest from the mobile one by an intermediary unctuous liquid.

But in the automatic body of animals all that proceeds more skilfully or, I should say, more divinely. There indeed even the humour which is supplied and where it is supplied reveal a skill by far greater. The parts are arranged so that the liquid concealed in the vicinity as in a storeroom is secreted more scarcely or more copiously depending on the more or less intensive usage, without our noticing it. Then, after it has accomplished its task, it is carried once more to the parts through other paths. Thus the movements of the parts in the mouth are enhanced by the arrival of saliva. Thus the unctuous liquid expressed from the glands beneath the tunic by the food itself to be swallowed makes swallowing easier. To the same end all the tube of the intestines is smeared over its inner surface by viscous slime. It is for no other cause that a moderate abundance of certain humours is found in many other parts of our body.

But above all these phenomena appear most delicately in the eyes. There indeed peculiar vessels offer themselves to be considered. They bring humour serving for the movement of the eyelids and derive the same humour from there to somewhere else. Since I know they have not yet been noticed in this manner (I even think that the afferent ones have not yet been described despite an origin obvious to anybody. Their continuation, however, was not recognized by the great anatomists of this century which appears from their writings), I decided to describe them in the present work as I observed them together with the glands in different beasts, mainly in calves.

But *the glands which moisten the inner surface of the eyelids* are of two kinds. The illustrious Wharton calls one *lachrymal*, the other *innominate*. They belong both to the kind of conglomerate glands. It seems of course that the task of the conglomerate glands is to prepare a peculiar humour to be excreted through the efferent lymphatics. This was observed previously in the pancreas and in the lower jaw only but it is obvious now in several other parts. Besides those which have been mentioned for the mouth, other fleshy parts made of bundles of glands show it. Some are also observed beneath the internal as well as beneath the posterior tunic of the nose and beneath the tunic of the throat and in the superior aspect of the epiglottis whence they are found conspicuous in the cartilage itself to the inferior aspect of the passage. In these the presence of lymphatic vessels is very easy to show. The matter is not so obvious for the glandulous flesh which presents in the external acoustic meatus, between the cartilage and the skin since the colour of the cerumen seems to indicate another origin. But it is not surprising that these glands excrete a yellow juice since a similar juice is also excreted from the glands of the mouth, a fact which I saw in the head of an ox as well as in that of a sheep. Anyway I should think that the aqueous excretions which occur through the ear without any injury of the hearing originate from these glands when infected (which Platerus recalled who mentions in Vol. 3. l.2. c.7. a young girl who at intervals over a long time had had a discharge of some measures without any other lesion) since it could hardly have been possible through the tympanumwith intact hearing. You may say, if the channel which arrives to the nose from the ears is choked, the humour to be excreted in the nose has finally found a path in the ears. But, whatever there is, the said function of the glomerate [glands] undergoes day by day a favourable experience and this is confirmed by the following.

The other gland of the eye called the *innominate* is situated in the superior-external region. It appears big and almost round in this place. From there to the inferior aspect of the orbit it becomes thinner and thinner curving in a kind of narrow flap.

Its anterior part concedes free exit to the vessels which excrete there the secreted humour. They proceed from the gland itself through the intervals between the lobes into which the gland is delicately divided. From there the vessels proceeding forwards inside the inner tunic of the eyelids perforate the tunic in small holes at a short distance from the eyelashes. The orifices of the said streams are easily pulled in view. You will stretch a little the everted eyelid in the external angle. Soon indeed at a recess one half inch from the external canthus in the angle itself three, in the lower lid four, in the upper lid six and sometimes seven [orifices] gape. A thin piece of hair passing through discloses an easy passage to the gland itself. I found these [orifices] last year when, to the light of a candle on November 11, I exposed an eyelid of a sheep's eye pulled out from the orbit and denuded of its external tunics, to find out whether this is transparent. Soon indeed rivulets glittering with lymph revealed themselves.

The other gland called the *lachrymal,* is oblong and also concealed in the internal or larger canthus. Besides blood vessels and nerves which it has in common with all other glands, it is provided with a peculiar cartilage and with two vessels besides the efferent lymphatics. The *cartilage* comprises two parts. The *thinner* one appears to anybody who scrutinizes a calf's eye, wide, semilunar and tending to the nature of a membrane or rather degenerating into a membrane, provided with a thicker margin smooth on the side which touches the eye but rough with blunt asperities on the opposite side. It is narrower than the other one which is *thicker*. It is truly cartilaginous and white, narrower where it continues the middle of the margin. From there it expands little by little to a greater amplitude between the gland itself and the eye where, narrower again, it ends near the middle of the gland. The *excretory vessels* go by two. They originate inside the gland, surge to the sides of the cartilage and open out where the cartilage becomes narrower upwards, between cartilage and eye in small orifices not to be considered as some papilla but as one aperture. That is how I observed them for the first time on June 19 of this year, when I was enucleating an eye.

The *lachrymal points*, as I observed in sheep, calf and dog, although they appear double in both eyes near their larger angle where however you will have followed the carving in the bone to the lachrymal orifice, you will have noticed that they concur in one trunk. This trunk, where it lies above the osseous canal, leads forwards to the internal side of the rest of the bone, ends in one opening, quite conspicuous and not far from the outermost part of the nose.

Although I have only seen the said vessels in beasts, I have no doubt that they are present in man as well. Since indeed they are similar to that of a gland, they are set in a similar place, since the humour which is found beneath the eyelids is not different, it must thus be admitted that their vessel is of the same kind. The lachrymal points not less conspicuous in man than in beasts reveal that their continuation ought to be found also in man.

The humour proceeding from these glands and their vessels, which is observed between the eyelids and the globe of the eye flows out in the nose through the lachrymal points either scarcely only, it is then noticed by few people, or very profusely and then it comes under the name of tears. Nobody ignores that various opinions have been handed over by different people. This is an argument among others, and not the least, which shows of how little force the brain, even the keenest, is if it does not rely on sufficient experiments. What indeed has not been imagined on this subject by different authors? There were some who distinguished the substance of the tears from a less viscous humour and assigned to both a different origin. Platerus asserted that the latter exudes from the veins of the eyes and the former originates from the brain. Those who derive the tears from the brain do not agree with each other. Satisfied with the brain alone, they disagree over the paths: some imagine an anterior canal of the choane, others nerves, still others veins and I do not know what else. Some actually, besides the brain call other parts to help. Vesling thinks that tears flow together to a channel to the lachrymal points partly from the brain through the second hole of the cuneiform bone, partly from the top and sides of the head. Those who, excluding the brain, think of other afferents do not agree with each other. Some indeed deduce that they come from the nourishment of the eyes, others from the crystalline humour and from excrements of the vitreous humour, the illustrious *Schneider* from the proximal arteries, the very clever Descartes from vapours which get out of the eyes in more abundance than from any other part

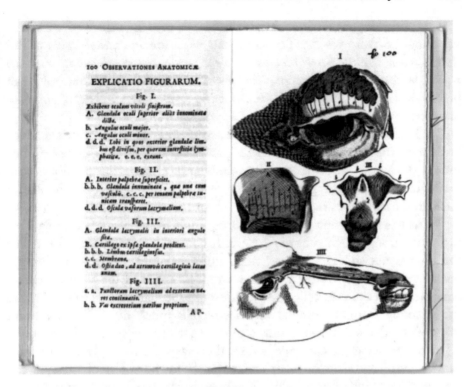

Fig. 2.5.2   Glands of the eye, facsimile edition 1951, 8°. Fig. I. Showing the left eye of a calf. A. Upper gland of the eye called innominate. b. Larger angle of the eye. c. Smaller angle of the eye. d.d.d. Lobes into which the anterior margin of the gland is divided and through the intervals of which lymphatics. e.e.e. emerge. Fig. II. A. Internal aspect of the lower eyelid. b.b.b. Innominate glands which appear through the thin tunics of the eyelid together with vessels c.c.c. d.d.d. Orifices of the lachrymal vessels. Fig. III. A. Lachrymal gland situated in the internal angle. B. Cartilage proceeding from the same gland. b.b.b. Cartilaginous margin. c.c. Membrane. d.d. Two orifices, one on each side of the cartilage. Fig. IV. a.a. Contuation of the lachrymal points to the tip of the nose. b.b. Excretory vessels proper to the nose.

because of the magnitude of the optic nerves and the multitude of small arteries. (Fig. 2.5.2)

But, although various fairly clever opinions seem *to be pressed upon* with great probability from these people, experience however demonstrates paths far different, more in agreement with the usual way of nature. As indeed what is common to all parts is completed by universal organs, so the peculiar necessitated by singular parts are prepared in peculiar places and are excreted through peculiar vessels. I thus think that it is fairly obvious that this humour which serves to the movement of the eyelids, secreted in glands from arterial blood, is delivered through the vessels thus described.

That function was attributed to these glands even by others who did not yet know 86 these vessels. The illustrious Wharton in *Adenographia* c. 26. indeed says: *It must be admitted that these supply the small moistures which trickle in tears although not in such abundance*, although he had said just before: *nobody has so far really demonstrated how these glands spit out these moistures or through which vessels they excrete them.*

But it seems dubious whether the tears must be derived from there since great men have estimated that these glands are not up to this task. They indeed did not believe that such an abundance of tears can possibly come forth from such small glands. And it would result that tears should be attributed even to beasts, which seems absurd to many people.

But as regards the first point, if the magnitude of the drops is compared to the time during which they are collected, no problem will appear here. For the time is not so short that as much humour could not flow in through several vessels as is required to form a drop. Nor was all the humour which emanates from the eyes collected previously in the glands. It is sufficient that the secretion from the arterial blood which is called in at each pulsation becomes accelerated in the glands, a point on which we shall soon say more. As far as the beasts are concerned, since they have also these glands and these vessels, more abundant humour emanating from the angles of the eyes can also produce a kind of tears (which is very often observed) which must deserve the name of tears as well as that which spreads out from the eyes of men, without emotion, as a result of abundance of substance alone, of irritation, or of a defect of the organ.

I thus reckon that tears are nothing else than a humour destined to irrigate the eye arriving more abundantly, so that, as the vessels observed in the glands of the eyes show obvious nearby paths, they have no need be deduced from other ones concealed and remote. I will not deny that, as humour trickles from the eyes into the nose through a peculiar channel, although the nose is not deprived of its own fleshy glands, that perhaps also a somewhat salty humour can be derived from the brain in the eyes through separate vessels. Since it is not my way to offer suspicions for sale rather than truths, I will leave that an open question and present what is evident to explain all the phenomena which are observed concerning the tears.

Anyway, when the blood is good and the organs are well and in order, humour facilitating the movement of the eyelids flows in normal quantity and is derived into the nose through appropriate channels. If one or the other is altered either by internal or by external causes, the outflow of humour also will change. As far as organs are concerned, and the vessels which supply the substance, and the glands which secrete it, and the excretory vessels which bring out what is secreted, can be considered. If the vessels bringing in the blood are not in normal quantitative proportion to those which carry it away, normal outflow of the humour cannot be expected. Indeed the easier blood passes from the arteries into the veins, the less it allows its components to be separated, for that which should be separated goes out together with the rest through more open paths. And through the passages intended for serum, the parts close to these passages will pass slowly only, the others flowing over freely. Thus the more uneasily, the more abundantly serum will be expressed through the simple and porous tunics of the capillaries present inside the glands. While it is whirled about in these straits, it does not flow in as freely as it flows out. Everything able to get out through other paths will enter these more quickly than it uses to so that the velocity of the passage makes up for the narrowness of the paths. Thus, although not more numerous than they normally are, the channels will be open, although these channels are not dilated (if they were so dilated as to offer passage at one time to more than one serous particle or smaller components of this particle, they would also admit the thicker particles of the blood), only the velocity—resulting from the proportional change in size of the vessels—of the particles passing through these channels is sufficient to produce great abundance of tears.

But to find the most immediate cause of this varied proportion, it must be seen first in what part the change occurs, and then what produces the change. Only three vessels are known so far destined to carry this substance, arteries, veins and excretory lymphatics but the excretory lymphatics are unable to undergo this change (constricted indeed, they would receive less than before and, dilated close to the course of the blood vessels, they would transmit other things than serum, but, dilated close to another duct, they would not receive a quantity of humour greater than normal, unless perhaps they were subjected to some tension was to be imagined which by suction or some other kind of force would adapt a quicker movement to the contents which must attract serum. Yet such options are hardly admitted). It remains the veins and the arteries to one or the other of which or to both something unusual happens. But since the whole change consists in that blood flows in more freely than it flows out, either the arteries will be dilated or the veins will be constricted or both. But since constriction properly occurs in vessels and other organs which have circular fibres, but not dilatation except by accident, the change is not to be looked for in the arteries but in the veins.

There is not one single place in the veins which can be constricted in such a way while tears flow out of themselves. It is either the part close to the heart or that at the extreme close to the arteries or that in between which is liable to this change. In the vicinity of the heart there is a muscular portion of the vena cava so that it is undisputably subjected to the command of the mind. Depending on the different ways it is moved, it produces two primary effects from which many peculiarities in the working of the mind proceed. Either it pours in the heart both the descending blood full of lymph and the ascending blood which, in angry people at least, in the opinion of everybody, is mixed with bile in abnormal mutual proportion, more scarcely or more abundantly than normal and thus changes the normal ratio; or it sends more of one and less of the other. From this, various symptoms occurring in different emotions could be clearly and distinctly deduced. This is not the place to present all these matters in more details. But although I shall not doubt that, in the abundance of tears, this part has a great role and not only provides the heart with more blood, which sighing, red of the cheeks and sometimes swelling of some parts of the face point out, but it also changes the ratio of the concurring humours which more abundant serum in the blood reveals. If not excreted through the appropriate channels, it induces headaches and other symptoms which may be observed if the tears are suppressed, reliable sign that the normal arrangement of the humours is changed. Since, however, the heart distributes blood evenly in all parts, in many weeping people hardly anything else than tears will be observed. Since often, in movements of the body, in anger or in other excitation of blood by ambient heat or by absorbed drugs, the arteries are stretched by the blood more than normally without appearance of tears except in great commotions in which, however, the feeling of pain concur to causing sadness, there must be a special place which concerns the glands of the eyes more than the other parts. In the same way, although the middle course of the vessels seems to undergo something as appears from the astonishing distortion of parts of the face which we observe in children and before the tears flow out and while they are flowing if, however, no change occurs inside the gland, one occurs in the vessels and no more than normal secretion of serum from the blood would occur in that part alone. Thus there remains the extreme parts of the veins inside the glands themselves of the eyes, which in one way or another are continued in the arteries and which can contract and, when narrowed, be the immediate cause of increased secretion in a well determined part.

This appears from what we said above about the proportional change in size of the vessels and also from what we proposed in a later discussion in *On glands of the mouth*, thesis 39.

The moved part having been described, it is easy to come upon the moving one. Since indeed the tears are the consequence of emotions such as sadness, joy, sometimes also anger, since it is in our power to suppress oncoming tears for a while, since it is easy to produce them at will for young ladies who have trained their eyes to weep, it is clear that that through which the mind provokes at will other movements in the body and stops them, i.e. animal [spirits], is also directed to these glands and move their parts above all since there are obvious nerves in the gland, through which this force is transmitted to the other parts and it would be possible to show that an agreement of the singular parts observed in the glands by the connection of the nerves with the veins makes sense.

Thus if we see a changed proportion in size of the blood vessels which alone is not sufficient to increase the secretion of tears unless there is also in the blood a substance able to produce tears. This substance will also be liberated from the surroundings of the blood parts. This is the reason why tears sometimes cease in long weeping and come back again after a drink. Hence it is also clear why sometimes they flow on order according to that sentence of Ovid:

If the tears fail you, they do not always come at the proper time,[1]

It is for the same reason also that children and young girls as well as old people, richer in emotions of this kind than middle age men, shed tears. But tears of blood have even been observed in extreme sorrow. This supports our opinion. If indeed the veins, not made narrower than the arteries, would concede free passage to blood, this would never leave spontaneously through the thin passages appropriated only for the particles of serum. Therefore, backflow of blood in the veins being impeded, it must have either dilated or burst the channels appropriate to excrete to open a path which it followed until, the boundary of the animal spirits in the said parts being removed, it will be allowed to flow again free through the normal path.

It seems to me that the tears which are called involuntary proceed from this cause, to which must be added those tears which erupt against the will in a great effort to move something and those which are provoked by smokes and acrid vapours.

For other tears which have been called spontaneous, the matter is easy. Either they have causes occurring in the eyes as if the glands are infected with the excretory vessels or if the passage of the humour in normal quantity into the nose is impeded, or they are due to faulty serum in the blood. It is no wonder if, as a result of the proportion of the elements which have more affinity with the tears than with sweat, urine, pancreatic juice, saliva, these

---

[1] *touch your eyes with a wet hand.* The (unquoted) continuation from Ovid, *The Art of Love,* Book I, part XVII.

elements are excreted in the glands of the eyes through channels corresponding in shape to their own parts. From this it would be easily possible, as far as the tears are concerned, to explain various symptoms which occur in diseases.

The illustrious *Bartholin* thus reports the paroxysms of a hysterical woman which stopped spontaneously only after an abundant flow of tears had occurred. In this instance, I should reckon that the humour, more tightly linked by the ties of the parts of the blood, by which it provoked various symptoms affecting movement, sensation and other functions in all the body, when finally free either by way of drugs or by internal movement of heterogeneous parts of the blood, since it would be more like tears than like other humours, has found an outlet in the tear ducts. But it cannot be determined with any certainty unless the nature of the disease is examined, why it happened that the fluid in this woman is closer to the nature of tears than that of the other humours although the proclivity of women to weep may have conferred something to it, since finally the same was not observed in others. I could add more on ophtalmia and its cure but since the paths now have been discovered the matters are quite clear, satisfied with having proposed in this paper what concerns the function of the glands of the eye and the sources of tears. I will leave a more detailed deduction of the matters which can be related to that until the time when Providence will permit me a complete description of the glands in which I will also show how the glands spread everywhere under the skin produce the sweat.

2.6   On Vessels of the Nose

Before removing the hand from the table, I cannot refrain from attaching as an appendix to the foregoing what appeared to me in the nose itself, while attempting more thoroughly at investigating the lachrymal ducts after the bones of the nose had been broken, since it displays quite the same structure as the envelope of the eyes.

Since the tunic surrounding the nostrils not only is found moist in dead men or animals but also in living animals it is seen to be irrigated by some perpetual moisture. Without a continuous stream of aqueous humour flowing from the nose it is necessary that there are paths which lay down this humour in the nose and others which carry away what has been laid in this place. The structure of the eyes and of the mouth had already drawn me to this opinion not only for the nose but also for the pericardium and other parts. But since anatomical arguments, besides analogy, also require personal inspection, I held that it only had to be reckoned among suspicions because either member of this proposition seemed to be expecting its own adversaries. Those who require nostrils dry, not wet, to perceive odours well would indeed question the former paths and quote the words of the great Hippocrates

OPH 6 vol. I, 93–97. "De narium vasis" must have been composed in December 1661, being written as an appendix to the preceding one, "De glandulis oculorum", which bears the date of December 6, 1661, and together with which it appeared in print about New Year 1662, as the last of the four treatises in *Nicolai Stenonis observationes anatomicæ* ... Lugd. Batav. 1662. Further annotations by Vilhelm Maar are found in the transcripts, www.extras.springer.com.

against those who merely doubt. One reads in the book *On flesh*[1]: "and when the cavities of the nose are dry, the brain most accurately smells dry things", and further: "when the nostrils are wet it is not possible to smell". If thus smelling required dryness of the organ, no vessels should be admitted there since not only they would be useless but they would even impede the sense. Those who contend that any humour of the nose, if there was such humour, can be either dried by the air passing by or rejected in the throat through open paths, would not admit easily the former paths different from the cavity overarching the throat. And thus although I could have objected to the former that the nose of the dogs, whose smelling power everybody praises, is humid, and the very need of humour to facilitate the passage of vaporous bodies in the pores to the extremities of the nerves; although I should oppose to the latter on the one hand this position of the nose in man as well as in animals which does not suffer that humour falls back in the throat, on the other hand the nature of the humour such that it cannot be dissipated in thin exhalations by the air alone driven inside the chest or by the air rejected from the lungs together with warm soot. Vulcan himself, if called to witness, demonstrates that. Nevertheless, I remained dubious until, besides the *aqueduct* already known previously, and besides the glandular fleshes of the *tunic* called *pituitary* by *Schneider*, I observed also *the continuation of the lachrymal point all the way to the nose*, and another *duct peculiar to the nose*, all the *vessels moistening the nose*, and altogether the *strait*, almost a *channel*, through which the humour contained in the nose is transmitted to the palate. Then it was clear indeed that, unless the words of *Hippocrates* meant immoderate humour, war had to be declared to experience itself.

The first canal irrigating the nose, or *aqueduct*, which carries humour there where the cavities of the ears are moistened, has been described by various illustrious gentlemen. Therefore I will add nothing here but, satisfied that I have added it to the others, I pass to the next ones.

As far as the second canal through which humour is derived from the eyes to the nose is concerned, since it has been described above on page 84, it will not retain us for a long time either. The famous *Vesling* thinks that this ends inside the tunics of the eyelids, to which the famous *Schneider* also adds his assent. In *De catarrhis*, l. 3. c. 9 p. 348, he says: *Before this author (meaning Vesling), we examined these matters and others of the same kind and taught at many places that these holes did not go further. An eyelid indeed is made of two membranes. This kind of hole runs between them to the extremity of the orbit and*

---

[1] *Liber de Carnibus.*

ends there. But I have observed together with the famous Mr. *van Horne* already twice that [this canal] extends all the way to the nose not only in animals but also in man. It is plainly amazing that the very skilled gentleman, *Fabricius d'Acquapendente* who observed a sinewy canal in the cavity itself of the os lachrymale did not see that this canal continues in the lachrymal point. He thus tells in *De ocul.* part 1. c. 13: *For this reason, very wise Nature arranged two paths to evacuate the tears. One is through both corners of the eyes, another through the nose. At both extremities of the tarsi, i.e. the upper one and the lower one, two small orifices are apparent in each of the two angles, so small and narrow that they are hardly visible in the bodies of live animals, but in the cadavers they escape completely sight. It is through them that tears, when they are poured out, leave though preferably through the internal angle which nature uses for the outlet of tears, as it is lower. But the nostrils are also a still wider path. At their apex, near the internal angle of the eyes, holes are carved in the bone so large that anybody wonders at them. If you look at them carefully, some considerable canals will appear to you, carved at that place for this function, to evacuate tears through the nose. But you will also see a sinewy canal from the eyes enter holes of the same kind, as if it were a canal fitted to derive tears in the nose.* I wonder, I say, that the illustrious gentleman, together with other defenders of the same opinion, divide things which Nature wanted to be joined.

In the third place, vessels which originate in the tunic itself of the nostrils must be examined. They are of two kinds. Some are rather short. They are hardly visible except by expressing humour the drops of which proceeding little by little show the orifices of the vessels. The others are longer. They deserve the name of true canal.

The shorter small vessels pour out that humour which *Schneider* observed in an anterior pituitary membrane. I cannot agree with the famous gentleman who ascribed that entire task to the membrane without making any mention of glands. But, beneath the membrane itself, glandular portions are scattered over the extremities of the vessels which are found there in great number. These glandular portions are smaller at the front of the nostrils. The more you move backwards, the larger they are. Some narrow excretory ducts which have their origin there release a humour separated from the arterial blood.

I have not yet seen the longer vessels in man but I noticed them often in sheep and in dogs. There are two, one in each of the two cavities of the nostrils where it has its origin and its end. In those glands indeed which are concealed and scattered beneath the tunic of the nostrils in sheep, many roots for the glands are observed above the region of the molars, arranged in a delicate order. They converge into a trunk from which a duct arises which, where it rises towards the back beneath the sinewy canal, reflected forwards from there, proceeds between the back and the small vessel continuing the lachrymal point, until, near the extremity of the nose where a cartilage arises above an otherwise flat surface, it discharges itself in a fold conspicuous inside the cartilaginous protuberance of the lateral walls. It is not difficult to see whether you are looking for the orifices or for the springs. When you look at these springs in their course, you have to break off the bones of the nose, keeping intact the membrane by which they are covered and a lymphatic running through the external side of the tunic comes forth in its magnificence. If you continue upwards, you will find a delicate offspring of small branches carrying out the task of roots. If you want to find the orifice, the lateral walls must be separated from each other about the crest of the nose. Then the same lateral wall must be divided above the cartilaginous excrescence (at the lower part of which the orifice through which the eyes discharge themselves gapes). You will at once see a kind of furrow. This is short but a probe introduced into it leads to the small vessel. Very often in sheep, blackness shows, even to those who look incidentally, as with a pointing finger, all the fosse jutting outside the duct, where almost all the orifices of the excretory lymphatics as well in the cheeks as beneath the tongue are often seen tainted. In dogs, its orifice discharges itself, not on the back of the cartilage, but beneath it, and the canal itself does not keep at the same distance from the top of the nose in its course.

The vessels by which the nostrils are irrigated having been considered, the paths by which the same humour is sent off elsewhere when it has fulfilled its task deserve to be examined. There are two of this kind which deserve to be examined here, one in each part. The gaping doors of the nose which are open to the to and fro movement of air have not to be referred to here, since the healthy state of life, in which nothing is ever discharged by this way, shows that that which is open to the outside is not destined to this task. On the other hand, those which, directed inside are above the throat, besides that their size and function have made them known to everybody, so that it would be superfluous to describe them here, also their very position obviously shows that they are not up to this task. The passage which opens from the nostrils in the palate is very short. It does not deserve to be called a canal, but an orifice. If you look for the place, you will come upon the extremity of the nose where a cartilage arising above the edge of the third maxillary bone retains the humour coming there so that this does not flow outside. In man you will find a round orifice inside the barrier close to the sides of the vomer bone. This orifice is fairly wide at its upper circumference but it soon becomes so narrow that it does not permit even the passage of a hair in the palate. If you inspect the ceiling of the palate, certainly in man, and also in dogs, you will see a tubercle rising up near the root of the front teeth. On its sides, if you have pressed the tunic only a little, one drop springing forth on either side will demonstrate exchanges of the mouth with the nose. But in the calf, as well as in the lamb, in which the front glands do not exist in the upper side of the mouth, a heart-shaped figure will indicate the same orifice to the investigators. The sides of this heart-shaped figure will open a path into the nostrils to an introduced probe. But it must be observed that, in these two subjects, the position of this heart-shaped figure varies. Its apex is directed to the throat in sheep, whereas it is the base which is directed to the throat in cows. Neither is the conformation of this orifice the same in different skulls. In man indeed, if you return the upper jaw, you will find the orifices which appear to be separated by the vomer bone in the cavity of the nose, as also the small ones, joining into one below the teeth. In animals, however, they remain divided and they are not round but extend into oblong clefts.

Having seen the paths by which the nostrils receive the humour and let out the humour which they have received, one should go further and investigate the nature of the humour itself, on one hand the vessels which bring its matter, on the other the way in which it is liberated from the embrace of the sanguineous parts. But these points have been examined previously somehow. Deeper knowledge of the humour itself and of its mode of secretion, however, requires more experiments. Therefore, I will put that off for another time until my possibilities and time will permit to experiment actually the various matters which I conceived in my mind.

From those things which have been said here and above, it clearly appears how carefully the most intelligent Creator of the animals made it carefully so that no refuses would befoul the royal seat, I mean the head. The eyes together with the cavity of the ears had to be moistened, the nostrils had to be irrigated, and the mouth and throat had to be lined with an unctuous humour. Nevertheless, neither the eyes nor the nose nor the throat, if you live normally, excrete anything. What are the tears indeed if not the sign of a movement of the soul? What is the mucus of the nostrils, what the spittle of the mouth if not the sign of a life deviating from the laws of good health? "For the food exceeds the work done." As soon as that humour has fulfilled its task, from the ears it is derived through the aqueduct, from the eyes it is derived into the nose through canals, and from there it is let out partly into the throat through huge orifices, partly into the cavity of the mouth through a narrow channel, and, descending from here, through the oesophagus into the stomach, and pursuing its course into the blood vessels through the chyle and lymph duct, it completes a peculiar circuit of its own kind.

2.7 Origin of Sweat from Glands. On the Insertion and Valvula of the Thoracic Duct and of the Lymphatics

To Thomas Bartholin in Copenhagen

What I had destined to you for a long time, famous Sir, finally, after more than six months, presents itself to be subjected to your judgement. Since the *Disputation*[1] I had decided immediately to prepare a new print to transmit the integrity of what I had proposed in short at that time. But soon being called away for a journey through the most illustrious Belgian cities and drawn away by different hindrances arriving from here and there, and finally being retained by the dissections themselves while any new observation was an introduction to the next one, I could not bring this matter to an end. And if narrow circumstances at home had not retained me, I should have proceeded further in the examination of glands, wishing to survey the vessels of them all and to investigate the paths of the lymph visible in various cavities of our body. I indeed have no doubt that the task of all the glands which the famous *Sylvius* calls conglomerate consists of moistening adjacent regions through

---

OPH 7, vol. I, 101–103: "Sudorum origo ex glandulis. De insertione & valvula lactei thoracici & lymphaticorum" accompanied the copy of the then just published *Observationes anatomicæ* ... Lugd. Batav. 1662, sent by Steno to Bartholin. The letter is found in "Th. Bartholini Epist Medicin. Cent. III". *Hafniæ* 1667. pp. 262–266, numbered Epist. LXV. Further annotations by Vilhelm Maar are found in the transcripts, www.extras.springer.com.

[1]See Fig. 1.2.4, p. 78.

© Springer-Verlag GmbH Germany, part of Springer Nature 2018
T. Kardel and P. Maquet, *Nicolaus Steno*,
https://doi.org/10.1007/978-3-662-55047-2_18

peculiar vessels. But I am also close to believing that all aqueous humours occurring in various cavities must be derived from similar glands. Does not sweat also seem to have the same origin? Several times indeed, in the hospital, in patients emaciated and completely exhausted by the long tyranny of a disease, I saw just underneath the skin, either in the abdomen or in other places, countless small glands connected to the skin through very thin fibrils which seemed to have to be considered as vessels. A few days ago the famous Sylvius, who during this trimester teaches the practice of medicine, gave me the opportunity of examining anything of my choice. In a subject in whom he had demonstrated the parts which served his purpose, after the skin had been opened in the completely swollen thighs, I saw glands which otherwise are usually hardly visible, so well visible that their origin, by which they must be reckoned, could not be concealed. I should willingly have completed that and other matters concerning mainly the nature of glands composed of several small acini, adding those other things which, while I do them, presented themselves together to me in other parts, not really new but questioned by many people today, if time and short supply had allowed it. Subjects for study indeed are not found so easily here and I dared not delay more the publishing as it has already been somewhat delayed. Certainly, this delay in addressing myself to you would have embarrassed me very much if this kindness, so rare in others, most familiar to you, of which your last letters testified, had not given me more courage. I rejoice and hold myself for happy to whom a good fate has given such a great gentleman as teacher. And if only the things which have been started by my insignificance would please you, I will judge that I have obtained the largest fruit of my labour.

On various things which we saw in the Anatomical Theatre and in the house of the famous Mr. *van Horne*, I will add nothing since I do not doubt that they were reported to you by the famous Borch. If I am not mistaken, I wrote last time something on the thoracic duct with the investigation of the insertion of which I was busy more than once. But recently I observed that its orifice in the vena cava is provided with a fairly large and delicate valvula opposing regurgitation of the blood. This same valvula also closed the orifice of the lymphatic duct descending from the neck. At first I had introduced a hair through the said lymphatic all the way to the vena cava itself, and, from there through the thoracic duct, trying to find a path into the same axillary vein, I finally hardly penetrated in it. Thus as I noticed the hairs being in the cava itself, I opened this vein and made apparent two small orifices very close to each other and covered by a common valvula. The junction of lymphatics near this division of the vein is amazing. But the thoracic duct itself deserves even greater admiration: a probe introduced into it leads into the vena cava and, if from there you deviate, it enables you to ascend the probe in the descending lymphatic all the way to the neck, as, however, a different way opens from the orifice of this, through the same lymphatic, also into the sanguineous canal.

In what concerns Mr. *Blaes*, I should never have even mentioned him, if he and his brother had treated me friendlily. But, in the theses, I did not write any word which could reveal an angered mind, although he acted so harshly against me in his letter that, if I had had to answer in similar terms, sarcasms would have had to be resorted to. But I preferred to present the naked truth provided with its explanations rather than, by lowering myself to unfriendly and harsh words, to seem to have put aside the respect which a private student owes his professor.

I thank you very much for taking pain of commending me to the promoters of science, and, among them, to the Magnificent Mister Chancellor. Since I have no means of being able to deserve your obliging mood, I will recommend you to the most holy Divine Power and I will harass Him with the most ardent prayers that He gives you with the new year new favours of complete happiness, and makes it so that you will return to heaven late and remain prosperous with us for a long time. I end with this wish and ask strenuously that you continue to find me more worthy of your favour. Do well and excuse my hasty writing.

Leiden, January 9, new style, 1662

Yours devotedly,

Niels Stensen

2.8 Why Tobacco Powder Makes the Eyes Clearer. Observation on a Milky Gelatin

To Thomas Bartholin in Copenhagen

The more your last letter delighted me with pleasure, the more the hindrances which prevented me from answering until now made me embarrassed. I hope nevertheless that my very close friend *Matthias* will perform the task of intermediary if this intercession is needed otherwise by the one whose kindness competes with his over the world highly praised erudition. By commending the writings of a disciple to your colleagues and even to our most clement King, you have shown in fact as well as in words that these writings have pleased you. It is not so much your eloquence in raising me— which makes me blush, conscious of my own insignificance—that made me happy, as your well-disposed feeling from which this most friendly judgement resulted. It is true praise to be praised by somebody who is praiseworthy, and who has deserved more titles for praise in medicine than you? Although I know that all that is feigned is very alien to your frankness, I knew, however, that in this matter you accommodate yourself to the tradition of the teachers. They praise the works of their disciples not because these are worth this commendation but to encourage them and to incite them to do more. Thus,

OPH 8, vol. I, 108–111. "Cur Nicotianæ pulvis oculos clariores reddat. De lactea gelatina observatio." This Letter from Steno to Th. Bartholin is dated Leiden May 21, 1662. It is found in Thomæ Bartholini *Epist. Medicin. Cent.* IV. Hafniæ 1667. pp. 1–10, numbered Epist. I. Further annotations by Vilhelm Maar are found in the transcripts, www.extras.springer.com.

although your dexterity at dissecting, which is an object of admiration to all, easily sees that my endeavours and labours have been devoted to an insignificant matter and neglected by nearly all people in the past that, however, so that a hope once conceived is sustained, they are stimulated to greater results by your praise. But your works which deserve preservation promise a better fate to my small vessel than that of *Wharton*. It seemed not enough to you to have praised me to our men, so that you made my name known to the universal literary community. So that your extreme benevolence would not leave anything untouched, you even did so that I would be assured the clemency of the King. Well, every time I turn all that in my mind, I cannot but have mixed feelings: should my soul perceive greater delight from this most happy message or greater pain by considering its own weakness? One solace only remains to me, that of hoping that heaven will return the favours which it is not in my power to return, heaven will recompense so many of your merits by utmost glory and all kind of happiness.

As far as Mr. *Blaes* is concerned, I frankly acknowledge that it makes me very sad when I hear from professors of this place what was reported to them by the older Blaes through the younger, not on account of the cause itself since so many evidences would immediately offer themselves to me that any judge, even elected at the choice of the plaintiff, could not have condemned this cause in good conscience; but on account of himself with whom I have to struggle about the thing. Therefore, as nothing is more opposed to my nature than to take the pen against him, nothing would be more agreeable than if the whole matter could be settled by your arbitration. Those who know my evidences against *Blaes* know that. They very often wonder that, although accused very seriously by him to my teachers and although being provoked in the most caustic letters, from such a crop of evidences I have pointed out one only supplied to me from his own writings. Neither his biting letters nor his wagons of abuses have appeared to me of such importance that, because of them, I should transgress the limits of moderation. Neither do I dwell on the reproaches of *Deusing* who will teach me moderation: I will actually learn this from his preface where he very clearly presents himself not only as the promoter of *Bils* but also as an imitator of his abuses.

Shortness of time did not allow me to reflect on, how does tobacco powder amend weakness of sight. Being devoted to others as it is fit, we have no time for ourselves. But if it must be deliberated on the spot and what springs forth under the hand of the writer is to be brought under your eyes, you will forgive if precipitation has led away from the truth in some points. Since heat has the property to diffuse minute and easily mobile parts whenever possible, and since it is very obvious that tobacco teems with most subtle particles, I do not doubt that the forces of the powder driven into the nose together with the most subtle effluvia, not only enter the blood by penetrating the conglomerate glands underneath the tunic of the nostrils through the excretory lymph ducts (which in the nose are as abundant as they are short), and, by striking the thick and viscous humours in the blood, make them prone to be sifted out, but also push themselves in upwards, brought between the eyelid and the eyeball through the canal continuing the lacrimal point. There, if they do not do anything else, they at least make more and more fluid the viscosity covering the cornea. For the limpidity of the humour lining the eyelids inside is not always such that sometimes intermingled thicker and more viscous particles do not adhere to the tunics of the eye, the thinner ones being removed. This is what occurred rather often to my father of blessed memory, who, after an excessive outflow of humour, quickly restored their pristine acuity to the eyes by bringing near them ginger root moistened with saliva. Actually, the subtle, sharp and warm particles of ginger provoked tears by stimulating also the membranes and, by cutting through and weakening the thick humour which had been drawn before the membranes, they opened a freer entry to the rays. What if one thought that the thinnest parts of these drugs penetrate the cornea itself and make the humours more suited to transmit the rays by thinning and purifying them? Many people have believed that in enchantment spirits are rising from the heart of the possessed through the eyes together with rays and penetrate into the eyes of the other person and attach themselves to the blood so as to produce there the same ailment. This I think I have read in the *Dialogues* by *Lilio Giraldi*, not to mention others whose names I have no time to search for. What about the fact that also the very intelligent *Descartes* wrote in his *De homine tractatu* that an outlet for the tears is open through the same membranes? If he thought that the cornea affords a passage to tears getting out, but the ancients thought that the cornea affords a passage to harmful spirits to get in, it would not be absurd to attribute the faculty of penetrating the cornea to the subtlest particles of drugs, above all since there is no membrane so dense that it is not perforated by almost infinitely many pores. I will, however, leave that, since neither this opinion on enchantment satisfies everybody nor, how clever it be, the way *Descartes* explains the tears relies on no necessity after the new rivulets of tears proceeding from both glands of the eye have been discovered. And no membranes are found which, although porous, would be suited to receive indiscriminately any very subtle substances. It will be sufficient that tobacco powder with its most efficient part either finds a path into the blood through the excretory lymph ducts of the nose, or rises to the external surface of the cornea through the prolongation of the orifices, since the whole matter can be explained in these two ways. And indeed, from this the cornea is cleaned, and a drug sent to the heart through veins and propelled from there through the arteries into all parts. It is also carried to the inside of the eye and through vessels either between the sclera and the choroid, or to the bottom of the eye, conspicuous at the centre itself of the retina. Tobacco powder not only purifies the substance itself of the transparent membrane but also makes more limpid the humours contained in the tunics. And what more could be desired from this powder than that the tunics as well as the humours be purified, the thick matters be thinned, the turbid ones be clarified? A proper shape and position of the crystalline lens is also required as well as a size of the pupils matching these shape and position, which must vary according to the more or less intensity of the light. But a drug will hardly modify the shape of the crystalline humour. Age does it more easily. Its position actually varies depending on the abundance of the vitreous body and movement of the ciliary processes. But the opening of the uvea[1] will obey the usage of looking at certain objects rather than the forces of certain drugs. Thus people accustomed to read very small writings and are used to moving the objects closer to their eyes do not distinguish more remote things. Inversely, seamen always looking into the distance hardly recognize smaller things suddenly presented. Hence it is said that myopes, if, having to make a journey, they look back at the place from where they left, finally they will correct the defect of nature. The uvea emulates the nature of a muscle. It obeys the order of the light rays rather than that of the will. I have indeed observed in myself as well as in others that, when a candle is brought near or the sun rays are let in, the uvea constricts, even if we resist. But why do I say so much on this? It indeed belongs to the disciples to submit their opinion to the judgment of their teachers so as to be led back on the right way if they have not followed up the explanations of the teachers. Similarly, I promise to myself, famous Sir, that either if I go astray from the truth or if I have not presented these matters clearly, warned by you, I will do better in the following. But before ending this, I will add some words about something observed in my own eyes. *Gassendi* recalls in *Vita Peireskii* that *Peiresc* has observed that his eyes keep images taken from objects for a long time, above all when the eyes are moistened from sleep, as he reports. Something similar happened also to me not long ago when, together with my friends, the noble *Gersdorfs*, in a pastime I observed through a biconvex lens collecting the rays of the sun, a variety of the image depending on the various position of the lens. That igneous disc of

---

[1]The iris, and probably not what is today defined as the uvea.

the sun had impressed its image so strongly on the eyes that, back in a darker room and looking at the wall in the shadow, this fire itself seemed to shine red on the wall, but fading away little by little. It is not only one representation of the sun which is left in the eye, but one can notice as many ardent globes as there are circles seeming to fly around in the air after having looked at the sun, if, one fading away, you have followed another, as long as the flame deceiving completely the sight has stood still on the wall. If, however, the wall has been illuminated with light, other colours will appear besides the golden red, resembling not a whole circle but the exterior rim of a circle or a ring. I should think that rays reflecting from the illuminated wall do that the representation impressed previously is not observed so distinctly as it does usually otherwise when we look at a place in the shadow, which does not reflect rays at all or reflects very few of them.

But I am afraid this is too much on the eye. Let me add only one or another observation which presented itself to me in the abdomen of a dog cut open not so long ago. The singular trunks of the mesaraic veins connected to the proximal trunk on both sides around the intestines allowed free backflow to the blood from one trunk into another, and this anastomosis of veins appeared rather as a perfect prolongation into the whole mesentery so that, however, all the intestines received their small vessels from these intermediary ducts. In the same dog, between the kidneys, towards the end of the receptacle very white gelatine had accumulated to the size of a goose egg, which, before the common tunic of the peritoneum had been taken away, I suspected, while wondering, to be the receptacle of the chyle. But soon, as the matter was stripped of all envelopes and did not flow down, I recognized my error. Once the glutinous humour was removed, I found the receptacle of the chyle at its place.

I could add more to this but then I should seem to transgress the limits of a letter or to abuse of your patience in reading my rudimentary efforts. I hope that my little book which you wanted has already been presented to you by *Matthias*, together with the treatise of *Roland Storm* on febrifuge powder. Do well, famous Sir, and continue to favour

Leiden, May 21, 1662 devoted to you,

Your disciple
Niels Stensen[2]

---

[2]Thomas Bartholin's reply: EP 8.

2.9 Anatomical Observations in Birds and Rabbits

To Thomas Bartholin in Copenhagen

When I published my few observations, I had decided to lay down the anatomical knife until more convenient times and to take up again the nearly cast away geometer's rod, so as not to appear to have wasted all time and labour if I deserted completely my research for which I spent many hours in the past and which I would have treated not as my primary, but as my unique work, if straitened circumstances at home had not so much convinced as forced me to prefer the useful to the pleasant. But hardly were my fingers, rid of blood, slightly besprinkled with this very pleasant powder that partly the fairly acid faces of famous gentlemen, partly their unfriendly writings that presented my opinion in a sense different frommine, denied me the happiness desired for a long time so that they imposed on me the necessity to answer and also to return to this bloody task.

I will say nothing of our famous *Blaes*. Because since that time he has limited himself to private letters, neither have I judged to have to stir anything publicly. I hope that this same kindness of yours which sent you in the middle of this dispute will make so that *Blaes*, even if his mind is perhaps a little upset, will not lose the reins and say things which will be prejudicial to

---

OPH 9, vol. I, 115–120: "Observationes anatomiciae in avibus & cuniculis" is found in Thomæ Bartholini *Epist. Med. Cent.* IV. *Hafniæ* 1667 pp. 103–113, numbered Epist. XXVI. Further annotations by Vilhelm Maar are found in the transcripts, www.extras.springer.com.

© Springer-Verlag GmbH Germany, part of Springer Nature 2018
T. Kardel and P. Maquet, *Nicolaus Steno*,
https://doi.org/10.1007/978-3-662-55047-2_20

himself (which I could show that he has already done in private if I was stirred by the desire of writing), and that I am not forced to submit the whole thing to the eyes of a fair reader and thus be kept busy in spending good hours in a matter concerning the integrity of my reputation, but contributing not at all to research, either public or private. I think that you know of what titles our *Deusing*, who has attacked the famous Mr. *van Horne* with vehemence, gratified me. As he yields to nobody in eloquence in assailing others, if he contended as well with mathematicians to demonstrate about the truth, he would be a terror for all those who do not take his party. But one can already say of his weapons what in the past *Timon* said of the thunderbolt of Jupiter in *Lucian*: "It seems to them that you brandish a firebrand which neither by fire nor by smoke frightens, but only is believed to hurt because it covers with soot". Actually, if nothing had to be done to prevent him from interpreting the silence of other people as meaning his victory, I should suffer him to revile me with impunity. Finally, fully to complete the number of intrigues, *Anton Everaerd*, a famous physician of Middelburg, has also joined. This, over-coming the others by his moderation since he dislikes outcries, also keeps off from reasoning but, however, not from experiment. Some days ago he published a treatise entitled *Lux e tenebris affulsa, ex viscerum monstrosi partus enucleatione*.[1] In this, after a description of a monster deprived of head and chest at first, he deplores that the liver is proclaimed wrongly by some people to be a useless weight since it is needed as well as the heart (he had indeed seen in this monster two small hearts attached on both sides of the liver like additions). He mentions me next and believes that he has demonstrated for the eyes of his associates that there are milky vessels laying down milky juice for the nourishment of the parts commonly called spermatic, a thing, which I had fought against both with reasoning and with experiments. But either my eyes deceive me or Nature gave to the Zeeland rabbits what it denied to the Dutch ones; I opened the abdomens of old rabbits, from the outside I looked at those of young ones, I examined the udders as well as the uterus of pregnant ones, but I could not yet be as perspicuous as to observe other milky vessels than the mesenteric ones and the thoracic, and those which, origi-nating from the glands of the udders, carry milk to the tit. But, in these, I never looked in vain for lymph, either in the liver, or in the neck or beneath the capsule, or in the groins. The famous gentlemen *Sylvius* and *Borch* are witnesses, who nearly always honoured my dissections by their presence. Not so long ago, I constricted all the vessels which run below the receptacle with

---

[1] Light shining from obscurity as a result of the detailed explanation of the viscera of a deformed offspring.

one ligature, thinking that, if there were there any milky vessels, they would present themselves with a sign. But, immediately after the ligature whatever was between the tie and the capsule faded away, whereas, beyond the tie, a huge number of vessels increased progressively more and more to greater mass. They contained nothing else other than the most limpid lymph. But if I added more about the matter the best known by you since the first discovery of the lymphatics, I could in no way avoid being called silly.

As you see, excellent Sir, as partly private letters of famous gentlemen, partly publications had drawn me back once more to practice of anatomy, so that repeated investigation of the same matter would not result in aversion, I judged that satisfaction was to be sought for in variety. As I turned to animals to this end, I saw in an aquatic raven which the Belgians call "skolfer" something which I had not yet observed in other animals. Above all I admired the very delicate insertions of the biliary and pancreatic ducts in the intestines. Bile indeed was poured out in the intestines through two distinct ducts, one originating from the gallbladder, the other from the liver. The pancreatic juice got out through as many canals at a place in the middle between the two biliary ducts so that the pancreatic ducts were at a distance of one finger-breadth from each other, and the two biliary ducts were each at a distance of half a fingerbreadth from the pancreatic ducts. All four orifices actually faced the pylorus. The distance from this to the closest biliary canal was more than a span. This delightful spectacle urged me to examine the other viscera of other birds. And look! By chance a duck presented itself to me. Its intestines also receive the biliary and the pancreatic ducts but with no distinct orifices as in the raven. There is one and the same access to the intestines for all four ducts. Whereas probes introduced through the pancreatic ducts face the pylorus, sent through the biliary ducts they reach an opposite position so that the hairs intersect each other in the same outlet, forming a cross. In dogs I saw two ducts going from the same pancreas to the intestines. But I have not yet observed more than one canal carrying bile into the intestines. Your father of pious memory remembered in *Anatomicae Institutiones* to have sometimes found a bifid common duct. But he did not mention whether the origin of each was different or the same. But to return to our raven, the gallbladder did not lie beside the liver but, extending right through the fat to a width of about three fingerbreadths, it resembled the trunk of the vena cava by its position and, at a superficial inspection, even by its colour, but it was larger in width inasmuch as it was not less than the thickness of my fifth finger. Where it joined the liver, three small branches of it were observed. One rising in the liver emitted an infinity of ramifications. A second one was smaller and was directed transversally towards the hepatic duct. The third, mentioned previously, descended towards the intestines. When the gallbladder was inflated, a path was open for the wind into the intestinal duct. In this subject I could not penetrate into anything else whether because there was no access into the liver or the hepatic duct, valvulas closing the way, or whether because thicker bile had occluded the passages.

As we examined this and other things, the movement of the heart growing weak presented a very elegant spectacle. The heart of this bird was indeed so vigorous that its pulsations did not lose the least either of their velocity or of their power until the second hour after the opening of the animal. So as not to be idle spectators, not only we had observed meanwhile the mentioned biliary vessels but also various other things such as the membranes functioning as diaphragm according to the opinion of *Harvey*, the kidneys with their ureters and the orifice of the ureters, the figure of the stomach, its position, several tubules about its left orifice, the most delightful green matter adhering to the inner surface of the stomach and the other things contained in the abdomen which the lack of time and the limited space of paper do not permit to list, and which, moreover, you have seen very often in other subjects.

Finally the heart failed. That we turned our eyes and mind to it, we owed to the famous *Sylvius* (he had arrived with the illustrious *Piso* and the famous physician of Danzig, Mr. *Gottwald*, although some urgent matters calling Mr. *Piso* back to Amsterdam did not allow the presence of the eminent gentleman by me for a long time) who said that he had rather often observed in dogs that it was not the heart but the part of the vena cava closest to the heart which dies last. This bird of ours also confirmed the truth of this observation. As the heart indeed was pulsating less often, a movement of the auricle as well as of the vena cava appeared, very conspicuously distinct from the movement of the heart. It continued in the descending branch of the vena cava where it joins the heart and in that part of the auricle which is nearest to the vena cava, until 12 noon, after the heart had been completely dead (the chest was indeed opened at nine o'clock in the morning). But, when the auricle finally also stopped, the vena cava alone was seen to move even after two o'clock in the afternoon. After the right ventricle of the heart had been resected and all the blood which was in the vena cava between the liver and the heart had poured out, a swelling rising upwards of the transverse fibres, one following the other, was seen nevertheless in the tunics of the vena cava collapsing on themselves. (Having then dissected the heart, I saw in the right ventricle one valvula only, fixed to the auricle and fleshy such as I observed later also in a *hen*). Having since then looked for the same in other subjects, I did not have the same success. I indeed observed in four rabbits which I opened since then that the movement of the vena cava was the very last of all but it did not happen to me to observe in any of them movement of the fibres of their vena cava. But, and this is peculiar to the rabbits, as in the vena cava of the bird the descending branch pulsated, in rabbits the ascending branches constricted themselves at intervals. The left one of the ascending branches in the rabbits leaves the vena cava close above the auricle and, from there, finally rises transversely directly to the base of the heart to the left side. The three branches of the vena cava thus converge to the right auricle, the descending one and both ascending ones or, if the vessels must be described according to the movement of the blood, both descending ones and the ascending one which is alone. The first of them in which I looked for the prime mover displayed only this: when the auricle failed with the heart, the vena cava pulsated five, often six times, before the auricle pulsated once. After the blood had been evacuated, everything stopped.

A few days later another rabbit was opened. In this, the pulsation of the right auricle slackening a little, movement of the vena cava was seen more distinctly now in the right branch, now in the left. This movement continued even when the tunics were cold to the touch and it did not stop after the apex of the heart had been resected, but when, after the first blood had been poured out, new blood propelled again from the liver had been supplied to it, it constricted its tunics at intervals with as much force as previously. Even until the third hour after the beginning of the dissection, as, after the ascending trunk of the vena cava had been ligated and all the blood had been expressed, the idle sides of the vessels had collapsed over each other, once the ligature was cut, the pristine movement returned again.

What we saw in a third rabbit was more clearly observed. At the fringe of the auricle opposed movements were noticed, at one moment from the inferior edge to the superior, at another moment contrarily from the superior to the inferior as if something was seen to crawl. As the fringe was plainly dead, about in the middle, some fibres, however, by their contraction, seemed to give signs of life. Here I did not resect the apex of the heart as I did in the former, previous rabbit. But, through a tiny hole admitting hardly the head of a pin, I let out the blood, as I had put ligatures on the branches of the vena cava, the ascending ones as well as the descending one, and immediately everything was immobile. But, as the unhurt vessels of the heart supplied new blood, movement returned soon, although fairly small. It, however, strengthened almost instantaneously after all ligatures had been removed. I had opened the animal at about six o'clock in the afternoon and nevertheless the vena cava pulsated after nine o'clock at night. At intervals of time, as a result of repeated ligatures, the vena cava had omitted many pulsations.

Once I had removed the ties for the last time, I saw very clearly and obviously, in the same edge where the superior branches of the vena cava meet the inferior one, the tunic contract almost instantaneously into a strait and, after the second thrust of this edge, the ascending branches themselves move.

In a fourth rabbit, I did not notice anything different from the others as far as the movement of the heart and vena cava is concerned.

This is, famous Sir, what I could observe on the part moving last of all. I have not dared yet even to reflect upon why and how these movements occur since other occupations refuse me the time which the difficulty of this matter requires and the weakness of my natural capacity persuades me that the matter is beyond my forces. If it would please your kindness to let me know the opinion of its supreme and divine judgement, this would add greatly to your pristine favours to me.

Descartes' Tractatus de Homine appeared these days published together with figures by *Florens Schuyl*, Senator of the renowned city of s'Hertogenbosch and professor of philosophy there. In this book there are some not inelegant figures. It is certain that they have proceeded from a clever brain. I doubt, however, whether such images can be seen in any brain. But for too long I retain a famous gentleman whose time is devoted to more serious cares. Do well, famous Sir, and continue to favour.

Leiden, August 26, 1662   Your devoted disciple

Niels Stensen

2.10 Observations Concerning Movement of the Heart, the Auricles and Vena Cava, Extracted from Dissections of Various Animals Carried Out Here and There

1. In a cat, as the peristaltic movement of the intestines and many and big swellings of the biliary vessels had retained us in their examination for a long time, returning to the heart, we found everything immobile. But hardly had I compressed it three or four times with my fingers, their movement returned immediately to the heart and to the vena cava, although very rare, but together and very distinct, so that, as it appeared clearly, the movement started in the vena cava, now from its upper part, then from its lower part, and ended near the heart.

OPH 10. vol. I, 123–127: The observations mentioned in "Ex variorum animalium sectionibus hinc inde factis excerptae observationes circa motum cordis" are either identical with those mentioned in the previous letter to Th. Bartholin or they are closely connected with the latter, being in all probability made almost at the same time, i.e. in Leiden in 1662. The text itself practically only consists of brief notes on each observation, and Steno has made no attempt to work them up into a whole; no doubt they were meant to form part of an exhaustive study on animal motion or more especially on the motion of the heart, which, however, he never accomplished. The very last part of the "Treatise", in language as well as in contents, differs from the other part, and has certainly been added by Steno at a far later period, viz. when from 1672 until 1674 he was *Anatomicus Regius* in Copenhagen and left the "Treatise" in the hands of Bartholin to be published by him in *Acta Medica et Philosophica Hafniensia*, where it is found in vol. II, 1675, as No. XXXXVI, p. 141. Further annotations by Vilhelm Maar are found in the transcripts, www.extras.springer.com.

© Springer-Verlag GmbH Germany, part of Springer Nature 2018
T. Kardel and P. Maquet, *Nicolaus Steno*,
https://doi.org/10.1007/978-3-662-55047-2_21

2. The animal, a dog, dies while vainly I attempt to drive air from the coeliac artery into the lymphatics of the liver. At once I open the chest, I free the heart, which is very swollen by air, from the pericardium, and at the same time I see that movement has returned.

3. To the same end, I had opened a dog in the presence of friends at four o'clock in the afternoon. As I had experienced the same luck with the lymph, with nothing achieved, I had dismissed them and, having been retained by other friends, I came back to the cadaver at six o'clock, I saw the ventricles of the heart very widely distended. By shaking them a little with the fingers, I observed that the auricle moved and at the same time the part of the heart close to the auricle contracted clearly. But what I wondered most at, back from dinner at eight o'clock I touched once the swollen subject and clearly saw the auricle pulsate quite often. I observed this until nine o'clock several times, always in the same way.

4. In another dog we saw every fifth or seventh pulsation of the auricle release one movement of the heart, about the meeting of the pulmonary artery with the heart. Right at the start, after any contraction of the heart, a protuberance of the tunic outside the flesh of the heart appeared. The heart immobile, when pricked about the right ventricle, pulsated at once. It remained immobile when twitched at any other place.

5. In a dog, I saw the apex of the heart cut off and set on the fingers, when twitched either with a nail or with a knife, contract itself so that, the sides retracting to each other, it fell from the finger. But I saw the same apex move in like manner by being turned over.

6. Cut out of its membranes, a foetus of dog would have breathed out quite often the air taken in before due age with repeated loud wailings. I then opened its chest and twisted it set on my fingers so that the movement of the heart and the arterial duct would appear. I saw the very delicate and transparent heart turn red when the apex receded from the base, then whiten like juiceless flesh when the apex moved to the base. The adjacent artery revealed blood not only by its swelling but also by its colour. The blood flew out at once as a result of an injury in the neighbourhood. At the beginning, for each pulsation of the auricle, the heart beat twice but afterwards only one movement of the heart corresponded to two pulsations of the auricle. The heart was then cut out and warmed up by the heat of the hands. Even if I pierced it through more than 20 times from all sides with a needle, it nevertheless continued its movement for a long time.

7. In rabbits, before I tackle movement, it must be known about the course of the vessels of which frequent mention will be made below that there

are three branches of the vena cava, one inferior penetrating the dia-phragm and two superior, one rising straight at the right side, the other from the right stem directed transversely to the left side near to the base of the heart, and thence rising up at the left side. After these premises the matters which concerned movement were the following. Whereas the heart with the auricles was immobile, I saw the superior vena cava continue its movement, even after the apex of the heart had been resected, even after the blood which was in the vein had been evacuated as long as new blood was supplied, even if everything was cold to the touching 3 h after the beginning of the dissection. But also, before the heart became rigid, I observed various movements in it depending on whether it was compressed variously with the fingers.

8. In another rabbit I saw the part of the vena cava attached to the auricle pulsate sometimes five and often six times before the auricle moved even once.

9. In a she-rabbit dissected in about August 1662 in the presence of the famous gentlemen Mr. *Sylvius* and Mr. *van Horne*, different things were observed concerning mainly the movement of the auricles and vena cava. While the right auricle pulsated more sparingly and slowly, a movement of the vena cava finally appeared in its right and in its left branch. The auricle moved once after two or three pulsations of the vena cava, but not always in the same way. Sometimes, indeed, the movement crawled in the margin of the auricle from its inferior edge to its superior edge. Sometimes the same margin seemed to be stricken in a plainly opposite way, from the superior edge to the inferior. Sometimes, while the margin was absolutely immobile, there was some contraction of the auricle about the middle plane. (In the heart of a dove I also saw movement beginning from one edge of the auricle reach the opposite edge). Then the three branches of the vena cava were ligated and all the blood which was within the ligature in the vena cava, the auricle and the right ventricle was evacuated through a small hole made with a little needle in the bottom of the right ventricle. All movement stopped at once. You would have said that everything was dead. But this immobility did not last long. New blood regurgitated from the veins of the heart and distending a little the tunics of the vena cava [that had] collapsed on each other, had produced new but small movement visible only in the vena cava alone. After removal of the ligatures had allowed blood to flow back to the heart, complete movement was restored as well in the vena cava as in the auricles. I had opened the rabbit at about six o'clock in the afternoon and, after half past eight, when the auricle itself had stopped living, the

vena cava still lived. I repeated the same experiment in the presence of other friends with success. Blood evacuated was taken in once more through the same injury to the heart. The usual movement returned, and this in a peculiar way. A pulse was indeed observed right in the edge where the superior branches of the vena cava meet the inferior one, and mainly on the left side of this edge, completely different from all movement of the other parts. For any second beating of this edge, a branch, now the right, then the left, moved once. But movement was seen to begin also in the right branch already languid, where it perforates the pericardium, and to continue towards the heart.

10. In foetuses of rabbit retrieved from the uterus, I noticed that the heart becomes longer after a contraction and if it is held up in a proper position it appears clearly that, as it becomes longer, it falls by its own weight. Depending indeed on how you change its position, it falls towards different sides. The heart being thus collapsed at first, the auricles pulsate and then the cone of the heart rises almost instantaneously forwards and upwards, the ventricle swells on the right side and becomes shorter, all the heart becomes harder to the touching and soon the pulmonary artery is seen to swell. The resected cone did not pulsate. The right side being resected, the left nevertheless pulsated. After the auricles and arteries had been resected, the heart itself pulsated in the hand when pricked. When the auricle moves, mainly the left, a depression is obvious in its middle.

11. In a young aquatic raven thrown out of its nest in a tree in our presence, as the movement of the heart began to become weak about 2 h after the opening, the pulsations of the auricles and vena cava could be clearly distinguished from it. All three movements were distinct from each other also in time. The heart finally being immobile, the part of the auricle closest to the vena cava survived a long time until, the auricle also being entirely immobile, only the vena cava pulsated below the heart, displaying two movements plainly distinct from each other. One was visible in the more remote part externally, the other on the internal side in the part closest to the heart. But what I wondered most at—which I could not observe in another subject since then—was that, the right ventricle of the heart having been resected and all the blood evacuated, a movement persisted nevertheless in the tunics themselves of the vein, already utterly collapsed on each other. In this movement one of the transverse fibres after another was a little raised, resembling a thread led transversally over the vein, and this while proceeding towards the heart which had not yet stopped its movement at two o'clock in the afternoon, whereas I had begun the dissection at nine in the morning.

12. The dissection of a hen, besides other fairly extraordinary things, showed some things concerning also the movement of the heart, which must not be hushed up here. Not to mention what is fairly ordinary about the movement of the heart and auricles, not only each of the two branches of the inferior vena cava as well as that part of the superior branch which runs transversely towards the right to the base of the heart moved very distinctly for a long time, but also those veins which run over the length of the heart from the apex towards the base gave so clear evidence of their movement that even the pulsation of the tunic seemed to begin at this place where it spreads out small roots towards the cone, and to continue from there all the way to the base of the heart.

13. The heart of a chick extracted from the egg on the seventh day from the beginning of the incubation, freed from all vessels, pulsated for a long time in my hand. Then, immobile as if it were dead, it was recalled to life by the breath of my mouth and this succeeded several times in the same heart.

14. In another chick, as I came back to the heart later after a rather long delay devoted to other parts, by warming up the whole body with the heat of my hand and by approaching simultaneously a warm finger to the heart, I restored to this its pristine movement. Then I called to life for a second time the dead heart by propelling into it new blood from the liver. Similarly, I restored the vital heat in the heart already dead of another by the heat of a finger. But I have also observed that the dead heart of a young rabbit returned to pulsating spontaneously more than 100 times by the heat of a finger. From what was said it results that:

1. Interrupted movement returns to twitching of the heart either by a solid thing such as a needle, a knife, a nail, or by exhalations or dilatation of the blood.
2. The movement is not that of all the substance but of individual fibres, and not only of whole fibres but also of fibres deprived of each of their two extremities.
3. The fibres do not contract over their entire course at the same time but little by little through their parts, beginning from one extremity and continuing towards the other, which is also seen in the peristaltic movement of the intestines.

Other things could also be concluded from this. They will be presented in more detail if God lets me complete what I undertook on the heart in the past. Meanwhile, others will be able to experiment as much as it will be given to the researchers to progress in the way the animal moves and on the cause of the movement. The following facts furnished to me a very valuable argument for trampling the arrogance of human intellect: 1. After so many centuries of work and speculation the philosophers have so far discovered nothing about the true cause and mode of animal movement. 2. Those who pronounced their opinion on this matter with great authority have all sold us false theories instead of demonstrations. I am far from promising a full explanation of everything. But God must be thanked who at least made obvious to all the falsity of those who are in error. He of old sagaciously said: *It is virtue, to fly vice; and the highest wisdom, to have lived free from folly.*[1]

---

[1] Horace, *Epistles* I, 1.

2.11   Vesicles in the Lung. Anatomy of a Pregnant Rabbit. Experiments in Lungs. The Milky Vessels of the Udders. Observations in a Swan, Etc.

To Thomas Bartholin in Copenhagen

If ever any slowness in answering joined to quickness of writing, then certainly this one needs to be excused: although blessed during the same time with two letters, I have dispatched none for months and, as it is not permitted to delay further my duty of answering, an abundance of evidences has accumulated. But your well-known affability promises easy indulgence for both.

A long desired microscope offered by a friend finally provided me with a further opportunity of reflecting upon the observations of *Malpighi* concerning the parenchyma of the lungs. Having read your letters addressed either to Mr. *Borch* as well as to me, I did not doubt that various vesicles could be shown in the lungs, and not only in one way. In dried-up lungs indeed, if all blood is washed off as *Malpighi* does first before the lungs to be dried are exposed, it is beyond doubt that several cavities are to be seen where

OPH 11, vol. I, 131–136: "De vesiculis in pulmone. Anatome cuniculi praegnantis. In pulmonibus lacteis experimenta. In cygno observationes, &c." is found in Thomæ Bartholini Epist. Medicin. Cent. IV. *Hafniæ* 1667, pp. 348–359, numbered *Epist.* LV. Further annotations by Vilhelm Maar are found in the transcripts, www.extras.springer.com.

the lobes were dissected. Many small branches, of the vein as well as of the two arteries,[1] are indeed dissected together which are all provided with obvious cavities as is clear to everybody. If, before they are dried, the bronchi of the lungs are distended with air, how many small cavities and of what size is it possible to see in the dissected specimens? As countless extreme subdivisions of the trachea are certainly scattered throughout the substance of the lungs, air distending them violently, they display an appearance of countless vesicles if they are divided. But it thus does not seem necessary to state that the parenchyma of the lungs is made of pure vesicles. From this it does not result that the extremities of the trachea solely end in vesicles not to speak about the extremities of the vein, artery and lymphatics. How much did the microscope reveal to me in the dry lung of a dog, besides the round orifices of divided vessels and myriads of very minute filaments running without any law, absolutely nothing was seen, no round small cavities covered everywhere by a membrane, nothing at all which could be called a vesicle. Neither did the intact surface of a dried lung display any sign of vesicles under the microscope. There were protuberances but these were so large and unequal, not bulging at their middle, nor circular in their perimeter so that they do not at all deserve to be called vesicles. But I am far from withdrawing my trust in the experience of the most intelligent gentleman because I did not succeed in one or two attempts, perhaps because I did not enter the same path. But this will suffice on Malpighi until it will be possible to examine the *Letters* themselves.

When opening a rabbit these days, referring other things to a better opportunity, I observed the following points about the lungs of foetuses (with four of which the rabbit was pregnant). The fourth foetus having been left in the uterus inside its membranes, I liberated and pulled out the other three from the enveloping membranes. I ligated also the umbilical cord of one and, look here, after a short delay the small animals moved their mouth and chest and finally breathed in air with great force so that all could easily acknowledge full respiration. Then by trying the force of their legs for themselves, after some efforts they became so strong that one of them by crawling here and there went finally to rest below the legs of the mother lying on the table, being less twitched by the cold of air there. I retrieved the strong animal from its warm asylum and, after having cut open the chest of the animal unwilling as much as his forces permitted, I found the lungs white and already spongious, although the lungs of the other one dead inside its membranes in the uterus were dense, full of blood and completely different from those of the

---

[1]pulmonary artery and trachea.

former, wondering that the change was made in such a short time. Of older I observed the same concerning respiration in a pregnant dog, while not giving attention to the lungs at that time. The foetuses, although the time of birth was not yet reached, were cut out and, exposed to air, gave back the inspired air with repeated wailings. But these things are frequent. Rare is what the very intelligent young man and very diligent in anatomical exercises, Mr. *Swammerdam*, showed to me, propelling air in the heart by way of the lungs. When the animal is dead, with a bladder he continues a reciprocating movement of the lungs simulating respiration until finally bubbles descending into the heart through branches of the pulmonary vein demonstrate the success of the experiment. Also worth to be mentioned seems what I noticed about the movement of the vena cava mentioned by the same author and then mentioned by Mr. *Padbrugge*, a young man also dealing with anatomical subjects with great ardour, and which was observed for the first time by the very intelligent Mr. *van der Lahr*. They saw that, if the jugular vein is displayed in vivo, as well as the vena cava running in the abdomen through the middle of the back, every time the animal has distended its thorax in inspiring, the vena cava in the neck and in the abdomen is evacuated so that, whereas it was red when distended, it whitens instantaneously, the tunicas being compressed, to return immediately to its round shape and purple colour. Those who saw that hitherto in various ways presented various explanations. The phenomenon, it seems to me, could admit such explanation if one could invoke fear of vacuum. If indeed, as people say, the lung distended by dilatation of the thorax almost sucks and attracts air, one could say that, because of the same widening of the thorax, the portion of the vena cava which is inside the thorax, dilated, attracts that blood which is found in the branches of the vena cava as well above as below the thorax so that the vena cava outside the thorax is emptied whereas the vena cava which is inside the thorax is filled. But since fear of vacuum seems almost completely overcome and run away into vacuum, I am afraid, the opinion of those who are used to explain similar phenomena through pushing alone is apparent to me. According to these theories, the thorax while dilated by the action of muscles drives away anything around itself where the resistance is less, at one and the same time it propels air into the lungs, viscera of the abdomen downwards and the blood which is in the vena cava outside the thorax, into the part of the vena cava which is inside the thorax. Therefore the vena cava inside the thorax is so filled up because the vena cava which is outside the thorax sends its content there. That phenomenon could perhaps be explained in this way if further experience confirms it and altogether the farther function of respiration in the circulation of blood would be discovered. But where did the lungs, where did respiration lead me away? I must return to the other important matters in your very learned letter.

Many observations attempted to show that some milky vessels spread out from the insides to the udders but, as much as I can gather, none has yet demonstrated this. Schenck has certainly seen nothing else than a small vessel originating in the glands themselves of the udders and running from there to the nipple. He himself bears most evident testimony to it when he says that milk flows out through the nipple by pressure of this same small vessel. Neither should I easily believe, if some small vessels proceeded from the insides, that these are to be continued directly to the nipple since it would result that parenchyma has been given to these glands in vain and, that in dissections, udders distended by milk could not be separated without dissection of these vessels, therefore not without any outflow of milk. Of these two points, experience contradicts the latter and reasoning denies trust in the former. But if it must be decided that there are two kinds of milky vessels, one ascending from the insides to the glands, the other carrying to the nipples milk received from these same glands through their own roots, the same famous gentleman will not be able to demonstrate by his experiment that the milky vessel he saw was of the first kind which is, however, the only one which is in question, since the milky vessels of the second kind which excrete their content outside the body is not easily questioned by anybody. As far as the observations of others are concerned, these darken the matter rather than illuminating it, since, depending on the variety of the authors, the description of these ducts makes questionable the reliability of the experiment and what anybody has seen at first with the eyes of his mind, he would believe to have seen it from then with the eyes of his body. One indeed leads milky vessels from the thoracic duct through the muscles of the thorax to the udders, another one from the receptacle through the abdomen above the muscles of the abdomen to the same place, another one from the uterus to the same part through the muscles of the abdomen and, nevertheless, all present experiment as witness. But what if one in dogs, another one in rabbits, and the third in man has seen what he asserts? What, if in any of these a triple canal is to be found to the udders? These and similar suspicions would perhaps be admitted if various evidences presenting by far more likeliness of truth were not opposed to them. But there will be more on this matter, if it pleases God, in the answer to *Everaerd* because of whom, to observe the same things, I did not spare expenses more than labour, bringing all variety of rabbits to dissection. But the more I try in this instance, the more the truth of your lymphatics shines forth.

What I wrote on the gallbladder observed in a raven was perhaps a little too obscure. I indeed said that it did not lie on the liver and, nevertheless, I said that it was attached to the liver. But, I thought these two were different things, because, in the other animals which it happened to me to open, the gallbladder was attached to the liver not only by way of some fibrils but it was lying on the liver with about its half, attached to the surface over its length by a small skin which they had in common. This, however, adhering to the liver with one branch only, free everywhere for the remainder of its body, showed itself to me like an oblong little bag. But neither does the observation of *M.A. Severini* on the liver adjacent to the heart seem to take place in all ravens, since in this one, although it was small and still without feathers at many places, the interval between the right auricle and the liver was of some fingerbreadths. At intervals various experiments presented themselves to me while doing or looking for something else. As well some other reasons provided me with an opportunity of examining the movement of the heart, which I remembered then. To that end I searched more carefully into the fibres of the heart. On these, with the help of divine grace, I will publish soon a small treatise with, joined to it, a reply to the statements of *Everaerd* and the abuses of *Deusing* and *Blaes*. Laying off your friendly intervention, the latter published at Christmas time the letters which he had sent to you more or less 2 years ago. In these, not to mention the very bitter things which he spewed out against me, he commits more than six anatomical errors concerning the description of the duct itself and the interpretation of *Ideae* and *Med. Gen.*, inasmuch indeed as not from my description he could gain knowledge of the true origin of this duct or of its orifice.

There is no doubt that an excessive abundance of bile does much for frequent evacuations of the bowels. It is less certain whether a deeper position of the gallbladder can contribute in anything to the evacuation. However this be, I always observed about the same position of the gallbladder, the same number of biliary as well as of pancreatic ducts, not only in this one but also in other birds.

I think that you have already heard from friends what was observed in a mouse, a dormouse, a cat and in various uteri of cows, such as on the insipid bile as well as on the uterine hernia found in the two groins of a dog.

The swan which I opened these days displayed various things. Above all I wondered that on each side of the tongue more or less 13 orifices were counted situated on the same straight line. Each of them, when compressed, emitted juice and a probe could be introduced fairly deeply in them. 1. All the gullet and also the gullet above the oesophagus showed countless small orifices from which juice was expressed without any difficulty, white in the gullet, aqueous in the oesophagus. 2. Fleshy fibres running from the extremities of the ribs into the membranes of the lungs make up perhaps for the absence of diaphragm. I would add the other things if I did not think that they have been made known by the friends who were present with me.

These days I discovered in a human cadaver the errors of *Blaes* concerning the superior salivary duct but also I noticed clearly in the eyes of the same human body the vessels which I observed in the eyes of animals. Some of them led from the external gland of the eye into the internal tunic of the upper eyelid, others from the lachrymal point into the nose. I could however not observe anything on the said lachrymal gland.

A head of horse, as well as various fairly excellent things, displayed in the brain also something which I had not observed in others, such as, in the noble[2] ventricle, a plexus similar to the choroid; in this as well as in the choroid itself, was present over the vessels, a substance very similar to small glands. The size of the pineal gland was conspicuous enough but its colour was blackish externally, internally grey marked by many dark spots so that every one may say that black bile does not leave intact even the seat of the soul. But this is perhaps normal in horses. I indeed found the same also in another horse head which I dissected since then. Certainly, the more I open brains, either of other animals or of birds of various kinds, the less the structure of the brain of animals thought out by the noble *Descartes*, most ingenious and very appropriate otherwise to explain animal actions, seems to fit animals. This would be easy to show by the presentation of other things observed in the brain, if these observations, as they were done mostly in the afternoon, not rarely at night, had not in themselves hitherto much of the night; if the same observations, interrupted because of ordinary occupations did not show many gaps in many places; if the favour of Pluto denied to me had not acted so far so that, out of those many things which I see to be done every day to this end, I shall have carried out very few. But these must be taken care of until leisure and means have led the yet bitter fruits to some maturity. Do well, famous Sir, and continue to favour,

Your devoted disciple
Niels Stensen

Hastily
Leiden, March 5, 1663.

---

[2]fourth ventricle.

2.12 Diversity of Lymphatic Ducts

The diversity observed in dogs near the junction of the lymphatic ducts with the vena cava on the left side of the neck is found either in the branches of insertions, I, which are many here, few there, or in the small rings, K, which are completely absent in some dogs, are present in a certain number in most, are fewer in others still, sometimes narrow, sometimes fairly wide, as it appears in the figures.

Figure 1 is due to the diligence of my best friend Mr. *Swammerdam*. I have prepared Figs. 2 and 3 which are a little different as far as the insertion is concerned. I found it superfluous to give particular explanations, since one same common explanation is enough.

Of vena cava,

A. *Trunk above the heart*
B. *Jugular branch*
C. *Axillary branch*
D. *Branch directed to the posterior aspect of the neck*

OPH 12, vol. I, 139–142: The investigations on which "Lymphaticorum varietas" is based were most likely made by Steno in collaboration with Swammerdam, some time in 1662 and 1663; one of the drawings is likewise due to *Swammerdam*. Steno had planned an exhaustive work on the lymphatic system, but never realized his plan. The text was, at any rate as far as the latter part is concerned, no doubt composed long after the time when the investigations were made, probably during Steno's stay in Copenhagen in 1672–1674. The "Treatise" appeared in "Thomæ Bartholini Acta Medica et Philosophica Hafniensia. vol. II" *Hafniæ* 1675 as No. XCVII, pp. 240–241. Further annotations by Vilhelm Maar are found in the transcripts, www.extras.springer.com.

© Springer-Verlag GmbH Germany, part of Springer Nature 2018
T. Kardel and P. Maquet, *Nicolaus Steno*,
https://doi.org/10.1007/978-3-662-55047-2_23

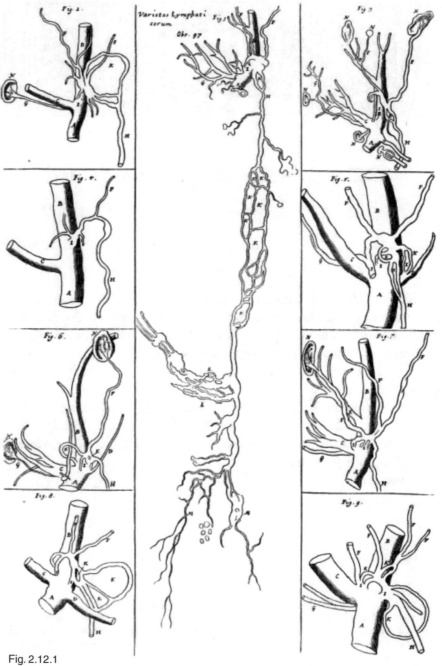

Fig. 2.12.1

E. *Branch directed to the inside of the neck*
F. *Lymphatic ducts of the neck*
G. *Lymphatic ducts of the left foreleg*
H. *Common trunk of the lymphatic ducts below the neck, called thoracic*
I. *Branches by which the lymphatic ducts continue into the vena cava*
K. *Rings of lymphatic ducts through which most often either vessels or nerves pass*
L. *Lymphatic ducts from the intestines and liver*
M. *Lymphatic ducts from the loins and hind legs*
N. *Conglobate glands*

Since vessels of the same kind admit such an important diversity in various individuals of the same species of animals, it easily appears that God, the creator of all things, among other matters to be acknowledged from the body of animals, wanted also to display to us these two attributes of divinity: he is neither carried away by chance since he does not move away from the regularity of a general conformation nor forced by any necessity since he arbitrarily changes particular conditions in individuals (Figure 2.12.1).

XIII

2.13   Prologue of an Apologia in Which It Is Demonstrated that the Blaesian Judge Is Ignorant in Anatomy and a Slave of His Sympathies

A book was published these days in Utrecht by *Nicolas Hoboken* entitled *The new salivary duct of Blaes drawn into the light*. If you consult the Preface, this promises to display in this book *matters which can enlighten the honest reader and make him know the first discovery not so long ago of a salivary duct out of the maxilla,* and this *so that he can judge from these matters by what right and to what purpose* Niels Stensen *recently attempted by all means to claim this*

From απολογια, speaking in defense through the systematic use of reason. (Wikipedia).

OPH 13, vol. I, 145–154: "Apologiae prodromus, qvo demonstratur, judicem blasianum & rei anatomicae imperitum esse, & affectuum suorum servum" is the last of Steno's writings relative to his controversy with Blaes about the priority of the discovery of the parotid duct. This pamphlet was caused by a little book, with the title of *Novus ductus salivalis Blasianus, in lucem protractus A.M. Nicolao Hoboken, Ultrajectino, Philosophæ & Medicinæ Doctore ac Practico,* Ultrajecti 1662, which Nicolaas Hoboken (1632–1678), a young friend of Blaes, published in his defence. In his pamphlet *Hoboken* stood forth as a judge in the quarrel between the two men. The judgement he pronounced was entirely in favour of *Blaes*; a decision at which he arrived all the more easily, as he did not enter into any investigations of his own in the matter in question. Hoboken's pamphlet consists of a preface, written by himself, an anagram on Steno's name (see below); a letter from Hoboken to Blaes; a letter from Blaes to Hoboken; three testimonials from former pupils of Blaes; a letter from Blaes to Th. Bartholin, dated Amsterdam July 16, 1661 (reprinted in "Thomæ Bartholini Epist. Medicin. Cent. III." *Hafniæ* 1667. Epist. XLIII, pp. 158–184); and last a letter from Hoboken to Blaes. Further annotations by Vilhelm Maar are found in the transcripts, www.extras.springer.com.

© Springer-Verlag GmbH Germany, part of Springer Nature 2018
T. Kardel and P. Maquet, *Nicolaus Steno,*
https://doi.org/10.1007/978-3-662-55047-2_24

discovery for himself and to deny it to Blaes. If you read this book you see that it has almost the shape of a trial in which *Blaes* is the prosecutor, I am the prosecuted and *Hoboken* is the judge. Indeed it was not enough for this *Hoboken* to have published a letter of *Blaes*. Healsodrew the censorial rod towards a man whom he does not know and hastened to cast down a judgement on a matter which is known to him as well as to *Blaes*.

Blaes (I am to demonstrate this in the "Apologia" itself with the words of Blaes), accuses me of *deceit, ingratitude, bad manners, injustice, blundering, foolishness, perfidy, incivility, inveracity, treachery, calumny, scoffing, malice, arrogance, perversity, audacity, shamelessness, impudence, fatuity and depravity.* What a list of titles! Those are the honourable words of Blaes! Who would have expected these words from a professor, or from any respectable gentleman, in a lawsuit not yet tried?

The words to which the mind of our judge *clang condemn me as being pushed by the stimuli of presumption, as being unmindful of myself, as infringing and violating the rules of obligation, the kindness and respect due to a friend, a host and even a more than very friendly teacher. They blame my efforts such as those who, even if they are very brilliant in the beginning, etc., in the end after the disguise and the mask have been removed, usually produce in an author remorse and shame for an insult, if he may be so frank.* To rise as high as he can, he publicly claims in an anagram *that I am a* usurpator and a traitor. Please recognize hence how free of any bias is the mind of the judge.

If you want to know the state of the controversy from which all this proceeds, it is as follows: *Had Blaes in private teaching or at another occasion shown to me the salivary duct which I have described in the* Anatomical observations, *or any part of it? Blaes* asserts this and *Hoboken*, accepting this simple assertion as a demonstration, pronounces his sentence. Thence this forbears another question: *Did Blaes know anything about this duct before I showed it to him?* Indeed *had he any distinct knowledge of it before my theses appeared?* As I said in *Anatomical observations on glands of the mouth*, thesis 13, it seems to me that, when I lived in his house, I showed a duct to Mr. *Blaes* who did not indicate what it was. Seven months later I asked my illustrious teachers in Leiden whether they had observed it before or had seen it described by others. The matter having been examined and even publicly disclosed in man, the illustrious Mr. *Sylvius* for the first time in the amphitheatre adjacent to the hospital and then again the illustrious Mr. *van Horne* in the great amphitheatre demonstrated it, at the same time commending my name, which was very kind of them. Hearing that, *Blaes* about one whole year after the discovery points out to my illustrious teachers thus mentioned that *he* had discovered this duct first and had shown it to me.

Soon he publishes the *Concepts of Medicina Generalis* in the preface of which he claims that this duct was discovered by himself. *Blaes* thus calculated for me who was planning a travel a future opportunity to investigate these matters. After I left Leiden, he mentioned in his teaching his duct as if he had discovered it (for the description of a *Blaes* duct, which is an imaginary salivary duct, will demonstrate that he never made any real demonstration. Indeed, the dissection which he carried out recently in Muiden, during which he looked for it in vain, as I hear, proves this). Hence when he heard that this was done in Leiden he had either to reveal himself or to accuse me. Shame prohibited him the former. Conscience protested against the latter. Finally shame prevailed but with what happy result the outcome will show. Myself, expecting nothing less than that from *Blaes*, even if other matters also presented themselves by which I could have demonstrated the opposite, I used this only as unique argument against him: If a whole year after the first discovery, in the book which he dedicated to the Great Consuls and Ordinary Treasurers of the municipality of Amsterdam, in which he publicly claims to be the first discoverer of this duct, in which he acknowledges not to present anything new besides this, and in which he describes the origin and end of the duct of *Wharton*, which he claims for himself; he describes neither the true origin nor outlet of the duct; hence he attributes to the gland from which this duct originates another obscure function; then it cannot be said that he knows the origin or the outlet of this duct nor that he has described this duct to somebody else and even less that he has discovered it. And yet *Blaes*, etc. Thus I undertook to prove this argument in my theses, leaving out all the other matters which occurred between *Blaes* and myself and which I could have presented in my favour. This is my malice, this is the calumny, this is the dishonesty, and this is the treachery. It is on this ground that I am accused by *Blaes* and condemned by *Hoboken*. If *Blaes* is a man, let him prove with any true argument that I have either done or said anything against him which I am not ready to defend in front of a legal magistrate. If the matters which I bring forwards seem perhaps too harsh to anybody, this must be assigned to the one who commits them, not to the one who reports them. If a mind conscious of the morally right would not find it dishonest to be disturbed by sayings devoid of reason, I should retort in the same way by right of retaliation. But if I did that, the readers would believe that they are among old women brawling on the market place. I will save up all my vindication for others, i.e. the legal magistrate. It will be sufficient here to present some points for my defence, nothing to my vengeance.

If our judge had been chosen by common agreement of both or at least worth the position which *Blaes* takes in Amsterdam, I should show here to him how extraordinary were the services which *Blaes* was not ashamed to blame me for, what *Bartholin* asked from *Blaes* in the first recommendation letter in my name; in what ways *Blaes* made himself my host, in what matter he had a disciple, with what skill he did that, so that under him, besides common and poor chemical operations, I learnt nothing. (Indeed, during more than 3 years, the illustrious gentlemen *Mr. Simon Paulli*, Mr. *Th. Bartholin*, Mr. *Bachmeister* and others, either candidates in medicine or students, had already shown me these matters in anatomy, not to mention the others I looked for in vain from *Blaes*). Meanwhile I should lose the opportunity of studying several matters in Leiden, not only in medicine but also in other arts necessary to a physician. From the end of the winter until the summer holidays, i.e. the best parts of a whole year, it would have been possible to learn from gentlemen famous all over the world for their experience and science. I should demonstrate all that, I say, and its details more than clearly to our judge, and altogether I should explain, after I had examined more thoroughly these artifices of *Blaes*, with what patience I have supported everything, that I always spoke friendlily of *Blaes* even when he started to undertake these things against me, all that by regard for his friends.

Hence I should continue, and 1. I should demonstrate that the duct which I described was not known to *Blaes* before 3 years counting from 29 June 1662, neither when I arrived in Amsterdam nor when he published his *Medicina Generalis*, nor when he wrote a letter to Mr. *Bartholin*, nor when he transmitted this letter to *Hoboken* last year. 2. I should make appear more clearly from this letter of *Blaes* that he does not understand either what *Wharton* says of the submandibular and parotid glands, or what *Blaes* himself wrote in summary about these glands, taken from Wharton, in his *Medicina Generalis*, nor how important is the distinction of the glands given by *Sylvius*, or what I presented about the salivary glands, or what anatomy itself teaches about them.

But, since the professorial dignity of *Blaes* requires another judge, since deferring the cause to the tribunal of him who has no expertise of the matter which is concerned and does not know how to refrain his feelings seems to oppose the laws of equity, I will hold back all these matters for a more appropriate place and time.

But returning to our judge, though I was dealt with by him unfriendlily and dishonestly, I will accuse him of nothing else than credulity since I think he was convinced not to have to expect anything wrong from a friend and that mistakes in anatomy could not be presented with such self-confidence by an anatomist. The precipitation of the sentence, whereas examination of the matter was omitted, is actually the consequence of his credulity or rather of the blind love of a friend.

If indeed he had examined the words of both and had appealed to an autopsy to account for the parts, he would not have said *that I had attributed to myself the* Blaes *duct, he would not have promised the readers to ascertain the discovery of the external maxillary duct from that letter of* Blaes, but he would rather *have observed the anatomical mistakes of* Blaes *and would have discovered the most obvious evidences of the absence of sincerity of* Blaes.

Imaginary Salivary Duct, or

the True Blaes Duct

As It Appears From the Writings of Blaes.

It originates from the internal submandibular glands (*Medicina Generalis* p. 63) or from that gland from which the lower one originates but not precisely from one point (*Duct. Blaes.* p. 31). By these glands he does not mean the external submandibular (*D.B.* p. 34) nor what he described under the name of parotids in *Medicina Generalis* (*D.B.* p. 39), but those which present from the parotid glands to the chin with variable magnitude and colour. (*M.G.* p. 23), or which are one and the same with the internal submandibular and cannot be distinguished from it unless we wish to divide one into two or three parts (*D.B.* p. 32). It is not above the mandible but it lies beneath it like, or even more than, the internal submandibular glands (*D.B.* p. 40) parts of which surround the ear (*ib.*). It fills the fairly deep cavity present beneath the ear (*D.B.* p. 52).

It carries away the saliva to the anterior glands of the mouth so that the saliva can be expressed hence with the movement of the tongue (*M.G.* p. 64, *D.B.* p. 45). There is not only one large opening provided with a papillary excrescence, but there are also others less conspicuous in which divisions from the same duct open (*D.B.* pp. 37, 48). This opening, at least it is supposed, is further away from the front teeth than the opening of the other duct (*D.B.* p. 42). It is suspected that the saliva is carried through the lymphatic vessels even from the *Blaes* parotid towards these anterior glands (*D.B.* p. 50). In the disputation which took place on July 8, 1661, which *D.B.* mentions on p. 36, he says: the inferior runs to the extremity of the mandible, around the lower aspect of the chin, to the side of the frenulum linguae; the superior runs to the side of the maxilla without determining whether it is inside the teeth or outside, whether in the cheeks or elsewhere.

In a letter to Mr. *Bartholin* written a month later, on p. 23, he actually mentions an exit to the cheeks.

As you may know, when he held the discussion, he had not seen my theses. He had read them before writing the letter as he himself acknowledges on p. 37 when he says: *This disputation took place on the same day*, i.e. July 8, *as the correction of Mr. Stensen, since I had not seen the Stensen discussion so that this discussion should perhaps not be considered as taken from mine.* A truly extraordinary admonition uncovering very neatly the tricks of *Blaes* by the acknowledgement of *Blaes* himself. Indeed, he says that a discussion took place during which he could not determine the issue, before he had seen my writings. He learned the following two points from my theses and figures: 1. the gland from which the duct originates fills the pit beneath the ear; 2. the duct, which he had said in his discussion to run to the side of the maxilla without determining it completely, ends in the cheeks. Sure, Mr. *Blaes*, p. 35, works to no purpose when he says that *what he brought to the discussion was not taken from the discussion of Steno but actually was observed from dissections of animals.*

I should indeed be sorry if he had acquired from my theses such gross errors, as he put there. Competent people will judge whether he could have drawn them from dissections of animals.

This is the origin; this is the end of *Blaes* duct. Do not question the truth of the observation. It was the third year after the discovery when *Blaes* dispatched his letter to the tribunal of *Hoboken* showing to him that he still maintained the opinion which he favoured when he sent the letter to *Bartholin*. In this letter, written one and a half years after the discovery, he says on p. 23 that *the matter is easily investigated: because they are easily found, what is said above as what is said below: they are short, dense and less apt to break. Soon he adds about me: if the matter is easy for the one who never saw it, why is it difficult for the one who frequently showed it to others;* then on p. 35 he says: *I undertook an investigation, even as often as possible, and I will say very truly that it is easy to find that either salivary duct takes its origin from the submandibular glands. Busy all the time with anatomical dissections in anatomical lectures, how could I have been so carefree as not to investigate the matters which I was the first among all to point out to the world, and not to have defended my name, above all since many people would attempt at taking that from me, whom they see becoming more famous with the help of God?*

Since it is an easy matter, time is long and the opportunity frequent and the causes are urging, we must believe that this duct of *Blaes* is a reality.

Let us see whether this is the one which I described.

Blaes' Duct	What I have observed
It originates from the internal maxillary.	It does not originate from the internal maxillary.
Or from the same gland, from which the lower originates.	Not from the same gland, from which the lower originates.
Not from those which *Blaes* in *Medicina Generalis*. calls parotid. But from those which, from the parotids to the chin, present with a varying kind of colour and magnitude.	But from those which *Blaes* in Medicina Generalis calls parotid. Not from those which, from the parotids to the chin, present with a varying kind of colour and magnitude.
This gland both *fills the rather deep cavity under the ear* and yet cannot be distinguished from the internal maxillary unless we will cut the one thing into two or three parts.	From the gland which *fills the rather deep cavity under the ear*, but very easily can be distinguished from the internal maxillary because they are two distinct glands, by the membrane between them, and also by the thickness, rather clearly separated from each other.
It does not lie at the cheeks but lies under them, more than the internal one under the maxillaries.	It lies at the cheeks and not under them.
It proceeds to the side of the upper cheek and ends there (because he says: the lower proceeds to the side of the frenulum linguae, the upper onto the side of the upper cheek).	It does not proceed to the side of the upper cheek, be it in man or in animals, because it finds its exit in the mouth.
It leads saliva to the foremost glands of the mouth to be expressed from there by the motion of the tongue.	It does not lead saliva to any glands of the mouth but secretes it immediately through its own outlet.
And there it does not have only one *large opening* surrounded by an outgrowth of papillaries, but also other more inconspicuous ones.	It has only one *large opening* surrounded by an outgrowth of papillaries in some, in many humans in the middle of the plane mouth without anything protruding; but apart from this no other openings.
Its outlet is in the cheeks.	*Its outlet is in the cheeks.*

Who does not recognize clearly from this with what right *Blaes* claims that I attribute to myself a duct which he discovered, examined so often and for so many urging causes? With what right did he accuse me in the presence of very famous professors of Leiden? With what right does he attempt to expose me to public ignominy? But who does not see from that that our judge has hurried up the sentence without examining the cause, since he said that I have attributed to myself the works of *Blaes*, and then dissipated himself in outcries? *The wasp of Aesop in Phaedra* was more cautious, who before pronouncing a sentence between bees and drones judged that the author had to be recognized *from the taste of the honey and the shape of the honey-comb*.

But let us see the certitude of the discovery which our judge promises us. Although it is very true that this *Blaes* duct is *Blaes'* invention and, before it originated from the mind of *Blaes*, was observed by nobody, since, however, it is not sufficient to know the truth, unless the causes from which we are induced to admit this truth are also known, let us see what were the causes which impelled *Hoboken* not only to believe himself that the *Blaes* duct was discovered by *Blaes* but also to attempt to convince others.

Certainly, by turning over the letter, you see that *Blaes* claims that he has made the discovery and has shown it to me, and that altogether by the testimonies of disciples written two and even 3 years after the lectures took place he proves that he has shown it to them. Since *Blaes* claims that he has made the discovery, since he proves by some written testimonies of his disciples that he has shown it for such a long time, we are thus sure that *Blaes* has made the discovery and we are thus sure that *Blaes* has shown it to me. It is as if these were mathematical demonstrations to which there is no reply. I will not explain here how they unfold their testimonies since I will do that in the answer itself. Only, to make it clear to everybody for how much his simple assertion is to be held I will present other demonstrations taken from the same letter of *Blaes*.

If, to defend his name, above all in the matter which he was the first to discover for the world, one must believe *Blaes* as an anatomist, as a respectable and truthful gentleman, when he says that the matter not only is trivial but has been often examined by him, then one must also believe *Blaes* when he explains to himself what he observed concerning his own discovery examined so often by himself because of urging causes. But one must not believe *Blaes* when he explains what he observed by himself concerning his own discovery examined so often by himself because of urging matters. Let thus the reader himself conclude with what name *Blaes* must be greeted. But that one must not believe in the observations of *Blaes*, all will admit who judge that what obviously opposes experience must be counted openly among false things. What is taken from the letter of *Blaes* himself is of this kind.

Anatomical Mistakes Committed by *Blaes*

in the Description of the *Blaes* Duct

P. 31 and 32. The conglomerate parotid glands cannot be distinguished from the internal submandibular unless one can divide them into two or three parts. Similarly the superior salivary duct as well as the inferior takes its origin from the internal submandibular glands, although not from one precise point.
P. 32. The internal submandibular, or a part of it, is what I call the conglomerate parotid gland.
P. 37 and 44. The external duct, besides one conspicuous opening at its extremity, has more other inconspicuous ones. They end extensions of the same duct.
P. 39. What he describes in *Medicina Generalis* as parotids are not those which *Wharton* described under the name parotids and which I called conglomerate parotids.

Similarly, what he describes in *Medicina Generalis* as parotids are the parotids which I call conglobate parotids.

P. 40. Some small parts of the conglobate parotid, even if you consider its inferior part, surround the ear.
Similarly, the conglomerate parotids are as subjacent as, or even more subjacent to the mandible than the internal submandibular glands.
P. 43 and 45. There are glands to which an external duct carries saliva which has to be expressed from there with the movement of the tongue.
P. 50. Some juice is sent from the conglobate parotids to the anterior glands more than to those to which a salivary duct extends.
P. 52. The parotids of *Blaes*, if they are conglobate, as he himself calls them above, have the same substance as the submandibular glands.

The falsity of all these points has been made known to friends, some of them also to the public, and I take on myself to demonstrate that so that no doubt remains to anybody. Let only *Blaes* come forward and demonstrate these points on one or the other side of a head either of a man or of an animal in the presence of judges fitting the dignity of a professor from Amsterdam, on the opposite side I will demonstrate the contrary.
This is the first example of *Blaes'* sincerity; listen then to another.
He says in *Blaes Duct* on p. 48 that his demonstration of the salivary duct before my arrival can be confirmed under oath. Thus it will be confirmed under oath that either its origin was demonstrated in the submandibular glands or its outlet in the anterior glands of the mouth: and there is neither such origin nor exit. Thus it will be confirmed under oath that *Blaes* demonstrated something which has been seen neither by *Blaes* nor by anybody else, since it does not exist. You who did not know *Blaes* hitherto, from

these two arguments, learn of which credibility this gentleman is, until you will have appraised his forlorn character in his apologia. I say forlorn. I will not indeed make it odious by any bitterness of epithets. I content myself with writing down the history alone of the facts so that it appears to everybody how worth are for me the lustre of the Athenaeum of Amsterdam, the dignity of a professorial person, and above all my love for modesty, so that, although so much burdened by him with reproaches, I will not return even one.

Since our judge can neither acknowledge the examples of the absence of sincerity of *Blaes*, nor notice his anatomical mistakes, nor see the difference between *Blaes'* duct and mine, it is clear that he attributed too much to his friend as being established in a public dignity and presenting his observations with so much trust, and that he did that deceived by his credulity so that if he brought to light a true *Blaes* salivary duct, i.e. an actually imaginary salivary duct, he would altogether disclose to the eyes of everybody that *he is ignorant in anatomy and that he is the slave of his sympathies.*

Let this suffice to demonstrate the precipitation of the sentence pronounced by our judge, based on prejudices until I shall have the leisure to present to the eyes of everybody the matters which *Blaes* and *Deusing* flung at me, examined with their own arguments, so that it appears more clearly how easily reasoning is carried away from its tracks and deviated where it gives way to sympathies.

Leiden, 1663

2.14 New Structure of Muscles and Heart

To Thomas Bartholin in Copenhagen.

I sent to *Niels Krag* the Prologue of the *Apologia*[1] to be presented to you. I hope this was taken care of properly. I received some copies of the *debates* published again by the famous *Sylvius* and of other publications on fevers, to be offered to friends. Among these friends there is none who deserves more my veneration than you, illustrious Sir. I trusted the documents to a friend who is going to Copenhagen these days to deliver them to you. For the rest, I am busy with a thorough examination of the heart and muscles, hoping, if the events leave me enough time for this work, to complete soon the structure of both with figures. Whether that which I observed in the muscles was noticed by others, your boundless reading will teach me. I will not mention the vessels which did not yet display much and this of small importance. However, as far as their fibres are concerned, I cannot admire their delicate structure enough. Anyone who studies any single fibre will see that in the middle part it is fleshy and at both extremities it is tendinous, a fact which is well known. But I have more rarely seen described the composition of the junctions. Thus, the fleshy portion does not extend in a straight line from one extremity of the muscle to

---

OPH 14, vol. I, 155–160: "Nova musculorum & cordis fabrica" is found in Thomæ Bartholini Epist. Medicin. Cent. IV. Epist. *Hafniae* 1667, pp. 414–421, numbered Epist. LXX. Further annotations by Vilhelm Maar are found in the transcripts, www.extras.springer.com. Facsimile of first edition and English translation in *Steno on muscles*, 58–75. Illustrations from REX.

---

[1]OPH 2.13, p. <533>.

© Springer-Verlag GmbH Germany, part of Springer Nature 2018
T. Kardel and P. Maquet, *Nicolaus Steno*,
https://doi.org/10.1007/978-3-662-55047-2_25

the other but it traverses the muscle between the broad expansions in such a way that the single fleshy parts run parallel. When making a section from one extremity of the muscle to the other along the direction of the fibres, this structure is clearly seen (Fig. 2.14.1):

A B. *The tendon extending over one surface of the muscle*
C D. *The other tendon extending over the opposite surface*
E E. *The fleshy part of the fibres between the two tendons*

Sometimes the structure of a whole and simple muscle is as shown here (Fig. 2.14.2):

G H. *One tendon extension*
I K. *The opposite tendon extension*
L L. *The fleshy belly*

There are also compound muscles, and not only of one kind. The ones which seem most elegant to me are those in which the fibres are disposed in the following manner (Fig. 2.14.3):

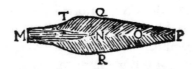

P. *One of the tendons which is split into two expansions, the upper one extends towards Q, the other, the lower one, extends towards R*
M. *The other tendon which distributes its fibres through the middle of the belly*
From N to D and farther between the middle and the two opposite external expansions the parts of the fleshy fibres are parallel

The membrane peculiar to every muscle must not be overlooked. The transverse orientation of its fibres and their distribution between the fleshy fibres seem to contribute not a little in the explanation of the movement. But about this and about the particular structure of some muscles more noteworthy, more will be said later on. But I wished to point out these matters to you incidently so that, if you observe something about them either by dissecting or by reading, it would be possible to the one delighted by your information on one hand to save work and costs and on the other hand, to orientate the explanation of the description according to your precepts.

As far as the substance of the heart is concerned, I will, so I believe, make it obviously demonstrated that nothing is found in the heart which is not found in a muscle and nothing is absent in the heart which is found in a muscle, if you consider what makes the essence of a muscle. This will appear obvious for the auricles as well. Hence it is obvious that there are no straight or circular or, as described by *Vesalius*, oblique fibres in the heart. All are of the same kind. They are simple such as those which are found in any other muscle. In the heart, however, they are different from some of the others only in that they do not run straight but, where they go down obliquely from the base towards the apex, they return upwards. I will demonstrate at autopsy, I hope, both extremities of all the fibres, together with the function of the eminences or muscles, the origin of the valvulas, their number and movement, and much of what concerns the vessels. Your opponents who reckon that the figures published in your *Anat. Reformat.* (edition of Leiden, 1651) on p. 245, showing us the systole and diastole of the heart are at the very opposite of obvious reasoning as of the diameter, will see from this that they have claimed a triumph before victory since the action resulting from the structure of the fibres is described most obviously therein. But the less accurate opinion on the parenchyma of the heart according to which people believed that the heart behaves like a bladder, thus that it dilates when it shortens and swells, and that it becomes narrower when it lengthens and collapses, misled them. The structure itself of the fibres proves what you demonstrate there with ligature, dissection and touching, and thus can explain almost everything which you present on p. 249 about the shape of the heart during its movements. However, it will also be obvious that the conus is not the densest, but its apex where the fibres return upwards, in the heart of an ox, is thinner than the tip of a needle. It was not yet possible to see whether the septum affords a passage. Following the small cavities, I found them closed. The fact that also the external walls have small cavities not smaller than the others as also that countless passage of fibres through the septum, seems to argue against it.

As you see, this is hardly sufficient for somebody who is quickening his writing as there is your letter to me to which I must answer, even if it is little, before I remove my hand from the table.

I had believed that the movement of the vena cava corresponding to inspiration was discovered by Mr *van der Lahr* but after that I learned that this discovery was due to three gentlemen, *Padbrugge, van der Lahr* and *Becker*, who attended together the dissection organized at that time. For what concerns the compression of the depressed adjacent parts during inspiration, this idea cannot find any place here since, in the neck and in the abdomen, the vein freed of the vicinity of all the otherwise incumbent parts is nevertheless evacuated so that its sides touch each other. Nor does the objection which you advance against compression appear to be compelling, since it is not required for the air to spread through the membranes of the vein in the thorax. It is only required for the blood present in veins outside the thorax to enter the dilated vein inside the thorax. The following facts actually seem almost to compel me to favour compression. I think it is beyond any controversy that when air is acted upon by force from something which pushes it, it is either condensed by this compressing object in the place where it is, or it is propelled to another place. Mechanics, however, teaches that a great force is necessary to condense it and experience shows that condensation never occurs unless all the ambient bodies are strong enough to resist the compressed air. If even a small part of the surroundings is unable to resist, air propelled by a pushing cause finds there a cleft through which it escapes. Thus if, while the ribs moving away from each other and the diaphragm moving downwards exert a force on air, these parts contained in the thorax yield to air by as much as the external surface compresses air, I should judge necessary that air does not condense but that it pushes where it finds a smaller resistance. Those parts which yield are those which are dilated by the distension of the thorax such as the lungs and the trunk of the vena cava present inside the thorax. Thus air fills the distracted lungs or rather air displaces the parts which do not resist. The blood which is outside the thorax, to distend the yielding membrane of the vena cava inside the thorax is thus propelled to the place which does not resist it. If my reasoning misled me here and led me away from the truth, I should wish to be returned to the right path by your explanations.

I am amazed that Mr *Hoboken* sent you his letter to defend *Blaes*. He would certainly have deliberated with himself and with his *Blaes* if he had been opposed to the printing. As actually *Blaes* and he himself demonstrate, those who have the least experience are the most presumptuous. I do not think that as many mistakes can be found in many big volumes as *Blaes* committed in the defence of what he found himself, errors which *Hoboken* endorsed by his approval. I should pity *Blaes* who could not refrain from talking and then from writing. By his writings he publicly displayed a character which does not fit a respectable gentleman.

As far as horses and mice are concerned, it was not possible to open either the skulls of the former or the uteri of the latter.

Thank you very much for the promised *Diatribe*. The famous *Borch* greets you. *Deusing*, as of old threatens the famous *Sylvius* of the worst things. But this most moderate gentleman smiles contemptuously at these fulminations. I cannot add more. Fare well and love,

Your disciple,

Leiden, April 30, 1663 Niels Stensen

2.15 Specimen of Observation on Muscles and Glands

To the most serene and powerful
Prince and Lord
FREDERICK THE THIRD King of Denmark, Norway, of the Wends
and the Goths,
Duke of Schlesvig–Holstein, Stormarn and Ditmarsk,
Count in Oldenburg and Delmenhorst,
My Very Clement Lord,
I very humbly offer what I happened to see in anatomy so far including
these few matters.
The very devoted undersigned,
NIELS STENSEN

2.15.1 The Muscles

At the beginning of last year, when, so as to satisfy the wish of illustrious friends I demonstrated in different private dissections in Leiden and Amsterdam what anatomy, thanks to divine favour, had disclosed to me

---

OPH 15, vol. I, 163–192: "De musculis & glandulis observationum specimen", *Hafniæ* 1664, appeared in the same year in Amsterdam and was reprinted in Leiden in 1683. The investigations, on which the first treatise is based, were made in Holland in the immediately preceding years (1661–1663). Further annotations by Vilhelm Maar are found in the transcripts, www.extras.springer.com.

© Springer-Verlag GmbH Germany, part of Springer Nature 2018
T. Kardel and P. Maquet, *Nicolaus Steno*,
https://doi.org/10.1007/978-3-662-55047-2_26

concerning the structure of the heart, I promised these friends to publish at the first opportunity the details illustrated by different figures. To carry out this the more happily, since some of them generously offered me their assistance, I wished also, as much as I could, to direct all endeavours to it so as to submit within the shortest time possible to the examination of everybody the same matter more developed, the foremost parts of which I noticed had pleased so important gentlemen. But despite all my efforts and although I can never praise enough the kindness which they bestowed on me and the help which I as a guest was offered in their hospitals, I was not so lucky to achieve my goals, partly because different obstacles diverted me from that at the most convenient time for dissection, but first and foremost because the research itself unexpectedly entailed the investigation of more matters. It is proper to scientific research that one cannot keep the various areas so isolated from each other as not to be obliged to bring many of them into one's consideration at the same time. The longer you stick to the details the more elements you lack in the whole and the more abundant material to investigate in the whole you stumble against (Fig. 2.15.1).

The first dissection of the heart revealed the following to me: 1. In the heart there is no parenchyma different from its fibres; 2. No fibre terminates in the cone but the different fibres twist around the cone and ascend at the side opposite to the one where they descended. The course of the fibres which I noticed to be the same in all, neither straight nor circular but only a little incurvated about their middle, occurred to me when I sought confirmation in other hearts since then. I thought that an easy way was opened here to achieve a distinct and clear knowledge of the heart inasmuch as I saw, thanks to these few observations, many questions which had opposed the most prestigious anatomists of our time and of the past to each other disappear spontaneously. But when I took the pen to undertake a more accurate description of every part, when I scrutinized more carefully the position and the arrangement of all the parts in the whole heart as well as in single fibres, day after day, when light appeared, brought in new darkness for the dissipation of which there is no other remedy than from the knowledge of the muscles. And so examination of the muscles as an approach to the investigation of the heart resulted in the coming to light of much else regarding the tongue, the oesophagus, and those muscles which raise and lower the ribs. Indeed during the dissections organized for the sake of my friends, some strange things came plainly to light. Consequently, since I was hindered not only by the very difficulties which occurred but was also held back partly by the need of other investigations, partly by my eagerness for knowledge, I could not reach the desired result as quickly as I had expected. When finally I had almost concluded the

Fig. 2.15.1   Title copperplate from "De musculis et glandulis" 1664

whole investigation, an unexpected event occurred which not only drew me away from my notes and dissections but at the same time removed all hope of returning to them for some time. So that no one shall believe that I break my word, and to avoid my action being ascribed to other motives so that I am groundlessly exposed to unfair suspicion, I shall in the present survey give my views on the heart and muscles to the extent my memory avails. As I do not have sufficient copies of the observations which were published some time ago for them even to be distributed only among my friends, let me add to the observations mentioned here, the matters which I regard more important on

A Specimen of Obfervations upon the Mufcles : Taken from that noble Anatomift Nicholas Steno.

THE Anatomical Difcoveries I had an Opportunity of making laft Year, concerning the Structure of the Heart, in my private Dif-fections made both at *Leyden* and *Amfterdam.* To oblige the Curiofity of many Perfons of Note, my particular Friends, I was then prevailed upon to engage my felf to pub-lifh them, together with their proper Fi-gures, with the firft Conveniency that oc-
5

cur'd. And fince feveral among them were fo generous as to offer me all the Affiftance they were capable to give or procure me for the carrying on fuch a Defign, I was very willing, as far as in me lay, to direct the chief of my Studies that Way, and accor-dingly proceeded.

In the Firft Diffection of the Heart, I dif-covered the following Things : Firft , that there is no *Parenchyma* belonging to the

Fig. 2.15.2 From the 1711 English translation, Kardel 1986, p. 105

the glands expanded here and there, and illustrated with new observations (Fig. 2.15.2).

That which may be called a mother not less fecund in evils than in errors, the all too negligent consideration of that which can and must be considered, often treats despicably or at least leaves sunken in obscurity the ones who are most worthy our notice and, on the other hand, often exalts and sets to the highest degree the ones that do not deserve it. The glands are an example of poor dealing, the heart a living image of exaltation. The heart has been made the residence of the innate heat, the throne of the soul, even the soul itself by some. The heart has been greeted as being the sun, the king whereas, if you examine it properly, you will find that it is nothing else than a *muscle*. I know that *Galen* has written in *On anatomical procedures* 1.7§.8, that *those who assert the heart to be a muscle understand nothing.* I also know the mockery by the great *Hofmann* of those who deduce the movement of the heart from its nervous plexus. He says in *De thorace*, line 1. Chap. 1.8: *I will rather laugh at them than refute them.* But had the mentioned famous gentlemen not been more prone to discuss than to discover the truth, if they had not preferred the principles which they themselves shaped to those which are deduced from the phenomena of nature, they would have been less rash in their judgment not only on this point but also on many others. Neither would they have expressed such an unfair criticism about those who held an opinion of which they perhaps could not be convinced. To have denied the truth would not matter much if they had denied without laugh and scorn. But even if they laugh and claim that those who, with the prince of physicians, *Hippocrates*, say that the heart is a muscle and think with the more modern authors that the nerves contribute to its movement, and the heart is actually a muscle, reasoning concurring with observations is obviously sufficiently convincing about this point. The famous gentlemen, Mr *Harvey* and Mr *Sylvius* who can never be sufficiently praised for their many experiments, had already taught that previously in part although the former had no clear knowledge of the fibres and other worries did not permit the latter to achieve the undertaken project. To make what was said more obvious, let me start by mentioning briefly these points so as to make the description of the muscle a little clearer and more distinct, I hope, than what has been done hitherto, *on one hand by making more certain the function of these muscles from the position of their fibres, on the other hand by outlining the true structure fitting all muscles.*

In the thorax, autopsy does not always tally with the writings of the anatomists. Therefore, it is not surprising that so many different opinions and so many controversies arise about this subject. To list all the controversies raised on account of the muscles which serve respiration being far beyond our present scope, in order to settle them, it will be sufficient to point out what anatomy brings out.

The anatomists disagree on the number of *intercostal muscles* and some would even not allow them any place among the muscles, only because the true course of their fibres was not sufficiently perceived. This is most apparent for the external layer where the anatomists have taken completely different muscles for one and the same. I admit that all carry out the same function but, notwithstanding this, they must be considered as distinct when their tendinous insertions are at different places and make different angles with the ribs, as these which descending from the transverse processes of the vertebrae into the upper aspect of the lower ribs seem to be properly named *elevators of the ribs*. Here, the tendon which is contiguous with the vertebrae appears to be diffused externally, but the one which inserts in the ribs expands on the opposite side over the inner surface. The very skilled *Spigelius* calls these the origin of the intercostal muscles. As far as I know, he is the only one who noticed the relation between the intercostal muscles and the vertebrae. But how can that be called the beginning of an intercostal muscle which is not at all intercostal since it joins, not a rib but a vertebra with a rib? Besides, at the level of the lower ribs there are hardly intercostal muscles unless this relates to the extremity of the lower tendon, like an apex, where inferiorly this tendon forms with the ribs an angle greater than 30°. Neither must this part of the external intercostal muscles be neglected which connects the bony extremity of the upper rib with the cartilage of the lower. What could be said about the length of the flesh and the insertion of the tendon easily appears to the observer from the general structure of the muscles.

The absence of observation of the course of its fibres has obscured the true function of the *lumbo-sacral muscle*. Much might be said about this but since I have not yet examined everything in different subjects, since I have not had access to a corpse on which I could make this investigation, I shall briefly recount what I saw in the few specimens that it has been possible to dissect. The anatomists ascribe the origin of this muscle to the os sacrum and the spinous processes of the lumbar vertebrae, whereas only a very small part of it has its origin there. The rest, which is by far the greater part, takes its origin from the ribs. The fibres do not run from the lower rib to that which is next above it but some of them span three or more intermediate ribs. Neither do the fibres which leave from one rib insert in one other rib alone but they extend some to three, others to five and even to seven ribs. In the same manner, each visible tendon, not satisfied with the fleshy part which is furnished by one rib, in some places gathers up fibres originating from five ribs. Notwithstanding this confusion, the fibres can be easily separated and shown in a subject of average bulk, by making a straight section so as to descend from the tendon backwards so that its different bellies are seen, and then by a lateral or transverse section from the space between the ribs, directed upwards so that the number of tendons of one belly appears. Each entire belly allows to several tendons everyone its share and each entire tendon receives its part from several bellies.

The muscle called *triangular* proceeds from the bony extremity of the true ribs to the mid-line of the sternum but the courses of its fibres are not the same. Whereas the upper ones form an acute angle with the sternum, the angles formed by the others which receive them is the less acute, the lower they are.

The description of the *diaphragm* abounds with many questions and falsities. Neither do all its fibres converge towards its centre nor is the beginning of its movement at the centre. One extremity of its fibres is in the vertebrae, the other is in the circumference through which it is inserted in the ribs and the xiphoid process of the sternum. The fibres of the tendons actually constitute a central membrane. They are interposed in between the fleshy parts like in a digastric muscle. Their course is not straight but varying, very nicely convoluted. Anatomists ascribe to the diaphragm two crura only inserted in one vertebra. I have never seen fewer than three and very often four. But always, as many as there were of them, they were inserted in the upper aspect of two vertebrae adjacent to each other and that about the middle of their upper aspect whence they seem to descend a little through the middle of the linea aspera of the vertebrae. Recently, in a dog, in which the tendon of the diaphragm presented with three parts, I observed the smaller left one to be inserted in the third lumbar vertebra, both the right ones which were each larger than the left to be continued in the third and fourth lumbar vertebrae. The anatomists have said that the course of the fleshy parts is like the radii of a circle but incongruous: if this is an adequate comparison for those which insert into the circle of the ribs, it is not true at all for the ones the tendons of which converge towards the back. Part of the diaphragm adheres to the rest through membranes, not everywhere through fibres. All its peripheral part between its crura and the ribs appears completely free if you separate the pleura and the peritoneum. Let me say with all the respect for the anatomists that I have not yet seen in their tables a picture of the diaphragm showing that they have made a true dissection of it. So far they have drawn neither the course of the fibres either in the fleshy part or in the tendinous membrane, nor the insertion of the tendons either in the ribs or in the vertebrae, nor the gap destined to the passage of the stomach, not to mention other most neat features which they could have noticed at the passage of the flesh into the tendons. But, God willing, I shall say more of these points and others elsewhere. What debates have not been written about its beginning and its end, even if the diaphragm does not draw one part to the other? Indeed, when it is contracted at its utmost, it is not straight. Nor is it provided with a pulley around which it moves (unless one wants to call the contents of the abdomen a pulley). Nor does any part to which it adheres, except for the xiphoid cartilage alone, move when the diaphragm moves: for who will deny but that the vertebrae are immovable? The muscles which raise the ribs actually maintain them rigid. There are many other points by which the absence of a more accurate examination of the muscles has obscured the description of respiration. But nothing is more surprising than the fact that the anatomists assigned for the movement of the ribs muscles such that, if the opposite extremity of these muscles were immobile, other muscles would contract so that the part in which they are inserted should be considered as immobile in relation to the ribs. This dismissed, let me describe the function of some muscles in respiration. The one who will not refuse to examine carefully the angles formed by the back, the ribs, the sternum and the muscles must find a demonstration of these muscles, perhaps not less certain than by mathematics.[1]

On the Use of Muscles in Respiration. The elevators of the ribs together with the external intercostal muscles, 1. move the ribs upwards and backwards; 2. they dilate the intercostal spaces; 3. they shorten the intervals between the

---

[1][i.e. will come up to the certainty of a mathematical demonstration].

cartilages and draw the cartilage of the lower rib towards the bony extremity of the upper rib.

The triangular muscle draws upwards the lower parts of the sternum towards the bony extremities of the true ribs. It shortens the distances between the cartilages and serves to dilate the spaces between the bones.

The internal intercostal muscles 1. almost pull down the sternum; 2. they move the cartilages away from each other; 3. they draw the upper ribs towards the lower and shorten the distances between them.

The muscle called lumbo-sacral moves the ribs away from each other as well as it draws them towards each other.

The brim of the diaphragm is not so much shortened as its convexity is straitened. All the lines indeed which one can imagine drawn from the vertebrae to the rest of the circumference, whether the diaphragm be loose or tensed, are incurved in some part or other, convex towards the thorax, concave about the abdomen. The less tense they are the more they are convex, the wider the abdomen, the narrower the thorax. The more contracted they are, the less the surface of the convex diaphragm, the wider the thorax, the narrower the abdomen. Thus during inspiration, the bottom of the thorax is depressed. It rises during expiration. I do not see what misled the otherwise so careful *Aranzius* who says he has observed in a mare, by looking and by touching, that the diaphragm is contracted in expiration.

First in a swan and later in a hen I have seen that the fleshy fibres proceed from the ribs to the pleura. They are unquestionably not inactive during respiration. But this is a passing comment.

Let us now proceed to the observations which concern the structure of the muscle itself.

The Structure of Muscles. The doubts which surround the theory of the muscles would almost deter me from undertaking a study of the heart. But happening to have at hand a dead rabbit, I took hold of a leg and separated its muscles to see whether there was any hope left of attaining any greater certainty. The first which I happened to cut open, after being cut off and divided with one straight section from an extremity to the other did present itself in the most simple figure of all that I have ever seen afterwards: opposite tendons indeed gathered at the extremities spread out where they had arrived at the fleshy belly so that one expanded through the upper surface of the middle belly and the other through its lower surface and grew more and more slender. The fleshy fibres moved in a straight course between these tendinous expansions, each of them continuing the tendinous fibres. Afterwards I examined many other muscles. All of them without exception fully confirmed my first observation. In all I found the same figure as in the first yet not everywhere in the same manner. For in some I observed it quite simple, in others, compound, and this in different ways. But hardly in anyone did I find it nicer than in the temporal muscle which, dissected with the knife following the direction of its fibres from the coronoid process of the mandible up to the skull, delighted the eyes with a pleasing spectacle. For the lower tendon ascending in the middle of the flesh is gradually thinned into a kind of sheet into which the fleshy parts adhere on both sides in the manner in which the little feathers adorn the wings of birds. The flesh which originates from the inner side of the tendon continues proximally to the skull. The opposite part has a tendon expanded over. In like manner it inserts into the skull by way of this tendon. I wonder that the anatomists who saw the tendon coming forth from the middle of the temporal muscle were not impelled to further examination by the very appropriate course of its fibres. There are several other muscles here and there to be investigated, which are remarkable by their structure.

Substance of the Tongue. Then I proceeded to the examination of the *tongue* where anatomists tell us that the extremities of the muscles end in a peculiar substance. But, after examination, I saw that this peculiar substance is muscular. This has also been shown by others and Nicholas Massa demonstrated in his *Libro anatomiæ introductorio*, from the different movements of the tongue, although he was compelled to say that this flesh is *not distinguishable but has to be ascertained by its actions only.* But what may seem impossible to us must not be called impossible straight away. On careful observation indeed, one can see distinctly all the series of fibres from the tip of the tongue to its base. There are only three series of fibres in all the body of the tongue. The external fibres close to the upper surface of the tongue have a course parallel to the length of the tongue. Of the others which occupy all the middle part there are two kinds. One descends from the upper surface of the tongue, the other runs from the middle towards the sides, in man. These two kinds are arranged into certain series meeting one another alternatively. Each of them contains hardly more than one fibre in their thickness as if it were layer over layer. The upper surface of the tongue itself keeps the tendons of the fleshy parts enclosed between the external tunic and the flesh. But this is sufficient to give a glimpse since this is not the place to discuss the opinions concerning the taste and the different movements of the tongue in the pronunciation of different letters.

The Muscular Tunic of the Oesophagus. Although in certain species the *oesophagus* is provided with many interwoven fibres, all of them can conveniently be reduced to two kinds, i.e. two spirals which, where they meet in their opposite course, intersect each other in such a way that one of them passes right beneath the other and then soon afterwards passes above it until the fibres meet again. I have not yet found out by experiment whether other fishes are provided with the muscle by which the ray contracts the upper orifice of its stomach. This delicate and unique muscle surrounds the orifice thus mentioned with many coils forming a spiral.

It would lead us too far to list all the muscles and to give an account of all the differences that occur in them. Let me only propose some certain conclusions drawn from all those which were examined.

Components of the Muscles.

I. In every muscle there are found arteries, veins, nerves, fibres, membranes. Some people add also lymphatic vessels which, although I have seen them very often around the muscles, I admit frankly not yet to have seen them in the muscles.

II. No muscle was met, the single fibres of which do not go forth into a tendon on both sides. All these tendons, on both sides or on one side only, either gather together into strong tendons, or spread into a membrane, or degenerate into bone rigidity. This occurs at the end as well as at the origin of the muscle.

III. The flesh is no parenchyma or stuffing but the same fibrils which, being tightly joined to each other form a tendon, and when being looser constitute the flesh. And in this sense, it is true that the tendon is one continuous body from the origin of the muscles to their end.

IV. It is not easy to find in any muscle a fibre which forms one straight line but each of them is divided into at least three lines which comprise two alternate angles. Neither are these three lines always straight, but often the flesh (as at least in the sphincters which nobody will easily deny to be counted among the muscles) and sometimes the tendon also (as when the fingers are flexed in a fist) form a curved line.

V. The three above-mentioned lines of the single fibres in the same muscle are not of the same length although all the fibres, when taken together, in most muscles show hardly any difference in length.

But, to make these matters better understood, a more particular naming of the parts must be presented. Any straight and simple muscle, like the other muscles, admits two sections along the course of the fibres, one straight and one transverse, not in a plane cutting everything transversely through the middle but led from one side to the other, so that the tendons remain intact with the flesh. So as to have distinct names for distinct things, let us call orders the straight series of fibres and ranks the transverse ones.

The fibres of the same order are in the same plane and form an oblique parallelogram or the figure of a rhomboid. Their two opposite parallels, where they form acute angles with the other parallels, extend outside the angles on both sides. And these parallels prolonged outside the figure correspond to the opposite tendons which go forth in different directions. The other parallels comprised between these and filling in all the area of the figure represent the position of the fleshy parts. Excuse me if, even when dealing with physics, I give mathematical names to physical and not mathematical lines. These lines indeed are not deprived of any width so that the single fleshy parts do not represent as many parallelograms and the two opposite collections of tendons do not represent two trapezia. But let us leave these niceties to mathematicians. It must first be noted here that in each of the two tendons, however thin they be, there are as many filaments lying upon each other as there are fleshy fibres in the muscle belly. This shows how much more compact the tendons are than the fleshy parts. 2. The inner filaments of the tendon are always shorter than the outer ones. Thence, in the same order, although all the fibres where they are fleshy are equal to each other, those of the tendinous parts on both sides maintain a certain inequality, which you may not improperly compare with the pan-flute[2] of the Ancients. Thus if you will consider more closely any order, you discover three figures, two opposite pan-flutes at the extremity and a parallelogram in the middle.

The rank of fibres generally is resolved into three figures not lying on the same plane, one made of fleshy parts which are rectangles and two made of tendons which are either rectangles, or triangles or trapezia depending on the different figures of the whole tendon. Hence it is easy to determine the quantity of each of the different fibres in the ranks. From ranks superimposed on each other so that their middle planes form more or less a parallelepiped, a muscle results where the structure of the ranks is made conspicuous from a series of fibres in the orders. Since there are as many ranks in a muscle as there are fibres in any order and in the uttermost ranks of the opposite tendons, the greatest variety is observed in the extension of the same rank; in the others, the closer they are to the middle, the more the inequality of the opposite tendons progressively decreases.

VI. Actually and conveniently, every muscle is divided into a belly and tendons, i.e. a middle and extremities. Controversies about the end and the beginning and about the ligament are never to be settled unless you determine a certain basis. If I were allowed to speak freely here, I should

---

[2]syringe.

not gaze at the flesh nor the tendons and I should not care about the insertion of the nerve, but I should say that the immobile extremity is the origin and the mobile extremity the end. Since the extremities sometimes exchange their role so that the extremity which was mobile becomes immobile in another position of the body, I should rather avoid these words and be satisfied with the designation of extremities since, properly speaking, none of them deserves the name of origin or end. A muscle indeed is not generated as if progressively proceeding from one part towards another, this being inconsistent with the motion of the blood. Neither is the immobile extremity always the same in all muscles. And if the point towards which the displacement of the other parts occurs is to be considered as the origin, then in all muscles the belly is to be considered as the origin inasmuch as, in various positions, the two extremities move towards the belly and the insertion of the vessels is in the belly. Thus, if a progressive generation of the parts is to be admitted, this origin must be looked for where the vessels are inserted. There are other questions of this kind on which much time and paper have been spent but, finally, when all is considered, they serve to adorn a pleasing discourse on probabilities but do not contribute at all to discovering the truth. Therefore, to cling to what is certain, let us retain the division of the muscle in a belly, or a body made of the middle flesh, and tendons, or the extremities. For it must not be thought that in a muscle the fleshy part of which inserts in bones without any intermediate tendon, such as part of the temporal muscle and several muscles of the limbs, one of the extremities is lacking. In the first generation of the parts, what are now bones, were they not tendons? And in birds tendons frequently turn into bones, where the fleshy parts continue into bone without interposition of tendons in the same way as the fleshy parts of the temporal muscle insert into the skull.

Thus in every muscle there are two tendons made of ranks of tendinous fibres which, as they proceed from the opposite parts and expand through the opposite half bellies become gradually more slender. These muscles are either simple and their expansions appear on the outside, or they are compound. In these, at one extremity, the tendon penetrates through the middle of the flesh and, at the other extremity, it surrounds the flesh on both sides so that this muscle seems almost to be made up of two muscles.

The length of the belly is not always a measure of the fleshy fibres since a very long belly often has very short fibres.

VII. A membrane not only surrounds the muscle with transverse fibres but also in the same way gradually insinuates itself between its different fibres.

VIII. The action of a muscle is a contraction but this does not necessarily result in shortening a straight line between two points at the extremities of the muscle, but always in a shortening of the different fibres of the muscle between these two points.

It is not the tendon which contracts but the flesh comprised between the tendinous expansions. Its contraction results in that two or more cross-sections of the belly come closer to each other. Thus the first instrument of the movement is not the tendon and we owe the movement neither to the origin nor to the end of the muscle.

How this contraction occurs is difficult to determine. Many attribute it to a filling in of the fibres, some to their emptying and some to both. I should be bold if I arbitrated between them. Therefore, I publicly proclaim that the causes and ways of action *are not obvious* and, since an explication through analogies greatly pleases many people, I will present here something which is not completely foreign to the purpose. Those who drive posts and palisades into the ground for foundations use an engine by which several men raise a ram[3] by way of cables, each of them pulled by a single hand. Not inconveniently this machine can be compared with a muscle divided transversely through the middle of the flesh. The ropes indeed represent the tendons. They are longer and longer depending on the distance the men are at. The weight hooked to the ropes represents the mobile part and the men themselves represent the fleshy fibres. By shortening while together they pull their ropes, the men indeed move the weight. Similarly, the contracting fleshy fibres, while they pull the fibres of the tendon move the mobile part. This being only a comparison, I shall insist no further on this.

The shorter the flesh in a long belly, the stronger the force of contraction since the number of fibres is greater.

IX. Whatever part of the body does not lack any part necessary to a muscle nor has another part which is denied to muscles, and which also has the structure suitable to a muscle, must be called a muscle. It is not required that it be subjected to the command of the will for in many parts muscles are never moved following the arbitration of the mind. This is proved in several muscles of the larynx and the tongue, as well as

---

[3] pile-driving apparatus.

in those belonging to the back. And what part does the soul play in animal movement? Does it not imitate by different turnings a movement in another animal until it finds the right movement for itself? This is too obvious to be discussed further since anyone can find within himself as many clear proofs of this matter as he may wish.

What the aforesaid structure of the muscles may contribute after further examination to a better understanding of some diseases and their care appears as in a nutshell, from what follows.

I said that the flesh is contained between tendinous expansions. Actually, not so long ago I saw in a turkey, in several muscles of the legs, that all the tendinous expansion was so free from the underlying flesh that it adhered to it only by some very thin fibrils which could be immediately torn by the slightest force and the tendon thus appeared like a membrane. The fleshy extremities themselves which previously had continued into this tendon were somewhat more whitish and firmer than the rest of the flesh and parted easily from this. At the same time they were surrounded by some moisture. I made this observation only once in a cock. What does actually prevent the same to happen sometimes also in human muscles so that this may sometimes be the cause of a weakening or of a complete absence of movement in a certain part, depending on whether the tendon parted completely or partly from the flesh? It has been believed so far that the fibres extend in the muscle in a straight line from one extremity to the other. This would result in that movement would be impaired by any transverse division but hardly by a longitudinal one. It is obvious that, if the middle of the muscles were divided transversely, all the fibres would be cut. But this structure of the muscle explains that a wound may be inflicted transversely through the middle of a muscle with a broad iron without any detriment to its movement, provided the iron remains between the same parallels without injuring the opposite expansions. On the other hand, all these straight divisions, as they are almost in infinite number, which do not remain between the same two orders of fibres, impair more or less the movement together with the fibres, depending on how the knife is directed.

What has been proposed here concerning the muscles, if it applies to the heart, is sufficient to demonstrate the initial proposition: *the heart is actually a muscle.* I will clarify that point in more detail elsewhere. Here I will present the main points in a few words.

The Heart's Constituting Parts.

I. In all the substance of the heart nothing else occurs than *arteries, veins, nerves, fibres, membranes.* But in a muscle nothing else than these items occur either. I will not mention fat and bones because not all hearts are provided with these and some muscles are provided with these. Whatever is said of the parenchyma as being distinct from fibres is due to the imagination rather than to the senses of anatomists. In some subjects I saw bright ducts running over the surface of the heart. Dissection taught me that they were not lymphatics but true nerves. A plexus of nerves described by *Fallopius* is confirmed by an autopsy as well as their distribution over the surface and their penetration inside the heart.

II. At examination of the fibres of the heart none occurred to me which was not fleshy in the middle and tendinous at both extremities. This is also common to all fibres of muscles.

In the heart as well as in any other muscle, the course of the fibrils is uniform: indeed all are bent like bows about the middle of the flesh, they make no angles except with the tendon at both extremities. Those who believe to know the ends of nature as if they were of her cabinet council have imagined straight, transverse and circular fibrils. They appear somewhat intricate but they are neither confused nor inextricable in so far as all those which occur in the left ventricle present with the same twist, as clearly demonstrated by the resolution of the cone in a star-like figure. Neither do their labyrinthine convolutions impede the one who proceeds carefully in the right ventricle. They would be more easily understood with figures than with words. I will, however, give some idea of them in a few words.

At a first glance, the outer surface of the heart, after being freed from its membrane and fibres, would seem to show that the course of all the fibres is on all sides the same, oblique from the basis to the cone. But then if you separate the fibres from each other, you will notice that very few of them reach the cone because, as soon as the first ones have descended from the basis under the next ones (let me here call first fibres the ones which are seen anteriorly at the orifice of the left ventricle where this meets the right ventricle), contorted immediately, they pass underneath the others towards the left. The next fibre always descends a little further than the previous one, before bending. When incurved it passes beneath the next one which in turn is deflected beneath the twisting first. The same is true for the course of almost all the fibres. It cannot be more conveniently explained than by imagining, or by building with your hands a parallelogram formed by fibres arranged in a parallel position, sufficiently long. You turn one of the shortest sides to which the extremities of the threads are attached so as to achieve a complete circle. Then indeed you will see the same picture of the threads twisting and ascending on each other. But if you imagine more parallelograms of the same kind and make them concur to form a parallelepiped and if you give to the other of the faces to which the fibres are fastened, the position which I said and if, besides, you set the same shorter faces previously opposite to each other, in the same plane so that the middle and looser course of the threads represents an arc, you will no longer complain of a confusing bundle of fibres in the heart. There is here a need of figures more clearly to demonstrate in particular each of these things. Because I am forced to postpone these figures to other times, it will not be appropriate to present here in details: how the course of the fibres changes in the walls of the ventricles; why interior prominences result from the concourse of external fibres coming from different places; what elegant figure a different dissection of the cone displays; how to represent the separation of the fleshy parts from each other into the opposite parts, where the tendons continue into the flesh; and several other points which present themselves spontaneously to the one who moves his hand with care and patience, and which in any case must be completely revealed for a description of the heart.

The middle or fleshy portion of the fibres constitutes the belly of our muscle and also the bellies of the heart. The substance of the heart, although obvious to the eyes and the hands of everybody, has bred as many controversies as any other part. People distinguish in it a basis and a cone which hitherto they thought to be very thick although its middle point or the centre which is left in the middle of the incurved fibres is so thin as to be hardly equal to a pin's head, since there is nothing there but the concourse of the internal and the external membrane.

The tendons which receive the extremities of the fleshy parts appear partly at the entrances, partly in prominences where they also form valves and go forth into the substance of the artery. (The veins indeed, inasmuch as they continue one of the tendons of the auricles, have nothing to do with the heart but through the auricles.) It is here that the origin and the end of the muscle must be looked for. The end is formed by the orifice of the pulmonary artery and the tricuspid valves of both ventricles together with that part of the aorta, that part which forms the mitral valve. The origin of the muscle is represented by the tendons which provide the remaining portion of the basis of the great artery.

I will not show here where the tendon expands either above or between the fleshy parts. Neither will I describe the fleshy valve in the right ventricle of birds. Nor will I enumerate the variety of the different parts, which I observed in different animals since I have no figure of them and I do not remember all of them.

A small bone is found in the heart, not only in large animals but I have very often seen it even in sheep. It is not present in all animals and, in those in which it is present, it is nothing else than a part of a tendon which is degenerating into the hardness of bone. Not only is the solidity of bone also observed in the middle of a tendon in the legs of birds, but also in any animal one of the tendons of many muscles turns into bone either totally or partly. This is easy to demonstrate in the skull as well as in the hands and feet. How should we assign to a small bone taken out from the heart of a stag other virtues than to another bone of the same animal (for I have no doubt that in different animals also the nature of the bones is somewhat different) when a part of this bone is made from the tendons of a muscle?

III. The proper membrane of the heart intersects the fibres of the heart in a transverse course of its fibres, and insinuates itself between them in the same way as the membrane of a muscle does.

Thus, since all the true properties of a muscle belong also to the heart and nothing is found in the heart but these properties, *the heart actually must be called a muscle, which it deserves because of tendon, flesh and nerve.*

This seems to be directly contrary to what the great *Hippocrates* said: *the heart muscle is certainly strong, not in its tendon but in its flesh.* However, to the one who compares these words in their order with ours, both will appear to be in agreement. For he says that the heart is a muscle the strength of which is not in the tendon but in the flesh. Thus, whatever the opinion of the author was, his words are not contrary to us.

If I did not believe that anybody will easily find by himself what questions this unique proposition resolves, nothing would be easier than to join here a long list of these questions some of which hitherto puzzled not only the Ancients but even *Harvey* himself and many other modern authors. Let me, however, mention some among the principal consequences which closely result from this proposition.

What the Substance of the Heart Is Not. If it is certain—and reasoning helped by the action of the senses makes it certain—that the heart lacks nothing which exists in a muscle, and that nothing which is denied to a muscle is found in the heart, *the heart is no longer a substance of its own kind and, therefore, it is neither the seat of a certain substance like fire, innate heat, the soul, nor the generator of a certain humour, like the blood, nor the producer of some spirits, e.g. vital spirits.*

If indeed the heart is the same as the muscles for all its contents, all which is contained in the substance of the heart and in the vessels is of the same nature as what is contained in a muscle, for the same thing delivered to each of them through the same vessels is received by the same substance. To this a last touch is added: the same action, a contraction of the flesh, is observed in both. Let others, quickerminded, see whether the immediate cause of this contraction can be different from the immediate cause of the contraction of muscles despite such similarity of all the organs and their favourable agreement.

But all this concerns the substance of the heart only, not the cavities. Let the advocates of the heart discuss whether something nobler than the blood contained in the vessels is to be found. As far as I see, I am certainly afraid that all about the dignity of the parts will be pure rhetoric. But be that as it will, I undertook to examine the substance only and to demonstrate from the course of the fibres all the phenomena which occur in the movement of the heart, for its own reasons. Although I have completed the greatest part of it, I am, however, compelled to adjourn the subject for another time since, as I warned at the beginning, deprived of papers and dissections for some time, I will hardly find leisure to return to them. Please accept that I thus presented these points early so that, if the industry of those who are in favour of the truth should find in them something to be improved, and if they would kindly let me know, it will be corrected in the work itself and that which is still lacking will be provided by longer thinking and experience.

On the Motion of the Heart. However, not to appear as having said absolutely nothing about the movement, let me add this proposition. When the fibres of the heart shorten, because their beginning and their end are about the basis, *the cone is necessarily drawn upwards.* Because the fibres descend from the basis towards the left to the cone, *the cone rises in an opposite way.* Because, in the right ventricle the fibres descend on the inside from the septum and rise obliquely towards the back through the external surface, *they raise the bottom of the right a little towards the basis, and therefore, the heart becomes shorter and also rounder on the right side.* Because the walls actually become shorter and thicker, *the cavities of the ventricles become also narrower. Thus the heart when round is not more capacious than when it is elongated and, therefore, the rotundity does not proceed from more dilatation of the blood, nor is there a diastole at this time.*

What is to be observed in the orifices of the heart with regard to the space and with regard to the valves, how the threads move, what the muscle of the auricles is, in how many ways I have seen the movement of the heart to stop and then to start again, indeed sometimes many hours after death, and several times at that, and what additionally concerns other points the details of which do not come to my mind, I will keep all this for a whole book. With this the muscles are dismissed and I proceed to the glands.

2.15.2   The Glands

As long as the nature of the glands was ignored, they were thought to be of little value, in a way not alien to the common habits of living people. Often many people are in the dark, not known outside their door, not because they are not worth a better fate but because their value is ignored. The glands thus owe much to this century which was keen in investigating their nature and happy to rehabilitate them.

Conglobatæ Glands. The purest nymphs of Bartholin removed all round or conglobate glands from among the dirt (according to the judgment of the Ancients about all the glands). They cleaned them from the sewage of the past centuries by an overflow of their lymph.

Conglomerate Glands. The conglomerate glands owe the origin of their restoration to *Wirsung*. But, as the first author paid with his death the glory of [having discovered the duct of] the pancreas, his followers were not very lucky either. *Deusing* indeed wanted to cheat *Wharton*'s industry by praising himself, and *Blaes* attempted to make the discovery of the external salivary duct disgraceful for its discoverer. But the other vessels will enjoy, I hope, a better fate than the one that met the first three.

What our Goddess, Anatomy, revealed to me about each of the two kinds, I said part in *Observationes* published 2 years ago and I will present the rest in the observations which I have not yet finished. I will make here a short mention of the main points only. These, however, concern either ducts not yet described in the previous observations or the illustration of ducts observed previously, or an improved knowledge of glands of both kinds.

The new ducts, at least for me, are:

I. *The external salivary duct*. Its origins are in the conglomerate parotid gland; its outlet is almost at the centre of the inside of the cheek in man. In between, the middle of the duct varies in the different animals. In man, dog, hare, rabbit, it is straight. In sheep it is incurved but its entire course is above the mandible. In the gender of cows it actually descends down to the inferior angle of the mandible, from there, where it runs to a fold carved in the mandible itself, it rises towards the cheek. The *Prologue of the apologia* published a year ago has clearly exposed in what consists the difference of this duct from the one described by *Blaes*, which is hardly to be found in any other than the author, except perhaps in the inhabitants of the moon or those crawling in the spaces between the worlds of

Epicurus[4]. Since the matter about this matter is very clear, I hope, and I promised friends, under a certain condition, not to proceed further, I will here only mention his evidences (as it is his only argument). This was written a long time after the thing was done and by his disciples. The only one among them who deserved to be trusted, acknowledged in the presence of two illustrious professors that he had committed an error in the assignation of the time and that he knew that mention had been made of a new duct but he could not remember with any certainty whether this duct was above or below and whether or not either mention or demonstration of another salivary duct had been made, besides this one.

I supported with an equal mind these and other grievances which would have made other people angry, even if the justice of my cause was and is such that I can at any time call to trial himself, *Hoboken* and his anagram. But my conscience and the opinion which honest people among whom I have always been have about myself, are sufficient for me. Let me behave so that, if it pleases them together with *Deusing* to insult me longer, I can refute them in front of a proper audience. He designated my theses as injurious pamphlets and, not to mention other matters, he was not ashamed to blame my age as a sign of an impotent mind. But let an old man learn from a youth what he taught the youth by his example: nothing disgraces an honest man more than addressing ignominious words to somebody without adding the true reasons, like somebody for whom either wine or anger or senility added to the volubility of the tongue what they took away from the proper use of the mind.

II. *The small vessels of the cheeks* compensate their smallness by their number. Their glands in man are well stuffed all around the lips and can be compared with grapes. The distribution of vessels and nerves to them increases the loveliness of the face. In beasts they fill all the plane of the cheeks. In them, i.e. the beasts, some are concentrated into one body, they lie over the length of the mandible and give ducts of average size. The orifices of these ducts, set on the same straight line, hardly appear unless some humour is expressed, even if, once they are discovered, a fairly thick bristle can be introduced into them. Such is the equality and flatness of the surface at those places. In some sheep, however, black colouring surrounds the orifices; others are more scattered, some lying on the flat tunica of the cheeks, some raising this tunica into rough and pointed nipples.

---

[4]i.e. empty space.

III. *The ones which are seen inside the gums, on the sides of the tongue are called sublingual glands* and are different from *Wharton*'s glands. In the tongue of some birds they are of a rare elegance. There are 10 or 12 orifices on the sides of the tongue itself near its basis, distributed on the same straight line. If the tongue is compressed, droplets of serous humour surge everywhere. In the swan the glands easily admitted a probe, fairly deeply, in proportion to the size of the tongue. The same is true also for the duck. There may even be more ducts in the aforesaid animals but, delighted by the very beautiful aspect of the ones thus described, I did not give attention to others. In a turkey not only did I see this straight series of outlets, but I even saw all the tunica, perforated, conceal the bodies of glands and the outlets of the ducts either between the basis of the tongue and the orifice of the larynx, or in the sides of the tongue itself, or beneath the tongue. Glands and ducts lie so widely scattered in the tongue of other animals that they do not easily permit the introduction of a hair in cows and animals of their kind, although the brightness of the contents shows the delicacy of the ducts which lead from the glands to their outlets in a parallel course and the expressed liquid reveals these orifices.

IV. *The entire tunica of the upper palate perforated by countless holes receives the ducts originating from the glomeruli of glands.* They are the more conspicuous the further you reach to the back. If you open the bill of birds, you notice there also sources of liquid in the upper palate. All which follows the white tunica close to the bill and *surrounds the cleft everywhere*, has the colour of the flesh. If you press it, it will reveal the outlets of the glands. Before anything else, two points must be noticed at the anterior extremity of this substance. Through these two points great abundance of a whitish humour is expressed. In the chicken, they even admit the introduction of a probe, and not only at the surface.

V. *Passage descending from the nose into the anterior aspect of the palate inside the teeth in those who have teeth.* If it pleases you, call it a strait, the comparison is suitable. This path is wider and more obvious in beasts than in man. It appears differently in different animals. This can be ascertained by comparing the severed mouths of a calf, a sheep, a dog, a rabbit. In birds it is replaced by a cleft dividing the posterior aspect of the palate.

VI. *Small canals perforating the cartilage of the epiglottis.* Originating from the caruncles lying over the epiglottis, they evacuate the humour transmitted through the cartilage in its back side projecting over the larynx.

I will not add to them those ducts which I observed in the oesophagus and in the gullet of various birds, since I learned that things of the same kind were observed previously in a chicken by an illustrious gentleman, formerly very dexterous worshipper of anatomy. In the swan the tunica of the oesophagus besprinkled with countless small points thrusted forwards droplets of trickling aqueous humour. In the gullet the tunica was actually thicker and rough and showed not much fewer sources of a white and viscous juice. After this, let me go up to the nose as if it were to a second floor.

VII. *Small ducts which irrigate the tunica of the nose.* Examination of the tunica separated from the bone and inspected from its back side teaches that its humour is not secreted from openings of arteries (nature either closing or opening them at will).

VIII. *This duct of the nose, which so far I have not yet seen except in sheep.* Its roots originating in the back of the nose gather into one branch into which a probe can be introduced. This branch proceeds through the sides of the nose towards its anterior aspect.

IX. *Drains, convergences of the lachrymal points in one duct on each side, deriving the humour of the eyelids into the cavities of the nose.* Here there are big differences in different animals. In man, small holes carved in the apices of the eyelashes send down small canals, not easy to demonstrate unless the tunicae are left intact after cautious fracture of the bones of the nose for soon, where they enter the bones (up to there they are separated by a thin membrane), they continue widened into a fold of the tunica of the nose. In the sheep, the hare, the calf and the rabbit, the outlets are not found in the eyelashes themselves but a little inside. They are to be sought for more deeply in birds where they are much larger than in any other species. A longer probe introduced in them comes out through the cleft of the palate. Here the membrane which distinguishes the holes is somewhat short.

X. *Ducts of the eyelids or lachrymal.* They are more in number but, with respect to their sources, they are reduced to two kinds: 1. the external ones lying beside the external angle of the eye have their origin in the gland which occupies this angle and which is called innominate by many people; 2. the internal ones which are commonly called lachrymal. Their source is a gland. Some animals have a nictating membrane. They have gaping orifices of ducts arriving inside through this membrane. In birds a probe can even be introduced into them.

XI. Let me add to this what I saw in a ray, with the help of the illustrious teacher *Simon Paulli.* The ray variegated with many black spots is covered by a skin sprinkled everywhere with holes from which droplets of viscous liquid proceed, if you press on the orifices. A hair can be introduced freely through those orifices. And these are ducts which I think nobody has noticed. If *Galen* had observed that many of them are visible without dissection, then he would have repeated this question in his *De usu partium* l.2 where one can read:[5] *Consequently, do you wonder or do you not believe that some of the points which are more obscure in anatomy have been ignored by those who did not even know what was apparent before anatomy existed?* But to whom does this reproach of the great anatomist apply? *Galen* ignored many points visible without dissection and I do not doubt that many such points remain which, perhaps too obvious, do not succeed in drawing our attention. It was well-said by *Galen* if he did not except any of the mortals. Since he made this remark only about his predecessors, he seems to have been an unfair censor of other people and not to have considered the very narrow limits of human knowledge at this point. What follows serves to illustrate ducts previously observed.

I. The lymphatics are illustrated both by *the general structure of the conglobate glands* and *by the true insertion of the lymphatics in veins.* It is usual for the conglobate glands which I had the opportunity to observe so far, that they comprise, besides arteries, veins and nerves, two kinds of lymphatics, some afferent, others efferent. In this they are different from the conglomerate glands which have efferent vessels only. The substance itself, everywhere full with the roots of the vessels, presents a conformation which cannot be better compared than with the body of the kidneys, to which some of the glands are very similar. I mean in shape, not in colour. Externally indeed they are globular. Internally they are not excavated into some kind of basin. However, if you attack the separation properly, you will find a cleft between the edges protruding everywhere. From this cleft the roots of an efferent lymphatic vessel come forth together with a membrane. In the outer surface, if you direct the blade of the knife cautiously, you will see branches coming from many different afferent lymphatic vessels enter the substance itself in many off-springs of their branches. The parotid conglobate gland in sheep and calves testifies for this by its delicate design. Three differences stand up between these glands and the lymphatics which carry their contents back into the blood. Some indeed originating from the primary sources, whatever these will finally be

---

[5]Galen, *Of the uses of the different parts of the human body*, XVII, lib. II c 3.

found to be, are sprinkled over the surface of the glands; some lead from the cavity of one gland to the hump of another; and others surge from the cavity of these glands up to the vena cava.

The true insertion of the lymphatics in the veins, as it was eluded to in the investigation of many people, did not deceive anybody more than the noble *Bils*. He seems to have been the first to discover the true way of dissecting it but he seems also to have been the first to suffer then being diverted to absurd opinions. This will be published more extensively elsewhere, where the obvious cause of his errors will also be given, already pointed out previously in my *Observations*, and the variety of the insertion will appear in different figures. Here let it be sufficient to have said in a few words that the insertion of the lymphatics in the veins is to be demonstrated about the confine of the axillary and the jugular vein, on both sides, and in fact on the right the insertion of those which come from the right anterior parts in brute animals, such as from the right bag of the pleura, the right front leg and the right side of the head; on the left, besides the vessels of the left side analogous to the aforesaid ones, the insertion also of the thoracic duct which receives all the lower lymphatics and joins with the upper ones before the insertion. The common concourse of all the confluences on the left side does not form a receptacle and is not provided with valves. Neither does it, contrary to what *Bils* wanted to persuade us, discharge its contents into the vena cava through one outlet alone. I have indeed never found out less than two, very often three and sometimes four of them and, to make sure for the onlookers that chyle and lymph are poured into the blood through this path, I filled the vessels thus concurring now with lymph, now with milk, by simply compressing them, then, by a different movement, I opened the path to the blood without hurting either the vessels or the valves. But these points mentioned in a few words previously will be treated more extensively elsewhere together with the manners of discovering the valves of the lymphatics. The ring or labyrinth of Bils is not visible in all subjects. Nor, where it is present, does it always appear with the same shape. Or it is nothing else than a deviation of a lymphatic provoked by arteries and veins or nerves passing across.

II. An observation made on birds that also clarifies the function of the liver, serves to illustrate the *duct of Wirsung*. In the different kinds of birds which I could open, I indeed saw a double pancreatic duct meeting a similarly double biliary duct (one of which was running from the gallbladder which did not lie against the liver, the other was running from the liver). The insertion of the four ducts occurs in three different ways: either all converge into one outlet, or one of the pancreatic ducts penetrates below the common orifice with its biliary counterpart so that the intestine is accessible through two orifices only, or the outlet conceded to every duct is such that four different paths following each other on the same straight line open to a probe in the intestine. Recently, I saw in a turkey a biliary duct single where it left the liver, split twice during its course towards the intestine which thus received the bile from three vessels.

These points, I think, throw considerable light on the utility of both humours and confirm their meeting, which is illustrated also by the peculiar insertion in birds of the pancreatic duct into the biliary duct at midway, before arriving in the intestine. I see afterwards that this insertion has already been described and drawn by the most diligent *Highmore*.

III. About the *duct of Wharton* I will say one thing which eluded me too when I published my *Observations*: he writes that his duct runs from the adjacent gland (the conglobate gland adjacent to the conglomerate one) into the inferior maxillary (conglomerate) gland itself which initially I also thought to be present there, relying on the observation of *Wharton*. But, later, after having dissected further, I observed that it penetrates between small glomeruli of this maxillary gland into the other conglobate gland and thus must be counted among the lymphatics of *Bartholin*.

The Immediate Origin and Motion of Lymphs. From these observations and from similar observations made by others and by ourselves, part of the following was brought forward by experiments, part resulted from reasoning:

I. All lymphatics have to do with glands. I should say that the glands must be considered as the first spring-water of all lymphatics. But the first origin of the lymphatic vessels of *Bartholin* is that more uncertain that it admits several likely explanations. One thing at least remains certain: in the body no lymphatic has been observed which does not either originate from a gland or lead into a gland.

II. All lymphatics continuing conglomerate glands pour their contents into certain cavities such as the eyes, the ears, the nose, the mouth, the oesophagus, the larynx, the gullet of birds, etc.

III. The ones which are attached to the conglobate glands carry back their contents into the vena cava, either directly or by way of other conglobate glands, and indeed not only about the left axillary vein to which all the thoracic duct generally joins itself, but also in the right axillary vein where I should wish *Bils* to show us the distribution of his trickling liquid to the right lymphatics of the head, or the concourse of the upper lymphatics of the right side with the thoracic duct.

IV. Thus, all the glands themselves are viscera through which the lymphatics scatter their roots.

And reasoning convinces us that all this must be considered as true since it is proved by the testimony of the senses. On the other hand, the following propositions must be considered only as probabilities. They are arrived at by induction even if this induction is not complete for all the points.

α.

At any place of the animal body some aqueous humour is normally found smearing the surface where this surface is perforated by the outlets of lymphatics originating from a conglomerate gland. From this, one could deduce: 1. the origin of the *pericardial watery liquid* is the same, which an observation on the thymus by an illustrious friend seems to confirm. 2. The sources of *sweat* thus are not different. More about them later in the examination of sweat. In this examination, by induction from all the parts which can constitute the skin, by comparison of sweat with other humours and by examination of the causes which move these humours, it will be possible to demonstrate that nothing more probable is said from what has been observed so far concerning the skin than that this is a glandular substance different from the others only by the fact that, besides capillary vessels, it has also roots of hair scattered through its body. The vessels perforating the skin of rays confirm this here. It is from them that I judge the true cause of the slipperiness of fishes is to be expected. I believe that the serum of those which people tell to have been generated without skin was more viscous than that it evanesced at the first contact with air or that their skin was moistened by continuous sweat, as a result of the abundance of humour in the body, or of the size of the paths, or because of an obstacle to the passage of the serum through other small glands. 3. *Humour* is secreted *in the ventricles of the brain* in a similar way. This seems to be enhanced by the glandular substance observed previously by other people in the choroid plexus. 4. The humour which smears the parts of *the thorax and the abdomen* seems to be analogous in its origin. 5. Different sieves are not to be sought for the *humour from which the foetus appears and is fed.* The animals indeed in which there are cotyledons display spongy fleshy parts adhering to the womb as well as placental roots easily extractable from the fleshy parts without bleeding. In the cat, where a wide ring surrounds the egg, I separated the egg with its ring from the womb without bleeding and the substance of the ring was actually glandular. In the same way, glandular prominences made rough also the surface itself of the womb to which it adhered. Inside the amnion of some animals there are also white corpuscles which contain blood vessels. But these points, as I said, are to be counted among the cohorts of suspected matters.

β.

The lymph of conglomerate glands, after pouring out from the vessels into certain cavities does not remain there but is carried elsewhere inside the body through other paths, thus from the ears and the eyes into the nose, from the nose into the mouth, from the mouth into the oesophagus, etc. Hence it appears that paths must also be sought for through which the pericardium, the brain, the thorax, the abdomen are evacuated. One may even suspect the same for the watery liquid of the eye. Among the matters to be sought, I set the paths which evacuate the humour of the brain. This matter indeed has not yet been ascertained, although nothing is heard more often among physicians than that humours trickle down from the brain in the subjacent parts. Frequent dissection of the brain convinces one that there are always humours in the brain but through what paths these are secreted has not been so easy to find out so far. In a horse, as I liberated the pituitary gland from the countless plexus in which it is intertwined, I saw that I had in veins an open path for my probe. Therefore, I came to suspect that the humour of the brain percolated through the pituitary gland is received in the veins without the intervention of other vessels, but then I could not repeat the same dissection to the same end through other investigations.

Indirect Origin of Lymphs. From the orifices we followed the lymphatics to their sources which, as they raise many questions macroscopically, also abound in uncertainties microscopically. The following opinion deduced from what was observed hitherto does not seem deprived of probability. According to this opinion, the *glands are sieves*, built in such a way that, whilst blood passes from the arteries to the veins through the capillaries, the serous parts liberated from their surrounding of bloody elements are expelled by the force of the heat through pores fitting them, into lymphatic capillaries. This is concurrently directed by nerves attached to the last veins, subjected to the will of the mind elicited by the perception of the senses.

And this was said about the nature of glands in general. To this the following points should be added:

Tears. I. Tears appear to consist of a humour normally smearing the eyelids, carried from the glands through obvious paths. Their quantity is increased by emotion in the same way as the presence of delicate food provokes salivation in the one who has appetite so that there is no need to resort to a juice of the nerves, a serum of the brain, a direct exudation from the arteries. I dare not yet assert whether the pores which I observed several times in the cornea with admiration transmit a certain part of the aqueous humour: I assuredly saw humour coming out through the pores but is it not to be assigned to the substance of the tunica itself or is it to be derived from the included water, I could not determine it easily before further examination since it was never tried except in doing something else.

Gastric Humour. II. Since serous humour always flows down in the stomach from glands present in the ears, eyes, nose, all the mouth and the oesophagus, and the inner crust of the stomach, if compressed, exudes some humour but otherwise is harder than any other digested substance, and, without showing any tracts of fibres, looks like a kind of true parenchyma, who will doubt that these sources permanently supply serous humours to the stomach? And to whom is it not evident altogether that something else is found in the stomach besides heat alone? If we proceed to the function, chemistry is to be summoned to the discussion. But let me not transgress here the limits of anatomy.

Milk. III. I will hold for established that milk in the udders is secreted from the blood not otherwise than any other humour in other conglomerate glands, until those who draw it out from elsewhere have actually described their paths and have confirmed them with reliable facts. The illustrious *Anton Everaerts* (who does not seem to have considered that he put forwards a unique explanation only about the milk of the udders and that he has given up about more frequent observation of the breast, that I speak actually only about the milky ducts to the udders when I say that he observed them only once) always calls for an autopsy. I separated udders distended by milk from the subjacent muscles as carefully as possible but, although the udders were swollen with milk as well as the receptacle of chyle, I could, however, not see any passage from the latter to the former. But to have declaimed their assertions only is not to have refuted them. One draws off the chyle to the udders through the thorax from the thoracic duct, another from the receptacle through the abdomen, a third one from the womb through the muscles of the abdomen, and all boast of autopsy.

Catarrh. IV. The one who would say that glands are the sources of catarrhs will easily explain the symptom since the head abounds in glands everywhere. In the last volume of *De Catarrhis*, the illustrious *Schneider* wishes to make believe that he has discovered them. However, one finds that he contradicts himself in the third volume, where he claims that the sources of catarrhs are membranes and describes how nature opens the ends of the vessels in these membranes and, once the liquid has been secreted, closes them again, without any mention of glands. To correct that, he adds in the last volume that nature, when closing the vessels, uses the lid of the glands. But the illustrious gentleman does not seem to have considered that the flow is continuous and that an increase in quantity of the flowing and a change in quality provoke diseases so that there is no place for the invented lids, and the glands are not lids but sieves.

And let this dissertation about the muscles and the glands be enough as some examples of observations.

You see here explained quite a few points neglected so far, some even listed by the most trained anatomists among the impossibilities. To what was said it would perhaps be possible to add other matters of no less importance. But, since to present what I promised requires much time and since these matters also suppose a mind free of other worries, I will not mention them by name, satisfied to have shown at least with this specimen that not all which can be said was said in our century and that our eyes are not so dull that much of which many people would despair to know, could not be clearly demonstrated even by their action. If, to the one who wishes to imitate this, the first subjects do not seem to tally immediately with what was announced, let the second point be considered and the third be added before blaming for being false which may not be so. These matters were not observed one day only, nor were they seen in one subject only, nor were they done in such a youthful manner, howsoever *Deusing* goes on to scoff at a young man, that they have not had as onlookers, gentlemen and old men with a great name and well skilled in this art, who do not refuse their agreement with what was seen.

2.16 A Letter on the Anatomy of the Ray

To

The most excellent

WILLEM PISO

Very successful doctor in medicine,

Very illustrious Sir,

The memory of this day when, spontaneously approving the endeavours of a youth to investigate in Leiden with the very famous *Sylvius*, you were the first to show me the way to acquire your manifest approval, has not yet left my memory. Although the antiquity counts the ravens among the inauspicious birds, for me indeed, no dissection was ever done more auspiciously than that of an aquatic raven, since nothing would ever meet the expectations and wishes with greater profit. How indeed could I have offered to myself that in the future a gentleman very much praised by all, who had already made his name known for ever by the most desirable and diligent marks of genius and power of judgment, would trouble himself in such a way that not only the training exercise of a youth would be deemed worthy of his presence but that

OPH 16, vol. I, 195–207: The dissections of two rays, on which "De anatome rajae epistola" is based, must have taken place in Copenhagen shortly after Steno's return in the spring of 1664. Further annotations by Vilhelm Maar are found in the transcripts, www.extras.springer.com.

579

he would liberally offer his love and favour with the kindest words? I have very often admired this singular kindness since, moreover, I noticed that the first achievements had so pleased you that you wanted also to attend our other experiments in Amsterdam. So that I should not seem to ignore so many evidences of a great favour, I judged that I had officially to testify that I was grateful, at least with a little letter. Thus, since I should wish also to join to it other matter of writing so that it would not be insignificant, and I have so far looked in vain for such matter, here is finally a very opportune letter presented on a matter which you mentioned at your house where I had the honour to be. It indeed pleased the very famous gentleman Mr. *Simon Paulli*, physician of the King and prelate of Aarhus, my teacher respectable as a father, to admit me to the dissection of a ray on last March 21 of this year. There, as different matters nice to be seen and pleasant at least to us by their ordinance had forced our admiration for him, I believed that I should not more properly apologize for my long silence than by compensating the delay of the letter by the novelty of these matters.

There were at our disposal two rays both of the same sex but of different size, although the smaller one was taken by some people for a male because the prolapse of the intestine through the vulva was deceiving. Hanged from scales, the smaller exceeded 10 lb whereas the larger did not weigh less than 30 lb. The width of the larger one was four and a half spans and exceeded the length or the straight line from the tip of the mouth to the beginning of the tail by one span. The inequalities of the diagonals denied the animal a square shape. Neither had it a rhombic figure since the sides which were equal to each other were not parallel but concurred to form the same angle.

To those who looked at it from beneath the surface, everywhere besprinkled with countless black spots (I should have said globules of black glass if they had not been flat), just as many striae continuing these spots were displayed, although they were less blackish, because they were more deeply situated. Not only did probing with a bristle through the hole that perforates the spots disclose that these striae are hollow like vessels but also the humour contained in these were clearly indicated, and the expression of humour resembling the vitreous body of the eye through the orifices of the spots obviously demonstrated that actually they are vessels excreting through the surface of the body a humour similar to the lymph secreted in the conglomerate glands of other animals, in colour as well as in consistence. Those who trust their senses will easily concede to me that this is that humour which makes the ray slippery. I should not deserve the trust of all if I said that I arrived at this opinion not only as a result of this phenomenon but also of several others, since I think that all other slippery fishes have vessels running through the skin, if not completely similar, at least analogous to these in some way and, therefore, any small spots scattered over the surface of fishes must be suspected to have this function. When these are absent, other small, untinged points must actually be sought for. I should even incur the disapproval of those who would accuse me of being entirely addicted to a vice familiar to the anatomists inasmuch as a general conclusion is drawn immediately from one observation alone. I should have wondered that a liquid as glittering as the purest lymph is carried inside black ducts and excreted through black orifices, if I had not seen that very often in the mouth and nose of sheep. And this was obvious to anybody even without dissection just as was also the various position itself of the striae, which are very nice in various places.

Some tendinous fibres leave from the muscles lying over the abdomen to the skin or at least to a membrane tightly attached to the skin, so that they cause the hands great difficulty of dissection to the one who separates the skin spread over these muscles. They thus delight much the eyes. Intersecting each other alternately in a transverse course, they form a very pretty intertwining or, if you prefer, a pattern not dissimilar from that which microscopes disclose in webs made of a texture not so delicate. If you have pursued dissecting further and very accurately to the sides, you will notice that other fleshy fibres also rise to the skin among tendinous fibres. In the same way, ascending tendons of muscles situated on opposite sides intersect also each other in the middle cartilage of the pubis. These phenomena were only observed incidentally, not less than the muscles themselves of the abdomen. However, they have not been presented less satisfactorily since I saw in them something which, if I am right, I remember having observed in other muscles which I happened to dissect: i.e. fleshy fibres that end in a tendon on both sides and very short fleshy fibres to be observed in the long belly of a muscle compensating for their lack of length by their number.

In the abdomen, having been opened crosswise, the very nice position of the liver turned our eyes towards it alone. The liver itself was yellow verging to pale. It appears at the top of the abdomen, below the sternum, from the right to the left, occupying the space with its very thin base. From this, three lobes like as many tongues or imitating the prongs of the trident of Neptune, extended downwards over a length of one span. One of the lobes descended through the middle of the belly between the pylorus and the stomach, whereas the other two reached the two sides of the abdomen. Between these lobes two intervals gaped. The left one was filled by the width of the stomach. The right one contained the gallbladder above and the pylorus with part of the intestine below.

Since actually the left part of the hypochondrium did not contain the whole liver and the right side could not have made a claim for the spleen, which is of a bright red, since the transverse width of the upper abdomen was filled by the liver, the spleen had chosen for itself a middle position beneath the middle lobe of the liver between the two orifices of the stomach, or, to speak more properly of this subject, between the large cavity of the stomach and the pylorus. It was attached there to the angle of the stomach and membranes, and to blood vessels running through membranes.

From the other side of the pylorus, the angle which the reverted intestine makes downwards, offered the pancreas a convenient seat. There, the pancreas was provided with its duct, not less than in other kinds of animals.

From the mouth to the anus, by which way the intakes are led, the route, four spans long and more, is curved at very few places: first about the end of the stomach where, made narrower, it rises to the pylorus; from there, a little above the pylorus where the distorted intestine descends towards the right; thirdly in a place corresponding to the right groin of other animals where, striving to the left about the spine, it ends in a trace of the rectum.

You will look in vain for a tongue in the mouth where everything is flat and neither excrescence nor emergence protrudes. Continuing the mouth, without oesophagus, the stomach seems to make one and the same cavity with the mouth in dead animals. This cavity is fairly large for the size of the whole animal so that the large aperture of the mouth admitting easily two fists together can give a measure of bodies to be swallowed. However, I should believe that the fleshy fibres, thick and very strong which in the throat surround the upper orifice (like a spiral or a snail), while they function in live animals as a sphincter, separate the stomach from the mouth over a short distance. But, in my opinion, this part could not be called oesophagus since, whenever I divided tubes of oesophagus, all were made of two spirals inter-secting in opposite courses, not to mention other common prerequisites for an oesophagus. I shall tell how it actually is so that you will get an idea of for how many dead animals which had concealed viscera within their viscera one animal is alive. The larger one had devoured a one-half-span-long small cod (which we call "Dorsch") with a meagre sole not much smaller. From the smaller one we drove out, besides a turbot, nine sea-crabs of middle size, the whole spine of—I do not know what—fish, certainly not smaller than the cod mentioned above. I do not wonder that the soft flesh of the ingested fishes inside these confines, as if it were decayed flesh, became dissolved little by little, but the cartilages became softer, the shells themselves of the crabs were crushed (the minute particles of these shells tainted here and there in red the thick juice in the vicinity of the pylorus hinted less than they demonstrated).

But I should greatly wonder if this occurred from heat alone without the intervention of any humour. Certainly, the crust coating the inner surface of the stomach (the substance of which left after cooking becomes firmer and harder and does not display any trace of certain fibres) did not seem to me more similar to anything other than glandular substance. Spread with some viscous juice in live animals, this crust, if well compressed, exudes anew a new juice of the same kind; but this is not the place to discuss about these matters. As a strong sphincter surrounded the right orifice of the stomach, so the left one was tightly closed so that a finger could not be introduced in it except by force.

Beyond the pylorus, about a fold of the intestine, two ducts were opening between tunics inside the large intestine, neither at the same place nor on the same straight line but opposite to each other at the extremities of a diameter. One was the pancreatic duct, the other the biliary duct. This is called intestinal here and it continues into the cystic and hepatic ducts.

The inner surface of the intestine was also coated with a crust, but not as thick as in the stomach. If you reverse it towards the non-divided, proximally below the pylorus you will see that an elevated figure almost fringed resembles a snail shell. It descends down to the region of the coecum, not different from what I very often demonstrated to friends in the coecum of rabbits. There are three series of fibres in this snail. The two outer ones rise obliquely from the tunica of the intestine to its extreme orifice. The middle one travels along the length but neither this proceeds in a straight way. Not far from the end of this cochlear layer, an anterior fringe almost nervous appears. This is also white and resolves into filaments which, I think, are nothing else than its fibres collected at this place and twisted downwards. Do not think that this body is insignificant, it is more than two finger breadths long and eight finger breadths wide folding the intestines in circles, as I said. Thus the circles of the snail replace in a very short intestine the circles which in other animals form a very long intestine. The liquid contained in these intestines, like a thick juice, by its colour, disclosed the admixture of bile. In the coecum, which was very short and similar in substance to a parenchyma, nothing was found except a viscous humidity, concealed in the inner surface, rough with many small cavities. Therefore, a bright light seems to have been thrown on the function of a part obscure so far, as it appears presently from that which the famous *Simon Paulli* inserted by modifying it copiously, God willing, in his short reflection *Quadripartitum*[1] on this intestine. In this, various descriptions,

---

[1] *Botanicum.*

anatomical as well as practical, will nicely teach that a part unknown hitherto and ascribed to the classes of the dull and idle things which we say have been made only as a pretty mark, is neither dull nor idle but is of another class of more necessary function than just to make a pretty mark.

Here there is almost no mesentery since the intestine, otherwise free, receives only in very few places vessels proceeding a little inside a membrane. Hence its easy prolapse which, as I said at the beginning, made the fishers believe that it was a penis.

About the bottom of the abdomen on the sides of the spine, there was a rubicund body, one span long, one finger breadth wide but less thick. However, being divided and split, it was similar in all to that body which occupies an analogous position in birds. *Harvey* in *On the Generation of Animals* calls it kidneys in all birds, in snakes, in oviparous quadrupeds and even in fishes, and he claims it escaped the notice of *Aristotle* and other philosophers.

After the organs devoted to nutrition had been removed, the organs of generation appeared. There was a big difference between them in this pair. The younger seemed never to have conceived yet. Full of eggs, the older by the size of its passages, bore witness of frequent deliveries.

This animal, among the oviparous, was provided with two distinct organs to generate and exclude eggs, think of one on each side, arranged with great elaboration of the parts and built with no less skill.

In the ovary, besides eggs which were many of diverse size, and besides the membranes covering the eggs, and the blood vessels running through the membranes, and various intertwinings of fibres connecting all, the substance on which they lie must also be noticed, which at first glance you would take for the pancreas for its colour and softness. When dissected, you will actually find it full of a thick liquid with bloody striae giving a white colour. I wonder, however, whether it was molten parenchyma since, after this juice had been expressed, a more solid substance of the same colour, provided with many vessels, remained nevertheless, lying beside the membrane. This substance looked nothing other than glandular.

Although it is called uterus (its part indeed in which conception and conformation occur as well as the nutrition of the conformed foetus deserves this appellation), it can hardly be considered as a uterus because from the ovary to the white body not yet removed, it resembles a kidney: from its medial cavity which is almost equal to its whole width, a fairly wide and thick neck descends towards the drain of the uterus; above, it tends to the dia-phragm its horn made up of very thin substance very easy to break, the true boundary of which, on either side, would not present itself to the researchers.

By inflating the horn, we followed each of them up to the diaphragm where the air found a free exit.

The body itself of the uterus, so called here only for the sake of distinguishing, when freed and dissected from its coating tunica, is made of two substances, one external white, and one internal grey. They are provided with almost the same difference in colours as that which distinguishes the medulla of the brain from its cortex. But the whole body which is white is not of the same substance: you will find indeed that many fibres extending from the inner tunica to the outer one, longer in the middle, progressively decrease towards the sides and, besides the fibres, mainly on the sides, there is another substance not distinct from the fibres, similar to a glandular substance. The diverse necks of the diverse uteri do not join in one orifice but, on both sides, the uteri, each provided with its own orifice, carry down their contents into a common receptacle provided with its orifice. This receptacle is otherwise called cloaca in the birds although a membrane separates this from the intestine as a dividing wall.

In the younger one, the lowest part which was the substratum of the eggs was made conspicuous by its white softness, displaying on both sides many bubbles of the same magnitude, bright almost like water, opposite to each other about their middle. In this animal actually only a very thin tubule of membrane communicated with the horns and necks of the uterus. From there, a way out into the common cavity was not more open to a probe or to insufflations than from the cavity above. This obviously indicated that the bar of virginity was not yet violated.

On the beach, the limits of sand extend furthest when that of the ocean turn over at ebb. Among other things which the retiring water leaves on the dry sand, frequent small shells are found that display the shape of an oblong rectangle. Their longer sides, like in a case, extend on both sides beyond the surface of the parallelogram. It is usual for researchers to find them empty. It is actually rare to find in their swollen belly a substance different from air or sea water or sand, as happened to the very illustrious *Simon Paulli*. This gentleman saw two shells similar to those described above, turgid with a viscous humour, resembling almost the substance which flows from decaying mussels. After the content had been scraped out from the inner surface of the shells which concealed mussels, the shells had a dark blue colour inside. The people who retrieved them asserted for certain that these were also included in the stomach of a ray. Those who gave us the description of animals illustrated with figures depicted a shell similar to the ones above, with a bunch of eggs lying over it. They claim that the eggs are eggs of a ray and that the shell should be called the uterus of a ray. Let me say that I disagree, and why this seems unlikely to me. I wonder that *Rondelet, Gesner, Jonston* and so many gentlemen have pronounced such a simple and absolute judgment on a matter so lightly examined and have not given more careful attention to the function of the uterus in the generation of animals. What animal, I ask, rejects an excluded uterus with the delivery? And how many uteri would be needed for so many eggs which, deprived of place, are produced in excess on both sides by the ovaries of the rays? Among the anatomists, there are, I must say, very experienced gentlemen to whom an egg has been presented as a uterus which, if also the shell thus were an egg of this ray, would also deserve the name of uterus according to their opinion. Neither to be misled nor to seem to mislead others, let us scrutinize without prejudice all the parts of the egg as well as of the uterus so that it appears whether those who said that the egg is an exposed uterus have considered everything rightly.

The parts of an egg either change into the substance of the chick such as the albumen and the yolk, or surround the chick, such as the envelopes. Some of these adhere to the animal itself. Such are the amnion which encloses the chick floating in a liquid, the small bag of yolk attached to the intestines, and the chorion enveloping the whole. Others are free from vessels everywhere. There are three: a thin membrane thus which closely surrounds the chorion; another one thicker, which is white, around the first; and finally the external cortex which makes everything more secure by its hardness.

These are, indeed, the parts which we accounted as distinct in an egg. The comparison thus carried out will show what similar to these a uterus turgid with a foetus has. In the uterus also humours from which the foetus rises from are collected. There are also membranes in continuity with the foetus which surround the humours and the foetus. There are three at least in the beasts as I have separated the whole allantoids not only in a calf but also in a sheep, even if *Harvey* in *Exerc.* 68[2] says that he did not find it in a sheep. Since actually these are not those which are called uterus in viviparous animals, it is sufficiently clear that, because of them, this name uterus must not be attributed to an egg. What does thus remain? Certainly nobody would call uterus the shell together with the two membranes free from any community with vessels, which alone are left, unless one also said that the only function of the uterus is to contain the foetus without having, moreover, any commu-nication either with the mother or with the foetus. This would be far below the dignity of a very skilful part since the uterus alone does for the viviparous

---

[2]W. Harvey, *On the Generation of Animals.*

animals all that the oviparous animals expect from the ovary, the oviduct and the egg.

The structure of this substance indeed is such that the arteries of the mother in certain places transmit there all the humour which the eggs receive from the chalice of the ovary and from the thickened but soft and whitish body of the oviduct. The body itself of the uterus not only warrants the same function as the three external envelopes of the egg but also, receiving from the mother life as well as sensation and movement, it keeps the foetus warm with moderate heat while bearing it. Also, at the proper time of delivery, it mostly contributes to its proceeding to the light. This movement of the uterus is very obvious in vivisection of animals. Add that every uterus, while bearing a foetus, carries the egg: although it yields to the egg exposed in this place since, before the foetus is formed, it does not have all that is necessary for the formation and growth of the foetus but, this growing progressively, it receives new increments nutrition from the mother. It is, however, similar in that it has no communication with the vessels of the mother, contrary to what most people have believed so far.

Since thus what is common to uterus and to egg is not that thanks to which the uterus holds its name, what actually remains in the egg[3] is nothing else that some dead envelopes which have no life neither from the mother nor from the foetus. Since, moreover, the uterus itself provides the foetus with all that which the chick receives from the ovary, the oviduct and the egg, it indeed warms up inside itself the true egg and brings the foetus to the light in due time, the egg could not be called uterus except completely improperly.

From this it is obvious that the said small bag, even if it had been connected with an egg of ray, must, however, not be called uterus. It remains to investigate whether there are certain phenomena from which it could be deduced that it had been an egg of ray. If we had fish ponds which would restrain the freedom of rays to move about, an autopsy would shortly settle this controversy. Since we have none, we shall deserve indulgence if we admit nothing else than what it is blameless to collect from data. You will not easily convince somebody comparing the ovaries and oviducts of a ray with organs which must be considered the same in a terrestrial species, that the egg which, while it sticks to the ovary, is round, when it has tumbled down, will take in addition four thin and soft outgrowths, even if they are long. Its inside, which is close to the ovary is not much different from a square and the cavity of the body would perhaps bring about the square and flat shape of the shell but

---

[3]Read this as *ovarian follicle*. The egg itself was not yet identified by microscopy.

whence should we deduce that they are outgrowths of the same shell? Neither do the descriptions agree with each other. Making believe that the clustering of eggs clings to that which the picture of the authors demonstrates, what else will you gather from that other than that a fish, perhaps a ray, has poured out eggs over this body? And what has been retrieved from the stomach does not at all prove anything, since the gaping aperture of the mouth consumes anything it meets. The mother, grave of her own offspring, could have swallowed the egg expelled from the uterus, but this actually does not have any kind of truth, since the same evidence allows saying that some kind of zoophyte has been absorbed by the ray. As long indeed as either a similar bag has not been found in the oviduct of a ray or an offspring of a ray has not proceeded from such a cover, since I am not afraid of tearing away the authority of great gentlemen, I will defer these matters to be examined to the judgment of others. It will be enough to have demonstrated that the said bag, if even it had been an egg of a ray, should not have been called a uterus.

And thus leaving the abdomen, a way was opened with the knife to the innermost of life, where, distinct from the membrane coating the entire cavity, the pericardium was looked for. It reproduced the shape of the whole heart but inversely. Indeed, matching this flat head at its top, it has been accepted as a tail a fleshy portion separating the orifice from the base of the aorta, and actually matched nicely the protrusive angles of the sides with its angles. The heart had one ventricle and there was one auricle only situated between the heart and the gullet. The passage of the vena cava commands to give this the name auricle. Resolution by cooking showed that the two bodies are made of fleshy fibres. At this place where the artery divides into branches, a yellow gland is concealed near the tunica of the orifice.

Traces of lungs were here neither clearer nor more obscure than in other fishes. If you carefully inspect five gaping clefts on each side inside the mouth, you will see the hiding places of gills shaped so that the water taken into the mouth, when it has been sent forth through these inlets, cannot return through them because outside skin descending on any hole covers them.

Now what else do the countless orders of blood vessels which supply the gills and make them magnificent than to subject the blood which they contain to a modification from the ambient medium. This happens either by sending out something from the blood or by receiving in the blood external matters, or by achieving both by one and the same action. I beseech those who consider the thing itself rather than its concepts or its appellations to see whether the thing does not finally fall back to the same, either what is received from the ambient medium in the same cavity is sent out by the same way by which it was sent in, or that what has been ingested in one way is sent away through another. But, in the passage it touches the extremities of the blood vessels which surround the bronchi in the lungs and protrude bare in the gills. And indeed it is certain at least in breathing that it is required that the ambient medium, be it water or air, always renewed be brought to the extremities of the vessels. Those who deny lungs to fishes seem to be right since the different organs deserve different names. But it does not result from this that fishes lack a part by which they carry out a similar function. People say that it is not wrong to use cautiously the name of respiration if we are mindful of the deviation of its meanings, as water cannot properly be called airy spirit. But, since I should believe that the vessels of terrestrial animals do not retrieve the thick air which surrounds us more than the vessels of the fishes drink in water, who knows whether water does not do for them what air does for us, by yielding to them thinner bodies contained in its embraces, and which are the spirits of something, if they yield something otherwise. Indeed if they receive refuse only, the matter is easy and liable to no controversy.

The second ventricle denied to the heart raises the greatest difficulty. But this difficulty will not be such for us if the different points are considered properly otherwise. We shall concede that all the blood is not carried through the gills. Nor does all the blood pass through the kidneys, the liver and the other viscera but, nevertheless, all these serve to the mass of blood. But more precisely: not to mention the embryos to which something can be restored, I do not forget that, even in grownups, if we believe the anatomists, a route sometimes open from the right to the left auricle is testified to have been found. In these people all the blood is not transmitted through the lungs. But so as not to seem to proceed from the abnormal to the normal, I ask that this quantity of blood which is transmitted through the gills should be considered and it will appear that this quantity is sufficient so that, concurring with the rest of blood from there in the auricle, to a suitable proportion it reduces everything easily.

Since, thus, the gills in fishes carry out the same function for the blood as the lungs do in other animals, although the latter are different from the former in some respects, let us not bring into the discussion the black substance attached to the spine, or let us not have recourse to other subterfuges. Nothing is as necessary to maintain the life of animals as the alternate acceding of the ambient medium to and its receding from blood vessels whether this occurs by receiving the ambient medium inside (and this either through the bronchi like in quadrupeds and in birds, or through the cavities of bladders like in frogs) or by bathing the protruding vessels of the gills in the medium outside. It finally amounts to the same thing since the function of the lungs is carried out in any case. Thus it is not because they are deprived of lungs that the fishes must be called mute but because they have everted lungs so that they can be affected by the ambient medium, even if they do not reject through a pipe this medium received in a cavity. As far indeed as the sound emitted by animals is concerned, we see that what is needed is only that the stream of sounds passing through straits be squeezed out in its passage through obstacles besetting the narrow place. The very fitting lid of the pupils would almost have escaped my sight. *Apelles* will not have depicted more nicely the disc of the sun sparkling in every direction with its shining rays than it was done partly in the eye of the ray by that which animates all colours, nature. The order even of the rays which surrounded the middle circle was very elaborate and the golden brightness of the whole disc glittered. It lacked only that the rays would not set free the whole crown, leaving an arc by which the lid was fastened above to the iris, not less bright. And indeed it was actually part of the uvea so that in its backside it was tainted as usual in a tawny colour. The nicety of the structure is admirable but its function is even more admirable. What the pupils could not have expected from the eyelids (there is plainly no upper one and the lower one is hardly noticeable), this organ described here provided to it, now admitting the rays, then keeping them out depending on the diverse influence of internal or external impulses. Nature to so great an extent is not tied to certain organs to act but is able to carry out a function with diverse instruments. This is demonstrated here by the uvea, the lungs, and the intestines. For whom would it, therefore, not be very sure that the more successfully and fully reason may have deduced a priori diverse modes of function of organs, the less reliable will it be in a certain case, unless the senses have determined the matter through experiments.

It seems that I have presented much on a matter not that great and I should have given more complete arguments and added much more if, while I had the knife in my hands, those same things which the pen succeeding the knife wanted had occurred to me. My very illustrious teacher indeed had not the leisure to go down to all details and I who seldom dissect fishes had never had the possibility of examining the viscera of rays before the kindness of the same very famous gentleman urged me to do so. Not to mention all his other paternal favours to me which I always remember with gratefulness, I owe him much for this reason only that he has given me the opportunity of observing matters which, if they may not please common people, will at least not displease you. This I gathered from your very learned and altogether very pleasant discourses on your journey in Brazil, which I will always remember with delight. Hence I perceived clearly where you have been stirred by the desire of examining small wonders of this Earth, and how much you spent to survey natural things since, during the time you remained there, you could not fully satisfy your desires because of other occupations. Your curiosity does not know any boundaries and would not devote itself so much to rarities if you did not judge that the rarest things are hidden among the most common ones. My eagerness to send these matters to you thus cannot be blamed. It would have been my duty to provide all the details for this veneration to stand for your sight without blush. But although an urgent journey opposed such desires, I would not, however, withhold them till they reach maturity, since I should expect to myself the same goodness from honest people which you, your kindness, certainly presented to me, you who look for matters not for words, and where actually they are even inferior to expected values look for the will, not for the effort. The very illustrious *Simon Paulli* to the affability of whom all these matters are due, bids me to greet you with the most courteous words. Our very famous *Bartholin* (who has often made honourable mention of you) would have bidden the same if his stay in his estate had not prevented us to see him. Do well, very famous Sir, and greetings.

Copenhagen, in the year 1664, April 24.

Yours sincerely,
Niels Stensen

2.17   On the Passage of Yolk into the Intestines of the Chick

A Letter Addressed to PAUL BARBETTE,
Most Renowned and Experienced Physician

Many are the rare discoveries this age has produced in the field of anatomy, and if these were to be accounted monsters, anatomy would not yield first place to Africa in its numerous brood of prodigies. And why should you not when they are so closely related to the monstrous? There are some that merit exhibition to everyone because their novelty excites wonder, and there is no dearth of those displaying real abnormity in that they violate the ordinary laws of nature, but very often it is the impetuosity of the observer that invents a monstrosity when there is nothing abnormal in reality. So, too, there are many things that the Africans themselves wonder about, doubtless only because they consider that everyone ought to be shown that which no one has

OPH 17, vol. I, 211– 218: "De vitelli in intestina pulli transitu epistola". The observation on which this treatise is based must have been made after Steno's return, to Copenhagen in the spring of 1664, and before the June 12 (old style), the date of the treatise. It was published together with the two previous Treatises (XV and XVI) in *Nicolai Stenonis de musculis & glandulis observationum specimen … Hafniae 1664*. Further annotations by Vilhelm Maar are found in the transcripts, www.extras.springer.com. OPH 17 was translated and annotated by Margaret Tallmadge May in *Journal of History of Medicine and Allied Sciences* 1950; 5, pp. 119–143, with a postscrip, 338). Translation reproduced with the permission by the publisher.

© Springer-Verlag GmbH Germany, part of Springer Nature 2018
T. Kardel and P. Maquet, *Nicolaus Steno*,
https://doi.org/10.1007/978-3-662-55047-2_28

seen before. Sometimes, however, chance links animals of different species as they gather at the common drinking places and produces hybrids, just as nature in her operations turns aside from her accustomed path sometimes in one way, sometimes in another, and brings forth offspring which to the Africans seem monstrous. But when the supposed deformity of the thing which appears throws them at first glance into instant flight, quaking with terror, they then imagine a monster a thousand times more deformed than what they actually saw (which was perhaps nothing short of monstrous). This is the way with morbid and superstitious persons; I cannot think what terms for specters and portents they do not use to describe any meaningless noise, the delusive shapes and colours of the clouds, often even their own shadows, and other trivial causes of childish terror, and they have no trouble in persuading the credulous that they have truly seen prodigies when, in fact, the actual appearance was far different. Now just as fright deceives the Africans and the credulous, so does joy trick anatomists when some observation which at first sight is unusual, favouring their preconceived opinion or offering a point of departure for a new line of thought, impels them to redouble their "eureka" and leaves no room for pursuing a more careful investigation. There are sometimes still other impulses of the mind, here to be passed over in silence, that have caused many to father monstrosities in their writings. It would be invidious to illustrate my meaning with modern examples, while one who brought forward examples culled from the Ancients would seem to be denying them their due respect. Nor is there need to seek elsewhere for a fault which is easy for any one of us to detect in himself. For I do not believe that anyone who is not too slight or too casual a student of anatomy will deny that just as he does occasionally observe some new things and some that are extraordinary, so he not infrequently persuades himself that he has seen what he has not.

And when those who are striving to gain an understanding of animals through experimentation are carried away by the rash impulses of their minds as if driven by a sort of whirlwind, what wonder is it that many anatomical monstrosities are hawked abroad which are never to be found by others, no matter how assiduously they may pursue the search? Unless the mind is tranquil, it will by no means be free to apply itself to a close examination of facts which can and ought to be closely examined, and unless every least detail is noted in so far as the minuteness of the object or its intricate diversity allows, the pathway to error is downhill and very easy. Therefore, just as in all dissection there is need of great and heedful attention, lest of its own accord the straying mind should distract the far too agile eyes from the object before them, so the greatest caution must be exercised when anything is encountered in variance with the commonly accepted opinion, lest one should be deceived by outward appearances and shout one's triumph before victory. For my part, I have been badly frightened by other men's records, and in any unusual observation I would not dare to put any trust in my own eyes unless, after very frequent and repeated examinations of the same thing, they also had as witnesses of the observations the eyes of others having expert knowledge of the point in question. And if, accordingly, you should come upon anything resembling a monster in these pages which you are now about to read, you may confidently believe that it has been examined by the eyes of men most highly trained in anatomy.

I myself began by putting it down as a monstrosity when I observed *that in the chick the nutriment, hitherto believed to be received from the yolk by the umbilical vessels and conveyed to the portal vein, is taken from the tunica of the selfsame yolk through a duct of its own directly to the middle portion of the intestinal canal.* It has been ridiculous to many for the chick to be nourished by mouth; how much more ridiculous for it to be nourished by the way of intestines themselves, especially since even the eyes of the mind have never imagined that this occurs in any condition of life? Now many have thought and still think that while the animal is living in the fluids without air, nutrition is accomplished by means of the umbilical veins; others teach that the liquid in which the foetus floats is conveyed into its body only through the mouth. In the perfect animal, the mouth is known as the entrance for different kinds of sustenance where not only solid and liquid foods descend through the oesophagus, which is the common opinion, but also, equally with all the other spiracles of the body, airy spirit is drawn in by the lungs, as certain chemists do not so much conjecture as maintain. There are stories of many sick persons who have prolonged life for some days by odours alone, and of not a few who have derived benefit from nourishing clysters. But to whom would it occur even in a dream that there is any condition of life in which nutriment is poured into the midst of the intestines?

Therefore, the stranger and more paradoxical it seemed, the more carefully I have thought it should be studied, until every doubt and every fear should be removed by observations always consistent in every subject I examined, so that I might avoid hawking abroad fiction for truth, and for the normal, something monstrous. These observations are as follows:

In the egg, both albumen and yolk have their own membranes, in both cases provided with blood vessels; of these membranes, the one surrounding the albumen is destroyed when the egg hatches, but the other is taken inside the abdomen and is found in the chick for a long time, even later than the 30th day after hatching according to the observations of *Harvey*. *Aristotle* calls the vessels of each membrane umbilical cords and thinks that the one running to the yolk is attached to the small intestine; but whether the yolk hangs down outside the abdomen, or whether by the gradual drawing together of the abdominal wall it is retracted and lies hid within, it has its own duct by means of which it is attached to the intestines. Blood vessels passing over the middle portion of the intestine crowd against it on both sides, and so it happens that by enveloping this third duct the blood vessels have hindered its discovery. The tunica of the yolk resembles a little sac with a large cavity but a very narrow mouth, from which a little vessel leads to the intestines at a point about half way between their beginning and end; it is scarcely as thick as the blood vessels, and moreover is the color of a ligament, so that you would think it a nerve, were it not that the passage of liquid proves it pervious. The way in which it enters the intestines does not differ from the insertion of the pancreatic duct, except that the latter is attached to the intestines on the same side as the mesentery, whereas the yolk finds an open passage on the opposite side. By this passageway the yolk is poured into the intestines after it has been liquefied and converted into a clear fluid, the containing sac diminishing gradually along with its contents until finally, when all the juice has been used up, there will adhere to the intestine an elongate hollow tubercle, which is the only trace of the yolk permanently remaining, so that by means of it the place where nutriment was taken in through the intestines can be demonstrated even in adults. But in order to be so much the more certain of the truth of the matter, I laid open the intestine and saw that when I squeezed the sac, at first a somewhat thin liquid, followed later by a thicker one, ran out through the little vessel I have just described; and since this always happened in the same way whenever I tried it, I finally came to the conclusion that this passage is the one by which the yolk enters the body of the chick, and that the process called chylification actually takes place in the intestines as well as in the stomach. But to make this whole matter more fully evident, I will briefly set forth the circumstances which led me to an investigation of it.

Uncertainty over the function of the liver has long kept me in doubt, and even to this very hour it has not been granted me either to observe experimentally or to learn from any other source facts sufficient to settle that question beyond all doubt. Since those who refer everything to reason alone, because they are so clever, seem to proclaim each his own opinion in all but dictatorial fashion, and since those who are poles apart from one another in the way they solve the problem are alike confident in their proofs, anyone who is avoiding mental bias and is not an adherent of any school sect will not easily be able to divine which of the many opinions he may safely adopt as the more certain. But those who have entirely devoted themselves to experimentation are so far from removing every uncertainty that they have made the whole problem much more perplexing, while they either heedlessly disregard the rules prescribed by that method of searching out truth or knowingly and wittingly disregard them for the sake of an opinion laid hold on once and for all, which they fasten upon and cling to forever as to a rock, referring everything to it, both what they have seen and what they have not. In no other matter is this tendency more clearly evident than when the question of the route of the chyle is dealt with. After the true route of the chyle and lymph had at length been established, although experiment seemed to have removed the logical difficulties, and the diligence of those copying the same procedure to have exposed the falsity of many of the experiments, the following considerations nevertheless remained to make uncertain the mind striving after truth. First, it was not clear that all the chyle is conveyed by this route, since observation showed only that part of the ingested nutriment filtered through the tunicae of the intestines it sent through the lacteals up to the vena cava; and now who will vouch for it that all the remainder passes into excrements? Who has convicted of error all those who believe that the office of the veins is not only to carry back the blood transfused from the arteries, but also to convey to the heart along with the blood any substance fit to enter their orifices? And if you allow that this happens in the rest of the body, what reason is there for denying something similar to the intestines? This uncertainty is accompanied by another of no less importance, namely, that an investigation of this matter has not been made in every kind of animal. Now if, therefore, animals are to be found which have not been provided with lacteals, or if a time of life is to be observed when the blood vessels convey straight to the liver the nutriment taken in at their extremities, why should like conditions be granted to other animals also, especially as long as nothing has been found to prove the contrary? How momentous this doubt—to pass over the many others I could adduce derived from this source—should be considered, although few have remarked upon it, the single example of *Harvey* will show. He refused to acknowledge lacteal vessels because he believed that in birds the yolk is carried through the umbilical vessels to the liver. Hence in *Exerc.* 21[1] he says: "And if the argument based on the course of the umbilical veins (which I have said end in one or two trunks at the porta of the liver) holds good, the chick already then is nourished by the attraction of nutriment from the yolk through the umbilical vessels in exactly the same way as it is later nourished by the chyle brought from the intestines through

---

[1]William Harvey, *On the Generation of Animals.*

the mesenteric veins. For in both cases the vessels end at the porta of the liver, to which they convey the nutriment that each has attracted. Thus there is no need to resort to lacteal veins in the mesentery (which are nowhere found in birds)." These words have caused me no little trouble since I decided that if the veins discharge this office in the chick, the privilege ought not lightly be withdrawn from them in other animals or at another time of life. Hence it is that I have never ventured to oppose the liver, although I have never been one of its partisans; for the authority of *Aristotle* and *Harvey* weighed so heavily with me that I allowed as an established fact that idea about the transmission of the yolk into the liver through the umbilical vessels, especially since I saw that it was not inconsistent with reason.

But since in their accounts I failed to find what happens after the yolk has been exhausted from the sac, and certain other points of no great importance, indeed, but nevertheless contributory to a full review of the facts, I took from a brooding hen one of her eggs, whose shell had just been broken by the chick making haste to hatch.

When the clothings had been stripped off, the swollen abdomen betrayed the presence of the yolk lying hid within. I will not recount the facts, more agreeable to be noted by observation than by reading, about the way in which the abdomen is enveloped by the clothings, but leaving all that, I will say that while in simple faith I was searching for the passages followed by the yolk from the sac to the liver, the passage I was not seeking from the same sac into the intestines offered itself unbidden to me for discovery. Thus the dicta of other men are too often oracles for us until a more careful examination reveals how far removed they are from the truth. I had read the words of *Harvey* confirming the opinion of *Aristotle*, and I saw that reason favored them both, with the result that not even the least suggestion of doubt occurred to me until the whole structure was, as it were, overturned in a moment by my own observation, which was the direct cause of rendering the throne of the liver now all the more shaky by the withdrawal of this prop also. For since any foetus takes by mouth the liquid in which it floats, and the chick takes the albumen by mouth and the yolk in its intestines, I see no evidence which proves conclusively that part of the humours are conveyed to the liver through the umbilical vessels rather than to the heart, although we are by no means without facts which give apparent support to the idea. Now I am not actuated by partisanship for the organs, nor have I ever declared war on the liver, by I have asked of its defenders nothing more than that they should cite experiments to bolster the inventions of their reason. Hence, although we might, now that we have clear and open passages for the nutriment, set the following limits to the function of the umbilical vessels, that by their heat and perchance by whatever liquid eventually seeps through the ends of the vessels, they merely make the humours suitable for nourishing the body, but convey no humours into the body, nevertheless, while the dispute over what the mouths of the veins receive is yet unsettled—whether it is only blood, or in addition to blood, any suitable substance applied to the pores—to proceed to such a point of boldness would seem an unsafe thing to do.

Moreover, it is an inquiry of no little importance when we investigate the way in which any substance applied to a part of the body transmits its particle into the body. There are veins which can receive them, and in many animals lymphatics to which some have assigned the same function are not lacking; but if indeed you mean that there is penetration through the pores of the skin, membranes, muscles, and bones to the inmost viscera, you must in this event mean: either the material is subtle, as in bodies actually hot or cold, just as in magnetic substances too, or the parts themselves are so relaxed by the force of disease that true passage occurs, unless, perhaps, the substance is so corrosive that it would force a way through everything in its path, even though it were iron. I will admit that the bile has been seen to stain the parts in its vicinity with a yellow colour although the gall bladder was unbroken, that fluid has been observed to seep through the bladder of the ox, and that dust from the cut stones has been found inside a bladder hung up in a sculptor's workshop, although it could not have easily entered through the tightly closed orifice of the bladder; but a condition which is to be ascribed to fibres of membranes either relaxed by disease or dried out does not apply to a living and healthy body where everything is replete with spirit and moisture. I see no reason, therefore, why heart strengthening poultices are placed over the region of the heart other than the course of the veins from that part of the skin to the heart is shorter than if you were to apply the same poultice to the hands, seeing that otherwise anyone ordering a purgative dose to be held in a sick man's hands would have no greater difficulty in causing a bowel movement than the one who smeared an ointment with the same faculty on the navel. Thus, neither in one place nor in the other does any penetration at all take place through the pores of solids, but only a mixture with the blood of particles received by the vessels which produces the desired effect. Cantharides[2] behave in the same way and true poisons not differently. But it would be a long task to unravel every problem involved in this general question. Since, therefore, substances applied to the outside of the body are received by the mouths and the veins and follow their channels, why should

---

[2]Blister beetles.

not the fluids contained in the membranes of the uterus also offer a part of themselves to the umbilical vessels? This is plausible reasoning but it is not applicable in every case. For who would believe that the urine in the bladder is resorbed by the extremities of the veins when it has only just been secreted in the kidneys from the blood or, whatever is in fact the source, at least from other useful parts, a fact which otherwise anyone is altogether bound to concede, if he grants to all veins a structure such that they receive besides blood other substances applied to them, provided only that these are fit to enter? For why does urine have greater penetrative force when it is applied externally and exhibits its strength there? With equal reason we might reach the same conclusions in the case of bile and other residues. But if here you object that nature abhors her own excrements in disgust, you will accomplish little, since it would detract greatly from her wisdom if she who deals prudently with the bladder is caught being so improvident with the skin that in the skin she allows to be mingled with the blood what in the bladder she is unwilling to receive in the veins. No if, on the other hand, you call to your aid the fact that phlegm coats the inside of the bladder, you would seem to be bringing forward a more appropriate reason. But if you indeed admit the presence in the urinary and the gall bladders, or elsewhere, of obstacles preventing the junction with the blood of substances which enter freely in another part of the body, who will give us such full knowledge and understanding of the inner surface of the foetal membranes that he dares boldly to proclaim his opinion? Now just as the situation regarding the umbilical vessels is doubtful, so the truth is not sufficiently established concerning the branches of the portal vein, both for reasons I have just given, and because I have not as yet seen either lacteals, lymphatic vessels, or conglobate glands in the mesentery of birds or fishes; whether this has come about through faulty observation, or because of the absence of the structures in these animals, is a question I do not yet wish to decide. But if now it were established with certainty that there are animals whose intestines entrust all the chyle to the red veins of the mesentery for transport, you could not deny the liver a part of the chyle in other animals (and by chyle I mean whatever is separated from the ingested nutriment and forced out through the tunicae of the intestines into either lacteals or red vessels) unless you were to show that either the innermost villous tunica of the intestines, whose villi all face the center of this coiled tube, or some other tunica, differently formed in different animals, commit to blood vessels in one animal what in another it has destined entirely to the lacteals. But let us leave unsettled that which is doubtful until a future day shall yield experiments that either restore a part of its former function to the liver or show that it is entirely unworthy of the honour of the chyle. And yet it would add little to its dignity even if it did permit the free passage of all the chyle, since nowadays, indeed, most people have ceased to believe that the solid organs (which either set the blood in motion or act as sieves for what is to be strained out of it) contribute anything toward producing or perfecting it. But where have I been led by this discussion of the liver? I wanted to make known to you the route followed by the nutriment of the chick, a route to me new and all but abnormal, and at the same time to point out the circumstances which led me to discover it. This did not admit a sufficiently simple explanation without a consideration of the liver also, seeing that the umbilical vessels surround the yolk when it is contained in its own sac, and when it has been taken into the intestines the branches of the portal vein touch upon it, both these vessels having communication with the liver. Farewell, most illustrious Sir, and hold in esteem yours,

Copenhagen, June 12, 1664 (old style)

Niels Stensen

2.18   The Discourse on the Anatomy of the Brain

OPH 18, vol. II, 3–35: *Discours sur l'anatomie du cerveau* was delivered at Paris in 1665. English translation by Paul Maquet. Annotation based on Vilhelm Maar, OPH II, 313–317.

Vilhelm Maar: When going to Italy Steno left the manuscript in Paris, and, in 1669, it was published by a printer named Robert de Ninville,. Steno's relative Jacob Winslev (Jacques Bénigne Winslow, 1669–1760) reprinted the whole *Discourse* in his *Exposition anatomique de la structure du corps humain*, Paris 1732, and in all subsequent editions, including one German, seven editions in Italian from 1743, and one in English by G. Douglas in 1763. In 1671 a Latin translation appeared in Leiden, a translation of which was made by Guido Fanoisius, a physician of that town. In 1903 a Danish translation by Vilhelm Maar appeared in Copenhagen. *A Dutch translation can be found in J.G. Vugs *Leven en werk van Niels Stensen*, Universitaire pers Leiden 1968. A facsimile reprint of the *Discours* with (the second) English translation by Alexander J. Pollock, also with a German translation by Adolf Pilz, with comments on Steno's anatomical diagrams by Adolf Faller, and with introduction and annotations by Gustav Scherz, see *Nicolaus Steno's lecture on the anatomy of the brain*, Copenhagen: Nyt Nordisk Forlag Arnold Busck, 1965.

The *Discourse* was reviewed in the *Transactions of the Royal Society*, London, on September 20, 1669, see the reprint in Niels Stensen, *Foredrag om hjernens anatomi*, 1997, pp. 61–66.

© Springer-Verlag GmbH Germany, part of Springer Nature 2018
T. Kardel and P. Maquet, *Nicolaus Steno*,
https://doi.org/10.1007/978-3-662-55047-2_29

DISCOVRS

SVR L'ANATOMIE
DV CERVEAV.

A MESSIEVRS DE
l'Aſſemblée de chez Monſieur
Theuenot.

 ESSIEVRS,

Au lieu de vous promettre de con-
tenter voſtre curioſité, touchant l'Anatomie du Cerveau ; ie vous fais icy

To the Gentlemen at an Assembly at Mr. Thévenot's

Gentlemen,

Instead of promising to satisfy your curiosity in what concerns the anatomy of the brain, I do confess here sincerely and publicly that I know nothing of this matter. I should wish with all my heart to be the only one forced to speak so, for I could take advantage with time of the knowledge of others and it would be very lucky for the human kind if this part which is the most delicate of all and which is liable to very frequent and very dangerous diseases was known as well as many philosophers and anatomists imagine. Few of them take example of the simplicity of Mr. *Sylvius*,[1] who speaks of this matter only with uncertainty although he has worked on it more than anyone I know. The number of those who take pains to do nothing is unquestionably the largest. These people who are so prompt to assert anything will give you the description of the brain and the arrangement of its parts with the same self-confidence as if they had been present when this marvelous machine was built and as if they had penetrated all the plans of its great Architect. Although the number of these assertions is large and I must not be accountable for the feeling of others, I remain convinced that those who are looking for (II,4) solid science will find nothing that could satisfy them in all that has been written on the brain. This is certainly the main organ of our soul and the tool with which the soul carries out admirable tasks. Our soul believes to have so penetrated everything which is outside itself that there is nothing in the world which could set bounds to its knowledge. However, when it has entered its own house, it cannot describe it and does no longer know itself. It will suffice to have seen the dissection of the great mass of matter which the brain is made of to have good reason to complain of such ignorance. You see on the surface diversities which deserve admiration but when you go as far as penetrating inside you do not see anything. All you can say is that there are two different substances, one greyish and the other white, that the white substance is continuous with nerves which are distributed all over the body, that the greyish substance is used as bark for the white substance in some places and in other places it separates the white filaments from each other.

If we are asked, Gentlemen, what these substances are, how the nerves join in the white substance, up to where the extremities of the nerves penetrate it, we must acknowledge our ignorance if we do not want to increase the number of those who prefer the admiration of people rather than sincerity. Saying indeed that the white substance is nothing else than a uniform body as would be wax, in which there is no concealed artifice, would be having too low consideration for the most beautiful masterpiece of nature. We know for sure that where there are fibres in the body, everywhere they display a certain

---

[1]As early as in 1641 Frans de la Boe Sylvius had written additions and notes to the section on the brain in the first edition undertaken by Thomas Bartholin of the Anatomy of his father, Casper Bartholin the elder's, *Institvtiones Anatomicæ* ... Lvg. Batavorvm 1641. In this edition the fissure of Sylvius, for instance, is described and pictured for the first time. The parts due to him are marked F. S. - Moreover Sylvius was the author of the dissertation, published in 1660, *De Spirituum Animalium in Cerebro, Cerebelloque Confectione* ..., which constitutes No. IV of *Disputationum Medicarum Pars Prima*. Amstelodami 1663. When Steno later on in his *Discourse* alludes to Sylvius' criticism af Descartes' doctrines, he probably refers to some verbal remarks of Sylvius'; perhaps also to various opinions expressed in the two dissertations, published in 1663, *De Febribus*.

behaviour towards each other, more or less compound, depending on the operations to which they are destined. If the substance is fibrous everywhere, as indeed it seems at certain places, it must be recognized that these fibres must be artfully arranged since all the variety of our feelings and of our movements depend on them. We admire the artifice of the fibres in each muscle. How much more must we admire them in the brain where these fibres enclosed in such a small space carry out each its operation, without any confusion and without any disorder.

The ventricles or cavities of the brain are not better known than its substance. Those who accommodate there the spirits believe (II,5) to be as right as those who destine them to receive excrements,[2] but both are fairly puzzled when the source of these excrements or of those spirits must be determined. These can also come from the vessels which are seen in the brain as from the substance itself of the brain, and it is not easier to indicate their way out.

Among those who set the spirits in the cavities of the ventricles of the brain, some have them passing from the anterior ventricles to the posterior ones to find there the entrances to the nerves[3,4,5,6]; others believe that the extremities of the nerves are in the anterior cavities. Some claim that the excrements of the brain are in these ventricles because they see there something similar; the same people find that there is as much slope in the brain to let them down into the medulla as there is to lead them into the funnel called *infundibulum*: but let us suppose that everything gets into the funnel, you can let them out through the sinuosities of the dura mater and there is some

---

[2]Among those, who supposed that the ventricles of the brain contained animal spirits, may be mentioned e.g. Vesalius. See *De Humani corporis fabrica Libri septem*. Basileae 1543. lib. VII c. I pp. 622–623. An example of the other view, mentioned by Steno, is Bartholin's *Institvtiones Anatomicæ*, lib. III c. VI (p. 279 in the edition of 1641): *Vsus cavitatum vel ventriculorum cerebri est, esse excrementorum conceptacula* ... and c. VI p. 270: *GLANDULA PITUITARIA ita ab usu dicta, quod excrementa cerebri ex ventriculis per infundibulum suscipiat* ... Even Willis, in 1664, thinks that the excrements of the brain are conveyed down into the infundibulum and there discharged by means of the pituitary gland. See *Cerebri Anatome: Cui Accessit Nervorvm Descriptio Et Usus*. Studio Thomæ Willis ... Londini 1664. c. XII p. 146.

[3]In Bartholin's *Institvtiones Anatomicæ* ... Lvg. Batavorvm 1641, lib. III c. III p. 259 another view is maintained: ... *quartus* ... *ventriculus; in quo nos statuimus veram esse animalis spiritus generationem. Et hæc medulla oblongata est omnium nervorum, quicunque ibi sint, principium & origo; cerebrum verò ne unius quidem nervi initium; contra quam communis fert opinio.*

[4]See Bartholin's *Institvtiones Anatomicæ* ... Lvg. Batavorvm. 1641. lib. III c. VI p. 274: *Cæterùm hîc observabis facilem esse humori in dictis ventriculis contento exitum & secundum spinalis medullæ ductum descensum.*

[5]Steno here possibly refers to *De Spirituum Animalium in Cerebro, Cerebelloque Confectione, per Nervos Distributione, atque Usu Vario*. Respondente Gabriele Ypelaer, Amstel. 4. Februarij 1660 (see note to vol. II p. 3. l. 10 from bottom).

[6]See e.g. *Cerebri Anatome* ... Studio Thomæ Willis ... Londini 1664. c. XI p. 134.

reason to believe that they find passages which lead them directly into the eyes, into the nose and into the mouth.

There is even less certainty about the subject of animal spirits. Are they blood? Would they not be a peculiar substance separated from chyle in the mesentery glands? Would the serosities not be their sources? Some people compare them to wine spirit and one can wonder whether they would not be the matter itself of light? After all, the dissection which we usually carry out cannot throw light on any of these problems.

If the substance of the brain is little known, as just said, the true manner of dissecting it is not better known. I do not mean that kind of dissection in which the brain is cut in slices. It has been known for a long time that it does not throw much light on anatomy. The other dissection carried out by unwinding the creases is a little more artistic but it shows us the outside only of what we want to know and this very imperfectly.

The third manner which adds to the unwinding of the creases a separation of the greyish body from the white substance goes a little further. However, it does not reach beyond the surface of the medulla. (II,6)

Various combinations of these three manners of dissecting are made and others could be added: longitudinal sections, cross-sections.

Personally I assert that the correct dissection would be to follow the threads of the nerves through the substance of the brain to find out where they pass through and where they arrive at. It is true that this manner meets so many difficulties that I do not know whether it could ever be completed without very special preparations. The substance of the brain is so soft and the fibres so delicate that one can hardly be touch without breaking them. Thus, since anatomy has not yet reached the level of perfection of being able to carry out the correct dissection of the brain, let us have no more illusions. We must rather sincerely acknowledge our ignorance in order not to cheat ourselves first and then the others, by promising to show the true structure of the brain.

It would be boring to list here all the opinions and all the disputations which can be delivered on the subject of the brain. The books are full of them. I will mention only the principal errors which persist in the mind of several anatomists and which can be proved false by anatomy itself. They will be limited to the following. Among those who claim to know, some show you separate parts in the brain which constitute the continuation of the same substance, others want to persuade you through the administration of anatomy that the parts touch each other without being attached to each other although they obviously join through threads or through vessels. Some attribute to the parts the position which they think necessary for the system that they have imagined, without considering that nature has situated them in the completely opposite way. Some will demonstrate to you the pia mater where it is not and do not recognize the dura mater where this is obvious.

If need be, they would even pretend the substance of the brain to be a membrane. I have too high an opinion of the men of letters in general to believe that they do that in order to deceive others. The principles which they have established and the manner of dissecting to which they submit themselves do not allow them to do otherwise. (II,7) All anatomists would demonstrate the same in the same way if all of them used the same method. No wonder that their systems keep up so poorly.

The Ancients were so preoccupied by the subject of the ventricles that they took the anterior ventricles for the seat of common sense and destined the posterior ones to memory so that judgment, as they say, being accommodated in that in the middle can easily make its reflections on the ideas which come from one and the other ventricle.[7] There is nothing else to do than bidding those who, with the Ancients, support this opinion, to give us the reasons that would compel us to believe them, for I can assure you that of all that which has been asserted hitherto to establish this opinion, there is nothing convincing. And this nice vaulted cavity to the third ventricle in which they had set the seat of judgment and the throne of the soul is not even there. You can see how to judge the rest of their system.

Mr. *Willis* gives us a quite peculiar system. He accommodates common sense in the *corpus striatum* or striated body, imagination in the *corpus callosum* and memory in the cortex or in the greyish substance which envelopes the white one.[8] But there would be much to say if all these hypotheses were to be examined in detail. He describes the corpus striatum as if there were two kinds of striae, the ones ascending, the others descending, and,[9] however, if you separate the grey body from the white substance, you will see that these striae are all of the same nature, i.e. they are part of the white substance of the corpus callosum, which goes to the medulla of the back separated in various lamellae through the extremity of the greyish substance.

How can he be so assured to make us believe that these three operations occur in the three bodies which he destines to them? Who can tell us whether the nervous fibres start in the corpus striatum or whether they rather pass through the corpus callosum up to the cortex or to the greyish substance? Assuredly, the corpus callosum is so unknown to us that as long as one has some mind, one can say anything about it.

---

[7] See e.g. Vesalii … *de Humani corporis fabrica Libri septem*. Basileae 1543. lib. VII c. I p. 623.

[8] See Willis, *Cerebri Anatome*, c. XI p. 136.

[9] See Willis, *Cerebri Anatome*, c. II pp. 29–30 and plate VIII facing p. 167.

As far as Mr. *Descartes* is concerned, he knew too well (II,8) the short-comings of the description that we have of man to explain his true structure. Therefore, he does not undertake to do that in his *Traité de l'Homme* but he explains to us a machine that would do everything men are able to do.[10] Some of his friends explain themselves here a little differently from him. Nevertheless, one sees at the beginning of his work that he meant it so, and in this sense it can reasonably be said that Mr. *Descartes* has outdone the other philosophers in this treatise which I just mentioned. Nobody else has explained mechanically all the actions of man and mainly those of the brain. The others describe man himself. Mr. *Descartes* speaks only of a machine, which, however, lets us see the insufficiency of what the others teach and lets us know a method of looking for the functions of the other parts of the body as evidently as he demonstrates the parts of the machine of his man, which nobody has done before him.

Mr. *Descartes* thus must not be condemned if his system of the brain is not strictly in agreement with experience. The excellence of his mind which principally appears in his *Traité de l'Homme* covers the errors of his hypotheses. We see that very skilled anatomists such as *Vesalius*[11] and others could not avoid making similar errors.

If these great gentlemen who spent most of their lives in dissections have been forgiven their errors, why would you be less indulgent to Mr. *Descartes* who has spent his time very happily on other research? The respect which, with everybody, I think to owe to minds of this kind would have prevented me from speaking of the shortcomings of this treatise. I should have been content to admire it with some other people as the description of a nice machine all of his invention, if I had not met many people who take it quite differently and who want to present it as an exact report of that which is deeply concealed in the recesses of the human body. Since these people do not agree with the very evident demonstrations of Mr. *Sylvius*,[12] who has often showed that the description of Mr. *Descartes* is not in agreement with the dissection of the bodies which it describes, I must, without reporting here all this system, (II,9) point out some places where, I am sure, if they want, they will see clearly and acknowledge a big difference between the machine

---

[10]See *L'Homme* De René Descartes *Et Vn Traitté De La Formation Dv Foetvs Dv Mesme Avthevr. Auec les Remarques de Lovys De La Forge* ... A Paris ... 1664. I: p. 1: *Ces hommes seront composez comme nous, d'vne Ame & d'vn Corps* ... II: *Ie suppose que le Corps n'est autre chose qu'vne statuë ou machine de Terre, que Dieu forme tout exprés pour la rendre la plus semblable à nous qu'il est possible* ...

[11]See *De Humani corporis fabrica Libri septem.* lib. VII c. I pp. 622–24.

[12]See above note 2.

which Mr. *Descartes* has imagined and that which we see when we make the anatomy of human bodies.

The pineal gland has recently been the subject of the most important questions on the anatomy of the brain but, before considering the fact and solving the question of where this gland is, I must first show the opinion of Mr. *Descartes* on this subject and that with his own words. Here are various passages in which he talks of it and which are confirmed by other pages of his treatise, which one can see at the end of this dissertation.

There is a ratio of the surface area of the gland[13] and *the inner surface of the brain, Q.*

In the cavities of the brain the pores are directly in front of those of the small gland. F.

The spirits flow from everywhere in the gland into the cavities of the brain. D.

The gland can serve to the actions despite the fact that it is now inclined to one side, now to the other. L.

The small pipes of the surface of the cavities are always facing the gland and can easily turn towards the different points of this gland E.

There is thus no doubt that he believed the pineal gland to be entirely in the cavities of the brain.

One must not stop at what Mr. *Descartes* says in some places, that it is located at the entrances of the cavities for this is not contrary to what he says elsewhere since, as a result of its size, the gland can, according to his opinion, occupy the place which is near the entrance to the cavities,[14] or some other place in the cavities, and still be inside, as he says in all the other passages.

Let us now see whether this opinion is consistent with experience.

It is true that the base of the gland directly touches the passage from the third ventricle to the fourth, as you see indicated in the figure. But the posterior aspect of the gland, i.e. half of it, is so much outside the cavities that it is very easy to satisfy the attendants on this point. To this end (II,10) there is nothing else to do than to remove the cerebellum or small brain and one of the protuberances of one of the tubercles of the third pair,[15] or both if you want, without touching the ventricles for, this having been done skilfully, you

---

[13] [Neither Maar nor Scherz] have been able to find this and the following quotation from Descartes. The meaning of the capital letters added to each quotation is not obvious.

[14] See *L' Homme*, e.g. XIV p. 11.

[15] *Les tubercules du troisiéme paire* are the *corpora qvadrigemina*. The third nerve was still at that time a common denomination of the oculomotor nerve, as the olfactory nerve was not considered one of the nerves of the brain, though already Casper Bartholin the elder in his *Anatomicæ Institutiones* of 1611 (see the reprint of Rostock, 1622, lib. III c.II p. 346) probably before anyone else, looked upon this formation as a nerve.

will see the posterior aspect of the gland completely bare without that any passage appears there through which air or some liquid could enter the ventricles.

Now to become enlightened on the situation of its anterior aspect and to show that this is not in the lateral cavities, one has only to consider them after they have been opened, whether the method of Mr. *Sylvius* has been used to open them or that of the Ancients,[16] for the thickness of the substance of the brain will always be seen between the gland and the lateral cavities. This truth can always be demonstrated without dividing the substance of the brain, by separating from its base the part that contains the said cavities for, so doing you will find the gland so much outside these cavities that it cannot face them, being prevented from doing so by the ties which keep this part of the brain joined with its base. The Ancients knew that the part of the brain commonly called the vault or the fornix is not continuous with the base of the brain but it supports its folded substance and forms a third cavity below. It is true that by pushing air forcefully into the entrance of the cleft of the tubercles of the second pair,[17] air raising the vault breaks the threads which link it to its base, and makes a larger cavity to appear. Hence it has been imagined that, when the spirits swell the cavities, the vault rises and the surface of the gland faces from all sides the surface of the cavities.

I say that this has been imagined since, although the vault rises as I just said, only the anterior surface of the gland can face the lateral cavities. For the rest, whatever the preparation, one will never achieve that the posterior aspect faces the posterior ventricles. But if you do not force the brain by breaking the skull, or by blowing air forcefully between its parts, or by using some other form of violence, you will find nothing in this third ventricle the middle of which is very narrow and which is filled only by the large vein which makes (II,11) the fourth sinus[18] and by the glandulous bodies which accompany this large vein.

I acknowledge that, behind the cleft and exactly below its posterior hole, there is a cavity which is coated both in front and on the side by the part of the choroid plexus which rises towards the fourth sinus, and which is closed in the back by the pineal gland the anterior aspect of which is entirely continued and when the fornix, or the vault has been removed this cavity remains completely beneath the first and looks much like an inverted horn.

---

[16]See, about this, a little section written by Sylvius in Bartholin's *Institvtiones Anatomicæ* ... Lvg. Batavorvm 1641. lib. III c. VI pp. 279-284.

[17]*Les tubercules du deuxiéme paire* are the *corpora mamillaria*.

[18]i.e. *sinus rectus*.

As far as what Mr. *Descartes* says that the gland can serve to the actions although it sometimes inclines on one side, sometimes on the other, experience assures us that the gland is quite unable to do that for it is so squeezed between all the parts of the brain and everywhere so attached to these parts that you cannot give it any movement without forcing it and without breaking the ties which keep it attached. For as regards this situation, it is easy to show the opposite of what Mr. *Descartes* says about it for it is not perpendicular to the brain, it is not turned forwards as some of the most skilful ones believe, but its tip always faces the cerebellum or small brain and forms with the base an angle of about 45°.

The connection of the gland with the brain by way of arteries[19] is not true either for the circumference of the base of the gland holds to the substance of the brain or, better said, the substance of the gland continues into the brain, which is quite contrary to what he says in article H.

The hypothesis of the arteries gathered around the gland and rising to the great channel is a matter of no little consequence for the system of Mr. *Descartes* since the separation of the spirits and their movement depends on it.[20] However, if you believe your eyes, you will find that it is only a collection of veins which come from the corpus callosum, from the inside substance of the brain, from the choroid plexus, from several places of the base of the brain and from the gland itself; that they are veins and not arteries and that they carry blood back to the heart, whereas the arteries carry it to the brain. Some people believed that (II,12) Mr. *Descartes* wanted to continue the nerves to the gland but that has not been his opinion.

The friends of Mr. *Descartes* who take his man for a machine, I have no doubt, will be good enough to believe that I do not speak here against his machine the artifice of which I admire, but for those who undertake to demonstrate that the man of Mr. *Descartes* is made like the other men: the experience of anatomy will let them see that such undertaking cannot succeed.

One will say that they believe also to be based on experience and on anatomy. I answer that when dissecting the brain there is nothing more ordinary than to commit faults without noticing them. That will appear clearly in the course of this dissertation. I had thought of reporting the other systems of the brain by which it has been attempted to explain the animal

---

[19]See *L'Homme.* LXXIV p. 77: … *& qu'elle [la glande pinéale] n'est pas toute ioiente & vnie à la substance du cerueau, mais seulement attachée à de petites arteres.*

[20]Ibid. XIV p.11: … *elles [les arteres] montent tout droit, & se vont rendre dans ce grand vaisseau qui est comme vn Euripe, dont toute la superficie exterieure de ce cerueau est arrosée.*

actions, the source and the components of the liquids of the brain. But since then, I have considered that this was an undertaking which demanded more application and more leisure than the aim of my journey allows.

Dissections or preparations being liable to so many errors and anatomists having been prone hitherto to building systems and adapting to them the softness of these parts, there is no wondering whether the figures made according to them are not exact. But the faults of dissection are not the only cause of what is lacking for their exactness. The sketcher sometimes adds his ignorance of his art; the difficulty of giving their parts in a drawing their relief and their perspective, and the difficulties of letting him know what is to be observed most carefully are always invoked to excuse him. The best figures of the brain that we have had so far are those which Mr. *Willis* gave us. Nonetheless some faults here and there must be noticed in them and many things should be added to make them perfect. In the third figure[21] he represents the upper gland, in other words the pineal gland, as a round ball. If it were without point, as represented in his figure, it could not be said that its point faces the front rather than the back. Nor do you see anything of the substance of the brain, which is in front of the base of the gland and which passes beyond from one side of the brain to the other and, looking at the figure, you would think that there (II,13) is nothing in front. Behind the gland a space appears between the bodies of the third pair of tubercles,[22] which is met at the base of the brain. This space appears completely different when seen in reality. The thin expansion of the white substance of the brain which continues with the middle of the cerebellum and which right there is very thick is not represented, nor the true origin of the pathetic nerves which get out from this same expansion. He also shows as separated the bodies of the second pair of tubercles[23] whereas they are usually together. The under aspect of the vault seems there to be all of the same substance whereas one finds there bumps and a very elegant structure. The cross-section of the corpus striatum actually shows rays but these are very different from, what Fig. 8 of Mr. *Willis* displays. Seeing it, you would imagine that these white rays are continuous with the anterior part of this corpus striatum, whereas the anterior part of this corpus striatum is made of greyish substance which, passing between the white rays, is such in this manner of dissecting that it does not seem to hold to nor to be joined to any other body.

---

[21]Willis, *Cerebri Anatome*, plate facing p.49.

[22]i.e. *corpora qvadrigemina anteriora*. See Willis, *Cerebri Anatome*, plate I facing p. 25.

[23]i.e. *thalamus opticus*. See Willis, *Cerebri Anatome*, plate I facing p. 25.

In the second figure,[24] the *infundibulum* or funnel has nothing to do with nature: the nerves which move the eyes are straight whereas they should be curved. You do not see the true origin of the threads which get out from the base of the brain to form these nerves. The *Varolius*[25] point could be better and more distinctly expressed. Also, the anterior roots of the vault which you see in Figs. 7 and 8 are not separated as these figures show but they touch above where they form an acute angle.

The line marked G.G.G. in Fig. 7 seems to be an uninterrupted line although that which is shown between the roots of the vault has no connection with the extremities. In the same figure, the pineal gland holds to the substance of the brain through two small funnels. I will not speak of the figures of *Vesalius*,[26] *Casserius*,[27] etc., for since the last and most accurate ones are so far from the perfection which they could have, one can imagine what to think of the others.

I have seen three figures only of *Varolius*. They express very poorly the nicest remarks which were ever (II,14) made on the brain. I do not know whether the figures of the first edition, that of Padua in 1572, are better than those which I saw and which are of Frankfurt[28] 1591 and which are also in the anatomy of *Bauhin*. Among those of Mr. *Bartholin* there are three which represent dissections carried out in the manner of dissecting the brain which Mr. *Sylvius* gave us in which the author himself warns the reader of some faults.[29] But without considering various others which are in these figures in general, I will only say that there are few figures in which the true situation of the gland and the true duct of the third ventricle are found. We have none either which shows well the choroid plexus or which represents the

---

[24]All of the editions have, erroneously, troisiéme instead of deuxiéme.

[25]The original edition has *point* instead of *pont*.

[26]See *De Humani corporis fabrica Libri septem*. lib. VII p. 605 ff.

[27]See *Ivlii Casserii Placentini … Tabvlæ Anatomicæ LXXIIX, Omnes nouæ nec ante hac visæ. Daniel Bvcretivs … XX, que deerant suppleuit et omnium explicationes addidit*. Venetiis 1627. lib. X p. 85 ff.

[28]Constantii Varolii … *De neruis opticis, multisque aliis præter communem opinionem in humano capite obseruatis…* [Patavii 1573] is reprinted in Constantii Varolii … *Anatomiæ, Siue De Resolvtione Corporis Hvmani Ad Cæsarem Mediouillanum* Libri IIII.. Francofvrti 1591. p. 119 to the end. See p. 150, 152 and 153. The figures are most likely the same both in the original edition and in the reprint, but the editor has not seen the original edition. The first two of the three figures, though somewhat altered, are also found in Caspari Bauhini Basileensis *Theatrvm Anatomicum Novis figuris æneis illustratum et in lucem emissum opera* & sumptibus Theodory de Brÿ … Francofurti at Moenum 1605. lib. III plate XVIII p. 687. [Maar] has not seen the original edition of Basle 1591.

[29]See Bartholin's *Institvtiones Anatomicae* … Lvg. Batavorvm 1641. The figures of the brain are taken from Vesalius, with the exception of the above, mentioned three figures, which are due to Sylvius (pp. 257, 261 and 277). Sylvius himself criticises these figures, e.g. lib. III c. II p. 250 the Fig. 4 on page 257.

ramification of the veins contained in the lateral cavities, the distribution of the arteries, the concourse of several veins which form the fourth sinus, the glandulous bodies which are there in fair number.

Gentlemen, you have thus seen how the dissection of the brain has been done hitherto, how little light has been drawn from it and how unreliably the figures show the parts which they should represent. From that, please judge how unreliable the explanations based on such poor foundations are. It even happened that those who undertook to provide these explanations—by I do not know what folly which was met in most of those who wrote on arts— used very obscure terms, metaphors and so inappropriate comparisons that they confuse the mind almost equally of those who know the matter and those who want to learn it. Besides, most of these terms are so low and so unworthy of the most noble material part of man that I am as surprised by the trouble of the mind of the one who was the first to use them as by the patience of all the others who, for such a long time, have always used them. Was it necessary to use words such as *nates*, *testes*, *anus*, *vulva*, *and penis* since they are so poorly related to the parts which they mean in the anatomy of the brain? Indeed, they resemble them so little that what one calls *nates* is called *testes* by another.[30]

The third ventricle is an equivocal term. The Ancients designated by this term[31] a cavity below the *fornix* of the vault. They believed that (II,15) this vault was separated from the base of the brain and they represented it as if it was set on three legs to sustain the body of the brain which rests on them.

Mr. *Sylvius* takes for the third ventricle a canal that lies in the substance of the base of the brain between the funnel[32] and the passage which leads beneath the posterior pair of tubercles of the brain to the fourth ventricle.[33] Some people when dissecting separate the bodies of the second pair of tubercles and take for the third ventricle the whole space between these two bodies, which they have created by separating them so that the third ventricle is sometimes the cleft which is above, sometimes the canal below, and others

---

[30]See e.g. Bartholin's *Institvsiones Anatomicæ* ... Lvg. Batavorvm 1641 lib. III c. VI p.278: *meatus in se geminos habet [ventriculus tertius] .... alterum posteriorem, membranâ vestitum; quod foramen quidam anum vocant. Vulva is: fovea inter eminentias [corpora qvadrigemina] oblonga. Penis is: glandula pinealis*, and *nates* and *testes* are *corpora qvadrigemina anteriora & posteriora*. The latter term, which is now generally used, is due to Winslow, who endeavoured to eradicate the older terms by substituting the name *tubercules quadrijumeaux* (Exposit. Anat. Paris 1732. p. 621 § 79). The old terms, however, curiously enough still survive in most French Anatomies.

[31]See Galenus, *De Usu Partium* (Περί χρείας τω εν ανθρώπου σώματι μορίων) lib. VIII 10–12.

[32]infundibulum.

[33]See Bartholin's *Institvtiones Anatomicæ* ... Lvg. Batavorvm 1641 lib. III c. VI p. 271 and pp. 275–278. - *Les deux paires posterieures des tubercules* are the *corpora qvadrigemina*.

assert that it is the space between the canal and the cleft created by the rupture of the bodies which I just described. Here are thus three very different kinds of third ventricle. Only the second of them is true in reality for the first and the third result entirely from the preparation.

A fourth meaning could be added if the small cleft below the vault were taken for a passage from the two anterior ventricles to the fourth. But it is very small and so filled by the vessels and the glandulous bodies of the choroid plexus that I doubt that there is there some communication between the anterior ventricles and the posterior ones since, according to the designation by Mr. *Sylvius*, the third ventricle is large enough for this purpose. Thus the situation of this canal of Mr. *Sylvius* is so appropriate to this function that, if you want something to go from the lateral ventricles to the fourth ventricle, nothing can go before the funnel and this canal are filled first.

We count two glands in the brain but we do not know whether one or the other has anything in common with the glands, except for the shape, and even this shape when carefully examined is not always found to be consistent with that of the glands. The upper or pineal gland does not resemble a fir cone [pomme de pin] in all animals, nor in man himself. The lower gland is called pituitary although there is no assurance at all that it acts on the phlegm.

The choroid plexus represents a lacework of vessels (II,16) but you see there easily veins distinct from the arteries and you can as easily follow the distribution of the former and that of the latter separately. The name of "vault" suggests a vaulted cavity which, however, is not there at all when you look for it properly. The corpus callosum, according to the common usage, means the white substance of the brain that appears when the lateral parts are separated. But it is true that this part is entirely similar to the rest of the white substance of the brain and thus one does not see any reason to give a particular name to a part of this substance.

There are two ways only to arrive at the knowledge of a machine, one that the master who made it discloses us its artifice, the other to take it to pieces to the last spring and to examine all these separately and together.

These are the true means of knowing the artifice of a machine and, however, most people have believed that they had better guessed it rather than seeing it by examining it closely with one's senses. They were content with observing its movements and, on these observations alone, they have built systems which they presented as truths when they have believed that they were able thus to explain all the effects which had come to their knowledge. They did not consider that one thing can be explained in different ways and that only the senses can assure us that the idea which we have formed about it is consistent with nature. The brain being indeed a machine, we must not hope to find its artifice through other ways than those which are used to find the artifice of the other machines. It thus remains to do what we would do for any other machine; I mean to dismantle it piece by piece and to consider what these can do separately and together. In this research one can say with good reason that the number of those who have showed the eagerness of true curiosity in this matter is very small. In all centuries chemistry has had private people and princes who built laboratories for its purpose, but few people have displayed such eagerness for anatomy. It has not depended on princes. Several of them have been curious of such important knowledge and built magnificent theatres destined to dissections which they have (II,17) even sometimes honoured by their presence. But those who carry out the dissections have always wanted to appear expert in this science. None of them has been willing to confess how many things remained to be learnt and, to conceal their ignorance, they contented themselves with demonstrations of what the Ancients had written.

The anatomists would be justified to complain about me if I did not explain more here, to show that they are not completely in the wrong which I seem to accuse them of when I say that they do not sufficiently apply themselves to anatomical researches.

Those who deal with such researches are usually physicians or surgeons. Both are compelled to attend to their patients and as soon as they have acquired some knowledge and some fame, they are unable to spend the necessary time on researches. But they should not undertake to heal a body the structure of which they do not know, i.e. they should not venture to rebuild a machine the pieces of which they would not know.

The others who do not see patients and who have no other job than the profession of anatomy in schools do not believe to be more compelled to carry out research than the physicians and surgeons, for the goal of their profession is to teach those who want to practice medicine or surgery the description of the human body which the Ancients left us. When that which is in their writings has been clearly demonstrated by the one side and distinctly understood by the other, both think to have accomplished their duty. The boundaries between these two professions have been so poorly marked that true knowledge of the machine of the human body, which was most necessary, is neglected as not being in the province of anatomy nor of that of the physician nor of the surgeon. I say that to make researches which would teach us the truth requires a man entirely, a man who has nothing else to do. Even the one who makes profession of anatomy is not suited to that since he is compelled to carry out public demonstrations that prevent him from engaging in this application for the reasons which I have mentioned and for others which I will now present.

1. Properly to examine each part requires time and such concentration of mind that all other work and all other thoughts must be dropped to accomplish this task. (II:18) Practice makes that impossible for physicians and surgeons as do anatomical presentations for those who make of that a profession. Sometimes it takes years to discover that which can then be demonstrated to others in less than an hour. I have no doubt that Mr. *Pecquet*[34] spent much time before carrying the chyle from the mesentery to the subclavian artery and I may not be believed if I say how much difficulty I met before being able to show the true insertion of this same lymphatic duct of Mr. *Pecquet*, the figure of which had been given by Bils.[35] By now a half hour or one hour is enough to prepare and demonstrate one and the other together.

2. Although the anatomists open thousands of corpses in the schools, it is pure luck if they discover something. They are compelled to demonstrate the parts according to the Ancients and, to this end, they must even follow a certain method. Researches, on the contrary, do not admit any method. They must be tried in all possible ways.

All the other parts must be divided to show the one which they are asked for. On the contrary, researches demand that the least part should not be divided before having been examined first. If this method was followed in the schools, the attendants would consider the one who is dissecting as an ignorant. They would be right in complaining of the time spent because of him since, often, after having searched for a long time, he would not find that which he had undertaken to show them. You thus see that those who have professed anatomy hitherto have not been compelled to search and even that they could not have succeeded so that it is not their fault if anatomy has not progressed for so many centuries.

---

[34]Pecquet in 1649 found the thoracic duct in a dog (Joannis Pecqveti ... *Experimenta Nova Anatomica, Qvibus Incognitvm Hactenvs Chyli Receptaculum, & ab eo per Thoracem in ramos usque Subclavios Vasa Lactea deteguntur* ... Parisiis 1651. The thoracic duct was found independently by van Horne, that published the discovery in 1652. Thomas Bartholin was the first who demonstrated the duct in man: *De Lacteis Thoracicis in homine brutisq; nuperrimè observatis* ... Hafniæ 1652. This was followed by his discovery of the lymphatic vessels: *Vasa Lymphatica, Nuper Hafniæ in Animantibus inventa, Et Hepatis exseqviæ.* Hafniæ 1653.

[35]The figure mentioned is found on p. 18 of *Epistola Apologetica* ... Roterodami 1661. See vol. I p. 230. col. 1, foot-note.

This science, to speak in general, has thus been dealt with little success and the researches on the brain particularly have even less succeeded, having not been undertaken with the necessary diligence because of the difficulties in dissecting this part. Let us now consider of what it consists and whether some of those who have tried have managed the way the matter deserves.

Mr. Bils has applied himself in anatomy without having studied what the Ancients have written. But I have no doubt (II,19) that he would have furthered his knowledge if, after having seen what good the Ancients had done, he had spent his time and his zeal in new researches. It must be admitted that so nice experiments are reported in the writings of our predecessors that we should have risked to ignore them if they had not warned us. It even sometimes happened that they told us truths which our contemporaries have not recognized because they did not examine them with sufficient application. On the other hand it is true that what the Ancients and the moderns taught us about the brain is so full of disputes that there are as many snags and disputes, as many doubts and controversies as there are books of anatomy. Nevertheless, one can make great profit of their work and even draw great advantages from their mistakes. I speak of the authors who have worked by themselves. As far as the others are concerned who have worked only on the works of others, one can read them only as amusement and it is not always useless to do it, but they would have deserve more merit and their studies would have helped those who work if they had made an exact report of what the anatomists have written about the brain or if they had developed according to the laws of analysis all the manners of explaining mechanically the animal actions or if they had written an exact catalogue of all the propositions which they found. Among these propositions they should have carefully distinguished those which are based on fact and experiment from the others which are only reasoning. But so far nobody has done so. Therefore, only those who worked by themselves must be considered.

The first matter to be considered there is the description of the parts, in which one must determine what is true and certain to be able to distinguish that from propositions which are false or uncertain. It is not even enough to be able to enlighten oneself. The evidence of the demonstration must force others to agree. Otherwise the number of controversies would increase instead of decrease. Every anatomist who has applied himself to dissecting the brain demonstrates by experiment what he tells of it. The softness of its substance (II,20) obeys him so well that his hands unwillingly form the parts as the mind has imagined them before and the attendant often seeing two contrary experiments made on the same part finds himself well puzzled, not knowing which he must accept as true and finally he denies both to get out of trouble.

Therefore, in order to prevent this inconvenience, it is absolutely necessary, as I said, to pursue in all dissections a convincing certitude. I admit that this is difficult but I also know that this is not completely impossible. Do not believe, Gentlemen, from what I just said, that nothing is certain in anatomy and that those who practice it form with impunity the parts at their leisure without being liable to be convinced. You may question whether the parts which are showed to you separately were not joined previously but it would be impossible to show them joined together if they had not been so normally. To become free of this doubt and to make sure that the parts showed to you have not been joined together one must only examine them in the state in which they are normally without forcing them at all but let those whom one wants to convince do everything possible to demonstrate them joined. One can arrive at the same certitude in other circumstances and particularly when dealing with the situation of the parts as long as one does not touch anything before having examined it previously and even as long as one expresses at each moment what one is touching. To this end, one must not only be heedful of the part which is being dealt with but also make reflection on all the operations which have been performed before arriving at the part. Such operations may have altered this same part for, by dealing with the external parts, you often change the internal ones without noticing it and, when you happen to discover them you may believe that they are such as they appear to you and you forget that you have yourself significantly changed their situation and their attachment to other parts. I will report here an example in an anatomical matter, the most famous in this century. Those who deny[36] the continuation of the pineal gland with the substance of the brain and the attachment of the vault to the base of the brain would not talk of a fact (II,21) with so much self-confidence if they did not believe to have clarified the matter by experiments carried out with all the necessary attention. In their experiments they must not have considered the changes which occur when the outside has been removed and that, by so doing, the ties which attach the skull to the dura mater are torn. When raising the upper part of the skull I have seen that the middle of the dura mater remained attached to it even after I had opened it sufficiently to introduce three fingers between the separated parts of the skull. How does this raising of the dura mater occur unless the internal parts which are attached to it suffer from this violence? The pineal gland holds to the fourth *sinus* which is attached to the *sinus falcis* so that you cannot raise the dura mater whatever little at this place without forcing the pineal gland. The

---

[36]Cf. *L'Homme.* LXXIV p. 77.

same *sinus falcis*[37] receives all the veins which pass between the vault and the base of the skull and keep these two parts joined together. There is a fairly firm connection between the upper aspect of the brain and the dura mater by way of the reins and, when you raise the dura mater the upper substance of the brain which is attached to them follows and the fourth *sinus* being pulled upwards causes that the connection between the vault and the base breaks. At the beginning I have often been mistaken and I could not understand why these attachments were not always felt. But later, seeing in horses, sheep, cats where the part of the dura mater which separates the cerebellum from the brain is changed into bone, I broke many of the internal parts by making evulsions of this bony part; I began to recognize the cause of this mistake and I learnt that separating the skull properly was not an operation with little consequence. One always makes a circular section in the human skull to remove the upper segment but if another section were to be made in this segment at right angles with the first, the skull would be opened more easily without forcing the brain much. It must indeed be admitted that the chisel, the saw and the pincers can never be handled without concussion or shaking. A small circular saw could be made which would provoke no big shaking, above all if it were rotated about an axle prepared in a certain manner and set between two pointed bars. Such a saw (II,22) could be used for various other purposes that can occur in the separation of the skull. But if there were some liquor which could dissolve bone in a short time or soften it, nothing more convenient could be wanted and that would be the best of all manners of separating the skull.

It is not enough to pay exact attention at every moment, the manners of dissecting must also be changed; being as they are as many evidences of the truth of your operation. They can equally satisfy yourself and convince others.

This will seem strange to those who believe there are set Laws according to which the dissection of every part must be done and who claim that anatomical management inherited from the Ancients must be totally observed without either changing anything or adding anything. I should admit that the Ancients could have given inviolable rules for the dissection of each part if they had had a perfect knowledge of these parts. But, as they were as little enlightened as those of this century, and in various peculiarities even less than we are, they would have been as unable as we are of prescribing the true manner of dissecting in which there will be nothing constant nor fixed until more discoveries have been made. It is, however, necessary, you will say, to

---

[37] i.e. the *sinus longitudinalis inferior.*

use some method for dissecting the parts according to what is known of these now. I will easily agree. It is right to use the method of the Ancients for want of a better one but not as an accomplished thing.[38] The main cause which has maintained many anatomists in their errors and which prevented them from going further than the Ancients in their dissections was that they believed that everything had been so well noticed that nothing else remained to be researched by the moderns, and they took the old rule of dissection for inviolable laws. They did not do anything in their whole life than demonstrating the same parts with one, the same method. Instead of that, anatomy must not be submitted to any rule and must change as many times as dissections begin. Hence this has the advantage that if it does not always discover something new, it recognizes at least that it has been mistaken in what was seen previously, (II,23) mainly when there is some dispute, for then the attendants must be free of prescribing the laws of the dissection.

It is true that this manner of dissecting does not allow a great show and one cannot play the scientist and at the same time admit one's own ignorance. For myself, I prefer admitting mine than reciting with authority opinions the falsity of which will be demonstrated by others some time later. We have seen great anatomists who fell into that trap and we still see others who imagine that people will have more faith in their obstinacy than in what they see with their own eyes. I leave this vanity to those who feed on it. I try to obey the laws of philosophy[39] which teach us to search the truth while questioning our certitude and not to be content before having been confirmed by the evidence of the demonstration. I cannot give you more obvious evidence of the necessity of changing the dissections than the two following. It is a sure experiment that when one has blown into the beginning of the cleft which is below the vault, one finds the vault separated from the base and a fairly large cavity in between. The same happens when the skull is removed forcefully, as I said above. This is so obvious that those who carry out and those who attend this operation believe that nothing more certain can be done. If one begins to doubt, there is no other way of freeing oneself of this doubt than to attempt to demonstrate this cavity by other means. If, indeed, this cavity is normally there, you will always find it whatever the way you look for it. But if, through another kind of dissection, you find that the cavity is not there and that the parts between which the cavity should be are bound together without any

---

[38]Winslow, *Exposit. Anatom.* has *assurée*.

[39]Steno evidently refers to Descartes; and Guido Fanoisius is obviously aware of this, when he translates: *Legibus stare Philosophiæ satagens, quæ scrutari veritatem d o c e t, de ipsius certitudine dubitando* ... See *Dissertatio De Cerebri Anatome ... Operâ & studio* Guidonis Fanoisii ... Lugd. Batav. 1671. p. 56.

space in between, you must then be convinced of the error of the former demonstration and you will clearly see that the force of the air which had been blown in has provoked this appearance.

If the human brain is dissected in the manner of *Varolius* and *Willis*,[40] after having been removed from the skull, you will normally see the bodies of the second pair of tubercles separated in the middle from the white substance which is in front of the gland and which will be ruptured later. (II,24) When the same dissection is carried out with the brain remaining in the skull, one and the other are seen intact and it is then easy to notice, by comparing the two sections, that the cause of the first mistake was the gravity of the lateral parts which break those in the middle.

After having drawn a true and very accurate draught of the parts of the brain, discovered the mistakes and their causes and decided the true manner of demonstrating these parts with all the necessary precaution, it must be attempted to explain what has been seen by correct and exact diagrams, for it would be better to have none than to have false or imperfect ones. The portrait is used when the original is not there, in order to keep the memory of the latter. There are even people who never see these parts but in painting. Their aversion of blood prevents them from satisfying their curiosity by inspecting subjects and nature so that, if the figures are not such as they should be, they give false ideas to those who use them to learn anatomy and trouble the others who use them to help their memory.

That is why all possible means must be used to have exact pictures to which a good sketcher is as necessary as a good anatomist. Application and a particular study are also necessary to take measurements properly and to see how the dissection must be carried out and how the parts must be arranged distinctly to express all that which is to be seen in the brain. Some difficulty is met that is peculiar to this part when one wants to make a drawing of it. For the other parts, indeed, it is enough to prepare them once to achieve their picture. The brain, on the contrary, when being prepared caves in before its diagram has been completed, so that one must draw from several brains in order to achieve one single figure. Having not been considered, perhaps, this may well be the cause that in anatomy there are no more imperfect figures than those of the brain.

I did not say anything so far of the functions of the parts nor of the actions called animal because it is impossible to explain the movements occurring through a machine if the artifice of its parts is not known. Reasonable people

---

[40]See Willis, *Cerebri Anatome*, c. I p. 5.

must find these peremptory anatomists very pleasant when, after having talked (II,25) of the function of the parts the structure of which they do not know, they bring forth for reason of the functions which they attribute to them, that God and nature do nothing in vain. But they are mistaken in their application here of this general maxim and what God, according to the temerity of their judgment, has destined to one end is found later to have been made for another end. It is thus better here also to confess one's ignorance, to be more reluctant to decide, and not to undertake so lightly to explain such a difficult thing on simple conjectures.

What I have said so far is only the smallest part of what I believe must be done in order to acquire some knowledge of the brain for, to this end, it would be necessary to dissect as many heads as there are different species of animals and different states in every species. In the foetus of animals one sees how the brain develops and what would not have been seen in an intact brain, and in its whole, will be seen in the brains which have been altered by some disease.

In live animals all the things that may cause some alterations of the actions of the brain must be considered, whether they come from the outside like liquors, injuries, medicines, or causes that are internal such as the diseases of which medicine counts a great number. There is still this reason for working on the brain of animals that we treat these as we wish. One trepans them and performs on them operations of surgery to learn the way to carry out these operations. Why not perform these same operations to see whether the brain has some movement and whether, by applying certain drugs on the dura mater, on the substance of the brain or in the ventricles, it would be possible to learn some particular effects?

Various trials could also be carried out without opening the skull, different drugs be applied externally on it, others mixed with the food, others injected in the vessels, to learn from that what can disturb the animal actions and what is best to restore these when they are disturbed.

The brain is different in different species of animals. This is another reason to examine them all. The brain of birds and fishes is very different from that of man and, in animals with a brain the closest to ours, (II,26) I never saw one in which I did not find some very obvious difference.

Such a difference, whatever it may be, always throws some light on the researches and may teach us that which is absolutely necessary. There are animals in which fibres appear more clearly than in man. The parts which in man are mixed and joined together are sometimes found distinct and separated in other animals. In others, the substance is found to be more or less firm, the size unequal and the situation different.

I will not dwell more here on this matter since I am convinced that everybody will easily admit that we owe to the dissection of animals almost all the new discoveries of this century and that some parts would never have been recognized in the brain of man if they had not been noticed in that of animals.

What we have seen so far, Gentlemen, on the insufficiency of the systems of the brain, on the shortcomings of the method which has been followed to dissect and to know it, on the infinity of researches which should be undertaken on men and on animals and this in the different states in which they should be examined, on how little light we find in the writings of our predecessors and on all the attention necessary when working on such delicate pieces, must undeceive those who keep to what they find in the books of the Ancients. We shall always remain in a miserable ignorance if we content ourselves with the little light they left us and if the men most prone to make these researches do not join their works, their industry and their studies to arrive at some knowledge of the truth which must be the main goal of those who reason on the subject and who honestly study. (II,27)

End

Passages from the writings of Mr. Descartes[41]
which confirm that which was asserted on page [OPH II,9]
and following of this dissertation.

Page 11. For it must be known that the others who bring them from the heart after they have divided them into an infinity of little branches and have composed these little tissues which are spread like tapestries over the bottom of the cavities of the brain, gather about a certain small gland, a, located at about the middle of the substance of the brain, b, at the very entrance of its cavities, and which have at this place a large number of small holes through which the subtlest parts of the blood which they contain can flow into this gland but which are so narrow that they do not permit the bigger parts to pass. It must also be known that these arteries do not stop there, c, but that, having gathered several in one, they rise straight and go to this big vessel, which is like a Euripus from which all the external surface of the brain is sprinkled.

Page 12. The gland must be imagined as an abundant source from which the smallest and most agitated parts of the blood flow simultaneously everywhere in the cavities of the brain.

---

[41]The quotations from *Descartes* (OPH, II, 27–28) do not appear in the Latin translation of this Treatise. In the present Edition they are reprinted *verbatim* from *L'Homme*, Ed. 1664, as both the original edition of this Treatise of *Steno's* and the reprints in *Exposit. Anatom.* have misprints, which are fatal to the sense.

For elucidation of this tricky part of the Discourse consult: Stensen, *Discours*, ed. Andrault, 127, note 1.

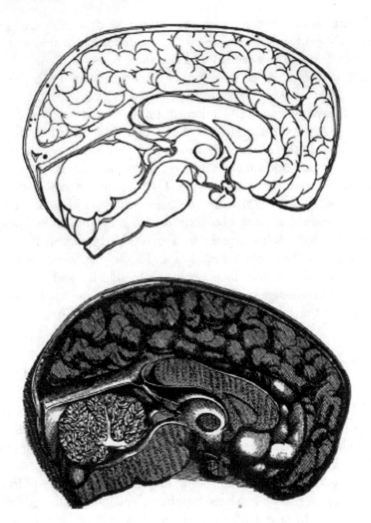

Fig. 2.18.1 Sagittal section of a human brain in two displays

63. Imagine the surface facing the cavities as a fairly thick compressed network all the meshes of which are as many small pipes through which the animal spirits can enter, e, and which, facing the gland from which these spirits originate, can easily turn here and there towards the various parts of this gland.

65. The spirits do not only stop in one space but, as fast as they enter the cavities of the brain through the holes of the small gland, f, they tend first to those of the little pipes which more directly opposite.

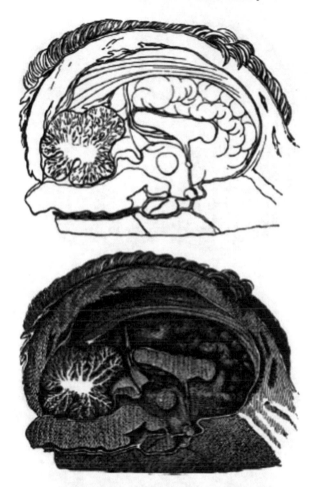

Fig. 2.18.2   Stensen's drawing of Descartes' concept of the brain in two displays

72. By explaining how the figures are drawn in the spirits over the surface of the gland, g, he determines fairly evidently the ratio which he makes between the internal surface of the brain and the surface of the gland.

77. Besides that, consider that the gland is made of a soft material, h, and that it is not at all joined and united with the substance of the brain, i, but is only attached to the small arteries (the skins of which are fairly loose and flexible) k, and supported, as in scales, by the force of the blood which the heat of the heart pushes towards it, l, so that very little is necessary to determine it to incline or to bend more or less now to one side, now to the other, and to do so that, while bending, it directs the spirits which get out of it to certain places of the brain rather than to others. *And a little later,* if the

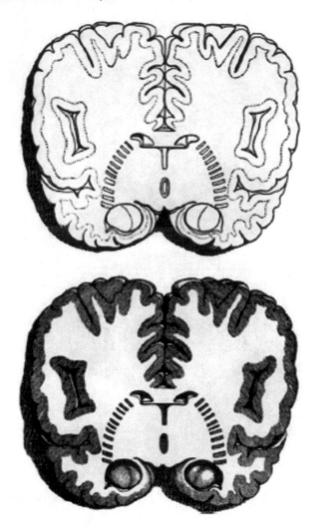

Fig. 2.18.3   Coronal section of human brain in two displays

spirits were of exactly the same strength, etc. m, they would support the gland upright and immobile at the centre of the head.

78. The spirits thus getting out, more particularly from certain places of the surface of this gland than from others, can have sufficient strength to turn the little pipes of the internal surface of the brain in which they will penetrate, n, towards the places from which they are getting out unless they find them already turned.

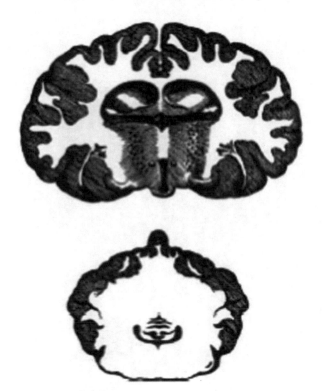

Fig. 2.18.4 Two coronal sections through posterior regions of human brain

For elucidation of this tricky part of the Discourse consult: Stensen, *Discours*, ed. Andrault, 127, note 1. (Figs. 2.18.1, 2.18.2, 2.18.3 and 2.18.4)

2.19 Observations on Egg and Chick

The first thing which appeared when the shell was broken at its big end was the tunica which coats the entire inner surface of the shell. It serves as a common envelope for all the parts contained in the egg. It is rough externally, smooth internally, and, towards the big end of the egg, it forms a fairly large cavity. Next to this, a second tunica has a smooth external surface and closely covers the albumen. When it is broached, the albumen flows out.

Closer to the middle of the egg than to the extremities, one sees chalazae adhering on either side to the tunica of the yolk by a pedicle made of a small white nerve. Thanks to these strings the examiner can pivot and move the albumen in all directions. At about the same distance from each of these chalazae, a white circle is seen with a whitish centre, and surrounded by another circle like the halo of the moon.

On the opposite side of the yolk there were several other circles with their centre marked by a spot, not white like the first one but rather reddish.

OPH 19, vol. II, 39–47: The research on which *In ovo & pullo obervationes* is based is most likely what Steno carried out when collaborating with Swammerdam during their stay in the house of Thévenot at Issy, in 1665, and which is mentioned by Swammerdam in the prefatory dedication to Thévenot in *Johannis Swarnmerdam … Tractatus physico-anatomico-medicus de respiratione usuque pulmonum*, Lugduni Batavorum, 1667. This is closely connected to earlier research by Steno, which was the basis of his treatise *De vitelli in intestina pulli transitu epistola*, dated June 12 (old style), 1664, OPH XVII.

Whether the text in its present form was written by Steno as early as 1665 or only shortly before the treatise was to be printed is an open question. The treatise was first published in *Thomae Bartholini Acta Hafniensia II*, Hafniae, 1675, where it is found as No. XXXIV, pp. 81–92. In Holger Jacobaeus' manuscript (see the Introduction to OPH XXXVI) the treatise is found in a somewhat abbreviated form, written in the third and not in the first person. This copy may have been made from Steno's manuscript, before the latter was printed.

© Springer-Verlag GmbH Germany, part of Springer Nature 2018
T. Kardel and P. Maquet, *Nicolaus Steno*,
https://doi.org/10.1007/978-3-662-55047-2_30

In them several concentric stripes were observed. I reckoned up to five of these circles in one egg, and three in another, but they were unequal.

I set the albumen in plain water and noticed that it did not mix completely with the water but that the largest part sank to the bottom in such a way, however, that each drop left in the water, from the surface to the bottom, a track marked by kinds of distinct fibres which became twisted delicately almost like small ropes. Then I threw the yolk in the water. It sank to the bottom. By shaking, the albumen was resolved into filaments and the yolk mixed intimately with the water. It tainted the water with its small corpuscles which, however, fell down to the bottom soon afterwards without the water becoming clearer again.

In a hardboiled egg one sees several fibres arranged as rays, near the centre at the place where the chalazae attach to the yolk. At mid-distance between the chalazae there is a white circle. In the middle of the yolk there is something almost white.

Day 1

After 12 h incubation, the small white spot seemed to have increased a little and different parts could be distinguished in it. This spot changed position depending on the position of the egg, and always rose upwards as if this part of the yolk was lighter than all the other parts contained in the shell. I observed the same change in position also on the following days, the second, third, fifth, sixth, seventh and eighth.

Day 2

The next day several circles were reckoned in the white spot mentioned above. This spot was not only much larger than the day before but also protruded above the surface of the remainder of the egg in the same way as the transparent part of the cornea forms a bulge on the eyeball. There was an uneven white point at the centre.

In an egg opened towards the end of the second day, I distinctly saw the whole white spot and, near this, a bluish oblong figure as if it were the body of the chick with its membrane, of the size which it could have on the second day of incubation. About this figure, several reticulate small blood filaments formed a one finger breadth large circle. The whole was surrounded by several whitish circles of uneven curvature which formed kinds of waves. The outer circle had a diameter of more than two and a half finger-breadths. The membrane could not be touched without being immediately torn and humours flowing out of it. This proves that there was already a beginning of decay.

After 47 h of incubation, there were two thick and irregular yellow circles around the whitish point.

Day 3

The cavity was not yet considerable. The yolk, however, was very close beneath the tunica and, in that part of it, the outline of the chick was visible, namely small blood vessels which seemed smaller when the yolk still was in the egg but formed a wider contour when the yolk had been taken out of the egg. From the middle of the almost pulsating centre, two vessels extended almost to a straight line, from there returned to the opposite side and, being then reflected from the big end of the egg, joined each other. Beside these two vessels, two others extended towards the upper part, forming an oval figure. They threw out to the sides several ramifications which formed a ring with their convolutions. Some vessels were seen proceeding below as well. But the most remarkable was a small skin, quite transparent ⑧ and oblong.

In another egg I found that the yolk had completely changed colour. Besides various yellowish striae, there was also such a bright space ⊂⊃, resembling water. But I could not decide what it was.

In another egg, after the 71st hour, there was an oblong bloody fringe, the longest diameter of which was about two finger breadths long and the small diameter was less than one and a half finger breadths long. The vessels which left from the centre situated beneath the chick extended over the long diameter. In the part where the head of the chick was, the vessels ran straight without subdividing towards the fringe whereas in the opposite part they subdivided into several ramifications before continuing into the fringe. The vessels which followed the direction of the long diameter were larger and formed larger ramifications. The head of the chick was more than a third of the whole body and the remainder of the body extended in length resembling a small worm. In the middle, a small blood vessel was seen. In the region of the heart there was another blood vessel twisted into a spiral.

In another egg, a white point with a white circle seemed to hide away within the blood vessels. These parts were not as bulging as the day before. The yolk was paler. One of the chalazae seemed to be continuous with a small transparent canal contained in the tunica of the yolk. I assume that this was nothing else than a part of albumen less fluid than the rest.

Day 4

On the fourth day, the rudiment of the chick was less mobile and it was not found at the big end of the egg but almost midway between the big and the small end. I even observed in two other eggs that the chick was at the very small end. This shows that the position of the macula is not certainly at the big end of the egg and that it does not always move towards the cavity at the big end.

In another egg opened towards the end of the fourth day, a transparent, somewhat yellowish, vesicle was seen at the anterior part of the body near the tail. After the amnion had been removed, the vesicle still contained its humour, was nearer the tail and was more aqueous than previously. In another egg this same vesicle seemed to be full of a crystalline liquid.

In the body of the chick, a globular part could be distinguished from an oblong part. In the globular part a circle appeared, which I took for the rudiment of an eye.

The distribution of the aorta was obvious, descending along the back as well as turning outside near the head. Red vessels could be seen which extended from the centre to the periphery although the arteries could not be distinguished from the veins. At the place where a white point with a white circle had been observed the day before, today one could see the oblong part of the chick. Beneath the body of the chick, inside the tunica of the yolk, there was a part which appeared white.

Day 5

The fringe of blood vessels had become larger and anastomosed at different places with blood vessels surging from the centre. The fringe no longer formed a re-entrant curve but the continuity of the circumference was interrupted at several places whereas I had seen one interruption alone previously. All these vessels seemed to ascend from the parts lying beneath the chick, i.e. they spread through the tunica of the yolk on which the chick lay so that their continuity with the chick is not so easily seen without turning the chick over and pulling it at little from its position. We noticed that the vesicle which we had seen full of liquid the day before was adorned with blood vessels. The tunica of the yolk was provided with its chalazae as previously. In the chick the dark circle of the eyes made the pupil more obvious. Two kinds of vesicles were observed in the head. At first I took the anterior one for the brain and the posterior one for the cerebellum but observations on the following days revealed my mistake. The tip of the bill was also conspicuous. A small whitish elongated body above the heart marked the wings. The feet were more obvious. A kind of tail prolonged below the feet seemed to be the rump. The movement of the auricles was distinct from that of the heart. The trunk of the aorta was fairly visible as well as the vessels which were seen externally along the back. Perhaps was it the canal of the medullary cord?

Day 6

This vesicle which had begun to appear on the fourth day was already much increased. It adhered to the umbilical vessels. The chalazae had come quite closer to each other. The tunica enveloping the yolk was easily distinguished from the other tunicae. The substance of the heart was whitish, while that of the auricles was red. In the eye, the black tunica, besides the hole of the pupil, had a small gaping cleft, extending from the pupil downwards. In the anterior part of the head two oblong eminences lay near each other. The posterior eminence which I had taken for the cerebellum was fairly considerable and remote from the anterior eminences. In the space between them, another kind of eminence was visible, but very small.

Day 7

The chalazae were quite close to each other like the day before. The albumen was quite viscous. The liquid observed in the vesicle of the fourth day was yellowish. The cleft in the black tunica of the eye extended as far as the optic nerve. Inside the black circle in the space which I had believed to be the pupil, another smaller circle which formed the true pupil had become conspicuous. Examining the tunica of the eye more closely, I indeed saw that the black tunica had been the choroid which appeared through the transparent cornea, so that what I had taken in my previous observations for the pupil had been all that part of the eye which together comprises the iris and the pupil. The crystalline humour was quite small. In the head the substance of the brain was white and fairly firm. These two anterior eminences appeared excavated. White striae were seen on the posterior eminence which was quite big. The inferior part of the bill was split. The substance of the heart appeared truly fleshy. The stomach with a small portion of intestine, the liver and the kidneys were distinct inside. The inferior part of the abdomen had begun to close. The anterior part of the whole body down to the wings was quite long and wide. Before the end of the seventh day the bloody fringe had occupied half the egg. In the foetus itself the feet, the wings and the rump were distinctly recognized. The head alone was as big as the remainder of the body. The eyes were large and protruding. The bill was not yet seen to protrude beyond the eyes but the opening of the mouth was considerable. The posterior tubercle in the brain was also conspicuous. The crystalline humour appeared completed whether it adhered to the vitreous body or whether it was detached from it. It had a diameter of about the size of the latter. The albumen was yellowish and fell to the bottom.

In another egg the viscera were enclosed. The heart was quite big. The ventricles and auricles were clearly recognizable. The liver was not red but rather yellowish. The stomach and the intestines were conspicuous. The vertebrae of the spine were wide in the lumbar region. All along the spine transverse lines displayed a delicate figure. The feet were divided into digits. The extremity of the wing was wide. The bill was not yet protruding beyond the eyes. Its lower extremity was bent into the cavity of the mouth and dull. The tongue was short but thick. The eyes had a considerable magnitude. That part of the dark tunica nearest the middle was blacker than in the rest.

Day 8

I saw the membrane called chorion expanded over almost the entire egg. After the albumen and the yolk had been taken out through a hole at the inferior part of the egg, and the chorion had been removed, the amnion with a liquid as clear and as limpid as crystal appeared to our eyes. The foetus floated in this liquid. Its head was proportionally bigger than the remainder of the body. The eyes were quite large. Its mouth was open. The tongue could be seen in it. The tail or rump was extraordinarily long. The heart distinctly moved with the auricles. Even after having been resected and separated from all its vessels, it kept moving in my hand for a long time. And when finally it had stopped throbbing and looked quite dead, I made it throb again by breathing on it. This is an experiment which I have repeated several times to the astonishment of the attendants. The brain no had longer an aqueous consistence but its consistence was stickier. The cerebellum was quite protuberant. The crystalline humour was very firm and the ribs were very delicate. What I had taken during the first days for the cerebellum was the continuation of the optic nerves. The humour around brain and cerebellum was obvious. The inferior and lateral parts of the brain, previously apart, were then joined. The tongue was conspicuous. The neck was covered by its skin. The cavity of the thorax was still quite open. The feet were divided into digits.

Day 9

The fourth ventricle of the brain was very wide. Part of the sternum, the lungs, the testicles and the ureters were beginning to form. I broke a shell and dropped all of the egg onto a plate. The bloody fringe of the chorion was delicate. In its middle the chick looked like a small cloud. At the place of the heart there was a part made in a half circle which received and drove forth blood through a kind of circulation. I dropped another egg onto a plate in the same manner. All the parts of this egg were larger. I had at first opened this at the small end and I had seen that the chorion did not yet surround all the yolk. The chorion seemed to be double. Part of it enveloped the yolk, the other the foetus. In the foetus the eyes were larger than the days before. The wings and the feet were quite distinct and some obvious movement was even observed in them. The liver was yellowish. All these parts were by then covered by their teguments. The head was bigger than the body. All the vessels appeared so to say prominently on the membrane.

Day 10

In an egg I found neither chick nor any trace of a chick. Nevertheless I removed the egg from its shell. I saw in the intermediate space between the two chalazae something white which was not arranged in circles. In another egg which I had opened at its small end, I found the albumen quite viscous and verging somewhat to yellow or rather this colour resulted from the yolk seen through the albumen. Blood vessels were scattered between the external tunica of the egg and the albumen but quite slender. The ones which were prominent on the tunica, between the albumen and the yolk, were very delicate. After having put the whole egg into a vessel we saw, inside a crystal clear liquid which had some bitterness, an almost monstrous foetus moving the feet and wings. Its head and its eyes were indeed of an outsized magnitude. Four distinct parts could be seen in the brain. The brain itself was fairly firm. When it was opened, a hole was seen in the middle of the retracted parts. The eyes were not yet entirely covered by the eyelids. The retina was conspicuous by its whiteness. The vitreous body was not different from the crystalline humour of other animals as far as its consistence was concerned. It had the size of a pea. A very little lens adhered to its upper surface. The bill was completely formed. A white point was noticed at its tip. The auditory cavity being examined very carefully showed a small blood vessel. The blood vessels of the neck were well visible. Skin covered all the parts and was rough. Traces of feathers were even seen on the back. The small ossicle which lies near the other bones was obvious in the wings. In the feet blood vessels were also clearly visible on either side of the digits, the joints were fairly protuberant. The rump was quite large. Inside the foetus the lungs were seen. There were two lobes of the liver which was yellowish. One of the lobes lay to the left of the stomach. The stomach was white and extended considerably downwards. The three coeca and the kidneys were observed. The heart was already plainly dead but I revived it by warming it up in my hand. After that, when it had again stopped moving, I compressed the liver and I saw that the vena cava which lay between the heart and the liver began again to move. I could even observe a fairly distinct valvula at this place. The auricles appeared only as a very thin membrane at some places of which kinds of small white strings were seen.

Day 11

In an egg which I had opened at its big end, the cavity was fairly considerable. But the external tunica was so dense that nothing could be seen through. This being broken, the second tunica was very thin. It enveloped the albumen in which there were several vessels. Through the albumen we saw the chick lying on the yolk in a cavity fitting it as if it had found a bed. It was full of life and moved the head, the feet and the wings. At intervals it made efforts as if to hop in the middle of the liquid. The glutinous albumen had occupied the big end. I wonder whether this did not result from the egg having been shaken. After the shell had been broken little by little and after the membrane had been detached with a stylet, I turned it over and dropped all the egg on my hand at first, then on a plate so that all the yolk was still in its membrane and that the other humours had not flowed out of their tunicae either. After I had examined the distribution of the vessels, dissected the chorion and the amnion, and divided the umbilicus, I put the foetus itself on my hand. The body of the foetus having been thus warmed up revived the vital fire of the heart which was still throbbing on 1 h later. On both sides in the thorax, some matter as transparent as crystal was seen, similar to that which I had already observed on the right side in the region of the stomach. That was unquestionably the liquid of the amnion which had penetrated through the trachea and the bronchi. Near the orifice of the stomach a green point was seen. The spleen was clearly visible. The lobes of the liver still verged somewhat to yellow. The disproportion between the eyes and the other parts of the body had decreased a little although being still far from the normal proportion. In the middle of the four parts of the brain there was something white, perhaps the pineal gland.

Day 13

In the chick this was the figure of the eye. The small circle represents the lens; the crescent, the nictitating membrane, and the outer circumference which is oblong, the rim of the eyelids which progressively grow during the next days. The gap in the black tunica of the eye was closed except at the place where it touched the optic nerve. The filaments which this nerve sends to the lens were black. The cerebellum, although quite small, was clearly visible. The two chalazae were still obvious. The vessels of the mesentery were delicate. The abdomen was almost closed except for a passage for the umbilical vessels and the intestines. The canal from the yolk to the intestine was quite obvious. The lungs were whitish. At different places there were small black feathers bulging through pores of the skin.

Day 18

The skin did not yet cover the entire abdomen and, beside the fringe of the true skin which was rough and sprinkled with a quantity of tubercles arranged in order, one could also see the margin of another envelope which was quite smooth and level. The anterior and posterior parts of the brain seemed to come closer together and, in the intermediate space between them, something with looked like the pineal gland was seen. In the stomach there was some dark green substance. The chalazae were still visible as well as the bloody fringe conspicuous since the third day of incubation and which did not yet enclose the whole yolk. The eyelids were still open.

In another egg, the stomach contained a white substance which looked like curdled milk. It was mixed with some greenish substance in the vicinity of the oesophagus. There was also much bile in the gallbladder of the same colour with fairly bitter taste.

Day 20

All the albumen being consumed left only a small white part which I should easily take for the chalazae and which was as if it were the centre about which the bloody circle which enclosed all the yolk had condensed. The continuation of the small bag of yolk with the intestine between two blood vessels was visible. But the most remarkable were the yellowish vessels spreading their branches like those of a tree, and the trunk facing the intestine. They were discovered in great number inside the small bag which had been dissected, adhering to the bag through capillaries only. These vessels were beautifully shown after a long opening was made in the upper aspect of the bag still attached to the intestines so that part of the yolk could be poured out. Then, indeed, these vessels spreading their many ramifications appeared arranged approximately like a fern and attached to the internal walls of the bag. The liver had still a yellowish pale colour so that the blood could draw its colour from the liver. By then the eyelids were shut in the two chicks.

2.20 Dissection of an Embryo Monster for Parisians

The bones on the right side of the palate with the upper lip were cleft so that they made the right cavity of the nose continuous with that of the mouth. The mother attributed that to the fact that she was fond of eating rabbits. But was it not the humour descending from the nose in the mouth through the twin holes situated in the fold of the teeth, being excessive either in abundance or in acrimony on the right side, which changed the right aperture into a cleft when the movement of the humours opened to itself their paths and cavities in a material not yet firm enough?

All the fingers of the left hand but the thumb were attached to each other. The middle finger which is normally the longest was the shortest. Perhaps the stronger connection of the extreme fingers thwarted the growth of the middle one.

The sternum was entirely cartilaginous. It was coherent above over its small part and the rest was split in two parts. Each of them closed the intercostal spaces on its side. They kept viscera protruding outside the body thus open. The heart, the liver, the stomach, the spleen which adhered to the stomach, almost all the intestines and the right kidney were bare outside the cavities of the thorax and the abdomen thus open. What if the suspicion brought forwards concerning the nose is applied also to the bellies? The cause, which could indeed change the orifice of the palate into a cleft, could also produce

---

OPH 20, vol. II, 51–53: "Embryo monstro affinis parisiis dissectus" is found in *Thomae Bartholini Acta Hafniensia Ann.* 1671, 1672, Hafniae 1673, as No. CX, pp. 200–203. In the heading Steno is styled as *Anatomicus Regius*, the position he is considered to have held during his stay in Copenhagen. The treatise, together with the following one, was probably composed to be published in *Acta Hafniensia* on the basis of notes made by Steno during his stay in Paris in 1665 and during the earlier part of his residence in Florence from 1665 until 1668. See, moreover, the notes to the next treatise, OPH XXI.

© Springer-Verlag GmbH Germany, part of Springer Nature 2018
T. Kardel and P. Maquet, *Nicolaus Steno*,
https://doi.org/10.1007/978-3-662-55047-2_31

an opening of the thorax and abdomen. There is unquestionably humour in the thorax and abdomen. It must be sought for whether even in the presence of mucous envelopes this abundance or acrimony of this humour could break these envelopes and open a way for itself also in the amnion. Only the lungs had kept their position in the thorax. Several fissures could be seen in the kidneys. The adrenals were large and almost triangular.

Those whom the considerable size of the clitoris had prevented from further examination had believed that it was a male. But the external aspect of a female sex as well as the presence of a uterus in the abdomen proved that it was a female. And so that nobody would believe it was hermaphrodite, the nymphet covering the upper part of the clitoris left its lower part bare in such a way that not only the prepuce covering the glans penis but also the skin which normally envelops the whole penis was totally absent. I will not mention the testicles[1] which were two and were concealed in the abdomen as is usual in females.

What deserved above all the attention and admiration was an unusual formation of the arteries proceeding from the heart. I discerned at first that the pulmonary artery seemed to indicate that it was much narrower than the aorta. Therefore, I cut it open from the right ventricle of the heart to the substance itself of both lungs and I reckoned that the canal which extends from the pulmonary artery to the aorta, otherwise conspicuous in every foetus, was completely absent here. After I had actually opened the right ventricle, from there a probe pushed upwards in the aorta close to the septum found an orifice opening as far as into the aorta as easily as had it passed from the left ventricle into the same aorta. There were thus three orifices in the right ventricle, one from the auricle and two others into arteries, and the same canal of the aorta common for the two ventricles formed a double orifice thanks to the septum of the heart. In the auricles there was nothing different from the normal conformation of a foetus. But, although the structure was unusual in the arteries, the effect resulting from this was, however, normal and, indeed, this is the movement of blood in every foetus: just as the vena cava discharges itself in both auricles,[2] so does the right ventricle empty in both arteries, and as the left ventricle receives blood from both auricles, so does the aorta from both ventricles at the same time. Thus either blood at first enters the pulmonary artery, coming from the right ventricle through the proper canal[3] from there to be sent into the aorta, or the canal of the aorta

---

[1]Ancient term for ovaries.

[2]Through foramen ovale.

[3]Ductus arteriosus.

which straddles the same right ventricle, the movement of the blood remains the same from the right ventricle into both arteries. I have nothing to say about the cause of this phenomenon. If indeed you say that the pulmonary artery has remained distinct from the aorta in the open thorax but otherwise in a closed thorax receives blood from the right ventricle to be sent into the aorta, I see two objections: It cannot be taken for granted that such an opening of the thorax always results in a similar structure of the arteries and, even if this was observed, it would be uncertain how a thorax opened in such a way would contribute to a change in the structure of the arteries. It appears clearly in infants that, while the pulmonary vessels are little by little dilated by respiration, the arterial canal changes with time into a ligament so that this is patent in the foetus for this cause alone that all the blood leaving the right ventricle cannot pass through the pulmonary vessels. But I do not find with any certainty why in the present case it did not enter the pulmonary artery but opened for itself a direct passage into the aorta. But whatever the cause, I think it is obvious evidence of the wisdom of nature when the same effect results not in the same way, yet in an ingenious way. What this foetus shows by sending part of the blood from the right ventricle into the big artery provides an example of how the solution of a problem inside an animal has not always been solved in the same way.

2.21 The Uterus of a Hare Dissolving Its Own Foetus

A tumour bigger than a fist in its right horn when dissected, removed white matter, tenacious, similar to soap, diffusing something bloody with pungent odour but without any stench of decay. When this matter was dissolved in water a hairy ball appeared bristling with small bony spikes protruding here and there. The hair having been removed and the bones which adhered to nothing having been unfolded, this ball appeared to be a baby hare, fairly large. All its hind parts were consumed except for the hair, part of the spine with the bones of the feet and a part of the liver. But all the anterior parts such as the diaphragm, the heart, the lungs, the ribs with their muscles and the shoulders had kept their normal colour. In the head, moreover, neither the eyes nor the brain were decayed yet. The internal tunicae of the uterus which had been in immediate contact with this matter did not show any sign of decay. It was thus certain and obvious that all resolution of a retained foetus does not occur through decay of the foetus and corruption of the surrounding tissues, and that nature is provided with a variety of all kinds of solvents which it uses at choice to remove solid encumbrances. Nor should I doubt that some softer parts dissolved in this way inside the uterus have spread out through the openings of the veins new nutritious humour for the blood, and all these parts would have had to be dissolved with time and to be returned into the blood of the mother, but if hair and bones had escaped the force of dissolving other [parts] and bones, would have had to be removed from the uterus, the viscosity of the

OPH 21, vol. II: 57–60: "Uterus leporis proprium foetum resolventis" is found in *Thomae Bartholini Acta Hafniensia Ann.* 1671, 1672, Hafniae, 1673, as No. CXI pp. 203–207.

surrounding matter being progressively changed into more fluidity, if the mother feeding on her foetus had escaped the hands of the hunter.

I saw that indeed in a snake of Florence to which two live little sparrows had been given to swallow, in my presence, one seized by the head protruded in the throat under the compression of the sinuous curves of the body. The other caught by the tail had been ingested in the belly in the same way. The volume of the sparrows exceeding the thickness of the snake, two knots stood up indicating their arrival in the belly. After some days in a casket in which I kept the snakes under close guard, two kinds of excrements were found. One part was white, friable like chalk. The other part was made of the feathers of the sparrows gathered within a ball. It was as if the animal, cold to touch, had drawn out of the whole sparrows by the force of a strong solvent, as could be judged by the eyes and the hands, every juice which it could withdraw to make food, and had sifted out unchanged what resisted the efficacy of this solvent. But this is common to the venter of all animals as they dissolve some parts of what they have swallowed and reject other parts intact depending on the various natures of the liquid dissolving solids and on whether this liquid contains more or less viscous matter. But actually I never found a dissolving liquid of this kind in a uterus except for this one time and I do not remember that it was observed by others. Perhaps, in animals which take a quantity of nutriments fitting nature, the opportunity of a liquid inducing decay is not so easy to meet. It may be that something is retained outside its place or outside due time. I indeed remember having opened also a dog, the intestines of which contained very small balls of lead inside their tunicae, sticking there for a long time already without the least sign of decay, and large wounds very often heal spontaneously without previous decay not only in animals but also in men abiding by the laws of temperance. From this observation it is possible to draw at least a good reason of hope for the parturients who, after their efforts, retain all the remnants of dead foetus, if they do not otherwise overflow with putrid humours from another cause. It is not a light comfort in a desperate situation to be able to raise a hope of salvation by one reliable example, when wearing off of the forces is feared as a result of hopeless pain. But to come back to the industry of intelligent nature, it appeared in the first example that nature does not carry out the same movement of blood by the same paths. The other example shows that nature does not always achieve dissolution of a solid present outside due time by fluids of the same kind. I will add an example of nature urging itself at carrying out unusual achievements, by way of human industry. This is testified by what follows.

A Goose's Big Liver and Milk

While Jews indeed, in order to obtain more fat, feed geese with particular care, the abundance of fodder increases the volume not only of fat but also of the liver, and plainly modifies the colour either of the liver or of the serum. I saw that in Florence, among other rare wonders of nature for which the love of the Medici princes in promoting studies offered me a daily opportunity with the utmost goodness when their liberality quietly provided for my needs not only sufficiently but also pleasantly, profoundly and honorifically. In view of investigating the cause of the increased liver, at this time I opened two geese, one frugally fed the blood of which contained little serum, and this quite aqueous, and the liver of which, actually very small, was yellowish. The other had been fattened up by copious fodder for many days. Its blood when let out of an artery was of a bright red. But soon, as it progressively cooled, its colour changed until, above clotted red blood at the bottom of the vessel, white milk floated without sign of aqueous serum. The serous parts of this milk were separated from the caseous parts by addition of vinegar. Nor did I wonder any longer that the milk of women is concealed in their arterial blood without any true mixing. A phlebotomy carried out in men shortly after a very copious meal is also in agreement with this observation, and not infrequently milky serum is retrieved which is holden for slime by some people. This experiment further confirms the opinion of those who teach that the serum of blood is nothing else than chyle more or less dissolved in the blood after shorter or longer delay, and that milk is nothing else than recent serum secreted in the udders after one or two circuits. Concerning the liver, these are the consequences: the colour of the blood does not come from the liver but rather the colour of the liver results from the colour prevailing in the blood, namely if red liquid predominates, the liver is red; but if the liquid is serous, the liver takes the colour of the serous liquid; it becomes yellowish when the serum is yellow as a consequence of too much bile; it is actually whitish when the serum always remains white by swallowing of frequent new fodder and no time is allowed to it for further resolution like in the present instance. In what concerns the monstrous increase of the liver, I will see in a next examination how far it will be possible to me to strive to the truth. Whilst, from being small and yellowish, it becomes white and big, as external causes, one observes that movement of the body is hindered and more abundant fodder is given than normally, as internal causes, one finds that milky serum and fat are amazingly increased. As far as bile is concerned I do not remember having investigated its quantity or its taste. But it is also clear that, if movement is allowed again and feeding is diminished, the liver recovers its previous size. In this instance the parenchyma of the liver demonstrates also a common point with fat: as they increase and decrease together, when the sudation that movement usually provokes is denied, the matter of chyle is also increased. Since more fodder is ingested than digested, with the arrival of new matter, the new bile always becomes duller. But it is clear that the more pungent bile is, the more nutrition of the animal is hindered. In the formation of a foetus, I also remember having found at intervals a humour almost insipid in the gall-bladder, at a period when the parts grow rapidly. Thus blood full of milk or chyle, when it passes in the liver from the hepatic artery and the vena porta into the vena cava, leaves certain parts of chyle to be set down on the parenchyma of the liver from which its whiteness and sweet taste which are taken away again and again by the blood made more pungent in its continuous circulation without arrival of new chyle. Chemical analysis could discover whether there is, between the substance which increases the parenchyma of the liver and that which increases the quantity of fat, another difference than the consequence of the presence of salt in the liver because of the bile and the absence of salt of bile in the fat. Here it will be enough to have hinted at the explanation which nature gives to those who investigate the increase of fat and the simultaneous increase of the liver and, moreover at the artifice used to achieve what nature does not usually perform by itself.

2.22 Specimen of Elements of Myology

2.22.1 Most Serene Grand Duke

While I publicly acknowledge that I do not at all deserve the great proofs of your favour, that does not diminish your very accurate judgment nor constitute any risk of my striving after honours. The favours of the greats disclose the generosity of the giver rather than the merits of the receiver. As the providence of the Divine Power is not overturned because the less worthy experiences good fortune, so the sagacity of princes is not diminished by the fact that they shower great favours not only on those who deserve them.

I do, indeed, interpret it as a great manifestation of your favour that in Italy, in Florence, in the palace blooming with arts, O Prince, most renowned among men of letters the world over for your firm grasp of affairs, you have chosen to give me, a man from the North who scarcely rises above mediocrity in talent, some share in those hours in which you relax your mind, when,

---

The book "Elementorum myologiæ specimen" was published in Florence in 1667, Fig. 2.22.1. It is in three parts, OPH 22, 23 and 24, with a common preface, "Most Serene Grand Duke". The annotated English translation of OPH 22 by Sister M. Emmanuel Collins, O.S.F. (1900–2002), Paul Maquet and Troels Kardel was published in *Steno on Muscles*, (SOM, 77–225). We are grateful for permission to reproduce this text from the publisher, the American Philosophical Society, Philadelphia. The annotations by Vilhelm Maar are found in the transcripts, www.extras.springer.com. The foot-notes to the translation are from SOM, 229–230. Stensen uses *italics* in large sections of this work, in general when he describes arguments and definitions, and he uses ordinary font for anatomical descriptions like those related to Plates I–III and for his considerations. Some illustrations are scanned copies from the printer's MS in the the Royal Library, Copenhagen, with free web-access.

© Springer-Verlag GmbH Germany, part of Springer Nature 2018
T. Kardel and P. Maquet, *Nicolaus Steno*,
https://doi.org/10.1007/978-3-662-55047-2_33

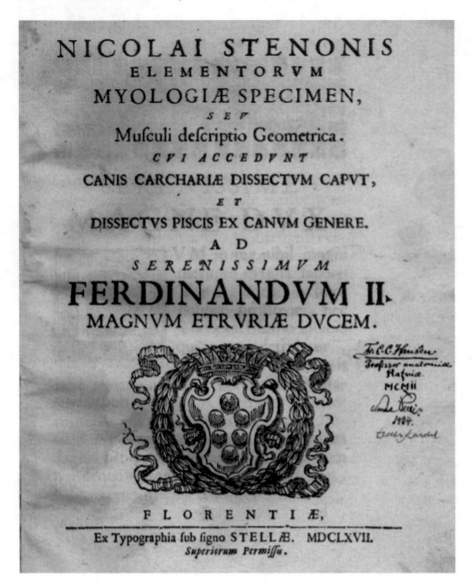

Fig. 2.22.1 Frontispiece, the *Elementorum Myologiae Specimen*, Florence 1667

wearied by official responsibilities, you take that delight in the mysteries of art and nature which others would seek in games and pleasure. Moreover, you have given me hospitable reception in the city and have allotted me everything which might possibly be of service for my studies and experiments. You have freely accommodated me beyond my expectation. I pass over the many and greater kindnesses far beyond my merits by which from day to day you displayed your benevolence toward me.

If I had the gift of eloquence, I would show my gratitude by recalling your fame, by congratulating you on anatomy and the other arts. I would demonstrate that you are, not only the beneficent Maecenas of all, but even also the most expert critic. But I am not versed in the exercise of speaking. The masters of that art recognize that this matter is beyond their abilities, even if they should bring together into one all the things which are separately proclaimed by different people about the greatest benefactors and wisest princes. Since I have no other token of my gratitude, it behoves me to imitate the plants whose fruits speak in silent eloquence of the mildness of the air and of the generosity of the earth. The various animals which you allowed me to dissect have contributed not a little to the description of anatomy. I now humbly present, dedicated to you, this *Specimen of Elements of Myology* with two descriptions of anatomical dissections appended.

I wished to demonstrate in this dissertation that unless myology becomes a part of mathematics, the parts of muscles cannot be distinctly designated nor can their movement be successfully studied. And why should we not give to the muscles what astronomers give to the sky, what geographers to the earth, and, to take an example from microcosm, what writers on optics concede to the eyes? These writers treat natural things mathematically so that they may be more clearly understood. The structure of the muscles requires almost necessarily that they be explained mathematically. But why do I claim for the muscles alone what is due to the entire body? Our body is an organism composed of a thousand organs. Whoever thinks that its true understanding can be sought without Mathematical assistance must also think that there is matter without extension, and body without figure.[1] Nor is there any other cause of the many errors which have foully defiled the description of the human body than that anatomy has hitherto disdained the laws of mathe matics. For, while ignorant of the rule of the legitimate prince, in its, may I not say, blind judgment, anatomy has governed all things; it has thrust on us the dubious for the certain, the false for the true, the unknown for the known. Anatomy, finally, has brought the matter to such a point that nothing remains more unknown to man than man himself.

How fortunate it would have been for us, how fortunate for all human kind, if our ancestors had decreed that those who spent a lifetime in the study of anatomy would pass on to posterity only that which is well established.

---

[1] This and the following sentence is quoted directly in Phil. Trans. R. Soc. 2 (1667/68): 627.

Our knowledge would not be so wide but it would be less hazardous. And if medicine, based on these established principles, did not relieve the sick of their pain, it would not add to this pain. Now we have enormous books on anatomy and medicine. Nevertheless, we drag the wretched sufferer among thousand tortures, we even drive him to a thirsty death through a thousand torments. Our greatest misfortune is that often when we deem to be helpful, then indeed we are most harmful. While deploring this common misery, I expose its cause, but I do not promise any remedy of my own. I hope that in time one will be discovered by others. And why would it not be permitted to hope for great things, if anatomy was transformed so that experimental knowledge would rely only on well established facts and reason accepted only what has been demonstrated; in other words, if Anatomy used the language of Mathematics. But this work remains for others, abler in mind and hand than I am. Limiting myself to a short incursion in muscles, I wished to make an attempt. I expose there their true structure by a new method and I demonstrate that the mode of contraction through inflation of spirits such as proposed by the majority hitherto is built on a very shaky foundation.

But if in this new bold undertaking I shall have given little satisfaction to your expectation, this knowledge is still in the seedling stage and expects increments at least, if not maturity, from further investigation and study. This is of a good omen since this tender knowledge delights in being treated gently by your hands first. Farewell, Most Serene Prince! Live long unharmed and reign happily for many years to come.

2.22.2   Specimen of Elements of Myology[2]

SINCE experience revealed a new structure of the motor fibre, the structure of the muscles, so far not considered by anybody in the different muscles already known, begins to be recognized, but a way is also open for an investigation of those muscles, which in the past nobody could prepare properly nor explain clearly. Almost three years ago I published a *Specimen* on the muscles of the heart, tongue, oesophagus, and some other organs, but without illustrations. I will now illustrate it with figures, partly of already

---

[2]SOM, 91 ff. Proclus (410–485 A.D.) defined Element as follows: "There are in the whole geometry certain leading theorems, bearing to those which follow the relation of a principle, all-pervading, and furnishing proofs of many properties. Such theorems are called by the name of elements; and their function may be compared to that of the letters of the alphabet in relation to language." Quoted from the introduction to Euclid, vol. 1, p. 114. Concerning Stensen's references to Euclid, see 'Quotations from Euclid, "Elements" used by Stensen in the geometrical description of muscles', SOM, 1994, p. 243.

known, partly of unknown muscles. These illustrations are shown here with the aim of making clear that the geometrical *structure of the muscles* which I am to propose is not just an artefact, but is derived from experience. This thus will be a small *Specimen of Elements of Myology.*

If it does not displease the public, on some occasion I intend to give a complete description of the muscles according to these new principles. There is no muscle of which there cannot be said anything specific. Of many only very little has been said so far. But also the true structure of the bones, as yet studied by no one, will altogether emerge easily and clearly from these observations, since the tendons are inserted in the bones.

I have often desired to undertake this work. But I would never pretend that what pleases me should be accepted by all others. According to an old saying, love makes people blind to their own progeny, and it is a frequent experience that what displeases other people is often what pleases authors most.

The structure of muscles in antiquity. Hitherto philosophers as well as anatomists have explained the structure of the muscles according to the illustration (Fig. 2.22.2):

At present I shall say nothing about this figure, except that what is best known to authors is totally unknown to nature. It seems to me most trustworthy to represent the structure of muscles the way I found it in many simple muscles, and I hope to demonstrate it in all compound muscles.

The new muscle structure. Relying on this basis I represent a muscle as *a collection of motor fibres arranged so that the flesh in the middle forms an oblique parallelepiped and the tendons form two opposite tetragonal prisms* (Fig. 2.22.3).

I can imagine that a number of people will stop after these introductory remarks and decide that this new muscle structure is just a new chimera. But I hope that these people will be kind enough to wait until they have read the

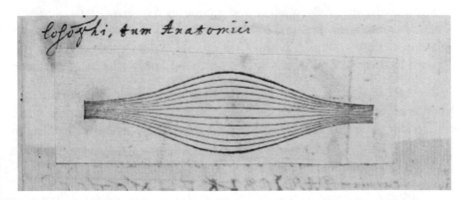

Fig. 2.22.2 The structure of muscles in antiquity. From the printers manuscript, www.kb.dk/rex

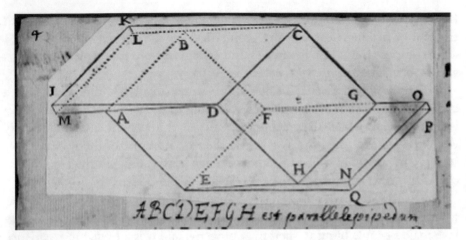

Fig. 2.22.3 The new muscle structure. A B C D E F G H is the flesh parallelepiped; D A M I C B L K and E H N Q F G O P are the two tetragonal tendon prisms

entire dissertation before expressing their opinion. They will indeed realize that I follow the track of nature closely, presenting nothing unnecessary.

In order to understand the composition of the muscles more clearly, I will put together a list of explanations of all the terms used, in the form of definitions, as the geometricians usually do and I shall begin with the motor fibre.

2.22.3   Definitions[3]

1. *A motor fibre* [fibra motrix] *is a definite structure of very tiny fibrils, immediately joining together lengthwise. Its intermediate part differs from its extremities in consistency, thickness and colour, and is separated from the intermediate part of the adjacent motor fibres by transverse fibrils continuing into the aponeurosis.* Figure 1 in Plate I represents a motor fibre. B C is the intermediate part, and A B and C D are the extreme parts. What is proposed in this description can be demonstrated in any muscle of our body.

   Why it is called motor. I name this fibre motor, because it seems to me to be the true organ of the movement of an animal. The entire muscle which hitherto was given this name is nothing else than a well-determined collection

---

[3]The 44 "definitions" are structurally determined, mostly based on anatomical observations. Microscopes were still insufficient to identify the muscle fibres.

of such fibres as its analysis shows to the senses. The muscle does not act unless its single fibres act, as dissecting living animals demonstrates.

When I name the motor fibre the true organ of the movement of an animal, I do not determine whether that part that shortens is shortened in itself, or whether it is crimped by fibrils going transversely through it, or whether the contraction is carried out in any other way. Whatever of these occurs, the motor fibre rather than the muscle is what must be termed the animal's organ of movement.

How to understand that fibrils lie in immediate contact. When I say that fibrils are joined immediately together, their close contact excludes all fluid in between. These same fibres would be continuous rather than contiguous. But we see such a phenomenon in mechanics, when two cubes, or any other two bodies, are said to be superimposed in whatever way immediately one upon the other, even if ambient fluid is not totally excluded. Thus two plane mirrors, which are moistened, are said to be superimposed immediately one on the other, even though water in between separates their surfaces. Whatever small amount of water may be in between, they are said to be in immediate contact because nothing solid is in between.

2. *THE FLESH is the intermediate part of the motor fibre surrounded by transverse membranous fibrils. The flesh is soft, of a certain width and thickness, and it has different colours in different animals, reddish in many, grey in others, and even white in some.* I remember having seen both red and white muscles in the same paw of a rabbit. I am here speaking of raw flesh, since most meat whitens when cooked.

3. *THE STRUCTURE OF THE FLESH could be represented by a parallelepiped* A B C D, *whose end surfaces* A C *and* B D, *i.e. those which continue into tendons, form oblique angles with the transverse surfaces* E D *and* A F,

whereas the lateral surfaces E B *and* C F *form right angles with both the end surfaces* A C *and* B D, *and with the transverse surfaces* E D *and* A F. In order to show their surfaces more clearly, I have drawn them larger than nature.

Three objections concerning the structure of the flesh

Three objections may arise here: First, I say that the extremities of the flesh, which continue into the tendons, are plane; Second, I supposed that the lateral surfaces form right angles with the other four surfaces; Third, I have determined that the transverse surfaces are inclined on the end surfaces.

Answer to all concerns together.

I could give one answer to meet all three objections, just by saying that I describe here one regular fibre as a norm for all other fibres, which would be no less permissible for me than for all the others, who explain problems in complex and extraordinary things by means of ordinary and simple examples. But in order not to be considered as presenting something without giving a reason, I will answer each of the three objections separately by reporting the experiment which is the foundation of what I say.

Answer to the first objection concerning the end surfaces.

Thus, concerning the end surfaces of the flesh, I remember having seen once in the leg of a turkey-cock,[4] after I do not know what disease, that the flesh had parted from the tendon expansion; the end surface of the flesh appeared to me to be flat, as they are seen when the middle of the flesh has been cut transversely. But also in boiled meat, when the tendon parts from the meat, the end surfaces of the flesh appear flat.

Answer to the second objection concerning the lateral surfaces.

With regard to the lateral surfaces, I admit that the flesh is so extremely delicate that it is not possible to determine with any certainty its location in relation to the transverse surfaces. But, since in lumps of meat we see at many places the lateral surfaces as perpendicular to the transverse surfaces, it is not without reason then to assume the same relation of the surfaces in the single parts constituting these lumps of meat. The end surfaces of many muscles are clearly demonstrated to be at right angles with the lateral surfaces, so it can be easily admitted that the lateral surfaces form right angles with the other four surfaces.

Answer to the third objection concerning the oblique angles.

As far as the oblique angles are concerned, experiments obviously show that the end surfaces form oblique angles with the transverse surfaces. This appears at the extremities of the flesh freed from tendons, as well as in the middle of the flesh when cut through a plane parallel to the end surfaces.

However, I do not wish to impose on the reader the impression that I have examined all muscles in all animals and that I believe for certain that the position of the surfaces in relation to each other is everywhere such as I have described. I claim only with certitude to have found this position of the surfaces in many cases. The demonstration of this simple and regular structure of the flesh is based on these observations. It is thus not without reason that I propose it as a model for all others.

How are the surfaces of the flesh?

4. *According to the definition of the parallelepiped of the lateral surfaces* E B *and* C F *have oblique angles, while the transverse surfaces* E D *and the end surfaces* A C *and* B D *are rectangles.*

---

[4]Similar observation mentioned in *De musculis ...*, OPH XV.

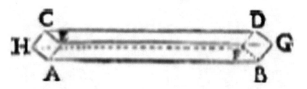

Fig. 2.22.4 See definition 4

In order to define the edges of these surfaces I propose:
Lateral edges of the flesh.

5. I call *FLESH EDGES* those edges shared by the transversal surfaces, such as A B and its parallels.

Tendon edges of the flesh.

6. *THE TENDON EDGES* are those edges shared by the end surfaces, such as E A and its parallels.

Transverse edges of the flesh.

7. *THE TRANSVERSE EDGES* are those edges shared by the transverse surfaces and the end surfaces, such as E C and its parallels.

Length of the flesh.

8. *THE LENGTH OF THE FLESH* is the line between the end surfaces and parallel to the lateral edges of the flesh and thus equal to these, such as A B.

Width of the flesh.

9. *THE WIDTH OF THE FLESH* is the distance between the lateral surfaces and thus equal to the transverse edge, B F.

Thickness of the flesh.

10. *THE THICKNESS OF THE FLESH* is the distance between the transverse surfaces.

Tendon.

11. *THE TENDON* is the extreme part of the motor fibre; it is hard and white. Since both ends of a motor fibre have the same consistency and colour, I will call both of them tendons, while I call ligament the remaining of the tendons which joins two bones.

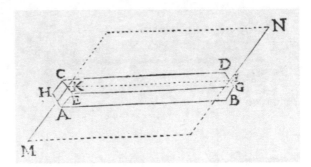

Fig. 2.22.5 See definition 12

Plane of the motor fibre.

12. *The two opposite tendons* K M *and* I N *in the same motor fibre* M K, I N *are in the same plane* M N, *which is parallel to the lateral surfaces* A G *and* H D *and divides the transverse surfaces* C G *and* H B *in two. This plane* M N *can be called* THE MOTOR FIBRE PLANE (Fig. 2.22.5)

Inflected motor fibre.

13. *A motor fibre is said to be* INFLECTED, *when in the motor fibre plane* M N, *the tendons* K M *and* I N *are inclined on the flesh* H G *at obtuse angles* M K I *and* K I N, *whereas the end surfaces* C A *and* D B *form acute angles* E A B *and* B G E *with the transverse surfaces* C G *and* H B. This is also shown in Plate I, Fig. 2, in which the tendons A B and C D are inflected towards the flesh B C, both tendons forming obtuse alternate angles A B C and D C B with the flesh.

Rectilinear motor fibre.

14. *A motor fibre is said to be* RECTILINEAR *when the flesh of the motor fibre forms a straight-line.* Curvilinear fibre, the flesh of which is bent into curves will not be dealt with here, since the definitions given here apply only to the rectilinear muscle.

Equal motor fibres.

15. EQUAL[5] *MOTOR FIBRES are those which are equal to each other in all respects.*

Equally equal motor fibres.

---

[5] *Equal* denotes equal in content.

16. *EQUALLY EQUAL*[6] *MOTOR FIBRES are fibres, the flesh of which is equal, and the corresponding tendons of which are equal.*

Unequally equal motor fibre.

17. *UNEQUALLY EQUAL MOTOR FIBRES are fibres the flesh of which is equal and the tendons taken together are equal, while the individual tendons of one muscle compared with the individual tendons of the other are unequal,* as in Plate I, Fig. 3, which shows nine unequally equal fibres.[7]

Unequally equal motor fibres arranged according to the increment of the tendons.

18. *UNEQUALLY EQUAL MOTOR FIBRES ARE SAID TO BE ARRANGED ACCORDING TO THE INCREMENT OF THE TENDONS when the single tendons which are on the same side exceed each other by an equal amount* as in Plate I, Fig. 3. All the tendons in the space B A E F as well as those in the space D C G H on the other side increase by equal increments.

The Order

19. *An ORDER*[8] *is a series of rectilinear motor fibres unequally equal inflected at the same angle and arranged according to the increment of the tendons in such a way that flesh parts are set upon flesh parts at the transverse surfaces, and the tendons set upon tendons directly so that all the tendon edges of the flesh pat-ts form the same two straight-lines.*

To understand better the compound nature of the Order, I displayed in Plate I three more illustrations, Figs. 3, 4, and 5. Figure 3 shows the rectilinear, unequally equal motor fibres, arranged according to the increment of the tendons. Figure 4 shows the same motor fibres equally inflected on both sides, but not yet united. Figure 5 shows the same, now with the fibres united, in which, however, spaces have been left between the flesh parts in the illustration for more clarity. F B and G C are the two straight-lines at each end of the flesh.

---

[6]*Equally equal* denotes equal in shape.

[7]In *De musculis* … (OPH 15) Stensen compared the *Unequally equal* arrangement of tendons with the shape of Panpipes, *Syrinx.*

[8]In the 1684 translation of Thomas Willis, *De motu musculari,* by Samuel Pordage [Ordo] was translated as ORDER and [Versus] as TURNINGS. In the 1712 translation of *De musculis,* (OPH 15) [Ordo] was translated as RANK, and [Versus] as LINE or ROW. ORDER and RANK were preferred for this translation.

I could have described the parallelepiped of flesh and the attached prisms of the opposite tendons in a solid [three-dimensional Order. However, I do not intend to introduce anything here just to impress and every thing must have a purpose. For this reason I describe those matters alone which will necessarily be used in what follows.

Plane of the Order. Plate I, Fig. 5.

20. *From the definition of the Order it follows that the planes of all the motor fibres coincide with the same plane* A H, *which I call THE PLANE OF THE ORDER, in which the opposite tendons form two scalene trapezia* A B F E *and* H G C D, *while the flesh parts form an oblique parallelogram* F C *comprehended by the flesh edge* F G *and the straight-line* G C *formed by the tendon edges of all the flesh parts. I, therefore, call this line the TENDON EDGE.* Depending on the larger or smaller number of constituent motor fibres forming an Order, the flesh parallelogram may acquire the form of a rhombus or of a rhomboid.

Height of the Order.

21. *THE HEIGHT OF AN ORDER is the distance between the tendon edges* G C *and* F B.

Thickness of the Order.

22. *THE THICKNESS OF AN ORDER is the distance between the flesh edges* F G *and* B C, *which, according to the definition of the Ordet; is equal to the sum of the constituent motor fibres' thicknesses.*

Similar Orders.

23. *SIMILAR ORDERS are Orders the parallelograms of which are similar.*

Homogeneous edges of Orders.

24. When comparing two Orders, THE EDGES of flesh are said to be *HOMOGENEOUS, and the EDGES of the tendons are said to be HOMOGENEOUS, when considering edges which are in similar positions.*

Compound Order.

25. *An ORDER is said to be COMPOUND when the surfaces of several Orders are mutually joined to each other in the same plane.*

The Rank.

26. *A RANK is a series of equally equal rectilinear motor fibres inflected at the same angle. The planes of the motor fibres are parallel. The flesh parts are set immediately upon flesh parts along the lateral surfaces of the motor fibres in such a way that the transverse edges of all flesh parts are on the same straight-lines.*

How many and what surfaces are in a Rank. Plate I. Fig. 6.

27. *From its composition it is clear that in the Rank* I D *there are three distinct rectangular surfaces the middle one of which* K C *is THE RECTANGLE OF FLESH PARTS, while the end surfaces* I B *and* L D *are THE RECTANGLES OF THE TENDONS.*

The rectangle of flesh parts.

28. THE RECTANGLE OF FLESH PARTS K C is delineated by a flesh edge B C and a straight-line C L made of the transverse edges of all flesh parts. I call this straight-line THE TRANSVERSE EDGE.

Width of the Rank.

29. *THE WIDTH OF A RANK is the distance between the flesh edges* B C *and* K L, *which, according to the definition of the Rank, is equal to the sum of the widths of all the single flesh parts.*

Unequally equal Ranks.

30. *UNEQUALLY EQUAL RANKS are such ranks in which the rectangles of flesh parts are equal, while the sum of the two tendon rectangles of one rank is equal to the sum of the two tendon rectangles of the other rank, but the individual tendon rectangles of one rank are unequal to those of the other rank.*

There can be two descriptions of a muscle depending on whether this is considered as made of orders or of ranks.

31. *A simple rectilinear muscle is a body made muscle of several equal orders, similar, parallel and set immediately upon each other, so that all orders are congruent with all other orders. (See Table I, Fig. IV).*
32. *Or, a muscle is a body made of unequally equal, similar Ranks arranged according to the increment of the tendons, with flesh parts set immediately upon flesh parts and tendons upon tendons in such a way that all ranks are congruent with all other ranks. (Table I, Fig. V).*

33. *From the definitions it appears that whatever the composition of the muscle, there is in it one parallelepiped of flesh parts and two tetragonal tendon prisms as represented above.* Since, however, the thickness of the tendons is not needed to explain the movement of the muscle, and that part of the tendon protruding beyond the flesh (which I call the extruding part of the tendon continuations) rarely produces any noticeable difference, it is better to consider only those tendon surfaces in which the end surfaces of the flesh parts are. Thus, to consider the movement of a muscle it is not necessary to choose more than three pairs of surfaces. They are named according to the description which I used for the flesh. They are the end surfaces, the transverse surfaces, and the lateral surfaces.

34. *THE END SURFACES, made of the end surfaces of the flesh parts in immediate contact are rectangles delineated by the transverse edge and the tendon edge of the Orders.*

Each of the two is called BASE OF THE MUSCLE.

35. *THE TRANSVERSE SURFACES are the flesh surfaces of the outermost Ranks in the muscle and, like them, rectangular, being delineated by a transverse edge and a flesh edge.*

36. *THE LATERAL SURFACES are formed by the surfaces of the outermost Orders in the muscle, and, like them, they form oblique parallelograms delineated by a tendon edge and a flesh edge.*

From this it appears that the edges of the muscle can be expressed in the terms already explained, which are:

37. *THE TRANSVERSE EDGE is the same as the transverse edge of the Ranks.*
38. *THE TENDON EDGE is the same as the tendon edge of the Orders.*
39. *THE FLESH EDGE is the edge shared by the Ranks and the Orders.*

The three dimensions of this parallelepiped are thus expressed in the three straight-lines, which can be called the height, the width, and the thickness.

40. *The height of the muscle is the distance between the end surfaces and is equal to the height of the orders.*
41. *The width of the muscle is the distance between the lateral surfaces and is equal to the transverse edge.*
42. *The thickness of the muscle is the distance between the transverse surfaces and is equal to the thickness of the orders.*
43. *The length of the muscle is the straight-line between the outer extremities of the opposite tendons; if for a muscle you have only the parallelepiped, the*

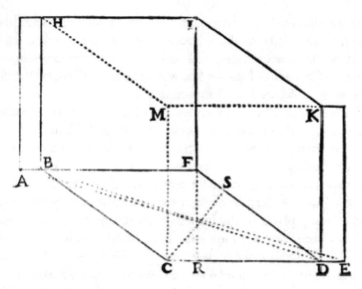

Fig. 2.22.6   See definition 44

length is the distance between the transverse edges at their maximum dis-tance, thus it is equal to *the longer diagonal in the parallelogram of the order.*

44. *The muscle is said* to *contract when its length becomes shorter* (Fig. 2.22.6).

In the present figure:

B I and C K are the end surfaces;

B M and F K are the transverse surfaces; B D and H K are the lateral surfaces;

K D and its parallels are the transverse edges; G D and its parallels are the tendon edges; B C and its parallels are the flesh edges;

F R is the distance between B F and C D, or the height of the muscle;

D K is the distance between I K and F D, or the width of the muscle; G S is the distance between B G and F D, or the thickness of the muscle; and A E or B D is the length of the muscle.

Use of the new muscle structure to demonstrate that muscles can swell during contraction even without the arrival of new substance.

These descriptions include different matters which should rather be placed under propositions to be demonstrated according to the structure of the flesh, the order, the rank, and the muscle. But they are given [as definitions], as I am not going to propose all the elements of myology, but only to provide those which are sufficient for a clear understanding of the muscle structure.

Since I believe that in the above presentation I have accomplished this, it now becomes my task briefly to show how this structure is used to explain the movement of the muscles; not that I am giving the true explanation of the movement, which I admit I do not know, but rather I will show that what has been presented by others is far from certain.

When a muscle contracts, the amount of fluid substance contained in it either increases, or diminishes, or remains unchanged. Thus, there are three different ways to explain the movement of the muscles and to represent their mechanics according to our observations.

Many have supposed it to be an unquestionable fact that, during contraction, the mass of the muscle increases, because an obvious swelling is perceived at many places in the human body during the contraction. These people then reason, that the muscle is like a bladder which becomes shorter the more it is filled. To make it clear that a swelling alone is not sufficient to make one assert that new material arrives into the muscle during contraction; I will demonstrate that in every muscle that contracts there is a swelling, even if the contracted muscle remains equal to the non-contracted muscle. To this end, in what follows, it is necessary to make some suppositions and some demonstrations.

2.22.4 Suppositions[9]

1. *When a muscle contracts, the tendon edges do not change.*
2. *When a muscle contracts, the individual parts of flesh contiguous in the same plane remain such.*
3. *When a muscle contracts, the individual parts all along the flesh move and shorten equally.*
4. *When the flesh shortens, its width remains the same.*
5. *A contracted muscle is equal to the same muscle non-contracted.*

Which of these suppositions are certain and which are uncertain? Among these suppositions the first three are certain, as I will demonstrate experimentally elsewhere. As far as the fourth supposition is concerned it appears that the width of the flesh parts does not increase. Whether it diminishes remains questionable. If it were certain that the width diminishes, a swelling in the contracted muscle would manifest itself more clearly. But even if the width is

---

[9]The suppositions are determined from macroscopical observations of the movements of muscles, SOM, 123 ff.

not diminished I will show that contraction will result in a swelling. I made the fifth supposition, not because I believe it is certain, but, if it is admitted, I demonstrate that a swelling will occur in the individual muscles when they contract. There are some who take the fifth supposition for granted, saying that the width of the muscle increases as much as the muscle's length decreases, following the same principle as when an oblong rectangle is changed into an equal quadrate. But, so far, neither have I seen my fifth supposition truly proven by anyone, nor does an explanation with rectangles agree with nature, not to mention several other things which might be added.

Lemma I[10] *On the parallelepiped of muscles. When a muscle contracts it does not cease to be a parallelepiped. When the single parts of flesh are in contraction, all parts change equally (supp. 3)*[11] *over the whole length of the muscle. They remain in the same planes (supp. 2) contiguous to themselves. Those parts which were straight and equal (def. 19 and 31) before contraction will remain so after contraction. In the same way their opposite surfaces which were parallel (def. 33) before contraction will remain parallel after contraction. Thus the muscle parallelepiped made of single flesh parallelepipeds does not cease to be a parallelepiped during contraction.*

Corollary. Since a contracted muscle is a parallelepiped, the opposite angles, the opposite surfaces, and the opposite edges will be equal. Hence, since the extremities of the muscle (def. 43) are its two opposite transverse surfaces most distant from each other, it will not matter with respect to the muscle whichever extremity meets the greater resistance during the movement, and accordingly becomes the immobile base.

Lemma II *On the width of the muscle. When a muscle contracts, its width does not change.*

Let the muscle be a parallelepiped C H I D *in which* H B *is the width and* D K *the immobile end. Let it contract to form the contracted muscle* C O Q D, *the width of which is* O N *(Fig. 2.22.7).*

I claim that the straight-line O N *in the muscle* C O Q D *is equal to the straight-line* H B *in the muscle* C H I D.

---

[10]The 6 lemmas are geometrical deductions of the muscle structure (determined by the definitions) set into motion (determined by the suppositions). According to Proclus, lemma is "a proposition which is assumed for the construction of something else: thus it is a common remark that a proof has been made out of such and such lemmas. But the special meaning of lemma in geometry is a proposition requiring confirmation." Quoted from the introduction to Euclid, vol. 1, p. 133.

[11]References in brackets refer to Euclid, here changed to Heiberg's definitive text numbers. See SOM, 243.

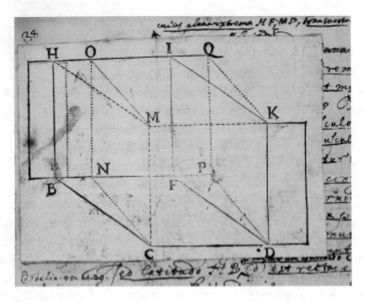

Fig. 2.22.7 See Lemma II and IV. From the printer's manuscript in Royal Library, Copenhagen

When the muscle contracts (supp. 2) *the single parts of flesh in the same surfaces remain contiguous. But the width* (supp. 4) *of the single parts of flesh does not change. Thus, the straight-line made of* (axioma I.2) *the contiguous widths of several parts of flesh will not change. But the width* H B *in the non-contracted muscle* C H I D (def. 42 and 29) *is the straight-line made of the widths of the several parts of the flesh. Therefore, it will be equal to the straight-line* O N *in the contracted muscle.*

Lemma III *On the width of the muscle. When a muscle contracts, its base remains the same.*

Let the base of the muscle be the surface A D *with the tendon edges* A B *and* C D, *the transverse edges* A C *and* B D, *and the diagonals* A D *and* B C (Fig. 2.22.8).

I claim that the base A D *of the muscle remains the same when the muscle contracts as it was before the contraction.*

The base of the muscle, or in other words (def. 34), *one of the muscle's end surfaces, is a rectangle delineated by the tendon edge* A B *and the transverse edge* B D. *But the tendon edges do not change* (supp. 3), *and the transverse edges* (def. 33) *are equal to the width of the muscle, which likewise* (Lemma II) *does not change. As a consequence, the edges of the base remain the same in the contracted muscle as they were in the non-contracted muscle. But the diagonals* A D *and* B C *also remain equal. Since*

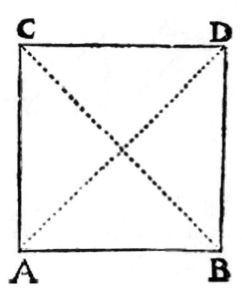

Fig. 2.22.8 See Lemma III

indeed the base of the muscle (def. 34) is a rectangle made of the end rectangles of the parts of flesh immediately adjacent, its diagonals are equal to each other. Therefore in both diagonals there will be the same position and number of extremities of flesh parts. Therefore, since composites made of equal (supp. 3) parts similarly arranged change equally, the diagonals in the contracted muscle will also remain equal. Since the edges of the base A B *and* B D *are the same in the contracted and in the non-contracted muscle, and the diagonals of the base remain equal in the contracted muscle, i.e., the base of the contracted muscle remains a rectangle, it is obvious that the base does not change, when the muscle contracts.*

Lemma IV[12] *On the height of the muscle. The height of a contracted muscle is equal to the height of the non-contracted muscle.*

Let C H I D *be a muscle in which the end surfaces are* H F *and* M D, *the base is* M D, *and the immobile extremity is* D K, *and let the contracted muscle be* C O Q D *in which the end surfaces are* O P *and* M D, *and the base is* M D.

I claim that the height of the muscle C H I D *is equal to the height*[13] *of the muscle* C O Q D.

---

[12]See the figure in Lemma II.

[13]1667 ed.: *altitudinem.* OPH: *latitudinem* is an error.

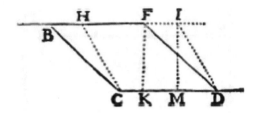

Fig. 2.22.9   See Lemma V

When the muscle contracts (Lemma I), *it does not cease to be a parallelepiped; and its base* M D *is the same in the contracted* (Lemma III) *as it was in the non-contracted muscle; also the contracted muscle* C O Q D (supp. 5) *is equal to the non-contracted muscle* C H I D. *But equal parallelepipeds with the same base* (XI. 29) *have the same height. Thus, the height of the contracted muscle is equal to the height of the non-contracted muscle.*

Lemma V *On the lateral surface. The lateral surface in a contracted muscle is equal to the lateral surface in the non-contracted muscle* (Fig. 2.22.9).

Let the lateral surface in a non-contracted muscle be a parallelogram B D, *in which the tendon edges are* B F *and* C D, *and the height is* F K. *This muscle contracts. The lateral surface of the contracted muscle becomes* H D *in which the tendon edges are* H I *and* C D, *and the height is* I M.

I claim that the lateral surface H D *is equal to the lateral surface* B D.

The height of the lateral surface or (def. 36), *which is the same as the height of the order in the muscle* (def. 40), *is equal to the height of the muscle. Thus,* F K, *the height of the lateral surface* B D *is equal to the height of the non-contracted muscle, and* I M, *the height of the lateral surface* H D, *is equal to the height of the same muscle when contracted. But, when a muscle contracts,* (Lemma IV) *its height remains the same. Thus,* F K *is equal to* I M. *But the tendinous edges* (supp. 1) *do not change either; thus,* C D, *the tendon edge of the lateral surface* B D, *is equal* to C D, *the tendon edge of the lateral surface* H D. *Since* B D *and* H D *are parallelograms of equal height and have the same base,* (I. 35) *they are equal. Consequently, the lateral surface in the contracted muscle will be equal to the lateral surface in the non-contracted muscle.*

Lemma VI *On acute angles. When a muscle contracts, its acute angles widen.*

Let the lateral surface in a non-contracted muscle be B D *in which the tendon edges are* B F *and* C D, *the flesh edges are* B C *and* F D, *the acute angles are* F D

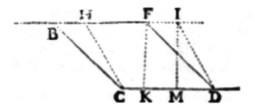

Fig. 2.22.10 See Lemma VI

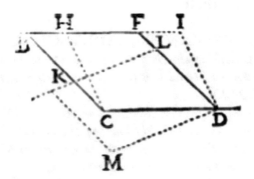

Fig. 2.22.11 See Lemma VI

C *and* F B C, *the immobile extremity is* D, *and let the muscle contract. Now the lateral plane becomes* H D, *the tendon edges are* H I *and* C D, *the flesh edges are* H C *and* I D, *and the acute angles are* I D C *and* I H C.

I claim that the angles C H I *and* I D C *are larger than the angles* C B F *and* F D C.

Since the lateral surfaces B D *and* H D *(Lemma V) are equal, and since their base* C D *(supp. 1) is not changed, and the surfaces are on the same side (they are indeed in the same muscle), they are (I. 35) between the same parallel lines* B I *and* C D *(Fig. 2.22.10).*

Therefore, in the triangle C B H *(I. 16) the external angle* C H I *is greater than the inner and opposite angle* C B H; *but the angle (I. 34)* I D C *is equal to the angle* C H I, *and the angle* F D C *is equal to the angle* C B F, *or* C B H, *since they are opposite angles of the same parallelogram. Consequently the angles* C H I *and* I D C *are larger than the angles* C B F *and* F B C. *Thus, when the muscle contracts, its acute angles widen.*

When the acute angles in the muscle widen, either the flesh edge, or the tendon edge is immobile, depending on which of them meets the greater resistance; the one which is thus immobile is called the IMMOBILE EDGE *(Fig. 2.22.11).*

Thus, in the order B F, D C, *if the edge* F D *meets less resistance than the edge* C D, *then the edge* C D *will be immobile, and the contracted muscle will be* H I D C. *If, actually, the edge* C D *meets less resistance, then the edge* F D *will be the immobile edge, and the contracted muscle will be* K L D M.

After these premises, we will now arrive at the proposition itself.

2.22.5 Proposition

EVERY MUSCLE SWELLS WHEN CONTRACTING

Since swelling is nothing else than an increase of one or several dimensions, increase of the thickness of a muscle is the same as the occurrence of a swelling in the muscle (Fig. 2.22.12).

Let the lateral surface of a non-contracted muscle be B D *in which the flesh edges are* F D *and* B G, *and the thickness of the muscle is* C R. *The muscle contracts. The lateral surface of the contracted muscle becomes* H D *in which the flesh edges are* H C *and* I D *and its thickness is* C S.

I claim that the straight-line C S *is longer than the straight-line* C R.

The lateral surface H D, *in the contracted muscle* (Lemma V) *is equal to the lateral surface* B D *in the non-contracted muscle, and the rectangle* C R × F D (I. 35) *is equal to the parallelogram* B D. *They have indeed the same base* F D *and they are between the same parallels* B C *and* F D *since* C R, *or the thickness* (def. 22) *is equal to the distance between the flesh edges. For the same reason, the rectangle* C S × I D *is equal to the parallelogram* H D.

Therefore, since (axioma I.1) *two elements which are equal to a third are equal to each other, the rectangle* C R × F D *will be equal to the rectangle* C S × I D. *But* (VI.14) *in equal rectangles the sides are inversely proportional; thus,* F D/D

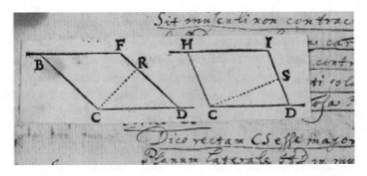

Fig. 2.22.12 See Proposition. From the printer's manuscript in Royal Library, Copenhagen

I = C S/C R. *But* F D *is longer than* D I *since the flesh of a non-contracted fibre is longer than that of the contracted fibre. Thus,* C S *is longer than* C R, *i.e. the thicknesses of an Order in a contracted muscle is greater than the thickness of the Order in the non-contracted muscle. But the thickness of a muscle and the thickness of the Orders of the same muscle are equal. Thus, the thickness of the contracted muscle is greater than the thickness of the non-contracted muscle. Therefore, every muscle swells when contracting.*

I could have demonstrated the same according to Euclid XI. 34 by considering the muscle as a solid structure with one of the transverse surfaces as the base and the distance between the transverse surfaces as the height. But it is against the rules of the method to use a solid[14] *structure when a plane structure is sufficient.*

Because of the varying form and location of the muscles, this swelling will not be observed in the same manner in each case. I shall disclose here in a few words the principal variations that may occur.

On the various swelling in long muscles depending on their different locations.

First, let be the muscle B D *in which the flesh edges* F D *and* B C *are so much longer than the tendon edges* D C *and* F B, *that even when the muscle is maximally contracted, for example* B L, *the shorter of the diagonals* P M *still form the obtuse angles* M P B *and* P M L with the opposite tendon edges P B *and* M L (Fig. 2.22.13).

In this muscle the lateral surface B D *is supposed to be perpendicular to the part on which the muscie is supported, the extremity to lie immobile* B *and the immobile edge* B C *to lie close and parallel to the same part* B D. *From what precedes it appears that there will be an equal swelling over the entire space occupied by the edge* P L, *since this swelling is nothing else than that created when* (a) *the thickness of the distance between the flesh edges in the Order increases. The measure of this swelling is the excess of length of M N over the straight-line* M I. (a) def. 22.

In the same muscle if the immobile edge B F *is supported to be directed towards the part on which the muscle is supported, the swelling will not lie even as it was in the preceding case. The further you move a finger over the transverse surface* B C *from the fixed extremity towards the mobile extremity, the bigger you will feel the swelling.*

Since not only B C *recedes from* F D [forming B M], *but even* F D *withdraws from the surface of that part on which the muscle is supported* [forming F L], *thus the maximum swelling is the excess of length of* M I *over the straight-line* N I (Fig. 2.22.14).

---

[14]three-dimensional.

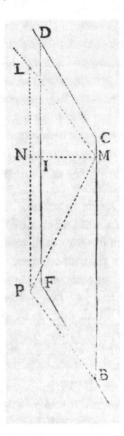

Fig. 2.22.13 See below

In the same muscle the lateral surface B D *is supposed to be parallel to the part on which the muscle is supported. Since during contraction the width of the muscle* (a) *remains the same, no swelling will be observed above the supporting part in this instance.*

This shows that equal muscles can be placed in different ways so that in one a swelling is felt, in another an even larger swelling, and in a third no swelling is felt at all (Fig. 2.22.15).

(a) supp. 4.

On the variable and swelling in short muscles depending on their location.

Second, in the muscle B D, *the flesh edges* B C F D *are so much shorter than the tendon edges* B F *and* C D, *that even the shorter of the diagonals* F C *in the non-contracted muscle forms acute angles* B F C *and* F C D *with the tendon edges* B F *and* C D.

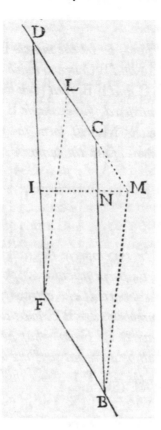

Fig. 2.22.14   See above

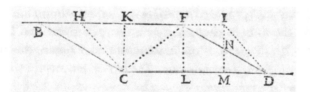

Fig. 2.22.15   See above

In this muscle the immobile extremity is supposed to be D. *The immobile edge* D C *is close and parallel to that part on which the muscle is supported. From what has been previously demonstrated, it is clear, that in all that space occupied by the tendon edge* B F *no swelling is to be observed, since the distance between the tendon edges* (a) *is the height, which always remains the same* (b). *But in the space* F D *an uneven swelling is felt, since* (c) *the angle* F D C *widens. As the extremity*

F D *proceeds from* F *to* I, *when one puts a finger on the edge* F D, *between* F *and* N, *one feels the swelling, which raises the finger and progresses under it.* (a) def. 21. (b) Lemma IV. (c) Lemma VI.

Even though one may get an idea of the compound muscies from the simple ones, I will nevertheless give one example of the compound (Fig. 2.22.16).

On the swelling of a compound muscle.

Third, a muscle A E *is made of the simple muscles* A C E I *and* A G E I. *The tendon edges are* C A, E I, *and* G A, *the flesh edges are* E G, I A, *and* E C, and the acute angles are at A *and* E. *The straight-line* C G *will be longer in the contracted muscle than it was in the non-contracted muscle.*

Since the acute angles widen during contraction of the muscle (a), *the angle* C A I *is wider in the contracted than in the non-contracted muscle. And for the same reason the angle* G A I *also widens. Thus,* [ergo] *the angle* C A G *made of the angles* C A I *and* G A l *is wider in the contracted than in was in the noncontracted muscle. But since the tendon edges* A C *and* A G (b) *are the same in the contracted as they were in the non-contracted muscle and an angle formed by equal sides has been demonstrated to be larger in the contracted muscle than it was in the non-contracted muscle, the base* C G *will also be larger in the contracted than it was in the non-contracted muscle. In this instance a swelling will appear in the middle.* (a) Lemma VI. (b) supp. I.

I THUS THINK IT IS AMPLY DEMONSTRATED IN EVERY MUSCLE THAT WHEN IT CONTRACTS SWELLING OCCURS, EVEN IF NO NEW SUBSTANCE ENTERS THE MUSCLE.[15] This is what I had promised to demonstrate, in part to make it clear that, whatever clever arguments are proposed from several sides about an influx of new substance into the muscle, they are by no means proven, and in part to show the usefulness of the new muscle structure to explain the movement of

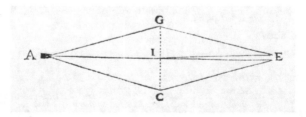

Fig. 2.22.16   See above

---

[15]The general conclusion emphasized here in capital letters, followed by the aim of the study.

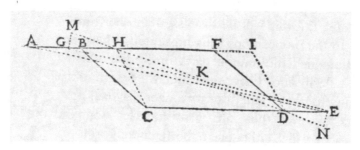

Fig. 2.22.17   See below

muscles. If I demonstrated that any part of flesh in a motor fibre cut trans-versely contracts in the same proportion as did the whole flesh even after arteries, veins, and nerves have been removed, then there would remain nothing of their theory. Moreover, the structure they propose is unlike any natural structure, and the reasoning by which they inferred the arrival of new substance is not certain, nor is the arrival of new substance itself likely. But this will be discussed another time.

In the propositions thus presented, I have not mentioned the tendon parts protruding outside the flesh, since I have measured the length of the muscle by the length of its longer diagonal only. As, however, there are many muscles from the flesh of which either one or both tendons protrude outside the flesh, one might wonder whether those tendon parts, which were parallel before contraction, would extend into one straight-line in continuation with the longer diagonal, or whether the change of position of the tendons could restore the length of the muscle which the contraction itself had shortened.

I remember that this problem was presented to me once in a learned assembly. Therefore, to show that the one who raised this objection had a good reason to wonder, and altogether to make clear what must be said about the whole matter, I decided to add the following consideration.

A muscle A E is supposed in which the tendon edges F A and C E continue outside the flesh parallelogram. B A and D E are the protruding parts of the tendons, A E is the length of the muscle, and E its immobile extremity. The muscle contracts. The contracted muscle is G E. The total tendon edges are G I and C E, their protruding parts are G H and D E. The length of the muscle is G E, and its longer diagonal is H D.

Let the line H D on both sides continue ad infinitum *(a), and mark H M equal to H G (b), and D N equal to D E. (a) post. I.2. (b) I.4 (Fig. 2.22.17).*

Then please compare the straight-line M N and the straight-line G E. It appears that the straight-line M N will always be longer than the straight-line G E.

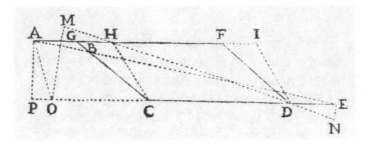

Fig. 2.22.18   See below

The straight-line H G (c) *is equal to the straight-line* H M. *In the triangle* G H K *the sum of the sides* H G *and* H K (d) *is larger than the third side* G K. *Thus,* [ergo] *the straight-line* M K, *being equal to the sum of the two mentioned sides* [H G + H K], *is longer than the straight-line* G K. *It is demonstrated in the same way that the straight-line* K N *is longer than the straight-line* K E. *Thus,* [ergo] *since the single parts of the straight-line* M N *are longer than the single parts of the straight-line* G E, M N *itself will be longer than the straight-line* G E. *Therefore, the same contracted muscle is longer if the protruding parts of the tendons coincide with the longer of the diagonals than if both tendons remain parallel to each other.* (c) from the construction. (d) I. 20.

But actually if one wants to compare the straight-line M N *with the straight-line* A E, (e) C D *is drawn* ad infinitum *from* C. D O *becomes equal to* D M (f). *A perpendicular* A P *is drawn from the point* A (g) *to the continuation of the straight-line* D C.

(e) post I.2. (f) I.4. (g) I.12.

There may be three different cases: the perpendicular A P *can fall either on the extremity of the straight-line* O E, *or inside or outside this extremity.*

Whether falling on the extremity itself of O E *or outside this extremity,* M N *will always be smaller than* A E (Fig. 2.22.18).

Let P E *be equal to* MN. *In the trigle* A P E, *the angle* A P E (a) *is a right angle. Thus,* [ergo] (b) *it is equal to the sum of the other two angles* P A E + P E A. *Therefore,* [ergo] *it is larger than* P A E *alone. Since actually in the same triangle* (c) *a larger angle spans a longer side,* A E *will be longer than* P E *or its equal* M N. *If the perpendicular falls outside the point* O, *the outer angle* A O E (d) *in the triangle* A P O *will be larger than the inner and opposite angle* A P O, *which angle I have already demonstrated to be larger than the angle* P A E. *Thus,* [ergo] [the angle A O E] *is much larger than the angle* O A E. *Therefore the straight-line* O E (e) *or* M N *is shorter than the straight-line* A E.

(a) from the construction. (b) I.32. (c) and (e) I.18. (d) I. 16.

If the perpendicular A P fell inside O towards C, *it might happen that the angle* E O A *would be equal to the angle* E A O. *Therefore, the straight-line* E O *or* M N *would be equal to the straight-line* A E. *It might even happen that the angle* E O A *would become larger than the angle* E A O *and, consequently,* M N *larger than* A E. *But since the contraction can never be so slight that a situation like this could possibly occur, it is needless to dwell longer on this situation. Especially since experience indicates that the muscles themselves are in continuation with the bones, they are packed against one another; and are constricted by membranous envelopes in such a way that the protruding parts of the tendons at any time can hardly be extended into a straight-line common with the longer diagonal.*

Now that some Specimens of Elements of Myology have been presented, I have to demonstrate that they are certain through examples taken from Nature itself. I shall do this rather by showing figures of different muscles (Plate I, II, III) than through explanation, since the whole matter is so evident that, without any explanation, inspection alone is sufficient.

Most of the figures show the surface of the Orders in which different single edges and different angles are displayed at the magnitude at which I have measured them in cadavers.

The tendons are drawn like trapezia without making any distinction as to their fibres because, on one hand, they are not needed to explain the movement of the muscle and, on the other hand, all of them could not be represented accurately in such a small space.

Before proceeding to the figures of muscles, let me repeat briefly in Plate I points proposed at different places above which to me seem necessary to understand more easily at a glance the motor fibre, the Order, the Rank, and the structure of the muscle.

2.22.6   Explanation of the Plates

Now that some *Specimens of Elements of Myology* have been presented, I have to demonstrate that they are certain through examples taken from nature itself. I shall do this rather by showing figures of different muscles than through explanation, since the whole matter is so evident that, without any explanation, inspection alone is sufficient.

Most of the figures show the surface of the orders in which different single edges and different angles are displayed at the magnitude at which I have measured them in cadavers.

The tendons are drawn like trapezia without making any distinction as to their fibres because, on one hand, they are not needed to explain the movement of the muscle and, on the other hand, all of them could not be represented accurately in such a small space.

Before proceeding to the figures of muscles, let me repeat briefly in Plate I points proposed at different places above which to me seem necessary to understand more easily at a glance the motor fibre, the order, the rank, and the structure of the muscle.

Plate I (Fig. 2.22.19)

Figure I. A B C D is a rectilinear motor fibre; B C is the flesh; A B and C D are the tendons.

Figure II. A B C D is a rectilinear inflected motor fibre; A B C and B C D are the alternate obtuse angles.

Figure III shows nine unequally equal [Stensen's def. 17] rectilinear fibres arranged according to the increment of the tendons [Stensen's def. 18].

Figure IV shows the same nine motor fibres inflected.

Figure V shows the same nine motor fibres united in the way they appear in an order, in which B C F G is the flesh parallelogram, and A B F E and H G C D are the two tendon trapezia. Between the flesh parts empty spaces are left to show all more distinctly.

Figure VI shows equally equal motor fibres located as they appear in a rank. There, B C L K, is the flesh rectangle, and A B K I and L C D M are the two tendon rectangles. To create perspective according to the rules, it has not been possible to draw all the fibres at the same magnitude, because the different surfaces are not on the same plane.

Figure VII shows a muscle made of nine Orders as described in the definition of the muscle. A D is the first order followed by eight parallel orders.

Plate II (Fig. 2.22.20)

Figure I. A H is the plane of the orders in the interior part of the calf muscle, in which A B F E and H G C D are the tendon trapezia and B F G C the flesh parallelogram. In this parallelogram F B and G C are the tendon edges, and F G and B C are the flesh edges. A is the upper extremity, H is the lower extremity, and G D is the inner surface which is contiguous to the soleus muscle. The truth of what we proposed above concerning the orders appears from a simple inspection.

This is a simple muscle of the kind of those in which the shorter diagonal always forms acute angles with the tendon edges. Therefore, what has been

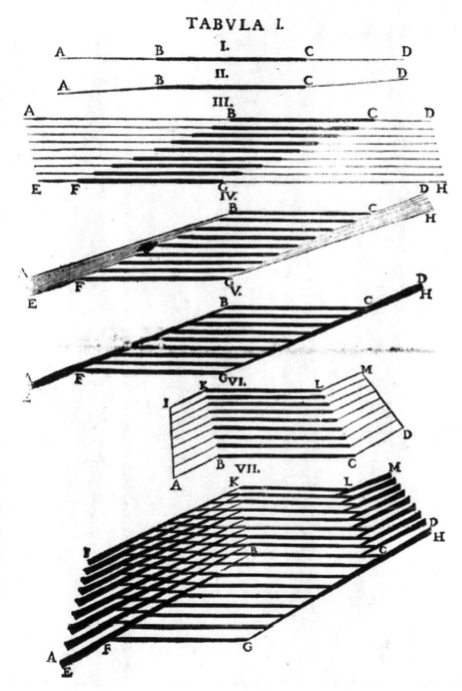

TABVLA I.

Fig. 2.22.19 Stensen's Plate I, Fig. I–VII. Inner structure of pennate muscle

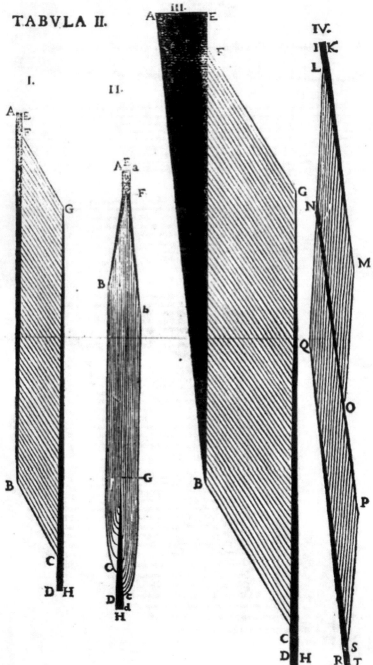

TABVLA II.

◄**Fig. 2.22.20** Plate II, Fig. I–IV. Muscle structure as seen in longitudinal section of muscles, from left: gastrocnemius, biceps brachii, semimembranosus, semitendinosus muscle

said above in the second example about the swelling of the muscle can apply to it.

Biceps muscle. Figure II. E H is the plane of the orders in the biceps muscle of the arm, in which E is the upper extremity, and H the lower extremity. In this muscle two things must be noticed which differ from the previous muscle. (1) The muscle is composed of two simple muscles A H and a H put together in such a way that the lower tendons D C G H and d c G H constitute one tendon hidden inside the flesh, whereas the upper tendons A E F B and a E F b contain flesh inside themselves. (2) The length of the flesh parts here is such that the shorter diagonals always form obtuse angles with the opposite tendons. When this muscle contracts, an obvious swelling is felt, since the angle B E b widens and the distance between the flesh edges B C and b c, i.e. the thickness of the muscle, increases. Add that the extremity H is raised. Therefore, the angle formed by the straight-line H E and the humerus widens. These concurrent causes can produce an obvious swelling even without the arrival of any new substance.

Figure III. A H is the semimembranous muscle the structure of which differs from that of the gastrocnemius muscle in that the upper tendon trapezium A B F E is not similar to the lower tendon trapezium H G C D, since the motor fibres are not equally inflected on the two sides. For this reason the tendons of many fibres raised above the end surface of the flesh parts form a membrane, thus giving the muscle its name the semimembranous muscle.

Figure IV. K R, the so-called semitendinous muscle is so named because its lower extremity R T has tendons united in the shape of a cylinder. Here the explanation of this arrangement is that the upper tendon of one muscle is united with the lower tendon of another muscle in such a way that the flesh parts of the two muscles remain separated from each other by a tendinous curtain N O.

Plate III (Figs. 2.22.21)
Deltoid muscle.

Figure I shows the delicate structure of the deltoid muscle in which 12 single muscles are counted. They are united and arranged in the same manner as the two muscles which I have described in the biceps muscle. The empty looking spaces above and below are filled up with flesh. But with this resection another part of the tendons is resected into which this flesh extends. I wished to show here those parts of flesh alone the extremity of which would be conspicuous.

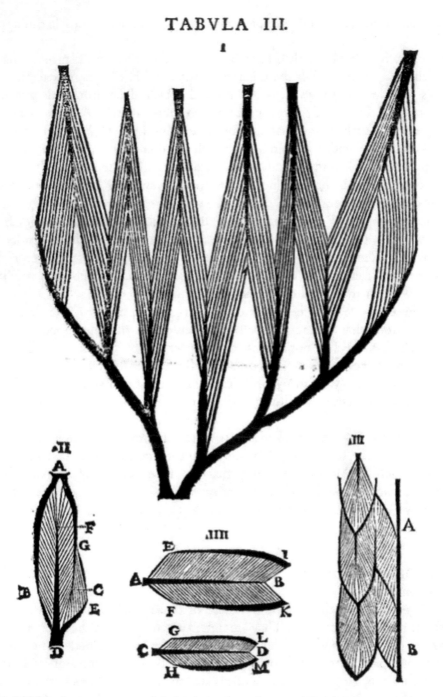

Fig. 2.22.21 Inner structure of skeletal muscles: I. human deltoid, II. human masseter, III. fish, IIII. adductor and abductor in lobster claw

Masseter muscle.

Figure II shows the plane of an order in that part of the masseter muscle which connects the angle of the mandible and the anterior aspect of the zygomatic bone. A is the upper end which continues into the zygomatic bone, D is the lower end which continues into the mandible. The side ABD faces the mandible.

This order is made of three different orders, the first of which is D B A F, the second D C A F, and the third D C G E. The flesh edges are B D, F A, D C, and E G, and the tendon edges are B A, D F, C A, and D E.

Explanation of the swelling observed in the contraction of the masseter muscle. Anybody can easily feel a swelling on himself during the contraction of the [masseter] muscle by clenching one's teeth. From what was said above, this swelling is easily explained without the arrival of new material.

Since the angle B D E *is made of the three acute angles of the three muscles* B D F, F D C, *and* C D E, *anyone of which* (Lemma VI) *widens during contraction of the muscle, the angle* B D E *widens also. But since the flesh* edge B D *facing the mandible meets resistance, the edge* D E *has to move.* This is obvious: if you put a finger on your mandible close to its angle, you will feel the swelling descend towards your finger. This results from the fact that the edge D E recedes from the edge D B.

But the angles B A F *and* F A C (Lemma VI) *widen in the same way since they are acute angles in the muscle. Thus, about the zygomatic bone into which the extremity A continues there will also be a swelling. About the middle of the masseter muscle where the flesh edge* E G *of the smallest muscle ends, i.e. in the space* E G, *there will be swelling in the same way, since the acute angle* E G C *increases.*

Thus it appears that a swelling must be felt at the extremities and at the middle of this muscle. And if somebody likes to consider further the swelling due to the acute angles of the other two parts of the masseter muscle set behind this exterior part and the other located in a curve behind the inner part, he will no longer be surprised that although the substance in the muscle is unchanged, an obvious swelling will occur.

Muscles of fish.

Figure III. Hardly anyone has mentioned the muscles of fish. Their delicate pattern most obviously confirms our observations. I could bring forward several examples supporting this point of view, such as that of the muscles corresponding to our intercostal muscles or that of the muscles moving the eyes, the gills, the fins, etc. But since I still have not satisfied all of my own curiosity so that in some cases I still have to investigate further, it will be enough to present here one example, drawn from the shark, *Canis carcharia.*

The plane here is perpendicular to that plane in which all the vertebral spinous processes are located, and parallel to the line of the extremities of all the spinous processes. The straight-line A B is a section commonly shared by the former plane and the plane of the spinous processes. The elegance of the structure appears at a simple inspection (Fig. 2.22.21).

The adductor and abductor muscles of the lobster's claw. Figure IV. I wanted also to consider the example of the lobster among the crustaceans.

To this end I pulled out the muscles of the major claw, which open and close the claw. In accordance with the usage of anatomists, I shall call them the abductor and the adductor muscles. The larger figure shows a compound order of fibres in an adductor muscle. There, A B, the tendon inside the flesh, is inserted into the mobile part. E I and F K are the outer tendons adhering to the immobile part. The smaller figure displays a compound order of fibres in an abductor muscle. There, C D, the tendon inside the flesh, is inserted into the mobile part. G L and H M, the outer tendons, continue into the immobile part.

Of what does the power of the claws consist? Fig. 2.22.22

I do not doubt that readers will be pleased to know of what the great power of the claws which break fairly hard bodies consists. Therefore, I wanted to show the following figure in which I represent the mobile part of the claw divided by a plane perpendicular to the lateral surfaces of the muscles and parallel to their end surfaces or rather coinciding with the end surfaces sited within the flesh. The uninterrupted lines represent the position of the part when the adductor muscle contracts. The dotted lines show the position which the same part takes, when the abductor muscle contracts.

B A C K is the mobile part of the claw, A K is the length of this part, B A C is its basis, A is the centre of movement, C is the extremity into which the abductor muscle is inserted, B is the extremity in which the adductor muscle is inserted, C E is the length of the tendon edge in the abductor muscle, D B is the length of the tendon edge in the abductor muscle, F G is the width of the adductor muscle, and H I is the width of the abductor muscle. Since the adductor muscle both is wider and has longer tendon edges than the abductor muscle, it has more flesh parts. For that reason alone it would be stronger than the abductor muscle. Moreover, the insertion of the adductor muscle is more remote from the centre of the movement than that of the abductor muscle, since A B is longer than A G. The adductor thus is stronger for two reasons: more flesh parts and a more efficient insertion.

But all the necessary examinations have not yet been carried out to determine the ratio of the forces in these muscles, and it remains to be determined in each how much the flesh parts decrease in length, considering

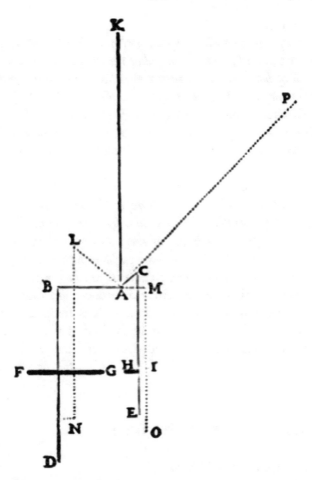

Fig. 2.22.22   Mechanics of lobster claw

both the whole length of the flesh and the distance travelled by the extremity of the flesh during a contraction. Should this indeed be explained more extensively, other premises would be given, which would require more time and would be different from the present description.

What to take into consideration to explain the relative forces of the muscles.

To make it clear to everybody that it is not so easy, as it has been believed so far to examine properly the forces of muscles, and that they cannot be wrung out in agreement with nature from principles different from those which I proposed here, let me show what must be considered in the muscles to find out their relative forces, beyond their insertion, which was so far about the only subject of interest.

In simple muscles.

If any two simple muscles are unequal, they must be reduced to two rectilinear muscles of the same width, by obtaining that the transverse edges are the same in the end surfaces of both. Thence, in these muscles of the same width, the tendon edges of one muscle must first be compared with the tendon edges of the other muscle since these reveal the number of flesh parts. Then the flesh edges must be compared with the flesh edges, to ascertain how much they shorten their length in contraction, and to ascertain the distance travelled by their extremity during contraction.

In compound muscles.

If two different compound muscles are compared, they must be reduced into two simple muscles of the same width. This is done by averaging the length of the flesh parts of different lengths in each one of the muscles (if there are differences in length) and by reducing the different end surfaces into one surface of a given width. When this has been done, the two muscles can be compared as I described.

It should then be obvious that since differences are found either in the structure of the flesh or in the end surfaces, it is necessary to analyze the muscles by our method and to measure in the body itself the flesh edges, the tendon edges, the transverse edges, and the other elements which were proposed here.

2.22.7   On Muscles of the Back

On muscles almost unknown hitherto.

In order to introduce here something concerning muscles of our body only confusedly known hitherto, not to say completely unknown, let me present the spine muscles reduced to a certain order.

Anyone who compared the writings of anatomists on the muscles of the back, and wished to imitate their methods of handling these muscles, will easily recognize that the different anatomists are as much in disagreement with each other as they are all far-off from the truth. I could confirm this by the acknowledgment of several anatomists, but I shall use only one who stands out particularly to me to count for all, and that one is Fallopius, when he says: "The muscles are so diverse and complicated that it is no wonder, if the anatomical authors do not agree. To be honest, [the muscles of the back] look to me like an undigested mess, or a confusing chaos of muscles, for which I need a teacher who clearly dissects them before my eyes and reduces their parts to a certain number and order."

I CALL VERTEBRAL MUSCLE a muscle continuing at both extremities into vertebrae. It is either straight or oblique.

A STRAIGHT VERTEBRAL MUSCLE is a muscle both extremities of which are on the same line parallel to the spinal cord. Therefore, the position of the muscle itself is parallel to the cord. Either the muscle is *MEDIAN, and both its extremities continue into the spinous processes or spines of the vertebrae; or the muscle is LATERAL and its extremities continue into the lateral or transverse processes of the vertebrae.* The spinal ones face medially whereas the transverse ones face laterally. But besides the muscles described so far there are others both median and lateral muscles. These same names could be given to them, but since the word transverse describes a transverse position of the muscle rather than a connection with the transverse processes, it seemed to me less confusing to call them lateral, since the transverse processes are lateral.

AN OBLIQUE VERTEBRAL MUSCLE is a muscle the extremities of which are on two lines parallel to the spinal cord. Therefore, the position of the muscle itself is oblique to the spinal cord, either *RECEDING FROM THE MIDDLE, in which case its upper extremity is in the transverse processes:* the lower part of the splenius muscle and also the lower part of the longus muscle in the neck belong to this kind, or *the vertebral muscle is VERGING TOWARDS THE MIDDLE, in which case its lower extremity is in the transverse processes.* The second part of the longus muscle in the neck and all the muscles which lie directly over the vertebrae on both sides, from the second spinous process in the neck down to the os sacrum, belong to these. The name semispinatus is given to these muscles not incongruously, although, besides the semispinati of others, that of the os sacrum and others also belong to this kind.[16]

From what has been said, it is clear that the vertebral muscles can be described in a few exact terms without any ambiguity. Thus their four kinds are the MEDIAN, THE LATERAL, THOSE VERGING TOWARDS THE MIDDLE, AND THOSE RECEDING FROM THE MIDDLE.

It is common to almost all these muscles that one upper vertebra receives muscles from several lower vertebrae, and that one lower vertebra sends muscles to several upper vertebrae. I demonstrated elsewhere such an arrangement of the muscles even in the ribs when I described the lumbosacral region.

But in order to show even more clearly, what I proposed here about the vertebral muscles, I decided to illustrate this with the two figures herewith (Fig. 2.22.23).

---

[16]The meaning of this sentence is not clear to the translators.

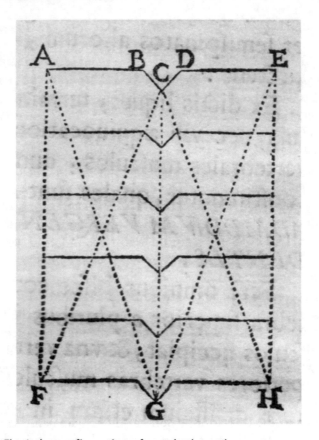

Fig. 2.22.23 *Fig. I,* the configuration of vertebral muscles

Figure I shows the various positions of the vertebral muscles. A B, D E and the other lines similarly positioned represent the transverse processes.

The angle B C D, and the other similar angles represent the vertebral bodies and the spinous processes.

Since the angle B C D represent a spinous process, B C and C D are the sides and C is the tip of the spinous process.

A F, C G, and E H are lines parallel to the spinal cord, I call the muscles situated along these lines the straight vertebral muscles. Those on the line C G are median vertebral muscles, and those on the lines A F and E H are lateral vertebral muscles. G A, G E, F C, and H C are lines the extremities of which are on two different lines parallel to the spinal cord, i.e., the extremities of F C and G A are on the lines A F and C G, whereas those of G E and H C are on the lines C G and E H. I call the muscles in this position oblique.

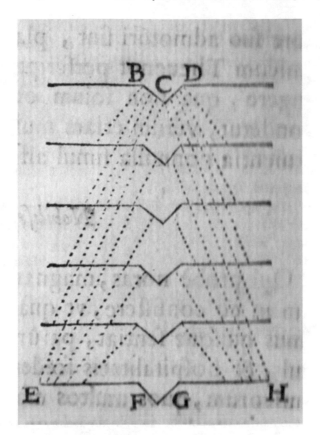

Fig. 2.22.24 *Fig. II,* the configuration of vertebral muscles

Among them F C and H C are those verging towards the middle, and G A and G E are those receding from the middle.

Fig. II. It is seen in those vertebral muscles verging towards the middle, why those coming from one spinous process B C D have opposite extremities inserting into several transverse processes of lower vertebræ and, conversely muscles originating from the transverse processes of one vertebra E F and G H have their opposite extremities inserting into spinous processes of several upper vertebræ (Fig. 2.22.24).

Those who have understood this figure will easily understand the position of all semispinatus muscles which occupy the grooves directly on the exterior sides of the vertebrae from the second cervical vertebra down to the os sacrum. There is only one difference: in some places the muscles span several vertebrae whereas in other places they span fewer vertebrae. Such a preparation does not require much cunning. One has only to separate the muscles by introducing the knife with a light hand in the spaces between the spinous processes downwards, and in the spaces between the transverse processes upwards. The true position of the other vertebral muscles could be displayed in the same way, but I shall reserve that for the Myology itself. Presently it has only been my purpose to bring to the public's judgment an example of elements of Myology, while providing certainty on these matters through figures taken from Nature. The same works indicate how some muscles known superficially only, for which I discovered the intrinsic structure, and others known confusedly only so far, not to say completely unknown, can be prepared properly and described clearly.

Since, however, I have no doubt that the prejudiced critics of my works will champ their gums as they usually do, I have decided to append here a letter to my noble friend Thévenot in which not only their objections are refuted, but also some data are presented for a better understanding of the muscles.

2.22.8   Most Noble Man[17]

It is demonstrated that movement in the heart and in the muscles takes place the way it appears.

Since I have met several people who, although having seen that both the substance and the composition of the fibres in the heart are like those of the fibres of the muscles, nevertheless cannot bring themselves to acknowledge that the heart is a muscle, because they do not dare to depart from the authority of *Galen, Descartes*, and others. For their benefit, I will demonstrate the following proposition: *The structure of the motor fibre in the heart and in the muscle is the same: thus the phenomena of movement in the motor fibre which are manifest to our senses and are seen in the muscle are the same in the heart.* The phenomenon in each of them will be made clear by induction.

Which part of the muscle contracts.

I. When a muscle contracts, its different motor fibres shorten.

This becomes evident when the aponeurosis of a muscle is divided and the fibres are separated from each other. Thus, fibres liberated from mutual contact become shorter for the same reason as before, when they were closely joined together inside the aponeurosis of the muscle as if they were inside a common envelope. For, as I observed very often, even when most of the fibres

---

[17]SOM, 183 ff. is an appended letter to Melchisédec Thévenot, benefactor during Stensen's sojourn in Paris, with detailed reply on the criticism he had met on heart and muscle.

are resected, the others continue to move for some time. The same is obviously observed in intact muscles such as in the panniculus carnosus, in the diaphragm, in abdominal muscles, and, in fact, in whichever muscle is made visible after the skin has been peeled off, the abdomen opened, the shoulder resected, or after other dissections have been carried out, they are revealed to the light. Very often in such circumstances not all the fibres are seen to move at the same time, but individual fibres move separately at different times. The amazing movement so often seen in the dying heart is not different, for here also the fibres that previously moved simultaneously move at different times, thereby causing this phenomenon.

Which part of the motor fibre contracts.

II. When a motor fibre shortens, the flesh part alone shortens.

I could confirm this by other different experiments, but at present it should be sufficient to present one which can serve for many other propositions.

After having divided into two a muscle with a simple structure from one extremity to the other along the length of the fibres, without touching the tendons, you divide the transverse flesh in one of the parts. You will observe that the divided fleshy part will become much shorter than the intact fleshy part, while the length of the tendons actually remains the same.

Similarly, when dividing the ventricles of a living heart, the movement of the fleshy threads is obvious, while there is no movement of the tricuspid valves.

What can be observed in the flesh when it contracts.

III. When the flesh of a motor fibre shortens, it also hardens, and its surface which, before contraction was smooth, becomes rough.

Whoever will test with his own fingers and eyes the contracted diaphragm and the contracted heart of the same animal, will obviously find out the truth of this proposition.

What happens to the flesh after contraction

IV. The flesh of a motor fibre again relaxes to a certain length after contraction.

Others have shown this about the muscles long ago. In the heart, nothing is more evident than this since, after the blood has been drained and the auricles resected, the heart does not stop beating for quite a long time, now growing shorter by drawing into a cone at the base, then becoming longer by relaxing again.

Whether it is all the flesh or single parts of it which contract.

V. Whichever part of the flesh is divided transversely, becomes both shorter and harder to an equal degree, and relaxes again to a certain length as does intact flesh.

I have often demonstrated this in the muscles of the shoulder and of the neck when I explored the junction of the thoracic duct and lymphatic vessels in the region of the throat and shoulder in half dead dogs. An easy way to carry out this experiment is by cutting across the flesh with a scissor at three or four places in a muscle with fairly long flesh. The flesh parts between the divisions repeat their contraction several times and become hard to the touch and rough to the sight. Not infrequently I have succeeded in seeing the same in the chest cavity of birds. The apex of the heart, cut off the base and set on a finger contracts repeatedly so vigorously that it leaps from the fingertip. Those who have not observed that the same phenomenon is shared by the other muscles, attribute it to a heart's pulsatile faculty (as they call it). When it thus jumps off the finger one feels its manifest hardness and does not see the usual smoothness.

Whether the vessels contribute immediately to the contraction of the motor fibre.

VI. A motor fibre contracts as well when the arteries, veins and nerves are resected as when they are not resected.

There is no one who has not noticed the trembling of the panniculus carnosus in recently slaughtered animals, from which both heart and head have been removed. The sternum after being divided with the cartilaginous part of the ribs has been casually observed by some people to jolt at intervals. Some believe that this pulsatile force is transmitted from the heart, without noticing that this action comes from the intercostal muscles themselves. My friend Swammerdam has often demonstrated to us that a frog swims quite a long time with its heart cut out. In a tortoise, movement has been noticed even in the feet and tail twenty four hours after the head had been cut off. This movement continued some time after the heart had been excised. In Leiden, while intent on testing the experiments of Bils when, after the scapula had been raised in moribund dogs, I divided the plexus of nerves spreading out to the left front limb, I saw muscles convulse, not only when I divided their intact nerves, but also when I pressed strongly, or divided the parts of the divided nerves adhering to muscles, the same reaction that my friend Swammerdam had previously observed in frogs.

What is surprising, therefore, is that the heart, extracted from the body, moves with repeated beats, although neither new blood flows into its vessels, nor does new spirit enter it through the nerves from the brain.

Since, therefore, not only the structure of the motor fibre is as I observed it to be three years ago, but a movement obvious to the sense is observed in the same way in the heart as in what are commonly called muscles, I hope to have demonstrated clearly what the Ancients said, namely, that the heart is a muscle.

What I have proposed here concerning the movement of muscles, refers to the change which occurs in these muscles when they contract, without suggesting any cause from which that movement proceeds. Hence, I add nothing concerning the power of the will, since I believe that it is evident that every voluntary movement is carried out by muscles, but that not every movement executed by a muscle is voluntary, so that it fits to a muscle alone, but not all, nor at all times, to be an organ of voluntary movement.

Finally I must also answer those who claim that all myology in all details has been established completely by anatomy long ago as much as it is possible, and that therefore my work is but that of an idle man.

Answer to those who consider the new work of Anatomists concerning the muscles as ridiculous and superfluous. Now if the opinion of these people was destined to diminish my own endeavours only, it would be a matter of little consequence, and I would not waste any effort in commending my own work. A more important issue is at stake. Thus, not to appear themselves as being ignorant, they try to obstruct the path of the ones who investigate further. Thus, as much as they can, they validate forever their ignorance wrapped in a veil of science, at the greatest detriment to both truth and health. I am not here elaborating my own defence to demonstrate the usefulness of my work, but so that others about to tackle the same subjects successfully do not let themselves be deterred from beginning.

For my part, I acknowledge that both ancient and modern authors have accomplished much in this field. *Galen, Vesalius, Fallopius, Spigelius, Placentinus, Aquapendente, Riolan,* and other great men have dealt with this problem with such skill that one could easily deduce that the whole issue had already reached the Pillars of Hercules, which the Ancients believed highly imprudent to pass. But it is not disgraceful to the Ancients that their posterity did penetrate beyond these pillars, and one will not rightfully reproach the great high priests of anatomy not to have investigated all the arcanes of nature. I know the reverence I owe to my teachers, and I am sure that, if brought back to life, they gazed upon the works of present day anatomists, their verdict would be far kinder than that pronounced by our censors.

As it is evident that much about the muscles still remains unknown let me review briefly as in a quick survey what is common to all muscles and then what is peculiar to some.

What is common to all muscles may be reduced either to their solid or to their fluid part. If I say that both were completely unknown to the Ancients, and not only to them, may the shades of the great men of the past forgive me.

It is not my wish to repeat what I have said elsewhere about the muscles to show that the Ancients did not know what is obvious to the eye of anyone any time meat is brought on the table. I honestly knew that quite often what is most evident does not receive the least attention; nor do I have any doubt that, hidden among what I myself have discovered, there are things which, perhaps, are simpler and more obvious than what I have seen. Therefore, if I were to act like a charlatan in making these remarks, I should rightly expect retaliation.

Chemical analysis of the muscle. Therefore, I pass on to other matters. As far as I know nobody has associated chemistry and anatomy in such a way as to show clearly and distinctly—not by following the precepts of an art, but by tracking the footprints of Nature—in what flesh, tendon, and bone are the same or in what they differ. My very famous teacher Sylvius has successfully studied the humors of our body, and if I remember well, I have even heard him lecture on the substance of tendons and bones according to these same principles. But even though this very famous gentleman achieved much in this field of study, he daily impressed on his students the fact that he had not been able to achieve everything, lest he seem to prefer his own glory to the public interest. To this end, what he thought he himself did not yet truly understand, he proposed as hypotheses and suggestions, by which he stimulated others to investigate and at the same time offered material for further research.

Anatomical analysis of the muscle. But just as no one has yet set up a true chemical analysis of the motor fibre, so nobody has attempted a true dissection of the muscle, which is conducted through the various planes according to the laws of mechanics.

Separation of one muscle from another. What about the fact that rules for true separation of one muscle from another have not even been established anywhere? Thus, the actual extremity of many muscles remains unknown. Frequently several muscles have been considered as one muscle even though they are either used in moving different parts, or in performing different movements of the same part.

The terms of Myology. I face the confusion generated by words like beginning, end, connecting link, stretching part,[18] and other terms aimed at describing parts of the muscles. This confusion is an obvious indication that the matter itself is not well known.

Concerning the fluid part of muscle our knowledge is so uncertain as to be non-existent.

How manifold is the fluid of muscle?

---

[18]Stensen indicates the ancient meanings of *ligament* connecting link, and *tendon* stretching part.

It is certain that there is fluid in the fibrils of which the motor fibre is made, between these fibrils, between the motor fibres themselves, in the membranous fibres of the muscle, and between these same membranous fibres. It is also certain that these fluids are not all of the same kind. Whether they are distinct in their separate locations and whether they are also different in their material properties is not so certain.

Which known fluid does the muscle fluid correspond to?

It is not known to which of those fluids, which we consider known to us, any of these fluids is similar. Many call them animal spirits, the subtler part of the blood, its vapor, juice of the nerves. But these are mere words, nothing proved by experiment.

Those who proceed further introduce salty and sulphurous parts, or something analogous to spirit of wine, which may be true perhaps, but are neither certain nor sufficiently distinct. Experience teaches us that exhausted strength may be restored by drinking spirit of wine, but whether to ascribe this to the humor which we call spirit itself or to another substance which makes the spirit fluid, or perhaps to another, cause closely linked to Spirit, who will determine?

What is the movement of this fluid?

Just as the substance of this fluid is unknown to us, its movement is uncertain, since it has not yet been established, either through reasoning or through experiments, whence it comes, by what means it proceeds, or where, on departing, it escapes.

From where does it come? One can recognize the fluid surrounding the arteries, the nerves, and opposite muscles as its source.

Where does it go? The veins, the pores of the body, the bones perhaps, and the nerves may offer an exit. In the muscle itself it can be carried in a simple movement from the middle towards the extremities, from the extremities towards the middle, from one extremity towards the other; but it can also move in a less simple displacement.

In what way does the fluid of the contracted muscle differ from that of the non-contracted muscle?

There remains another problem no less momentous and not yet solved: namely, in what does the movement of the fluid in a muscle differ when this contracts, from the movement of the fluid in the same muscle when this is at rest, non-contracted? Is its quantity changed, or does it remain the same? If some new fluid arrives, is this incoming fluid of the same nature, or different from the one which was previously there? Does the fluid move because the solid part contracts, or does the contraction of the solid part proceed from the movement of the fluid?

What else in the muscle has to be examined with scrutiny.

But in order to investigate successfully what is lacking in the knowledge of the muscles, the solid and the fluid part of the nerves must be examined at the same time, and even the composition of the blood one must be known. Their examination cannot be carried out properly unless the nature of fluids and the way in which objects affect our senses are also studied.

It cannot any longer be ignored how incomplete our knowledge of what is common to all muscles has been and still is, and how great an area of investigation lies waiting for those who are not shy of work. Nor would I discover a less extensive field of investigation if I enumerated those matters peculiar to individual muscles and hitherto unknown. But rather than explaining it, let me hint at it in a few words.

Different structure of different muscles.

There are almost as many structures as there are pairs of muscles. I am not surprised that they have not been observed by those who do not look beneath the muscle's surface. But it surprises me that those who have drawn the muscles often have been more exact than those who described them in words, and that the industry with which painters have approached Nature's skill had not the power to urge them to admire such a work of art, which is the first step to investigation.

Movement of individual muscles.

I am not going to review the errors which are frequently committed concerning the movement of parts by individual muscles. It is evident to everyone that the true extremities of these muscles are unknown, and their actual movement is either completely unknown or known only by chance.

All of the following matters thus have been ignored hitherto and some still are: the substance of the solid part in the motor fibre and its shape; the substance of the fluid part and its movement both during and outside contraction; an analysis and the structure of all the muscles; the separation from each other of many of them and the movement of some of them. Since, I say, all these matters have been unknown hitherto and most of them remain unknown even now, it is clear enough that our censors have no right to proclaim that Myology is complete in all its aspects, however much labor remains if somebody undertook a description of the muscles illustrated by accurate figures.

But some will say: no one will deny that much still lies hidden, but most will deny that what they themselves do not know, either can be known or is worth knowing.

It is shown that many among the matters which are unknown on muscles can be made known.

There is no need of an elaborate answer to demonstrate that many of these problems can be solved. It is evident from the examples of my observations and of the elements that so far only a small number of them have yet been exposed, and, if these alone were pursued, the true extremities, the true structures, and the true movements of all muscles could be displayed as certainly as Geometry is accustomed to prove its propositions. Whether all remaining issues will be discovered so easily, however, is deservedly questionable. But even so, this doubt itself, explored with assurance, can make known for certain, matters which so far were unknown, and, if it cannot determine the true mode of contraction, it can at least distinguish what is certain in it from what is uncertain. Nobody should ignore how much this alone should be valued.

No one who loves truth and cherishes health will ever deny that the matters which remain to be investigated at the price of much work deserve to be known.

It is shown that these matters deserve to be known.

There is no need to borrow arguments from Orators, to prove that sweating on this work is not the fact of a man who spends his leisure time, as my critics were not ashamed to assert, often in the presence of famous men. The elegance of the artefact alone deserves an endeavour in research even a thousand fold greater, since it reveals such great evidence of Nature's intelligence. Add that it deals with the motor fibre: with the part that moves the limbs, that breathes in the air, that moves the blood, in short, on which the signs of life and death depend. But who will call idle the desire to explore the nature of a part which is so far almost unknown, and who sees what can be accomplished in this exploration? But such considerations escape our critics.

Since you have often personally taken my side, you as well as myself readily remember the objection: What's the use of wanting to know? or, What practical application does it have? So by repeating their questions again and again, adorned with various figures of speech, they try to make those who remain ceaselessly alert to new discoveries, appear ridiculous, I may even say troublesome.

How much medical practice owes to the experiments of anatomists of this century may be shown in. greater detail elsewhere. One example alone will suffice. Anatomy has disclosed innumerable errors committed in the etiology and revealed errors in most of the reasons given for the application of medical remedies. Instead of responding here, I ask them to search their conscience to determine on what solid bases rest all what they in their brashness so glibly deliver to explain apoplexy, paralysis, spasms, convulsions, prostration, syncope and other symptoms of the movement of animal. On what foundation do they rely in deciding what remedies to apply to the cure of these same evils, not in applying them to the paralysis, nor to the convulsion, but to the paralytic, and to the convulsed person?

If they face the fact that their diagnosis is mere words, their cure but guess work, they will admit, even reluctantly, that it serves a purpose to discover what is true and certain in anatomy. Do they not object that everything has remained the same down the centuries? The answer is in short: they all look for remedies and for that part to which the remedies apply, but few attempt to understand. But, just like the construction of a machine built by someone else must be precisely understood by the one who must restore the movement of this damaged machine, similarly the nature of the blood, of the nerve fibre, and of the motor fibre must be investigated as far as human zeal permits by the one who wishes to cure not only by luck the symptoms affecting the natural movement.

In myology much that can be known remains hidden to us, and since it is not only in the interest of truth but also of good health that these points be known, it is clear to anyone how little justified our critics are in ridiculing the new experiments of the anatomists and in claiming their labours to be the occupation of an idle man.

These are the matters that I decided to bring to light in order that my friends might know what answer to give to those who speak not too kindly about my work.

Farewell, most noble man, and do continue to appreciate me.

2.23 A Carcharodon-Head Dissected

I have no doubt that a long and uninterrupted account of observations on muscles would be distasteful to the reader; accordingly, since variety is the spice of life, I have decided to add to what has gone before material which should provide an opportunity to review various isolated observations. And I could have wished for no other item more suitable than that which presented itself to me as I was dissecting the head of a shark. Actually the most Serene Grand Duke of Tuscany, on being notified from Livorno that an unusually large fish had been caught some miles from the port, ordered that the head should be brought to Florence and handed over to me for dissection.[1] This allowed me to observe a number of things that clarify our knowl-

---

OPH 24, vol. II, 149–155: *Canis carchariae dissectum caput* is the second of the three treatises, which Steno published jointly in Florence in 1667. The translation of Canis Carchariae is by Alex J. Pollock, of the Notes by Mrs. Maria Grandt, Lecturer, Copenhagen, and James G. Gordon. The Latin text of the printed book was compared critically with the printer's manuscript of the Royal Danish library. Notes are from Scherz, GP 123 ff., that relates to the translation pages, GP 73 ff. Additions by the editors are marked *.

---

[1] *Lorenzo Magalotti,* in his letter of October 26th, 1666 to the Archbishop of Siena, Ascanio II Piccolomini describes the capture of the Canis Carchariae. Adding a drawing of the head he tells that a French fishing-boat in the previous week had observed the head of a Lamia some miles from Leghorn, between Gorgona and Meloria, "accompanied by thousands of stupid fishes", the reason why the fishers called the animal a sea-owl. Finally the shark was pulled ashore with a coil, tied to a tree and killed by blows. The enormous belly with its bowels was thrown into the sea, the liver was rated to 300, and the fish itself to 3500 lb (N. St. *Epistolae* 922 f.—Scherz, *Vom Wege* 75 f. One pound then was 339, 54 g). As C. Dati wrote to Ottavio Falconieri many curious people were present at the dissection, cf. Mercati, *Metallotheca* etc. p. xxxiv.

© Springer-Verlag GmbH Germany, part of Springer Nature 2018
T. Kardel and P. Maquet, *Nicolaus Steno*,
https://doi.org/10.1007/978-3-662-55047-2_34

edge of the parts of animals, which I wished to set down here illustrated by various other observations.

At first sight it became clear that this was the head of *Canis carchariae*, since everything zoologists say about that kind of fish was applicable to this specimen.[2] I will not argue whether the name *Lamia* fits the same fish, since it is agreed that the term Lamia should be assigned properly to a flat fish, even though in many places the name Lamia is given to this fish of ours.[3]

The diagram that is shown under the name of Lamia, I owe to the kind favour of a very learned friend, *Carlo Dati*,[4] who when he saw that an engraving of the lacerated head was less than satisfactory for the reader's purpose, bearing in mind the Metallotheca Vaticana[5] by *Michaele Mercati* of Miniato, is granted me, from the various bronze plates that belong to it and are in his possession, the use of those which illustrate the head and teeth of the Lamia (Fig. 2.23.1) and the larger tongue stones (Fig. 2.23.2).

---

[2] A description of this Charcharodon Rondeletii, now called carcharodon Carcharias, Linné 1758 is by T. Jeffery Parker, Professor of Biology in the University of Otago, New Zealand: Notes on Carcharodon Rondeletii. In: *Proc. Zool. Soc.* 1887. Nr. III pp. 27–40 with 5 plates.

[3] Plinius 9. *Nat. Hist.* 40 (78): Planorum piscium alterum est genus... quos bovis, lamiae, aquilae, ranae nominibus graeci appellant.

[4] *Carlo Roberto Dati* (1619–1675), the author of *Vite dei pittori antichi* (Firenze 1667) and member of the Accademia del Cimento. He was one of the most learned philologists of Italy and from his twenty first year a member of the Accademia della Crusca, and as a pupil of Galileo also much interested in natural science (OPH 2, 323).

[5] *Michele Mercati* (1541–1593), born in San Miniato (Toscana), doctor of philosophy and medicine and a great naturalist. Nominated by Pius V as prefect to the Vatican Gardens and under Sextus V put in charge of the Museo Vaticano, especially its collections of minerals and fossils, and Archiater. The manuscript of his *Michaelis Mercati Samminiatensis Metallotheca Opus Posthumum... Opera & studio Joannis Mariae Lancisii... Illustratum.* Romae 1717 was in Steno's time in the hands of C. Dati. (Capparoni, Profili 1, 53 ss.). The plates borrowed by Steno are found on pp. 332–333.—In the MS. the text here runs... *spectantes, meis usibus eam concessit, qua, Lamiae (Ut ille appellat) caput et dentes expressos vides.* The quotation (from Ut ea quae..., glossopetrae variant) in chapter LXIX, pp. 333–334 (followed by chapters on Glossopetrae mediae and parvae) has supposita instead of supposititia. Mercati's work, the source of Stensen's Plates was not published until 1717, see copy in transcripts.

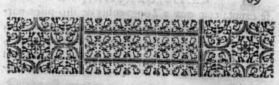

69

CANIS CARCHARIÆ
DISSECTVM CAPVT.

ON dubito , quin Lectori faftidiofa
fuerit longa , & non interrupta , rerum mufculos fpectantium expofitio;
quocirca , cum in varietate oblecta-
tionem animus inueniat , materiam
priori fubiungendam iudicaui , quę varijs , nec cohę-
rentibus inter fe , obferuationibus recenfendis occa-
fionem præberet . Nec commodiorem aliam defi-
derare potui , quam quæ mihi fefe obtulit , Caput
Canis Carchariæ diffecanti . Scilicet Sereniffimus
Magnus Etruriæ Dux , cùm Liburno ipfi fignifica-
tum effet , ad aliquot milliarium a Portu diftantiam
captum infolitæ magnitudinis pifcem effe , caput
illius Florentiam apportari iuffit , mihiq; tradi dif-
fecandum . Licuit mihi in eo nonnulla videre , animalium partibus cognofcendis lucem haud obfcu-
ram afferentia , quæ varijs alijs obferuationibus il-
uftrata hic exponere volui .

He gave permission also to add here the following extract from the said
Mercati's manuscript containing much that is well worth knowing and a wealth
of varied instruction about soils, salts, oily fluids, stones, bodies of idiomorphic
shapes, and so on; this manuscript would have remained buried in eternal
darkness, had not the very learned *Dati's* skill brought it out of the underworld
and provided an opportunity for it to be exposed to the light of day:

I have always striven so that those matters with which I deal should be as
intelligible as possible; and it was my resolve not only to teach facts but also to
remove errors and conjectures, to relegate them from their place in the established
order, or at least to make fewer errors; thus, for this reason, I am compelled to
interrupt the sequence at this stage. For I observe that the large tongue stones and
the teeth of the Lamia are confused even by the careful. This error has arisen from
their similarity which is so great that anyone who did not know their origin would
suspect nothing, anyone who has not compared their characteristics on both sides,

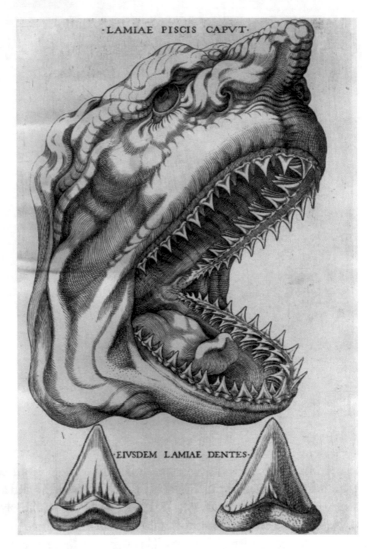

Fig. 2.23.1 Etching by M. Mercati of the head and teeth from a giant shark. (This Plate follows in number Tabula I-III on muscles in OPH XXII)

would not distinguish between them. I feel certain that the reader will not be ungrateful for this digression, especially to a wonder of nature, namely the monstrous appearance of a huge beast (Fig. 2.23.1). Rondelet[6] has described the

---

[6]See *Gulielmi Rondeleti. .. Libri De Piscibus Marinis, in quibus verae Piscium effigies expressae sunt...* Lugduni 1554. lib. XIII, pp. 390–393, especially p. 391. A great deal of Rondelet's text has been cut out. The MS. has *contenti*, yet both Mercati and Rondelet *contecti*.

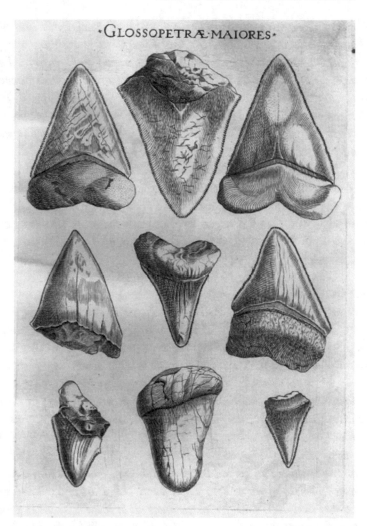

·GLOSSOPETRÆ·MAIORES·

Fig. 2.23.2   Mercatl's etching brought by Stensen ot large tongue stones, glossopetrae majores

Lamia accurately: it has a huge head, a very large mouth opening, bearing on the top and bottom very sharp and very hard teeth of a triangular shape; there are six rows of teeth; the teeth of the first row project from the mouth and are curved at the front; the teeth of the second row are straight; the teeth of the third, fourth, fifth, and sixth rows bend back in the mouth, for the most part placed on each side of the jaws, in soft spongy flesh. The fish is carnivorous, very ravenous, and man-eating, for it devours even whole dead bodies, as has been shown by dissection. *Some have been captured at Marseille and Nice (says the same writer) in the bellies of which have been found a man armoured with a cuirasse. I was pleased to have the opportunity to depict the head of a Lamia, in my possession, in an engraving. There is really little to distinguish between teeth and tongue stones. Tongue stones are mainly thicker; teeth are thinner, and shine less, so that we can arrive at a decision between bone and stone by inspecting the material of the tongue stones (Fig. 2.23.2). Also, the teeth have one unvarying colour, white, or yellow with age; tongue stones vary in colour.*[7]

And these then are the things *Mercati* relates about the Lamia. I might have added to these various other observations relevant to the present study, collected from various writers, but since these are readily available to those who own or can visit libraries, I proceed to those observations which I believe are not common knowledge, with particular reference to the skin, the eyes, the brain and the teeth.

Vessels of the Skin

There were many pores in the skin between the snout and the eye region, in the upper part of the head as well as on its sides, through which exuded a viscous white fluid, like jelly, each time that part of the skin was compressed. One could have said it was like a pomegranate cut through the middle, since, when the skin is cut at that spot, the middle vessels continuous with the aforesaid pores are cut: as such moreover was both the colour and sheen of that stagnant fluid. When the skin was cut away from there, a depression occurred in the cranial cartilage, thus leaving a fair sized space between the skin and the cranium; the inner extremity of this space was filled with the kind of vessels referred to, being itself surrounded by white material through which was spread a vessel, perhaps a blood vessel, which sent out a small branch to each of the vessels. These vessels, three or four inches long, were as thick as a medium sized goose feather; almost all of them lay close together under the skin, but, for penetrating the skin, some had their own special canals, many in close contact with each other, opened out into only one common passage in the skin itself, even when the openings were less close together at the point where they reached the skin.[8]

---

[7] A description of the ampullary tubes and the system of mucous canals discovered in 1664 by Stensen in selachian fishes and described and carried on by Stefano Lorenzini in the *Torpedo* (1678). The author pays great homage to Stensen, his teacher, and to Holger Jakobsen (Jacobæus).

[8] Here Steno describes the ampullary tubes and the system of mucous canals discovered by him in 1664 in selachian fishes, and later described and carried on by his pupil Stefano Lorenzini in the Torpedo (1678); this treatise pays great homage to Steno, his teacher, and Holger Jacobsen (Jacobæus) (1650–1701). (Cf. R. Spärck in Scherz, *Indice* 88 s. and Cole, *A History* 374 ss.). The last sentence *perventum esset* is in St.'s hand, instead of some crossed out formulations.

This confirms greatly the opinion that I expressed in the *Anatomy of the Ray*[9] concerning the origin of the oily fluid that besmears the surface of fishes; this fluid is just as necessary to the fish in helping them move more easily through water as it is necessary to coat ships that must plough through the sea with grease and other fats.

As it is the vessels in the skins of fish that we deal with here, perhaps I may at the same time describe a passage that I observed in the skin of an eel, in the presence of my very dear friend, *Francesco Redi*.[10] There was a cavity in the skin itself, behind the head, around the beginning of the backbone, from which the passage came out on each side; when this passage reached a region almost halfway between the belly and the spine, it continued from there straight to the tail along the middle of the sides of the fish, in the region where a line shows on the outside extending from head to tail. There were three other cavities on each side, flanking the large cavity, spaced at some distance from each other. The part of the passage between the individual cavities was coated internally with a cartilaginous substance, so that the whole course of the said passage to the tail, on both sides, was composed of many quite hard tubules between which stood out certain papillae that were perforated in such a way that air blown into them could come out easily, while it was difficult to insert a probe. When the middle cavity behind the head was blown into, the cavities lying between it and the middle lateral line swelled up, but the air escaped only through the papillae that were positioned along the middle lateral line. Besides the smaller papillae just described, there were others, quite large but less numerous, situated above the middle lateral line near the head but standing out from the line itself towards the tail. And this indeed is the description of a new passage first observed by me in the skin of the eel. Since then, I have observed the same in other fish, together with larger cavities behind the head; and I need little to persuade me that a similar passage will be found in all fish where papillae are found on the middle lateral line. What might be the function of so perforated a passage I shall not yet

---

[9] As to the Rajae Anatome and Steno's first discovery of the mucous canals, see OPH I, 196.

[10] *Francesco Redi* (1626–1697), an eminent physician in Florence and physician of the court from November 28th 1666, was also a great naturalist and poet (Cf. his *Bacco in Toscana*). He was a member of the Accademia del Cimento and worked energetically for the great Italian dictionary of the Crusca-Academy. Both his *Esperienze intorno alla generazione degli insetti* (Firenze 1668), where he succesfully attacked the general belief in *generatio spontanea*, and his most important publication *Osservazioni intorno agli animali viventi che si trovano negli animali viventi* (Firenze 1684) on various animals prove him to be an observer and discoverer of high rank. He collaborated with Steno, and was congenial to him also through his modesty and religious uplift.

venture to decide, since I have found it empty, without even a trace of slimy fluid present.[11]

But whatever the reason for this, at all events one can deduce from the observations made here that the skin of the fish has many excretory vessels, whatever the kind of material that separates from the said vessels; which seems to establish to no small extent my opinion about the human skin, in which I have stated that it has its own excretory vessels originating in glands, whether these glands lie just under the skin, or are woven into it, or are arranged both ways. Though, to be sure, imperceptible transpiration, recorded by the Ancients, has been confirmed by the experiments of Santorio,[12] a question remains both as to the parts from which this material comes and the means by which it separates. Some believe that the said material flows from the contents of the head, chest, and intestines, indeed from every part of the whole body lying under the skin, to volatilize through the pores of the skin and its intermediate parts. Others have the opinion that the material is carried through the arteries to the skin, to find open pores in the skin through which it passes outwards. I admit willingly that there is certain material that penetrates even the harder parts of our bodies at all times, and that the said tenuous material can bring the majority of the attenuated parts of our body with it, even from the innermost recesses; chemistry provides us with a clear example of this in the calcination of deer-horn. Nevertheless, since it seems most probable to me, I consider it valid that the denser bodies which pass out in this transpiration, either in the form of vapour or of fluid, are carried through the arteries to the glands of the skin, being expelled from there through special vessels.

Substance of the Skin

The quite dense and hard skin received so many and so large muscle tendons that it seemed to be in its entirety a web of these tendons.

On this occasion, I began experiments on the same subject with other fish, and there I saw that most of the tendons very clearly continued into the skin: This reminded one of an elegant dissection of the tendon fibres of the ray as a result of which I believed at the time that the fibres reached out either into the skin itself or into a membrane joined very tightly to the skin[13];

---

[11]Here Stensen describes the lateral system of canals declaring himself ignorant of the purpose of these canals.

[12]See Ars *Sanctorii Sanctorii... De statica Medicina...* Venetiis, 1634. Santorio published his doctrine of *transpiratio insensibilis* in 1614. It was quoted by Stensen in the Chaos-MS (fol. 56ᵛ).

[13]Cf. OPH, I, 101. 186. 196.

now, indeed, I hold that they are undoubtedly intermingled with the skin itself. In like manner, I have discovered often, in examining snakes and reptiles that the muscle tendons of these animals are joined to the skin. What may I say of the hedgehog? Here, not only is the remarkable skin muscle (called by others the *panniculus carnosus*) joined to the skin, but also each quill sheath has its own motile fibre, which item I shall demonstrate more fully if I am able at some time to throw light in an orderly way on the rest of my anatomical observations, which are due to the love of the most Serene Grand Duke for natural science.

Thus the abdominal muscles of the human body have several tendons joined to the skin, especially in the part of the *linea alba* above the *umbilicus;* the same can also be seen at the edges of the *linea alba* if the skin is dissected a few inches from the middle, cutting parallel to the *linea alba* and separating the skin with its substrate of fat from the underlying muscles; for this separation could be carried out in such a way that the true path of the tendon fibres through the middle of the fat is seen to join the skin where they run from the middle to the sides in the skin. The same may be observed also in the elbow, where the tendon fibres, spreading to all parts, are incorporated in the skin, a space being left in the middle where they adhere less to be skin. I say nothing of the *palmaris* muscle in the hand, the *plantaris* muscle in the foot, and various muscles spread through the whale body, joined to the skin at one end, hitherto known by the name of *panniculus carnosus.*

From all that has been written, it is quite clear that the tendons run together to make up the substance of skin: this is likewise confirmed by the fact that the skin, in the same way as the tendons, melts when heated into a gelatinous, or rather a glutinous substance. I admit that I did not pay attention to all of this when I gave my opinion on skin elsewhere, as I thought that the confused mixture of fibres in the skin was explained sufficiently by assuming them to be the ends of vessels (by which I understood veins, arteries, and nerves).[14] But now, since I have noted in the shark that the skin vessels have their gland beneath the skin, since an innumerable bundle of tendon fibres were seen to have reached the skin, I take it as certain that the greater part of the skin is a web of fibres of arteries, veins, nerves and tendons. Regarding human glands, as mentioned above, I am not certain whether they are to be found in the skin, under the skin, or in both places.

---

[14]At Stensen's time the nerves were looked upon as hollow, an idea attacked by him in his myology, where he demonstrated geometrically that the swelling of muscle-contraction was not to be explained by a fluid led from the brain to the muscle by the canals of the nerves. See OPH. II, 76.

It is clear from the proven insertion of tendons in the skin what should be determined about the *membrana adiposa*, the *membrana carnosa*, and the *membrana musculorum communis*. Most people have thought them to be mutual wrappings for the body: but the *panniculus carnosus* does not cover the whole body, since the kind of skin-muscle which should form a basis for this *panniculus* appears only in certain places; the *membrana adiposa* is nothing more than an accumulation of fat placed between the aforesaid tendon fibres running from the muscles to the skin; the said *membrana musculorum communis* is not a single membrane but a membranous expansion of the tendons issuing almost everywhere from different muscles, from which fibres disappear into the skin.

From this can be seen what should be our opinion about many other membranes, namely that they are not always coverings which adhere strongly to neighbouring parts on immediate contact, but indeed very often continuations of these as a result of a transition of fibres from coverings into neighbouring parts, or from neighbouring parts into coverings, which the hollow plates in the cranium of an embryo, and the jagged fringe of the *pericranium* and *dura mater* show sufficiently elegantly. The same is likewise clear in the middle tendon of the diaphragm, held joined to the *vena cava*, the *pleura*, *peritoneum*, and other parts: so that in consequence, no one can fail to observe from the aforesaid by what right the *dura mater* may be held to be the common source of the remaining membranes.

Cartilage

There was cartilage instead of bone, whose substance was outwardly quite hard and opaque but which was internally soft, transparent, and filled with blood vessels. The sight of those purple trees in this transparent substance was unusually beautiful, yielding not a fraction to the blades of grass, flies, and other bodies confined in amber and crystal that delight lovers of the things of nature.

One can perhaps doubt whether the cartilage in cartilaginous animals, like the bones in other animals, is composed for the most part from the ends of motile fibres: at least it is clear that the harder part of those cartilages that receive the most numerous tendons is almost completely composed of those same tendons. Nor would I reach any other conclusion from reasoning than that its central transparent substance is of a tendinous nature, since we observe that tendons themselves are reduced to a similar substance solely by heating.

A cartilaginous cylinder taken to be the optic nerve (Fig. 2.23.3/Tab. V, Fig. I.).

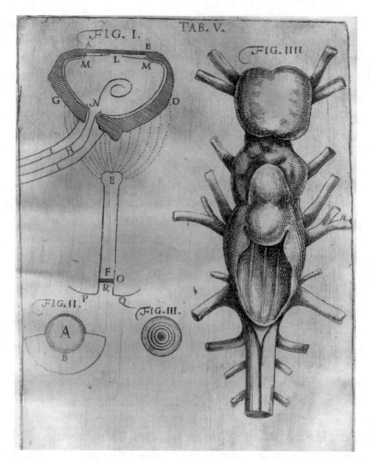

Fig. 2.23.3   Tab. V: Brain and eye from giant shark

Nor was the eye lacking in reward for those who dissected it.[15] *Between the sclerotic and the inner part of the orbit of the eye ran a cartilaginous cylinder E F, one end of which, F, was joined to a cylindrical prominence from the orbit of the eye P Q rising through the intermediate softer matter O; the other end, E, stretching out to the eye, was not joined to the eye itself but ended about two inches from the eye in a larger, uneven top; the head of the cylinder was connected to the eye through a soft, fibrous and somewhat white substance, so large in bulk that it occupied the whole of the rear part of the eye, G D.*

---

[15]Here Stensen correctly describes the stem which, in the *plagiostomes*, fixes the eye ball in the orbit.

I am of the opinion that here the wisdom of nature shines forth elegantly; since a small eye in a large cavity would move about in an uncontrolled fashion, unless some fulcrum acted as support, and the antagonistic muscles controlling counter-movements would be ineffective unless the less mobile base was attached everywhere to the readily mobile eyeball. This is the cylinder which zoologists have taken to be the optic nerve, deceived by the similarity in shape and by the position of the straight muscles; for just as they surround the optic nerve in other animals, so in this animal they surround the cartilaginous cylinder.

The optic nerve, H, at first glance, seems to be fairly thick, but on removal of the many membranes I, I, in which it was concealed, it was hardly the size of the human optic nerve. Nor was it as solid as the human nerve; allowing that other writers record this as due to the change which occurred in the parts in collapsing after death, nevertheless it does not seem to me that this reason holds here, since the optic nerve in other animals usually retains its solidity and firmness for many days after death.

The insertion of the optic nerve in the sclerotic or cornea was halfway between the internal angle of the eye and the very centre of the eye, opposite the carti-laginous cylinder; and it had its own obvious opening where the solid substance of the sclerotic was pierced.

The muscles of the eye. There were also six eye muscles albeit their position was quite varied, for the straight muscles that usually arise around the spot where the optic nerve leaves the cranium were observed here around the point where the cartilaginous cylinder joins with the orbit. But the uppermost of the oblique muscles, which otherwise are carried in a straight line from their source right to the trochlea, ran diagonally just like the inferior oblique muscle. Thus the eye of the shark is also an example of the fact that nature has not confined herself to the same law everywhere, but can often obtain the same effect in a different way.

The Sclerotic Membrane

The front translucent part of the sclerotic membrane, A B, known as the cornea, was flat in this case, the remaining part was really hard and similar to the cartilages in the same fish.

So also in birds, it is found that a large part of the sclerotic consists of bone. Thus it is possible to suggest that a large part of the sclerotic membrane is composed of mobile fibres of tendons, since not only is it a construction of the truly tendinous *dura mater* but, in truth, it contains also tendons both of those muscles that are inserted into it from the exterior eye muscle and of the uvea muscle or muscles that are interwoven with it, and since, in various animals, its consistency is similar to that of other parts of the same animal that are likewise composed of tendinous fibres. The membrane demonstrated by anatomists in dissections as being composed of muscle tendons of the eye is probably due only to the mode of preparation.

Space L, the diameter of the pupil, is rather small. LM represents the uvea. The substance of the crystalline lens. Fig. 2.23.3/Fig. I.

The crystalline lens, confined in its own sheath, consisted of three substances. The inmost part occupying the centre and surrounding parts, was hard and composed of lamellae, which, when whole, were seen to be transparent just like crystal, but which became white and opaque on the instant of being dissected, the outer part of the crystalline lens next to the membrane flowed just like water, the remaining part, lying in the middle region between the centre and the sheath was also of a medium consistency, resembling glue in its stickiness. The solid ball surrounded by sticky fluid moved around freely in water.

I have often observed in the crystalline lens of living fish that the substance nearest the centre was hardened and composed of lamellae in the same way as an unbroken lens in the eyes of boiled fish is found to be both hard and composed of lamellae. Such hardness of the lens does not occur in the eyes of animals that we class as warm-blooded, though these eyes, when boiled in hot water and hung out to dry, seem also to be composed of lamellae. Thus, a few weeks ago, I saw that the lens of a blind woman's eye was hard and composed of lamellae after death, and I believe that the lens which the very famous Guglielmo Riva, most skilled in anatomy, observed in hardened condition in a blind person, would have been composed in the same way.[16]

Anyone who has observed the gradual solidification of crystalline fluid that occurs in the eyes of living fish, in the pearls of oysters, and in the teeth of cartilaginous fish, all of which takes place in the middle of cold fluids and cold places, recognizes readily that when investigating fluids and solids, there are other things to be considered apart from heat and dryness.

---

[16]*Giovanni Guglielmo Riva* (1627–1677), famous surgeon at the Hospital of S. Maria della Consolazione in Rome; in 1664 he accompanied Cardinal *Flavio Chigi* to France and was archiater of *Clement IX* from 1667. Stensen met him first (May 1666) in the Villa Ludovisi together with *M. Malpighi*, became his cordial friend and also belonged to his private academy in Via della Pedacchia, the most active centre of anatomy in Rome, teaching it and practising also on corpses. Riva gave a first true and graphic description of the chyliferous system, left a manuscript *De Latice in animante*, i.e. on fluid which must have interested Stensen exceedingly, and was also known for his experiments with transfusion. His interest in the crystalline fluid of the eye is shown in a letter on an observation *de restitutione humorum oculi*. His pupil G. M. Lancisi (1654–1720) published Mercati's *Metallotheca* (Mieli, *Gli Scienziati Italiàni* 213–219).

These variations in the crystalline substance seem to confirm the opinion of the most talented Philippeau,[17] who also, since he had undoubtedly observed the same in fishes, was convinced that the ciliary processes attached all around the crystalline lens could, by shortening, reduce the convexity of the lens according to the ability of the confined fluid to resist their pull, and that for the same reason, the shape of the crystalline lens (which in the human is believed to be composed of two hyperbolas) can be varied according to various distances of the object in view.

The membrane of the crystalline lens (Fig. 2.23.3/Fig. II).

The convexity of the lens is enveloped at the front by the membrane of the vitreous fluid, so that this membrane becomes in truth common to the vitreous fluid B and the crystalline A. It is clear that those who believe that the crystalline fluid is enveloped by a membrane at the front only, have taken part of the vitreous membrane as belonging to the membrane of the crystalline lens.

Shape of the Crystalline Lens

The shape of the crystalline lens, confined within its own membrane, apparently round while it held fast to the ciliary processes and was hidden within the vitreous membrane, seemed to be composed from two spherical segments, the lower of which had a circumference greater than the upper. I do not know whether I may be permitted to suggest that this observation is similar to what has been recorded about the crystalline fluid in certain fish, where the posterior segment jutted out on all sides with the anterior segment.

I add nothing here about the muscle of the uvea, since I have not yet satisfied myself about it. This much I will point out here: it differed greatly from the uvea muscle in other fish observed by me in Paris, and approached in size the uvea of horses and oxen that I observed in the same city. But the place for this, as for the wartlike vessels of the *choroides*, the lymph ducts of the eye, and also the junction between the transparent cornea and the sclerotic, will be when I set down my experiments completed in Paris at the house of my dear friend Thévenot.[18]

---

[17]Haller writes in *Bibliotheca Anatomica* (Tiguri [Zürich] 1774–1777) 1, 493, when mentioning this passage: *Cl. Phelypeaux hypothesin probat, processus ciliares ejus lentis convexitatem deprimere* (Ibid 1, 513). Haller mentions *Vincentii Phelipeaux de praecipius actionibus automatice in homine Lovan,* 1662 (OPH v. 2, 325). Stensen met Phelipeaux in Bourdelot Academy in summer 1665.

[18]These experiments were never published and the notes referring to them have in all probability been lost. Perhaps Winslow had them (Cf. *L'Autobiographie de I. B. Winslow* publ. par V. Maar. Paris 1912. p. 85 ss.). The *transparentis* after *corneae* in the MS. is from Stensen's hand.

Cavities in the Cranium Behind the Orbit of the Eyes

Behind the eye socket, in the cranium, was a cavity containing a capsule filled with grey matter. Whether this cavity functioned in place of ears is open to doubt.

The brain (Fig. 2.23.3/Fig. IIII).

The mass of the brain was quite small, being scarcely three ounces in the fish which weighed more than three thousand pounds.[19]

Yet, it seems to be characteristic of these fish that a huge body-mass is animated through a very small brain, assuming, for the present, that all manifestations of life emanate from the brain alone and that one does not attribute the origin of part of these manifestations to the spinal cord. Since the arguments against this proposition are not inconsiderable, I shall examine them more closely.

1. The nerves from the brains of fish are so few that they do not seem sufficient for animal functions. In the shark, which is under discussion here, no more nerves emanate from the brain than are able to pass at one time through the area shown in diagram, this area is made up from cross-sections of all brain nerves and a cross-section of the spinal cord directly beneath the fourth ventricle. Who could be readily convinced that all motion and sense perception of this massive animal could be governed by such a small number of nerve fibres? Not to mention the fact that if one were to eliminate from this surface specific nerves concerned with vision and other senses, this area would be very minute. But we will consider only movement for the present. The number of muscle fibres serving motion in this animal was so great that an area put together from cross-sections of individual fibres would amount to many Florentine square feet. It is easy for everyone to see, on consideration, how many muscle fibres, of the length shown here, could be contained in such a huge fish. Thus, anyone will realize that it would be almost miraculous if so many muscle fibres were brought into movement by so few nerves. In man, this evidence is not so obvious because the number of muscle fibres is far smaller here and the number of nerves is much greater.

2. It may be seen that more nerves pass outwards from the spinal cord than can pass through a cross-section of the spinal cord directly beneath the fourth ventricle of the brain. Thus, since the other ends of all nerves would not seem to be in the brain, one cannot, obviously, lead all animal activity back to the brain.

---

[19]Three ounces mean about 93 gramme.

The number of nerves, in the fish, which proceed from the spinal cord, is so large that an area formed by the junction of their cross-sections would be much larger than the area of cross-section of the spinal cord itself. It may also be verified, in these fish, that the thickness of the spinal cord is maintained throughout its length, although it should become gradually thinner towards the tail, indeed it sends out even more nerves, the further it is from the brain. Nor is there a lack of proof, in other living creatures which, over and above a spine, have jointed limbs, to confirm that the spinal cord grows much thicker around the lumbar region. Indeed, in birds, there is a diamond-shaped cavity in the same region.[20] All these things seem to verify that the majority of nerve endings are to be found not in the brain but in the spinal cord.

Since, then, more nerves begin in the spinal cord than can run from the brain down through the beginning of the said cord and since the number of muscle fibres exceeds quite disproportionately the number of nerve fibres going out from a single brain, one may have valid doubts whether all animal functions are to be looked for in the brain alone.

I know that it may be objected against the mentioned difficulty that the one nerve fibre could serve the movement of several muscle fibres. Nor do I ignore the objection that might be made against the first difficulty, that the spinal cord, in as much as its thickness is decreased by the emission of nerve fibres, could enlarge itself by a corresponding amount or more by building up new grey matter. But from lack of certainty in experiments these assertions must here be considered doubtful.

I might have used another argument to call in question the hitherto accepted function of the brain, for I have observed, on making a ligature in the descending aorta, without previous cutting, that voluntary movement of all the parts below ceased in proportion to the tightness of the band, while it was restored similarly as the knot was loosened. I have been making such observations for several years and have demonstrated them in various places, particularly in Florence, where the dog survived without suffering any damage to its mobility, after the ligature was removed. Since this experiment requires completion by other methods, as yet untried, I shall add nothing more about it here.[21]

---

[20]Stensen was the first to mention this *Cavitas rhomboidales*; the name he gave it, is the one still in use.

[21]This is known as Stensen's experiment, demonstrated at several locations during his travels. Not mentioned in the short report is that the investigator may have used a large curved needle to insert a band from one side of the thoraco-lumbar region crossing the body in front of the aorta to exit on the other side of the animal. When tied over the back of the animal, the aorta was compressed towards the spinal column causing immediate paralysis lasting as long as the band was kept tight. Ischaemia of the spinal medulla causes such immediate paralysis while ischaemia of the muscles of the limbs does not. (TK).

I have presented many more controversial points concerning the brain in the discourse which I left in Paris, in the hands of a friend.

Connection of the Optic Nerves
The optic nerves did not run on opposite sides, as in the case of other fish where the nerves are separated from each other, but, instead, as in the case of man, were joined very tightly together as one nerve.

The Jaws
The jaws and teeth were made available for examination by turning the head over. I do not wonder that whole men have been found in the belly of these fish, since the jaws in this shark could have let a whole man pass through them without difficulty. *The transverse diameter of the mouth joining one corner of the jaws to the other, was one Florentine yard long, the other diameter, perpendicular to the first, in other words right from the middle of the upper jaw to the middle of the lower jaw was four fifths of a Florentine yard long.*[22] I shall not delay the reader by describing the shape of the teeth, since the first plate shows these drawn from life.

The Number of Teeth
As to the number of teeth, I am not surprised that only six rows were observed by Mercati, since the head that he saw was dried up; the remaining rows, with a softer base, had fallen out or had been lost in another way as the head dried. Nor am I able to say anything with certainty about the number of teeth owing to the curiosity of the men who first handled this head, for they had torn out most of the teeth and in same places had cut away the gums with the teeth so that not even a trace of the teeth was left: at least I can state that *there were 13 rows of teeth in the middle of the lower jaw, the curved inner and lowest ones were so hidden in the soft, spongy flesh of the gums that they did not come into view unless the gums were cut away.* I cannot comprehend the use to which nature has destined teeth that are curved to such an extent, since, being buried in the flesh, they could on no account serve for the chewing of food. Which may explain the fact that animals of this kind swallow everything in their path without chewing it, as witness the findings of whole men in their stomach, if we can believe the accounts. The first row serves to hold the prey fast so that it does not escape, and perhaps to tear larger bodies into pieces so that they may pass into the belly, but regarding the others, I do not see the

---

[22]In the MS.: *qvarta*. A *braccio fiorentino* was 58 cm long.

purpose for which they were made, unless materials may be said to be made inevitably.[23]

The Substance of the Teeth

The teeth showed three different kinds of consistency: some indeed were completely soft; you might call these flesh, if on other occasions, anything soft found in the body is to be taken as flesh; many were completely hard; some were partly soft with their remainder hard, and of these some had the whole outer crust hard, others had only a hard tip. Moreover, the order of succession was such that the four lower rows were soft, though they were not different from the others in their triangular shape and jagged edges. These had recently hardened near the apex and over a small part of the planes, but those which were in the next highest rows were completely covered into a hard crust; all the rest were hard right through. If those whose inner substance was still soft were broken, it was clear that this substance was dense and made of elegantly arranged fibres.

In the same way I have observed with younger Indian swine that the whole inner part of the teeth was soft, but the outer part of the teeth which rises above the sockets, was hard, while the remaining part was the more soft, the deeper it lay in the sockets; the same may be seen with the teeth of calves or other young animals. Over and above, in the teeth of Indian swine, I have observed that the individual molars seemed to be composed as it were of several smaller ones, though it became evident, when the sockets were broken, that each consisted of one tooth; their apparently multiple shape is due to the complexity of the crust which is very similar to a capital **❷** letter in Latin script, as may be seen from the attached drawing representing a cross-section of a molar.

The area that is drawn here in black was filled with a substance like tartar, which usually grows on the outside of teeth. It is perhaps like the structure of the teeth of a hedgehog shown to me on an earlier occasion by the very able anatomist Tilmann.

What was said hitherto proves clearly, first, that the substance of teeth initially soft, becomes hard as time passes, a feature common to teeth and other bones; secondly, that the teeth do not become completely hard all at once, but the first signs of hardening substance appear at the apex; in this they differ from other bones, where the first hard knots almost always appear in the

---

[23]As early as in the treatise *Historia dissecti piscis ex canum genere* Stensen recognizes the real use of the shark-teeth: *Interiores ordines sursum inversi gingivis firmiter erant infixi; ut adeoqve ex tam ubere dentium numero minima pars usul esse possit, nisi delapsis èxterioribus interiores sensim succederent* (OPH 2, 150). In *Prodromus De solido* there is no doubt any more as to the use of the teeth-rows (OPH. 2, 216 s.).

middle; thirdly, that the substance of the teeth is a double one, an outer one and an inner one that cannot be called a membrane. In this too, teeth differ from other bones, because the latter have a marrow, but the former enclose a fibrous body.

Three things have to be considered here: first, whether the teeth are continuous with the rest of the body, or are indeed only in contact with it; secondly, whether the crust comes from a fluid which is secreted from the fibrous body, or is indeed a continuation of the same; thirdly, what is involved in the difference between soft and hard teeth. But more experiment will be necessary, before anything certain can be determined about this; it would surely be highly desirable, because ignorance of their nature hitherto has meant that the cure of almost all sicknesses affecting teeth is left only to chance. Who can stop tooth decay once it has begun? Who can lighten their pains? Who can give a clear explanation of the symptoms of teething, or cure it as desired? But if we had a clear idea of their substance, and if we but could make comparisons with other substances, then I do not doubt that we might find a better cure for so many sicknesses, and that the number of those who complain of being toothless would be much less.

2.23.1 Concerning Tongue Stones

The controversy to be decided in regard to the larger tongue stones is whether they are the teeth of *Canis carchariae* or stones produced by the earth.[24] To be sure, some would have it that bodies dug from the earth bearing a resemblance to parts of animals are the remains of animals that were formerly in those places and are now decayed; others believe them to have been produced in the same places without animals being involved. I do not yet have the knowledge of this matter to pass judgment on it here; and though my travels have taken me through various places of this kind, nevertheless, I do not dare to guarantee that what I shall observe in the rest of my journey will be similar to what I have observed up to now; chiefly, since I have not yet seen what my very famous teacher Bartholin observed in his journey to Malta.[25]

Thus just as in legal affairs, one takes the part of the plaintiff and the other submits himself to the decision of the judge, so I produce, from what has

---

[24]Due to cautious method, perceptible also in several expressions of the MS., Stensen—though at that time probably quite sure of the organic origin of fossils—retained his statement until by all means of observation and conclusion, there could be no doubt left.

[25]Thomas Bartholin visited Malta in 1644, (cf. Garboe, *Thomas Bartholin* p. 1, 65 ff.).

been observed in the past, the proofs of those who reckon those bodies to be of animal origin, setting down perhaps at another time the reasons for contrary opinions, and looking always for a true judgment from more learned men.

Digression on Bodies Resembling Parts of Animals that Are Dug from the Earth

Therefore, I begin to set down, methodically, the present digression on the origin of bodies, resembling parts of animals, dug from the earth, and regarding the earth itself, with the desire that things I pronounce as uncertain will be held to be indeed uncertain. But lest the reader be led to expect many new ideas and because of this expectation complain that he has been deceived, I wish to warn him beforehand that some of the propositions have been made already by others; that many are owed to the observation of my teachers; there will be very few to which I have not been an eyewitness.

2.23.2   Historia

Observed facts about these soils and bodies.

1. The soil from which bodies resembling parts of aquatic animals are dug is in certain places rather hard, like tufa and other kinds of stone; in other places it is rather soft like clay or sand.
2. The said soil, whether rather soft or rather hard, is almost everywhere compacted, and is resistant to not too violent pressure.
3. In various places, I have seen that the said soil is composed of layers superimposed on each other at an angle to the horizon.
4. I have observed in clayey soil, that these layers, which differ in colour from each other, are split apart in several places, and that all the fissures which are filled with material of one colour are almost perpendicular to the layers themselves.
5. In those soils that I have been able to observe up to now, bodies of different kinds have been concealed in the same soil, sometimes in the harder, and sometimes the softer sort.
6. I have observed that the number of these bodies in clay is quite large in the surface but quite small in the soil itself.
7. In the same clay, I have observed that the deeper one goes into the soil, the more fragile are the said bodies; indeed, some of them crumble into

powder at the slightest touch; almost all of those that were in the surface could be reduced to whitish powder without much effort.

8. In rocky ground, I observed both that these bodies are much more abundant and that they have the same consistency all through the rock, and also that they were attached to the rock as if they were embedded in lime or gypsum.

9. Whether they are dug out of harder or softer soil, bodies resembling different parts of aquatic animals are not only very like each other but are also very like the animal parts to which they correspond; there is no difference of any kind in the course of the ridges, in the texture of the lamellae, in the curvature and windings of the cavities, and in the joints and hinges of bivalves.

10. The said bodies may be either rather hard like stone or less hard so that they may be reduced to powder without difficulty.

11. Very many oyster shells are found in some regions, deformed and hardened into one lump; sometimes also, broken scallops and mussels are dug up; some people have seen, in the same place, many tongue stones clinging as it were to the same matrix; these were not all of the same size nor were they all complete.

The following conjectures, based on the observations presented offer some glimpse of the truth.

2.23.3 Conjecture 1

Whether the soil today produces these bodies. Soil from which bodies resembling parts of animals are dug does not seem to produce these bodies today.

In the case of soft soil, there is little likelihood that the soil produces these bodies, but rather that it destroys them, since the deeper they are buried, the softer these bodies are (Observation I.) and the less they can withstand a touch. Nor should anyone believe that their greater softness arise from the fact that they are not yet fully developed; things that are soft while they are being formed keep their parts together with some glue-like material (as may be observed in the fresh bark of young pine or almonds), but these bodies are lacking in every kind of glue-like material and disintegrate into dust, so that their softness seems to proof of decay, not growth. It is no argument against this that their numbers seem to increase on the surface of the soil, (Observation I.) for this is due to the rain that washes away the soil between them.

On the contrary, when the substance of those (Observation 7.) that are on the surface is rubbed to dust without much effort, this rather proves that their decay, begun in the soil, has been interrupted by the intervention of rain.

One may conclude that they are not produced in our time in hard ground from the fact (Observation 8.) that they are found all through rock with the same consistency, and that they are surrounded on all sides by the hard material, for if any such bodies were produced anew today in these rather hard soils, the surroundings ought to give way to them during their growth, and the bodies themselves would no doubt show differences from those produced long ago.

Thus, since no bodies seem to be produced anew in harder soil, and since in many regions softer soil probably destroys these bodies, we may suspect not without reason, that soil from which bodies resembling parts of animals are dug does not produce these bodies today.

Conjecture 2

The said soil does not seem to have been firm when the bodies referred to were produced in it.

Whether the soil in question has always been of the same firmness. Bodies that expand by slow growth can certainly raise heavy objects resting upon them and may widen fissures in rocks; this is proved by tree roots in hard ground, in walls, and in cliffs. Nevertheless, while these bodies make a suitable space for themselves, frequently they cannot avoid being hampered by the resistance of the harder obstacle, which is exactly what happens to young tree roots that become twisted and compressed in countless ways in harder ground, so that they assume shapes different from roots found in softer ground. But the bodies that we are dealing with here are in fact always of the same shape (Observation 9) whether they are dug up from softer ground, hewn from rocks, or taken out of animals[26]; it would seem then that since these bodies do not appear to be produced today (Conjecture 1) in the places where they are found, and since things that grow in from soil are found to be strangely deformed, but these are everywhere alike, the soil would not have been firm when the bodies referred to were produced in it.

---

[26] *Oluf Borch*, Stensen's teacher and friend in Copenhagen, published a treatise on stone-building in cavities in rocks as well as in living organisms (*De generatione lapidum in macro-et microcosmo*. In: *Acta med. et phil. Haffn*. V. (1680). The MS. as *intueamur*.

Conjecture 3

Nor can there be strong opposition to the belief that the said soil was once covered with water.

Whether it may have been covered with water. This may have occurred in two ways, according to whether we assume that this piece of ground always had the same situation, or that it has changed its situation some time.

Regarding the first assumption, we learn from Holy Scripture that all things, both at the beginning of creation, and at the time of the Flood, were covered with water. Tertullian[27] writes elegantly about this: "A change occurred in the whole world when it was covered with all the waters; even now, sea shells of mussel and whelk range over the mountains seeking to prove to Plato[28] that the very peaks have been under water". No weight should be attached to the arguments set out by people of the opposite opinion when they say that bodies of this kind ought to be found everywhere if they owe their existence to the waters covering all places, or at least, that such bodies when found, should not be found only in high places. An easy answer may be given to both objections: since not all kinds of water carry everything, and what is strange about the fact that these bodies which are concealed in lowlands, covered by fresh soil, appear uncovered in high places, when we observe flat zones at the foot of mountains being filled with, as it were, the scourings of mountains scraped away by the violence of rainfall?

And if anyone should believe that portions of soil in places from which the said bodies have been dug have changed their situation at some time, he cannot be held to think anything that is contrary to reason or experience. Indeed, when we consider the fissures in the layers that are filled with material of one colour (Observation 4) whereas the layers themselves are of varied hues, it seems indeed quite likely that this piece of ground, shaken violently by a gigantic movement, broke on falling back, and so reached its new situation. It would be easy, to show how great are the changes in soil caused frequently by earth movement, from various examples, if the evidence of Tacitus alone were not sufficient[29]:

"During the same year, 12 towns in Asia Minor were laid waste by an earthquake in the night, whereby the catastrophe became even more unforeseen and calamitous. And the usual resource on such occasions—to take refuge in the open places—was of no use, since people were swallowed up in the yawning earth. Huge mountains are said to have been levelled to the

---

[27] See Tertullian, *De Pallio*,c.2.

[28] See Plato, *Critias 11B*.

[29] Tacitus, *Annales II*, 46.

ground; the flat ground is said to have risen into steep mountains, and fire broke out among the ruins".

Thus, since both the configuration of the ground itself and examples from other places indicate that this soil once had another situation, since it seems (Conjecture 2) that the said soil was once less firm, what is to prevent us from ascribing this softness to the waters, and what is more, to believe that the soil, before it changed its site, was covered with waters, whether the waters were exposed to the open air or were covered by the earth's crust?

Conjecture 4

There seems also to be no objection to the belief that the said soil was at some time in the past mixed with water.

We suggested in the preceding proposition that this soil might at one time have been covered with waters; now, we go a stage further, to prove that the said soil may have been mixed with waters.

That clay and sand are mixed with strongly agitated water is so obvious from the headlong course of torrents through such soils, and from the agitation of waters by the wind, that no further explanation is needed. Nor is it difficult to prove that sand, clay, tufa, and all sorts of solid bodies may be concealed in stagnant water, even the most limpid water. Solid bodies may be concealed in water in two ways: they may be concealed as powder, or their elements may be concealed in it. Solid powder may mix with water by itself, as all kinds of salts and vitriols illustrate, or it may unite with the water through the intervention of a third substance: thus minerals are dissolved in waters under the action of acids, oily substances by the aid of alkaline salts, whereby the salt gives to the oil and the acid to water the heaviness by which the oil is pressed down into the water and the minerals are lifted upwards into the same water.

The elements of a solid body may also be concealed in water in two ways: for either the solid elements themselves, in total or in part, are found therein, or particular substances are present in the water that assume a different form from it and are transformed into solid. For this reason, most people believe that mineral waters contain the elements of the minerals, and from this principle is derived the source of that radical solvent of metals, with which people work anxiously to extract mercury and sulphur from single metals.[30]

---

[30]The Arabian alchemists *Jabir and Al-Razi* in the eighth and ninth century taught a new theory, the germ of which is contained in Aristotle's *Meteorology*, viz. that metals are composed of mercury and sulphur and are generated in the earth from these. (I. R. Partington, *A short History of Chemistry*. 2nd Ed. London 1951, p. 29 f.).

Places from which the dissolved solids may have flowed. These then are the ways in which solids can appear under the appearance of water, and no great effort is required to find the places from which these solids joined the waters covering our lands.

The bosom of the earth conceals solids and fluids of all kinds: neither the juices that flow through the secret courses of the earth nor the exhalations that meander through these places can leave intact solids that nature destined them to dissolve, if they come into contact somewhere else. Moreover, juices flowing all the time from the veins of the earth into the waters, both those exposed to the atmosphere and those covered by the crust of the earth, spread the solids dissolved below ground through the substance of the water. But again, all kinds of bodies that are given off into the atmosphere from water, earth, plants and animals, and there combined in a wonderful way, will be able to reach the said waters in the form of rain or in some other way that eludes our senses. So, for example, animals of various kinds, indigenous to water, during their lives deposit the excretions of their bodies into water, and, when they die, are, as it were, totally dissolved in the waters.

Thus, since solids of all kinds may be mixed with the waters, since the places from which these solids could have joined the waters are obvious, why should we be astonished that either powders or the elements of clay, sand, tufa, and other stones should be mixed unseen, with the same water? Nor need anyone believe that the juices which dissolve these hard bodies ought to be acid to the taste, and therefore unable to sustain animals' life. I have seen my most amiable teacher Borch dissolve a very hard pebble in ordinary water; why then should we not grant to nature what we cannot deny to art?

Conjecture 5

I cannot see anything to prevent us from regarding the said soil as a sediment gradually accumulated from water.

Whether it may be taken for a sediment of water. We have just seen (Conjecture 4.) that there is no obstacle to the belief that this soil was mixed with water; moreover, it is clearly obvious (Observations 3 and 4.) that in various places it is composed of layers of different colours superimposed on each other; in addition, in those places where all the soil is of the same colour, it is nevertheless possible to distinguish the difference between layers. Thus, the layers themselves lead us to believe that the soil is a sediment from water, and indeed the difference between layers at least indicates that the said soil was deposited gradually, if it does not prove it completely.

I shall now make clear the ways in which sediments could have been deposited, so that these matters may in fact be more readily understood.

The ways in which sediments in waters accumulate. If we believe that the water under discussion could receive muddy water, either from the ocean or from torrents it is certain that the bodies which make the water muddy ought to sink to the bottom when the violent motion ceases. Nor do we need to seek diligently for examples of this type since both the beds of rivers and their estuaries give sure proof of it. One thing should be noted here: the bodies that make the water muddy are not all of the same weight; thus it follows that, as the water gradually calms down, first the heavier particles then the less heavy ones settle down; the lightest particles, however, float longer in the vicinity of the bottom before becoming attached to it. It is clear, in consequence, that frequently different layers will be found in the same sediment.

The ways in which solid bodies hidden in water may be secreted. But if we believe that limpid water has united with solid bodies transferred to it from the atmosphere, from soil, and from animals, it is not difficult in this case either to find different ways by which solid bodies contained in the limpid waters might have been secreted there. I shall explain briefly here the most important of these.

1. If "subtle matter" agitating the particles of a fluid[31] does not penetrate the fluid with the same force at all times, the solid particles that were moved uniformly with the fluid particles by its greater force will become detached from the grasp of the fluid when the said force ceases. Thus, blood, except when it is warm, is not completely fluid or uniformly red; on being cooled, it separates into parts that differ in colour and consistency. So, frequently, a clear urine loses its transparency with its warmth, recovering it when placed on a fire. By the same reckoning, warm juices coming out of the earth, or warm exhalations from the same earth, mixed with the waters, could deposit the more solid of the finely divided materials brought with them, when their warmth ceases. Nor does this warmth need always to have been considerable and unfavourable to animal life dwelling in these waters; it will have been sufficient if the "subtle matter" agitating fluids coming from some other region should move with more than its customary speed.

2. If the more volatile parts of a fluid have been able to evaporate slowly, the ratio of dissolved matter to solvent being changed, a quantity of dissolved solid ought to settle out in proportion to the quantity of solvent given off.

---

[31]In his *Prodromus De solido…* Stensen speaks more extensively of the meaning of this and related terms (OPH 2, 190 ff.).

This is illustrated frequently from experience of fluids, both of those that contain powered particles of solid and those that conceal the elements of solids. In the first case, crystals of salt separate from salt water; in the second, tartar separates from wine. So many examples may be observed daily that sediments accumulate in an identical way in all waters, that I judge it superfluous to gather them together here.

3. If we believe that different fluids can flow together, either simultaneously or at different times, from different places, we should admit readily that solids may have been precipitated from fluids brought from one place by some kind of fluid brought from elsewhere. Thus substances dissolved in acids may be precipitated by addition of salts, and substances dissolved by the action of salts may be precipitated by the addition of acids, since acids and salts unite more easily with each other than with other solids. It is possible also for solids dissolved in acids to be precipitated in other ways, as we see with metals, where one metal dissolved in acid may be precipitated by the addition of another metal. Similarly, tinctures extracted with alcohol[32] may separate out when water is poured into them. But again, if two fluids are poured together, a solid may congeal immediately. Thus, in Paris, in the Academy at the house of my great friend *Thévenot*, I have seen *Borel*,[33] greatly skilled in chemistry, pour together two quite clear liquids which immediately became so solid that not even a drop left the glass container when it was inverted. Why then, may we not suspect that rain falling from the atmosphere at various times, and of varying composition, or perhaps juices and vapours of different kinds from the earth, when mixed with the waters, may sometimes precipitate bodies dissolved in them and at other times dissolve bodies that have been precipitated from them? This is evident in urine collected from one and the same person at different times, since a solid deposit laid down during the first days, and adhering most firmly to the base of the container, is very often dissolved, during subsequent days, by fresh urine from the same person, only to collect afresh, soon after, from the second urine. What varieties of diet

---

[32] *Spiritus volatilis* means volatile matter (OPH 2, 327), probably alcohol.

[33] *Pierre Borel* (1620–1689) promoted to M. D. in Montpellier came to Paris in 1653, where shortly afterwards he became *Conseiller et Médicin Ordinaire du Roi*. Stensen had already as a student in Copenhagen read Borel's *Historiarum et observationum medicophysicarum centuriae* IV (Parisiis 1656) and made extracts from them, also f. inst. concerning fossils; at Thévenot's lie often met Borel and attended his chemical experiments together with Borch (Epistolae 1, Ss., 2, 907.). The experiment Steno here remembers is reported by Borch from the meeting of the Thévenot academy on the 9th December 1664. Borel then took a little water, added to it a little sulphuric acid—according to Borch's supposition—and a little butter of antimony of a watery consistency; this mixture coagulated instantaneously, and the vessel could be turned upside down without a drop running out (J. Nordstrøm, *Swammerdamiana* 63).

accomplish in the humours of the microcosm, so alteration in the sun and moon and various other changes could produce in the humours of the earth. *Gassendi*, the "glory of France", supports this assertion with the clearest of examples in his learned work in which he explains the origin of stones.[34]

4. Those who contend that anything whatsoever may come from anything whatsoever, or, at any rate, admit that the smallest items in nature are subject to a variety of changes, could explain this matter in an other way; it is possible, however, for us to imagine clearly and plainly enough that "subtle matter", as it moves through water and air, changes the various fluid parts, which gradually take on a new form, into solid bodies. I have seen, at the home of our friend *Borch*, a white earth come from the purest water, a tasteless earth from alkaline salt, and, from the atmosphere, a salt that is unaffected by fire: these may be explained in different ways according to various first principles.

Such are the various ways in which solids may be precipitated from a fluid, nay more, fluid from fluid (as may easily be shown of those fluids which form the atmosphere); if the layers in our soil have not been formed in all these ways, it is certain that they could have been formed in such ways.

But whatever the exact way in which solids are separated from fluids, they appear either in the form of powder, as in the case of metals precipitated from acids, or as coagulated material, whether it be softer, as in blood where it is fibrous, in milk where it is cheesy, in May dew[35] and rain water where it is a viscous sediment or whether it be harder, like tartar in wine, crystals in salt water, and stony crusts in various springs.[36] It is clear from this that crusts could have hardened out of the most transparent waters, crusts of varying consistency, crammed full indeed with minerals of various kinds.[37]

How well then everything fits together! How unanimously they come together in agreement. We find the position of the soil suited to its having

---

[34]See, moreover, Petri Gassendi... *Opera Omnia*... Lugduni 1658, vol. II. lib. III *De Lapidibus ac Metallis* c. I. p. 112. In Stensen's Chaos-Manuscript of 1659 many extracts of Gassendi's works may be found. (See Scherz, N.-St. und G. Galilei 16f.).

[35]May-dew was used for certain alchemical purposes, but may be mentioned here by Stensen because the air in spring is full of vapours and small particles emanating from plants and soil, absorbed at this time of the year by the dew falling. The *sedimentum viscosum*, which Steno says is formed ja dew and rain-water, is probably only the sediment produced in course of time by the growth of algae and other microorganisms (OPH 2, 327 and Garboe, *Ædelstene* 124).

[36]It means a crust of carbonate or lime. (OPH 2, 327).

[37]*The paragraph, *But ... kinds.* is written on a separate small piece of paper. This paragraph may well give the reason for Stensen's exclamation that follows: *Qvam bene itaqve conveniunt omnia!.*

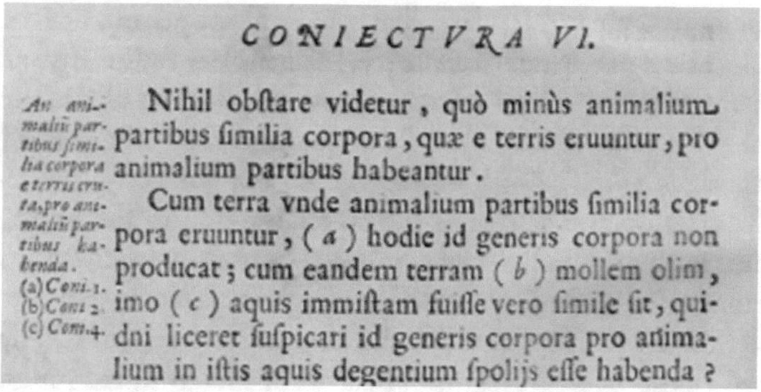

CONIECTVRA VI.

An ani- Nihil obftare videtur, quò minùs animalium
malii par-
tibus fimi- partibus fimilia corpora, quæ e terris eruuntur, pro
lia corpora
e terris eru- animalium partibus habeantur.
ta, pro ani- Cum terra vnde animalium partibus fimilia cor-
malii par-
tibus ha- pora eruuntur, (*a*) hodie id generis corpora non
benda.
(a)*Coni.*1. producat; cum eandem terram (*b*) mollem olim,
(b)*Coni.*2. imo (*c*) aquis immiftam fuiffe vero fimile fit, qui-
(c)*Coni.*4. dni liceret fufpicari id generis corpora pro animalium in iftis aquis degentium fpolijs effe habenda ?

Fig. 2.23.4 On the biological origin of fossils, **Conclusion**

been able to hold waters; we know that both powdered soil and the elements of the said soil could have been mixed with the waters; we do not ignore the ways in which they could have both entered and separated from those waters, nay rather we pay close attention to the variety of layers in the soil itself. Why then is it impossible for this soil to have been a sediment from water?

Let those for whom it is not enough go into underground grottos from which stones were once quarried, and they will observe new rock forming in place of the rock that was removed, nay more, they will perceive stone icicles, formed from bodies secreted by atmospheric fluid, hanging from the vaults: these icicles, hollow inside and made of many cylindrical lamellae, receive neither water nor rock from the vaults, this is not only indicated but also proved by the structure of the lamellae (Fig. 2.23.4).

2.23.4 Conjecture 6

Whether bodies dug from the ground and resembling parts of animals should be considered parts of animals. There seems to be no objection to the opinion that bodies dug from the ground which resemble parts of animals should be considered to have been parts of animals. Since the soil from which bodies resembling parts of animals are dug does not produce this kind of body today (Conjecture 1), since it is likely that, the said soil was once soft (Conjecture 2), nay more looks as if it was in truth mixed with waters (Conjecture 4), why not allow us to surmise that bodies of this kind are the remains of animals that lived in those waters? Indeed, if it is agreed to examine their position in the soil, it does not seem that they could have collected in this way, unless they may be said to have gathered together gradually with sediment from the water. Nor is it against our ideas that they are found in such numbers in harder soil. There is no difficulty to be found in that for anyone who has examined in detail the way in which new rock is formed in the subterranean galleries of the earth, where stones were quarried formerly.

On the position of these bodies in the ground. For whether a cream-like crust of stone hardens on the surface of the water, sinking to the bottom when it has become heavier, or particles of stones are produced evenly throughout the water, settling out gradually, the sediment grows only at a slow rate, thus, only those things which are already adhering to the bottom, whether they be dead animals, skins of dead creatures, or live animals unsuited for locomotion, will be covered over by new sediment; the rest of the living animals, striving above the said sediment, fills the waters with numerous progeny before a new sediment is laid down there. In addition, is may be stated that (1) the stagnant water in these subterranean galleries always preserves the animals that were once produced there, contrary to what happens with running water, (2) shellfish and similar kinds of animals do not prey on their own kind, other aquatic animals do and so consume each other, and (3) the shells of shellfish are rarely consumed while other aquatic animals almost completely dissolved in water. All this evidence seems to me to carry no little weight in establishing my theory, especially since nothing can be readily brought against it from the shape and substance of those bodies.

On the shape of the bodies. With regard to the shape of the bodies of which we speak, (Observation 9) since this corresponds exactly to parts of animals, the similarity of forms seems to suggest a similarity of origin; indeed it is difficult to believe that such great conformity should be observed in any other basis, whatsoever you might propose for their manufacture. And herewith is the clearest proof of this. Who does not acknowledge that hexagonally shaped rock crystals, cubes of pyrites, crystals of salt from experiments in chemistry, and countless other bodies precipitated from fluid, have shapes that are far more regular than the shapes of scallops, bivalves, whelks, and the rest. Nevertheless, we observe in these simple bodies sometimes the apex of a truncated corner, sometimes several bodies adhering to each other without order, sometimes planes that differ from each other in size and position, and a variety of other ways in which they diverge from the customary shape. How much greater and more numerous should be the defects observed in bodies possessing a much more composite shape, and in those that are copies of animal parts. But if in certain places many oyster shells are found hardened into one mass (Observation 11), this is not different from what happens in the sea, since from the sea too may be drawn huge masses of oysters of different size, that adhere together in a wonderful way. If certain mussel shells are found broken across the middle, the edge of the fragment itself provides evidence that another part was once attached to it; indeed this is often found close to the first. But if several tongue stones of various size, not all of them complete, are observed sometimes to stick together, as if in the same matrix, the same is noted in the jaw of a living animal where neither are all the teeth of the same size nor are the teeth arranged in the inner rows completely hardened. Thus, since defects that occur very frequently in the simplest bodies are rarely found in most of the composite bodies, since no defects are observed in those composite bodies which are not found in exactly the same way in animal parts, since the said bodies, no matter where they are dug out, are both very like each other and very like the parts of animals, it is easy to show that the shape of those bodies is no obstacle to our considering them to be parts of animals.

On the substance of these bodies. When we pass on to the substance of these bodies, it is not contrary to our opinions either. For whether, like stone, it is hard and heavy, or, like calcined bodies, it is light and easily reduced to powder, nothing is shown by this that could not have happened with animal parts of this kind. We observe that the more solid parts taken from animals are made up of two different materials: one, which is converted to a fluid by the action of a more "subtle" fluid, becomes visible as an exhalation or a liquid; the other, being resistant to the motion of the more subtle fluid, keeps its complete shape for a reasonably long time, until at length, after a very long delay, it is broken down into powder. Thus, all sorts of bones and horns exposed to an open fire, stags' antlers and the rest, calcined scientifically, as they say, lose most of their fluid materials, nevertheless keeping their pristine shape and, as far as can be seen, their size. However, I dare not affirm that their size is not diminished. Indeed, after the animal juices have been expelled, the pores in those bodies could be filled with an other fluid of the same volume, but it could also be that these same pores decrease in size when the more solid parts collapse together. Thus, I have seen solid metals change their size under variations in heat and cold, without changing shape; as a favour to me, my very dear friend *Lorenzo Magalotti* demonstrated this with a copper bracelet,[38] and this can be shown any day to all interested in the things of nature.

We owe these experiments to chemistry, but I do not doubt that nature operates in a similar way in the bosom of the earth. While the collected sediment hardened together with the said bodies over a long period of years,

---

[38]See *Saggi de naturali esperienze fatte nel' Accademia del Cimento... e descritte dal Segretario di essa Accademia*, 1666, 182 ff. Magalotti was the secretary of the Academy.

the subtle fluid could not have left the same bodies intact, but must have, according to the nature of the surrounding soil, either extracted animal juice from them, or added a mineral juice to them, or introduced a mineral juice after the animal juice was removed, or, unless we are unwilling to find change in the smallest things in nature, transformed the animal juice into mineral. And thus, I reckon that I have shown sufficiently clearly indeed that neither in the soil from which bodies resembling parts of animals are dug nor in those bodies themselves is it easy to find anything which is an obstacle to the belief that those same bodies may be regarded as the parts of animals.

While I show that my opinion has the semblance of truth, I do not maintain that holders of contrary views are wrong. The same phenomenon can be explained in many ways; indeed nature in her operations achieves the same end in various ways. Thus it would be imprudent to recognize only one method out of them all as true and condemn all the rest as erroneous. Many and great are the men who believe that the said bodies have been produced without the action of animals. For, omitting those who are well known, *Mercati* of San Miniato, whom I have mentioned above, holds this opinion, as also *Antonio Nardi*[39] in his "Scenes of Tuscany", the manuscript of which book, containing many problems in physics and mathematics, is owned by my most renowned friend, *Francesco Redi*, in the service of His Grace the Grand Duke of Tuscany. These men have their reasons too, which are so much the less to be rejected the more numerous are the admirable works of nature that excite our astonishment afresh each day.

On the larger tongue stones. Returning to the proposition, having completed this digression, I shall fit some of what I have stated to the larger tongue stones. That they are teeth of the shark is shown by their shape, since they are closely matched, planes against planes, sides against sides, base against base. If we believe the accounts, new islands have sprung up from the depths of the sea,[40] and who knows Malta's origins? Perhaps at one time when this land lay under the sea, a place where sharks lurked, their teeth were buried frequently in the muddy seabed, but now they are found in the middle of the island owing to a change in the position of the seabed caused by a rapid conflagration of underground exhalations. Nor does any difficulty arise from the great number of tongue stones produced from this island. More than two

---

[39]*Antonio Nardi* was a pupil of *Galileo*. His *Scene Toscane* is a scientific miscellany in nine parts. The MS. which has never been published is in the Biblioteca Nazionale Centrale in Florence, in MS. Galilei, vol. 130.

[40]Cf. Plato, *Timæus*, 24 E ff. and Plinius, *Naturales historiae libri* XXXVI, lib. II, c. 85 ff.

hundred teeth may be counted in the said fish, to which other new teeth are added day by day.

Thus since the bodies resembling parts of animals that are dug from the ground can be considered to be parts of animals, since the shape of tongue stones resembles the teeth of a shark as one egg resembles another, since neither their number nor their position in the earth argues against it, it seems to me that those who assert that large tongue stones are the teeth of a shark are not far from the truth.

This discourse was ready for printing when *Manfredo Settala*,[41] Canon of Milan, whom everyone knows for his unique knowledge of the things of nature and his indefatigable zeal for enriching his museum, told me on a visit to this place, that there were many things among the rarer pieces of his collection which quite clearly favour my conjectures, which is gratifying to my ears, in as much as I am aware of how much weight they gain from this man's assent.

---

[41] *Manfredo Settala* (1600–1680), a great traveller and linguist, but especially well known on account of his great mechanical ability, made many ingenious pieces of apparatus and models, as well as several fine microscopes. He held a canonship in Milan, in which town he also laid the foundation of his large collection of paintings, antiqvities, natural curiosities and models of maehines. See *Museum Septalanium Manfredi Septalae… Pauli Maria Terzagi… geniali laconismo descriplum.* Dertonae 1664… Cf. Epistolae 1, 33 s.

2.24 Description of the Dissection of a Dogfish

Hardly had I ended my previous discourse, when the most serene Grand Duke sent me another dogfish[1] to be dissected in Pisa. I wish to add its description briefly related to the previous ones since it presents many connections with them.

Is it the so-called Salvian's pig? The dogfish, otherwise similar to Salvian's pig, is different from this in that its dorsal fins are deprived of stings. One could say: perhaps stings are provided to the male gender only while denied to the female since the fish object of the present description was of the latter sex.

Fins, number and position. There were two pairs of fins under the belly, one under the diaphragm near the end of the gills, the other on the sides of the anus. There were also two fins on the back, the posterior one very close to the tail, the anterior one stood midway between the snout and the aforesaid posterior fin. But also the tail made of two fins represented a scalene triangle the upper side of which was twice as long as the lower one.

Gills. Five pairs of gills concealed under the skin had their openings, but small, carved in the skin. These openings release water swallowed by the mouth. Behind the eyes an orifice which others have instead of an ear opened in the cavity of the mouth so that it is obvious that water passes through this way also. I could not observe whether there is a way analogous to the auditory passage, from this orifice into the intricate cavity carved in the skull.

OPH 24, vol. II, 149–155: *Historia dissecti piscis ex canum* is the third, of the three treatises that Steno published jointly in Florence in 1667.

---

[1]Maar: *Scymnus lichia* (Bonap.) shark.

© Springer-Verlag GmbH Germany, part of Springer Nature 2018
T. Kardel and P. Maquet, *Nicolaus Steno*,
https://doi.org/10.1007/978-3-662-55047-2_35

Excretory vessels of the skin. The snout, flat at its upper part, was perforated on all sides by many openings. When compressed these openings poured out a viscous humour. Since I had observed similar vessels in the heads of a giant shark and of a smooth lamprey, and in the skin of a ray, I no longer doubted that this unctuous humour coating the surface of the fishes must originate from these sources. This shows the industry of ingenious nature which oiled the surface of fishes to let them split more easily the opposing waters. Nature actually wished that copious sources of this unctuous humour be in the snout, by which the force itself of the opposing water compressing the skin would express thence the unctuous humour.

Teeth of the lower jaw. The teeth were flat in the lower jaw, round in the upper jaw, sharp in both. One would hardly describe the shape and position of the flat ones as nicely as nature shaped and set them. Therefore, omitting their description, I mention their number and consistence in a few words. There were eight rows of them, each provided with nineteen teeth. The upper row stood right up, the others were inclined inwards on the jaw. The two lower rows had not hardened in any of their parts although they had a shape by which they were very similar to the others, and the sides were equally provided with teeth. This proximal row had only the apex hard. The fourth from the bottom had kept the base alone soft. In all a difference of the inner substance from the outer crust was obvious.

Teeth of the upper jaw. Round teeth equipped the upper jaw. There were six rows of them and in each row there were seventeen teeth. The inner rows of the inverted gum above were firmly implanted so that, of such an abundant number of teeth, a very small part can be used, except if the inner ones succeed little by little to the fallen outer ones.

Tongue. The cartilage of the tongue receiving many fleshy parts of muscular fibres seemed to be made for its largest part of the tendons of these fleshy parts.

Muscles of the eyes. The obliquus superior muscle of the eyes was not trochlear but had an origin and a course similar to those of the obliquus inferior. Two by two, the recti muscles had their origin in a cartilaginous cylinder which the four recti together surrounded in the same ratio as in the dogfish and which, since then, I have observed in the smooth lamprey. Perhaps this is also common to other fishes.

Lens. The substance of the lens was triple, the middle was hard and made of lamellae; adherent to this from everywhere, another part was very glutinous; the third one, close to the tunic, was completely aqueous. But this is also the case for most other fishes. This is in favour of those who believe that the convexity of the lens changes thanks to the ciliary processes, depending on the distance from the objects.

Brain. Figure 1 shows the conformation of the brain. A is the fourth ventricle. B is the cerebellum. C, C are two bodies in continuity with the optic nerves, hollow inside, what is seen also in the brain of birds: these cavities were continued into the fourth ventricle. D, D are the optic nerves. Their connection was visible without inverting the brain. E represents the mamillary bodies. F represents the nervous filaments diffusing into the tunic of the nose from the mamillary bodies. G is a membranous hemisphere full of very thin lamellae membranous in like manner, parallel to each other and perpendicular to the axis H I. This shows how nature, everywhere similar to itself, wished to carry out through these small membranes in this fish what it achieves in us and in many other animals by wrapping up numerous lamellae of cancellous bone in membranes. Indeed, where corpuscles diffused through ambient fluid would affect more easily the perception of this part, nature achieves this by a number of lamellae so that the expansion of a large surface occurs in an exiguous space and in such a way that the ambient fluid contained in this space and divided variously between the lamellae touches a huge surface of the body. The larger the surface which is touched by the ambient fluid, the more corpuscles of the ambient fluid can act together on this part. I do not dispute here whether odours are perceived through a humid medium. Something must be assigned to this action. Whatever the name given, let it not be denied that corpuscles diffused through a humid medium (able to twitch the membranes or nerves of this part) exert their forces where they touch this part.

Weight of the brain. To find the ratio of the brain to the remainder of the body, I used scales. It appeared that the brain hardly weighs three tenths whereas the eviscerated fish weighed more than fifteen pounds and together with its viscera weighed twenty six pounds.

Peritoneum. When the abdomen was cut open, the peritoneum immediately revealed a delicate texture of tendinous fibres some of which represented a warp, others a weft. It was obvious that those fibres had to originate from the muscles lying over the peritoneum.

Liver. The weight of the liver exceeded five pounds. It had two lobes equal in length. One occupied the right side of the abdomen; the other, the left side. Its medial part connecting the two lobes beneath the diaphragm concealed the gall-bladder. From there, the passage of the biliary duct to the intestine was in the substance itself of the liver. One can suspect that, in animals which we believe to be deprived of a gall-bladder, this may perhaps be hidden in the substance of the liver and elude the industry of the researchers.

Contents of the stomach. The stomach contained three small fishes unhurt by the teeth.

Their surface was excoriated at some places. I should rather say that it was gnawed at by the corrosion of some dissolving humour.

Intestine. After a short circular course beneath the pylorus, by which circular course the pancreas and the spleen were wrapped up, the intestine continued in a straight course down to the anus. Two things were remarkable in this intestine.

1. The muscular tunica, half an inch thick, was made of spiral fleshy fibres.
2. A membranous spiral filled its whole cavity. The intestine itself was one span long. The fringe of the spiral along its axis was five spans long. The other fringe adhering to the intestine was eight spans long. The excrements in this spiral had a yellow colour. The coecum adjacent to the anus was very small and coated by a red crust.

Mesentery. Here there was no mesentery since all the part of the intestine extending in a straight direction was free from all part. In the vicinity of the stomach there was a portion of some membrane and a small vessel adhered fairly strongly to the intestine itself.

Ureters. The ureters of the kidneys could not escape being seen; such was their amplitude. In the median womb a tubercle remarkably stood out, the apex of which was perforated by the mouths of the ureters.

Ovaries. There was an ovary on each side which contained many eggs different from each others in magnitude, colour and shape (Fig. 2.24.1). They were there large, of middling size, very small. They were white, aqueous, and yellowish. They were round, oblong, unequal with many tubercles. The ovaries resembled two oblong small bags and I could not see through which passage the eggs depart from there to enter the oviduct.

Oviduct. The oviduct in like manner was double although the outlet into the fundament is the same for both and the mouth of the funnel in the abdomen is the same, as appears in Fig. 2 in which A is the mouth of the funnel and B is the outlet of the oviduct. Each of the two oviducts where it had arrived at the diaphragm beneath the liver, and had ascended between the liver and the diaphragm, reflected downwards, was attached to the midline of the abdomen (which is called *linea alba* in other animals) in such a way that the mouth of the funnel could capture eggs floating in the abdomen. I believe that the oviducts of the ray were conformed in a similar way. I had cut away their upper orifice together with the liver. This resulted in that air found an outlet beneath the diaphragm, as I have shown in the *Anatomy of the ray.* In the same *Anatomy of the ray*, to follow the common opinion I said about the

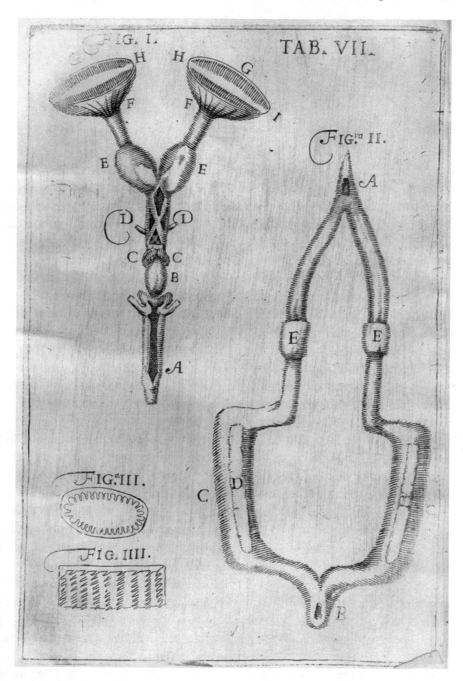

Fig. 2.24.1 Tab VII. Reproductive organs in a shark. See text

> non ampliùs dubito , quin mulierum teftes ouario
> analogi fint , quocunque demum modo ex teftibus
> in vterum , fiue ipfa oua , fiue ouis contenta mate-
> ria tranfmittatur : vt alibi ex profeffo oftendam , fi
> quando dabitur partium genitalium analogiam ex-
> ponere ,

Fig. 2.24.2 Concluding sentence on female ovaries not being "female testicles"

uterus: this does for the viviparous all that which the oviparous expect from the ovary, oviduct, and egg. Thus actually, since I have seen that the testicles of the viviparous contain in themselves the eggs, since I have observed that their uterus open in the abdomen in like manner as the oviduct, I no longer doubt that the testicles of the females are analogous to an ovary whatever the way either the eggs themselves or the matter contained in the eggs are transmitted from the testicles into the uterus (Fig. 2.24.2).

I will show publicly elsewhere, if and when there will be an opportunity to explain the analogy of the genital parts and to correct this error by which people believe that the genitals of females are analogous to the genitals of males.

Contents of the oviduct. The lower part of the oviducts, C, C, presented with a swelling above the cylinder of the intestine. When this swelling was cut open, white liquid flew out on which some bodies, long and not very thick, floated. They are indicated by the letter D in Fig. 2.

I was not perspicuous enough to see anything in these bodies contained in the oviduct except two tunicae, the outer one of which was rough and thick, the inner one smooth and full of crystalline humour. I indeed should believe that these bodies must be considered as eggs in which there was not yet any sign of a foetus; that therefore the outer membrane was the chorion, the inner one, the amnion; that it actually had been the function of the milky humour, progressively dried up more and more, and thinned, to penetrate the said membranes in order to provide the foetus with material for its formation and feeding. In the same way the eggs of frogs and of many fishes, surrounded by viscous humour, are seen to increase progressively even if they are no longer attached to the body of the mother.

Mammae [shell glands?]. At the place where the oviducts were narrower, a small ring denser and larger was seen in them (letter E in Fig. 2) the analogous of which I saw in a ray. But I did not see anything similar in birds and snakes which have oviducts in like manner, nor did I know its function. The zoographers call it mamma.

Substance of the oviducts. To look at the substance of the oviducts more thoroughly, I examined them after they were boiled. I saw something fairly pretty as appears in Figs. 3 and 4. Figure 3 shows a cross-section of an oviduct and Fig. 4 the inner surface of the same oviduct. Here I observe three things.

1. The substance itself of the tunic has appeared fibrous to me, which is observed in like manner in other animals.

2. All the inner surface is full of nipples arranged in parallel lines extending over the length of the oviduct. It is perhaps through these nipples that the milky humour which is to be changed into the substance of the embryo is secreted; it is perhaps through these nipples that the egg increasing with time is attached to the oviduct.

3. The amount of milky humour left in the oviduct has been hardened by boiling. This proves that there is a great affinity of this humour with white of egg. On the surface of this hardened milk, as I had taken away the tunic of the oviduct and its nipples, a parallel structure of nipples appeared most elegantly for the same reason that shapeless wax poured in plaster moulds, when hardened, represents the shape of the mould once the plaster has been removed. So that one actually should not assign this to imagination rather than to experience, I call upon the testimony of *Vincenzio Viviani,* mathematician of the most serene Grand Duke, who was present as a keen observer of these facts and of others contained in the present book.

2.25 Anatomical Observations Concerning the Eggs of Viviparous Animals

To confirm and light up observations of friends on the reproduction of animals from an egg, let me add to their works what divine generosity pointed out to me from the dissections of various animals concerning the eggs of the viviparous. By egg[1] I mean not only the round vesicles full of humour which constitute a great part of the testicles but also the chorion with all its contents. I use terms which are usual for most people, the testicles of females meaning the ovaries, the uterine tubes and horns, and the uterus meaning the oviducts. The ovaries of the female testicle provide the principle of the egg. The oviduct, i.e. the horns of the uterus with the uterine tubes, provide what is required for the perfect growth of the foetus.

OPH 25, vol. II, 159–166: *Observationes anatomicae spectantes ova viviparorum*. In the former treatise *Historia dissecti piscis ex canum genere*, OPH XXIV, Steno, in describing the dissection of a shark, for the first time maintained, that the testes of women were analogous to the ovaria of oviparous animals and ought to be called by that name. Steno also said that he hoped to take up this subject again, when he had made some more investigations. These notes, for the most part based upon dissections in the following years, from and including the year 1667, although a few of them date as far back as to the stay in Holland, are what Steno published in this Treatise; he probably felt that he would not succeed in completing the intended and more extensive work on this subject. The manuscript may have been handed over to Thomas Bartholin, while Steno was living in Copenhagen as *Anatomicus Regius*, but the treatise was published in *Acta Hafniensia*. vol. II. Hafniae 1675 (as No. LXXXVIII, pp. 210–218) not until Steno had left Denmark, and not till three years after the publication of *Regneri De Graaf de mulierum organis generationi inservientibus tractatus novus*, Lvgdvni Batav. 1672, and of *Johannis Swammerdami… Miracvlvm natvrae sive uteri muliebris fabrica…*, Lugduni Batavorum 1672. The first small section of the treatise was, no doubt, added as a kind of introduction to the observations, which are only connected by their subject being the same. In that section Steno also refers to the observations of de Graaf and Swammerdam, in the words *amicorum observationes*. See, furthermore, Gosch, *Udsigt over Danmarks zoologiske literatur*, II 1. Kjøbenhavn 1872, pp. 235–237.

---

[1] i.e. egg follicle.

© Springer-Verlag GmbH Germany, part of Springer Nature 2018
T. Kardel and P. Maquet, *Nicolaus Steno*,
https://doi.org/10.1007/978-3-662-55047-2_36

In a Cow's Uterus

There are two glandular parts in the cotyledons. They are like flesh. One is attached to the chorion and is implanted on the other part of the cotyledons by countless blunt extreme ends like as many roots. This part, if compressed, emits whitish juice which adheres to the membrane of the uterus. The glandular substance can very easily be separated from the membranes of the uterus except where it receives vessels. When these are divided, red points testify that they are blood vessels.

The allantois inside the chorion extended on both sides into the horns. One of these was twice longer than the other. Insipid and white water retrieved from there soon laid down a white sediment.

Many round sorts of vesicles were seen in the amnion. Its water was very abundant, salty and greenish.

In the foetus the umbilical arteries were much wider than the iliac arteries from which they come out. Outside the umbilicus the umbilical vein was double. The spleen was fairly delicate, oblong, and black. The stomach was distended by transparent humour. The lungs, although dense, however appeared to be made rarer when blown. When the abdomen was incised a copious quantity of blood flowed out.

In a Second Cow's Uterus

The testicles were full of various vesicles which contained yellow water. After having taken out the foetus with all its envelopes, I separated the amnion as well as its humour which it contained so that the humour of the allantois did not flow out. On its surface the amnion had many small glands almost white or small cavities scattered here and there. It actually contained salty humour. The inner surface of the amnion was full of white small firm points fairly extended each of which receives obvious blood vessels. They seem more numerous about the umbilicus. At the boundary of the amnion with the skin a kind of ring is observed.

In the choroid, besides larger glands constituting the other part of the cotyledons, small glands are scattered over the entire outer tunica. They are resolved into minute parts similar to the tiniest sand but quite abundant. Some filaments seemed to reach each of them. They were unquestionably narrow capillaries of blood vessels. Shall I not believe that other glands corresponding to these are found in the inner tunica of the uterus as is observed in the cotyledons? Certainly over the inner surface of the uterus there were many yellow granules, also many red vessels but, beneath the tunica, not a few corpuscles similar to conglomerate glands were also observed.

About the extremity of the allantois there is a kind of knot hanging down outside the chorion.

In the abdomen of the foetus there was much blood effused from vessels. The liver was nibbled almost everywhere, the gall-bladder was white and the bile itself was white, none the less bitter.

The spinal cord was thicker about the loins and the neck, thinner in the middle of its tract.

In a Third Cow's Uterus
In the vagina as well as in the cervix, fairly sticky matter was observed, similar to the white of egg which became more tenacious after some days of incubation. The foetus occupied one horn. In the other there was nothing else than the continuation of the allantois with the chorion, extending almost to the extremity in which there was white and turbid humour. Aqueous vesicles were seen in the testicles. Two others were attached to the gaping duct of the uterine tubes into the uterus. The inflated urinary bladder swells up the allantois.

In a Fourth Cow's Uterus
The foetus was as large as a small dog. The diameter of the larger cotyledons was four finger breadths long. In the portions of these attached to the uterus large gaps remained after the roots of the portions attached to the chorion had been removed. Although the colour of the roots verged on the red because of the blood vessels which they contain, they were pulled out without effusion of blood. This shows that the blood vessels of the mother are not continued into the blood vessels of the foetus. In the foetus itself, after dissection, the taste of the bile was not bitter, its colour was similar to washed flesh.

In a Fifth Cow's Uterus
The chorion not only filled the horns with its extremities but had also a fairly long filament on both sides extending almost over the whole uterine tube with white matter hardened into a globular mass. Besides the humour of the

amnion and that of the allantois, a third one appeared to be present enclosed in a peculiar membrane.

White spots were seen over the outer surface of the chorion as if they were some crusts which would be attached to the uterus.

After the uterus had been cooked, not only the cotyledons but also all the inner surface of the uterus was porous. Perhaps the difference between the cotyledons and the remaining surface of the uterus is that more copious matter exuding in the former progressively produces a bigger mass of parenchyma by connecting tubercles. Scantier crust only covers the surface remaining outside these cotyledons.

The fleshy parts of motive fibres in the substance of the uterus very often intersected each other. In these fibres I could determine nothing ordained except for a triple duct. They were directed towards the testicles, the wings, the horns and the tube. Many also seemed to end in its outer membrane.

The orifice of the uterine tube facing the testicle is similar to the mouth of the oviduct which faces the ovary in birds. Fleshy fibres spread out outside the straits of the tubes. They pour out all around but do not all reach the extreme brim of the expanded orifice.

After cooking the testicles, I examined the eggs which were many in one, fewer in the other, all hardened like albumen. Where fewer were found, there were many particles glittering with vivid red and some substance fairly large, yellowish, through the middle of which a filament was led up to the surface of the testicle. At its extreme end a point was seen as if it were a hole.

In a Sheep's Uterus

There was nothing else than protuberances. When dissected these showed obvious vessels. There were, however, protuberances covered everywhere by a membrane. In the middle of the flat surface of the protuberances some almost red colour was seen but no quantity of humour worth noticing.

In Another Sheep's Uterus

The cervix of the uterus was quite narrow and made harsh by various small eminences here and there and was impassable by a probe. The outermost surface of the chorion tainted in various whites as if in small lines which, although they were quite irregular in many places, nevertheless in others extending along their length seemed to include some spaces almost round in their interstices.

The number of cotyledons was high and their position was not definite. In the middle, vessels of the chorion were attached to them. By compressing them in the confines of the uterus, whole glands attached to the chorion came forth from small cavities adhering to the uterus. A substance similar to these glands was found in the small cavities themselves.

Then with a hand raised, I tackled the separation of the chorion in which I happily succeeded although the amnion parted from the centre of the cotyledons rather more uneasily. In the allantois I recognized some very thin vessels. The water of the allantois was almost tasteless, that of the amnion a little salty.

When the foetus was dissected, water flowed out from the abdomen. The spleen was small and almost round. The gall-bladder was attached to the liver on which lay another very thin vesicle movable hither and thither. In each of the two dissected vesicles the humour was white and hardly bitter to the taste. In the stomach the water was limpid and hardly tasted salt. In the smaller intestines green colour was observed. In the larger intestines there were filthy excrements.

In a Third Sheep's Uterus
The uterus was divided in two parts by an obvious intermediate septum. In the chorion thus pulled out I counted hundreds of cotyledons. Three foetuses were covered by a common chorion but they were separated from each other by an amnion and an allantois.

I opened one of them and, after having inflated the urinary bladder, I showed that a path opens in the proper allantois. Among the foetuses one was black since an obvious small tuft of black hair was preparing its exit through the skin. All had already a beard. The bitterness of the bile was not worth noticing.

In a Bitch Which Had Died by Itself
There were several tumours in the left groin. After the skin was dissected there, the peritoneum appeared to be dilated beneath the extremity of the obliquus internus muscle like a bag to the magnitude of a fist so that, however, it seemed that the second process of the peritoneum had a petty

addition of this shape. A portion of the intestine constituted the smaller part of the tumour. The larger part was formed by the left horn of the uterus with

membranes intertwined with much fat.

These membranes had a singular shape. On the right side there was also a tubercle which when dissected contained the membranous ligaments of the uterus all with their normal colour. But on the left side the outer and inner aspects of the tumour were corrupted by gangrene at many places.

From the testicles of the uterus up to the lower rib a kind of long and straight muscle seemed to rise. As I attempted at separating this from the subjacent parts, I separated the diaphragm together. A very delicate vessel, I do not know of what kind, descended to it straight from the diaphragm and was divided into ramifications.

In the Uterus of a Live Doe-Rabbit

I saw a movement of the uterus which stretched itself out, contracted, became longer even if it seemed to me that I had prevented the afflux of blood in these parts by ligating the aorta at about the middle of the spine. I also saw that the horns moved but I mostly wondered that near the bladder fibres destined to feeling moved manifestly. They did not belong to the horns nor to the uterus but were like some ligaments.

In Another Doe-Rabbit's Uterus

Besides many vesicles full of aqueous humour in the substance of the testicles, three other vesicles covered by a common membrane were near the orifice of the other uterine tube. One alone similar to this one was found near the orifice of the other uterine tube. The tubes themselves were quite open and expanded.

In a She-Bear

The testicles consisted of several whitish granules like ovaries of fishes. The open extremities of the uterine tubes enclosed these from everywhere so that no more than a very small hole was open from there in the abdomen. Once this hole was dilated, the testicles fell out spontaneously.

In a Hedgehog

Two kinds of motive fibres are directed towards the uterine tubes. One has its opposite extremity above, beneath the kidney. I wonder whether it is in the first rib or only in the peritoneum. The other has its opposite extremity below towards the groin. This is called the round ligament. Its extremity continued into the uterine tube seems to scatter its fibres through all the horn up to the concourse of this with the opposite horn. The upper fibres towards the extremities of the horns seem to collect in a white knot from which fibres delicately scattered leave towards the bow-like course of a blood vessel nourishing the eggs. It is not certain whether these fibres are continued into the motive fibres of the uterine tubes. They can act either together and thus keep the tube as if it were tensed or apart and then pull either upwards or downwards. Are they not set in motion in the conception, or in the delivery, or outside these periods, or in all these periods? About the extremities of the horns there is a vesicle inside which one sees on one hand eggs, on the other hand a cartilaginous duct bent into a circle, not closed on itself, however. Where there is no circle, there is a small opening. From a vesicle in the abdomen in the vicinity of this opening the uterine tube opens inside a vesicle near which there is a delicate white tubercle which perhaps prevents the eggs from falling outside the vesicle into the abdomen, and makes their entering the tubes easier. All the eggs are arranged in order around a blood vessel forming a circle to which they freely adhere through small vessels like in the ovaries of birds. Here more blood vessels are directed towards the uterine tubes, the horns and the uterus than to the ovary. Thus this happens because more matter arrives to the egg at these places than when it is attached to the ovary.

Situated beneath fibres to move the skin by an admirable artifice, the udders occupy the region of the chest and of the abdomen on both sides. On each side they constitute one glandular body although they form five equal nipples.

Between the muscular fibres of the skin and the other muscles as well as between the conglomerate milky glands, there were more conglomerate glands turning red together beneath the milky glands in the chest.

I wondered that the urinary bladder was so swollen that, bigger than a chicken egg, it pushed the intestines upwards and, close to the gall-bladder, it took from this a green taint since, however, all urine flowed out under slight compression after the death.

In a Doe-Hare

There were two substances in the testicles, both grey but one verging more to the dark, the other to the white. In this substance many round spots were seen, very similar to those which in chickens constitute the beginnings of eggs at the base of the ovary. The open uterine tubes lay on the testicles.

In a Salamander

On both sides there was an ovary full of eggs. An oblong yellowish body adhered to this ovary. In the same way there was an oviduct on each side. They were not joined at any extremity. In them there were eggs concealed in some humour.

In a Tortoise

Many eggs adhered to the fringe of a membrane. The smallest of them were white, the middle ones turning from white to yellow, and the biggest ones glittered with a saffron yellow colour. In the different eggs like a gap the containing membrane gaped in a round space. This membrane otherwise in chickens leaves an oblong line deprived of vessels. Beneath the membranes of the ovary on both sides a white body was adjacent to the vena cava.

The two oviducts join in a common orifice. In each oviduct two eggs were adherent, coated by a white and already hard cortex.

In Women

The first in whom I remembered having dissected the testicles was said to be a quinquegenaria. There were many vesicles inside her testicles. The cervix of the uterus when compressed exuded serous humour. In the round ligament of the uterus fibres similar to motive fibres seemed to leave towards the pubis.

In two others as old as the first, I saw these peculiarities: in one the testicles, besides some calculous granules, contained two calculi, one round but very small, the other bigger but different from many tubercles; in the other woman, besides other usual things, there was one calculus similar to the tuberous calculus of the first, not by its substance inasmuch as this was normal, but by its shape.

In a younger woman, eggs were numerous and round, not all of the same size, but all difficult to separate because of the solidity of the substance in which they were contained.

When examining more closely the round ligament of the uterus in another, I found blood vessels, nerves and fleshy fibres continuing in the substance of the uterus. The extreme edges of the uterine tube abounded in blood vessels over all their length, running forth almost like in the gills of fishes. Inside the uterine tubes fibres run forth over their length. Blowing from a tube into the uterus found an open passage.

2.26 Observations Concerning the Eggs of Viviparous Animals

Hound-Fish

In a *Galeus lævis*, a fish of Italy, three foetuses were contained in each oviduct, all of the same size. They did not withdraw from each other as usually happens in other animals but, set near each other, they turned their heads forwards.

Each foetus had its membrane which could be taken for an amnion since, like an amnion, it closely surrounded the foetus floating in limpid humour. It was, however, different from an amnion in that it was attached to the placenta, which would be characteristic of a chorion. I wonder whether the tunica enclosing the entire foetus together in each oviduct was a chorion or whether it was actually the inner membrane of the oviduct. In one foetus, substance similar to egg adhered to the lowest part of the amnion. From the tunica made of this substance, a rough line extended to the amnion. This line was not dissimilar from a vessel which spontaneously folded up as described, once the liquid of the amnion had been sucked out. ⋀⋁⋀ There was one placenta only, and this one small, for each foetus, rubicund, adhered to the oviduct around its inferior orifice, and the overspread membrane formed a cavity.

OPH 26, vol. II, 169–179: The research on which *Ova viviparorum spectantes observationes* is based were, like those which form the basis of the previous treatise, OPH XXV, made in direct continuation of the observations which Steno had made on sharks and which had been published in 1667 (XXIII and XXIV). This treatise was first published in *Th. Bartholini Acta Hafniens.* vol. II., Hafniae, 1675, as No. LXXXIX, pp. 219–232 and its last passage from *Qvanta divinae sapientiae* (vol. I p. 178) to the end, was obviously added at this later period.

© Springer-Verlag GmbH Germany, part of Springer Nature 2018
T. Kardel and P. Maquet, *Nicolaus Steno*,
https://doi.org/10.1007/978-3-662-55047-2_37

The umbilical vessels penetrated the abdomen of the foetus through a cleft below the region of the diaphragm between the two anterior wings. I observed their course by following in one of them bubbles of air distinct from the accompanying liquids. These bubbles, pushed further, entered the intestine. In another foetus, the intestine being inflated, while I shook the foetus violently, I soon opened a way for the air right on into the placenta. This showed that a small vessel which is not a blood vessel is comprised in the umbilical vessels. One of its extremities continued into the snail-like intestine hidden inside the abdomen. The other adhered to the placenta where the upper surface of the placenta overspread in a thin membrane forms a cavity. It is evident that the liquid contained in this small vessel is transmitted from the placenta into the intestine since the orifice by which it gapes in the intestine is similar to the orifice of the pancreatic duct through which the pancreas discharges its contents into the intestine. If the passage had been from the intestine into the placenta, the foul humour of the intestine would have displayed a manifest sign of that in the canal itself. The same small vessel was known by *Belon* as I learnt from *Aldrovandi* but, if I am not mistaken, it was conformed otherwise. He indeed says that a small bag is contained in the abdomen which has to be filled with an nutriment like yolk, brought there via this path. This small bag must be different and from the stomach and from the intestines since he describes in the same fish an empty stomach and an intestine full of excrements. From the structure of the vessels it appears that in this kind of fishes the nutriment is poured forth in the intestine from the cavity of the placenta, as it is from the yolk in birds, as long as nutriment is supplied to the foetus from the humours of the mother. But similar water seen in the surrounding stomach shows that the same foetus is nourished also through the mouth.

For more clarity's sake I decided to add a figure of this canal.

At the sides of the anus there were two nipples in which blood serum similar to that which was contained in the abdomen was secreted. Looking for their duct I found that this is nothing else than holes through which liquids effused in the abdomen could pour out. Who knows whether other animals have not something analogous to the said orifices? Certainly the abdomen is hardly ever found empty of humour even if it is opened in a live animal, and unless one says that this humour transpires through the substance in the middle of the body, one must admit that there are paths deriving it in the bladder, the intestines or in other places.

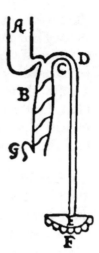

A. Stomach.
B. Snail-like intestine.
C. Insertion of the canal in the intestine.
D. Canal itself.
E. Insertion of the canal in the receptacle of the placenta.
F. Surface of the placenta which adheres to the oviduct.
G. Coecum.

The cavity which, concealed behind the eyes in the cartilaginous skull, undoubtedly serves to hearing. It was very beautiful. Three circles had been carved in the skull. One was almost horizontal, the other two were perpendicular to the first and to each other. This anfractuous duct contained a cartilaginous canal which, expanded twice or thrice in the shape of a round receptacle, received in these places fibrils from the auditory nerves. Another canal softer than this solid canal was enclosed. They were both full of the most limpid water. I will deserve indulgence if I present nothing here on the function and mode of action of this canal by which the fishes hear since I do not even know how we hear. About the structure let me add that the canal is twisted in a triple circle which was cartilaginous in this fish being carved in solid bone in men and quadrupeds. In birds the bony canal is surrounded everywhere by cancellous bone. The anatomists call that cavity the labyrinth. At the same place of the skull there was another larger cavity from which soft and white substance was pulled out. This substance hardened in air and did not suffer any damage from fire. It is very likely that calculi from this matter become petrified in the head of fishes. In the *Description of a Dogfish* I described what else I observed in the present fish, i.e. vessels preparing unctuous humour for the skin, nostrils, muscles of the eyes and a cartilaginous cylinder distinct from the optic nerves, the softness of the teeth growing up from below the other teeth if one is allowed to call other teeth a roughness of the jaws which resembled a file. In the brain there was hardly any difference except for the volume which here, in relation to the size of the body, was much bigger than it is usually in fishes. Neither were the small fishes collected in its stomach different from the food of another fish: I wondered that some of these small fishes were deprived of skin, fins and tail and even of scales as if somebody had attacked them deliberately. This was the clearest testimony of the efficacy of the dissolving fluid. It showed very reliably that this had not been achieved by the roughness of the jaws. The eggs adhered to the ovaries externally. They were not enclosed in a small bag like in another fish. The oviduct, for the rest similar to another oviduct, had a peculiarity: above the body which people call udders it was full of blood serum. Perhaps for this animal also a flow of lochia follows delivery.

Spiny Dogfish[1]

In the *spiny dogfish* the back is armed with two stings but unequal. The larger is the one which is closer to the tail and the fin nearest the tail takes up each of the two stings. The tail similar to a scalene triangle has a base longer than its lower side, shorter than its upper side. In the base a quite deep fold is carved. In the downwards part of the body two pairs of fins unfold themselves.

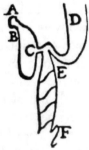

The liver is divided into two lobes. The oblong gall-bladder ends in a sharp tip. The bodies of the oviducts with their funnel and their lower orifice are similar to those which I observed when dissecting a dogfish. The open gap of the funnel showed many wrinkles of the fibres, from which I should very easily believe that it is muscular. In both ovaries there were many minute eggs containing milky humour. There were also four other eggs certainly as big as, if not bigger than, chicken eggs. All were yellow like yolk except for one which was quite white like milk and was rough from several wrinkles. Each ovary had a peculiar membrane containing all the eggs as if they were enclosed in a purse. Of the biggest eggs, three were contained in the right side, the fourth on the left side. In one of the oviducts there was one foetus, in the other, two. The heads of all three were turned towards the orifice of the uterus whereto the caudal extremity was also bent back, the brevity of the

---

[1] *Spina Piscis, Acanthias vulgaris.*

place preventing the fish from extending straight. I wondered that the humour in which the foetus floated was contained in the tunicae only of the oviducts, nothing being conspicuous there, neither peculiar membrane nor placenta. Soon, by examining more closely these small fishes, I observed beneath the region of the heart a whitish tubercle A, hollow but empty. When the abdomen of the small fishes was dissected, an oblong vesicle C came forth continuing into the described tubercle through a small canal B. The upper part of the snail-like intestine F was attached to the middle of the vesicle itself, the fairly large orifice E appearing by which the vesicle gaped in the intestine. When the vesicle was opened, humour from white to yellowish flowed out. The intestine already tainted by bile was full of this humour. In their stomach D the humour was similar to the humour contained in the oviduct, in which they floated, so that this description showed us also that the fishes, like the birds, receive their nutriment sometimes through the mouth, sometimes through the intestines.

Blood vessels delicately swinging in sinuous folds ran over the length of the oviducts on their inner surface. The alternate angles of their curves could be raised above the tunica.

Torpedo[2]

In a torpedo very small ovaries adhered to the posterior aspect of the liver on both sides. One of their extremities was close to the funnel. There was a great number of eggs. The bigger of them glittered with yellow colour. The smaller were diaphanous like water.

In another bigger torpedo, one of the russet ones, which weighed nine pounds, the ovaries were stuffed with many eggs. Most of these were either white or yellowish, few were quite aqueous. Beneath the ovaries, which lay beside the liver, there was a cavity which I should think is a portion of the vena cava coming forth outside the liver. There were six eggs in the right oviduct, eight in the left. The eggs did not adhere either to the oviduct or to each other. All floated in a humour which resembled partly the vitreous body of the eye, partly the most limpid water. The colour of the eggs in the oviduct was an intense green at some places. Others were a pale green. Their consistency was soft like fresh cheese. Their shape, no longer round but flattened

---

[2]Electric Ray.

by their reciprocal compression due to their mutual contact, was not far from the shape of a torpedo.

I should think that the foetuses are formed and grow in the oviduct of the torpedo although they adhere nowhere to the oviduct. A soft, white, relatively thick body was contained between the inner and the outer tunica of the stomach. Inside the stomach certain nipples protruded. They seem to be vessels of dissolving humour. Many vessels besmearing the surface of the torpedo with unctuous humour, arranged in the most beautiful order, extended from the head to the posterior aspect, by which the contained humour is more easily expressed. Worth noticing were fibres thicker than the thickest feathers of geese, soft and white, which, situated on both sides perpendicularly between the superior and the inferior tunica of the body, were separated from each other by transverse fibres and received obvious nerves arriving transversely so that I should not hesitate to say that they are motive fibres. They occupied a position which is between the lateral fins, and the gills and the head down to the region of the abdomen. In almost this shape, the anterior ones were shorter than the posterior ones. Cooked, they displayed very soft flesh. On a live animal this place was indeed soft to touch. But if at the same time that one touched this place, the animal contracted itself, the contraction was felt in the extremities of the fingers with a kind of pricking pain rising through the arm and which continued for some time. When the animal was quiet nothing was felt. In the same way, nothing was felt when the animal was dead.

The substance of the lens was triple. This is the figure of the uvea[3] which is able to close the pupil. There was a hard cartilaginous cylinder, also adhering firmly to the globe of the eye.

Argentina[4]

In the Argentina the ratio of one eye to the brain was 19 to 2. The abdomen was divided into two cavities. One of them was close to the dorsal spine and contained the kidneys and a long bladder of air containing inside several bodies similar to those which are seen in the vesicle of air in eels. The other cavity contained the stomach, the spleen, the liver, two oblong and white bodies which contained a considerable cavity opening in the rectum. Several animalcules were hidden inside the abdomen about the end of the rectum, similar to the gaping shell-fishes described by *Fabius Columna* except that they were deprived of testicles.

---

[3]Membrane of the eye comprising the iris, the ciliary body and the choroid.

[4]*Argentina sphyræna.*

Vipers

Vipers have two ovaries. Once I observed that the right one was higher than the left one. I did not trouble myself about whether this was so also in the others. In this viper, if I remember well, there were three eggs on the left side, five on the right. Something from the ovaries seemed to be a peculiar membrane in the form of a cylinder or of a portion resected from the intestine, closed on both sides to the observation. Eggs adhered to its inside.

The two oviducts also met in a common oblong orifice, situated between the rectum and the back. Towards the inside, they remain distinct by their orifices which are quite small and close to the extremity of the ovaries. In the right oviduct of one viper the orifice seemed to be expanded inside into a small bag which surrounded the extremity of the ovary. In the chicken a double oviduct has been found once although there was only one ovary on the left side.

She-Wolf

In a she-wolf the testicles contained many eggs, each of them conspicuous by its membranes and blood vessels just as it is usual in ovaries of birds. All were close to each other and were surrounded by some substance not completely membranous but approaching the nature of glands like the base from where the eggs hang from the ovary in birds. The cavity in which the testicles were enclosed was considerable and opened in the abdomen only through a small opening. The orifice of the uterine tube extended above the testicle. Through this orifice a whitish humour was found not only over all the length of the tube but also in the horn of the uterus about the inner orifice. The remaining part of the uterus being empty, the orifice of the uterine tube into the uterus was carved in the middle of a nipple. This is reliable evidence that humours pass uneasily from the uterus into the tube, but easily from the tube into the uterus. In the vagina there was a considerable tubercle in which the radii of a pentagonal figure were seen. The orifice of the uterus was found at the concurring point of the radii.

Hind

In a hind one-inch-high cotyledons presented with very abundant small canals, different in number as well as in position from those which were seen in a cow's uterus. As I separated the chorion, from there I took away roots fastened to the said cotyledons of the chorion without effusion of one drop of blood. Two substances appeared to the one who examined more closely the different roots, a whitish and transparent external one, like gelatine which easily suffered to be scraped away, a red internal one, extended over the length of the roots, continuing in blood vessels which penetrated the body of the foetus itself which by then was already completed with all its limbs and about one cubit long. I managed to separate the chorion so as to show the amnion and the allantois distinct from each other, without effusion of humours, and this more precisely to find the ratio of the humours. The ratio of the humours contained in the amnion to that contained in the allantois was of 30 to 13. The limpid and slightly salty humour of the amnion whitened when spirit of vinegar was poured in. The same spirit of vinegar changed from white to yellow the humour of the allantois which was sweet and white, with subsequent manifest precipitation. Rectified spirit of wine returned the colour of the liquid of the amnion which spirit of vinegar had modified and impeded precipitation every time it had been mixed with this liquid before spirit of vinegar was poured in. In the allantois oblong and whitish solid bodies were found, similar to curdled milk but more tenaciously compact almost as a polypus of the heart usually is. There was a path open to air from the bladder into the allantois. Very abundant white icicles hanging from it added a delicate decoration to the inner surface of the amnion. They were of different lengths; some of them had a length of a finger breadth. Blowing in the uterine tube made the uterus swell. The testicles were full of eggs. Motive fibres extended from the uterus or from its tubes as far as to the testicles. An oblong vesicle turgid with the most limpid humour clung to the membranes outside the testicle.

Sow
The uterus of a wild sow contained four foetuses. Among these the one which was the nearest the cervix had a bifurcated extremity of the chorion, one part of it rising in one horn, the other part falling towards the orifice of the uterus. There were four membranes for every foetus. The *first* was a chorion which was all wrinkled in creases similar to rings and found in the inner tunica of the uterus other creases which matched them. Infinity of white oval spots of the size of aniseed, which in some places touched each other and elsewhere were at a very small distance from each other, were disseminated over all the surface of the chorion. Something red was prominent in the middle of all the spots. Its spots in the uterus corresponded to the spots of the chorion and were similar to those in position and in size. At some places the spots of the chorion quite firmly adhered to the spots of the uterus so that I hardly

doubted that they ought to be considered as cotyledons. The remainder of the chorion as well as of the uterus was a rough surface with countless small cavities. The *second* membrane was the allantois the humour of which was scarce and filthy with yellow excrements. The *third* membrane was the amnion the humour of which was also scarce but limpid and a little gelatinous, full of yellow globules adhering to each other. The *fourth* membrane tightly surrounded from everywhere the foetus then already coated with hair, beautifully enveloping also the different limbs, the nails, the tails, and the ears. Only the mouth with the holes of the nostrils, the circumference of the umbilicus and the genitals was left bare. Very long hair protruded from the same membrane for the eyelashes and the eyebrows. From a new experiment it appeared that the foetus in the uterus is fed through the mouth since not only a humour similar to the humour of the amnion was found in their stomach but also solid globules similar in size, shape and colour to those which were in the amnion were observed there. Globules which remained in the rectum, very similar to the afore-mentioned ones, proved that the latter had been excrements of the rectum. What shall we admire more than the fact that excrements of the skin serve to nutrition if the excrements themselves of the anus are returned to the stomach? Obviously what is secreted from the internal fluids into the external fluid through the capillary vessels in the first days after the formation travels the same circle every time it is received anew in the body, as either the viscosity of the juice or the weakness of the heat leave there much crude things and as it is to be resolved in subtler parts with time, until it is made more pungent. Then it will be put aside sooner or later in the external fluid or the liquid of the amnion. It is rejected by the animal and stimulates the animal to get out that which affects the feeling too much.

She-Ass

The testicles of an ass were as big as a chicken egg being shaped like a kidney. We counted more than 20 eggs in the cavity of one of them, equal to and more numerous than those of a wolf, but spherical. I separated them intact from the testicles and from each other. ⬭ After one of the testicles which hid them away very deeply had been cooked, the eggs acquired the colour, consistence and taste of cooked albumen except that slight pungency was also perceived. The other eggs closest to the uterine tube which stood out as if they were on a bare surface remained fluid. ▽ The remainder of the substance of the testicle appeared to me to be fibrous as the nipples of the kidneys are, if you exclude a red triangular body which contained many small eggs. In the vicinity of the superficial eggs (if one is allowed to call so those which almost protrude in the surface) there were certain petty additions not dissimilar from the cups of eggs in chickens, which contract on themselves after the eggs have fallen out. The orifice of the uterine tube was above a part of the testicle where bare eggs stood up, and its capacity was smaller than that of a wolf.

She-Mule

I examined also the testicles removed from two mules. In the first the testicles were quite small but provided with abundant blood vessels. When they were dissected, no trace of eggs was found in them. The uterine tube was fairly long and folded into many turns. The external orifice close to the testicles was fairly open, the internal one was so tightly shut that it would not even allow air to pass into the uterus, although in a horn of the uterus a nipple was manifest which in other animals is found gaping. The inner tunica of the uterus was rough with many wrinkles but wider and inclined on the surface of the uterus. The orifice of the uterus was less tight although annular protu-berances which usually close this orifice were not absent.

In the second mule the testicles were big, like in an ass. Their concave part contained few small eggs, **O** and, in one of them, an egg of this magnitude, full of yellowish liquid. **( )** In the other there was an oblong dark red body with a structure very similar to that of a conglomerate gland. Its extremity extending towards the convex part of the testicle contained a cavity. The other extremity protruded outside the concave part of the testicle. All this body was free on all sides, adhering to the testicle by neither vessels nor filaments. The uterus, red inside, was swollen and all its vessels were turgid with blood.

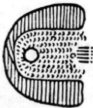

So a mule will be able to give birth without a miracle if there is a way ahead for the eggs in the testicles. Though it would not be without miracle, if, according to what was said, at the same time some important event would happen in politics and the birth-giving of a mule. But several reasons of the sterility in these animals can be given such as the absence of eggs or their too deep positions inside the testicles, or a substance unable of conceiving eggs,

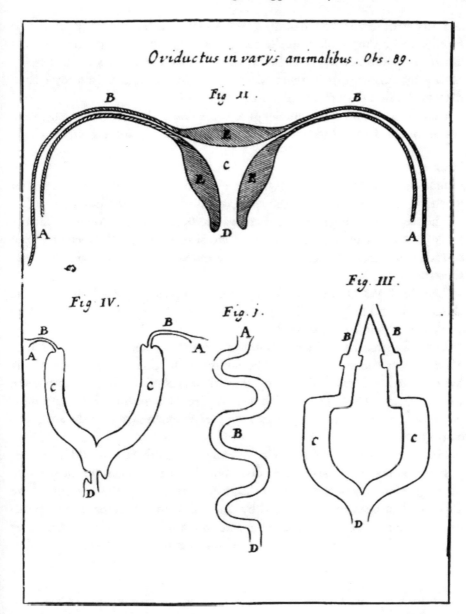

Fig. 2.26.1   Fig. I Oviduct of a chicken single for both orifice A, D and for the cavity B. Fig. II Oviduct of a woman, double BB for the internal orifice A, A and the part of the cavity close to this, single for the remainder of the cavity C and the external orifice D. In this figure, the cavity C as well as the thickness of the substance E, E, E and the length of the canals are showed with their true dimension. Fig. III Oviduct of the Galeus lævis, simple for both orifices A, D, double for the whole cavity BC, BC. Fig. IV Oviduct of a she-wolf, double for the internal orifice A, A and almost all the cavity BC, BC, single for the external orifice D

and many others reckoned among the usual causes of sterility. If it was possible to find more often in the testicles a body similar to the one which I described in the last mule, I should suspect that there is a kind of females among the quadrupeds in the testicles of which the eggs are already coated then in the first days of the placenta.

I do not mention the guinea-pigs and the she-badger in which eggs were also recorded. In the badger these are many, similar to minute grains.

Deer

In four deer I examined the eggs of the testicles. The first one was young, not pregnant. There were many whitish eggs full of transparent humour.

The second was pregnant but there was no beginning of a foetus. Only the inner parts of the uterus were swollen. One of its testicles contained an egg of considerable size. Part of the testicle seemed to be made of glandular substances.

The third was old and carried a foetus completely formed although the chorion adhered only very slightly to the uterus. The eggs of the testicles were smaller here than in the second.

In the fourth each of the two horns of the uterus enclosed whitish humour which, when cooked, hardened like albumen. In this way the eggs of the testicles hardened and whitened together when cooked. In these testicles there were some cavities empty of all humour. Others were full of a diaphanous humour like in the horn, and here white parenchymatous substance was conspicuous.

How much evidence of the divine wisdom and goodness it would have been possible to draw from the different dissections if the time spent for the pleasure of the present object and to vain discussion concerning the honour of the discovery had been totally devoted by the author to contemplate such beautiful and skilful objects! Because we assent remaining at the lowest matters, we shut off ourselves to enter the path to higher matters (Fig. 2.26.1).

2.27 The Prodromus to a Dissertation on a Solid Naturally Contained Within a Solid

OPH XXVII, *DE SOLIDO INTRA SOLIDUM NATURALITER CONTENTO DISSERTATIONIS PRODROMUS* was published in Florence in 1669. The *Prodromus* was translated into English by Henry Oldenburg in 1671 and by John Garrett Winter in 1916. It was translated into Danish by August Krogh and Vilhelm Maar in 1902. Scherz's English edition in *Steno—Geological Papers*, 133 ff., was translated by Alex J. Pollock, partly assisted by James G. Gordon.

The editors of the present edition have changed "solids … within solids" in the book title of Scherz' English edition into "a solid … within a solid", which is similar to the translation by John Garret Winter (Wi), since *De solido intra solidum* is singular case. Meanwhile, the "Prodromus on solids" remains a valid expression concerning two items, the containing and the contained solid.

© Springer-Verlag GmbH Germany, part of Springer Nature 2018
T. Kardel and P. Maquet, *Nicolaus Steno*,
https://doi.org/10.1007/978-3-662-55047-2_38

Dedicated to the Most Serene

Ferdinand II

Grand Duke of Tuscany

Florence.

From the Printing Shop under the Sign of the STAR. 1669 With permission of
the Superiors (Fig. 2.27.1)

2.27.1 Most Serene Grand Duke,[1]

While travellers in unknown territories hasten over rough mountain tracks
towards a city on a mountain top, it often happens that they judge the city, at
first sight, to be close to them; constantly, numerous twists and turnings
along the route delay their hope of arrival to the point of weariness, for they
see only the nearest peaks; in fact, those things hidden by the said peaks, the
heights of hills, the depths of valleys, or the levels of plains, whatever they
may be, far exceed their conjectures, and they, deceiving themselves, estimate
the intervening distances from their own desires.

Nor is it in anyway different for those who move towards true knowledge
through experiments, for as soon as certain tokens of an unknown truth have
become clear to them, they believe that the whole matter will be revealed at
once; they can never estimate properly the time that is required to unravel the
closely linked series of difficulties that, emerging gradually, as if from hiding,

---

[1]Scherz: Almost all of our Latin text has been taken from the first edition of *Nicolai Stenonis De solido
intra solidum naturaliter contento Dissertationis Prodromus* (Florentiae 1669). This edition (size 18,5 x
24,5) contains the treatise itself starting at Serenissime Magne Dux on page 1 to *singula observavi* on page
76; on pages 77 and 78 the approbation, and on an inserted last leaf some corrections also, after which
follows the plate sheet and the *Explicatio Figurarum*, both size 24 x 34,5 cm. This text is also published
in the edition of Maar in OPH 2, 183-224. To-day we know also of the printer's manuscript in Florence
(Bibl. Naz. Centr., Gal. 291, fol. 1^r to 26^v), on whose first page is the title, and beneath, written
by Viviani in person: Questo fù stampato sotto la mia cura (see Fig. 1.6.4). Viviani has also written
the whole of the text: the text itself takes 45 pages, after which follows on one of the last pages
the *Explicatio Figurarum*, though without the figures, and the approbations are missing as well. It is a neat
copy almost without corrections, at any rate with none by Niels Stensen himself, and its text corresponds
almost completely to the printed edition. Footnotes refer to a couple of deviations; the much more
numerous breaks between the various paragraphs in our book are quite naturally taken from the printed
edition. The translation of the Prodromus is by Alex J. Pollock, of the Notes by Maria Grant,
both for some part assisted by James G. Gordon. * Scherz' sources are not checked nor included
in the Bibliography of this edition.

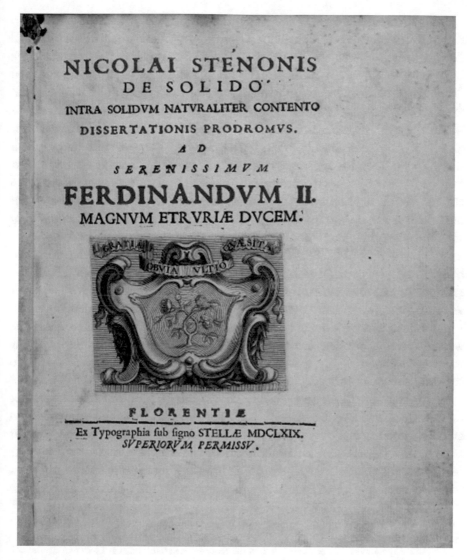

Fig. 2.27.1   The Prodromus on a solid contained within a Solid, Florence 1669

impede those who are hastening to their goal, by constantly throwing fresh obstacles into their path. Early work merely reveals certain common and generally known difficulties; what is included in these, now falsehoods to be overturned, now truths to be established, now dark areas to be lit, now unknown facts to be brought forth, will rarely be detected by anyone before he has been led to them by the thread of his research.

Democritus made use of a good example with a well in which,[2] since both the number and size of the hidden springs leave the quantity of material flowing into the well in doubt, a proper estimate of the work and time involved in emptying it could scarcely be made except by draining the well dry. Therefore, do not wonder, Most Serene Prince, that I should have said for a whole year, and what is more, almost daily, that the investigation, for which the teeth of the Carcharias shark provided the opportunity,[3] was near to completion. For, having on some occasions observed soils from which shells and other such marine deposits had been dug, having observed that those soils were sediments from a turbid sea and that is was possible to calculate in each place how many times the sea had been turbid there, not only did I make up my mind too hastily, but I also, in addition, gave daring assurances to others that a complete scrutiny on the spot would be a very brief task. But since then, as I investigated more closely[4] both each place and each body, these gave rise to a succession of doubts, indissolubly connected, which assailed me day by day, so that I saw myself more often brought back as it were to the starting gate when I believed I was near to the finishing posts. I might liken these doubts to the heads of the Lernean Hydra, since when one was destroyed, countless others grew out from beneath; at all events, I saw that I was wandering in the kind of labyrinth where the nearer one comes to the exit, the greater the circles in which one walks.

But I shall not linger in excusing my slowness, since you have abundant knowledge, from long experience,[5] of how complex is a matter involving a chain of experiments: but an excuse would indeed be required on the account that, having succeeded with a large part of the said task, I am seeking your permission to return to my homeland, interrupting everything, to pursue a study of anatomy, did I not know that you will not be displeased by the

---

[2]Cf. Diogenes Laertius, *Pyrrhon* (IX p. 72): "We know nothing, as a matter of fact, the truth lives in the depths". In the age of Niels Stensen the bottom of the sea, was often interpreted as a well or a cistern, Wi: [John Garret Winter 1916].

[3]This shark, of about 3.500 lbs. was at the time called Lamia or Carcharodon Rondeletii owing to the description in G. Rondeletii. *Libri de Piscibus Marinis.* (Lugduni 1554) lib. XIII pp. 390-393 (OPH 2, p. 115 ff. EP p. 922 f.). Stensen dissected its head in Florence and wrote the preceding treatise on it: *Canis Carchariae dissectum caput.*

[4]As to the numerous occurrences of marine sediments from the latter and earlier Tertiary period and petrifacts of those ages in Tuscany, compare the detailed descriptions in T. T. Viaggi, in the Elsa-valley 1, p. 201 f. 222; near Livorno 2, p. 461; Volterra 3, p. 11 ff., 5 p. 257 ff.; Valdarno 8 p. 53 ff.; 10 p. 237 ff., and also of Rodolico, *Toscana* p. 60 ff. 79 ff.

[5]Grand Duke Ferdinando II in person took part in the works of the Platonic Academy of 1628 and its successor, the Accademia del Cimento. He invented and constructed instruments and had the first large map of the moon drawn by means of the Galilean telescope (Alfr. v. Reumont, *Geschichte Toscanas seit dem Ende des florentinischen Freistaats,* 1-2 Gotha 1876/77, 1. p. 557.

obedience of a subject for another prince that you would be pleased to have from your own subjects on a similar occasion.[6] My expectation of your willingness in this matter is made more certain by the exceptional kindness with which, in pledging generous assistance towards the development of my studies, you wished to leave me complete freedom to digress however often the occasion might arise. Thus, since I can no longer dare to hope for the necessary time in which to complete the tasks begun, in meeting my promises, I do what by common custom is arranged for those in financial debt; they, when they lack the means to pay their debt in full, pay what they have, so that they will not lose their business; and since I am not able to complete all that was to be shown to you, I shall show you the chief items of what has been completed, so that I may not seem to have broken my word.

I should have been by no means unwilling to put off everything until I was at liberty to perfect the details, after my return to my native land, but for the expectation that the same situation will arise there as has been my experience up till now, insofar as new tasks have always stood in the way of the completion of initial undertakings: while I was intent on finding the number of glands in the whole body,[7] the wonderful structure of the heart dragged me off into an investigation of it[8]; the deaths of my relatives interrupted studies begun on the heart.[9] To take me away from a detailed account of the muscles, a shark of prodigious size was thrown up by your seas,[10] and now that I am wholly dedicated to my present experiments, I am called to other things by he whose command I must obey, under natural laws, and whose will I am urged to obey by the great favours that he has shown to me and my family.[11]

---

[6]An invitation for his return to Denmark, however, had not yet been issued officially on Apr. 20th 1669, in Venice, or after his return from Hungary on Oct. 27th (EP 1 p. 27, 194, 207 ff.).

[7]When at the beginning of 1662, in the Observationes anatomicae, Stensen published his discoveries about the glandular system, he wrote to Thomas Bartholin on January 9th, 1662: "... if my straitened circumstances at home had not prevented me, I should have continued my investigations of the glands. I wanted, you know, to investigate all their vessels and trace the tracts of the Lymph which is apparent in the various cavities of our body (OPH 1 p. 101 ff.).

[8]On August 26th, 1662 Stensen writes enthusiastically about his observation of a chicken's heart. Soon follows a series of similar observations, and on April 30th, 1663 the aims and methods of his investigation of the heart were clear to him (OPH I p. 118, 123 ff. 158).

[9]Stensen's stepfather Johan Stichman was buried in St. Nicolaj on October 29th, 1663 and in June, 1664 Stensen was to bury his mother there too. Soon after, he left Copenhagen (EP 2 p. 901 ff. OPH I p. 168).

[10]At the time of the huge shark's capture preceding Stensen's geological period he was on the point of editing his *Elements of Myology* and commencing a debate about it with his contemporaries (OPH 2 p. 61 ff.).

[11]The court was regularly supplied by the jeweller Sten Pedersen and Johan Stichman. Stensen's brother in law, Jørgen Carstensen, was clerk of the royal treasury and his family must have felt deeply obliged to the Royal family (EP 2 p. 900 ff. - Scherz, *Vom Wege*, p. 20 f., 36 ff.).

I do not wish to ask anxiously for what purpose all these things happen; I may perhaps attribute to myself things which are due to a higher cause. If prolonged meditation would have added something, as it were of my own, to discoveries not my own, it is certain that if I had remained closely concerned with developing one discovery for a long time, I should have debarred myself from the opportunity of discovering the rest.[12] Not knowing, therefore, what other experiments and studies await me elsewhere, I thought it best to set down here those things concerning a *solid naturally enclosed within a solid,*[13] as a token of my gratitude for the favours I have received, and to provide an opportunity for others, who from their own wish have the use and enjoyment of leisure, to cultivate studies in physics and geography with greater profit.

With regard to the production of a solid naturally enclosed within a solid, I shall first outline briefly the method of the Dissertation, and then describe succinctly the more important matters that occur in it.

I had divided the dissertation itself into four parts,[14] **the first of which,** acting in place of an introduction, shows that the investigation of marine objects found far from the sea is old, pleasant, and useful, but that its true explanation, less doubtful in earlier times, has become exceedingly uncertain in most recent times.[15] Then, having set down the reasons why later investigators departed from the belief of the Ancients, and why, although many excellent books by a large number of writers may be consulted, no one has up to now found a new answer to the controversy, I show, returning at length to you, that following many other things, in part newly discovered, in part freed from old doubts under your patronage, to you is also owed our expectation that the finishing touch will soon be given to this investigation.

In the second part a general problem[16] is solved, upon which depends the explanation of individual difficulties, that is, *given a substance endowed with a*

---

[12]The following utterance reveals, how strongly N. St.'s faith in Divine Providence determined even his research work (cf. his well-known prayer: Sine cujus nutu nec de capite capillus... OTH 2, 542): "Various incidents forced me, when on the advice of my friends I considered everything well prepared and easily discharged, and quite unexpected obstacles suddenly upset the plan I had conceived, whereas other things, which neither I nor anybody else would have expected, suddenly at the same time offered a possibility so that I could not have any reasonable doubts that God was behind (OTH I p. 393).

[13]Mark that Stensen here, even before the words geognosy and geology were common at the end of the 18th century, as shortly and precisely as possible characterises the science and history of the earth by the two principal subjects of enclosing and enclosed bodies.

[14]There are no enhancement of Stensen's division in four parts in the original edition; cf. the table of contents in the Introduction.

[15]Cf. with this the Introduction p. 21 and the notes on Italian authors in Rodolico, *L'Esplorazione* 43 ff.

[16]Here the method of actualism which, through v. Hoff and Lyell, prevailed a century later was first hinted as a principle of geological research. About occasional earlier references in antiquity, as well as at the dawn of our age, to the importance of the forces still at work for an understanding of the past, Kurd v.

certain shape, and produced according to the laws of nature, to find in the substance itself clues disclosing the place and manner of its production. Here, before going on to expound the solution to the problem, I strive to set down the meaning of all the terms in it, so that no school of philosophers may be left in doubt and dispute about them. **The third part** I have allocated to the examination of particular solids enclosed within a solid, according to the laws discovered in solving the problem.

The fourth part describes various conditions of Tuscany not dealt with by historians and writers on things of nature, and proposes a kind of universal deluge that is not rejected by the laws of natural movements.[17]

Indeed, I had begun to develop these in Italian, not only because I knew you would be pleased in consequence, but also to reveal to the illustrious academy[18] which has enrolled me among its members that just as I am least worthy of such an honour, so I am most eager to give evidence of the effort with which I am striving to attain some knowledge of the Tuscan language. Nor am I distressed that the necessity is imposed on me of postponing the said writing, for as my immediate journey promises me fuller knowledge of matters that will be useful in elucidation of the investigation, so the delay offers me more fruitful progress in my study of the language.

With regard to those matters explained by the aforesaid method, it would take a long time to write out all observations and the conclusions drawn from them, therefore sometimes I refer to observations, sometimes to conclusions, whatever may seem best to indicate the chief matters as briefly and as lucidly as possible.

There seem to me two main reasons underlying the fact that in the solution of natural problems not only are many doubts left undecided but also most often the doubts multiply with the number of writers.

The first is that few take it on themselves to examine all those difficulties without whose resolution the solution of the investigation itself is left marred

---

(Footnote 16 continued)

Bülow says: "All this fades before the effort of N. Stensen. He was the first to proceed really geologically i. e. historically in an inductive way, and besides, conscious of his method and therefore "actualistically" (Der Weg der Aktualismus in England, Frankreich und Deutschland. In: *Berichte der Geologischen Gesellschaft* 1960, p. 160 ff. - cf. also Oedum, p. 53 ff., and Hölder, Geologie, p. 481 ff. - See on the principle of actualism also H. Murawski, *Geolog. Wörterbuch*. 5. Aufl. Stuttg. 1963. p. 4 and A. A. G. Schieferdecker, *Geological Nomenclature* (Goringham 1959) Nr. 2664).

[17]In the text we find the usual Latin term: *diluvium universale*, the universal flood as it is described in Gen. 2, 17-20.

[18]The Accademia della Crusca founded in 1582 for the cultivation of the Italian Language and the publication of a Comprehensive Italian Dictionary. Stensen officially became a member on July 16th 1668 (EP I p. 19).

and imperfect. The present investigation is an obvious example of this point. Only one such difficulty troubled the Ancients, namely the way in which marine objects had been left in places far from the sea, nor was the question ever asked whether similar bodies had been produced in places other than the sea.[19] In more recent times the difficulty of the Ancients received less emphasis since almost everyone was busy inquiring into the origin of the said bodies.[20] Those who ascribed them to the sea did their best to show that bodies of such type could not have been produced in any other way, those who attributed them to the land denied that the sea could have covered these places; they all joined in extolling little known powers of nature as capable of producing anything whatsoever.[21] And it may be that a third opinion has sufficient in it to be acceptable, that some of the said bodies may be attributed to the land and some to the sea; yet almost everywhere there is deep silence about the doubt of the Ancients,[22] except that some mention floods and a succession of years of unknown duration, but only incidentally, and, as it were, in dealing with something else. Thus, in order to satisfy the laws of analysis to the best of my ability, I wove and unravelled the web of this investigation many many times, and examined its individual parts until there seemed to me to be left no further difficulty in the reading of authors, nor in the objections of friends, nor in the inspection of sites, that I had not either salved, or about which I had at least decided, from what I had learned up till now, how far a solution was possible.

The first question was whether the *glossopetrae* of Malta had been at one time the teeth of sharks; it was immediately obvious that this was similar to the general question of whether bodies resembling marine bodies which are found far from the sea were produced in the sea in past times.[23] But since

---

[19]In his *Geographia* Strabo argues about the discovery of shells in places far from the sea (C. 49, 50 I, 3, 4).

[20]Possibly Stensen thinks here of the then representatives of the Flood-theory (cf. Zimmermann p. 192 f., and Chapter II *Geological Science in Classical Times*, and Fr. D. Adams, *The Birth and Development of the Geological Sciences'* New York 1954, p. 8 ff.).

[21]Cf. Aristotle's theories on original procreation and its consequences, further Zimmermann 51 f., Hölder p. 359.

[22]Perhaps a reference to G. Agricola (1494-1553). Cf. W. Zimmermann, *Evolution.* Freiburg 1953, p. 187 f. or to B. Palissy († 1590), *Discourses,* p. 16 ff.

[23]Stensen's professor Thomas Bartholin had visited Malta in 1644 and called the sediments of the Cretaceous and the Tertiary ages "a lump of fossils". He does not want to say anything final about the glossopetrae, fossilised shark-teeth also called tonguestones, birds' tongues, swallow tongues, Lamiodontes, and snakes' tongues (OPH 2, 326): "whether they are snakes' tongues as the natives believe, through the curse of the Apostle (Paul) changed into hard stone, or if they were produced by creative nature as a stone concretion in the rock, which Gesner and Boethius tend to believe, or whether in accordance to what Plinius writes, we ought to consider them a kind of lightning-arrows, or on the other hand whether they are fish-teeth which they resemble and which were left behind on the continent after the universal or partial floods." (*Historiarum Anatomicarum et Medicarum rariorum Centuria* V—VI

other bodies are also found on land that resemble those which grow in fresh waters, in the air, and in other fluids, if we grant the earth the power to produce these bodies, we cannot deny it the ability of producing the rest. Thus, it was necessary to extend the investigation to all those bodies dug from the earth that are seen to resemble those bodies that elsewhere are observed growing in fluid; but many other bodies are also found in rocks, endowed with a certain shape; if it be said that they were produced by the power of the place, it is necessary to admit that all the rest were produced by the same power, so at length, I saw that the point had been reached where every solid naturally enclosed within a solid should be examined to determine if it was produced in the same place as it was found, that is, the nature of the place where it is found should be examined and then the nature of the place where it was produced. But indeed, no one will readily determine the place of production who is ignorant of the method of production, and all discussion on the method of production is futile unless we have some certain knowledge of the nature of matter; from this it is clear how many problems must be solved to satisfy one line of inquiry.

The second reason, one which nourishes doubts, seems to me to be that in considering the natural world those things which cannot be determined with certainty are not kept separate from those that can be so determined; as a result, the principal schools of philosophy are reduced to two classes; some, indeed, are prevented by scruples from putting faith even in demonstrations, for fear that the same error exists in them that they often detect in other declarations; others, on the contrary, would by no means show themselves constrained to hold as certain only those things in which people of sound mind and sound perception could express belief, they being of the opinion that all those things are true that seem to them admirable and ingenious. Indeed, the advocates of experiments have rarely had the restraint either to avoid rejecting entirely even the most certain principles of nature or to avoid considering their own self contrived principles as proved.[24] Thus, to avoid this reef also,[25] I decided to press with all my might in physics for what Seneca often urges strongly regarding moral precepts; he states that the best moral precepts are those which are in common use, widely accepted, and

---

(Footnote 23 continued)

Hafniae 1661, Cent. VI, Hist. I p. 193 ff.). Bartholin brought a collection of these tongue-stones with him, which N. St. must certainly have known. (A. Garboe, *Thomas Bartholin* 1, 59 ff.).

[24]It is obvious to think of Descartes in this connection. Yet Stensen as late as 1665 decisively rejected the purely speculative use of his method and his machine-theory of biology (N. Steno's *Lecture on the Anatomy of the Brain* ed. by G. Scherz. Copenhagen 1965 p. 170 ff. further OTH 1, 389 f.).

[25]The MS. and the Ed. Florence 1669 have *evitare*.

which are jointly proclaimed by all from every school, peripatetics, academics, stoics, and cynics[26]; and indeed those principles of nature could not but be best that are in common use, widely accepted, and are considered admissible by all from every school, whether those who are eager for novelty in everything or those who are devoted to the teaching of the past.

Thus, I do not determine whether particles of a natural substance[27] can or cannot undergo change, as its shape can, whether there are or are not minute empty spaces, whether in those particles, in addition to the ability to occupy space and the property of hardness, there may not be something else unknown to us; for these statements are not widely accepted, and it is a feeble argument to deny that there is anything else in a certain thing because I do not observe anything else in it.[28]

But, in truth, I do assert without hesitation that:

1. A natural body is an aggregate of imperceptible particles through which may pass the forces emanating from a magnet, from fire, and sometimes also from light, whatever may be the passageway by which access is found, whether between the particles, in the particles themselves, or both.[29]

---

[26]Particularly as an ethicist on a purely humanistic basis, Seneca influenced both the Christian Middle Ages and the Renaissance through his 124 *Epistolae morales* as well as through his *Naturales quaestiones*. In 117 he says: "Usually we pay much attention to the conviction of all mankind and with us the concordance of everybody is considered a proof of the truth". This concordance is derived from human nature, but also from doctrines consecrated by pious faith (O. Willmann, *Geschichte des Idealismus* 1-3. Braunschweig 1907. 1, 575 f.). In his *Epistolae* (29.11) is a passage the beginning of which reminds us of N. St.'s: *ex omni domo conciamabunt, Peripatetici, Academici, Stoici, Cynici…*

[27]This paragraph on the particulae, which has a rather abrupt beginning was a necessary chapter in a treatise on the nature and the structure of bodies and the important part played by fluids in nature. In the 17th century the interest in atomic theories was encouraged above all by Pierre Gassendi (1592-1665) and his *Animadversiones in decimum Librum Diogenis Laertii* (1649), in which he revived the atomism of Democritos - though modified. N. St.'s notes in the Chaos-MS. contain many excerpts of this work. (fol. 70-74. Vgl. EP 2. p. 907 ff.). Later Descartes had advanced a corpuscuiar-theory in the 3rd part of his *Principia Philosophiae* of 1644, where he distinguished between three kinds of substances. The tendency of the age was to replace philosophical atomism by scientific one, to which Stensen evidently quite consciously confined himself (cf. A. G. M. van Melsen, *Atom gestern und heute*. Freiburg/München 1957. p. 115 ff. 138). Mark also that Stensen speaks about *particulae*, not about *atoms* (Democritos), not about *minima* (Aristotle), and not about *corpuscula* (Descartes).

[28]The Latin text: *et debile argumentum est, quo nego esse aliud qvid in re quadam, quod ego aliud quid ibi non observo*, both in the MS. and Ed. 1669.

[29]The interparticle passages, Stensen imagined, were filled with a fluid that was according to Mieleitner an imperceptible substance similar to the universal ether postulated by Mieleitner and others in the 19th century, the existence of which can only be indirectly inferred. Similarly, Caverni regrets that Fracassati and Guglielmini have not followed Stensen's ideas concerning the power which makes the small particles crystallize. Stensen thought of an ethereal fluid and not like Willis of some vague "spirit", which by magnetic power makes the substance crystallize. Stensen's detailed and comprehensive statement is said to stand at the beginning of modern crystallography (R. Caverni, *Storia del Metodo Sperimentale in Italia* (Firenze 1893) 3. p. 612 f.).

2. A solid differs from a fluid in that in a fluid the imperceptible particles are in continual motion, mutually receding from each other, but in a solid, though the imperceptible particles may sometimes move, nevertheless they scarcely ever move apart from each other as long as that solid remains solid and intact.

3. While a solid body is being formed its particles change their position from one place to another.

4. So far we know of nothing in the nature of matter to explain the origin of motion and clarify our understanding of it; but the determination of natural motion[30] can be altered by three causes:

   1) By movement of the fluid that permeates all bodies, and we state that things that are produced in this way are produced according to nature.

   2) By the movement of living beings, and we state that many of the things that are done by man in this way are artificial.

   3) By the first and unknown cause of motion: and even pagan people believed that there was something Divine in those things that come about in this way. Certainly to deny this cause the power of producing results contrary to the usual course of nature is the same as denying man the power to change course of rivers, of struggling with sails against the winds, of kindling fire in places where without him fire would never be kindled, of extinguishing a light which would not otherwise vanish unless its fuel supply ceased, of grafting the shoot of one plant on to the branch of another, of serving up summer fruits in mid-winter, of producing ice in the very heat of summer, and a thousand other things of this kind opposed to the usual laws of nature. For if we ourselves, who are ignorant of the structure of both our own bodies and the bodies of others, alter the determination of natural motions each day, why should not He be able to alter their determination who not only knows the whole of our structure and that of all things, but also brought them into being. Moreover, to be ready to admire the genius of man as a free agent in things made by human skill, and to deny a free mover to things produced by nature would indeed seem to me to show great simplicity in pretended subtlety, since when man has performed most skilful things, he cannot, except through a fog, make out what he has done, which instrument he used, or through which cause the said instrument moves.

---

[30]Here Mieleitner (Mi) translates *determinatio* as destination probably destination of essence or existence, with direction. Cf. 2. Ed., p. 47.

I shall set forth these details, demonstrated by experiments and arguments, at greater length in the dissertation itself, so that it may be evident that no philosopher exists who does not say the same thing, though not always in the same words, or if he has spoken differently, nevertheless admits those points from which these details necessarily follow. For what I have stated about matter holds everywhere, whether matter is considered to consist of atoms,[31] or of particles which may change in a thousand ways,[32] or of the four elements,[33] or of as many different chemical elements as are needed to meet the variety of opinions among chemists.[34] And indeed what I have proposed about the determination of motion agrees with every mover, whether you call the mover the form, or properties emanating from the form, or the Idea, or common "subtle matter", or special "subtle matter", or a particular soul, or a world soul, or the immediate influence of God.[35]

Following this, I explain the various modes of speech that are commonly accepted through usage, by which we explain in various ways the diverse production of different bodies, and sometimes of the same bodies; for whatever contributes anything to any body does so either as its location, its material, or its mover: hence, when like produces like, there is conferred on the produced object location, material, and movement of production, just as the tiny plant enclosed within the seed of some plant received from the other plant both the material within which it was produced and the particulate

---

[31]The atoms of antiquity (essentially different from our chemical ones) were minute, similar, indivisible and unchangeable bodies which only differed in size and weight, could move and influence each other by pushing. As to details cf. Melsen, *Atom gestern und heute*, p. 30 ff.

[32]Probably the Aristotelian idea of minute parts - minima naturalia - which with each substance differ as to specific and substantial shape (Melsen p. 65 ff. Cf. also the opinions before Aristotle p. 24ff. and afterwards 313).

[33]The doctrine about the four elements dates as far back as Empedocles from Sicily (5th century B. C.), who regarded fire, air, water and earth as qualitatively different elements, and traces growth and change back to the blending or separating of the smallest parts of these substances. These four elements however are not so much the real substances as the qualities characteristic of them (Melsen p. 33 f.).

[34]Robert Boyle (1626-1691) in the *Sceptical Chemist* (1661) and *Origin of Forms and qualities* (1666) aimed at scientific explanation, and for principles seems to adopt substance, motion, and elements with invisible, smallest particles. The chemists of the 16th and 17th centuries (like Paracelsus, Glauber, van Helmont) tended to different explanations of the elements which, however, were not elements as we understand them to day: each element was a substance complex in itself, which had to be present at the construction of all other substances (Melsen p. 147, and M. Boas, *Robert Boyle and Seventeenth Century Chemistry* (Cambridge 1958 p. 75 ff.), and I. R. Partington, *A History of Chemistry*, Vol. 2, London1961).

[35]While in the passages above N. S. sums up old and rather new aspects of what substance is, here he mentions different hypotheses about influences on matter. He may think here of Plato's doctrine of ideas and Aristotle's morphology (Melsen p. 40 ff.). Descartes' corpuscular theory knows about a "minute substance" (Melsen p. 158). J. Kepler, in his *Harmonice mundi* is convinced of the idea of a spirit of the earth, which since the creation determines the image of the universe in mathematical proportions (Max Caspar, Johannes Kepler, 2. Auflage, Stuttgart 1950, p. 331 ff.).

motion which gave it its shape; the same is true of animals enclosed within the egg of similar animals. While the particular form or soul produces something, the movement of particles in the production of that body is determined by some particular mover, whether this be the mover of another similar body or something else similar to this mover.

Things said to be produced by the sun derive the motion of their particles from the rays of the sun, and by the same reasoning, those that are attributed to the influences of the stars could derive the motion of their particles from the stars; for since it is certain that our eyes are stimulated by the light of heavenly bodies, it will likewise be beyond dispute that the rest of matter might be influenced by them in the same way.

Things produced by the earth derive from the earth nothing else than the location in which they are produced and the material furnished to them through pores in that place.

Things produced by nature derive the motion of their particles from the motion of a penetrating fluid, whether this fluid comes from the sun, or from fire contained in terrestrial matter, or from some other cause not known to us, as, for example, the operation of the soul, and so on.

Thus, he who attributes the production of anything to nature names the general mover found in the production of everything; he who calls the sun to play this role defines the same mover in a more restricted way; he who names the soul or particular form puts forward a more limited cause than the remainder; but no knowledge is to be found by duly weighing the answers of all concerned, since nature, the rays of the sun, the soul, and the particular form are things known only by name. But since, apart from the mover in the production of bodies, the material and the location ought also to be considered, it is clear in consequence that not only is this answer more obscure than the object of our research but it is in every way incomplete; when shellfish found in the earth are said to be produced by nature so also those that grow in the sea are the work of nature; indeed nature produces everything, since the penetrating fluid is involved in the production of all things, but it could be said rightly that nature produces nothing, since that fluid by itself accomplishes nothing—its determination is dependent on the material to be moved and the location. Man provides an example for us: he can accomplish anything if the necessary things are all to hand, but he never accomplishes anything if these are lacking.

He who ascribes the production of anything to the earth names indeed the location, but since the earth bestows location at least in part to all the things of earth, the location certainly by itself does not explain the production of a body; the same can be said of earth as I said about nature, namely, that all things formed in the earth are produced by the earth, and, also, that of those things that are formed in the earth none is produced by the earth. These same few points set down above are sufficient to resolve all doubts in our proposed investigation; I have wished to express these here in the three propositions that follow.

2.27.2   Three Propositions

I

If a solid body is enclosed on all sides by another solid body, the first of the two to harden was that one which, when both touch, transferred its own surface characteristics to the surface of the other.[36] From this it follows that:

1. In the case of those solids, whether of earth or rock, which completely surround and contain crystals,[37] selenites, marcasites, plants and their parts, animal bones and shells, and other bodies of this kind that are endowed with smooth surface, these same bodies had already hardened at the time when the material of the earths and rocks containing them was still fluid; and so clearly it could hardly be that those earths or rocks produced the bodies contained in them, since they would not even have been in existence at the time when the said bodies were produced there.
2. If at any time a crystal is partly enclosed by a crystal, a selenite by a selenite, a marcasite by a marcasite, then at a time when these contained bodies were already hard, part of the containing body was still fluid.
3. In those earths and rocks containing crystalline and petrified shells, veins of marble, of lapis lazuli, of silver, mercury, antimony, cinnabar, copper, and other minerals of this kind, the containing bodies were already hardened

---

[36]Mi [Karl Mieleitner] translates: Wenn ein fester Körper... weicher bei der gegenseitigen Berührung auf seiner Oberfläche die Eigenschaften der Oberfläche des andern Körpers ausdrückt. Here we follow the translations of Ma [Vilhelm Maar, 1910] and Mo [Gius. Montalenti], *Stenone, Prodromo di una dissertazione ...*, Roma 1928] which not only facilitate a reasonable understanding of the sentence but also correspond more closely to the text (not *in sua superficie*, but *sua superficie*), so the word *exprimit*, expresses here might be better understood as "produces". Mo continues: "Not only the shape of the body belongs to the quality of the surface, but also the accidental, physical characteristic of its surface such as smoothness, radiance, brilliance, mother-of-pearl-likeness, etc. It is obvious, however that on examining the surfaces of two bodies in contact with each other that both may equally be called the counter part of the other. If the criterion suggested by Stensen shall be of any value at least the characteristics of one of the bodies must be known. This criterion is very important; even in our time we are supported by it when we want to ascertain the succession of fossilization in the various parts of a rock".

[37]By crystal Stensen almost always means rock crystal (quartz). Selenites are gypsum crystals or calcite crystals (calcspar). By marcasite, the ancients' mostly mean pyrites, the most frequent compound of iron and sulphur (Scherz, *Vom Wege* 143, 157). Pyrites and marcasite are, by the way chemically identical, though structurally different.

when the material of the contained bodies was still fluid, and so marcasites were the first to be produced, then the rocks in which the marcasites are enclosed, and next the veins of minerals that fill the fissures in the rocks.[38]

II

If a solid body resembles another solid body in all respects, not only in the state of its surface but also in the internal arrangement of the parts and particles, it will resemble it also in the method and place of production, excepting those conditions of place which are often found in any region and provide neither advantage nor disadvantage to the production of the body. Whence it follows that:

1. The strata of the earth agree, in location and manner of production, with those strata that are deposited from turbid water.
2. The crystals of mountains agree, in manner and place of production, with crystals of nitre, although it is not on that account necessary that the fluid in which they were produced should have been aqueous.[39]
3. Those bodies that are dug from the earth and which resemble in every way the parts of plants and animals were produced in the same manner and place as the parts of plants and animals themselves. But so that no new doubts may arise from an uncertain interpretation of the term place, I shall deal with this difficulty.

I understand by the term "place" the material whose surface is in immediate contact with the surface of the body which is described as being in that place; however the said material admits of a variety of differences, for it is either:

1. Completely solid, completely fluid, or part solid and part fluid.
2. Completely perceptible by itself, or partly perceptible by itself and partly by means of tests.
3. Completely in contact with the body that it contains or even partly continuous with the said body.
4. Always unchanged or undergoes change scarcely observably; thus the place in which a plant is produced is the material like that plant in which the

---

[38]Mi confirms this succession of time in the origin of pyrites in the sediments about them. Whereas Mo adds: Evidently N. S. thinks of the Tuscan pyrite and calcite pyrite (Boccheggiano), which probably trace their origin back to the segregation of volcanic stones from magma. The chronological conception of the hardening of magma occupies rather an important part of modern rock science and these early hints of Stensen's are very valuable.

[39]Of course Stensen did not know the chemical difference between quartz (SiO_2 silicon dioxide) and salpetre ($NaNO_3$ sodium nitrate) (Wi).

seeding takes shape; thus the place in which a plant grows is all that material whose own surface is in immediate contact with the whole surface of the plant, consisting at times of earth and air, at times of earth and water, at times of earth, water, and air, at times only of rock and air, in the way in which I have often observed roots of small plants, totally bare of soil cover, in underground regions, clinging completely to the surface of tufa; thus the place where the orange grows after its blossom has fallen is partly the peduncle continuous with it and partly the air in contact with it; thus the place where the initial development of an animal occurs is partly the amniotic fluid in contact with it and partly the umbilical vessels, diffused through the chorion, which are continuous with the animal.

III

If a solid body was produced according to the laws of nature, it was produced from a fluid.[40]

In the production of a solid body both its first outlines[41] and its growth should be considered; but just as I acknowledge freely that the outline of most bodies is not only doubtful but completely unknown to me, so I consider, without any hesitation that almost all of the following statements about their growth are true.

A body grows by the addition of new particles, secreted from an external fluid, to its own particles; this addition, moreover, is either made directly from external fluid or through the medium of one or more internal fluids.

Additions made directly to a solid from an external fluid sometimes fall to the bottom because of their own weight, as in the case of sediments; sometimes the additions are made from a penetrating fluid that directs material to the solid on all sides, as in the case of incrustations, or only to certain parts of the surface of the solid, as in the case of those bodies that show thread-like forms, branches, and angular bodies.[42] There it must be noted in passing that the said processes continue sometimes until they fill the whole of some space, whereby replenishments occur which are at times simple, at times formed from crusts, at times from sediments, at times from angular bodies, at times from various components intermixed in a variety of ways.

---

[40]On reading his treatise, we easily understand how Stensen came to this extremely penetrating view (Mi). But note also his previous biological research following his first discoveries about the glandular system.

[41]Mi translates *prima lineamenta* as circumference. K. v. Bülow suggests something like "first germ" or "first shape".

[42]In the text *angulata corpora*. Here crystals are meant in the modern sense. (Mi and Wi.).

Those particles that are added to a solid by an intermediary internal fluid either take on the shape of fibres (in so far as they are partly added through open pores along the length of the extended fibril and are partly distributed by the permeating fluid through the interstices of the fibrils into the shape of a new fibril) or form simple replenishments; animals and plants are formed in these two ways. As I am less versed in the anatomy of plants, I make no decision on whether several internal fluids are present in plants; in the case of animals, it is certain that various internal fluids are present, which I shall attempt to bring into a definite order.

Besides the "subtle fluid" permeating all matter, at least three kinds of fluid are observed in animals, the first of which is external, the second internal and common, and the third internal and specific to each part. By the term "external fluid", I mean that fluid in animals which not only surrounds their outer visible surface in the fashion of the atmosphere but also touches all the remaining surfaces of the body connected with the said outer surfaces by means of the larger foramina, such as the complete surface of the wind-pipe that comes into contact with inspired air,[43] the entire surface of the alimentary canal, by which I mean the mouth, the oesophagus, the stomach and the intestines, the complete surface of the bladder and the urethra, the entire surface that communicates with the uterus, at least during the years of puberty, the complete surface of all the excretory vessels in continuity from the capillaries even to the orifices that discharge their contents into the ears, the eyelids, the nose, the eyes, the alimentary canal, the bladder, the urethra, the uterus and the skin, a separate enumeration of which would show that many are truly external which are commonly taken to be internal, nay more, to be in most parts, and so:

1. Most of the worms and stones produced inside our body are produced in the external fluid.
2. Many parts are necessary to certain animals because they are present and not because the animal cannot exist without them.

I call the fluid which comes into contact with those surfaces an external fluid because it is in communication with the surrounding fluid by means of canals without intermediate capillary vessels, that is without cribration; the result is that while the cavities containing the said fluid may at times be closed, nevertheless whenever they are opened, all the fluid retained by them is discharged indiscriminately.

---

[43] The *Arteria aspera* (in the text) is now called the trachea, windpipe.

I call that fluid internal which is not joined with the external fluid except by means of the intermediary filters of the capillary vessels, and so never discharges all its components naturally into the external fluid without their separation.

The internal common fluid is that which is contained in veins, arteries and lymphatic vessels, at least those joining the conglobate glands[44] and veins. I call this fluid common because it is distributed to all parts of the body. I make no statement about the other common fluid which is contained in nerve material, since it is less understood.[45]

A specific internal fluid is that kind that flows around the capillary vessels of the common fluid and varies with the variety of places; for it is of one type in bloody parenchyma,[46] of another type in non-bloody parenchyma, of another around muscle fibres, of another in the ovaries, of another in the material of the uterus,[47] and another in other places; for the belief that the ends of veins and arteries terminate in a particle of the body of the least possible size, to distribute heat and nourishment at that place, is in accordance neither with reason nor experiment, but there are cavities everywhere in which parts secreted from the blood mix with the fluid of that place,[48] to be added next to the solid parts, just as particles worn from the solid parts return to the said cavities to be restored again to the blood, by whose operation they are carried back to the external fluid. The fluid of these cavities agrees in many ways with the teachings of the great Hippocrates[49] concerning air taken into the body.

Although I may be unable to determine why different fluids are separated from the same blood in different places, nevertheless, I hope that little

---

[44] *Glandulae conglobatae* are equal to the lymph vessels of to-day, contrary to the glandulae conglomeratae, the true and proper glands. The terms stem from Stensen's professor Fr. Sylvius, his pupil however, clearly ascribes to them an anatomical-physiological conception (OPH 1. p. 20 XIII and p. 183 ff.).

[45] Here Stensen thinks of the succus nervosus, the delicate inner substance of the nerves, which his age saw as a juice, in continuation of the idea of spiritus animales, an inspiration which Descartes also accepted, but which Stensen fought against in his Myology. Cf. E. Bastholm, *The History of Muscle Physiology* (Copenhagen 1950, p. 135 ff. 144 f.).

[46] The name of *parenchyma* was used by Erasistratos about the special substance of the lung, liver, kidneys, and the spleen and this because of the view which Galen also shared, according to which the substance was different from that of the muscles and was formed by the blood, which coming from the veins, coagulated in the said organs (OPH 2, 335).

[47] As Stensen occupies himself with the *fibra motrix* in his *Elementorum myologiae*, he also deals with the ideas of egg, oviduct, and uterus, and he first recognised that the *testes muliebres* are equal to the ovaries of the viviparous (OPH 2, 69. 152. 330 f.).

[48] The view - derived in various forms from antiquity - on the importance of the blood for feeding and keeping warm. Cf. on Hippocrates and Erasistratos, K. E. Rothschuh, *Geschichte der Physiologie* (1953) p. 8, 12 f.

[49] Hippocrates' book *De Flatibus* teaches that all diseases have the same cause, i. e. gas in the body, and that differences in the diseases originate from the place that has been attacked (Cf. OPH 2, 335).

remains to be determined about this matter, in so much as it is certain that it depends not on the blood[50] but on the places themselves; consideration of this problem involves these three items:

1. Consideration of the capillary vessels of the internal common fluid, the sole concern of those who ascribe everything to cribration through various pores, among which number I too was counted for quite some time.[51]
2. Consideration of a specific internal fluid, the sole concern of those who attribute a special ferment to each part; an opinion which might be partly true, though the term ferment rests on a comparison drawn from a very special process.[52]
3. Consideration of the individual parts of a solid, the chief adherents to which are those who, in attributing to each part its form, show that they recognize there something specific to the part, but unknown to us, which, according to the knowledge that we have gained up to now about matter,[53] can be nothing else than the porous surface of that solid and "subtle" fluid permeating the pores.

I should wander too far from the subject if I were to apply the above statements to an explanation of those things which occur in our body every day,[54] and cannot be explained otherwise. It will be sufficient to have hinted here that particles separated from the external fluid in various ways are carried into the internal common fluid, being sieved in the process, whence, having been secreted likewise in various ways and being transmitted into the specific internal fluid through a fresh sieving process, they are added to the solid parts either in the form of fibres or of parenchyma, as determined by an unknown

---

[50]As early as 1661 - contrary to the opinion of the ancients - Stensen calls the blood the housekeeper of the whole body, which supplies everything for the nourishment and production of the special fluids, and he also deals with the objections against this idea. In *De musculis et glandulis*, 1664, he provides a summary of what he has determined about glands and the lymph, and their relation to the blood (OPH 1, p. 34 ff., 198 ff.). Here, Montalenti points out the extraordinary importance of this paragraph III for mineralogy as well as for physiology: "Modern physiology which has concentrated on this fact hinted at by N. Stensen, has not got much further than Stensen as to the problem: Why are different fluids segregated by the same blood in different places".

[51]Cf. Stensen's Disputation *De glandulis oris* etc. of 1661 (OPH 1, 18 ff., 50 f.) on the opinion of the ancients that the glands are fungi.

[52]The iatrochemists J. B. van Helmont (1577-1644) and Stensen's teacher Fr. Sylvius saw chemical processes in the manifestations of life, which they called fermentation. They are begun by various ferments dispersed in the body (Francisci Delboe Sylvii *Opera Medica* (Amstel. 1680) p. 166 ff. 702 f.).

[53]Cf. note 33.

[54]Cf. note 50.

property of any individual part included in the consideration of the three statements made above.

Therefore, if one wishes to reduce solids enclosed naturally within solids to definite classes, by the above method, some of them will be found to have been produced by apposition from an external fluid, this refers either to sediments such as the strata of the earth, or to incrustations such as agate, onyx, chalcedony, eaglestone,[55] bezoar,[56] etc., or to filaments such as ami-anthus,[57] feathery alum,[58] various kinds of thread-like forms I have observed in the fissures of rocks, or to dendrites such as those plant-like shapes observed in the clefts in rocks; only superficial are certain ramifications observed by me in an agate which had trunks resting on the surface of the exterior lamella but whose branches were spread through the material of the inner lamella; or to angular bodies such as rock-crystals, the angular bodies of iron and copper, cubes of marcasite,[59] diamonds, amethysts, etc., or to replenishments such as every kind of variegated marble, granites, dendrites,[60] petrified mussels, crystalline materials, metallic plants,[61] and many bodies of this kind filling the places of bodies that have been destroyed.

Other solids are produced through apposition from an internal fluid; this refers either to simple replenishments such as fat, the callus joining broken

---

[55]Aetites, Eaglestones is the name given chiefly to clay and limonite concretions which enclose a body in their hollow insides and therefore rattle when shaken. It was believed that the eagle brought them to its nest to ease egg-laying, and miraculous power was ascribed to the stone particularly as an aid during pregnancy and birth (the big stone was "pregnant" with the small one). Plinius in his *Natural History* X 12 (3) XXX, 1230 (14) relates their miraculous powers (OPH 2, 335, Wi). Stensen's collection contained several pieces (*Vom Wege* 162). They probably stem from the rich deposits at Levane in the Arno valley, described in T. T. Viaggi 8, 352 f.

[56]The *Bezoarstone*, probably a Persian word, was a round ash-grey stone, consisting of layers, in size between a hazelnut and an eyeball. Inside the stone, to which was often attributed magical power, is sometimes found a grain of sand, a shell or a piece of coal (*Universel D'Histoire Naturelle...* Par M. Valmont Bomare... III Ed., Lyon 1791, V, 222. Cf. OPH 2, 335).

[57]The term of *Amiant* was used also for asbestos. Gf. Scherz, *Vom Wege* p. 168.

[58]*Alumen plumosum*, feathery alum; sometimes applied to a silky-fibrous, yellowish-white blending of sulphate alum, earth, iron-oxide and water, and besides to two different kinds of asbestos; cf. OPH 2, 336, and Scherz, Vom Wege 211.

[59]The original edition, as well as the manuscript has *ubi* instead of *cubi* marcasitae. As in earlier mineralogy, Stensen means Pyrites when speaking about marcasite cubes. Marcasite is of the same chemical composition as Pyrites (FeS), but is crystallized in orthorhombic shape, whereas pyrites-crystals follow the cubic system (Wi). Cf. the numerous marcasite pieces in Stensen's collection. *Vom Wege* 165 ff., Rodolico, *Toscana* 344.

[60]*Dendrites, Dentroitides* is mainly the name for minerals which exist in nature in the shape of ramified bodies. The MS. has *Dentiotides*.

[61]Ref. metallic plants cf. A. Kircher's "mercurial-tree" and other metallic trees in his *Mundus subterraneus*. II, lib. IX c. XXIV. Cf. Note 68, 69, 131.

bones, the cartilaginous substance joining severed tendons, the fluids that are the chief constituents of the substance of the viscera, the marrow of both plants and animals, or to fibrous parts such as the fibrous portions of plants, in animals indeed the nerve and muscle fibres, which are all solid bodies and for the most part are naturally enclosed within solids.[62]

Therefore, if at least all solids have grown from a fluid, if bodies mutually alike in all respects were also produced in the same way, if of two solids mutually in contact the first to harden was that which impresses its surface characteristics on the other surface, it will be easy, given the solid and its location, to make a definite statement about the place of its production. And this indeed is a general consideration of a *solid enclosed by a solid.*

2.27.3   Incrustations

I proceed to the more specific examination of those solids dug from the earth that have given rise to many disputes, especially incrustations, sediments, angular bodies, shells of marine animals, objects with the shapes of mussels, and of plants. To incrustations belong all the kinds of stones composed of layers, the two surfaces of which are indeed parallel but are not extended in the same plane. The place where incrustations are formed is the whole common boundary between fluid and solid, so that the shape of the layers or crusts corresponds to the shape of the place, and it is easy to determine which of them hardened first and which last; for if the place was concave, the outer crusts were formed first, if convex, the inner crust were first to form; if the place was uneven owing to various larger protuberances, once the smaller spaces were filled with the first formed layers, new layers were produced in the larger spaces: from this it is easy to explain all the varieties of form which are observed in sections of similar rocks, whether they show the round veins of a tree cut transversely, or resemble the sinuous winding of a serpent, or follow curves of some other type in an irregular fashion. Nor is it surprising that agates and other kinds of incrustations are observed to be rough like ordinary stones on the outer surface, since the outer surface of the outer layer depicts the roughness of the place; in torrents, however, incrustations of this kind are often found away from the place of production because the material of the place has been scattered by the bursting of the strata.

---

[62]In the first case undoubtedly, secretions are referred to; in the second concretions (Wi).

Regarding the way in which particles of the layers that are added to the solid are separated from the fluid, it is at least certain that:

1. Buoyancy or gravity are not involved.
2. The said particles are added to surfaces of all kinds, since surfaces that are smooth, rough, flat, curved, and composed of many planes inclined in various ways are found overlaid with crusts.
3. The motion of the fluid is of no hindrance to them.

I leave undecided, meanwhile, whether the substance under consideration that flows from the solid is different from the substance that agitates the parts of the fluid, or whether something else is to be investigated.

Differences in layers at the same place can be produced either by the diversity of particles leaving the fluid in succession, as this fluid is gradually dissipated more and more, or by different fluids being conveyed there at different times: so it happens that sometimes the same arrangement of layers is repeated in the same place, and often evident signs exist showing the ingress of new material. But all the material of the layers seems to be of a finer substance emitted by the rocks, as the following will reveal further.

2.27.4   The Strata of the Earth

Strata of the earth are related to fluid deposits because:

1. The pulverized matter of the strata could not have been reduced to that form unless, having been mixed with some fluid, it was extracted from that fluid by its own weight and was spread out by the motion of the said superincumbent fluid.
2. The larger bodies contained in these same strata obey for the most part the laws of gravity, not only with respect to the position of any individual body but also to the relative positions of different bodies to each other.
3. The pulverized matter in the strata has so adapted itself to bodies contained in it that it has not only filled the most minute cavities in each contained body but has taken on the smoothness and lustre of the said body on that part of its surface which touches it, even though the coarse nature of the pulverized material seems incapable of such smoothness and lustre.

Sediments then are formed when the contents of a fluid sink under their own weight regardless of whether these contents have been conveyed there from elsewhere or have been secreted gradually from particles of the fluid itself, either in its upper surface or from all the particles of fluid.

Although there are close similarities between crusts and sediments, they can still be distinguished easily because the upper surface of crusts is parallel to the lower one, no matter how uneven this has become as a result of larger projections, while the upper surface of sediments is parallel to the horizon or deviates only slightly from the horizontal. Thus in rivers, the alternating green, yellow and red mineral crusts do not remove the unevenness of the stony bottom, while a sand or clay sediment does in fact make everything level; because of this, I was able to distinguish easily between crusts and sediments in different composite strata of the earth.

The following can be determined about the materials of strata:

1. If all particles in a stony stratum are observed to be of the same nature and of fine size, it cannot reasonably be denied that this stratum was produced at the time of Creation from a fluid that then covered all things; Descartes, too, accounts for the origin of the earth's strata in this way.[63]

2. If, in a certain stratum, fragments of another stratum or the parts of plants and animals are found, it is certain that the said stratum must not be counted among the strata that settled out of the first fluid at the time of Creation.

3. If, in a certain stratum, we discover traces of sea salt, the remains of marine animals, timbers of ships and substances similar to the sea bed, it is certain that the sea was at one time in that place, however it came to be there, whether by its own overflow or by the upheaval of mountains.

4. If, in a certain stratum, we find a great abundance of rushes, grass, pine cones, trunks and tree branches and similar objects, we surmise correctly that the said material was swept thither by a river in flood or by a torrential outbreak.

---

[63]Descartes who is regarded as a supporter of Vulcanism, says in *Principia philosophiae* p. IV par XXXII, *De Terra*, that the solar like global mass of the earth gradually separated into shells of various kinds, a nucleus of glowing fire, a metal shell, a water shell and a stone shell, the latter of which, on bursting, created valleys and mountains. Here Stensen has in view the varieties of stones which compose the so-called fundamental rock of modern geologists, i. e. almost all gneiss and other crystalline slates; he, too, considers these forms "the oldest on earth", says Mi. Hölder also adds (p. 24) "a partly subconscious distinction between primary and secondary rock". Cf. here also Targioni Tozzetti's distinction of *primitivi* and *secondari* (Rodolico, *Toscana* 50 ff.). No doubt, Stensen's dynamic geology knows both exogene powers, such as sedimentation, erosion, and crumbling, and endogene ones, namely changes in the layers, tectonic forms, and volcanic symptoms) though without drawing further conclusions from these observations than his methodical caution permits. He can neither be counted among the neptunists nor - even less - among the vulcanists of later ages, abounding with hypotheses.

5. If, in a certain stratum, pieces of charcoal, ashes, pumice stone, bitumen and calcined bodies are observed, it is certain that fire occurred in the neighbourhood of the fluid, the more so if the entire stratum is composed of ash and charcoal only; such I have seen outside the city of Rome,[64] where the material of burnt bricks is dug.

6. If the material of all the strata in one place is the same, it is certain that the fluid did not absorb fluids of a different kind flowing in from various places at different times.

7. If the material of strata in the same place be different, either different kinds of fluid from different places streamed through that spot at different times (whether the cause be a change of winds or a more violent downpour of rains in certain localities) or the material in the same sediment was of varying gravity, so that first the heavier particles and then the lighter ones sought the bottom; changes in weather might have caused this diversity, especially in regions where a similar inequality of soil is observed.

8. If stony strata are found between earthy strata, it is certain either that a spring or petrifying waters[65] existed in the neighbourhood or that occasional eruptions of subterranean vapours took place, or that a fluid, having receded from the sediment that had been deposited, returned again when the upper crust had become hardened by the heat of the sun.

The following can be considered certain about the position of strata:

1. At the time when a given stratum was being formed, there was beneath it another substance that prevented the pulverized materials from sinking further; consequently, when the lowest stratum was being formed, either there was another solid substance underneath it or some fluid existed there which was not only different in nature from the fluid above it but was also heavier than the solid sediment from the fluid above it.

2. When an upper stratum was being formed, the lower stratum had already gained the consistency of a solid.

3. When any given stratum was being formed, it was either encompassed at its edges by another solid substance or it covered the whole globe of the

---

[64]The place outside Rome is no doubt the Monte Mario with its layers of clay from the Pliocene Age. Stensen probably observed a contact between two unconnected groups of stones which are found in the Seven Hills: the clay-layers mentioned and the vulcanic tuff with ashes, pumice, etc., so frequent in the neighbourhood of Rome. (We owe this note to prof. Fr. Rodolico).

[65]Here Stensen probably thinks of the hot springs producing calcareous sinter which he had been able to observe in Italy. (Mi). Cf. M. Messini - G. C. di Lollo, *Acque minerali del mondo* (Rome 1957), where the mineral springs of Italy and their mineral content are described, particularly in pp. 83 ff. and 145 ff.

earth. Hence, it follows that wherever bared edges of strata are seen, either a continuation of the same strata must be looked for or another solid substance must be found that kept the material of the strata from being dispersed.

4. When any given stratum was being formed, all the matter resting on it was fluid and, therefore, when the lowest stratum was being formed, none of the upper strata existed.

As far as form is concerned, it is certain that when any given stratum was being produced, its lower surface and its edges corresponded with the surfaces of lower and lateral bodies but that its upper surface was as far as possible parallel to the horizon; all strata, therefore, except the lowest, were bounded by two horizontal planes. Hence, it follows that strata which are either perpendicular to the horizon or inclined to it were at one time parallel to the horizon.

Nor are my statements contradicted by change in the position of strata and their exposed edges such as are observed today in many places, since obvious traces of fire and flood are to be found in the neighbourhood of those places. For just as water in breaking down earthy material, bears it down slopes not only on the earth's surface but also in the earth's cavities, so fire, in breaking down those solids that oppose it, not only drives out their lighter particles but sometimes hurls out very heavy masses; the result is that precipices, channels and hollows form on the earth's surface while, in the bowels of the earth, subterranean passages and caverns are produced, in consequence of which the earth's strata can alter position in two ways.

The first way is the violent upheaval of strata, whether this be due mainly to a sudden flare of subterranean gases or to a violent explosion of air caused by other great subsidence nearby. This upward thrust of strata is followed by the dispersal of earthy material as dust and the shattering of rock material into pebbles and rough fragments.

The second way is a spontaneous slipping or subsidence of the upper strata after they have begun to crack because of the withdrawal of the underlying substance or foundation; in consequence the broken strata take up different positions according to the variety of cavities and cracks. While some remain parallel to the horizontal, others become vertical; many make oblique angles with the horizon and not a few are twisted into curves because of the tenacity of their material; this change can take place either in all the strata overlying a cavity or in certain lower strata, the upper strata being left unbroken.

Alteration in the position of strata affords an easy explanation of various fairly difficult problems.[66] Herein may be found a reason for the unevenness of the earth's surface that gives rise to so many controversies, as are found in mountains, valleys, natural reservoirs, elevated plains and low-lying plains; but, passing over the rest, I shall deal quickly here with certain points concerning mountains.

2.27.5   The Origin of Mountains

It is clear that alteration in the position of strata is the main source of mountain formation from the fact that in any given range of mountains the following may be observed:

1. Vast level areas on the summits of some mountains.
2. Many strata parallel to the horizon.
3. Various strata on the sides of mountains inclined at different angles to the horizontal.
4. Broken strata, on the opposite slopes of hills showing absolute agreement in form and material.
5. Exposed edges of strata.
6. Fragments of broken strata at the foot of the same range partly piled into hills and partly scattered over the adjoining terrain.
7. The clearest signs of subterranean fire, either in the rocks of the mountains themselves or in their neighbourhood, just as many springs are found around hills that are composed of earthy strata.

---

[66]As to the importance of Stensen's doctrine about strata, there shall be quoted, though abbreviated, the opinions of some modern geologists, W. H. Hobbs state: "In geology we owe to Stensen the first distinct establishment of some of those great principles which to-day we regard as axioms... that rocks stem from sediments... that the sediments of seawater can be distinguished from those of fresh water and of volcanic products... that the order of the strata piled on top of each other determines the age of the formation. The originally horizontal position of the sediment formations is to-day considered one of the great fundamental principles of geology." (Wi. 172 ff.) Hölder says: "These observations (as to the material of the strata) were the first to give us information about the shifts of the stones and their causes., i. e. a Fazieskunde". In connection with this, Stensen arrives at the law about sediments, consciously understanding and distinctly formulating that a lower layer will always be older than the following higher one. But the law about layers is at once expanded, the slanting positions of the layers being explained by the influence of water and fire. Here St. first mentions the "bowlike curvature" i. e. the folding of the layers which he could observe in the Apennines. This can hardly mean the recognition of fold mountains but N. S.'s treatise is the first to point at appearances which are later termed tectonic (p. 25 f.). Mark the difference in Stensen's view, pointed out by Targioni Tozzetti, between *monti* and *colline*, mountains and hills. Hubert G. Schenk says:" The use of paleontology in industry was dependent on two fundamental principles: the law of layers and the law that fossils are different according to their geological ages. The former he ascribes to N. S., the latter to G. Soulavier (p. 1752 ff.).

Here, it must be observed, in passing, that hills formed of earthy strata have, for the most part, larger fragments of stony strata as their foundation; these, in many places, keep the earthy strata that are laid upon them from being swept away by the currents of neighbouring rivers and torrents; indeed, they often protect entire regions against the violence of the ocean, as the spreading reefs of Brazil[67] and exposed rocky shores everywhere bear witness.

Mountains can also be formed in other ways, such as by the eruption of fires that belch forth ashes and stones together with sulphur and bitumen, and also by the violence of rains and torrents whereby stony strata, already split by changes in heat and cold, are tumbled headlong, while the earthy strata, cracking under great heat, are broken down into various parts; from which it is clear that there are two chief classes of mountains and hills.

The first of these, those that are composed of strata, consists of two species, since rocky strata abound in some and earthy strata in others; the second class consists of mountains that rise, without order or arrangement, from fragments of strata and from parts that have been worn away. Hence, the following can easily be shown:

1. All the mountains of this era did not exist from the beginning of things.
2. Mountains do not grow in the way plants do.[68]
3. The rocks of mountains have nothing in common with the bones of animals, apart from a certain similarity in hardness, since they agree neither in material nor in manner of production nor in construction nor function; if one may be allowed to make any pronouncement on a subject so little known as the function of things.

---

[67]Willem Piso (1611-1676), to whom as a friend Stensen dedicated *De anatome rajae epistola* of 1664 (cf. OPH 1, p. 195, 248), had published an account of a journey to Brazil, where he was physician in ordinary to Count Jan Maurits of Nassau: namely *De Medicina Brasiliensi Libri quatuor* (Lugd. Bat. et Amstelodami 1648) and later *De Indiae utriusque Re naturali et medica libri quatuordecim* (Amst. 1658). W. H. Hobbs says that the shifts in the positions of layers are according to Stensen the main cause of the mountains. And he distinguishes rather clearly between three of the most important types of mountains, i. e. block- or fold mountains, vulcanic mountains and erosion-mountains. The importance of earthquakes for the origin of the mountains is close to contemporary views (Wi 173).

[68]Vegetative growth was frequently ascribed to minerals and mountains in the 17th century (cf. note 140). Fabroni writes about the journey of G. Montanari and Boni 1657, both of them friends of Stensen: "They also examined if metals grow like plants, i. e. by means of a fluid circulating in the ground". (OPH 2, 338). As to a widely spread belief in the vegetative growth of minerals as late as in the 17th century, cf. Pazzini, Le Pietre 40 ff., who also points out the fact that many functions are common to the organic and inorganic world.

4. The arrangement of mountain crests or chains, as some prefer to call them, along certain zones of the earth, is in accordance neither with reason nor experience.[69]

5. Mountains can be destroyed; fields can be carried from one side of a high road across to the other[70]; mountain peaks can be raised and lowered; the earth can be opened and closed again; and other phenomena of this sort occur which are considered to be myths by those who, in their reading of history, wish to escape the reputation of being credulous.[71]

Passage Ways for Things Issuing from the Earth

The same alteration in the position of strata provides a passage way for materials flowing from the earth, as in the case of:

1. Mountain springs whose waters are separate from the atmosphere, within mountain caverns, whether these come from subterranean waters or are waters condensed from the upper atmosphere into the place and thrust forth, which I believe is very common, since I have seen an abundance of water dripping from many caverns where every part of both roof and floor was solid.[72]

2. Winds rushing out of mountains, whether these are formed by air expanded by heat or various fluids belonging to the air give rise to them as a result of their boiling up from mutual collisions.

3. Fetid exhalations, fiery or frigid ebullitions, etc. Nor, moreover, is there any difficulty in the fact that cold and dry places boil up without any indication of heat whenever water is poured over them; that a hot spring gushes forth at the side of the coldest spring; that by earth movement a hot spring may turn into a cold one, and rivers change course; that valleys enclosed on all sides discharge water gathered as rain into lower regions;

---

[69]Athanasius Kircher (1602-1680), often a clever observer and often a child of his own time too, in his treatises based on speculation, fancy and analogy, was of the opinion that the chains of mountains (by which he did not mean mountains ranges in our sense), followed a straight line from east to west, from south to north, that they continued at the bottom of the sea and formed the back bone of the globe (*Mundus subterraneus* in XII Libros digestus, Amstel, 1665 Tom I. c. IX). As an undergraduate in Copenhagen, Stensen wrote out excerpts from his work: *Magnes sive de arte magnetica* (1640) and *Ars magna lucis* (1646), and later he was in correspondence with him, but not about geological questions.

[70]In the Latin text of MS. and Ed. 1669: ... ex uno latere in alterum per mediam viam publicam. Cf. on the agency of wind on mountains, Wi p. 232 f.

[71]Cf. regarding Stensen's district of research in Tuscany, Preller, *Italian Mountain Geology*.

[72]Mi says "strangely enough Stensen does not think of the possibility that the water may percolate the stone but believes that the water of the stalactite caves is condensed vapour".

that rivers flowing under the ground return to the surface in places; that in laying foundations, architects sometimes find that all their labour is to no useful purpose, when they encounter quicksand[73] as it is called; that in certain places when wells are dug, at first water is found near the surface of the earth, then after the earth has been excavated to a depth of several arms' length, fresh waters are found which, on being provided with an outlet, spring up to a height beyond the level of the water found originally; that whole fields with trees and buildings subside gradually, or are destroyed unexpectedly, so that large lakes now exist where once there were cities; that a plain is a danger to its inhabitants because of this kind of disaster, unless they are sure that the plain is founded on rock; that at intervals chasms open up, releasing pestilential gas, which are stopped up again when materials of any kind are heaped upon them.

The Origin of Variegated Stones and Repositories of Minerals
The same alteration in the position of strata has given rise to variegated stones of every kind, and at the same time provided a repository for most minerals, whether this occurred in cracks in the strata or in fissures found in their material, dry indeed but not hard, either between layers or in clefts; whether in the interstices between the upper and lower strata, after the collapse of the lower strata, or in places left empty by the decay of bodies contained therein; consequently it may be shown:

1. The minute and all but superstitious divisions of veins utilized by miners are built on the slightest foundation, nay clearly on no foundation at all; so that the divination of an abundance of minerals by means of roots and twigs is as dubious with regard to metals as the opinion of certain Chinese is ridiculous with regard to the head and tail of the dragon that they use in discovering a favourable position for burial places on mountains.[74]

---

[73]Stensen speaks about *arena viva*, living sand (Mi).

[74]This custom is mentioned by A. Kircher in *China Illustrate* (Amstelodami 1667); p. 135, refers to the book by the two Jesuit missionaries and explorers of China: P. Nik. Trigault and P. Matth. Ricci (OPH 2, 337). Stensen's sceptical view on the value of the divining-rod is a typical demonstration of his methodical sobriety. The divining-rod, in antiquity confined to legends and myths, was in the Middle Ages sometimes surrounded with religious ceremonies, and sometimes considered to be superstition or piece of devilry; in the 17th century it was commonly used, and the diviner belonged to the staff of mines (though then also sceptical voices were heard such as those of A. Kircher, C. Schott and P. Borel), and Descartes' and Gassendi's atomic theories were used for explanation. In the 18th and 19th century the interest in rods almost disappeared; in our time some research was done as to the eventual chemical, physical or physiological background, yet the practical geologist does not seem interested in it (Cf., C v. Klinckowström und R. v. Maltzahn, *Handbuch der Wünschelrute in Geschichte, Wissenschaft und Anwendung.* München 1931, and H. Ødum, *Ønskekvisten.* København 1932).

2. The majority of the minerals with which human effort is engaged did not exist from the beginning of things.

3. Many things may be detected in the examination of rocks that are attempted in vain in the examination of the minerals themselves; since it is more than probable that all the minerals filling either the clefts or wide spaces in the rocks obtained[75] their material from vapour forced out from the rocks themselves, whether this occurred before the strata changed their position, as I believe happened in the mountains of Peru,[76] or when the strata had already changed their position; so that a new metal can grow up in place of the exhausted metal, as is believed rather than known about the mining of iron in Elba,[77] since mining implements and idols found there were surrounded not by iron but by earth.

And I judged these items concerning the strata of the earth to be worth more careful examination not only because the strata themselves are solids naturally enclosed by solids but also because almost all the bodies that give rise to the problem proposed are contained in them.

2.27.6 Concerning Crystal

With regard to the formation of crystal,[78] I shall not attempt to determine how its first outline is produced; it is at least beyond dispute that things I have happened to read in the works of other authors on this subject are not relevant; for experience is not in accordance either with irradiations or with

---

[75]Mi: That is, *in nuce*, the lateral secretion theory of the 19th century. Mo: I think that this is an obvious reference to the pneumatolytic formation of minerals.

[76]Here Stensen probably thinks of a description of discovering gold and silver in the mountains of Peru in: *Historia Natural Y Moral de Las Indias … Commuesta par el Padre Joseph de Acosta* (Sevilla 1590) lib. IV, cc. IV—V. This description is found condensed in another book, which Stensen knew: *Le Mercvre Indien, ou le Tresor des Indes* (Paris 1667) by P. de Posnel. I, 1, cc. 1-111 (OPH 2, p. 337). Various short notes (Chaos MS. fol. 76ʳ80ʳ), by Stensen, probably, written down around October 1668, bear witness to several books of scientific interest which he must have studied at that time, e. g. East India, Bermuda, Guinea, Sumatra and Ethiopia.

[77]Cf. the Introduction p. 23 about N. St.'s visit to Elba. According to Simonin, *La Toscana et la mer Tyrrhénienne* (Paris 1868), p. 321, new valuable iron-layers were formed on Elba after they had been depleted in antiquity (OPH 2, 337). Wi: The belief that iron can grow out again and replace itself in time is explained by the limonite finds on the instruments mentioned in the text. About the wealth of the iron-ore layers on Elba cf. Ermenegildo Pini's explanation of 1777 in Rodolico, *Toscana*, p. 95 ff. and Ida Valeton. *Eisenglanz und Mohrenköpfe* (Merian, *Elba* Heft IV/XIX p. 16 f.).

[78]Stensen says *crystallus* meaning rock-crystal; very rarely, he says *crystallus montium*. Crystal in our general sense he calls *corpus angulatum*; in the translation: angular body (Mi).

the resemblance of the shape of particles to the shape of the whole,[79] or with the perfection of a hexagonal and the joint flow of parts towards the same centre, or other ideas of this kind[80]; this will be clear from various propositions that I shall bring forward and that are proved elsewhere by the most obvious experiments. But so that there may be no room left for doubt, it is of advantage to explain beforehand the terms that I use in naming the parts of a crystal.

A crystal consists of two hexagonal pyramids and an intermediate column that is also hexagonal. Here I call *terminal solid angles*[81] those angles that form the apices of the pyramids but those that are formed by the union of the pyramids with the column I refer to as *intermediate solid angles*; in the same way, I refer to the planes of the pyramids as *terminal planes* and the planes of the column as *intermediate planes*. The *plane of the base* is a section perpendicular to all the intermediate planes. *A plane of the axis* is a section in which lies the axis of the crystal, made up of the axes of the pyramids and the axis of the column.

There may still be doubt about the place in which the first hardening of the crystal begins, whether it be between fluid and fluid or between fluid and solid or in fact in a fluid by itself;[82] but the place at which the crystal grows, once formed, is solid at the region where the crystal is supported, whether that place be of stone or another already formed crystal. The remaining portion is fluid, excepting the obstructions that can be presented to it by roughness of rock or by other already formed crystals.

---

[79]The composition of crystals from very tiny particles similar to or like a crystal was accepted by Guglielmini later on and theoretically proved. H. Tertsch though, called him "less thorough and careful in the mere and sober observations of the facts than Stensen" (*Der Schlüssel zum Aufbau der Materie*, Wien 1947 p. 19 f.).

[80]There were different theories at the time about the origin of the rock crystal. Cf. *Musaeum Septalianum Manfredi Septalae... Pauli Mariae Terzagi... Geniali Laconismo Descriptum* (Dertonae 1664), Cento primus p. 241 ff. and Index p. 302: "Whether it originates from frozen water or condensed sugar or from humid slimy soil or from water and a crystal germ or like the stones by means of the heat of the sky condensed out of pure water" (OPH 2, p. 337 f.).

[81]Here Mi translates *angulos solidos extremos* with "outer solid angles" and *angulos solidos intermedios* with the "middle solid angles"; here too he sees a reference to Stensen's recognition of the constancy of the crystal-angle. A. Johnsen, (Die Geschichte einer kristallmorphologischen Erkenntnis. In: *Sitzungsberichte d. preuss. Akad. d. Wiss.* 1932 (p. 404-415) made the correction that *angulus solidus* does not mean a solid or constant angle, but only an angle of a body, or a "spatial" angle i. e. a corner. H. Seifert (p. 32 ff.) shares this opinion. K. v. Bülow, Mi and Johnson in common with many others, think that Stensen's quartz crystals with a double pyramid had come from the neighbourhood of Porretta. Mark here the strictly geometrical description.

[82]Stones which are not formed by sedimentation from water, in which group crystals are found, are called rock (*saxum*) by Stensen.

I shall not attempt to determine whether the surrounding fluid is aqueous; it does not matter what is said about water enclosed within crystals, since it is certain that, together with water, air is contained therein, and many crystals are found that enclose air alone; but, if the crystal had indeed grown in an aqueous fluid, all the enclosed spaces would be filled on all sides with water since it is consistently observed that water held in this way never disappears during any number of centuries.

Cavities in rocks, produced in diverse ways as stated above, afford such a location for crystals; the fact that entire hills consist of earthy material crammed full of crystals is no disproof, considering that mountains of rock suited to formation of crystals are found in the vicinity of the said hills, and from these hills of earthy material are dug out large unburied rocks that have been torn from the neighbouring mountains; some of these show fissures filled with the substance of marble, just as fissures in strata from mountains made of rock are filled. Moreover, the same cause that hurled fragments of strata torn from the neighbouring mountains on to the hills can have likewise disseminated over the same hills crystals that had been shaken out of cavities in the said strata.

The following propositions show what can be established about the place on a crystal to which new crystalline material is being added:

I. A crystal grows while new crystalline material is added to the exterior planes of the already formed crystal, so that there is no room at all here for the opinion of those who assert that crystals grow vegetatively,[83] that they draw their nourishment from the side on which they are attached to the matrix, so that the particles thus received from the fluid of the rock and transmitted into the fluid of the crystal are added internally to the crystal particles.

II. This new crystalline material is not added to all planes but mainly to planes of the apex only or to terminal planes, with the result that:

1. The intermediate or quadrilateral planes are formed out of the bases of terminal planes and, hence, the said intermediate planes of some crystals are larger, in others are smaller, and, in others are completely absent.

2. The intermediate planes are almost always striated but the terminal planes retain traces of the material added to them.

---

[83]Cf. note 61.

III. Crystalline material is not added to all terminal planes simultaneously nor in equal quantity. Hence, it is that:

1. The axis of the pyramids does not always continue in the same straight line as the axis of the column.
2. Terminal planes are rarely of the same size; in consequence, there is inequality in the intermediate planes.
3. Terminal planes are not always triangular just as all intermediate planes are not always quadrilateral.
4. The terminal solid angle is broken down into several solid angles, this being also the case, frequently, with solid intermediate angles.[84]

IV. Crystalline material does not always cover an entire plane but bare portions are left sometimes near to the angles, sometimes towards the edges, and sometimes in the centre of the plane. Hence, it is that:

1. The same plane, commonly so-called, does not have all its parts located in the same plane but in different planes extending above it in various ways.
2. A plane, commonly so-called, in many places is seen to be not flat but humped.
3. Inequalities arise in intermediate planes like the steps of stairs.

V. Crystalline material, added to planes upon the same planes, is spread out by the enveloping fluid and hardens gradually with the result that:

1. The surface of the crystal becomes more smooth the more slowly the added material has hardened, and is left completely rough if the said material has hardened before it spread out sufficiently.
2. The way in which crystalline material was added to the crystal can be distinguished since, where it has hardened suddenly, it reveals a surface full of small humps resembling variolar pustules, just as small drops of oily fluid are wont to float on an aqueous fluid; sometimes, if it has hardened somewhat more slowly, it shows also sunken three-sided pyramids. The twisted borders of the descending material show now the place to which the fluid material was being added, now the place to which it was being extended, now the arrangement of the material added, i.e. which came first and which last. And in this way certain unevenness always show in the crystals

---

[84]What is meant by this breaking down is shown by diagram 6 of the plate. Cf. besides the detailed description by Seifert p. 34 ff.

of mountains; nor have I ever seen a crystal whose still unbroken surfaces have the smoothness that the fractured sides of the same crystal show after it has been broken apart, however diffuse writers on natural topics become in extolling the smoothness of crystal extracted from mountains.

3. If they came in contact with the surface of the crystal before it hardened certain intrusive solid bodies are bound to the crystal itself as if they had been smeared with some kind of glue.

4. It (crystalline material) seems sometimes to flow down over neighbouring planes.

5. When certain places on those planes have been left untouched by added crystalline material, new crystalline matter approaches and spreading over these places forms cavities there, sometimes producing several layers, sometimes enclosing part of the external fluid, which, at one time is pure air, at another is air and water.

VI. The external fluid receives crystalline material from the substance of the harder stratum, so that:

1. Rocks of different types, emitting different fluids, produce crystals of different colours.

2. Sometimes, in the same place, the first crystals are the darker, sometimes the last; but, in the same crystal, the parts first hardened are sometimes darker than those hardened last.

3. When oysters, mussels and other bodies have decomposed in the earth, the empty spaces left by them are filled with crystals.

VII. The movement of crystalline materials to a region where the planes of the already-formed crystal are fixed does not arise from some common cause of motion in the surrounding fluid but varies in any given crystal; so, in reality, it depends on the movement of the tenuous fluid that flows from the already-formed crystal;[85] the result is that:

1. Crystalline material is added, in the same place, to planes that make different angles with the horizontal.

2. Crystals of different shapes are produced in the same fluid. I leave it to wiser minds to examine whether the said fluid is that by the aid of

---

[85]According to V. M. Goldschmidt, the structure of the crystals in question in this paragraph, depends on the number, the position and the size of the elements, i. e. the size of the ionic and molecular radii, the structure of the electron shells and the polarisation qualities i. e. the nuclear field of force. (*Universitas Litterarum*, Berlin 1955, p. 127). The numbers V—VII of the last three propositions are taken over from the first English translation by H. Oldenburg.

which refraction is produced, or if it is in fact some quite different fluid. The length of the threads of iron filings that rise around the poles of a magnet,[86] not only when the filings are in direct contact with the magnets but also when they are separated from it by an intervening sheet of paper, proves conclusively that the efficiency of a penetrating fluid is indeed great; when the magnet is moved under the paper in various ways, at one time filings of this sort, while one end remains stationary, describe on the paper at the other end all the arcs that can be drawn in a hemisphere; at another time they all advance from place to place like armed soldiers; again, when deflected by the nearness of another magnet, they form an arch just as if the individual parts of the filings had been glued together and had coalesced into a solid body.[87]

I should suppose that the small drops that form in a receiver from material forced out of a retort cohere mutually in a similar fashion, under the action of a permeating fluid; at first, they cling together on the inside of the upper part of the receiver, but later, when a number of drops have come together in the same bend, they fall downwards and form various globular threads that on occasion cling with their end parts to the sides of the receiver and sometimes unite with other threads. I should believe that threads of this kind that I have noted sometimes in the aqueous humour of the eye consist of globules formed in a similar way, nor should I believe otherwise than that threads and branches have been produced in the fluid by accretion from without. But whatever the case may be regarding these items, we must take a double movement into account in the growth of a crystal: one, the movement whereby it comes about that crystalline material is added to certain parts of the crystal and not to others, a movement that I fancy must be attributed to the tenuous permeating fluid, and that is illustrated by the magnetic example I have given; the other, the movement whereby new crystalline material, added to the crystal, is spread out over a plane; this movement must be derived from the surrounding fluid just as, when the iron threads have risen up above the magnet, whatever is struck off one owing to movement of the air is added to another. To this movement of surrounding fluid I should

---

[86]The 1669 edition has *poros* (OPH 2, p. 338). The printer's MS. had originally *polos*, which however, by the erasing of the upper part of the letter l was changed into *poros*.

[87]Maar thinks this to have been an experiment in the circle of the Accademia del Cimento (2, p. 338). Mi sees it as a text book example and points at a similar experiment by Robert Boyle: *The Effects of Languid Motion Consider'd* (edition of P. Shaw, London 1725, vol. I, p. 477 f.).

attribute the fact that not only in a crystal but also in many other angular bodies any given opposite planes are parallel.

It might be possible to prove from the previous discourse that the efficient cause of a crystal is not extreme cold,[88] that it is not only the ashes burned out by the force of fire that turn into glass,[89] that the force of fire alone is not the producer of glass, that not all crystals were produced in the beginning of things but that they are being produced even now, that it is not a task beyond the power of man to detect the formation of glass without the violence of fire, provided one undertakes a careful analysis of rocks in whose cavities the best crystals are formed; for it is certain that just as a crystal has formed from a fluid, so that same crystal can be dissolved in a fluid, provided one knows how to imitate nature's true solvent.[90]

It is no disproof that certain solid bodies, once their dissolving fluid or solvent has been removed, can be dissolved no more by the same or a similar solvent, for this occurs in bodies from which the entire solvent has been broken down by the force of fire; but crystal, and all angular bodies that dissolve in the midst of a fluid or harden in the midst of a solvent, can never come out in such a pure state that no particles of the solvent are left within the parts of the angular body; and upon this fact depends the main cause of variation by which crystal differs from glass not only in refraction but also in other properties, since, in glass, no parts of the dissolving fluid are present, as they have been driven forth by the violence of fire; for the fluid in which a crystal is hardened bears the same relation to the crystal as ordinary water bears to salts; this could be proved easily by recording the characteristics that are common to the formation of salts and the formation of crystal.

But, since I should be wandering too far from the proposition if I were to refer to all these things here, I shall mention one experiment only which seemed very beautiful to me. In various places in the same stone, lamellae separated from each other were filled with crystals, of which some were watery, others very clear,

---

[88]Since olden times the opinion was widespread that quartz originates from ice during extreme frigidity. The name is identical with the Greek word for ice (Mi).

[89]Mi thinks that Stensen incorrectly regards glass and quartz to be identical or at least chemically very closely related substances. Hobbs on the other hand says: It cannot be expected of Stensen - great though he was in his day to have discovered the important fact of the orientation of the crystal-molecules, but actually he pointed out the strange peculiarity of refraction which distinguished the crystal from amorphous substances like glass.

[90]The term menstruum which Stensen uses, means a dissolving liquid. Cf. a. o. R. Boyle, *Experiments and Observations upon colours* 2, p. 96: That gold dissolved in Aqua regis transfers its colour to the menstruum, is a general observation. The solution of mercury in Aqua fortis, does not usually seem to transfer a special colouring to the menstruum. (Wi). Aqua fortis (nitric acid) and Aqua regia (a mixture of nitric acid and hydrochloric acid) were also produced by van Helmont, and the goldsmith of the time used such acids practically as tests for gold and silver (M. Boas, *Robert Boyle*, p. 128).

some white, many coloured like amethyst, mingled together without any mixing of the colours, all exactly as experiments made here with salts show: that vitriol and alum dissolved in the same water have, after removal of part of the water, each formed separately without any mixing of their parts.

Angular Bodies of Iron

The angular bodies of iron[91] that I have happened to observe are classified into three types[92]; the first of these is flat, thicker in the middle, thinning out gradually towards the ends where it terminates in a solid angle; the second has 12 planes and the third 24. Sometimes from the second class an angular body is derived made up of six planes, resembling two three-sided pyramids formed together by their bases so that the base angles of one bisect the other's basal edges.[93,94]

The second and third classes of angular bodies of iron resemble crystal in regard to:

1. The place of production, since the place where iron forms is partly solid, partly fluid, and is a cavity in a rock.
2. The place to which material is added, since in iron also it is not added to all the planes but to certain planes only, and not always to all of these, nor always at the same time, but sometimes to one, sometimes to another, sometimes to the ends, sometimes to the middle.
3. The place from which the material of the iron comes, since this matter also seems to have flowed out of the pores of a more solid body.
4. The way in which the said material is directed towards the solid by the operation of the permeating fluid, and is spread out over the plane and smoothed by the motion of the surrounding fluid.

---

[91] Steno says *ferri corpus* - an iron-body or only ferrum - iron; one translation thinks that by iron body or iron is meant haematite (Mi), and he translates the headlines angulata ferri corpora with iron crystals. Maar explains this: Haematite crystals from Elba, this haematite is a fibrous deviation of common haematite.

[92] Now Steno says *genera*, now *species*, which is always literally translated, actually only different types of formation are meant.

[93] The sentence *ut unius basis .... latera bifarian secent* is put in the Ms. over the erased: *Ut in planis piramidum unius piramidis latera alterius piramidis bases medias secent*, fol. 17ᵛ.

[94] Mo says: *Haematite* or iron oxide (Fe_2O_3) is found in nature, actually either as segments in which the prevailing development of the base (111) makes the surface of the rhombohedron (the first type) almost indistinguishable, or a combination of two rhombohedra, the straight fundamental (100) and the straight obtuse one (211) (the second type), or, from a third aspect as a result of the combination of the shape mentioned with the hexagonal pyramid of second order (311). The six-sided shape of the second type double is the single rhombohedron (100). The two latter types are more frequent in the haematite of Elba, the former in the Alps and in volcanic areas.

They differ in material and shape; while the material of crystal is transparent, the material of iron is opaque[95]; the shape of crystal is made up of 18 planes, of which the twelve terminal planes are polished but the 6 intermediate planes are striated; on the second type of iron, however, 12 planes are counted, of which the 6 terminal planes are striated and the 6 intermediate planes are polished, and in the third type of iron 24 planes are counted, of which 6 terminal planes are striated and the intermediate 18 planes are polished; sometimes 6 other glittering planes are placed between the striated terminal planes, resembling the truncated sides of triangular pyramids.[96] It seemed to me to be worth consideration that a cube truncated at its extremity can show all the planes of the third type of an angular body of iron; for there are there six pentagonal planes coincident with the planes of the cube, bisecting each side of the cube's planes at the four angles; all of the remaining planes are found at the corners of the cube when they are truncated in a certain way.[97]

There is another thing about these same angular bodies of iron that is no less to be wondered at: in the second class of angular bodies of iron, the terminal planes, striated and pentagonal, change as time passes into three-sided planes; but the intermediate planes, three-sided and polished, become pentagonal, having two right angles close to each other,[98,99] but between the individual planes of a pair of pentagonal planes whose right angles touch, two triangles, or two three-sided planes, are formed, also polished, whose bases coincide with the perpendicular sides of the pentagons, so that in consequence the second class of iron is changed into the third. I am convinced, moreover, that a body of 24 planes may be made in the same way from a body of 12 planes, for the following reasons:

1. In the same mass of iron bodies, almost all the thinner ones have only 12 planes, while the thicker bodies have 24.[100]

---

[95]Here the substance of rock-crystal, is explicitly called *materia crystalli*, and the substance of haematite, matenia ferri, (Mi).

[96]They are the surfaces of another rhombohedron (M).

[97]While Mi and Mo call the comparison not very felicitous, Wi explains it: The polyhedric angles. Evidently Stensen thinks of the relation of the rhombohedron to the cube.

[98]Obviously Stensen imagines that the different modifications of haematite are the results of the respective developments of new crystalline shapes (Wi). Here Mi speaks about corners.

[99][void]

[100]In the Latin text congeries, i. e. mass, pile; no doubt the Elba groups of crystals are meant (Mi).

2. In certain bodies with 12 planes, the beginnings of triangular planes are observed; these are accessory and, if continued, make up a body of 24 planes.

I have sometimes observed in triangular planes a smoothness so perfect that not even the least unevenness was evident to my eyes, which is something that I have never happened to observe in any crystal; in other specimens, I have seen smaller circular planes laid on top of larger circular planes, the topmost, for the most part, being closest to the vertex of the triangle, so that it is permissible to wonder whether pentagonal planes are formed from the bases of triangular planes, since traces of striae appear in them, parallel to the bases.

It may be inferred from pieces of copper ore preserved by you among other curiosities of nature that, in the case of copper minerals, angular bodies are formed in the same way as was described for crystal and iron; but since all the spaces of the bodies are filled with an abundance of material, it is difficult to study the complete shape of the bodies. Nor is it different in any way from the angular bodies of silver sent to you from Germany.[101]

Concerning the Diamond

The same is inferred about the place and method of production of a diamond from its structure; as was inferred for crystal, namely,

1. They are produced in a fluid enclosed in the cavities of rocks, even though a famous writer on Indian matters attempts to convince us that diamonds grow again, after a certain length of time, in earth from which they were once dug.[102]
2. They are produced from the fluid by the addition of diamond materials.
3. The actions of both the permeating tenuous fluid and the surrounding fluid are to be considered in their production.

---

[101]As to the copper - and silver - ores in the Grand Duke's collections, which stemmed from the Tyrol, Saxony, Norway etc., and which Stensen completed during his travels, cf. Scherz, Indice (List of natural specimen), especially p. 266 where Targioni-Tozzetti's report on a list of more than 22 silver ores from Saxony is mentioned.

[102]Indian diamonds are often found embedded in limestone. When the diamonds, visible on the surface have been won and the limestone is left, it will crumble in the course of the year, and thus new diamonds will be visible, therefore the Indians believe that the diamond grows out again. The famous writer may be Garcias ab Horto: *Anomatum et simplicium aliquot medicamentorum apud Indos nascentium historia* (Antverpiae 1574) 1, 177. Cf. Garboe, *Ædelstene* 23 f.

Moreover, diamonds have various shapes; since some have 8 planes, others 9, others 18, and others 24 planes, most of which were striated, some also smooth.[103]

Although some were angular, nevertheless some surfaces are curved rather than flat.

Concerning Marcasites

The material of marcasites assumes various shapes[104]; sometimes it encrusts the surface of a place; sometimes it condenses into bodies of many planes; sometimes it forms rectangular parallelepipeds, which in common parlance we shall call cubes, through equality of all the planes is infrequently observed.

Since it has been possible for me to observe various things about cubes of marcasites, both regarding the cubes themselves and the places in which they are found, I shall discuss only these; their production, however, is different from that of crystals.

1. In time; since cubes of marcasites were produced before the production of the strata in which they are contained, but crystals were formed subsequent to the production of strata.
2. In place of production; for crystal, at least as it formed, was resting on a solid body, and consequently was contained partly in a solid place and partly in a fluid; but cubes of marcasites seem to have formed between two fluids, since no traces show even in the larger cubes of cohesion with another body, although small cubes are often found which, while growing, adhered to each other in the surface of the fluid.

We are taught, moreover, by the most substantial proofs of the great Galileo[105] that heavier bodies of this kind can remain on the top surface of a fluid while one of their surfaces is in immediate contact with an overlying and lighter fluid of another sort; the aqueous nature of one of the fluids referred to is shown by the material of the strata that is deposited from the said fluid.

---

[103]No regular diamond-shape but probably caused by the disappearance of certain planes.

[104]In the text De marcasitis which mostly means Pyrites, Steno uses the plural, as the translation does.

[105]In the Discorso: Al Serenissimo Don Cosimo Gran Duca di Toscana Intorno alle cose, che stanno in su l'acqua, o che in quella si muovono. Di Galileo Galilei. (Firenze 1612. Le Opere di G. G. Ed. Nat. 1V (1894) pp. 63-141), Galileo examines the floating of specifically heavier bodies on a specifically lighter fluid. Today these phenomena are explained by surface tension which appears to produce an elastic film on the surface, a thought which Galileo approached when he reached the conclusion that a solid body can float on a lighter fluid, if one of the surfaces of the bodies touches a liquid above it, which is lighter than the first. Stensen's remark about "great Galileo" was at the time a courageous acknowledgment of the physicist and probably of his own friendship with Viviani as well.

3. In the method and place of apposition, since the material of a marcasite is added to all the planes of the cubes, as opposed to what we have stated happens with crystals, this is clearly shown by the uniformity of all the surfaces of the cubes that I have myself cut from rocks; all the planes of these cubes had striae parallel to two sides, in such a way indeed that the striae in opposite planes ran the same way but adjacent planes showed different directions of striae. It follows from the direction of the striae that the surrounding fluid was arranged around every cube by a threefold movement; of these motions one was vertical, the other two parallel to the horizon but at right angles to each other.[106] Nor is it difficult to explain the manner of this triple movement, for while the fluid attempts to move away from the centre of the earth, that vertical movement is impeded by the base of the cube, to that the said fluid is deflected towards the narrower sides (since the impetus of the ascending fluids is stronger across the wider sides, consequently it allows no passage this way) and in this way two pairs of planes are marked by traces of striae; the third pair of planes received its striae from the part of the fluid that passes between the cube and fluid rebounding from the base of the cube.

4. In perfection of shape, for in crystals there is scarcely one indeed to be found in whose shape there is nothing lacking; cubes of marcasites, however, rarely have anything missing; nor is it difficult to reason this out, for, since in a crystal all the solid angles, with the exception of the terminal angles, are obtuse, and crystalline material is added separately to each of their planes, a given plane becomes less perfect, spoiling the shape of neighbouring planes, as more material is added to it alone: in cubes of marcasite, however, since all the solid angles are right angles, although one plane only has new material added to it, the said plane always retains the same size, the neighbouring planes keeping their shape.

Various other things are observed in cubes of marcasites, such as cubes enclosed within cubes, material of marcasite covered by a transparent material that encloses another marcasite, and other things of this sort that I reserve for the dissertation itself. There are also angular bodies broken down into

---

[106]The triglyphic striae of the surfaces of the Pyrites cubes which are signs of the low degree of physical symmetry which a pyrites-cube possesses, compared with the regular cube. The surface striae of the prism of the quartz crystals which Stensen found on them are actually produced by the rhombohedral surfaces which touch the prism (Mo). Wi: Stensen was probably the first to observe that the striae on the surfaces of Pyrites cubes are usually arranged in parallel in accordance with the three edges that cross each other.

lamellae, just as rhomboidal selenites are rhomboidal bodies that are broken down into other rhomboidal bodies[107]; and various other bodies which, although differing from crystal in many respects, nevertheless all agree with one another in this that they were formed in a fluid and from a fluid; which is also true of talc, that most famous of chemical substances,[108] so that there is a minimum of error in those who believe that the solid body of talc can be resolved into a fluid body, since it is beyond argument that it was formed from a fluid; but, however, there is no doubt, that those who attempt to extract this favour from it by the torture of fire are very wide of the mark; for talc, being accustomed to being dealt with more gently by nature, becomes indignant at such savagery in lovers of beauty, and, in revenge, yields to vulcan that portion of its own dissolution that it keeps enclosed within itself.

If an accurate investigation of angular bodies were begun, not only of their composition but also of their dissolution, we should soon acquire a sure knowledge of the variety of motion with which the particles of both the tenuous and surrounding fluids are agitated; just as this part of Physics is necessary to all for a true explanation of the workings of nature so also there are few who have engaged in it.

2.27.7   Shells of Molluscs

Among solids naturally enclosed by a solid, no group occurs more frequently, nor arouses greater doubt than the shells of molluscs. Therefore, I shall speak at somewhat greater length about these, considering first shells taken from the sea and then those dug from mountains.

Shells of every kind that at some period of time have enclosed a living creature reveal the following characteristics to our senses:

---

[107]This probably means *calcite* or either its cleavage particles, as well as dolomite (Mi).

[108]Mi says it is *gypsum*. Steno considers its water to be mechanically enclosed. But by tale he probably means the snow-white or yellowish leafy and soft magnesium silicate which is often used as a means of smoothing or as a cosmetic powder. Even Agricola and Boëtius de Boot in *Gemmarum et Lapidum Historia* connect our conception of tale with this, i. e. magnesium-silicate (Hinze II, 1, p. 815 ff. ef. Scherz, *Vom Wege*, 148 f.) R. Boyle says in *The Usefulness of Philosophy* (Ed. P. Shaw, London 1725, vol. I., p. 67): "A trustworthy person, a pupil of Cornelius Drebel, however, assured me that he could build a kind of furnace in which by the sheer power of fire he would be able to make Venetian tale fluid, which, I must admit, I cannot bring about by means of fire in a glass-flask. The tale generally used for cosmetics is so difficult of combustion that outstanding chemists consider all the products of combustion from tale to be false." According to Wi, K. v. Bülow says: "Tale was not to be smelted by the means at the disposal of the age".

1. The complete shells are divided into "testulae"[109]; the testulae are again divided into filaments, and these filaments are reduced to two types, differing from each other in colour, substance and location.[110]
2. The upper and lower surfaces in the testulae are nothing else than the extremities of filaments, but the lateral surface consists of the sides of these same filaments located at the edge of the testula.
3. The inner surface of the shell itself is identical to the inner surface of the inmost or largest testula, but the outer surface is composed of the outer surface of the smallest testula and the surface of all edges of the intermediate testulae.

The following points can be demonstrated clearly concerning the way in which shells are formed on animals:

1. The substance of the filaments is like the perspiration of animals in that it is a fluid excreted through the animal's outer surface.
2. The shape of the filaments can be produced in two ways, either by the very pores of the animal through which they are excreted or by the fact that the surface of the growing animal, becoming larger than the surface of the testulae that have hardened long ago, separates from this surface and so partly draws the viscous fluid contained between the two surfaces into filaments (a process that is commonly met in viscous fluids), and partly adds to it by excretion of fresh fluid because no other substance can enter between the two said surfaces.
3. Differences in the filaments depend on a difference in the pores by which the animal surface is perforated and upon a difference in the material that is excreted through the pores; for animals of this kind possess a twofold substance in their surface, of which one is harder and the other softer, both being fibrous; a more careful examination of these throws no little light on the investigation of bones.
4. All the testulae, excepting the outermost or smallest, were produced between the outer shell and the body of the animal itself and have thus not

---

[109]* Winter translates "testulae" as "subdivisions", Krogh and Maar in note 10 as "scales" of the shell. [keep this foot-note]

[110]The shells of most molluscs that are provided with them consist of an outer skin of organic substance, a prismatic layer consisting of polygonal small prisms, standing vertically on the surface of the shell, one placed on the top of the other, and of a layer of mother-of-pearl consisting of many lamellae, which are parallel with the surface. The two latter layers mainly consist of calcium carbonate, mixed with a little organic substance. All of them are secretions of the coating. In § 3 Stensen thinks of the two last layers and in § 4 he assumes the existence of this thin skin (Mo). Cf. with this paragraph of the *Prodromus* Spärck in *Indice*, 87 ff.

taken their shape independently but as a consequence of their location; the result is that, in oysters, the motion of the animal and the amount of substance often give rise to some variation in form.

Regarding the outermost testula, there can be some doubt whether the surrounding fluid touched the outer surface or whether it was protected by some kind of membrane; I should think, however, that only the latter view is correct because:

1. The filaments of all the remaining testulae were untouched by the surrounding fluid at the time they were formed.
2. We see in prickly cockles *chama hirsuta* that a membrane or something like a skin covers the outside of the shells. But the problem is about matter that is all but imperceptible, and it can be stated that the filaments of the first testulae had already hardened within the egg, since experience proves that oysters and other testacea arise from eggs and not from decaying matter.[111]

From what has been stated, it is easy to explain the following:

1. All the variety of hues and prickles which arouses the wonder of many in the case not only of native but also of foreign shells; for this diversity has no other origin than the edge of the animal enclosed in the shell. And indeed this edge, growing and expanding gradually from something exceedingly small, leaves its imprint on each margin of the testulae; since the said margins either form from the fluid that is excreted from the outer edge of the animal or are in fact the animal's outer edges growing anew, perhaps, like the teeth of the shark; in the place of an earlier edge and, like those same teeth, being thrust gradually outwards.
2. The formation of pearls; not only those that, adhering to the shells, are not quite round in form, but also those that, after the openings of the pores in the animal's surface have closed, acquire a round shape within the pores themselves; for the difference between the integuments of pearls and the testulae in shells of pearl-bearing molluscs is merely that the filaments of the shells are located as it were in the same plane but the integuments of pearls have their filaments distributed over the same spherical surface. An

---

[111]The spontaneous generation of Aristotle (*De Animalibus Historia*, lib. V. c. 15 and *De generatione animalium* 23, III, 9, 10 and 11) was conquered in this very age through thorough investigations by Harvey as well as by Stensen's friends, Johannis Swammerdam's *Historia insectorum generalis* (Utrecht 1669), and principally Francesco Redi, *Esperienze intorno alla generazione degli insetti. Da Lui scritto in una lettera all Ill. mo Signor Carlo Dati* (Firenze 1668. cf. the edition by A. Pazzini, Roma 1945, and EP 1, 21ff.). One of these friends was Malpighi, to whom we owe the doctrine: *Omne vivum ex ovo*.

elegant example of this was provided by one pearl among those that I broke at your command[112]; this pearl, although externally a glittering white, enclosed within itself a black body that resembled a grain of pepper both in colour and size, in which the placing of the filaments, with one end tending towards the centre, was very clear, and the rows of spheres of the same filaments could be distinguished. At the same time, I saw[113] that:

1. The unequal excrescences on various pearls are nothing but various very small pearls enclosed in the same common crusts.
2. Many pearls of a yellow hue are tinged with yellow pigment not only in the outermost surface of the sphere but in all the inner spheres; and so it is no longer possible to doubt that the colour must be attributed to the changing fluids of the animal, and that he who seeks to wash it clean washes an Ethiopian,[114] unless either the colour has been acquired, for instance, from wearing natural pearls on the neck, or the pearl was yellow in the outermost sphere only, as might be the case if the fluids of the animal had changed when the inner spheres were being formed.

There emerges clearly from these facts the error of those who, without knowing their nature attempt the clever imitation of pearls, for hardly anyone could attempt this feat successfully unless another Lucullus should replenish his aquaria with pearl-bearing molluscs[115] and either seek in the animals themselves the means of multiplying them or learn from this the difficulty of imitating the works of nature. I would not deny that it is possible to make, artificially, globules formed of various integuments but I should consider it indeed most difficult to arrange these same integuments from a series of

---

[112]The paragraph *Elegans hujus rei... sphaerae formabantur* is not found in the MS. but a sign seems to point to a note, fol. 20ʳ, now probably lost.

[113]Cf. OPH 2,339: The shell of the mussel is made up of three layers, the outer or "epidermic layer", the "prism" layer, consisting of short rods or fibres, and the inner layer of mother-of-pearl, consisting of thin plates or scales. In the pearl which Steno has examined, the layer of mother-of-pearl has probably only been very imperfectly developed while the "prism" layer has been uncommonly thick, so that he has chiefly noticed the fibres or rods of the latter, which all conveyed toward the centre. See cf. K. Möbius, *Die echten Perlen* (Hamburg 1858).

[114]This refers to a fable by Aesop (6th century B. C) who relates the following: A man bought an Ethiopian regarding the latter's genuine colour as the consequences of the neglect of his former owner. The new owner took him into his house and applied all kinds of soaps, trying to scrub him white by means of many different baths. However, he was not able to change the colour and nearly fell ill through his toil. What is characteristic, remains what is was. (Halm, *Fabulae Aesopicae*. Leipzig 1875, XIII. - OPH 2, 339 and Wi).

[115]About these fishponds Pliny says in *Natural history* VIII, 211(52) a, ix, 170 (54), that Lucullus had them built near Naples at heavy expense. A mountain was removed and a villa built, in whose pond the saltwater could hold an abundance of saltwater fish (OPH 2, 339 and Wi.).

filaments added one to the other, upon which process the natural lustre of pearls depends.[116]

Shells that lie buried in the earth may be reduced to three classes:

The first class consists of those that are as like the shells just described as an egg resembles an egg, since the shells themselves are resolved into testulae, the testulae into filaments and there is the same diversity and position of filaments. Even if marine testacea had never been observed, examination of the shell itself would prove that these shells were parts of animals that lived at one time in a fluid, as will appear from the example of bivalve mussels.

At the time when bivalve mussels were formed, the material contained within the mussel:

1. Had a smooth surface pierced with innumerable pores and had two varieties of pore.
2. Was a pliable substance less hard than the shell itself.
3. Communicated on one side with surrounding material; on the other side had no communication with it.
4. Withdrew gradually from the side where communication with outer material was denied towards the side where it had free communication with the said material.
5. Was able to open itself at intervals according to the size of the angle allowed by the hinges of the shells.
6. Grew from a small to a large size.
7. Transmitted through its own substance the material that made up the testulae of the shell.

Regarding the outer material surrounding mussels:

1. If it were not completely fluid, at least its power of resistance was less than the power of expansion of the substance contained by the shells.
2. It contained a fluid substance suited to the formation of filaments of the testulae of the shell; all these conditions of both the inner and outer regions, demonstrated by arguments and drawings in the Dissertation itself, show sufficiently that there was an animal within the shells and a fluid outside.

The second class consists of those shells that are in other respects like those just described but differ from them only in colour and weight; while some are lighter than usual, others are heavier than usual because they have their pores

---

[116]In the MS. the following after *difficillimum* has been erased: *Ne dicam supra vires humanas judicaverim*: I do not say that I could judge it beyond the power of man.

filled with added fluid; I say no more about those whose pores have been enlarged by ejection of lighter parts because they are nothing else but the shells of animals that are either petrified or calcined.

The third class consists of shells that are similar to those just described in shape only, for the rest differing completely, since neither testulae nor filaments are found in them, much less a variety of filaments. Some of these are composed of air, others, either black or yellow in colour, of stone; others of marble; others of crystal and still others of other material; I account for the formation of all of these in the following way:[117]

When the penetrating force of the fluids dissolved the substance of a shell, these fluids were either drained away by the earth, leaving empty spaces in the shells (these I call shells composed of air),[118] or were changed by the addition of new material, filling the said spaces in the shells with crystals or marble or stone, according to the variety of material; the most beautiful kind of marble, that known as Nephiri,[119] has its origin in this source; being nothing else than a deposit of the sea, filled with shells of every description, in which a stony material has replaced the decomposed substance of the shells.

The brevity of my plan does not allow me to give an account of all the things that I have considered worthy of note, in particular the kinds of shells dug from the earth; wherefore, passing by other matters, I shall refer here only to the following:

1. A pearl-bearing shell, found in Tuscany, in which the pearl was clinging to the shell itself.
2. The larger portion of a marine pinna, where the colour of the byssus[120] persisted in the earthy material that had filled the shell after decomposition of the byssus.

---

[117]The first kind of fossil shellfish are those which have gone through a proper fossilization, i. e. have been preserved in the layers of the earth. In the process of petrification others replaced the organic parts by mineral substances. Conches belong to a third kind which were first enclosed in a layer, then dissolved and only left their shapes behind which either remained empty or were filled with mineral substances. Modern paleontology also keeps to these distinctions of Stensen's. (Mo). Cf. also Rome, in *Indice* 93 ff.

[118]The text has *testas aereas*, which probably means "porous shells" (Wi).

[119]The term *nephiri* (also in the printer's MS.) is unknown and is declared to be a misprint by Maar, Mi and Mo (which it is not). Maar suggests *nephritic* thinking of nephritic marble now called breccia (OPH 2, 339 f.). Targioni-Tozzetti knows of a *Pietra Nefritica* on buildings in Pisa. Nephrite (jade), a feltspar like actinolite, from the hornblende group known by Aldrovandi as *Lapis Indicus* could certainly not contain conches (*Indice* 218).

[120]*Byssus*, the exquisite and expensive fabrics of the Phoenicians of the Mediterranean were manufactured from the same fluid secretion (musselbeard, musselsilk), which many mussels produce in order to fasten themselves to their substratum (Scherz, Indice 218). Cf. also note 17 of the *Ornaments*.

3. There are shells of oysters, of remarkable size, in which are found several oblong worm-eaten cavities[121] in all respects like those that are inhabited by a certain type of shellfish in the rocks of Ancona, Naples and Sicily.[122] Unless they were formed by insects building a nest from mud (which I can scarcely believe since the substance of the middle portion of the rock, in which no cavities are found, is identical to the substance of the cavities, these being gathered wholly about the surface regions), these cavities have been eaten out by worms, since this is both confirmed by the surface of the cavity and proved by the discovery in many cavities of a body, made up of thicker filaments, that conforms to the cavity itself in size and shape.[123]

Assuredly, the cavities were made neither by nor around shellfish, since testaceae of this kind lack organs for gnawing and no cavity corresponds to the shape of their shells. Nor is it surprising that rocks exposed to the sea afford a resting place, in the said cavities, for the eggs of shellfish cast up by the sea, for I have not yet observed any of these cavities lacking an obvious outlet. But if it is stated that the cavities were made by a petrifying fluid that hardened around certain bodies, some cavities would have been found enveloped on all sides by the same material, and lacking an opening.

4. A shell, partly destroyed internally, in which a marble incrustation, covered by various balanoids,[124] had replaced the substance eaten away; thus it is possible to conclude with certainty that the shell had been left upon land by the sea, then carried down into the sea, covered again by a new deposit and abandoned by the sea.
5. Very small eggs and helical shells which can hardly be distinguished except with the aid of the microscope.
6. Pectines, helical shells and bivalve molluscs, not covered with crystal but wholly crystalline in their substance.
7. Various tubes of sea worms.

---

[121]The MS. as well as the Ed. Florence 1669 has *qvos*.

[122]These oblong cavities are evidently produced by the Lithophagus (Lithodomus). These bivalves of the Mytilidae family perforate shells of the lamellibranches Melina, Ostrea and Pectin and form a bottle shaped cavity. Cf. K. A. v. Zittel, *Grundzüge der Paläontologie*, 3. Aufl., 1. Abt. (München 1910, p. 322 and Fig. 632 c) (Wi).

[123]They are cavities indeed, perforated by Phoiades, molluscs with a reddish shell and rough as a file, which serves to gnaw the stones. (Mo).

[124]Shellfish (Cirripedes), provided with a chalk shell shaped like a little flowerbud, which lives fastened to conches and rocks (Mo).

Other Parts of Animals

What has been said about shells should be said also about parts of animals and animals themselves buried in the earth, among which may be reckoned the teeth of sharks, the teeth of eagle-fish,[125] the backbones of fishes, and all kinds of whole fish, the crania, horns, teeth, femurs and other bones of terrestrial animals, since all of these either resemble true animal parts in every way, or differ from them only in colour and weight, or have nothing in common with them excepting their outward shape alone.

Great difficulty is caused by the countless number of teeth carried away each year from the island of Malta,[126] since scarcely a single ship calls there without bringing back some proof of this marvel. But to this difficulty I can find no other answer than:

1. That there are more than 600 teeth to each shark, and that new teeth seem to grow throughout the whole of the shark's lifespan.
2. That the sea, whipped up by the winds, tends to push bodies in its path towards a given place and to heap them up there.
3. That sharks gather in packs and so many sharks teeth could be left in the same place.
4. That in lumps of earth brought here from Malta, besides various teeth from various sharks, are found also various shellfish, so that if the number of teeth persuades us to ascribe their production to the earth, the construction of the teeth and their abundance in each animal, the similarity of the earth to the sea bed, and the other marine bodies found in the same place favour the opposite opinion.

For others, difficulty arises from the size of the femurs, crania, teeth, and other bones that are dug from the earth; but there is not much either in this objection that unusual size should suggest a method beyond the powers of nature, since:

1. In our age, men of very large stature have been observed.
2. It is certain that there existed at one time men of gigantic size.[127]

---

[125]In the text: *piscis aquilae*, refers to a shark-family, whose scientific name is Myliobatidae, popularly known as eagle shark, devilfish or sea devil. Their teeth are flat grinding teeth for crushing hard material. (Wi).

[126]Contrast his cautious way of presenting the problem in *Canis Carchariae* (OPH 2, 127 ff.) with the zoological and palaeontological reasons for the right view as we have put them here.

[127]Wi says: The belief in the existence of giants was supported by the finds of fossil animal bones and was widespread, cf. E. B. Tylor, *Researches into the Early History of Mankind* (London 1856), p. 314-317; *Primitive Culture*, 4th edition (London 1903), 1, 387.

3. Often the bones of other animals are mistaken for the bones of human beings.
4. To attribute to nature the production of truly fibrous bones is on a par with saying that nature can produce the hand of a man without the remainder of the man.

There are those to whom the length of time seems to destroy the force of the remaining arguments, since there are no recollections in any age to confirm that floods have risen to the places where many marine bodies are found today, if the universal deluge is excepted, from which time it is estimated that 4000 years have elapsed up to the present;[128] nor does it seem in accordance with reason that a part of an animal body has resisted so many years of wear, when we observe often over the space of a few years the complete destruction of the same bodies. But it is easy to reply to this disbelief, since the outcome depends entirely on the variety of soil; for I have observed strata from a certain type of clay which dissolved all bodies enclosed by it, owing to the thinness of its juice; I have observed many other strata of sand which preserved whole all that lodged therein. By such experiment it may be possible to come to a knowledge of that juice, which dissolves solid bodies; but the following argument provides to our satisfaction that it is certain that the production of many shellfish found today is to be reckoned as coincident with the ages of the universal deluge. It is certain that before the foundations of the city of Rome were laid, the city of Volterra was already powerful; but shellfish of every kind are found in the huge stones that are found in certain places there (the remains of the oldest walls), and not so very long ago a stone filled with striated shellfish was hewn from the middle of the forum; and so it is certain that shellfish found today in the said stones were already produced at the time when the walls of Volterra were erected.[129] And lest some say that only shells, converted to stone, or enclosed by stone, have suffered no damage from the gnawing of time, the whole hill on which the oldest of the Etruscan cities is built rises from marine sediments, laid on top

---

[128]According to O. H. Schindewolf, Stensen with his law of stratification created one of the most important bases for chronological order in palaeontology, but at the time, the background for an absolute dating was missing (*Grundlagen und Methoden der paläontologischen Chronologie.* (Berlin 1950). 3. Auflage p. 41). Hölder says: "During the 17th and 18th centuries the doctrine about the flood included in the first outlines of the history of our world was in the foreground for some time because the biblical account of it and the geological finds seemed to overlap". (*Geologie* etc. p. 130, Leibniz and John Woodward thought so too).

[129]The stretches of massive walls inside the Porta all' Arco of Volterra consist of yellow limestone mixed with conches called *panchinae* (Cf. Dennis, *Cities and Cemeteries of Etruria*, London 1878, 2, 144). (Wi). As to the huge masses of fossil mussels in and around Volterra cf. TT. Viaggi 3, 11ff.

of other, parallel to the horizon, in which there are many non-stony strata which abound in true molluscs that have suffered no change in any way, and so it is possible to say with certainty that the unchanged molluscs that are extracted from them today were produced more than 3000 years ago. From the founding of the city to the present day we reckon more than 2420 years have elapsed; and who will not grant that many centuries have elapsed since the first men transferred their homes there until it grew to the size that flourished at the time of the founding of the city? If we add to these centuries the time which elapsed between the laying down of the first sediment of the hill of Volterra and the withdrawal of the sea from the same hill, when strangers flocked to it, we shall easily go back to the time of the universal deluge.

The same authority of history forbids doubts that the huge bones that are dug from the fields of Arezzo have resisted the ravages of 1900 years, for it is certain that:

1. The skulls of pack animals found there are not of animals of this clime, as neither are the huge femurs and very long scapulae found in the same plain.
2. Hannibal crossed thither before fighting the Romans at the Trasimene Lake.
3. In his army there were African pack animals and turret-bearing elephants of immense size.
4. While he was coming down from the mountains of Fiesole, a large section of the animals selected for carrying packs perished in the marshy regions owing to excessive floods.[130]

---

[130]According to Eutropius (III, 8) and Polybius (III, 42), Hannibal had 37 elephants when he entered Italy. Polybius on the other hand states that all except one perished on account of the great cold immediately after the battle of Trebia (218 B. C.) in the Po valley, and Livius (XXI, 36) also remarks that almost all of them suffered this fate. In the Chiana valley Grand Duke Ferdinand II had a whole skeleton unearthed in 1663, bones of which were kept in Florence. On the rich stores of fossils of elephants in the Arno-valley cf. TT. Viaggi 8, 415 ff. In the museum of Montevarchi, Valdarno, fossils are still found to-day, which Stensen may have seen. Cf. also Forsyth Major, Considerazioni sulla Fauna del Mammiferi pitocenici e postpliocenici della Toscana in: *Atti di Società Toscana di Scienze Naturali in Pisa* (1865) 1, 7–40, 223–245, (1877) p. 207–227. Mammalian Fauna of the Val d'Arno in: *The Quarterly Journal of the Geological Society of London*, vol. XLI (1885)1-8. Ch. Depéret, Evolution of Tertiary Mammals, and the Importance of their Migrations in: *The American Naturalist*, vol. XLII (1908) 109–114, 160–170, 303–307 (Wi). Cf. also I. Cocchi's and F. Nesti's books in: *Indice* 197–246.

5. The place from which the said bones are dug was built up from various strata that are filled with stones rolled down from the surrounding mountains by the force of torrents; so that the obvious agreement in all details can no longer be hidden from anyone who compares the nature of the place and of the bones with the historical record.

2.27.8 Plants

What has been said about animals and the parts of animals holds equally for plants and the parts of plants, whether dug from the strata of the earth or lying hidden in the substance of rock; for either they resemble true plants and parts of plants in every respect, which is a rare occurrence, or they differ from them only in colour and weight, which is a common occurrence, sometimes burnt to charcoal, sometimes impregnated with petrifying fluid[131] or they correspond only in their shape, there being a great abundance of such cases in various places. There cannot be the least doubt that the first two classes were once true plants, in that the structure of their bodies forces us to this conclusion, and the nature of the place from which they are dug is not inconsistent with it. Those who object, with the view that earth transferred to houses became wood in the course of time, can affirm this only about the surface of the earth enclosing the wood, where the earth, dried up with time and crumbled into dust, disclosed the wood within; nor do the metallic filaments found in the pores of the said wood weigh against us, since I myself have drawn from the earth a trunk, with knots on its branches and a bark which testified to its plant origin, whose fissures were filled with mineral material. From hence no small light might be thrown on the doctrine of minerals, if an investigation were made of wood and the location of the wood to find what they could have contributed to the production of minerals. Many things are called bitumen that are proved to be nothing but charcoal by the channels in the fibres and the ashes of parts that are burned.

With the third class of plants, or plant shapes inscribed on stone, a greater difficulty arises; since we observe shapes of this kind in hoar-frost, in the mercury tree,[132] in various volatile salts, in a white substance that is soluble in

---

[131]Mi: that is how Steno imagined the origin of the fossilized coal.

[132]A. Kircher wrote in *Mundus Subterraneus* (tom. II, lib. IX, c. XXIV, p. 431: "On metallic trees and their artificial production. 1. experiment: The tree of the philosophers grows before our eyes: Take half an ounce of purest or barrel-cleaned silver and 2 ounces of Aqua fortis and mercury, in which the mentioned silver is dissolved. Blend the two substances in a pan, pour a pound of common water on top and pat it

water,[133] which in glass vessels not only grows on the inner sides of the vessels but sometimes rises up from the middle of the base into the free air. But if everything is considered properly, nothing arises to oppose the opinions expressed; for the shapes of plants inscribed on rocks may be reckoned in two classes; some adhere only to the surfaces of cracks, which I should concede were produced easily without a true plant, though not without a fluid; others not only appear on the surface of cracks but spread their branches everywhere through the stony substance itself; whence it follows that at the time when the said plant was produced, whether it was made in the manner of other plants, or in the fashion of a mercury plant, the substance of the stone had not yet given up the nature of a fluid; this is confirmed further not only by the softer consistency of the stone itself but also by the angular bodies that occur frequently in the dendrites of Elba, such as form only in a free fluid. But what is the need for other proof when experience itself speaks? I have examined various moist places, some exposed to the light, some underground, where, because of water flowing by, a rock growing into moss and other plants was being covered with new mosses of another kind.

I have, previously, reviewed the chief bodies for which the place of discovery has caused many to be indecisive about their place of production; and I have, on the same occasion, indicated how from that which is perceived a definite conclusion may be drawn about what is imperceptible.[134]

(Footnote 132 continued)

down, and you will see it grow daily, both the trunk and the branches" (OPH 2, p. 340). About the various views on the petrified juice of R. Boyle and van Helmont, cf. Pazzini, Pietre 24 ff.

[133]Ammonium chloride, Sal-Ammoniac (Wi).

[134]Regarding this paragraph Becksmann points out (N. Steno p. 329 f.) Stensen's "great scientific achievement", in contrast to "other occasional clever utterances". Before the Prodromus there had not existed any real geological thinking. "For his work does not amount to just a small section of the history of the earth or the interpretation of a single geological document, but to the first account planned on generous lines of the surprising results of the geological development of a landscape up to the present configuration its surface has built up, and the first to be planned for this purpose only, on authentic doctrine and criticism of sources". Stensen is a tremendous "Autochthoner" (original thinker). The more admirable is his achievement, amounting to genius, since he was in no way inspired or encouraged by intellectual currents of his age, but, became the founder of geological thinking completely on his own, by means of clear, logical thinking and by following up problems which arose out of a comparison of recent living beings with fossils.

2.27.9   Different Changes that have Occurred in Tuscany

How the present state of anything discloses the past state of the same thing is made abundantly clear by the example of Tuscany,[135] above all others; obvious inequalities in the present surface contain within themselves clear indications of various changes, which I shall review in inverse order, working back from the most recent to the first.

1. At one time, the inclined plane A was in the same plane as the higher horizontal plane B, and the edge of the same plane A so raised, as also the edge of the higher horizontal plane C, were continued further, whether the lower horizontal plane D was in the same plane as the higher horizontal planes B and C, or another solid body existed there, supporting the exposed sides of the higher planes; or what is the same, everything was once level in the region where, today, rivers, swamps, sunken plains, precipices, and inclined planes between sandy hills are observed, and at that time all the waters, of both rains and springs, were inundating that plain or had opened up subterranean channels for themselves under the plain; at least there were cavities under the upper strata.
2. At the time when plane B A C was being formed, and the other planes underneath it, the whole plane B A C was covered with water; or what is the same thing, the sea was once raised above the sandy hills no matter how high.
3. Before the plane B A C formed, the planes F G I had the same position that they hold now, or, what is the same, before the strata of the sandy hills were formed deep valleys existed in the same places.
4. At one time the inclined plane I was in the same plane as the horizontal planes F and G, and the exposed sides of the planes I and G were either continued further, or another solid existed there supporting the said exposed sides when the planes referred to were formed; or, what is the same, where valleys appear today between the flat tops of the highest mountains, there was, at one time, one continuous plain, under which huge cavities were formed before the collapse of the upper strata.
5. When the plane F G was formed, an aqueous fluid lay over it; or, what is the same, at one time the flat tops of the highest mountains were covered with water.

---

[135]Cf. diagrams 20-25.

Thus, we recognize six distinct aspects of Tuscany,[136] two when it was fluid, two when flat and dry, two when it was uneven; what I demonstrate about Tuscany by induction from many places examined by me, so I confirm for the whole earth from the descriptions of many places set down by various writers. But lest anyone be afraid of the danger of novelty, I set down briefly the agreement between nature and scripture, reviewing the main difficulties that can be raised about individual aspects of the earth.

With regard to the first aspect of the earth, scripture and nature agree in this respect, that everything was covered with water; but of how and when it began, and how long it lasted as such, nature says nothing, while scripture speaks. That there was aqueous fluid, however, at a time when animals and plants had not yet appeared, and that the fluid covered everything, is proved conclusively by the strata of the higher mountains which are free from all heterogeneous material; the outline of these strata testifies to the presence of a fluid; their material bears witness to the absence of heterogeneous bodies; the similarity in materials and outlines of strata from different mountains that are widely separated proves indeed that the fluid was universal. If it should be said that solids of different kinds contained in these strata would be destroyed in the course of time, it is difficult to deny that a clear difference would be observed there between the material of the stratum and material that per-colated through the pores of the stratum to fill the spaces left by decayed bodies. But if in certain places other strata filled with various bodies are found above the strata formed by the first fluid, nothing follows from this except that new strata have been deposited by another fluid on the strata of the first fluid, the material of which could in like manner fill the ruins of the strata left by the first fluid; and so we must always return to this, that at the time when the strata of unmixed material, obvious in all mountains, were being formed, the rest of the strata did not yet exist, but everything was covered with a fluid devoid of plants, animals, and other solids. Since no one can deny that the strata are of the kind that could have been produced immediately by the First Mover, we recognize from this the obvious agreement between scripture and nature.

About when and how the second aspect of the earth, which was flat and dry, began, nature is likewise silent, while scripture speaks; moreover, nature's

---

[136]By facies Stensen evidently understands a geological phase of development. Our contemporary conception of facies stems from Gressly the Swiss geologist, and expresses the whole of the characteristics of a stratigraphical unity. (C. Chr. Beringer, *Geolog. Wörterbuch*, Stuttgart 1951, p. 30) As A. A. G. Schifferdecker, *Geological Nomenclature* (Goringhem 1959) says: "the same of the lithological and palaeontological characters exhibited by a deposit at a particular point" (No 2539).

assertion that such an aspect of the earth did exist at one time is confirmed by scripture when it teaches that waters gushing from one source overspread the whole earth.[137]

When the third aspect of the earth, which is believed to have been uneven, began, neither scripture nor nature determines; nature shows that the unevenness was of some magnitude; scripture, moreover, mentions mountains at the time of the deluge; as to the rest, neither scripture nor nature determine when those mountains, of which scripture makes mention, were produced, whether those mountains were the same as the mountains of today, whether at the beginning of the deluge there were valleys as deep as today, or whether new chasms were opened by new breaks in the strata to lower the surface of the rising waters.

The fourth aspect, when all was ocean, seems to cause more difficulty, although in truth it is not difficult. The production of hills from marine deposits testifies that the sea was higher than it is now, and this not only in Tuscany but also in very many places far enough from the sea, from which waters flow to the Mediterranean; indeed in those places from which waters flow into the ocean, nature does not contradict what scripture determines about how high the sea was, since:

1. Definite traces of the sea appear in places that are raised several hundreds of feet above sea level.
2. It cannot be denied that as all the solids of the earth were in the beginning of things covered by aqueous fluid, so they could have been covered again by aqueous fluid, since change is indeed continual in the things of nature, but nothing in nature is totally destroyed. But who has investigated the structure of the interior of the earth and will dare to deny the possible existence of huge spaces there, at times filled with aqueous fluid, at others filled with an aereal fluid?
3. It is completely uncertain what the depth of the valleys was at the beginning of the deluge; but reason persuades us that, in the first centuries of the world's existence, cavities were gnawn out by water and by fire, so that slighter collapses of strata followed from this; however, the highest mountains, of which scripture makes mention, were the highest of the

---

[137]Genesis 2, 10-14 (from the Vulgate): "A river arose in the place of joy and watered Paradise and then divided into four main rivers: One is called the Phison and it flows around the land of Hevilath, where gold arises. The gold of the country is very good and both bellim and lapis onychinus are found. The second river is called the Gehon; it flows around the whole of Ethiopia. The third river is called the Tigris; it flows east toward the Assyrians. But the fourth river is the Euphrates".

mountains then found, but not the highest of those observed in the present day.

4. If the activity of a living creature can bring it about that sometimes places flooded with waters are made dry by its decision, and sometimes are flooded with new waters, why should we not willingly concede to the First Mover of all things the same freedom and the same powers?[138] With regard to the time of the universal deluge, sacred History, reviewing everything in detail, is not opposed by secular history. The ancient cities of Tuscany, some of which are built on hills produced by the sea, were founded more than 3000 years ago; in Lydia, however, we come nearer to 4000, so that it is possible to reckon from this fact that the time at which the earth was abandoned by the sea is in accordance with the time of which scripture makes mention.[139] Regarding the manner in which the waters rose, we can put forward various agreements with the laws of nature. If it should be said that the centre of gravity of the earth does not always coincide with the centre of its figure, but sometimes moves away from one side, sometimes from the other, according to the formation of subterranean cavities in different places, it is possible to put forward a ready reason why the fluid that covered everything in the beginning of things left certain places dry, and returned again to occupy them.

The universal deluge may be explained with the same ease[140] if a sphere of waters, or at least huge reservoirs of water, are arranged around the fire in the middle of the earth; whence, without movement of the centre, the out-pouring of the enclosed waters could be derived, but the following method also seems quite easy to me; by which both a lesser depth of valleys and a sufficient quantity of water are obtained without considering the centre, either of the figure or of gravity. For if we should allow that:

---

[138]The MS. here has clearly *motori*.

[139]Cf. Scherz, *Das Feste im Festen*, p. 29 ff. Through his conception of the historical development of the sediments and mountains, Stensen broke through the rigid conception of a single finished act of creation. - And as a matter of fact the chronology of the 17th century hesitated between 3700 and 7000 years as to the length of time from the days of the creation till the birth of Christ. - On account of his own short period of research and with regard to the limited astronomical, chemical and paleontological knowledge of his time, he preferred, on one hand to stick to his own scientific perceptions, sober and fearless, and on the other hand to include the historical picture of sacred as well as the secular history, which his age offered him, an attitude which our contemporary geologists declare to be fully and scientifically justified.

[140]Cf. Kircher, *Mundus Subterraneus* 1, p. 70-71 and 230 ff.

1. Passages through which the sea penetrates into hollows in the earth to supply water to the sources of bubbling springs were blocked by the slipping of fragments of certain strata.
2. Water, undoubtedly enclosed by the bowels of the earth, was driven by the force of the well-known subterranean fire partly towards the springs and partly ejected into the atmosphere, through the pores of the earth that were not yet covered with water; then the water, not only that which is always present in the air but also that which was mixed with it by the method described above, fell in the form of rain.
3. The bottom of the sea was raised up by expansion of subterranean caverns.
4. The remaining cavities on the earth's surface were filled with earthy material eroded from higher places by the continuous rainfall.
5. The surface of the earth itself was less uneven since it was nearer in time to its original state. If we allow these points, then we allow nothing contrary to scripture, to reason, or to everyday experience.

Neither scripture nor nature makes clear what happened to the surface of the earth while it was covered with water; from nature, we can put forward only this, that deep valleys were produced at that time, because

1. The enlargement of the cavities through the force of subterranean fires provided a place for greater landslips.
2. A return passageway leading into the deeper parts of the earth had to be opened for the waters.
3. Today, in places far from the sea, deep valleys are found filled with many marine deposits.

With regard to the fifth aspect, in which huge plains were revealed, the earth being made dry again, nature demonstrates the existence of those plains while scripture does not contradict their existence; moreover, nothing certain can be determined about whether the sea receded completely and immediately, or whether, indeed, in the course of centuries, new chasms opened to provide an opportunity for the discovery of new regions, since scripture is silent, and the history of nations regarding the first ages after the deluge is regarded as doubtful by the nations themselves and is considered to be full of fables. What is certain, indeed, is that a great quantity of earth was carried down to the sea every year (as is readily obvious to any one who considers the breadth of rivers and their lengthy courses through inland regions, and the innumerable torrents; in short, all the downward slopes of the earth), so that the earth carried down by the rivers and added daily to the shores left new land fit for new settlements; this is confirmed by the judgment of the

Ancients by which they called whole regions the gifts of rivers having the same name as the regions,[141] and by the tradition of the Greeks, when they relate that men, descending gradually from the mountains, inhabited maritime regions that were sterile from excessive moisture and were made fertile in the course of time.[142]

The sixth aspect of the earth is obvious to the senses, in which the said plains were changed into various channels, valleys and precipices, mainly through erosion by the waters, and sometimes by fiery conflagrations; nor is it to be wondered at that there is no account by historians of the time when any given change occurred. For the history of the first centuries after the deluge is confused and doubtful among secular writers; indeed, with the passing of the centuries they took it on themselves to celebrate the deeds of illustrious men but not the miracles of nature. Nevertheless, we lack the records, mentioned by writers, of those who wrote the history of changes that occurred in various places; and in so far as the remaining authors, whose writings have been preserved, report almost every year among the marvels, earth movements, eruptions of fires from the earth, flooding by rivers and seas, it is easily shown that many and various changes have occurred in 4000 years[143]; so that those who point a finger at the many errors in the writings of the Ancients on the grounds that various things are found there which do not agree with present day geography, are much mistaken. I should not wish to attach a ready faith to the fables narrated by the Ancients, but many things are found therein in which I would not deny belief. For in their accounts I find many things whose falsity rather than truth seem to me doubtful, such as the separation of the Mediterranean Sea from the Western Ocean, the passage joining the Mediterranean Sea to the Red Sea,[144] the submersion of the island of Atlantis[145]; the description of various regions in the journeys of Bacchus,

---

[141]Herodotus (II, 5) calls Egypt a gift of the Nile, and Plato (*Timaeus* 22 D) knows about an Egyptian priest who says to Solon: "and from this disaster (i. e. temporary ruin) the Nile rescues and frees us, he who is our saviour who will never be dry". (Cf. Strabon, *Geography*, C. 36, I, 2, 29) (Wi).

[142]The Greek traditions on flood disaster are found in Plato: "None but the shepherds on the mountains were left, who finally descended into the valleys, founding new cultures, as Troy was founded by people descending from Mount Ida." (Cf. Plato, *Timaeus* 22 c. ss. *Critias* 111 d, The *Laros* III, 677 a—b. OPH 2, p. 340.— Wi 268 f.).

[143]To explain the evolution of the surface of the earth within the limits of the time which was accepted by the 17th century, Stensen is forced to accept the theory of violent catastrophes in nature (Wi).

[144]As to the connection with the Mediterranean cf. Plato, *Timaeus* p. 24 c. Ss., and Plinius, *Naturalis Historiae Libri*, XXXVII. lib. II, c. 85 ss. (OPH 2, 340).

[145]Plato says in *Timaeus*, p. 25. c—d: "But later when the violent earthquakes and flood disasters occurred all your military power collapsed; in a single disastrous night it sank and disappeared like the island of Atlantis. That is why in this area the sea is impassable and pathless even to-day, because a muddy bar obstructs the passage, caused by the immersed island". Cf. K. E. A. v. Hoff, *Geschichte der*

Triptolemus, Ulysses, Aeneas and others, may be true, although it does not correspond with things as they occur in the present day. I shall put forward in the dissertation itself clear proof of the many changes that have occurred in Tuscany, over its whole area between the Arno and the Tiber, and though it is not possible to fix the time at which each change occurred, I shall, nevertheless, put forward arguments from the history of Italy, so that no one will be left in any doubt.[146]

This is a succinct, not to say disordered, account of the chief things that I had resolved to set down in the dissertation itself, not only more distinctly, but also at greater length, with in addition, a description of the places where I have observed each item.[147]

2.27.10   Explanations of the Plate

Since the brevity of my hurried writing has left a number of things less clearly explained, especially where angular bodies and the strata of the earth are concerned, in order to provide some kind of remedy for that fault, I have decided to add here the following diagrams, chosen from others. The first thirteen diagrams, intended to explain angular bodies of crystal, are divided into two classes (Fig. 2.27.2).

The first class contains seven varieties of the plane in which the axis of the crystal lies. In diagrams 1, 2 and 3, the axes of the parts, from which the body of the crystal is composed, form a straight line; but there is an intermediate column, missing in diagram 1, appearing smaller in 2, and longer in 3. In diagram 4, the axes of the parts that make up the body of the crystal do not make up one straight line. Diagrams 5 and 6 are of the class, countless numbers of which I could present to prove that in the plane of the axis, both the number and length of the sides are changed in various ways without the

---

(Footnote 145 continued)

durch Überlieferung nachgewiesenen natürlichen Veränderungen der Erdoberfläche. 1-5. (Gotha 1822-1841). Especially 1, 102 ff.

[146]Here, at the end of his Prodromus, Steno declares that he would prove his statements in his Dissertatio by mentioning the individual items and places. Since he never wrote this Dissertation such a work as Fr. Rodolico, *L'Esplorazione naturalistica dell'Appennino* (Firenze 1963) may act as a compensation. The book not only describes these places by the help of photos and maps, but also tells of the geologists of the 18th century, describing them, cf. p. 49 ff., 151 ff. His *Indice dei toponimi Appenninici* (417) together with Steno's Indice may be a great help for everyone who wants to investigate the latter's research during his visits to Amiata and Radicofani, Valdarno and the environs of Firenze, l'Alpi Apuane, Carrara, Lucca, Pisa, Livorno etc. Cf. also Rodolico, *Toscana* etc.

[147]This final remark must always be kept in mind, when judging the Prodromus.

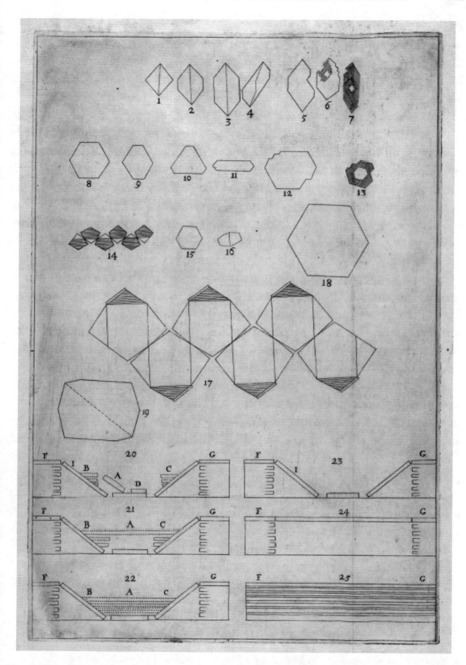

Fig. 2.27.2 Diagrams of angular bodies (1–19) and six distinct aspects of Tuscany (20–25)

angles being changed,[148] and that various cavities are left in the very middle of the crystal and various lamellae are formed. Diagram 7 shows how both the number and length of the sides, in the plane of the axis, are sometimes increased in various ways, sometimes decreased, by new crystalline material lain on top of the planes of the pyramids.

The second class contains six different bases of the plane. In diagrams 8, 9, 10 and 11, there are only six sides, with these differences, however, that in diagram 8 all the sides are equal, in diagrams 9 and 11 not all, but only the opposite sides are equal; in diagram 10, however, all opposite sides are unequal. In diagram 12, the plane of the base, which ought to be a hexagon, has 12 sides. Diagram 13 shows how sometimes the length and number of the sides in the plane of the base are changed in various ways as new crystalline material is added to the planes of the pyramids, without the angles being changed.[149]

The six following diagrams explain two different kinds of angular bodies of iron. Diagrams 14, 15, 16 serve to explain those angular bodies of iron that have twelve planes, and diagram 14, indeed shows all the 12 planes laid out in one plane, of which 6 are triangular and polished, the remaining 6 are pentagonal and striated. Diagram 15 is the plane of the base of the same body. Diagram 16 is the plane of the axis of the same body.[150]

Diagrams 17, 18 and 19 serve to explain those angular bodies of iron that have thirty planes, and diagram 17 indeed shows all those 30 planes laid out in one plane, of which 6 planes are pentagonal and polished, 12 triangular, also polished, 6 triangular and striated, 6 quadrilateral, oblong and polished. Diagram 18 is the plane of the base of the same body. Diagram 19 is the plane of the axis of the same body.[151]

The last six diagrams, while indicating how six distinct aspects of Tuscany may be inferred from its present appearance, at the same time serve to make more intelligible those things that we have stated about the strata of the earth. The dotted lines represent the sandy strata of the earth, so called from their main constituent, although various strata of stones and clay are mixed with them; the remaining lines represent rocky strata, likewise so called from their

---

[148] A distinct reference to the law of constancy of crystal angles (Mo).

[149] A second reference to the constancy of crystal angles. Hobbs calls Stensen the discoverer of the fundamental law of crystallography known as the law of constancy of interfacial angles (171).

[150] It is the combination of Rhombohedron (100) and striated pentagon with (211) smooth triangular bodies.

[151] The pentagons are the planes of (100), the striated triangles those of (211), the twelve smooth triangles belong to the pyramid (311), and the small quadrangles are the planes of the (211) obtuse rhombohedron (332). (Mi).

chief constituent, although sometimes other strata of softer material are situated among them. I have explained the letters of the diagrams in the dissertation itself, in the order in which the diagrams follow each other: here, I shall review briefly the order of change. Diagram 25 shows a vertical section of Tuscany at the time when the rocky strata were still complete and parallel to the horizon. Diagram 24 shows vast cavities eaten out either by the force of fires or waters, the upper strata being unbroken. Diagram 23 shows mountains and valleys produced by the disruption of the upper strata. Diagram 22 shows in the valleys new strata produced from the sea. Diagram 21 shows part of the lower strata of the new strata destroyed, the upper strata being unbroken. Diagram 20 shows hills and valleys produced there by the disruption of the upper sandy strata.

2.28   On a Calf with Hydrocephalus

To the Illustrious Grand Duke of Tuscany. Ferdinand II,
A Letter on a Calf with Hydrocephalus

In every kind of animals, any image firmly imprinted in the brain of the mother is sufficient to alter the course of the process which delineates the parts of the foetus. Sometimes, however, this alteration which is attributed to the imagination of the mother results from a peculiar disease of the foetus. It is certainly very difficult (not to say quite impossible) to find out which of these two is the true cause of the monstrosity, particularly if we must assess it from the outer parts without opening the inner parts in a foetus whose deformity is of an appearance normal for another species. The similarity suggests an effect stemming from a similar cause and there can be no reason to suspect a disease when all the parts appear healthy to us. I should consider the person to be a true Oedipus, who in such an instance could explain the phenomenon as it is, at a first glance, and who could set forth before the eyes of everyone that his opinion is in keeping with the truth. Some days ago I saw a calf whose head resembled that of dogs, in the round head of which either nature or artifice formed nostrils transversely flattened and opening along their medial longitudinal axis. Anyhow, I willingly acknowledge that I should

OPH 28, vol. II, 229–237: "De vitulo hydrocephalo" was originally written in Italian as a letter that has not been recovered. During his stay in Copenhagen Stensen caused it to be translated into Latin for publication in *Thomae Bartholini Acta Hafniensia Ann.* 1671 and 1672, Hafniae 1673, where it is found as No. CXXXI, pp. 249–262. The translator was Matthias Moth, who is styled *Medicinae Candidatus* (i.e. Bachelor of Medicine). Matthias Moth (c. 1647–1719), whose grandfather had been Physician to the Emperors Rudolph and Matthias, and whose father, a highly respected and able man, was Physician in Ordinary to Frederik III of Denmark, began by studying medicine abroad for several years, especially in Leiden. After his return, however, he gave up medicine, entered public service and with great ability discharged several high offices. DBL vol. XI, p. 485.

have attributed this condition to imagination rather than to a disease and I should not have recognized my mistake so readily if the illustrious Archduchess who had summoned me to see that monster had not given me the possibility of opening it. But, since, besides the causes underlying the deformity of the head, remarkable and unusual things were found in the structure of the brain, the illustrious Archduchess has thought it could not be a bad idea to make a written record of these things as they were discovered and to send it to you. This is what I will set upon in the following.

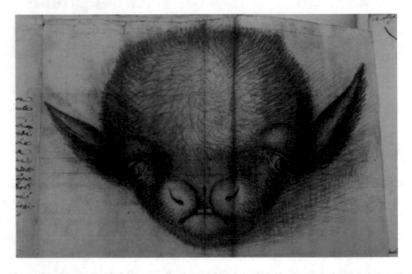

The illustration of the calf with hydrocephalus dissected by Stensen in Innsbruck is drawn in red chalk. No other such drawing exists from Stensen's hand. It may well have been provided by a court artist since it is found attached to the letter by Anna de' Medici, the Archduchess of Austria to Ferdinand II, her brother. FAS (For a colour version of this figure, please see the colour plates at the beginning of this book.)

Just as both the swelling of the head and the cleft in the nostrils constituted the external deformity in the present case, two unusual phenomena were also revealed after the skull was removed: there was, indeed, a great amount of water in the brain itself and there was a cyst beneath the skull at the root of the nostrils. The water tasted salty and had a reddish colour as is usual for all these serosities which have been retained in the inner cavities of animals for some length of time. This liquid weighed 4 lb. It had dilated the cavities of the brain to such an extent that much time was spent searching for the brain inside the brain itself. The skull was very similar to a hollow pumpkin. I was just about to have faith in stories I had not believed in previously, i.e. that people had been found without brain when, while separating the inner membranes from the skull, I felt in these membranes a body thicker than what could fit the membrane alone. In this body, however, I should not have found the parts of the brain in this instance, if the demonstrations of the brain which my teacher, the famous Mr. *Sylvius*, had frequently carried out in Holland in my presence had not guided me on this very difficult journey.

However, in order to provide a clear description of an abnormal state in a matter the normal state of which is not yet known clearly enough, it will be appropriate to begin by indicating the parts which I will consider in this brain and to give them the names which I think suitable for the understanding of everyone. The brain thus is divided into four parts one of which is a kind of base for the other parts and, therefore, is called the base. Above this, in front, there are two parts on the sides. They make up the second and the third part of the brain and are called the lateral parts. The fourth part is generally called the small brain or cerebellum. Because it leans upon the base behind the lateral parts, it is also called back part of the brain. On the upper side of the base one sees four pairs of tubercles in succession. These I call the first, second, third and fourth pairs of tubercles beginning from the front side of the base. Further, two small glands are found in the brain one of which is commonly called the pineal gland but which I call the upper small gland from the site which it occupies above the base whereas the other, called elsewhere the pituitary,[1] I call the lower small gland because of its situation beneath the base. Finally there are four cavities in the brain which, however, are nothing else than the different parts of one and the same cavity. The first two are called lateral due to their lateral position. The third cavity is situated in the base and the fourth extends between the base and the back part of the brain. I called thick membrane the membrane named dura mater by the Ancients and which is attached everywhere to the inner of the skull. I will call thin membrane the other membrane which directly coats the cerebral substance and which is named pia mater. The duplicated part of the thick membrane which, above, extends between the lateral parts of the brain is called the falx from its shape which looks like a scythe. I will use these words in the explanation of this rare case so that my description can be understood even by people unacquainted with this science and the matters which words are not sufficient to explain will be presented in illustrations.

But, now, back to the place from where I digressed. The dilatation of the whole increasing amount of water had exerted its force on the lateral parts of the brain and on the adjacent parts, and this in the following manner.

---

[1] Slime or phlegm.

Although the lateral parts should normally have been folded inwards at their extremities, over the second pair of tubercles (Fig. 2.28.1 Part 1), they were completely unfolded (Fig. 2.28.1 Part 2). Although they should have been united in the middle to the median parts which are called the corpus callosum (a), the septum lucidum (i) and the fornix (c), they were fully separated so that the falx (h), which is normally situated outside the cavity, protruded inside the cavity and prevented the extremities of the lateral parts of the brain from touching each other (Fig. 2.28.1 Part 4 kk). The extremities themselves of the lateral parts of the brain also which are usually attached everywhere to the second pair of tubercles were seen here at the sides of this pair only whereas the remainder was lifted from the base and expanded beneath the skull.

152   TH. BARTHOLINI

introrſum replicatæ ſupra ſecundum par tuberculorum,
part 1

erant omninò explicatæ (part 2.)

Cum in medio deberent uniri mediantibus partibus
part 3.

qvas (a) corpus calloſum, (i) ſeptum lucidum & (c) forni-
cem appellant, prorſus erant ſeparatæ
part 4.

ita ut falx(h), qvæ naturaliter extra cavitatem ſita eſt, intra
cavi-

Fig. 2.28.1

Although there should normally have been two cavities in the lateral parts of the brain (Fig. 2.28.1 Part 3 d, t) and, besides these cavities, a third one, according to the teaching of the Ancients (Fig. 2.28.1 Part 3 e), all that space was opened into one single cavity (Fig. 2.28.1 Part 4 g).

The cerebral substance of the lateral parts which otherwise is fairly thick had been thinned here by the pressure of the water, as appears in Fig. 2.28.2 which reproduces its exact measure. And, as a result of the very mass of the water, these convolutions of the brain which otherwise appear usually fairly numerous and deep had all disappeared. The only exception was the Sylvian fissure which, however, did not penetrate more deeply than is shown in Fig. 2.28.2 where (a) indicates the white substance and (b) the grey substance.

Fig. 2.28.2

The thin membrane which should penetrate in between the lateral parts and the base to form that membrane in their cavities, which is full of veins and small glands and is called the choroid plexus, was stretched out in the middle of the water and fastened above to the falx and below to the bordering area between the second and the third pair of tubercles.

That part of the thin membrane which should be turned obliquely backwards in its upper part was facing forwards here so that the vein of the falx, which is called the fourth sinus, made a very conspicuous angle with the veins of the thin membrane whereas they normally form a straight line. The upper small gland the base of which should have been united to the base of the brain at the place where the second and the third pair of tubercles concur, did not touch any part of the brain at all here but, immersed in water, with its back surface it adhered to the anterior aspect of the thin membrane which here, as mentioned above, was spread out at the place where two veins unite to form the vein called the fourth sinus in the falx. The front of this small gland was totally free and was not covered by this part of the thin membrane which usually forms there a kind of small bag at the front. The top of the small gland which should have faced backwards was turned towards the front. In its base there was a very conspicuous cavity from the sides of which some striae of grey substance extended downwards along the length of the thin membrane, one on each side.

These are the alterations which the mass of water had produced in the brain itself by distending the lateral cavities of the brain. But in the same brain there were also other things worth considering, the true cause of which is not easy to find since I do not recall having ever seen or read about anything like this.

There were minute tumours made up of grey substance over the white and fibrous substance in the inner surface of the lateral parts of the brain. In some places they were round and separated from each other like islands. In other places they were attached to each other and irregular. A larger group of these tumours was about four finger breadths away from the first pair of tubercles. When they were cut transversely a variety of patterns appeared which is shown in Figs. 2.28.3, 2.28.4 and 2.28.5. The grey cortex which surrounds the surfaces of white substance everywhere provides striking confirmation of the opinion of those who believe that the fluid which enters the filaments in the white substance is secreted in the grey substance.

Fig. 2.28.3

Fig. 2.28.4

Fig. 2.28.5

In the right lateral part, however, between the larger group of minute tumours just mentioned, and between the first pair of tubercles, in the vicinity of that group, there were such a great number of grey dots that you would have said they were the holes of some sieve.

Instead of the third pair of tubercles, there was one single tubercle provided with a lens-like cavity. Behind this there was a raised edge semi-circular in shape, grey in colour with many blood vessels which were parallel to each other.

The optic nerves were thinner outside the skull, much thicker about the chiasm but less hard than they usually are.

Moreover, there was a number of white filaments quite beautiful to behold which is represented in Fig. 2.28.6 where (a) indicates a white ligament transversely located where the third cerebral cavity extends down towards the lower small gland; (b) are two white filaments lying transversely over the fibrous substance (d) which proceeds between the second and the third pair of tubercles; (c) indicates many white filaments which extended from the lower part of the filament (b) obliquely towards the right side, over the grey substance of the third cavity, up to the left tubercle of the second pair which else would have been the continuation of the right optic nerve, and been folded above the base of the brain.

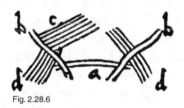

Fig. 2.28.6

Finally, these was nothing peculiar about the first, second and fourth pairs of tubercles, the spread out cerebral substance which is between the tubercles of the fourth pair and the median substance at the back of the brain (well enough known to *Galen* although it has hardly been observed by anyone after him), the hind part of the brain itself, the third and the fourth cavity, the network of small arteries in the duplicated thick membrane, called elsewhere rete mirabile, nor about the nerves originating from the base of the brain itself.

Two things above all seem to me to be astonishing: 1. that so great a force of the water which had distended the lateral parts of the brain, had left the tubercles of the second part attached in the middle through a white transverse ligament, grey substance being interposed behind, whereas elsewhere these tubercles are so easily disrupted that they are almost never found united during demonstrations; 2. that the lower small gland (destined for the majority of anatomists to the absorption of the serosities of the brain) was found here totally unaltered as far as the variety of its colours, or its magnitude, or the conspicuous cavity inside the gland are concerned, although a very large amount of this serous liquid floated around it.

The dilatation of the lateral cavities of the brain, containing 4 lb of water, could not have occurred unless the bones of the skull had given way to the water. Among others, the bones of the forehead had yielded more to the pressure so that, bulging from the sides laterally to the eyes, they hampered the vision, protruding in the middle, they formed there an angle with the bones of the nostrils like the one which is impressed by force in dogs to make them look prettier. Since these bones had been dilated immoderately, they could not unite above so that two finger breadths above the nostrils there was a gap unprotected by bones through the middle of the head, six inches long and two inches wide, almost rectangular in shape, whereas otherwise it appears rhomboidal in embryos. The base of the skull was unusually flat and the cavity carved in it to contain the lower small gland, appeared to be here almost completely flat, although it should have been shaped like the saddle of a horse.

An unusual thing in the skull (which could not have been produced by the water) was a protuberance sharp and high very similar to a styloid process situated in the right part of the skull. Not only was it covered by no part of the brain itself, but even the thick membrane was not spread over it.

So far I have been busy mainly reviewing the changes produced in the brain and the skull by the excessive amount of water. Now an investigation has to be undertaken concerning this water: what was it? when did it begin to stretch out the surrounding parts? and why did it attack the lateral parts more than the other ones?

Concerning the first question, about the nature of the water, its colour and its taste tell us that this is the same water which always enters the cavities of the brain when the animal is healthy, and is shortly afterwards expelled from them by the arrival of new water. The fact that the ways through which this water enters and leaves could not yet be determined exactly must not deter from this opinion. It is certain that similar water is found in the cavities of the brain whenever a skull is opened. It is very abundant in the brains of fishes even if you open the skull while the fishes are still alive. The same water is also found in other kinds of animals, above all when they have not yet acquired their final shape, as I have very often experienced in chicks from poultry, puppies, baby rabbits and others, dissected before hatching or delivery. There are indeed ways through which water could be carried away although it is not known which one of them serves to the mentioned task or whether they all contribute to it. There are many small glands in the brain, destined to separate the serum from the blood. Quite a few people attribute this function to the grey substance and I myself recall that lymphatic vessels have been found several times around the upper small gland, which may have been as many canals carrying water meant to moisten the cavities of the brain. But it will in no way be denied that the water, once it has made its way into the brain, finds its way out again although it can be rightly questioned whether this occurs through paths in the ears, the nose, the eyes and blood vessels or by other ways unknown to us. It is no new opinion that in certain fevers the brain is emptied through that canal which extends from the ears to the nostrils and it is told of many people that, after certain headaches, they feel a fairly copious stream of water running down through their nose, that this liquid is sometimes yellowish and that they soon perceive considerable relief. Moreover, if water is normally retained in the brain, which I have demonstrated above to be certain, then it is absolutely necessary that it makes its way out again, which is observed in every other cavity in animals.

But concerning the time from when this water began to dilate the lateral cavities of the brain, it can be taken for certain that, after its exit had been occluded, the water which subsequently came in gradually accumulated and began to dilate the surrounding parts within the interval between the completion of the formation of the base and hind part of the brain and the later folding of the lateral parts over the base, since both the base and the back part were found to be intact just as was normally constituted the vein of the falx which the Ancients called the fourth sinus even though the veins in the thin membrane continuing the former vein were found in an opposite position. This convinces me that both the falx as well as the base had already obtained their final shape before the accumulation of water took place there. Otherwise indeed, if the falx had not yet been completed and was still very tender, the extension of the vein in it would have had the same position as in the thin membrane.

Thirdly, it is sought why the water had extended the lateral cavities rather than those in the middle even though it was found inside all of them and they all together constitute one cavity only. If I measure these cavities, I find that the central ones are much narrower than the lateral ones. Therefore, the lateral cavities, because they are larger, were also able to take a greater amount of water and thus were exposed to a greater impetus of water. If I look at the cerebral substance which encloses the mentioned cavities, I find that the central cavities are made of firmer substance than the lateral ones. A greater amount of grey substance indeed is found in the lateral parts of the brain than in its base. But since the cerebral substance easily yields to pressure everywhere, the reasons which I have given here do not seem to be sufficient if they are not combined with the next two. The first one is that, during the formation of the brain, the base and the back part have already been completed when the lateral parts are still in the process of growing and do not yet occupy the central part of the base so that, the amount of water increasing at this time, the lateral parts not yet completed will have been prevented from uniting. The second reason is the one which I think is the most important of all, namely that the water will have exerted its force where it will have met less resistance. The other bones of the skull indeed give insertion to an infinite number of tendons which come from the temporal, cervical and dorsal muscles whereas the bones of the forehead have almost none. Therefore, they have got less force of resistance than the other bones of the skull.

If I have been somewhat prolix in considering the water, I will be more concise in describing the cyst. This was about three finger breadths long, one and a half finger breadths wide and one finger breadth deep and it was completely filled with thickened white matter. It had occupied so much space beneath the skull that it almost touched the bone over which the optic nerves form the chiasm. However, it was not connected to the brain. A thick membrane situated above the bone and another below prevented this. Outside no trace of the white matter was apparent but only some serosity came forth through that opening which was observed between the eyes near the roots of the nostrils.

It is not my task to say anything definite about the origin of this quite deeply seated cyst but, concerning the time of its origin, I think it must have arisen when the animal was not yet fully developed so that the serosities which came out from it could have formed a canal along the length of the nostrils which made these similar to the cleft nostrils of dogs. But if I may make a guess here I would assign a little more accurate time and say that this cyst has been formed before water had been enclosed in the brain, indeed that the canals meant to provide an exit to the water had been compressed and closed by the swelling of this cyst so that the cyst by its serosity will have been the cause of the cleft nostrils and that its swelling, by closing the exit of the water and by retaining this in the brain, will have been the cause of the swelling of the head. If somebody would spend some time considering these abnormalities found in the brain, he could certainly draw many conclusions very profitable to the knowledge of anatomy, as for example:

I. That one should question all those accounts in which it is said that the brain is either completely or partially dissolved in water. I myself would certainly have pronounced the same judgment about the present case if the observations made in our century had not taught me to find the parts of the brain even if they do not at all resemble the normal parts.

II. That one should also question those reports in which it is told that part of the brain has come out through the nose since, in these instances, there may have been a cyst like this one.

III. That union of the lateral parts of the brain by way of the corpus callosum, the septum lucidum and the fornix is not absolutely necessary for the animal's feeling and movement since this animal has lived for many weeks without them, so that those who build part of their teaching about the brain on this union can find reason from this case to question it.

IV. That it is not absolutely necessary for movement and feeling that the cavities be completely closed with cerebral substance since this animal lived for some time with these being open.

V. That the cerebral substance in its lateral parts and in the base can tolerate great pressure without complete loss of feeling and movement since here the cerebral substance underwent a pressure so great that the dural membrane and even the bones of the forehead themselves were forced to yield to its violence.

VI. That it is most likely that with time some vessel may be found at the position of the cyst that serves to discharge the brain's water.

Many other propositions could be drawn from the present case, concerning the substance, the structure and the functioning of the brain, as well as concerning the animal spirits and excrements. But in order not to come to bore you by writing a book instead of a letter, I will content myself with having demonstrated how the brain resists the greatest accidents although being the noblest and most delicate organ and how a disorder in the foetus itself can sometimes be the cause of these deformities that would be attributed to be produced by the imagination of the mother.

Innsbruck, June 1669.

2.29   Letter to the Grand Duke Cosimo III on the Grotto Above Gresta

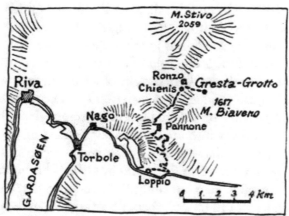

MOST SERENE PATRON,

Since the change in the weather spoiled all my hopes of seeing the frozen waters in the grotto above Gresta,[1] I returned to it after I had sent the last

---

Stensen's original letter in Italian does not seem to have been preserved. It must have served as a model for Fabroni, (*Lettere* 2, 318–321) and Manni (*Vita* 292–296) in their publications of the same in 1775. Maar adopted their text (OPH 29 & 30) as did Scherz in EP, 238–240. It is also to be found in Montalenti (117–119) and is the basis of this translation by Mrs. M. Rohde. Annotations by Scherz are from GP, 247 ff. The sketches indicate Stensen's routes to the grottos, see GP, illustration 105. The sketches from the grotto of Moncodeno in GP, 246 are not printed in OPH. Text from OPH.

---

[1]A grotto near Gresta Castle, north-east of Lake Garda, the property of the Counts of Castelbarco. The castle is situated in the Gresta valley above Loppio, near Pannone, and has been a ruin since its destruction, in 1701, by the Duke of Vendôme, a marshal of Louis XIV.

© Springer-Verlag GmbH Germany, part of Springer Nature 2018
T. Kardel and P. Maquet, *Nicolaus Steno*,
https://doi.org/10.1007/978-3-662-55047-2_40

letter to Your Highness, so that I should miss nothing that might help me to discover something new about it. I drew a plan of it,[2] as far as this was possible, given the unevenness of the ground, and I recorded several cross-sections, while keeping in mind the construction of the mountain above. While I was examining all parts of the grotto for this purpose, I noticed a wind coming up from the depths. After some exploration, I realized from the direction of the wind that the cold, which you feel in front of the grotto, was not coming from the ice in it, nor was the ice formed from a coldness which had become concentrated inside because of warmth outside. There was actually a current of very cold air coming through a cleft from remote cavities in the mountain. It caused the thermometer to drop at once to 13 °C, while the temperature of the water on top of the ice remained at 13½ °C, and of the air above the water at 14 °C. This is counting upwards from the bulb, and corresponds to 3 °C, 3½ °C, and 4 °C, if one does not count the first 10 °C.[3] In order to discover the source of this strong wind it suffices to consider it when it is at its strongest, i.e. when the sun is warmest, and to bear in mind the conditions in the deepest excavations when a mine is being dug. The only cold is the cold which enters from outside in summer and winter alike through the deepest clefts observed there. It would require a cold substance on the ground, such as water or ice, to cause a cold wind which would correspond to the coldness of its source. It is therefore highly probable that the rocks, which are continually warmed on both sides by the sun, melt the masses of snow and ice deposited the previous winter in the cavities, which are connected to the grotto by clefts. This melting process has two results: (1) cold air is sent out, and (2) the rocks on the floor of the grotto are covered with ice. It can be said, then, that the water in the grotto is frozen

---

[2]The plan has also been lost. However, we must note Viviani's mention of a drawing of a grotto in the mountains of Brescia, which he sent to the Grand Duke's steward on the 30th June 1671. (Florence BN, Gal. 269, fol. 237: *Disteso d'osservazioni de farsi da Lodovico Stoffiere Granduca Cosimo III sulla grotta delle montagne de Brescia, fatto da me e assegnatoli d'ordine di S.A.S. 30. giugno 1671*).

[3]Members of the Cimento Academy, working probably from instruments known before Galileo's death, were the first to produce a closed thermometer, which they described in their "Saggi", p. 2 ff. The earlier instruments were thermoscopes which showed only the change in temperature. The thermometer of the Cimento Academy resembled a present day one in shape, with a glass globe and a long tube. It was filled with ethyl alcohol and sealed to make it independent of the change in air pressure. In Tuscany, the highest and the lowest temperatures were taken as the fundamental points. The scale was however, divided into one hundred degrees, and each degree was marked by a white glass pearl which was burned in. It was only after the dissolution of the Academy that one of its members, Rinaldini, brought out the fundamental points as we know them to-day. Present day freezing point corresponded to 13.5° on this thermometer and 55° Celsius corresponded to 50°. (Caverni 1, 279 ss. Dannemann 2, 75-77. OPH 2, 343) and R. Vollmann, *Das Thermometer*. In: Ciba Zeitschrift. März 1944, p. 3301.

partly by cold air which blows over it, and partly by the coldness of the rocks beneath it.

As for the length of time which this freezing process takes, whether it is the beginning of summer, or the whole warm season, I must await a more accurate statement from monthly observations which Count Castelbarco[4] has said he is willing to make. This interest shown by Your Highness will enable us to come to a final decision about the century-long discussion on antiperistasis.[5]

I hear that there is a similar grotto above Lake Como. Since I am near it, and since Mr. Buondichi is trying to make the sightseeing easier for me by offering to be my companion, I thought it a good idea to take advantage of this opportunity. I hope in this way to give Your Highness more satisfaction by making as many observations as I can. The aforementioned Mr. Buondichi does me innumerable kindnesses daily, as does Count Alessandro Visconti although he has been ill until now; he sends me his carriage every day. Signor Manfredo Settala is always telling me how he feels beholden to you.[6] All these many special favours due to the kindness and protection which Your Highness has chosen to show me, only make me wish more than ever that God will give me enough ability to serve Your Highness as I should.

Your most humble and most grateful servant, Niels Stensen

---

[4]The Castelbarcos, a family from Valle Lagarina, who moved to Milan. In 1664 they obtained sovereignty over the districts around Ala, Avio, Mon and Bretonico, and were raised to the state of counts by the Emperor. Stensen's host was Count *Francesco de Castelbarco* (1626-1695), son of Scipione and Laura de Galvagni, married to Claudia Dorothea, Countess of Lodron (Francesco 1652, had sold the old castle and the jurisdiction over Castelbarco to Christian, Count of Lodron). Francesco's younger son, Giuseppe Scipione, married Constanza, the daughter of Cesare Visconti, in 1696; for this reason the branch of the family still living also has this name.

[5]The idea of *Antiperistasis* goes back to Aristotle and means the recollection and concentration of one's own forces to counteract the attack of an opposing force (Chauvin, *Lexicon Rationale*. OPH 2, 343 ff.). The theory of antiperistasis was used to explain the most widely differing phenomena, from fever with inflammation of the lungs to the observation that cellars are warm in winter and cold in summer. The members of the Cimento Academy did not find any confirmation of antiperistasis in their experiments and rejected it. (*Saggi*, 259).

[6]*Francesco Buondichi* a non-accredited Florentine resident in Milan from 1656-1692. A certain Count *Alessandro Visconti* († 1685) was cavalry captain in the service of Spain (EP 1, 240). *Manfredo Settala*, see note in OPH 23.

2.30   Letter to the Grand Duke Cosimo III on the Grottos of Moncodeno

Most Serene Grand Duke[1]

The grotto of Moncodeno[1] exceeded my greatest expectations. It presented characteristics which I had never read about in other writings, and which have never occurred to me on any other occasion. Here, too, I was able to confirm by observation the view I had begun to form by a process of reason in the grotto of Gresta. The most important characteristics are those to be seen in the ice formation, so completely different from anything I have ever seen. In some parts there is such similarity to a crystal, that I am not surprised at the

---

[1]The text of this letter is a faithful reproduction of the manuscript (Florence, Bibl. Nazionale, Posteriori 27. III, 12, now Gal. 287, 58ʳ to 61ʳ) of a letter written by Stensen himself almost without correction, and of the attached sketches of the grottos. Spelling, punctuation and paragraph division have been kept. This text was without doubt also used by *Fabroni* (Lettere 2, 321–327) and *Manni* (Vita 296–305), who printed it with practically no alteration—but without the sketches of the grottos and the letters referring to them. The same holds for Maar's publication (OPH 2, 245–248), for the somewhat abridged impression of Cermenati (La Ghiacciaia 55), and for that of Montalenti (120–126). The Epistolae publication (EP 1, 241–245) was the first to be based on Stensen's own letter and drawings. Translation by Mrs. M. Rohde and Annotations by Scherz from GP, 239–248.

---

[1]The *Moncodeno Grotto* on the north side of the Grigna Settentrionale (2410 m), also known as the Ghiacciaia di Moncodeno ca. 1600 m high, lies at the end of the Valle dei Mulini. In ancient times it was believed to be the source of the River Latte. Leonardo da Vinci describes it in *Codice Atlantico* (S. Saglio, Le Grigne, Milan 1937, 481).

opinion held by many that crystals are petrified ice. Here is a similarity, not only in transparency, but also in shape. If it were not for two experiences which I have had, I could easily be persuaded to the same opinion by such phenomena. The first experience is of a negative kind, viz. that I have not heard of a crystal being found in the ice of one of these grottos. The second is positive, viz. that you can find crystal in places where there is no ice at the end of the year, in fact, even in places where there is never ice.

But to get back to our grotto: ice is to be found there in the shape of columns in the middle of the grotto, in fact, wherever water drips continuously. This is partly along the rock mass on the side opposite the entrance; there is as much variety of form as there are different types of incrustation on those parts of the rock which are always wet, and partly on the floor of the grotto round the columns. Moreover, I did not find any water on the floor of the grotto, nor any ice on the horizontal surface. The incrustations on the rock mass, although exceptionally fine and delicate, were firmly attached to it, until the warmth of a hand or a flame dissolved them.

Some were in the shape of very small, clear drops, lined up against each other. Others were in the shape of small columns, continuing down from each other in a straight line (as a few of them may be seen at the letters f and c). Al the ones I saw were absolutely clear without the smallest air bubble, a thing which is very rare with ice.

The columns in the middle also consisted almost entirely of small columns like this, grouped in a circular formation around the central shaft. This gave the surface of the columns the appearance of bunches of grapes as one may see from the two sections, the one lengthwise, the other one perpendicular to the axis. Some of these columns looked as if they had been hollowed out lengthwise by a cylinder shaped body (A); others were hollow only at the top. On one of the columns the hollow did not resemble a cylinder, but was like some figure made up of several balls, one on top of the other (B).

Cross-section through the grotto

The position of these columns in the centre of the grotto afforded an unusual picture (as may be seen from the first and the second section through the grotto; the figures are made by eye and at the light of a candle, not with the exactness of a proper measure, as it was dangerous to walk on those uneven crusts of the ice.) No through draught was to be noticed here as in the grotto above Gresta. When I held a candle up to a cleft, wherever it was possible, the flame did not move. However, there was a very noticeable coldness there, and my feet went numb very quickly. The snow, which I presume must lie around the Gresta grotto, is piled up here in front of the entrance.

When I reached the grotto, exhausted by the walk past terrifyingly steep drops above and below the grotto, as well as by the strenuous climb, and overwhelmed by all the new impressions, I did not think of making many observations which might occur to me now. I would perhaps have made them there if the spot had been nearer some inhabited place, and not, as was the case, inhabited rather by goats and chamois than by people. All things considered, I think I have made observations in Gresta and Moncodeno which, after a few experiments with artificial ice, will solve some of the doubts about cold and warmth in subterranean parts. At least, I now see that the following conclusions can already be drawn from the exploration of the Moncodeno Grotto:

1. *The grotto is not warm when it is cold outside.* I did not get this information only from shepherds, who say the ice in the grotto is always there, or, as they put it, the ice has been there since the beginning of the world. I draw the same conclusion myself from the snow which would not be in there when it is warm outside, if it were warm inside when it was snowing outside.

2. *The ice forms there in summer, too.* There are also two reasons for this. The first reason is offered by the report of the same shepherds who lead their flocks up into the mountains in the very hot season. When there is no snow outside, they always take snow from the grotto, because they cannot get any other water for themselves and their sheep than that from the melted snow and ice. They assure me that the columns form again if removed. The incrustations of the ice give me ground for my second reason, although these are delicate, they are firmly attached to the rock.

d d d d Columns of ice in the middle of the grotto, some of them not more than a yard, others more wide than a man. A and B, see text above.

This would not be possible in a damp place if the coldness of the stone was not intense enough to freeze it at the same time.

3. *The water which freezes to ice there does not enter in great quantities, but almost imperceptibly, carried in by the wind rather than through the cleft in the rock.* I say this partly because you can hear the drops falling at intervals of a few minutes, and partly because a growth can be observed on these columns, which cannot possibly come from these drops. The drops tend more to hold the open indentation on the columns, and do not add to the growth on it. The formation of such a growth requires an even degree of dampness all round the column.

4. *The coldness in the grotto does not come from a concentration of the cold inside as a result of increased warmth outside. It comes rather from the coldness of the snow which lies near the entrance keeping the inner parts of the cave always cold.*[2] There is no ice-crust on the snow here, nor does the snow resemble frozen snow. The firmness of this snow recalls rather the snow found on mountain peaks and other places in summer, where the snow melts gradually and the water flowing underneath makes a way for itself. Similarly the snow here cannot turn to water or ice when it melts, and cannot flow away horizontally along the surface. It can happen in one and the same grotto that

---

[2]A telling observation directed against the idea of antiperistasis.

the snow near the entrance melts, while the water further away from the entrance turns to ice. The report of the shepherds throws some light on this point. They say that, when there is not much snow, there is a deep hollow between the rocks and the ice near the wood which serves as a ladder, and that if you throw a stone down, you can hear it rolling for a long time. The ice which forms the floor of the grotto is the ice they call "eternal ice" because it is there every year in the same form and because, in their opinion, it covers a vast surface. I have heard some people say that the River Latte gets some of its water from that melted snow. But however that may be, one thing is certain: if the sun shines on this side of the mountain all day, except for a few hours in the morning, it is not surprising that the snow and ice near the ladder (k) melt because of the heat from the rock between (a) and (k) in the first cross-section. This can easily be confirmed by observing how easily one can knock down the snow at the sides of the ladder (k) with a long stick. This would not be possible if the water from the snow turned to ice.

I could list other observations here, and the foregoing observations and thoughts could no doubt be more clearly and systematically arranged; however writing this down has taken so much more time than I expected that I beg Your Highness to excuse me for not re-arranging everything I have written. I am sorry, too, that I cannot go on to report on the irregularity of the rise and fall of the Plinian waters,[3] nor of the way the large grotto from which the River Latte[4] gushes forth all summer, dries up in winter. I hope to give Your Highness a personal account of these and of other peculiarities of the lake in a very short time, and I shall endeavour to take advantage of the first possible opportunity to travel to Bologna. It would be very ungrateful of me not to mention one point; I should like to recommend Mr. Francesco Buondichi to Your Highness for his services during the journey over the lake, and for the favours which he obtained from the landowners of the district. I should also like to mention the patience and good nature with which he procured every comfort for me,

---

[3]The name of a spring in the province of Como; it rises on the eastern slope of the Lario to the north of the town of Como and to the east of Torno. It is very well known, not only because of its lovely situation, but because of the mysterious appearance of its *intermittence*. In a description by Tristano Calco from the year 1493 it received the name of *Plinian Water*, because it was explored by both Plinys and described by the younger (*Dizionario corografico* 6, 282 f. and Saglio, Le Grigne p. 481 f.).

[4]This river, the smallest in Italy, (250 m long) on the east side of Lake Como, rises in a cave about one meter wide, 300 m above the Lario. From March until October or November, it gushes out of this cave, almost vertical to the rocks, foaming and turning as white as milk; in the warmest and wettest period its speed is 1 m^3 per second. The literature describing and explaining this phenomenon goes back to the sixteenth century, but it was not until research in 1921 that the reason was discovered, viz that the River Latte rises in a reservoir which has two outlets; one of these, the temporary one, is used only when the other cannot take all the water when the snow melts.

assisting me with his interest and kindness on the visits to the most difficult Alpine spots. Canon Settala recommends himself to Your Highness' protection. I myself wish Your Highness every contentment and greatness, and ask you at the same time to excuse the flaws of my hasty writing.

Your most humble and grateful servant

Milan, August 19, 1671.

Niels Stensen

Longitudinal section through the grotto[5]

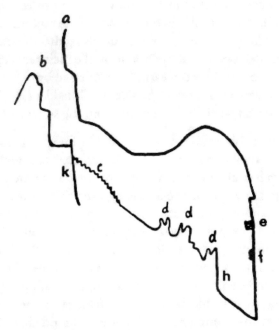

a. Slope of the mountain turned against northwest.

b. Entrance of the Grotto.

c. Steps formed in the snow.

d d d. Columns of ice.

e. Small columns of ice attached to the rock mass.

f. Small drops of ice attached to the rock mass.

dh. A very long side of a column six yards high.

k. A very long piece of wood that serves as a ladder, when there is no snow at the entrance, which is now almost wholly covered by snow; for this reason the shepherds say that there is a very deep cave between the ice and the rock mass

---

[5][See figure] These cross sections were meant to be only a provisional sketch. Stensen had ordered a complete and detailed plan from *Francesco Buondichi*, who promised the prompt delivery of the same, in a letter to *Viviani* from Milan on the 25th of May, 1672 (Florence BN, Gal. 164, fol. 191[rv]).

2.31 Invitation to the Preface Lecture

By

The Deacon

Of the Faculty of Medicine at the Royal University of Copenhagen

Thomas Bartholin

Royal archiater and honorary professor of medicine

Greets the students of anatomy.[1]

It is the enormous favour of the immortal God and of the very majestic King that the anatomy theatre, covered with dust and silent for some years, recovers. Indeed, since I gave up my task of dissecting and demonstrating, various hindrances intervened, either by injuries of wars or by a difficult imposed peace so that they withdrew the hands of them who were most trained in it. Meanwhile, although exhausted by work at night during so many years, tired of the work given to this kind of studies, and nearly consumed by various mockeries of a temerarious fate, with the indulgence of very clement kings I,

---

[1]Thanks to the assistance of Philip Endean S. J. I managed to translate this difficult text though the result is still my responsibility. In the list of animals dissected by Niels Stensen *Caprea sylvestris* appears. It might be a plant or a unicorn. At the advice of Dr. Scient. Elsebeth Hanken Thomsen, I translated it by goat. The rhetoric Latin, very difficult to translate, is by itself an homage to Niels Stensen. August Ziggelaar S. J

© Springer-Verlag GmbH Germany, part of Springer Nature 2018
T. Kardel and P. Maquet, *Nicolaus Steno*,
https://doi.org/10.1007/978-3-662-55047-2_42

already old, would have hidden in a corner of an estate. There with Aglaus Psophidius, more happy with least of work, envied by nobody, I hope, I would experience a minimum of any evil in life. However, I could not impose to my pen the rest allowed to the knife of dissection. Because I could not otherwise satisfy the wish of the public, I counterbalanced my duty by a promise having contemplated the work of others as from a vantage point, and I carefully avoided that my retirement should not be a burden or dishonour to the fatherland. Therefore, by turns and as far as time allowed, I took care of that which concerned our faculty of medicine, with so much diligence that not by my fault they should suffer damage, but I left intact this part of medicine from which I was excused either by function or health, or, in order to speak more openly and straightforward, by disgust of a skill, usually scorned and neglected by people. I wish that now may return the previous respect for anatomy and that with the right to return home dissections on animals may be resumed the more gladly and ardently as I once have taken care of this job. The one who before me turned well this boulder, Mr. SIMON PAULLI, after so many years of vacation, had always one desire that he might live to see Divine Anatomy either keep the fame which it had obtained among us or regain it. Under favourable omens every thing will succeed. The wishes were fulfilled. As waken up out of winter lair, I see with joy that the arena in which I so often have laboured, is trod. By the clemency of our very majestic King and Lord CHRISTIAN V, father of the fatherland, was called back to his homeland the most celebrated gentleman NIELS STENSEN, new Democritus of the century, He consoles the hope of scholars, he will witness to the fatherland that the fame obtained in the learned world by famous inventions and writings which respire bitten off nails, is not his private but public possession. In that intention, he started without envy, when scarcely he recently had set his foot in his native town, for the benefit of Asclepius' youngsters with lucky and ready hand to search the viscera of animals, in order to make visible every thing that was hidden. During the autumn of last year, though the weather was not enough favourable, he had publicly and privately dissected a human corps, two bears, a reindeer, a goat, hares, a cat, mice, a hedgehog, a squirrel, a dormouse, a monkey and other animals. Observations thereof I have put in the Acta Medica et Philosophica which are being printed. Not without exercise should pass the first months of the new year during these holidays, therefore he decided out of love for science and young people of the country, with the approval of the authorities and the agreement of the patron of the Academy SIR PETER REEDZ, Knight and the King's Great Counsellor to make in a humane corpse of female sex the experiment of his ability and doctrine in the Anatomical Theatre to the glory of God, the proficiency of Nature and the profit of the medical world. He will add to these acts and show for interested spectators of anatomy various things about the anatomy of a reindeer, living in northern countries. Our Mæcenas SIR PETER GRIFFENFELD granted it for public dissection. We praise these efforts for the good of the mortals, we praise the dexterity and eternal fame of the great man and wish the desired success and well deserved thanks for these efforts. We are drawn, not against our good will, to the intermitted studies and the glory of the growing anatomy, remembering Milo of Croton, who is said, when he already old saw athletes exercise, to have looked at his muscles and said with tears: But these are already dead. Every day the science of Anatomy makes progress and makes the veterans happy by grateful recording. Whenever prosectors make their hands bloody, then, before they clean them, new observations drip off. It is the praise of the century and the benignity of nature, which unfolds entirely for the interested ones. While it still was allowed by age, we lighted this lamp. Let now others with more success supply what will seem to be lacking to their perfection and splendour. Among these our Stensen, first of all prosectors by his ability of dissecting and easiness in detecting, has functioned with such successful genius in this respect, both at home and outside that he ravished to admiration of him the learned world. May Jupiter of the Olympus every day witness for Phidias, the Capitol and Dina of Ephesus for Mentor, to whom were consecrated art's vessels. For Stensen Europe's better theatres, words of Doctors, public fame arise, finally Man itself, whom with pleasure he will propose, so we know to be humans, born to be resolved, not to pride, not to slighting, finally not to contempt of God and nature. Whoever together with us does not disregard these secrets, let him approach our Democritus' cave, where he also will find gods, who nowhere will be more gracious. He will see that hands are not tired nor genius become deficient nor Nature become silent. For all who favour and are friends of the skill, first of all those who, dedicated to Aesculapius, wish and should know an accurate demonstration of the parts, our Anatomical Theatre will be open, under that rule and condicion that they hear and observe the incomparable Prosector with similar patience as they will see the laborious work being done, and receive the most famous Doctor, not to be judged by festivity but by action. He earnest so. According to the example of Hippocrates there is more care of wisdom than of gold. In order to inform those whose mind burns for love to know and to see, tomorrow's day will begin the Demonstration at two o'clock in the afternoon. Copenhagen, 28 January of the year 1673.

Under the Seal of the Medical Faculty.

DECANUS
FACULTATIS MEDICÆ IN REG. ACAD. HAFNIENSI.

THOMAS BARTHOLINUS

Archiater Regius
&
Medicinæ Profeſſ. Honorarius.

Anatomes Studioſis S.

 Ngens immortalis DEI & Regis Auguſtiſſimi eſt beneficium, qvod reviviſcere coeperit ſpiritumq; recipere Theatrum Anatomicum, per pluſculos annos, pulvere conſperſum & obnuteſcens. Enimverò ex qvo ſecandi demonſtrandiq; munere me abdicavi, varia intercurrerunt impedimenta, five bellorum injuriis five pace difficili injecta, ut manum operi ſubtraxerint illi qvi in eo fuere exercitatiſſimi. Ego interea, qvanqvam tot annorum lucubrationibus deſeſſus, laboribus huic ſtudiorum generi impenſis fatigatus, variiſq; temerariæ fortunæ ludibriis penè confectus, indulgentia Clementiſſimorum Regum jam Senior in angulum prædii me abdiderim; ut cum Aglao Pſophidio ibi felicior minimo labore, à nemine, qvod ſperem, invidendo, minimum in vita mali experirer, non potui tamen pennæ meæ imperare qvietem cultro Anatomico conceſſam. Qvia publico deſiderio aliter ſatisfacere non potui, voto debitum compenſavi, alie ruim labores velut ex ſpecula contemplatus, caviq; ſollicitè ne ocium meum patriæ eſſet oneri aut dedecori. Unde per vices & temporum ſpiramenta, qvæ ad Facultatem noſtram Medicam ſpectabant, curavi ea diligentia, ne mea culpa caperent detrimentum, ſed intaſtam liqvi hanc Medicinæ partem, à qva vel officii ratione excuſabar, vel valetudinis, vel, ut apertius ſine ænigmate loqvar, ſpretæ neglectæq; apud vulgus artis tædio. Redire jam priſtinam Anatomes reverentiam & poſtliminiò ſectiones animalium reſumi tanto alacrius ardentiuſq; cupio, qvanto majori cura hanc olim ſpartam exornavi. Qvi ante me hoc ſaxum volvit benè de Theatro noſtro meritus, cujus incunabula procuravit, D. SIMON PAULLI, poſt tot annorum vacationes unum ſemper in votis habuit, ut illo adhuc ſuperſtite Diva Anatome partam apud nos famam vel conſervaret, vel recuperaret. Auſpicato cuncta ſuccedent.   Impleta ſunt vota.   Ego ex veterno velut exci tatus, arenam calere lætus video, in qvo totiens deſudavi. Auguſtiſſimi Regis noſtri Dn. CHRISTIANI V Patris Patriæ indulgentiſſimi clementia ad patrios lares revocatus Vir celeberrimus, D. NICOLAUS STENONIUS novus ſeculi Democritus, Studioſorum ſpem ſolatur, patriæ teſtaturus acqviſitam præclaris inventis & ſcriptis demorſos ungues ſpiranti-fui in orbe eruditiſſimam, non proprium eſſe illi bonum, ſed publicum. Qvo animo ſine invidia cœpit cum vix pedem patriæ urbi nuper intuliſſet, in gratiam juvenum Aſclepiadeorum manu felici & prompta viſcera animalium ſcrutari, ut abdita qvæ vis propalaret. Cumq; menſes Autumnales prioris anni, qvanqvam aere non ſatis favente publicè privatimq; conſumpſerit in ſectione Hominis, Urſarum duarum Rangiferi, Capreæ ſylveſtris, leporum, felis, glirium, ſciuri, erinacei, cercopitheci & aliorum animalium, qvorum obſervationes in Actis Medicis & Philoſophicis, qvæ ſub prælo fervent, conſignavi; ne ſine exercitatione decurreret primi naſcentis anni, oportuniq; his ſacris menſes, conſtituit amore Artis & patriæ juventutis, faventibus Majoribus, & conſentiente Academiæ Patrono D. PETRO REEDZ, Eqvite & Magno Regis Cancellario, in humano fœminii ſexus cadavere, qvod oblatum fuit, majori cura & ordine artis ſuæ doctrinæq; felicitatem in Theatro Anatomico experiri, in Dei gloriam, Naturæ augmentum & commodum Reipublicæ Medicæ.   Junget eadem opera curioſiſq; Zootomiæ ſpectatoribus varia exhibebit ex Anatome Rangiferi Cervi, terris Borealibus proprii incolæ, qvem beneficio Mecænatis noſtri D. PETRI GRIFFENFELD, Eqvitis & Intimi Regis Conſiliarii, nactus eſt publicè diſſecandum. Qvibus conatibus bono mortalium dicatis, tanti Viri dexteritati famæq; æternæ applaudentes, optatum ſucceſſum dignamq; meritis gratiam apprecamur, qvi ad intermiſſa ſtudia & creſcentis Anatomes gloriam haud inviti trahimur, memores Milonis Crotoniatæ, qvi cum jam ſenex athletas ſe in curriculo exercentes videret, aſpexiſſe lacertos ſuos dicitur illacrymanſq; dixiſſe : *At hi qvidem jam mortui ſunt.* Augetur cottidie Anatomes ſcientia & emeritos grata recordatione reficit. Qvotieſcumq; manus cruentant Proſectores, anteqvam abluantur, novæ deſtillant obſervationes, Seculi ea laus eſt & Naturæ benignitas, qvæ le totam curioſis explicat. Nos, dum licuit per ætatem, hanc lampada accendimus. Alii jam majori felicitate ſuppleant qvod perfectioni ejus & ſplendori deeſſe videbitur. Inter qvos Stenonius noſter, Proſectorum nulli ſecundus ſecandi promtitudine & inveniendi facilitate, tam benigno genio his partibus domi foriſq; functus eſt, ut in admirationem ſui orbem eruditum rapuerit. Phidiæ Jupiter Olympias qvotidie teſtimonium perhibeat, Mentori Capitolinus & Diana Epheſia, qvibus fuere conſecrata artis vaſa ; Stenonio theatra Europæ meliora aſſurgunt, ora Doctorum, fama publica, ipſe deniq; Homo, qvem omnibus, qvibus volupe, proponet, ut homines nos eſſe ſciamus ad reſolutionem natos non ſuperbiam, non obrectationem, non deniq; Dei & Naturæ contemptum. Qvicunq; hæc arcana nobiſcum non faſtidit , ſpeluncam Democriti noſtri accedat , ubi Deos etiam inveniet, nuſqvam illis faventiores.   Non laſſari videbit manus , nec ingenium deficere, nec Naturam obmuteſcere. Patebit omnibus Artis fautoribus & amicis , in primis qvi Æſculapio conſecrati partium accuratam demonſtrationem deſiderant & noſſe debent, Theatrum noſtrum Anatomicum, ea lege & condicione, ut pari patientia Proſectorem incomparabilem audiant ſpectentq; , qva laborioſum opus peragi viderint, & ea benevolentia Clariſſimum Doctorem, non pompa ſed opere æſtimandum excipiant, qvam meretur ille, apud qvem, exemplo *Hippocratis*, major eſt ſapientiæ ratio, qvam auri. Ne neſcii ſint, qvibus ſciendi videndiq; amore inflammatur pectus, Demonſtrationis initium craſtinus dies dabit, hora poſt meridiem ſecunda. Hafniæ d, 28. Januarii Anno cɪɔ ɪɔc ʟxxiii.

Sub Facultatis Medicæ Sigillo.

Fig. 2.31.1 Thomas Bartholin's invitation to Steno's Preface Lecture, Copenhagen 1673. REX, KB 34:3, 79 2, folio

2.32 Preface to Anatomical Demonstrations in the Copenhagen Theatre in the Year 1673

That I stand in your presence, most worthy spectators of every rank, results from the generosity of the Creator towards his work, the favour of the King toward his subject, and my own expectation of the benevolent attention of all of you.

It has pleased God to disclose to me, although to one not seeking but rather reluctant, many things long denied before me to other anatomists far more worthy than myself. It has pleased the King to open today the anatomical theatre of his fathers, closed for many years, to show publicly the observations of others as well as mine. May it please you to pay attention not to the words and hands of the demonstrator, but to the revelation of the wonders of God in his works.

The visitors of a museum who want to see its curiosities hung and set about everywhere, indicated by the pointer or the rod of the custodian, do not feel offended if perhaps the pointer is rudely fashioned whereas an elaborately fashioned pointer will draw the attention of the spectators to itself. The anatomist is a pointer or a rod in the hand of God, pointing out the curiosities of the body like the guide of an exquisite museum. The anatomist himself sometimes deserves to be noticed because of the elegance of his

---

OPH 31 vol. II, 251–256: "Prooemium demonstrationum anatomicarum in Theatro Hafniensi anni 1673" is found in *Thomæ Bartholini Acta Hafniensia*, vol. II, Hafniæ 1675, as No. CXXXIV (misnamed CXXIV), pp. 359–366. Whether it was printed separately is not known. In Holger Jacobæus' manuscript, which is to be found in the Royal Library of Copenhagen, it is copied with a few omissions. Facsimile reprint of the *Prooemium* with English translation by *Sister M. Emmanuel Collins & Paul Maquet*, and with *Holger Jacobæus'* manuscript, see Chap. 2.33 and *T. Kardel* (ed.) *Steno—Life, Science, Philosophy*. Copenhagen 1994, pp. 112–146.

© Springer-Verlag GmbH Germany, part of Springer Nature 2018
T. Kardel and P. Maquet, *Nicolaus Steno*,
https://doi.org/10.1007/978-3-662-55047-2_43

diction and dissection. Such praise belongs to my teachers, my most famous predecessors in this place. Sometimes, however, the slip of his tongue and the clumsiness of his hands—this I acknowledge for myself—would rather offend than delight, if by itself the skillful structure did not rivet all the attention of the spectators.

But if at first glance the cadaver itself seems little attractive to some and even as a livid image of death looks horrifying to others, I will ask them all with insistence not to trust too easily their senses. The senses indeed deceive us as well when, in the Silenes of *Alcibiades*, they consider everything as contemptible and ridiculous, because the outward appearance is ridiculous and contemptible, as when they esteem highly a monkey dressed in purple because of the brilliance of the external colour. The world only promises more and greater things than it comes up with; nature comes up with more and greater things than it promises. Strictly speaking, both are deceitful, since with both, what lies hidden is different from what appears. It is, however, a pleasant error when one recognizes as the best and the most delightful what he had at first despised or feared as inferior or even most unpleasant. Diamonds, either struck from the rocks or retrieved from the mire at the foot of the mountains, at first sight present nothing but crudeness and filth; but, when a skillful craftsman has removed the ugly shell, they elate the discoverer by their brilliance and value. Inspection of the mines teaches us the same truth about other precious gems and even about gold, not to speak of pearls which can only be washed off from the stinking flesh of rotting oysters. All these examples show that a veil unpleasant to the senses very often conceals objects most pleasant to the senses (Fig. 2.32.1).

But it is not only a crude garment which conceals the most elegant beauties. Those works of nature, the outward appearance of which impels us to love them, disclose also such grace to the ones who examine their inside that the visible external elegance obviously gives only a slight idea of the hidden inner beauty. Whoever looks at a meadow from a distance in the loveliest time of the year will experience a most pleasant visual sensation from the mingling of delightful colours. But when then he stoops over the meadow itself to look more closely at the leaves and flowers of the individual plants, such a variety and beauty of shapes and colours unveil themselves that he is compelled to exclaim: from a distance they appear beautiful, but nearby they are far more beautiful! But if, indeed, he proceeds even further and examines in one single plant the inner structure of the different constituent parts, the goings and displacements of the fluids which perform all actions in the

Fig. 2.32.1 From Thomae Bartholini *Cista Medica Hafniensis*. Hafniae 1662

transition from seed to the mature plant producing new seed,[1] even if he becomes acquainted very little with all that as if it were through a mist, yet he sees enough to acknowledge that the pleasure experienced from what is known is nothing compared to that he would experience if he could know all that lies hidden. In his book *On old age*,[2] Cicero recognizes how much power even those things alone which are seen in the fields of the countryside have to delight the mind, although he observed only some of the least wonders which happen there as proofs of his assertion. How great the power and the influence of the human shape are on the minds of men all will acknowledge, who recall having looked closely at the loveliness of a figure with a mind not sufficiently prepared against enticements. And all this external appearance

---

[1] *Steno*, no doubt, is thinking of the investigations of *Malpighi* into the anatomy and biology of the plants, with which, being a personal friend of *Malpighi's*, he was well acquainted. The Preface of *Malpighi's Anatome Plantarum* is dated 1671. See, especially, the chapters *De Seminum Generatione* and *De Semine Vegetatione*.

[2] See *Cicero, Cato Major De Senectute*, cc. XV–XVI, especially XV 52.

which has such a powerful effect is, however, like the view from a distance of a meadow in blossom, nothing else. For, as in the field, only a small part of some flowers is seen, similarly in man nothing but the external appearance, and even very little of it, is seen. What else of the entire man does indeed one see than his face and hands? And of these how very small a part of their surface affects our senses? Frankly, he who knows the difference between the real and the apparent surface of the bodies or who at least applies the microscope on the skin,[3] will admit we see nothing of the human skin except its rougher tips, just as of the vegetation on a distant field we see only the highest tips of the ears. But actually if this hand, the pretty and well proportioned external aspect of which so often draws the entire attention of the observer, at the same time, transparent like a crystal, revealed also the colour of the tendons hidden there, rivalling pearls, and their artefact which surpasses every ingenuity, who would thence not promise the minds of the spectators a far greater pleasure? Yet if one is allowed to proceed into these same parts, the skin and the tendons, and to look closely at the most skilfully wrought textures of the fibres, the complexity of their course and the intricacy of their labyrinths of which we grasp very little and this only by conjecture since they escape every sense, who would cling any longer to the sensible perception of the external appearances alone and, from the attractiveness or the unpleasantness of that perception, judge the rest? Yes indeed, after having rejected all the errors of the senses, who would not repeat: beautiful is what appears to the senses without dissection; more beautiful what dissection draws forth from the hidden inside; yet by far the most beautiful is what, escaping the senses, is revealed by reasoning helped by what the senses perceive.

In order to turn souls away from harmful desires, moralists trace out all which is blameworthy in the object of love. The anatomist, however, asked for remedies for such desires, would not stoop to reproaches, but would raise the loving soul to nobler reasons of love, provided this soul is not completely unable to raise itself a little above the senses. If, however, longing for illicit pleasures has made him look sceptic rather than having made him truly sceptic (and I do not think that anybody is a sceptic except insofar as his love for illicit things finds an excuse for the vice in the obstinacy of doubt), even then we easily toss aside his complaints about the senses. He blames the senses for not displaying things as they are by themselves, but leaving us with either a false or, at best, an uncertain picture of everything. That complaint would indeed be legitimate if the judgment of things were committed to the

---

[3] *Steno* alludes to *Malpighi's* researches on the structure of the skin, which were published in 1665 (perhaps as early as 1664). Their title was: *De externo tactus organo anatomica observatio*. Neapoli.

senses. But this is not the case with ourselves and with our senses. It is not the function of the senses to display things as they are or to judge them, but to transmit to the reason those conditions of the things to be examined, which are sufficient for acquiring knowledge of things appropriate to man's chief end.

We have reason as a judge of the perceptible, thanks to which a reliable approach to the imperceptible is given us. Far be it for us no longer to be men and put ourselves below the beasts. Pursuing the most certain truth by pondering in frequent meditation, let us rather rise from ignorance to knowledge, from imperfection to perfection, and raise in ourselves thoughts worthy of man about his own true dignity. *If the smallest part of the human outer aspect is so beautiful and so greatly affects the observer, what beauties would we see, what pleasure would we experience, if we contemplated the entire structure of the body, if we gazed upon the soul which so many and so ingeniously constructed instruments obey, if we considered the dependence of all this on the cause who knows all that we do not know?* Beautiful is what we see, more beautiful what we know, but by far the most beautiful is what we do not know.[4]

So let us not dwell further on the senses, but with the mind's eyes, through the bodily eyes, as through a window of a very artistic palace, let us look out over this most delightful meadow in which there are as many flowers as there are parts and as many wonders as there are particles. Do not bring up with me the filth and stench of the cadavers which disturb so much the balance of the humours in some people that, even against their will, they keep them away or drive them from the anatomical theatre. This is the weakness of the body to which the mind, in its close union with it, is sometimes forced to yield. Yet, at the same time, those colours are not filthy, but ignorance is; bodies do not stink, but crimes do. The mind indeed deserves to be called, not a part of, but an image of the divine breath only if the mind feels offended by that alone which offends the divine breath from which it derives, and it rejoices in that alone by which the divine breath speaks to us of prudence, power, and goodness in a discourse, silent indeed, but surpassing all eloquence.

And this is the true goal of anatomy: Through the stunning work of art of the body to lift the spectators up to the dignity of the soul and, consequently, through the miracles of both, body and soul, to the knowledge and love of the Creator. Since indeed its object is the animal body, and particularly the human body in so far as this body must be dissected into parts which are accessible to the senses, such great and such obvious beauty cannot but arouse

---

[4] *Winsløv* uses this as a motto to his *Exposition Anatomique de la Structure du Corps Humain*, Paris 1732 and subsequent editions.

admiration and hence the wish to know what escapes the senses. This stimulates reason, from the consideration of the different parts and from the comparison of the single individual parts, to search after the author of such wonders. Reason acquires the more knowledge of this art, the less the presumption and the more complete the rooting out of prejudices with which it surveys that vast forest of experiences. If, indeed, no sensible man can look at a statue, a picture, a clock, machines and whatever well contrived machinery without immediately feeling driven to love the maker of such objects and to appreciate him, how should one gaze with attentive eyes at the structure of the human body which infinitely surpasses all human art, without feeling a forceful impulse to venerate and love its creator? The admirable rule of the divine providence for the creatures endowed with the ability to think is such that, at first, it rather overwhelms this ability with thousand delights through the different ways of perception, then it arouses the wish to seek out the true cause of these pleasures and finally, through discovering what was sought for, to find a way to recognize the giver in his gifts and to transfer all impulse to love from the gifts to the giver. Those who make it merely the servant for the prevention and healing of disease are in error, therefore, and treat Anatomy beneath its dignity. It does have a certain use there but not as much as we think since the recognition of an extraordinary state cannot extend beyond the knowledge of its natural state. Since the latter is yet quite limited, the former cannot advance its limits much further. True anatomy, however, which fits all observers, is a method by which God leads us first to the knowledge of the animal body, and thence of God Himself by way of the hand of the anatomist. For the anatomist should not arrogate to himself either his discoveries or his demonstrations. He is just a creature of God, engaged in the work of God, who not only watches on but also operates His own work. He cannot rightly attribute to himself apart from God anything except his own deficiencies and errors. Therefore, I want to ask all of you to praise with me the divine Goodness if you see something worth your expectation but to ascribe all errors of my tongue and hands to my impatience or a pride hidden for myself, which perhaps wishes more or greater or at least other things beyond the will of God and to which then is deservedly denied even that which I should have obtained easily otherwise. And so, with God as my guide I shall proceed to the anatomical demonstration of the present body. I will do everything to display to your eyes and to your minds what we know hitherto about the body through reliable researches or reasoning. The belief that it is sufficient for the anatomist to explain the parts prepared for the eyes and that the rest can be completed by the spectators through reading or thinking at home, is silly. I would gladly admit that assumption if there was nothing written on anatomy by our predecessors as being true which we in our time have recognized as erroneous, or if the mind engaged in the search for truth proceeded free from all prejudices. Now this is far from being so and, since nothing is more difficult to give up than prejudice, even what is carefully published today is not so deprived of errors that preconceived opinions do not leave their traces, and if I exempted myself from such prejudices, I should deserve the stamp of the most insolent arrogance. Indeed to defend this pursuit of truth to the best of my ability, conscious of the easy risk of error, and to avoid the mistakes committed by others, I will not keep to experiments alone nor bring forward arguments alone, but I will seek such a mixture of both that, by the reckoning of everyone, if not most, at least much will possess a demonstrative certainty. To this end from the general science of bodies I shall put forward only that which is common to all philosophers, even to the most opposed among them as I have explained elsewhere in one of my writings. I shall present the parts of the body not according to their different positions, but according to their substance and their functions, so as to be mindful of both brevity and clarity. I will be rather moderate in the refutation of the errors of others, remembering the words of a man as pious as he was wise: *The knowledge of the truth*, he says, *is sufficient to discern and overturn all errors, even those formerly unheard of, if they are only brought to light.*[5]

---

[5]*Augustine, *Epist.* 118.

2.33 "Niels Stensen's Public Dissection No. XVI of a Female Cadaver in Copenhagen's Anatomical Theatre 1673" as described by Holger Jakobsen (*Jacobæus*)

January 30 in the Afternoon

He separated the cuticle with the skin and all the teguments, which others name differently. He ascertained only three layers: the skin, the subcutaneous fat, and groups of motor fibres.

He said that there would be opportunity to deal with the skin among parts which are found around narrowings of the blood vessel.

That consideration of fat belongs more to the chemist than to the anatomist. What some bring forth about blood vessels which they think to be peculiar to the fat, is not yet that certain.

It is certain that in many parts fat is collected under the membrane around the vessels and the collection is sometimes drained out. Stensen said he

---

Jacobæus' report on Stensen's anatomical demonstrations on a woman's corpse, 29 January-8 February 1673 is preserved by the Royal Library, Copenhagen, NKS 309 aa 4°. It was published as OPH XXXVI in *Nicolai Stenonis Opera Philosophica*, vol. II, 300–307, by Vilhelm Maar. English translation by M. Emmanuel Collins and Paul Maquet in Kardel, *Steno, Life, Science, Philosophy*, 128–141. Photographic copy of Jacobæus' manuscript, ibid. 142–146. Our OPH XXXI b is commented in Part I, 302 and Fig. 1.9.4. There are no illustrations in this manuscript. However, on February 7 there is a reference in the text to Steno's illustration published in *Acta medica et philosophica* reproduced in OPH XXXIII displaying the circulation in a human dissection by Steno. Jacobæus' report on the Preface 29 January is close to Steno's account in the Preface and therefore not brought here. See Vilhelm Maar's annotations to Jacobæus' text in OPH v. II, 300–307.

Holger Jakobsen, or *Jacobæus* (1650–1701), was born in Aarhus, Denmark. He studied medicine and theology in Copenhagen. An educational excursion to Leiden in 1671 was interrupted by unrest and war. The following 2 years he studied with Niels Stensen in Copenhagen as described in Jacobæus' manuscript *Exercitatio anatomica*. In 1674 he became professor of history and geography at Copenhagen University, yet allowed to go for the grand educational tour until 1679, for a period visiting Stensen in Italy. He became later professor of medicine and rector of the university for several periods and a judge to the Supreme Court in 1691.

© Springer-Verlag GmbH Germany, part of Springer Nature 2018
T. Kardel and P. Maquet, *Nicolaus Steno*,
https://doi.org/10.1007/978-3-662-55047-2_44

himself did not know how the particles leave the blood and return back into it, possibly either through vessels of their own kind or excretory, such as in the kidneys, and in other parts, or indiscriminately through pores.

The groups of motor fibres constitute the third common tegument. We indeed see that cavities are formed from muscles, bones, and membranes, or from fleshy and tendinous fibres: motor fibres and that some muscles for some part of them provide for the cavity of the abdomen, for another part of them for the cavity of the breast.

He showed the viscera of the breast and the abdomen in their natural site.

[He showed] the muscles of the abdomen descending obliquely or external, ascending obliquely or internal, transverse and straight.

He said that the peritoneum is nothing but a continuation of the tendons of the muscles of the abdomen and diaphragm, as is the pleura, which on both sides by its cavity envelops the lungs and on the medial side of their cavities these form the mediastinum.

He explained the method of conceiving the site of the parts outside of the membranes: if one imagined each of the leaves of an Indian fig tree, tightly surrounded by a cloth, to continue down to the ground and there to be turned back upwards again to clothe all the leaves together in a single common envelope.

Concerning the peritoneum and the pleura, it must be noticed that there are different terms such as the mesenterium, the mesareum, the ligament by which the liver is suspended [the falciform ligament], the bat-wings [the broad ligament] of the uterus, their duplications and recesses; all designate various parts of the same membrane in so far as they extend differently from the contained parts; thus in the breast, the mediastinum is part of the pleura.

He showed the thymus in the breast.

The pericardium with its heart and water. In man the apex of the heart almost always inclines somewhat to the left, but the basis in a central position shows the auricles at the sides.

When several parts join together, to one of them is ascribed what belongs to another. 2. The vices of the fluids are very often attributed to the solids. Since the spleen, the intestine leaving the narrowings of the mesentery, the stomach, the colon, and the kidney are found close to each other, all pains in the left side are mostly ascribed to the spleen, although the spleen itself is often absolutely healthy. Thus, in the cadaver of an elderly, always believed to be splenetic, Stensen saw that the spleen was perfect but the left kidney was full of gravel. The very famous Sylvius confirmed that he had discovered a healthy spleen in many who had been considered as suffering from a splenic condition, but that the intestine adjacent to the spleen was almost always swollen. Colics are also confused with pain caused by a stone. The second mistake is that, while the uterus in hysterics, and in hypochondriacs, the viscera situated in the upper abdomen are being blamed, more often the defect is in the blood itself, in a humour receding from the blood, but not in the uterus or in the other viscera. What to say about the fact that symptoms of hysterics ascribed to the uterus for the most part are also common to males. Sure, the connection alone of the uterus with neighbouring organs suffices to show that symptoms in the upper abdomen, in the diaphragm and in the oesophagus must be traced back elsewhere than from the uterus.

January 31

He arranged the viscera drawn from another human subject on a large board in the order in which they follow each other.

The digestive tract is narrower in some places, wider in others. Dilatation appears mainly in the mouth, the stomach and the colon. The narrower canal between the mouth and the stomach is called the oesophagus. The dilatation starting from the oesophagus is the stomach, the beginning of which is the cardiac orifice, its end the pyloric orifice. Between the stomach and the colon, the canal from the pylorus to the orifice of the bile duct is the duodenum. The long canal from there is called the jejunum and then, at the end, the ileum. The caecum is the appendix connected to the beginning of the colon. The last part following the colon is the rectum.

To show the different conformations of the digestive tract in different animals, he compared the alimentary systems of a reindeer recently dissected, of a little fox, of a monkey, of a dormouse, of a hen, and of a ray fish.

In the reindeer the intertwining spirals of the thin and the thick intestines presented an elegant picture.

The caecum [is] double in birds. In ostriches, the tract has a delicate membranous spiral surging on the inner surface.

In the ray the caecum is single and as small as seen in some birds. In the marten and the bear there is none; in the monkey it seems to be a stretching of the thick outside the thin.

In man, although somewhat longer, yet it is almost always empty.

In the hare and the rabbit, it is rather long and interiorly provided with a spiral like in ostriches and filled with excrements, but never up to the top. On the contrary, around the end a visible parenchymatous substance appears from which a large portion of humour may be secreted from the blood into the caecum.

In the reindeer and the doe, it is altogether long and ample, without spiral, full of liquid excrements.

In the Islandic little fox and the dormouse it is fairly ample in proportion to the animal, twisted and containing excrements. The variety of the colon is also noteworthy.

In the marten and in bears no difference has been observed from the other intestines.

In man, while in some places it is constricted in length by motor fibres extended like bands, it is divided into different small chambers.

In dormice, the colon, arranged in an elegant manner, has inside several equal angles following each other along the length of the intestine where the spreading out of the legs which constitutes the angles faces the rectum.

Various things can be mentioned about the digestive tract.

Men become more quickly inebriated by small draughts than by full beakers; indeed abundant quantities swallowed at once flow down by their own weight; whatever does not touch the surface of the oesophagus or the stomach cannot transfuse its particles into the blood. But taken in small portions they transmit into the blood whatever spirit they contain by licking the surface of almost all parts of the mouth and of the oesophagus before descending to the stomach.

The digestive tract does not keep the same width from the mouth to the anus. Tt is found to be wider in some places and narrower in others.

The contents of the digestive tract fall into two categories, one, of the intakes, and the other of the excretions from the blood. If the fluids and solids taken in, after having been broken up and mixed outside the body were left to themselves, they would react on themselves and, in freeing particles, they would produce heat and bubbles sometimes simultaneously, whereas in the digestive tract they are added fluids secreted from the blood. These fluids have been prepared earlier from bodies of the same kind by way of the blood, and secreted through the sieves of glands and of the liver. This has two effects: their loosening proceeds more easily and only those which fit that individual are separated.

The entire duct from the mouth to the anus must be recognized as the place from which the distribution of the chyle begins, since spirituous liquors taken in by the mouth restore the strength of a tired man with pleasure almost instantaneously, and practitioners do not deny the efficacy of nourishing clysters nor the ability of mercurials in starting salivation.

Two things concur to the appetite, either a bitterness exciting the membranes, or pain from the constriction of the membranes because of emptiness. Perhaps true hunger occurs when both causes concur. It is certain that, when we abstain from food, the resolving fluid becomes more bitter the more frequent it carries out its circuit. If no new food arrives, the stomach is constricted after its contents have been expelled. All those would convince themselves that bitterness alone is not sufficient, who at certain hours after a meal vomited very bitter matter, without feeling hungry, unless perhaps one could answer that at the same time the inside of the stomach could have been coated by a crust of phlegm not allowing the cancelling of a bitterness of that sort.

We turn away from excrements, not because they are excreted by our body, for many animals live on excrements of other animals, and man likewise prepares not only the most pleasing odours but also food for himself from the excrements of other animals. He even mixes his food with his own excretions, such as saliva not yet spilled from the mouth, and he lives on his own excrements within the womb.

February 1

He next went to the dissection of the brain, since it quickly grows yellow and does not bear waiting.

Concerning the fibrous and white matter it is agreed that the impression of perceptible objects is transmitted to the mind through the extension of this matter into the nerves and that the decision of the will is transmitted from the mind to the motor fibres through the same fibrous matter. Both are proved by the most reliable experiments. When, after the nerve has been cut, or ligated, or otherwise impeded in whatever way, both feeling stops and the obedience due to the mind is lacking in all the parts toward which the same nerve sends branches, although the motor fibres are not denied movement.

All that is said about the animal spirits being distributed through the nerves is questionable. It is certain that, after intake of liquors like spirit of wine, a certain restoration of strength is felt. However, one does not know whether absorption of the liquor itself within the nerves occurs, or whether there is a larger amount of common fluid, or whether there is swifter movement, etc.

About the small particles at the base of the pineal gland, nothing else can be brought forward than what must be said about the small particles to be found almost everywhere in the rest of the body and is explained by the example of the building up of tartar on the teeth. There, indeed, the saliva clinging to the teeth gradually loses its more fluid parts, while thicker parts condense with time and harden. Or, to put forward a more common example, salt condenses in proportion to the evaporation of water from salt water. Tartar condenses at the bottom of a jar in proportion to the evaporation of spirit through the pores of the jar. Silver dust previously dispersed in aqua fortis condenses in proportion to the aqua fortis raised through an alembic. Similarly, both in the gallbladder, in the kidneys, in the small glands either of the tongue or of the rest of the body, and in the skin of gouty people, small stones condense in proportion to the evaporation of a thinner fluid.

Whatever the Ancients, whatever the Moderns imagine about the brain itself is by no means in agreement with the truth. After the error about the material part has been acknowledged, they will grow accustomed to philosophize more cautiously about the worthier part.

The brain can be divided into four parts, on both sides one lateral part, one posterior part, which is the cerebellum, the base from which the nerves and the spinal cord spring; 4 kinds of tubercles; the upper and the lower small gland [=the pineal body and the pituitary gland].

February 3

After the tunics of the eyes had been separated, he pointed out the aqueous, the vitreous, and the crystalline humours.

[He showed] the two muscles of the nictitating membrane, one, cone-shaped, the tendon of which is inserted in the membrane, the other, square, one side of which is in the outer membrane of the eye, the other having a perforated tendon.

One cannot deny a cause in the light moving it over almost infinite distances and carrying out its operation almost instantaneously, nor a cause of perception in the brain, nor a dependence of both moving and perceiving together, with the entire series of intermediate particles, either fluid, or solid, on the universal cause, which we call God.

What the perceiving cause or the mind can perceive by way of the narrow opening of the pupil surpasses all wonder: the vast space and hemisphere of the skies, distant bodies such as the heavenly bodies, countless objects such as the stars.

The nose separating the eyes brings about that we see many things at the same time with one eye that we do not see with the other, even that which we see with both eyes, we do not see with both eyes at the same place. This is proved by a simple experiment with a glass window. One place for the union of double vision is not needed within the brain, but one place in the brain corresponds to any part of the eye which perceives an object, and the mind perceives the single objects at their own places and without any confusion.

February 4

He examined the ear in which there are three small bones, the incus, the stapes and the malleus.

It is certain that movement is diffused gradually from the centre of percussion in every direction, a fact which he explains through the movement of vibration visible on the surface of water around a stone thrown into it. This is proved by experiments, that of the strings which produce a sound at the stroke of similar similarly tensed strings, and that of glass which at first vibrates totally to the human voice and finally shatters.

Two cavities communicating with the outer air, one outside through the external ear, the other inside through the nose; other cavities [are] more hidden, which are called the cochlea and the labyrinth. They seem rather to be different recesses of one single cavity. Three ossicles, nerves and muscles.

The labyrinth in birds and fishes is different. In the other animals indeed, it comprises circles carved in solid bone; in birds the circles are bony surrounded by cancellous bone; in fishes the circles are cartilaginous contained within cartilage of similar shape.

February 5

He showed the small glands of the eye, the passages from the eyes into the nose, from the nose into the mouth, the salivary glands, the tegument of the palate, of the lips, etc.

For a long time the glands were considered an obscure part of the body, until Wirsung, discovering a special duct in the pancreas, opened the way both to find the vessels and to recognize the function of the other glands. The glands are no longer to be considered either as a useless part of the body or as a part destined only for superfluous humours, but as a sieve separating from the blood a fluid necessary for the preservation of the individual. Hence through the whole digestive tract, there are such sieves, either very small and diversely scattered, or joined together into remarkable bodies.

The lymph nodes returning into the blood the humour recovered through vessels named after the famous Thomas Bartholin, have the same conformation as the kidneys, as for the concavity of the place from where the excretory vessels leave and as for the convexity of the opposite part.

Regarding the liquid which has to be returned to the blood through the vessels of these small glands, it is certain that within a few hours after eating, the vessels between the intestines and the left subclavian vein are full of a milky liquor, in the other vessels the liquor is watery.

The salivary glands, two at either side, the lower was described by Wharton, the excretory duct of which ends beneath the root of the tongue, and the upper gland was discovered by Steno.

The entire interior tegument, the tegument of the cheeks and lips is replete with minute glands which Prosector calls labial.

There are many minute glands in the oesophagus above the crop in birds.

February 6
He dealt with the reproductive organs, the uterine tubes, the seminal organs, the uterus, etc.

Ifl women, narrowings of the blood vessels are found in the seminal organs which must be called properly ovaries, as well as in the body and in the cervix of the uterus.

Women have an ovary, or vesicles (the liquor of which condenses into albumen when the vesicles are cooked) enclosed in membranes with narrowings of blood vessels and a substance of its own kind, from which the vesicles are automatically separated at the suitable time and extracted uninjured by means of skill.

The oviduct is a canal open towards the ovary and outward. In some it is single, in others double. Where it is double, in some it is divided about the middle.

Man's seed does not remain in the oviduct, but either the whole or the main part of it perhaps penetrates within the blood and [the nervous system?]. This also results from experiments of Harvey who found nothing in the uterus although he opened several fallow deer to this end.

The maternal blood never penetrates the membranes of the egg. This is demonstrated at first in the early days by the egg visible in the oviduct where the egg is found completely free, then about the time of birth when the placenta or cotyledons are divided into two parts, one adhering to the egg, the other to the uterus and no drop of blood is to be seen in any division.

The foetus is fed by and grows from a serous humour transmitted through the placenta from the mother's blood into the blood of the foetus and excreted in the outer surface of the foetus which transmits its parts within the amnion in a continuous circulation from the outer surface to the inner and vice versa, until finally the more bitter returns render the foetus restless by their stimulation, from which movement the uterus is stimulated to expel.

This very deed considered in itself, most worthy of all deeds of preservation, is also accompanied by the very great pleasure put there by the Creator in reward for those who carry it out properly. The harshest symptoms are set forth as a punishment for those who abuse it so that they finally take refuge in mercurials. To them Nature proclaims that if pleasure is not enough for arising love, pain is able to arise fear.

February 7
He dealt with the lymph nodes and blood circulation.

The total picture of the blood vessel is to be examined carefully according to the method at use among mathematicians. To this end let us imagine a canal with neither a beginning nor an end, i.e. returning back into itself, which we shall divide into 4 parts. In the first part [E, Pulmonary circulation], let us imagine the same canal split into several minute canals so that all the content must move from the ample space of the whole canal through the aforesaid narrowings equal in cross-section to the first whole channel. Let us leave the second [G-H, left side of heart] and the fourth part [B, right side of heart] intact and undivided. Let us divide the third [systematic circulation] part into 3 canals, two of which, however, must be provided in one place, the third [splancnic circulation] in two places, with narrowings similar to the narrowings of the first part. In the second and in the fourth part let us set double sluices in any part of the two so that, after the contents have been transmitted through them, the return to where they came from will not be open. Thus we have completed the true internal picture of all of the blood vessel. The narrowings of the first part correspond to the lungs [E]; the sluices through the second and fourth parts are in the heart with its valves. All this can be better recognized in the following figure [see OPH XXXIII].

February 8
He pointed out the muscles. He called the heart a muscle the fibres of which have their extremities in the base and their middle in the apex or about the apex.

The apex of the heart is thin so that there is nothing except inner and outer membranes concurring and touching each other.

The fibres of the tongue are threefold: straight, perpendicular, transverse.

2.34   Description of the Muscles of an Eagle

2.34.1   On the Muscles of the Head

1. With the skin of the head, the lower extremity of a muscle was cut off, the upper extremity of which was attached to the posterior aspect of the orbit and to the adjacent bone of the skull above the orbit.
2. The muscle which lowers the mandible is attached by its posterior extremity to the skull behind the ears. Its belly which is quite short passes immediately above the membrane which constitutes the inferior and posterior part of the acoustic meatus. Its anterior extremity attaches to the lower and posterior line of the mandible.
3. The strongest of all the muscles which move the mandible is the one which pulls it upwards and forwards. Its posterior extremity is attached to the inferior and posterior part of the mandible. Its anterior extremity is attached to the bone which corresponds to the posterior part of the vomer and to the bone which forms the anterior part of the palate. Flesh is attached immediately to this bone everywhere so that all this bone is nothing else than the ossified intermediate tendon of a composite muscle.
4. The muscle which corresponds to the temporal muscle has no tuberosity on which to be inserted as happens in the jaw of other animals. There is,

---

OPH 32, vol. II, 259–277: The dissection, upon which *Historia musculorum aquilae* is based, was made during Steno's stay in Copenhagen in 1672–1674, viz. on April 4, 1673. See the Appendix, OPH II, 308. The treatise was published in *Thomae Bartholini Acta Hafniensia.* vol. II. Hafniae 1675, as No. CXXVII, pp. 320–345.

© Springer-Verlag GmbH Germany, part of Springer Nature 2018
T. Kardel and P. Maquet, *Nicolaus Steno*,
https://doi.org/10.1007/978-3-662-55047-2_45

however, a tubercle a little in front of the joint. The muscle which corresponds to the masseter adheres to it.

5. The bone between the skull and the mandible has at least two muscles. The first of them has in the skull an extremity expanding from the middle of the base backwards and outwards. The other extremity expands over the length of the intermediate bone itself. The second muscle has one extremity in the lower aspect of the intermediate bone, and the other medially towards the posterior and inferior aspect of the mandible (Fig. 2.34.1).

6. Some muscular fibres extend from the posterior angle of the mandible on one side transversely beneath the trachea to the angle on the opposite side.

Fig. 2.34.1 An eagle dissected by Stensen. Drawing by his student Holger Jakobsen, REX

On the Muscles of the Eyes

Of the muscles of the eye the one which is met the first is the levator palpebrae superioris muscle. The superior oblique muscle lying beneath it is above a small gland lying in the anterior canthus. This gland resembles a muscle by its colour but is provided with a considerable excretory duct and passes through the nictitating membrane fairly close to the cornea.

Besides the levator palpebrae superioris muscle and the muscles of the nictitating membrane, seven other muscles are found.

The extremities of the oblique muscles are found in the anterior canthus, turned away by the eyeball, not much distant from each other so that there is there no trochlea and there is no use for one.

The nictitating membrane has two muscles of which the pyriformis alone would be sufficient to open the membrane but it would constrict altogether the optic nerve if it were not pulled at the same time by another quadrilateral muscle. One extremity of this quadrilateral is attached to the cornea while the other extremity attached nowhere forms a tendinous canal through which the tendon of the pyriformis muscle passes. The elastic force itself or the resultant of the membrane concurring with the convexity of the tunic of the cornea seems to be sufficient to open the nictitating membrane.

The admirable ingenuity of God appears more obvious every day from the comparison of diverse animals. Everybody admires the trochlea in the eyes of men and quadrupeds, and rightfully so. But it is beyond all admiration that, moving the eye without any trochlea in birds, he designed for the nictitating membrane a new kind of trochlea much more skilful. May the one who denies absence of restriction and sagacity to the universal cause learn at first to know mechanics and examine from this point of view the structures of the different animals and he must either lay aside man or admit an unrestricted and very sagacious agent.

On the Muscles of the Tongue, Hyoid Bone, Crop and Trachea

1. The first and lowest muscle has in its middle a tendon the opposite extremities of which extend to the sides of the mandible on both sides and behind the jaws.
2. In like manner it has its opposite extremities on both sides in the medial aspect of the mandible, extending from the middle of the jaw to the anterior angle, and it has a tendon in the middle between its extremities.

3. From about the middle of the first bone of the tongue, a muscle leaves to both sides. It extends almost to the middle of the branches of the second bone, branches corresponding to the greater cornua of the hyoid bone in other animals.

4. Coming from the posterior aspect of the first bone, a muscle ends below the tip of the tongue, curving the tip downwards.

5. Fleshy fibres extend from one branch of the first bone transversely into its other branch, beneath the second bone. They raise straight the tip of the tongue lowered down.

6. A muscle extends from the tips of the branches of the first bone on both sides to the second serrated row.

7. A lateral muscle of the hyoid bone has its anterior extremity towards the anterior part of the hyoid muscle above and in front of the joint. Its posterior extremity is in the posterior part of the mandible partly laterally, partly medially.

8. The geniohyoid muscle, bigger than the previous one, has one extremity in the posterior aspect of the hyoid bone, the other in the middle of the mandible.

9. On both sides beyond the joint there is a bony prong. From there and from another part of the second bone, a muscle extends backwards to the mandible.

10. Near the joint another muscle is seen, the opposite extremity of which is in the posterior and inferior aspect of the first cartilage.

1. Of the muscular fibres extending about the crop over the whole length of the neck, some have their posterior extremity on both sides near the region of the acromion wherefrom they rise and expand somewhat to the sides. Reflected backwards they finally have their anterior extremity in the skull in front of the extremity of the temporal, above the superior aspect of the acoustic meatus.

2. Other fibres have their lower extremity on both sides in the anterior and medial aspect of the branches of the bifurcated bone which is in front of the sternum. From there they rise and expand over the crop and, reflected backwards, either they surround the neck or they insert posteriorly in the skin. By their contraction all these fibres serve to evacuate the contents of the crop.

3. Two muscles ascending on the sides of the trachea have their lower extremity attached to the sternum. I had cut this extremity. From there they rise on both sides, attached to the trachea. I did not observe where their upper extremity reaches.

1. About the bifurcation of the trachea, in a comparison between the birds and the quadrupeds one discovers a nice sign of the ingenuity of the creator acting unrestrictedly. On account of the voice in the different birds He designed a different structure of the muscles for the bifurcation of the trachea, of which I have seen no trace exists in man and in the quadrupeds. In these, all the muscles of the voice joined the top of the trachea. In the eagle several pairs of muscles serving to the voice are found. The upper extremity of the first pair extends some fingerbreadths above the bifurcation, the lower one is laterally in the first cartilage. This pair seems to dilate the division site or bifurcation.
2. Other fibres are found externally in the branches themselves of the bifurcation. Their upper extremity is in the first cartilage of the branches, their lower one is distributed over the different lower cartilages. They thus serve to fasten together the cartilages of the branches. I did not examine other things.

2.34.2 On the Muscles of the Neck

Of the muscles that extend from the skull to the lateral aspect of the neck, the first has its upper expansion between the temporal muscle and the middle of the occiput, its lower extremity in the transverse processes of the fourth, fifth and sixth vertebrae.

The second, the digastric muscle, has its upper extremity near the middle of the occiput beneath the meeting of the previous pair, its lower extremity in the spinous process of the 13th vertebra. The first belly of this muscle is two and a half fingerbreadth long. The intermediate tendon is about three fingerbreadth long. The lower tendon receives nine muscles from nine different vertebrae. The upper four of them receive at their upper extremity as many muscles extending to lower vertebrae. Between the two digastric muscles of the neck there is another kind of muscles. Their lower extremity is in three spinous processes of the neck starting from that immediately above the spinous process in which the lower extremity of the digastric muscle is inserted.

The third muscle has its upper extremity starting on the lateral side of the first pair and expanding over the length of the lateral apophysis of the skull, its lower extremity in the middle of the neck over the second, the third and the fourth vertebra.

The fourth muscle has its upper extremity expanding beneath the upper extremity of the three previous ones, its lower extremity in the spinous processes of the second and third vertebrae.

The fifth muscle has its upper extremity near the middle of the occiput, its lower one in the upper aspect of the second vertebra.

The sixth muscle has its upper extremity on the side of the previous muscle, its lower one in the part of the first vertebra from the region of the spinous process towards the side.

Of the Muscles Extending from the Skull to the Inside of the Neck

1. Has a wide upper extremity expanding from the middle of the skull to its lateral process, a lower extremity in the middle of the vertebrae, starting from the first vertebra and continuing over several of the following vertebrae.
2. Has its upper extremity beneath the previous one, but more laterally, its lower extremity on the sides of the neck, starting in like manner from the processes of the first vertebra and continuing over several of the next vertebrae.

Of the Muscles Between the First Vertebra and the Others

1. On both sides of the tubercle which corresponds to the body of the vertebra, there is the upper extremity of a muscle the lower extremities of which extend between the lateral superior process and the site of the joint of the third, the 4th, 5th, and 6th vertebra where it begins to continue downwards through the apophysis of the vertebra. Hence it extends more laterally to the sides.
2. Beneath the previous one, in front, there is a small muscle. One of its extremities is in the body of the second vertebra, above, towards the middle, and also in the bony lamina of the first vertebra. The other extremity is in the lateral superior process of the third vertebra.

Behind

There is a pair of muscles the upper extremity of which is in the lower edge of the first vertebra, the lower extremity in the side of the spinous process of the second vertebra.

Laterally

There is a pair of muscles the upper extremity of which is on the sides of the first vertebra, the lower one in the upper aspect of the inferior articular processes (lateral inferior processes) of the second vertebra.

Of the Muscles Between the Second and Third Vertebra

In Front

1. It has its upper extremity attached to the inferior aspect of the spinous process and in the part of the transverse process facing the spinous process, its lower extremity in the upper aspect of the spinous process of the next vertebra.
2. It has its upper extremity in the posterior aspect of the transverse process of the second vertebra, its lower extremity in the next six vertebrae and finally in the meeting place with the lower extremity of the digastric muscle described above.
3. It has its upper extremity in the lateral aspect of the transverse process of the second vertebra, its lower extremity in the upper aspect of the lateral processes of the third vertebra.

Of the Muscles Between the Third Vertebra and the Next Vertebrae

In Front

1. It has its upper extremity in the sides of the spinous process, its lower extremity in the anterior extreme edges of the fifth vertebra including the waist of this vertebra.
2. It has its upper extremity in the transverse processes, its lower extremity in the edges of the fourth, fifth and sixth vertebra.
3. It has its upper extremity on the medial side, its lower extremity in the next vertebra almost on the side.

Behind

1. It has its upper extremity in the inferior aspect of the spinous process, its lower extremity in the superior aspect of the spinous process of the fourth vertebra.
2. It has its upper extremity behind, beneath the transverse process, its lower extremity in all the space between the spinous process and the upper aspect of the transverse process of the next vertebra.
3. It has its upper extremity outside the previous one, in the transverse process, its lower extremity in the sides of the sixth and seventh vertebra.

Of the Muscles Between the Fourth Vertebra and the Next Ones

In Front

1. It has its upper extremity in the opening beneath the spinous process, its lower extremity in the next two vertebrae a little away from the middle on the sides.
2. It has its upper extremity between the lateral process and the spinous process, its lower extremity above in the sides of the next vertebra.
3. It has its upper extremity in the transverse process of the fourth vertebra, its lower extremity in the next vertebrae, the fifth, sixth, seventh and the eighth.
4. It has its upper extremity in the inferior and lateral aspect of the transverse process, its lower extremity in the superior aspect of the transverse process of the next vertebra.

Behind

1. It has its upper extremity in the lower aspect of the spinous process of the fourth vertebra, its lower extremity in the upper aspect of the spinous process of the fifth vertebra.
2. It has its upper extremity in the posterior aspect of the inferior articular process, its lower extremity in the fifth, sixth, seventh, eighth and perhaps also ninth vertebra.

Of the Muscles Between the Fifth Vertebra and the Next Ones

In Front

1. It has its upper extremity in the superior aspect on the sides of the middle depression, its lower extremity in the superior aspect of the next vertebra.
2. It has its upper extremity in the apophysis and beneath the apophysis, its lower extremity in the sixth, seventh, eighth vertebrae.
3. It has its upper extremity between the apophysis and the transverse process, and in the lateral and inferior aspect of the same vertebra, its lower extremity above in the transverse process of the next vertebra.

Behind

One has its upper extremity along all the inferior angle beneath the prong which replaces the spinous process, its lower extremity partly in the middle of the next vertebra, partly in the transverse processes of the seventh and eight vertebrae, and in the lower extremity of the digastric muscle of the neck.

Of the Muscles Between the Sixth Vertebra and the Next Ones

In Front

There are three pairs like for the previous (vertebra) and they behave in the same way. The one that lies behind has also the same direction as the previous posterior muscle except, however, that it begins here (as do also the next four ones) to send fairly long fibres towards the tubercle of the 8th, the 9th, the 10th, the 11th and the 12th vertebra.

Other tasks prevented me from continuing the dissection of the muscles of the neck. The ones who are looking for God in the works of nature recognize the sagacious contriver, as anywhere else, in the various length of the neck, depending on the various species of animals. He provided with a shorter neck those in which he divided the extremities of the forelegs into digits mobile to all places of the body. But for those to which he denied digits, he wanted the extremity of the bill to act instead of digits, by moving the head further away from the trunk and by increasing the number of vertebrae of the neck.

2.34.3   On the Muscles Serving to the Common Cavity of Thorax and Abdomen

Seven true and two false ribs are reckoned. The latter are close to the neck, contrary to what exists in man in whom the false ribs are close to the loins. The first false rib is very short and has a muscle continuous with the last vertebra of the neck. The second false rib extends almost to the angle in the middle of the true ribs and has two muscles, one originating from the first thoracic vertebra, the other from the first false rib.

The true ribs are divided into two parts united by a joint. One of their parts is close to the sternum and corresponds to the cartilaginous part of the ribs in man, the other is attached to the vertebrae. Hence, there are three joints for any true rib: one by the rib with the sternum, another by the rib with the vertebra, the third between the parts of the rib. The ribs have two main movements: one towards the neck by which we breathe in, the other towards the loins by which we breathe out. The motor fibres serving to the former movement are inclined towards the neck, those serving to the latter are inclined towards the loins.

No muscular fibres are found in the part of the space between the last two ribs close to the sternum. But there is a muscle in the other part of the space towards the vertebrae. This muscle is inclined towards the lumbar region.

Perhaps other muscles attached to vertebrae and inclined towards the neck would have been visible beneath the ischium bone which had been removed from its place.

In the penultimate intercostal space, in the part close to the sternum, there are fibres of one kind only, all inclined towards the lumbar region. But, in the part of this space close to the vertebrae, the outer fibres are inclined towards the region of the neck, the inner ones towards the lumbar region. Moreover, the inner one of the intercostal muscles fills in all the intervals between the ribs and it is inclined towards the lumbar region. It pulls the ribs towards the loins and narrows the intercostal spaces so that it serves to expiration.

Inside the thorax one sees in the part of the ribs close to the sternum a considerable muscle one extremity of which is at the tip of the sternum in front of the first rib, the other is in the third, fourth and fifth ribs. This muscle widens the intercostal spaces and moves the sternum away from the spine so that it serves to inspiration. It is the antagonist of the previous one.

The obliquus externus which should rather be called transversus has an extremity starting in the part of the sternum about the last four ribs, continuing over the linea alba down to one and a half finger breadths from the os pubis. Its other extremity is in all the ribs (the more distant the ribs are from the first, the more the insertion recedes from the middle joint of the ribs towards the vertebrae) and in almost the whole margin of the ischium. The fibres more remote from the sternum look more oblique. This muscle narrows the spaces between the ribs and moves the sternum towards the spine. It thus serves expiration.

The half of the rectus abdominis muscle close to the pubic bone is completely tendinous. The other half for its most part is attached to the margin of the sternum and for the rest to the first rib.

Fibres extend from this first rib to the anterior margin of the ischium bone. Of those fibres, those which are closer to the linea alba are straight, those which are more remote from it become more and more oblique.

The internal obliques lie beneath the recti abdominis. They extend from the pubic bone obliquely towards the margin of the sternum.

The last four muscles thus described concur and serve to the excretion of the contents of the common cavity which in birds is formed by the thorax and abdomen, and which is divided into various cells by several membranes. They differ in that some of them pull the sternum straight towards the spine, others straight towards the pubis, others obliquely either towards the spine, or towards the pubis.

But in man also, the diaphragm, although forming two cavities, since it is mobile over its entire surface, does not prevent the muscles of the abdomen to serve less also to the excretion of air through the lungs or expiration.

2.34.4 On the Muscles of the Rump

1. A pair which is located in the middle has an extremity in the bone which corresponds to the sacrum, the other in the sides of the different spinous processes of the rump, beneath which other fleshy fibres are seen in the sides of the spinous processes.
2. There is an extremity in the bone which corresponds to the sacrum and in the different transverse processes of the rump, the other over two feathers towards the middle of the rump.
3. There is an extremity in the tips of the transverse processes of the rump, the other on the lateral side of the last feather.
4. There is an extremity towards the symphysis pubis, the other towards the last two feathers.
5. There is an extremity beneath the previous one, a little wider since it attaches not only to the pubic bone but also to the part of the ischium bone forming an angle with the pubic bone. The other extremity is towards the feathers lying about the middle.
6. There is an extremity in the posterior and internal margin of the ischium bone, the other towards the middle of the posterior bones of the rump.
7. There is an extremity beneath the transverse processes of the rump, the other, wider, towards the middle feathers.
8. Lying beneath the previous one, between the transverse anterior and middle processes, it has the same direction as the muscles of the vertebrae.

2.34.5 On the Muscles Connecting the Bones of the Wings with Each Other and with the Bones of the Trunk

The bones of the wing can be divided into seven groups: 1. of the scapula and clavicle, 2. of the humerus, 3. of the ulna and radius, 4. of the carpus, made of two ossicles, 5. of the antepenultimate bone, double, open in the middle with an ossicle which lies outside of it, 6. of the penultimate bone with an ossicle which lies inside of it, and 7. of the ultimate bone which is simple and unique. One counts here as many groups of bone as in the arm of man between the trunk and the tip of the thumb but there is one more in the other fingers of man.

A. Of the Muscles Which Connect the First Two Groups with the Trunk and with Each Other

1. One has an extremity partly in the flat area of the sternum, partly in the middle spine of the sternum close to the margin, partly in a bifurcated bone which is in front of the sternum. The other extremity is either in the linea aspera of the humerus or in another line extending from the medial aspect of the head of the humerus to the end of the linea aspera. This muscle is remarkable by its thickness.

2. Laterally and above the previous one there is a small muscle an extremity of which is in the part of the bifurcated bone lying in front of the sternum close to the top. The other extremity, in the pectoralis itself, is thin and wide, and gives a tendon from which, near the wing, a slender muscle leaves, formed in such a way that one of its tendons, bifid, surrounds the flesh while united to the flesh of the other.

3. There is an extremity partly in the angle which is in the sternum between its spine and its flat part, partly in a tendinous membrane between the anterior aspect of the sternum and the junction of the clavicle with the sternum; from there it passes underneath the ligament joining the clavicle with the scapula, and it inserts anteriorly at the top of the linea aspera of the humerus.

4. There is an extremity in the anterior and posterior aspects of the clavicle close to the sternum, the other at the top of the humerus medially.

5. Beneath the clavicle, one sees muscular flesh, an extremity of which is in the lower aspect of the clavicle, the other in the line of the sternum adjacent to the clavicle.

6. There is an extremity expanding over the fourth, fifth, sixth, seventh and eighth ribs (where three fleshy prongs are visible), the other beneath the scapula. It corresponds to the serratus major muscle.

7. There is an extremity in the external flat area of the scapula, occupying most of the scapula, the other in the cavity which is above in the humerus bone, or rather in the anterior margin of this cavity.

8. There is an extremity expanded in width first in the lower line of the scapula, from there obliquely to the vicinity of the clavicle. The other extremity is in the head of the humerus close to the clavicle. This muscle can be called musculus perforatus since the following.

9. Muscle has an extremity passing through the middle of this muscle up to the scapula bone. The other extremity is in the first of the true ribs.
10. There is an extremity in the anterior and superior aspect of the scapula, the other over the whole linea aspera of the humerus and certainly in the lateral aspect of the humerus.
11. There is an extremity where the clavicle meets the scapula, the other in the flat area which is below the linea aspera of the humerus.

B. Of the Muscles Which Connect the First Two Groups with the Third and the Others

1. There is an extremity expanding in a wide membrane, and attached partly to the acromial end of the clavicle, partly to the medial aspect of the head of the humerus. The other extremity is bifid inasmuch as it is inserted in the radius and in the ulna. Its structure is simple: tendinous expansions extend on either side nicely over the flesh.
2. There is an extremity partly in the inferior line of the scapula close to the humerus, partly in the lateral aspect of the humerus at a distance of two finger breadths below the scapulo-humeral joint.
3. There is a totally fleshy extremity in the upper aspect of the humerus, the other joins together with the distal extremity of the previous muscle into one tendinous membrane to which the small anconeus muscle is attached. This tendinous expansion inserts posteriorly in the radius and in the ulna.
4. A small muscle attached by an extremity to the second pectoralis muscle described above has its other extremity in the lateral aspect of the humerus close to the joint with the ulna.
5. There is a fleshy extremity attached to the tendon of the previous muscle. Both concur in forming a long tendon which inserts in a lateral process of the metacarpus.
6. 7. In the same place, in the vicinity of the bone, two small muscles have their first extremity, one tendinous, the other fleshy, in the aforesaid tendon. Their second extremity is in the medial aspect of an ossicle which is joined to the process of the metacarpus by an articulation.
8. There is an extremity in the lateral aspect of the head of the humerus, the other, all fleshy, in the ulna, not far from the humerus.
9. There is an extremity in the medial aspect of the humerus, the other in the medial side of the previous muscle.

10. There is an extremity in the medial aspect of the head of the humerus beneath the previous muscle, the other within the distal extremity of the previous muscle, but extending further over the radius.
11. There is an extremity in the cavity which lies between the two condyles of the humerus, the other in the ulna, close to the joint with the humerus.

C. In the Space Between the Bones of the Third Group, Laterally

1. There is an extremity in the medial aspect of the head of the humerus. Its other extremity passes through a sheath lying medially and on the lateral side of the joint is divided into two tendons one of which ends in an ossicle which is above the apophysis of the metacarpus, the other passes underneath the other tendon in the joint with the penultimate bone of the wing and inserts reflected obliquely behind in the posterior tip of this bone.
2. There is an extremity in the vicinity of the previous muscle but more posteriorly, perhaps also in a part of the ulna. Its other extremity passes through a sheath close to the angle of flexion and soon receives another tendon from the part which separates the sheath of this muscle from the sheath of the previous one, and so ends in the medial aspect of the metacarpus.
3. There is an extremity within the two previous ones. The other extremity, all fleshy, is in the part of the ulna adjacent to the carpus.
4. There is an extremity, all fleshy, partly in the radius, partly in the ulna, close to the humerus, and occupies the interstice between the two bones. The other extremity is in the apophysis of the metacarpus and follows the same course as the tendons of the first and second muscles lying about the ulna and radius.
5. There is an extremity, all fleshy, in the medial aspect of the radius, looking to the ulna, and occupies most of the radius but neither up to the joint with the humerus nor to the joint with the carpus. Its lower extremity passes through the sheath the most remote from the angle of flexion, follows the cavity of the joint where it leaves the sheath, and receives the tendon of a small but delicate and simple muscle the opposite extremity of which is on the lateral side of the joint. From there the extremity of the fifth muscle runs down to the next joint, where, with the formation of a sesamoid bone, it continues down to the last bone of the wing. In the lateral aspect of the angle between the bones of the third and of the fourth group, one finds a trochlea divided into three sheaths like three parallel grooves through which the tendons pass.

D. In the Space Between the Bones of the Third Group, Medially

1. There is a fleshy extremity over half the ulna in the vicinity of the humerus. The other one passes through a groove in the bone like a pulley, lying on the medial aspect of the joint with the carpus, and from there also over the medial aspect of the next joints down to the last bone.
2. There is an extremity, all fleshy, beginning where the first extremity of the previous muscle ends, and extending almost to the vicinity of the carpus. The other extremity passes through a groove in the bone, like a pulley, lying on the medial side of the carpus, laterally. Leaving from there obliquely, it inserts in the metacarpus close to the joint with the ulna.
3. There is an extremity in the medial aspect of the tendon which extends from the medial aspect of the head of the humerus down to the medial tubercle of the bone which is between the ulna and the carpus. I should think that tendinous fibres leave from this tendon, which from one side go forth to the feathers of the wings, from the other in the radius. The other extremity passes beneath the tubercle of the carpus or of the bone which lies between the humerus and the metacarpus, down to the penultimate bone of the wing.
4. There is an extremity in the lower end of the humerus, in the vicinity of the feathers, the other in the medial tubercle of the carpus or of the bone which lies between the metacarpus and the ulna. It seems to be tied to another muscle which is attached to feathers through fibres over its whole length.

E. About the Bones of the Fifth Group, in the Angle of Flexion, Laterally

1. There is an extremity in a tendinous membrane which is attached to the feathers of the ulna and to the anterior line of the metacarpus. The other one is double. Its wide part, close to the ulna, is in a flat cavity in the smaller of the two bones of this joint. A part ends into a thin tendon which here had been resected.
2. There is a trifid extremity. (Two ends of this extremity are separated by a ligament of a small bone lying between the ulna and the metacarpus. The upper extremity of the first muscle passes between these ends and the third one. The remainder of this upper extremity is attached to the margin itself of the same small bone over its half.). The other extremity is thin. It passes through a groove in the margin of the same small bone down to the tip of an ossicle lying on the anterior aspect of the bone of the sixth group where it receives fleshy fibres before attaching to the ossicle.

3. There is a fleshy extremity attached to the medial aspect of the larger bone of the fifth group. The other one passes through a groove in the extremity of the same bone, obliquely outwards to the posterior tip of the bone of the sixth group.

4. Lying between the two bones of the fifth group, a muscle has a fleshy extremity attached to the surface of both bones. The other extremity leaves upwards and passes between the bones of the fifth group and the feathers implanted over that bone, down to the last feather, to which feathers it seems to be attached. Perhaps this muscle contributes to the twisting of the feathers in the way which is required to close the interstices between them.

5. Beneath the previous one, in the same interstice between the bones, a muscle has an extremity attached for its most part to the medial aspect of the larger bone, from medially outwards, to the smaller bone for a smaller part. The other extremity passes outwards also through a groove in the bone of the sixth group between the extreme tips of the feathers and this bone itself, passing down to the medial tip of the last ossicle.

2.34.6 On the Muscles Uniting the Bones of the Legs to Each Other and to the Bones of the Trunk

A. About the Femur

1. It has a wide and tendinous upper extremity mostly in the upper and posterior margin of the bone corresponding to the ischium bone. Its narrow and tendinous lower extremity is in the upper and medial tip of the bone corresponding to the tibia.

2. It has a wide and tendinous upper extremity in the posterior margin of the bone corresponding to the ischium bone, from the end of the previous muscle to a tubercle at the greatest distance from the spine (from the anterior margin of the same bone, another tendon descends transversely through the medial surface of the confine which is between the flesh and the tendons). Its lower extremity joins with the lower tendon of the vastus externus muscle.

3. It has a wide upper extremity in the posterior margin of the aforesaid bone, starting from the tubercle where the previous muscle ends and continuing almost down to the bottom [of this bone]. Its lower extremity

passes through a very delicate trochlea the upper extremity of which is attached to the medial aspect of the femur, and the lower extremity of which is attached to the lateral condyle of the femur. In the middle of the trochlea a knot is seen as if it were the beginning of a sesamoid bone. Where the muscle comes in contact with the trochlea, it collects into a tendon fairly thick, shaped like a flattened cylinder, and forms a knot, beginning of a sesamoid bone. Below this knot it becomes thinner again and inserts in the tibia at a distance of two finger breadths below the head of the tibia, where a tubercle protrudes.

4. Lying beneath the previous one, another muscle has its anterior extremity in the middle of the posterior aspect of the femur, its posterior extremity between the muscles of the tail beneath the posterior extremity of the fifth muscle of the tail.

5. The upper extremity of a muscle analogous to the gluteus occupies all the margin of the bone corresponding to the ischium bone, beginning from the anterior aspect and then backwards through the superior aspect and descending along the posterior aspect down to the tubercle where the second of the muscles about the femur ends. Its structure is composed by two muscles, the lower tendon being enclosed inside the middle flesh. The outer fleshy fibres are shorter but the longer inner ones occupy a large part of the same bone. The lower extremity surrounds the external aspect of the greater trochanter almost in a circle.

6. Beneath this muscle, there is another one which occupies the anterior and middle margin of the ischiatic bone with its upper extremity. The lower extremity extends laterally below the greater trochanter transversely beneath the lower and transverse extremity of the seventh muscle but in an opposite direction.

7. A muscle has an anterior extremity above the posterior extremity of the previous one. Its posterior extremity actually continues through fleshy fibres of the lower surface of the ischium or of the hip. It pulls the femur backwards as the previous one pulls it forwards.

8. Beside the transverse tendon of this muscle in the thigh there is transversely another one, straight, of which I had resected the opposite extremity which no doubt was to continue to the posterior margin of the ischium.
Where they join the femur, the tendon of the seventh muscle is seen over the flesh, that of the eighth inside the flesh.

9. It has its anterior extremity beneath the tendon of the eighth muscle almost in the middle.

10. Fleshy fibres are found around a tendon of the greater trochanter like a purse. Their opposite extremity is in the periphery of the obturator foramen. The tendon itself passes through this hole inasmuch as it has its opposite extremity in the inner surface of the ischium bone.

11. A thin muscle has its anterior extremity in the inner margin of the iliac bone, not far from the acetabulum. Its posterior extremity is in the medial aspect of the femur, directed outwards.

12. It has its upper extremity in the inner margin of the hip behind the acetabulum. Its lower extremity is in the posterior aspect of the femur.

13. It has its upper extremity in the same place as the previous one but medially. Its lower extremity is in the femur inside the previous one where it is wider than the latter. There are two parts of a triceps muscle. Its tendons are attached partially to the tendons of the muscles lying in the posterior aspect of the lower leg.

14. It has its upper extremity in the lower aspect of the hip, its lower extremity in the medial aspect of the tibia. Its structure is simple with long fleshy fibres.

15. A thin and long muscle has its upper extremity in the anterior margin of the acetabulum, its lower extremity transversely over the anterior aspect of the knee, extending obliquely outwards.

16. It has its upper extremity, all fleshy, in the lateral and anterior aspect of the femur, over all the length of the bone so that, however, the tendon follows a straight line in the middle of the flesh. Its lower extremity is in all the anterior periphery of the tibia so that it almost surrounds the patella.

17. It has its upper extremity, all fleshy, in the medial aspect of the femur, over the length of the triceps muscle. Its lower extremity is in the anterior ridge of the tibia inwards.

B. About the Tibia and Fibula, Beginning from the Lateral and Posterior Aspect in the Vicinity of the Trochlear Muscle

1. It has its upper extremity in the upper and posterior top of the tibia below the trochlea its lower extremity expands beneath the tarsus extending almost to the vicinity of the digits. It constitutes part of this sheath which encloses there the tendons of muscles. Its structure is simple, its fleshy fibres of moderate length.

2. It has its upper extremity beneath the previous muscle, its lower extremity extending in a long and thin tendon which enters the cavity of a sheath and ends in the lateral aspect of the index finger. It flexes the first joint of

this digit obliquely outwards. Its structure is simple, its fleshy fibres longer than those of the previous muscles.

3. At its upper extremity is perforated by the trochlear tendon described above. In its lower extremity it becomes thicker about the joint of the shortest digit. Thus divided, it attaches to the first phalanx beyond its middle and to the second phalanx at its base. It is so dilated by the perforating tendons that it occupies all the width of the joint.

4. Two other muscles are attached above the previous one. The first which is the fourth in the group receives a tendon beneath the tarsus, descends obliquely from one side to the other through the region of the lateral malleolus. Meeting of its two tendons gives a common lower extremity which arrives at the digit next to the smallest one, where it begins dilating beneath the smallest joint and forms the first perforated muscle.

5. The other which is the fifth in the group has its lower extremity in the lateral aspect of the digit which is closest to the thumb.

6. It has its upper extremity in the lateral aspect of the tibia between the previous muscles and the anterior tubercle of the tibia. Its lower extremity is in the second phalanx of the digit next to the smallest one where it constitutes a first perforating or a second perforated muscle. Before entering the first perforated tendon, it proceeds to its lateral side. All these muscles, except for the first, are enclosed in a sheath part of which is constituted by the first in such a way that the common sheath is subdivided into particular sheaths.

7. The most part of the perforating tendons passes through a second remarkable sheath where they expand into a wide and hard body which has a bifid upper extremity continuing with one part on the lateral side of the femur beneath the upper extremity of the trochlea, with the other to the cavity of the femur below the knee, medially, so that it seems to occupy the place which the gastrocnemius normally occupies. Its structure is for its most part similar to the structure of the rectus femoris muscle in the thigh of a man.

8. A third sheath is remarkable. Another part of the perforating tendons passes through it. This part is attached to the fibula and to the tibia by its upper extremity, all fleshy, and occupies almost all the part. The lower extremities of 7. and 8. pass beneath the tarsus. When they arrive beyond the middle of the tarsus they seem to unite into one tendon by way of transverse fibres. They are then variously separated, united again, and divided once more. They constitute four perforating tendons departing to the different digits. The tendon adjacent to the thumb is thicker than the others.

9. It has two upper extremities, one in the medial aspect of the femur towards the joint, and the other in the medial aspect of the tibia. The lower extremity partly forms a sheath, partly passes through the sheath on the medial side, to the process of the tarsus, thus contributing to the formation of the membrane which includes all the tendons which fill the tarsal canal.

C. About the Tibia and Fibula, Anteriorly

1. It has two upper extremities, one tendinous in the top of the lateral aspect of the tibia, the other fleshy in the linea aspera of the fibula. There are also two lower extremities, one in the sheath itself beneath the tarsus, the other united with the first perforated tendon of the digit next to the smallest one.
2. The biggest among the anterior muscles has its upper extremity partly in the anterior and superior margin of the tibia, partly in the linea aspera of the tibia, descending from the head of the tibia down beyond its middle. Its lower extremity forms at first a knot, beginning of a sesamoid bone, where it passes through a sheath or trochlea. From there it inserts, for its larger part in the middle and superior concavity of the tarsus bone, for its smaller part in the lateral margin of this cavity.
3. Enclosed beneath the previous one, the extensor digitorum muscle occupies the most part of the anterior aspect of the upper end of the tibia with its upper extremity. With its lower extremity, however, it passes through the sheath lying medially, and it extends into the smallest digit and into the next two ones on their lateral side. The tendon of this extremity ascends inside the flesh where it has the delicate shape of a goose feather. This tendon also has a knot beneath the sheath.
4. It has its upper extremity more distally, partly in the tibia, partly in the fibula. Its lower extremity passes outwards through a small ring and proceeds obliquely towards the lateral process of the tarsus where it ends.

D. About the Bone Which Replaces the Bones of the Tarsus and Metatarsus, Beginning from the Upper Ones

1. It has three upper extremities, one in the medial aspect of the tarsus, another in the lateral aspect of the tarsus between the two tendons of the first muscle, a third between the eleventh and the twelfth muscle, so that it constitutes three bellies, the lower extremities of which join in one extensor pollicis tendon.
2. The abductor indicis lies beneath this muscle. It has its lower extremity in the inferior aspect of the first joint of the index. It has two bellies, one bigger and more proximal, the other small and close to the joint of the digit.

3. It has its fleshy upper extremity in the middle of the tarsus bone, its lower extremity in the first joint of the digit next to the smallest one.
4. It has its upper extremity towards the lateral margin of the tarsus, its lower extremity in the medial aspect of the smallest digit of which it is adductor.
5. Two muscles hide away in a bony canal beneath the tarsus. The one that lies outwards ends in the lateral aspect of the smallest digit.
6. The other lies inwards with two tendons one of which goes towards that part of the thumb looking to the inside of the foot, the other ends in the opposite side of the thumb. Both flex the first joint of the thumb.

This imperfect description of the muscles, perhaps not without its own errors, is not less arid to readers than their preparation has been delightful to observers; the most elegantly crafted structures very often met in them are indeed only obscurely described in words. But the flesh revealed to the eyes by the course of its fibres, the colour of the tendons, the proportion of the insertions and the distribution of its trochleae surpass all admiration. If it be God's will that I complete the myology of the many animals already begun, what now appears to be a sterile study will be abundantly fecund both for understanding the true causes of discrepancy in the outer shape of different animals and for bringing to light the mechanical artfulness, but most of all in order to defend the freedom to act of the universal cause against modern authors. By the acuteness of their arguments they seem to take away every freedom but too often they use themselves the utmost freedom in choosing the most exquisite delicious food. Thus in action they destroy what by words they try to build up with utmost effort.[1]

---

[1] A criticism of the conceit of those philosophers (Cartesians) who are able by sheer reasoning to deduce, by stringent human logics, how God must create living beings and no other way is possible. God has no other free choice than to follow the logically determined conclusions of the philosophers! The same argument was set forth in OPH I, p. 206: "For whom would it, therefore, not be very sure that the more successfully and fully reason may have deduced a priori diverse modes of function of organs, the less reliable will it be in a certain case, unless the senses have determined the matter through experiments." But here, in OP II, frugal Steno ends up with another moral: criticism of such fine gentlemen who at the table display their greed by allowing themselves the most exquisite delicious food. (AZ).

2.35 Circular Course of the Blood Made Clearer Through Separation of the Ventricles of the Heart

Captions

The primary common trunk of the blood system comprises:

A. the right auricle of the heart.
B. the right ventricle of the heart.
C. the trunk of the pulmonary artery.

The primary straits of the blood system in which:

DDD. Branches progressively divided more and more.
EEE. Straits themselves, invisible, which constitute the substance of the lungs for its most part and communicate with the roots of the excretory vessel of the lung,
Q, which is called trachea.
FFF. Roots joining progressively into a trunk.

The secondary common trunk of the blood system comprises:

G. The left auricle of the heart,
H. The left ventricle of the heart,

OPH 33, II, 281–282: "Receptaculi sangvinis circulus per ventriculorum cordis separationem ab invicem manifestior redditus". The figure was drawn by Steno after the dissection made by him in Copenhagen on February 7, 1673. It is described at length in Holger Jacobaeus' manuscript (see Chap. 2.33 <861>). The illustration appears with captions in *Thomae Bartholini ... Anatome quartum renovata ...*, Lugduni Batavorum, 1674, pp. 805–807.

© Springer-Verlag GmbH Germany, part of Springer Nature 2018
T. Kardel and P. Maquet, *Nicolaus Steno*,
https://doi.org/10.1007/978-3-662-55047-2_46

I. The trunk of an artery to be distributed in all the body upwards and downwards.

The secondary straits of the blood system in which:

KKK. Branches progressively divided more and more.

LLL. Straits themselves invisible, which constitute the substance of the kidneys, of the spleen, etc. for its most part, and communicate with the roots of the excretory vessels of the kidneys, RR, called ureters, and of other viscera.

MMM. Roots from the kidneys joining together with roots of other parts to form the primary common trunk.

NN. Roots from the spleen and other parts joining into the tertiary trunk of the blood system, which is not common to all parts, and which is called portal vein.

The tertiary trunk of the blood system,

O, which is called portal vein and is not common to all parts. >II 282

The tertiary straits of the blood system,

PPP, which form the substance of the liver for its most part with which the roots of the biliary vessels communicate. Hence, these straits join together into the primary common trunk.

2.36   Ornaments: Monuments, Signs, Arguments

Every ornament in church and palace, of ecclesiastical or secular ministers, is 1) a reminder of the curse on mankind, 2) a sign of the inward condition and ornament of the souls, and 3) a proof of coming beatitude.[1]

Every ornament, ecclesiastical or secular, is a reminder of the curse which rests on mankind. All will agree with this when they observe, 1. the hardships and

---

Translation from Latin by M. Rohde in Gustav Scherz, *Steno—Geological Papers*, 250–267. Notes by Gustav Scherz and his associates: Our Latin Text did not have the heading *Ornamenta* in the MS. of the Biblioteca Nazionale Centrale in Florence (Gal. 291, fol. 192r–203r - the volume also contains the Chaos MS.). This text corresponds in essentials to that of the Biblioteca Laurenziana (*Nicolai Stenonis Opera. Medic. Palat.* 36. Sermo XL p. 272–285). (L.MS.), which was published for the first time by V. Maar (*Oversigt over det Kgl. Danske Videnskabernes Selskabs Forhandlinger* 1910. p. 328–337), and which was also the basis for reproduction in *Opera Theologica* (2, 342–349). The copy not only dates from St.'s own period, but was also checked by him, corrected and implemented in parts by notes in the margin and between the lines. These additions throw also some light on the motive, place, and time of the writing (1675–1677), on St.'s mind, and his knowledge of language. We are grateful to J. and M. Götte of Berlin for a thorough revision of the Latin text which they also compared carefully with the L. MS. and the copy in the OTH. Peculiarities in our Ch. MS. will be apparent from its reproduction, but notes indicate discrepancies.

Translation reproduced by permission from the University Press of Southern Denmark. Read the manuscript in BNCF with errata by Stensen: http://teca.bncf.firenze.sbn.it/ImageViewer/servlet/ImageViewer?idr=BNCF0003561685#page/362/mode/2up.

---

[1]The classification 1, 2, 3 in box and 1, 2 below from St's hand.

© Springer-Verlag GmbH Germany, part of Springer Nature 2018
T. Kardel and P. Maquet, *Nicolaus Steno*,
https://doi.org/10.1007/978-3-662-55047-2_47

dangers of miners who mine gold and silver from the bowels of the earth, or of the skilled artisans who fashion them. To these men belong goldsmiths, coiners, metal workers who produce the finest wire,[2] and many others. Apart from the difficult work, there are also many dangers to life connected with their trades, e.g. mine collapses, vapours which can suffocate people in no time at all—or at least harm them for life because of the thick smoke from the lanterns,[3] subterranean air pressure,[4] the vapours from arsenic and quicksilver, from antimony and corrosive waters,[5] poor nourishment, and various other things which are harmful to the health of a human being. 2. All who have read of the misery of the pearl-fishers in India and those who wash pearls out of putrid shells, will also agree.[6] Those who have seen how much trouble

---

[2]Cf. the following impression of St's which could stem from his father's goldsmith shop: *Ex furnis, ubi minerale excoqvuntur, diffusa lux omnibus omnes etiam sanissimas facies Hippocraticas morti simillimas reddit, et quid de illorum luce dicam, qvae res sanctissimas ut perniciosissimas, impias et sacrilegas repraesentat?* (OTH 1, 271).

[3]St. had seen the iron-ore mines of Rio and the Monte Calamita on Elba, saltmines and lagoni around Volterra, quarries in Massa and Carrara, mines of salt, silver and emerald in the Tyrol and Salzburg, and gold and silver mines in Schemnitz and Kremnitz in the Slovakian ore mountains—probably also mines in Bohemia and Saxony.

[4]St. was familiar with Pascal's experiments with air pressure, and those of the Cimento Academy. He notes in his Chaos MS.: N.B. a continuation of new experiments *de pondere aëris.* Oxon. 1668 (Firenze. B. N. Gal. 291. fol. 76 v). Remember that it was not until a century later that H. Cavendish and C. W. Scheele described hydrogen and oxygen, and that the 17th century was still occupied with the investigation of the material composition of air (P. Walden, *Geschichte der Chemie*, Bonn 1950, 43 ff.).

[5]*Arsenicum*, here probably stands for Arsenic trioxide as it is seen in the rusting process of ores containing arsenic under the name of "Hüttenrauch"; it is also found in iron springs containing arsenic, and can be a deadly poison (In Italy in Levico, Roncegno; cf. also H. Bauer, *Geschichte der Chemie* (Leipzig 1905) 30 f. and T. M. Stiliman, *The Story of Alchemy and Early Chemistry* (New York 1960) 19 and 46. Also E. S. Dana, *Textbook of Mineralogy* (New York 1957) 370). Mercurius the Latin name for mercury, quicksilver. It is seldom found in its purest form in nature and is produced by the rusting of cinnabar (mercuric sulphide). Even very faint quicksilver vapours can be deadly poisonous (Cf. Stillman 44; Bauer 1, 30 f.). Antimonium stands for Antimony glance (Sb_2S_3), soluble in water and more or less poisonous; it was once used a good deal as medicine, but caused too much damage (Bauer 1, 56, *Dana Textbook* 411). Cf. also the pieces of arsenic-cobalt pyrites (276), cinnabar and mercury ores (44 x, 274, 275) and antimony (15, 273) in Scherz, *Indice.*

[6]Since Marco Polo's times many travellers had given reports on the West Indian pearl-fishing in the Gulf of Manar. St. was probably thinking of the extensive pearl-fishing carried out at Cape Komorin every spring by Parava fishermen from the lowest caste of the Sudro. Dressed in a scanty loin cloth, with a net for the mussels round their necks, and weighted down by a stone, these fishermen dived down into the water at the bare, sun-scorched beach; they gathered all the mussels within their reach within 1 or 2 min, and then they were pulled up again, to continue the work for 5 or 6 h. Some of the fishers bled through the mouth, nose and ears because they had held their breath for too long. Others lost consciousness and suffocated, and some disappeared completely in their struggle with the numerous sharks, or were pulled back into the boat torn to shreds. The mussels were then poured into large ditches, and, covered with flies, gave off a terrible stench before the pearls could be washed out of the rotting flesh. Moorish merchants and rulers cheated the poor divers out of the rewards of their labours and the lamentations of the womenfolk in Parava village over fathers, sons, husbands and brothers who never returned, were dreadful to bear. (G. Schurhammer, *Franz Xaver. Sein Leben und seine Zeit.* 2 (Freiburg 1963) 261 ff.).

goes into the making of silk will also agree with me.[7] The preparation of animal skins is just as wearisome, as is the polishing of stones or the perfection of any ornament.

All these examples bear out the judgment which the Divine Judge placed on Adam and his descendants: You will earn your bread by the sweat of your brow.[8] In order to earn their living, the hard-working poor must sacrifice their energy and life for what the idle rich use to adorn themselves. Many of the situations mentioned offer still another proof of the curse in so far as they usually originate in places not created by God, but after the malediction of earth, take shape in crevices, cracks, landslides or in subterranean caves produced in some other way. This holds for diamonds and all precious stones whose matter certainly was created at the beginning of time with the other material of the universe, and which was mixed with the other particles of solid and fluid bodies until, after the destruction of the earth it was secreted in old subterranean caves and took shape now to be used by human toil for its own purposes. What is to be said of pearls which are a sign of disease in mussels, and which would never had been formed if the divine curse on mankind had not contaminated everything else.[9] What is to be said of the small, almost

---

[7]St. may have seen how difficult it was to produce the larva out of the cocoon in the silk spinning-mills in Northern Italy. Two of his best friends, M. Malpighi and Jan Swammerdam, were particularly interested in the silk worm at this time (cf. EP 1, 30 f., Cole, *History* 177 f., 260 ff.). St.'s notes of 1668 show an interest in a work of Henry Bond in which he talks of mulberry trees for silk worms, and he also mentions a tract by a certain Mons. Isuard on "Silk and Silk Worms" (Florence B. N. Gal. 291, fol. 79^v).

[8]These quotations from the Scriptures (cf. Gen 3, 14, 17–19; 4, 11, 5, 29) are authorative on the Christian attitude to the problem of evil. The *sententia judicis, maledictio*, is the proclamation of temporal evil and eternal damnation as God's punishment for sin—especially original sin. The discussions about the nature and degree of the disturbed order of the universe and about conceptions such as paradise, freedom from suffering and death often varied and depended on the knowledge of the order of nature. The *carentia justitiae originalis* was regarded as a result of original sin as was the lack of order in nature caused by concupiscence and the upsetting of the natural order in man, which would also have existed without grace (the darkening of the mind and weakening of the will; cf. Thomas Aqu. S. th. I II g 82, a 5). In his treatise on "Original Sin, Sin and Repentance." (OTH 2, 453, ff.). St. refers to the opinions of Augustine and Bonaventure and to the theology of his own time. He realises of course that there is also innocent suffering, "so that the works of God will be revealed" (John 9, 3). He also knows of the healing force, wisdom and blessedness of suffering (Matth. 5, 10). On the other hand, in his treatise on the freedom of the will, he opposes the optimism of Leibniz who sees the creation as a necessary revelation of the being of God. Cf. Note 14.

[9]Ole Worm had already declared his support of B. de Boot's view that pearls are formed just like stones in the gall and bladder (*Gemmarum et lapidarum historia.* 3rd ed. Ludg. Bat., 168. A. Garboe. *Ædelstene,* 35).

invisible worms in the Kermes berries, the parasitic fruit of the plant-louse-bearing oak?[10] What should I say of the larger worms or the fly-bellies of the so-called Cocciniglia?[11] What of the purple snails of Byssus and several other phenomena.[12] An investigation of the single elements out of which all human ornament is made, dearly proves that they are nothing other than threadlike secretions from plants and minerals, the results of sickness or death, or of landslides, secretions and discharges from earth rubble. They are all mined from the earth with infinite pain and by risking lives, where the belly is the teacher, and they in reality are monuments reminding of the human curse.

The proper use of ecclesiastical and secular ornament is to point to the inner ornament of the soul, the striving after the perfection of the virtues of the true likeness of God. Spiritual and secular authorities are witness to this, and the qualities of the ornaments themselves prove it because they remind us very clearly of certain virtues. For spiritual authority, we need only read Moses on the sacred clothing of Aaron[13]; the psalms, the prophets, the songs about the adornments of Christ and the Church.[14] In the New Testament there are the presents of the Three Wise Kings,[15] the parable of the precious pearl[16] and the Prodigal Son, when he was received back into grace.[17] We find secular authority in the works of the poets and painters, not to mention

---

[10]The remark: "the small, almost invisible worms of the Kermes berries, the parasitic fruit of the plant-louse bearing stone-oak" refers to the insect-families of the gall-producing *Aphidae* and of the *Coccides* which swarm over the *quercus coccifera* (Scarlet-oak - Kermes-oak or pigmy-oak, an evergreen tree in Southern Europe). The bark was used in former times for tanning (Jul. v. Weisner, *Die Rohstoffe des Pflanzenreiches*. Bd. 1–3, Leipzig, 1914/21; 2 (1918) 177–221.

[11]*Cocciniglia* (Italian name), belongs to the family of *Coccidae (Homoptera)*, of which upwards of 800 species are known. Insects, usually minute, the male with one pair of wings, the female wingless- and usually so degraded in form that most of the external organs cannot be distinguished. Such Coccids, in their scale, an accumulation of excreted matter, are known as mealy-bugs or as "ground-pearls" (*The Cambridge Natural History*. Vol. VI. Insects by D. Sharp, 2, 592–599) and a dangerous pest on plants (Stefan von Kéler, *Entomologisches Wörterbuch*, Berlin 1963, 133 f.).

[12]The purple dye of the Ancients was got with great difficulty from the yellowish secretion of the snail species *Murex brandaris*, and was used particularly for royal robes. Byssus was first the fine transparent linen weave of the old Egyptians, got from a linen plant growing in the Nile Delta; second, mussel silk, also got with difficulty from a glandular secretion on the foot muscle of different mussels. (*Handb. der Naturwiss.* 6 (1912) 11; 9 (1913) 1232; cf. Scherz, *Indice* 44 B and C; 120 and p. 218, 230).

[13]*Exodus* 28, 2 ff. Bishop Epiphanius of Salamis (ca. 315–403) wrote: *De 12 gemmis Rationalis Summi Sacerdotis Hebraeorum liber ad Diodorum* in which the 12 precious stones in Aaron's shield are described allegorically and scientifically (printed too by Gesner, *De omni rerum fossilium genere*. Tiguri 1565).

[14]Cf. Ps. 45, Cant. 4 and 8. Cf. A. Dragsted, De ædle Stene og deres Mystik (Kjøbenhavn 1933) p. 162 ff.

[15]Matth. 2, 11.

[16]Matth. 13, 43 ff. Cf. Dragsted, p. 174 ff.

[17]Luk. 13, 22.

the historians, and the traces of Egyptian wisdom which have been preserved to our day.[18]

If we examine the qualities of the stones themselves, we find the most wonderful symbols of virtues.[19] The transparency of these objects as a matter of fact is a reminder that God and the saints can look into the soul, and see it with all its sins more easily than the mortal eye can look into the clearest stone with all its light and shade. This is a reminder of a thorough confession with the pain attached to the shame of revealing to a human ear what is always apparent to God. The hardness of the diamond signifies steadfastness, the green of other precious stones stands for the hope to gain a rich harvest of rewards in eternity for virtues which we nurture here. The fiery golden, deep red lustre is a symbol of the burning love for every good work which fills the eyes with the light of divine love and makes a person willing to offer up even blood to God in humiliation and pain. Only princes possess the carbuncle, just as this stone colours everything else with its light, so should the prince be aware that the morals of his subjects are determined by his own virtues and vices.

The whole globe is formed for the example of the king. The white lustre of the pearl is a symbol of purity. The feathers of birds and the skins of quadrupeds, once the emblems of heroes, should, like the skins of noble animals, spur the wearers on to great deeds, and offer the observer a proof of heroic deeds. Just as nobody can perceive these same ornaments of the body if the light from outside is absent, or the inner sight of the eyes, the outward light of divine grace and the inward spiritual health from the same divine grace must be present together to appreciate the real adornment of the spirit.

From the above we can see how unfortunate are those princes and great men who dazzle the eyes of their subjects with external splendours, but damage their own spirit in the eyes of God by vanity and sin. Of course, they rush for the approval of the people when they appear in majestic splendour,

---

[18]St. knew Athanasius Kircher in Rome, the most distinguished scholar of hieroglyphic and Egyptian literature in his time (EP 1, 109 f.).

[19]Many books on precious stones, in Classical Times and in Middle Ages, dealt with the symbolism of precious stones and interpreted their colours in the spirit of Christian perfection. This symbolism was still current when St. himself and Boyle gained a more thorough knowledge of minerals. J. G. Scheuchzer (1672–1733), Physica Sacra (I—IV) (Aug. Vind: 1731/35 1, 224 ff.) was still one of the main works in this field in the 18th century. In it the symbolism of stones is explained in the Razionale of the Jewish High Priest (A. Pazzini, *Le Pietre Preziose* 57 ff., 250 ff.). St. liked to see a reflection of spiritual beauty in the beauty of creation, cf. his *Proemium* (OPH 2, 249 ff.), the many exclamations on the beauty and fascination of glands (OPH 1, 18), of the ray's eye (1, 206) and of anatomy in general, in his scientific writings. Yet, it is remarkable that discussing the qualities of stones as symbols of virtue, Steno never speaks of their pretended psychical or medical qualities, the chief subject of old books of that kind (Cf. Garboe, *Ædelstene* 61ff.).

but their arrogance is in vain, partly because of the value of things which are actually worthless—rather the burden of the universal curse—and partly because the beauty of the objects was dug out of dirt, soiling and disfiguring the hands of the workers. In addition, the admiration of the spectators deserves to be pitied because they themselves, without being completely ignorant, are not able to attune their minds to what is really valuable and beautiful. Only those princes who strive before God and their subjects, to attain the spiritual qualities of which the external objects given to them by God are a reminder, are worthy of the name of prince. For this reason, the prince requires among other necessary knowledge, a knowledge of the symbolic meaning of visible objects so that he can look at himself from time to time and remember at once what some particular ornament means, and why it is worn on a particular occasion. The bishop has to say a prayer for each of the holy vestments he puts on, in which he begs God for the particular grace which they represent.[20] I don't see what argument men who encourage a frequent change of costume can offer to avoid giving the impression of instability, unless they want to introduce new incentives to virtue to the mind.

The frequent reminder of eternal bliss is another of the advantages of the contemplation of visible ornament. The word of God as handed down to us by his friends teaches us this. The Prophet Ezekiel reminds us of crystal, amber,[21] and the, sapphire when describing a vision of the divine glory. When describing Christ, the Apostle John refers to the golden belt, the fiery Aurichalch, the two-edged sword and the golden-edged candelabra.[22] In another vision he describes the heavenly mansions and mentions jasper,[23] sardony, emerald vision, white vestments, golden crowns, crystal of the golden dishes, perfumed garments,[24] white stoles, palms, a golden altar, the gleaming white byssus of a blood-stained white robe, the new town which is adorned like a bride for her groom, precious stones, jasper stone, crystal stone, the measure of the fine golden textiles, like pure glass, jasper, sapphire,

---

[20]To the privileges of a bishop belong especially the liturgical garments at episcopal functions, as for example the *garments for feet, hands, the Tunicella, the Cappa Magna, which are put on to the accompaniment of prayers*—most important are the mitre, ring and staff. *Codex Juris* can. 337; *Lex f. Theol. u. Kirche* 8 (1936) 373.—*Canon Missae ad usum episcoporum* etc. Ed. V. (Ratisbonae 1922) 11ff.

[21]*Elektron* stands either for amber, or a gold-silver mixture with a varying proportion. Ezechiel 1, 4 ff.

[22]In the Christian era, the terms *oricalcum* or *aurichalcum* undoubtedly refer to brass (Georgius Agricola, *De Re Metallica* ed. by H. Bl. Hoover and I. H. Hoover (London 1912) 409).—*Aurichalcite* a basic carbonate of zinc and copper, is found near Campiglia in Tuscany (Dana, *Textbook* 529).

[23]*Jasper* is impure opaque colored quartz (Dana, *Textbook* 473).

[24]Maar has *adornamentorum*. Sapphire may be a lapis lazuli. The term is also often used to indicate corundum gems of any color except red (Dana, *Textbook* 473).

chalcedony, emerald, sardonyx,[25] sard, chrysolite, beryl, topaz, chrysoprase, the hyacinth, the amethyst, and pearls.[26] We learn from St. Paul that all this should not be taken literally, but that it contains a higher meaning. He himself had seen similar things in-his visions, and he writes: "What eye has not seen, nor ear heard and what no human heart has ever known.[27]" We learn then in our search for the hidden meaning of the words that the beauty, everlastingness and clarity of eternal things are suggested by the words. This means that all the angels and saints and mortals and all their works will be revealed as in a crystal in a light, which allows neither darkness nor limits, since everything, even the purest gold is pure glass in comparison, whereas the transparency of gold signifies incorruptibility. The works will also be revealed as they are referred to under the names of clothes, when they are spoken of as being white because of the purity of purpose, and when they are compared to golden belts because of the splendour and unity of love. The following conclusion has general validity because we make the following observations from and with regard to all the ornaments and the temples and palaces of the spiritual and secular office-bearers. If the effects of the curse offer mortals so much beautiful material adornment as to capture their entire attention, what will the spiritual adornment be like which the eternal blessing won by the blood of the Son of God reserves for the blessed? This must suffice to show the threefold advantage of precious stones and all priestly ornament so that we might be moved to repent our sins, to practice virtue, and to long for the *visio beatifica*.

From the foregoing, it follows that none of the artificially produced objects of adornment would have been available to man, in fact would not even have been brought to light, if Adam had remained in a state of innocence. This leads to a new question, why God created such a wealth of material to no obvious purpose, why He gave man so many talents which were never to be used? A correct solution to this problem can be extended to all objects perceptible to the senses, and artificially constructed. Only one word need be changed and the beautiful replaced by the agreeable (or the pleasant). In the same way the variety of food, the whole harmony of musical instruments and

---

[25] *Sardonyx*, sardin like onyx in structure, but includes layers of cornelian (sard) along with others of different colors (Dana, *Textbook* 338, 473). On Sardin cf. note 36, and Dragsted, *De ædle Stene* 80ff.— Garboe, *Ædelstene* p. 37 ff.

[26] Apoc. 1, 12–16; 21, 15; 19–21 etc. Franciscus Rueus (De la Rue) wrote *De gemmis aliquot, iis praesertim quarum D. Johannes Apostolus in sua Apocalypsi meminit* (Francoforti 1608), printed by Gesner (cf. Note 21). *Hyacinth*, the orange, reddish and brownish transparent kind of the Zircon Group, used for gems (Dana, *Textbook* 610).

[27] 1. Cor. 2, 9. According to St. Paul it is usually ascendit.

all perfumes would have remained hidden not in the darkness of ignorance but in the realms of possibilities which would never be realised.

To do justice to this question we must admit that the greater part of earthly objects seems superfluous to us. This is the case:

1. in so far as we know most of them only superficially, as the history of plants and animals shows us,
2. in so far as we conclude from the discovery of new, previously unknown species that there are also fish in the deep sea, animals in distant deserts, plants on mountains and in woods whose species or name is not known to any man alive.
3. in so far as much is processed today which lay for centuries without any practical use, as for example the use of the magnet in sailing; other things which were once in use have now fallen into disuse e.g. purple, byssus, myrrh.[28]
4. the greater part of the materials from which art and handicraft take their working material, remains useless for us. How much gold and silver, how many precious metals and precious stones will remain hidden in the earth from the time of creation until the end of the world? How much wood in forests, or fruit from trees goes rotten without ever serving man or beast? How many millions of pearl-bearing mussels remain buried with their pearls in the sediment of the sea? Whoever considers the problem seriously, must admit that the greatest part of earthly things are, to our knowledge superfluous. However, no one can say whether they are really superfluous or useless because of this. it is one thing to consider such things within the bounds of our own intelligence, and another thing to examine them in relation to the total sequence of created things. It would be equally stupid to call useless something whose purpose we are ignorant of, and to deny the existence of things we have never seen. This would be the same as an apprentice who denied that his master had other instruments in his workshop than those which he saw in the first days, or that all the instruments whose purpose was unknown to him were superfluous.

If we turn to the Creator Himself, we can see the same wisdom on His part with regard to things with which we are familiar, and we realise that we cannot accuse him of stupidity in the face of the great number and variety of things, but must believe that there is not a grain of sand or dust missing in the whole universe, or one that serves no purpose or is superfluous. Therefore,

---

[28]*Myrrhinum*. The wine, spiced with a bitter resin-like Arabic plant extract, myrrha. (Forcellini, *Lexikon* 3 (1940) 325).

the most holy and wise King David has let Him say: "I knew all the birds in the sky, and the beauty of the fields is with me",[29] and another time, "He gives the individual stars their name",[30] although no mortal being can see what purpose any of them apart from the moon or the sun serves. The names which God gives to the stars are their qualities which are directed to the composition and preservation of the whole universe. We read, too in the Gospel that all the hairs of our head are counted[31] (which is of course not literally true, as if God had counted one hair after the other, but because he is, "everything existing in everyone",[32] and can see the number of them all as if it were by intuition. So it is far from true that the material for handcraft and art would serve no purpose if Adam had remained in a state of innocence. We may assume from the little that is fashioned for human purposes, that these things were not created for the ends to which we use them. Otherwise the Creator would have missed the purpose with the greater part of them. However, as we know from other cases that much that is determined for one end can be applied to other ends, the same must be ascertained of the objects of art and handcraft. Where others draw a conclusion less flattering for the Creator, we must conclude that the proof of his greatest wisdom is that He has created objects with their own purposes and functions within the order of the universe so that they can be used to various other ends. The tongue in the human body is used for selecting food, for chewing and swallowing, but also for revealing thoughts, at least since the fall. We might easily suppose that the mind had a more perfect way of expressing sensual impressions before the fall similar to that in which the Angels speak to the soul,[33] which the conversation with the serpent seems to indicate. We communicate to each other with our eyes, with our fingers, by writing and pictures, and who knows if there are not many other ways of communication hidden in the abyss of the possible. The mass of fire which penetrates everything is required to bring the particles of fluid into motion, and if it is collected round certain materials then it lights and nourishes them, and serves a thousand practical purposes. The magnetic force which continually holds the earth at a certain angle to the sky, proved to be suited for guiding ships.

---

[29]Ps. 49, 11.

[30]Ps. 146,11.

[31]Math. 10,30.

[32]1 Cor. 15,28.

[33]See Gen. 3,1f. Just as Hagar (Gen. 16, 7), Abraham (Gen. 22, 11), Tobias (3, 25ff.) or Isaias (6, 2) had visions of angels in the O.T.—so did Joseph (Matth. 1, 20 ff.), Zacharias (Luc. 1, 12) and many others in the N. T. Compare Gen. 3, 1 for the angel's words to the serpent.

With regard to the astonishing variety of talents which gradually reveal themselves according to the variety of the applications, nothing detrimental to the divine prudence can be concluded from this. It is true that in many cases many talents remain untried because they have never been needed, or because they were overlooked for a long time and not exploited until much later, or because, having been used for a long time, they were then left unused. This can be observed not only with the properties of the soul, which are destined for all types of work but also with the properties which were meant by God to serve theoretically or practically as a guide in the conduct of life. How many mortals remain at the level of animals all their life and use all their intellectual gifts to learn vice, to teach it and to practice it. Many stay like children their whole life long in the purely sensual sphere and never break through the barrier to the super-sensual. It must be remembered that the same faculties of perfection and will represent different human activities according to the difference of the objects to which they are directed. This difference is important so that a choice can be made and reward and punishment justify awarded. In a case of a shrine of costly stone I can use my ears, eyes and mind to establish the place of manufacture, the method of construction and the price and then reproduce the same work. In the same way I can learn from my spiritual leader the right way of using them according to God's will, following it in life, or I can from the prince of pride learn vain boasting and a stupid complacency about that ornament. If Adam had remained in his original state of obedience, all his activities would have been spiritual and he would not have directed his powers of perception and will to material things unless as spontaneous signs which God revealed. To understand this, we can think of the example of the letters of the alphabet which the scholar and linguist scarcely looks at because he is concentrating on the sense of the whole word. On the other hand, the man who understands neither the language nor the writing, gazes at the letters in admiration especially if they are artistically drawn. Thus, in the great book of the world, the material things are the letters which Adam, who was well versed in the language, scarcely saw since his whole intelligence was concerned only with their meaning. We draw this conclusion from the fact that he gave all the animals names in keeping with their nature,[34] and because he did not notice his nakedness until after the fall when he was dependent only on his senses.[35] There was no need for him to care for the needs of his body, since he didn't need any clothes or covering in

---

[34]Gen. 2,20.
[35]Gen. 3,7.

the mild climate; he used soft plants for a couch and the fruits of the choicest trees supplied him with wonderful food. He who considers the divine qualities in himself or another creature easily recognises in them as the causes the effects, i.e. the various forms of most of the sensual and super-sensual things, whenever he wants to lower himself from the loftier objects of contemplation to them as the lower ones.

Whence, although he had not soiled his hands with the blood and dirt of animals to turn their bowels into strings for musical instruments nor had he a fire to stretch iron into wire, nor had built canals from the three kingdoms of earthly things in a thousand ways with regard to size and shape, he would have had all sorts of ways of moving the air harmoniously, he would have had the whole variety of material necessary for making instruments of this type, he would have been able to see into the wide ocean of the possible if he had wanted to turn the eyes of his mind away from the contemplation of spiritual things to the pictures of sensual things. When man lost the close companionship of God after the fall, and the decree of the judge condemned him to physical work because of his various bodily needs, the powers of perception and will were very often taken up with the means which lead to the goal, and forgot the highest good because of the constant preoccupation with what was less good. It is a proof of God's lofty wisdom that He arranged the variety of objects from the beginning of creation without any stipulation. He foresaw the inclinations of individuals and the varying circumstances of opportunity and arranged the objects so they would not be superfluous to those who remained obedient to Him, and would offer those who had fallen away from Him the chance to choose from their own free will either to remain in a state of disobedience or to receive His ever ready grace, but always prepared to suffer the punishment for their transgression. I have put forward all these points so that, intoxicated by the artistic workmanship of beautiful objects, we do not make the results of the curse the object of our highest desires.

Bibliography

Abbreviated References

ADB	Allgemeine Deutsche Biographie, vol. 1–56. Leipzig 1875–1912
AGN	Algemene Geschiedenis der Nederlanden, vol. 7–9, Haarlem 1979/80
Acta Hafn.	Bartholin, Thomas, *Acta medica et philosophica Hafniensia*
BEM	Bartholin, Thomas, *Epistolarum Medicinalium*
BNBel	Biographie Nationale … de Belgique
BOD	(Borch, Ole), *Olai Borichii Itinerarium*
BU	Biographie Universelle (Michaud), Ancienne et Moderne, vol. 1–45. Nouvelle édition, Paris 1854–(65)
Chaos	(Stensen, Niels), Niels Stensen's *Chaos-manuscript*
DACL	*Dictionnaire d'archéologie chrétienne*, vol. 1–15, Paris 1930–1950
DBF	*Dictionnaire de Biographie Française*, vol. 1 ff., Paris 1932 ff
DBI	*Dizionari biografico degli Italiani*, vol. 1ff., Rome 1960 ff
DBL, 2. ed.	*Dansk Biografisk Leksikon*, grundlagt af C.F. Bricka, redigeret af Povl Engelstoft under medvirken af Svend Dahl, bd. 1–27, København 1933–1944
DBL, 3. ed.	*Dansk Biografisk Leksikon*, grundlagt af C.F. Bricka, redigeret af Sv. Cedergreen Bech, Bd. 1 ff., København 1979 ff
DNB	*Dictionary of National Biography*, eds. L. Stephen and S. Lee, vol. 1–22, London 1908–1909
DSB	*Dictionary of Scientific Biography*, ed. C.C. Gillispie, vol. 1–16, New York 1970–1980
DTHC	*Dictionnaire de Théologie Catholique contenant l'exposé des doctrines de la Théologie Catholique, leurs preuves et leur histoire*, vol. 1–15, 3rd ed., Paris 1930–1950
Enc. Ital.	*Enciclopedia Italiana di scienze, lettere ed arti*, vol. 1–16, Milano–Roma 1929–1939

© Springer-Verlag GmbH Germany, part of Springer Nature 2018
T. Kardel and P. Maquet, *Nicolaus Steno*,
https://doi.org/10.1007/978-3-662-55047-2

EP	(Stensen, Niels), *Nicolai Stenonis Epistolae*, see also Niels Stensen's Korrespondance
FL	Ehrencron-Müller, H., *Forfatterlexikon*
GP	(Stensen, Niels), *Steno. Geological Papers*
Holland–Danmar	Fabricius, K. et al. *Holland–Danmark* (details see below)
Indice	(Stensen, Niels), *Indice di cose naturali*
Kbh. Dipl.	Nielsen, O. (ed.) *Kjøbenhavns Diplomatarium*
Kbh. Univ. Matr.	Birket Smith, S. (ed.), *Kjøbenhavns Universitets Matrikel*
KU 1479–1979	Ellehøj, Svend (ed.), *Københavns Universitet 1479–1979*
Leibniz, SSB	Leibniz, Gottfried Wilhelm, *Sämtliche Schriften und Briefe*
LTHK	*Lexicon für Theologie und Kirche*
NBG	*Nouvelle Biographie Générale*, ed. Hoefer, Paris 1854–1866
NDB	*Neue deutsche Biographie*. Historichen Kommission bei der Bayerischen Akademie der Wissenschaften, vol. 1 ff. Berlin 1953 ff
NNBW	Molhuysen and Blok (eds.), *Nieuw Nederlandsch biographisch Woordenboek*
NSVO	(Stensen, Niels), *Niels Steensens (Stenonis) Værker i Oversættelse* (details see below)
OPH	(Stensen, Niels), *Nicolai Stenonis Opera Philosophica* (details see below)
OTH	(Stensen, Niels), *Nicolai Stenonis Opera Theologica* (details see below)
PHT	*Personalhistorisk Tidsskrift*
Pionier	Scherz, Gustav, *Pionier der Wissenschaft*
Positio	*Beatificationis et canonizationis servi Dei Nicolai Stenonis, Episcopi Titiopolitani (†1686) Positio*
Reisen	Scherz, Gustav, *Niels Stensens Reisen*
SHKG	*Schleswig-Holsteinische Kirchengeschichte*, Neumünster 1984
Sten. Cath.	*Stenoniana Catholica*

References

Adelmann, Howard B., *Marcello Malpighi and the Evolution of Embryology*, vol. 1–5, New York 1966.

Ahnfelt, Otto, *En skriftebok från Göinge härad på 1600-talet*. (Kirkehistoriske Samlinger), 4. R., vol. 5, København 1897–1899, pp. 344–345.

Allard, G. et al., *La science moderne (de 1450 à 1800)*, (Histoire Générale des Sciences), éd. René Taton, vol. 2., Paris 1958.

Andersen, Jesper Brandt, *Thomas Bartholin, Lægen og anatomen*, Copenhagen, FADL' Forlag 2017.

Andersen, Jesper and Niels W. Bruun, Steno-Fallots tetralogy og Bartholin-Pataus syndrom, *Fund og Forskning*, 54; 197–233, Copenhagen 2015.

Andersen, Vilhelm, *Tider og Typer af Dansk Aands Historie*, 1. Række: Humanisme, 1. Del: Erasmus, 1. Bog: Tiden indtil Holberg. København 1907.

Andrault, Raphaële and Mogens Laerke, eds., *Steno and the Philosophers*, Leiden: Brill, 2018.

Arbuthnot, John, *An Examination of Dr. Woodward's Account of the Deluge, &c., with a comparison between Steno's Philosophy and the Doctor's, in the Case of Marine Bodies dug out of the Earth*, London 1697.

Arnauld, Antoine, *Oeuvres*, vol. 14 (Paris 1778), photoprint, Bruxelles 1967.

Atti, Gaetano, *Notizie edite ed inedite della vita delle opera di Marcello Malpighi e di Lorenzo Bellini*, Bologna 1847.

Bailey, Edward Battersby, *James Hutton – The Founder of Modern Geology*, Amsterdam/London/New York 1967.

Bambacari, Cesare Nicola, *Descrizione delle azioni e virtù del' illustrissima signora Lavinia Felice Cenami Arnolfini*, Lucca 1715.

Bang, Gustav, *Kirkebogsstudier*, København 1906.

Barbensi, Gostavo, *Il pensiero scientifico in Toscana. Disegno storico dalle origine al 1859* (Biblioteca della rivista di storia delle scienze mediche e naturali, vol. XVII). Florence 1969.

Barfod, Immanuel, *Den falsterske Gejstligheds Personalhistorie*, Nykøbing Falster 1851.

Bartholin, Caspar, *De studio medico*, Hafniae 1628.

Bartholin, Thomas, *Vasa Lymphatica*, Hafniae 1653.

Bartholin, Thomas, *Acta medica et philosophica Hafniensia*. Vol. 1–5: 1671–1679, Hafniae 1673 – 1680. **Acta Hafn.**, www.books.google.com

Bartholin, Thomas, *Cista Medica Hafniensis eller Det medicinske fakultets brevkiste, som er fuld af forskelligartede råd, kure, sjældne tilfælde, biografier over københavnske læger og andre ting som vedrører anatomien, botanikken samt kemien. Dernæst følger 'En kortfattet beskrivelse af Anatomihuset i København' forfattet af Thomas Bartholin*. Ed. by Niels W. Bruun og Hans-Otto Loldrup, København 1982.

Bartholin, Thomas, *Domus Anatomica*. Translated by Niels W. Bruun, publ. by Hans-Otto Loldrup, København 2007.

© Springer-Verlag GmbH Germany, part of Springer Nature 2018
T. Kardel and P. Maquet, *Nicolaus Steno*,
https://doi.org/10.1007/978-3-662-55047-2

Bartholin, Thomas, *De Hepatis exautorati Desperata Causa cum praecipuis eruditae Europae Medicis Concertatio*, Hafniae 1666.

Bartholin, Thomas, *Epistolarum Medicinalium ... Centuria*, vol. 1–2, Hafniae 1667. **BEM.**

Bartholin, Thomas, *Universitetsprogram, 28. Jan. 1673*, Hafnia 1673. REX 34:3, 79–2.

Bartholin, Thomas, *On the Burning of His Library and On Medical Travel*, translated by Charles D. O'Malley, Lawrence, Kansas 1961.

Baruzi, Jean, *Leibniz et l'organisation religieuse de la terre*, Paris 1907.

Bastholm, Eyvind, *History of Muscle Physiology (Acta Historica Naturalium et Medicinalium vol. 7)*, Copenhagen 1950.

Baumann, E.D., *François de le Boë Sylvius*, Leiden 1949.

Baumgärtel, Hans, Alexander von Humboldt: Remarks on the Meaning of Hypothesis in his Geological Researches. Schneer, C.J. (ed.), *Toward a History of Geology*, Cambridge–London 1969, 19–35.

Beaumont, L. Elié de, Fragments géologiques tirés de Sténon, de Kazwini, de Strabon et du Boun-Dehesch. *Annales des sciences naturelles*, Paris 1832; 25, 337–377.

Beatificationis et canonizationis servi Dei Nicolai Stenonis, Episcopi Titiopolitani († 1686) Positio super introductione causae et super virtutibus exofficio concinnata (Sacra congregatio pro causis sanctorum officium historicum), vol. 38., Rome 1974. **Positio.**

Becker, T.A., (ed.) *Den ældste danske Archivregistratur*, vol. 4, København 1885.

Becker, Manfred P., *Die Ernennung von Johannes Alpen zum Generalvikar und Siegler durch Christoph Bernhard von Galen* (Studia Westfalica), Münster 1973, 53–75.

Becksmann, Ernst, N. Steno (1638–1686) und seine Stellung in der Geschichte der Geologie. *Zeitschrift der Deutschen Geologischen Gesellschaft*, Berlin 1939; 91, 329–336.

Bell, A.E., *Christiaan Huygens and the Development of Science in the Seventeenth Century*, London 1947.

Belloni, Luigi, *Übereinstimmungen zwischen Stensen und Malpighi auf dem Gebiet der Sekretion und der Inkrustation*, Scherz, *Dissertations on Steno as Geologist*, 1971, 140–148.

Bergendal, Jens, *Sørgeligste dog skyldigste Æris-Minde efter dend velædle/høylærde og af Kiøbenhafns Skole og Academie hoyemeriterede nu salige Mand Mag. Jørgen Eilersøn*. København 1686.

Bering Liisberg, H.C., *Christian den Fjerde og Guldsmedene*, København 1929.

Bernoulli, Johann, De motu musculorum, Dissertations, translated, edited and commented by Paul Maquet, August Ziggelaar and Troels Kardel, *Transactions American Philosophical Society* 87, Part 3, Philadelphia 1997, www.books.google.com

Berti, Enrico, *Il Palazzo Granducale di Livorno, Sede della Provincia*, da 'Liburni Civitas', anno IV, N. 11., Livorno 1931.

Bertrand, E. *Dictionnaire Universel des Fossiles Propres et des Fossiles Accidentels*, vol. 1–2, La Haye 1763.

Besaucèle, Louis B. de, *Les cartésiens d'Italie. Recherches sur l'influence de la philosophie de Descartes dans l'évolution de la Pensée Italienne aux XVIIᵉet XVIIIᵉsiècles, Thesis Faculté des Lettres d'Aix.*, Paris 1920.

Betschart, Ildefons, Stensen – Spinoza – Leibniz im fruchtbaren Gespräch *Salzburger Jahrbuch für Philosophie und Psychologie*, vol. II, München/Salzburg/Köln 1958, 135–151.

Bibliotheca Sanctorum, vol. 1–12, Rome 1961–1969.

Beukers, Harm, *The Sylvian Fissure*. In Koehler et al. (eds.) 2000, 51–55.

Bierbaum, Max and Adolf Faller, *Niels Stensen, Anatom, Geologe und Bischof, 1638–1686*. With a chapter by Anne-Liese Thomasen: Der Wandel des Stensenbildes. 2nd ed., Münster 1979.

Birch, Thomas, The History of the Royal Society of London for Improving of Natural Knowledge from Its First Rise, vol. 1–2. A Facsimile of the London edition 1756–1757. With a new Introduction by A. Rupert Hall and with an additional biographical note by Marie Boas Hall, *The Sources of Science*, No. 44., New York/London 1968.

Birket Smith, S. (ed.), *Kjøbenhavns Universitets Matrikel*, vol. 1: 1611–1667, Kjøbenhavn 1890. **Kbh. Univ. Matr.**

Bloch, Jørgen Carstens, *Den fyenske Gejstligheds Historie fra Reformationen til nærværende Tid*, vol. 1, Odense 1787.

Blondel, Laurent, *Vie des saints pour chaque jour de l'année*, Paris 1722.

Biographie Nationale ... de Belgique, l'Académie Royale des Sciences, des Lettres et des Beaux-Arts de Belgique, vol. 1–28, Bruxelles, pp. 1866–1944. **BNBel**.

Bøje, Chr. A., *Danske Guld- og Sølvsmedemærker før 1870*, 2nd rev. ed. by Bo Bramsen, vol. 1: København, København 1979.

Bonelli, M.L., The Accademia del Cimento and Niels Stensen, *Steno and Brain Research*, ed. by Scherz, Gustav, Analecta Medico-Historica, Oxford 1968; 3, 253–260.

(Borch, Ole), *Olai Borichii Itinerarium 1660–1665: The Journal of the Danish Polyhistor Ole Borch*, ed. by H.D. Schepelern, vol. 1–4, Copenhagen 1983. **BOD**.

Borelli, Giovanni Alfonso, *On the Movement of Animals, De motu animalium*, 1680, translated by Paul Maquet, Berlin: Springer 1989.

Børresen, Kari Elisabeth, Some approaches to Niels Stensen's theology, *Niccolò Stenone (1638– 1686) Anatomista, geologo, vescuvo*, eds. Ascani, Kermit, Skytte, Analecta Romana Instituti Danici, suppl. XXXI, 71. Rome 2002.

Borsig, Arnold von, *Die Toscana. Landschaft, Kunst und Leben im Bild. Einleitung und Erläuterungen von Ranuccio Bianchi-Bandinelli*, 5th ed., Wien–München 1964.

Børup, Marinus, Hvornaar blev Stensen født?, *Tilskueren, Maanedsskrift*, 1938; 55, 38–42.

Brand, P.W., Beach, R.B. and Thompson, D.E. Relative tension and potential excursion of muscles in the forearm and hand, *Journal of Hand Surgery* 1981; 6, 209–219.

Brasier, Martin, Deep questions about the nature of early-life signals: a commentary on Lister (1673) 'A description of certain stones figured like plants', *Philosophical Transactions of the Royal Society A 2015*; 373, 20140254.

Braubach, Max, *Wilhelm von Fürstenberg (1629–1704) und die französische Politik im Zeitalter Ludwigs XIV*, Bonn 1972.

Bricka, C.F., Fortegnelse over Danske og Norske, som ere immatrikulerede ved Leydens Universitet i det første Aarhundrede af dets Bestaaen (1575–1674), *PHT* 2, København 1881, 104–135, 193–210.

Bricka, C.F. and Fridericia, J.A. (eds.) *Kong Christian den Fjerdes egenhændige Breve*, vol. 1: 1589– 1625, Kjøbenhavn 1887–1889.

Brocchi, Gio Battista, *Conchiologia fossile subappenina con osservazioni geologiche sugli Appenini e sul suolo adlacente*, vol. 1, 1st ed. 1814, Milano 1843.

Brown, Harcourt, *Scientific Organization in Seventeenth Century France (1620–1680)*. 1st ed. 1934 New York 1967.

Brunsmand, Johannes, *Epistulae duae adversariae Nicolai Stenonii et Johannis Brunsmanni, ideo nunc editae, ut pateat, quam frivola ratione Stenonius lapsus sui turpitudinem honestare nitatur. Accessit historiae Francisci Spirae desperantis breviarium*. Hafniae 1680.

Bruun, Carl, *Kjøbenhavn. En illustreret Skildring af dets Historie, Mindesmærker og Institutioner*, vol. 1–3, Kjøbenhavn 1887–1901.

Bruun, Chr.V. (ed.) *Biblioteca Danica. Systematisk Fortegnelse over den danske Litteratur fra 1482– 1830*, vol. 1–4, København 1877–1902.

Bruun, Niels W., Fem nyfundne Niels Stensen-breve. *Fund og Forskning*, 2008; 47, 115–166.

Burke, John G., *The Work and Influence of Nicolaus Steno in Crystallography, Scherz*, Dissertations on Steno as Geologist, 1971, 163–174.

Bülow, Kurd von, *Stenos aktualistisch-geologische Arbeitsweise, Scherz, Dissertations on Steno as Geologist*, 1971, 149–162.

Bülow, Kurd von, Protogaea und Prodromus, *Akten des Internationalen Leibniz-Kongresses Hannover*, November 14–19, 1966. (Studia Leibnitiana Supplementa vol. II) Vol. II: Mathematik – Naturwissenschaften, Wiesbaden 1969, 197–208.

Bülow, Kurd von., Der Weg des Aktualismus in England, Frankreich und Deutschland, *Berichte der Geologischen Gesellschaft in der Deutschen Demokratischen Republik*, Berlin 1960, 160–174.

Carlquist, Gunnar, (ed.) *Lunds Stifts Herdaminne från reformationen til nyaste tid*. Ser. II Biografier; Bd. 5: Bara Ljunits och Herrestads kontrakt, Lund 1954.

Carøe, Kristian, *Studier til dansk Medicinalhistorie*, København–Kristiania 1912.

Caverni, Raffaelo, *Storia del metodo sperimentale in Italia*, vol. 1–6, Firenze 1891–1900.

Cawallin, Severin, *Lunds stifts Herdaminne, efter mestadels otryckta käller utarbetat*, vol. 1–5, Lund 1854–1858.

Cedergreen Bech, Sv., Kjersgaard, Erik and Danielsen, Jan (eds.), *Københavns Historie*, vol. 2: 1600–1728. Helge Gamrath: Residens- og Hovedstad, København 1980.

Chapelain, Jean, *Lettres de Jean Chapelain de l'Académie Française*. Tamizey de Larroque, vol. 1–2, Paris 1880–1883.

Chow R. S., M. K. Medri, D. C. Martin, R. N. Leekam, A. M. Agur, N. H. McKee, Sonographic studies of human soleus and gastrocnemius muscle architecture: gender variability. *European Journal of Applied Physiology*, 82; 236–244, 2000. https://doi.org/10.1007/s004210050677

Christensen, Axel E., Clausen, H.P., Ellehøj, S. and Mørch, S., *Danmarks Historie*, vol. 1 ff., Copenhagen 1977 ff.

Christensen, Carl, *Den danske Botaniks Historie med tilhørende Bibliografi*, vol. 1: Fra de ældste Tider til 1912. 1. Halvbind (indtil 1872), København 1924–1926.

Cioppi, Elisabetta and Stefano Dominici, *Genesi e sviluppo delle collezioni geologiche e paleontologiche | origin and development of the geological and paleontological collections*, in Monechi, Simonetta and Lorenzo Rook (ed.), *Il Museo di Storia Naturale dell'Università degli Studi di Firenze. Le collezioni geologiche e paleontologiche / The Museum of Natural History of the University of Florence. The Geological and Paleontological Collections*, ISBN 978-88-6453189-2, 2010 Firenze University Press.

Clarke, Edwin S., Brain anatomy before Steno. In Scherz, Gustav (ed.) *Steno and Brain Research*, Analecta Medico-Historica, Oxford 1968; 3, pp. 27–34.

Clausen, Helge, *The Written Word Is the Most Patient Missionary*, Copenhagen 2006.

Cole, Francis J., *A History of Comparative Anatomy from Aristotle to the Eighteenth Century*, London 1944.

Collas, Georges, Un poète protecteur des lettres au XVIIᵉ, see Jean Chapelain 1595–1674. *Étude historique et littéraire d'après des documents inédits*, Paris 1912.

Collection Académique, T. IVᵉ de la Partie Étrangère. Dijon 1757.

Congrès géologique international. Compte rendu de la 2ᵉ session, Bologne 1881, Bologne 1882.

Conring, Herman (1606–1681), ed. Michael Stolleis. *Beiträge zu Leben und Werk* (Symposium der Herzog August-Bibliothek [Wolfenbüttel 9–12 Dezember 1981]), Berlin 1983.

Corra, Giuseppe and Ferrari, Mario, Itinerari di Stenone nelle Prealpi Tridentine (Val di Gresta) e Lombarde (Alpe di Moncòdeno), Scherz, *Dissertations on Steno as Geologist*, 1971, 175–203.

Coturri, Enrico, Niccolò Stenone e l'influenza che ebbero su di lui e particolarmente sulla sua conversione al cattolicesimo i suoi soggiorni Toscani, *Stensen, Disputatio physica de Thermis*, 61–87, Montecatini 1966.

Coury, Charles, *L'Hôtel-Dieu de Paris. Treize siècles de soins, d'enseignement et de recherche*, Langres 1969.

Curley, Edwin (ed. and transl.), *The collected works of Spinoza*, vol. II, Princeton and Oxford, 2016.

Cutler, Alan, *The Seashell on the Mountaintop – a Story of Science, Sainthood, and the Humble Genius Who Discovered a New History of the Earth*. New York 2003.

Dalberg, C.A.S. and Plum, P.M., *Metropolitanskolen gennem 700 Aar*, København 1916.

Hermansen, Viktor, Roussel Aage and Steenberg, Jan, *Danmarks Kirker. København*, vol. 1, København 1945–1958. **Danmarks kirker.**

Dati, Carlo, *Lettere di Carlo Roberto Dati.* A cura di D. Moreni, Firenze 1825.

Descartes, René, *Oeuvres complètes*, Charles Adam and Paul Tannery, vol. 1–11, Paris 1897–1909, new edition 1964–1974.

Dewhurst K, Willis and Steno. In: Scherz, Gustav *Steno and Brain Research*, Analecta Medico-Historica, Oxford 1968; 3, pp. 43–48.

Diaz, Furio, Il Granducato di Toscana, 1: I Medici. In: da Giuseppe Galasso (ed.) *Storia d'Italia*, vol. 13, 1, Torino 1976.

Djørup, Frans, *Bidrag til Anatomiens Historie i Danmark*, Acta Historia Scientiarum Naturalium et Medicinalium, vol. 20, Odense 1972.

Dominici, S., and Cioppi, E., All is not lost: History from fossils and catalogues at the Museum of Natural History, University of Florence, in Rosenberg, G.D., and Clary, R.M., eds., Museums at the Forefront of the History and Philosophy of Geology: History Made, History in the Making: Geological Society of America Special Paper 535, https://doi.org/10.1130/2018.2535 (05) 2018, (in press).

D'Onofrio, Cesare, *Roma nel Seicento.* Fuori testo è riprodotta la pianta di Roma disegnata e incisa da G.B. Falda nel 1676. Fotografie di Cesare D'Onifrio, Firenze 1969.

Douglas, James, *Myographiae comparatae specimen, or, a comparative description of all the muscles in a man and in a quadruped*, London: Strachan 1707.

Dott, R.H. jr., James Hutton and the concept of a dynamic earth. Schneer, C.J. (ed.), *Toward a History of Geology*, Cambridge–London 1969, pp. 122–141.

Drake, Ellen Tan, *Robert Hooke and the Foundation of Geology: A Comparison of Steno and Hooke and the Hooke Imprint on the Huttonian Theory.* Diss. Oregon State University 1981.

Drake, Ellen Tan and Komar, Paul, D., A comparison of the geological contributions of Nicolaus Steno and Robert Hooke.*Journal of Geological Education*, 1981; 29, 127–134.

Dunin-Borkowski, Stanislaus von, *Spinoza*, Bd. 3: Aus den Tagen Spinozas, 2. Teil: Das neue Leben, Münster 1935.

Ehrencron-Müller, H., *Forfatterlexikon omfattende Danmark, Norge og Island indtil 1814*, vol. 1–12, København 1924–1935. **FL.**

Ellehøj, Svend (ed.), *Københavns Universitet 1479–1979*, Udgivet af Københavns Universitet ved 500 års jubilæet, Vol. 4, 5, 7, 13, København 1979–1980. **KU 1479–1979.**

Ellenberger, François. *History of Geology, Volume I. From Ancient Times to the First Half of the XVII Century.* First published as *Histoire de la Géologie*, Paris 1988. Translated by Kaula, R.K. Rotterdam, Brookfield 1996.

Eloy, N.F.J., *Dictionnaire historique de la Médecine ancienne et moderne*, vol. III, Mons 1778.

Emmen, Aquilinus, P. Franciscus de Hollandia O.F.M. (1650–1708), in saeculo Albertus Burgh *Archivum Franciscanum Historicum*, vol. 37, 1944, Firenze 1947, 202–306.

Escholier, Marc, *Port-Royal*, Paris 1965.

Eyles, V.A., The influence of Nicolaus Steno on the development of geological science in Britain. *Nicolaus Steno and his Indice*, Copenhagen 1958, pp. 167–188.

Fabricius, Knud, *Griffenfeld*, København 1910.

Fabricius, Knud, Hammerich, L.L. and Lorenzen, Vilhelm (eds.), *Holland-Danmark. Forbindelserne mellem de to Lande gennem Tiderne*, vol. 2, København 1945.

Fabricius, Knud, *Skaanes overgang fra Danmark til Sverige. Studier over nationalskiftet i de skaanske landskaber i de nærmeste slægtsled efter Brømsebro- og Roskildefreden*, vol. 1 (1645–1660), Lund–København 1955.

(Fabroni, A.), *Lettere inedite di uomini illustri*, vol. 1–2, Firenze 1773–1775.

Faller, Adolf, Niels Stensen und der Cartesianismus. In: Scherz, Gustav *Nicolaus Steno and his Indice*, Copenhagen 1958, pp. 140–166.

Faller, Adolf, Die Hirnschnitt-Zeichnungen in Stensens Discours sur l'anatomie du cerveau. *Analecta Medico-Historica* 1968; 3, 115–145.

Faller, Adolf, Wertschätzung von Stensens 'Discours sur l'anatomie du cerveau' im Verlaufe von 300 Jahren. *Veröffentlichungen der Schweizerischen Gesellschaft für Geschichte der Medizin und der Naturwissenschaften*, vol. 35, Aarau 1981.

Fauvelle, René, *Les étudiants en médecine de Paris sous le grand Roi. Essai sur leurs études, leur vie médicale et leur vie privée ainsi que sur la société bourgeoise dont ils faisaient partie.* Thèse, Paris 1899.

Fischer, Alfred G. and Garrison, Robert E., The role of the Mediterranean region in the development of sedimentary geology: a historical overview. *Sedimentology* 2009; 56, 3–41.

Fischer, Kuno, *Geschichte der neuen Philosophie*, vol. 2, *Spinozas Leben, Werke und Lehre*, 4th ed., Heidelberg 1898.

Fletcher, J.E., Medical men and medicine in the correspondence of Athanasius Kircher (1602–1680), *Janus* 1969; 56, 259–277.

Fontenelle, *Oeuvres*, vol. 6, new ed. Paris 1790.

Fosseyeux, Marcel, *Une administration parisienne sous l'ancien Régime. L'Hôtel-Dieu au XVIIᵉ et au XVIIIᵉ siécles.* Thèse, Paris 1912.

Francès, Madeleine, *Spinoza dans les pays Néerlandais de la seconde moitié du XVIIᵉ siècle.* Thèse, Paris 1937.

Franceschini, Pietro, Una misconosciuta priorità di Stenone: La scoperta della torsione a cifra 8 delle fibre muscolari del cuore, *Physis. Rivista internazionale di Storia delle Scienze, Firenze* 1972; 14, 431–440.

Franceschini, Pietro, L'Apparato motore nello studio di Borelli e di Stenone, *Rivista di Storia delle Scienze Mediche e Naturali, Firenze* 1951; 42, 1–15.

Frängsmyr, Tore, Steno and geological time, Scherz, *Dissertations on Steno as Geologist*, Odense 1971, pp. 204–212.

Fridericia, J.A., Studier over Kjøbenhavns Befolkningsforhold I det 17. Aarhundrede særlig omkring 1660, *Historisk Tidsskrift*, 6. R, Bd. 2, København 1889/1890, 219–263.

Fueter, Eduard, Geschichte der exakte Wissenschaften in der schweizerischen Aufklärung (1680–1780), *Veröffentlichungen der Schweizerischen Gesellschaft für Geschichte der Medizin und der Naturwissenschaften*, vol. 12, Aarau/Leipzig 1941.

Garboe, Axel, Thomas Bartholin. Et bidrag til dansk natur- og lægevidenskabs historie i det 17. århundrede, I–II, *Acta Historia Scientiarum Naturalium et Medicinalium*, vol. 5–6, København 1949/1950.

Garboe, Axel, Nicolaus Steno and scientific geology, *Nicolaus Steno and his Indice*, Copenhagen 1958, 99–119.

Garboe, Axel, *Geologiens Historie i Danmark: Fra myte til videnskab. Fra de ældste tider til 1835 (med Norge til 1814).*, København 1959.

Garboe, Axel, Niels Stensens (Stenos) geologiske arbejders skæbne. Et fragment af Dansk geologis historie, *Danmarks Geologiske Undersøgelser*, 4 R. 3 Bd. No. 4. København 1948.

Garboe, Axel, Niels Stensen's (Nicolaus Steno's) lost geological manuscript, *Meddelelser fra Dansk Geologisk Forening*, København 1960; 14, 243–246.

Garboe, Axel, Niels Stensens grotto letters (1671). An episode in the Life of the young Niels Stensen (Steno), *Hilsen til J. Christian Bay på Firsaarsdagen 12. Oktober 1951*, København 1951, 15–38.

Garboe, Axel, Præsten Jørgen v. Møinichen, *Fra Københavns Amt*, 8. R., København 1947, 3–20.

Garboe, Axel, Niels Stensen (Steno) set fra en sjællandsk Landsbypræstegaard (1672), *Lychnos* 1946/1947, Uppsala 1947, pp. 248–252.

Gaxotte, Pierre, *La France de Louis XIV*, Paris 1946, new ed. 1968.

Georges-Berthier, A., Le mécanisme Cartésien et la physiologie au XVIIe siècle, *Isis* 1914/19; 2, 37–89 and 1920/21; 3, 21–58.

Geurts, P.A.M., Niels Stensen en Albert Burgh, *Archief voor de geschiedenis van de katholieke kerk in Nederland*, Utrecht 1960; 2, 139–152.

Gosch, Christian Carl August, *Udsigt over Danmarks zoologiske Litteratur med en indledende Fremstilling af de videnskabelige Grundsætninger for Naturvidenskabens især Zoologiens Studium*, II. Afd. Bd. I, København 1873.

Gotfredsen, Edv., *Medicinens Historie*, 3rd ed. København 1973.

Göttler, Joseph, *Geschichte der Pädagogik in Grundlinien*, 3rd revised ed., Freiburg im Br. 1935.

Gould, Stephen J., The titular Bishop of Titiopolis, *Natural History* 1981; 90: 20–24.

(Grubbe), Sivert Grubbes Dagbog, *Danske Magazin indeholdende Bidrag til den Danske Histories og det Danske Sprogs Oplysning*, 4. R., 4. Bd., København 1878, pp. 78–81.

Guhrauer, G.E., *Gottfried Wilhelm, Freiherr von Leibnitz. Eine Biographie zu Leibnitzens Säkular-Feier*, vol. 1–2, Breslau 1846.

Haber, Francis C., Fossils and early cosmology. In: Bentley, Glass et al. (eds.) *Forerunners of Darwin 1745–1859*, Baltimore 1959, pp. 3–29.

Hackmann, W.D., The growth of science in the Netherlands in the seventeenth and early eighteenth centuries. In: Crosland, M. *The Emergence of Science in Western Europe*, New York 1976, pp. 89 109.

Hahn, Roger, *The Anatomy of a Scientific Institution: The Paris Academy of Sciences, 1666 to 1803*, Berkeley, cop., 1971, www.books.google.com

Hansen, Jens Morten, *Stregen i sandet, Bølgen på vandet*, Copenhagen 2000.

Hansen, Jens Morten, On the origin of natural history: Steno's modern, but forgotten theory of science, *Bulletin of the Geological Society of Denmark*, 2009; 57, 1–24.

Heim, Bruno B., Niels Steensens bispevåben, *Årsberetning for Niels Steensens Gymnasium*, København 1964, 48–52.

Heinekamp, Albert, (ed.) *Leibniz-Bibliographie. Die Literatur bis 1980*. Begründet von Kurt Müller, 2nd revised ed., Veröffentlichungen des Leibniz-Archivs, Bd. 10., Frankfurt am Main 1984.

Helweg, Ludwig, *Den danske Kirkes Historie efter Reformationen*, vol. 1, København 1851.

Hermansen, Viktor, Niels Stensens Mutter, *Stenoniana Catholica*, København 1960; 6, pp. 71–81.

Herning, Margrethe, Hjernesnittegningerne i foredrag om hjernens anatomi, *(Stensen), Stenos Foredrag om Hjernens Anatomi*. København 1997, 49–56.

Høeg, Eiler, *Licent. Med. Johann Valentin Wille (Johannes Valentinus Willius). Læge hos Abrahamstrups (Jægerspris) Ejer Overjægermester Vincentz Joachim Hahn i Aarene 1674 – 1676 samt kgl. Dansk Feltmedikus*, København 1934.

Hoffmann, Friedrich, Geschichte der Geognosie und Schilderung der vulkanischen Erscheinungen. Vorlesungen gehalten an der Universität zu Berlin in den Jahren 1834 und 1835, *Hinterlassene Werke von Friedrich Hoffmann*, vol. 2. Berlin 1838.

Hofmann, Jos. E., Nicolaus Mercator (Kauffman), sein Leben und Wirken, vorzugsweise als Mathematiker, *Akademie der Wissenschaften und der Literatur. Abhandlungen der Mathematisch-naturwissenschaftlichen Klasse*, Nr. 3. Wiesbaden 1950.

Hölder, Helmut, *Geologie und Paläontologe in Texten und ihre Geschichte*. München, 1960.

Hölder, Helmut, Leibniz' erdgeschichtliche Konzeptionen, *Studia Leibnitiana*. Sonderheft I: Systemprinzip und Vielheit in den Wissenschaften. Vorträge an der westfälischen Wilhelms-Universität Münster aus Anlass des 250. Todestage von Gottfried Wilhelm Leibniz, hrsg. von Udo Wilhelm Bargenda und Jürgen Blühdorn, Wiesbaden 1969, 105–125.

Hölder, Helmut, Stensens Bedeutung für die Begründung der Geologie und Paläontogie, Scherz, *Dissertations on Steno as Geologist*, 1971, 213–231.

Holzapfel, Helmut, Niels Stensen und die katholische Gemeinde in Kopenhagen, *Kirkehistoriske samlinger*, København 1981, 29–61.

Holzapfel, Helmut, *Unter nordischen Fahnen, die Militärseelsorge der Jesuiten in den nordischen Ländern im XVII. und XVIII. Jahrhundert*, Paderborn o. J. 1954.

Hoogewerff, G.J., De Twee reizen van Cosimo de'Medici prins van Toscane door de Nederlanden (1667–1669), *Journalen en Documenten*, Amsterdam 1919.

Hooke, Robert, Micrographia, Some Physiological Descriptions of Minute Bodies Made by Magnifying Glasses with Observations and Inquiries Thereupon, Observ. XVII. Of Petrify'd wood, and other Petrify'd bodies, London 1665.

Hooykaas, R., Book review / Gustav Scherz (ed.): Nicolaus Steno and his Indice, *Revue d'Histoire des Sciences et de Leurs Applications*. Paris 1959; 12, 181–186.

Hottinger, Johann Heinrich, Die Krystallologia (1698). In: Niggle, P. (ed.) *Veröffentlichungen der Schweizerischen Gesellschaft für Geschichte der Medizin und der Naturwissenschaften*, vol. 14. Aarau 1946.

Houtzager, H.L., Willem Piso, *Spiegel Historia*, Bussum 1977; 12, 390–393.

Hoy, Melissa G., Zajac, F.E. and Gordon, M.E., A musculoskeletal model of the human lower extremity, *Journal of Biomechanics* 1990; 23, 157–169. https://doi.org/10.1016/0021-9290(90) 90349-8

Humboldt, Alexandre de, *Essai géognostique sur les gisements des roches dans les deux hémisphères*, Paris 1823.

Humboldt, Alexander von, Kosmos. Entwurf einer physischen Weltbeschreibung, Bd. 1–4, *Gesammelte Werke, Bd. 1–4.* Stuttgart 1889.

Hunter, Michael, *The Social Basis and Changing Fortunes of an Early Scientific Institution: an Analysis of the Membership of the Royal Society, 1660–1685*, London 1976.

Hunter, Michael, *Science and Society in Restoration England.* Cambridge 1981.

Huxley, T.H., The rise and progress of palaeontology, *Nature* 1881; 13, 452–455.

Huygens, Christiaan, *Oeuvres complètes*, vol. 1–22, La Haye 1888–1950.

Iacobæus, Holger, *En lærd Families Liv og Livsvilkår i det sekstende og syttende Aarhundrede.* København 1931.

Imbert, Gaetano, *La vita fiorentina nel seicento. Secondo memorie sincrone (1644–1670)*, Firenze 1906.

Ingerslev, Vilhelm, *Danmarks Læger og Lægevæsen fra de ældste Tider indtil Aar 1800*, vol. I–II, København 1873.

Israel, Jonathan I., *Radical Enlightenment*, Oxford 2001.

Jahn, Melvin E., Some notes on Dr. Scheuchzer and on homo diluvii testis, Schneer. C.J. (ed), *Toward a History of Geology*, Oxford 1969, 70, 193–213.

Jakobsen, H., *Holger Jacobæus' Rejsebog (1671–1692)*, publ. by Maar, V. København 1910.

Jannaco, Carmine, Il Seicento. Con la collaborazione di Martino Capucci, *Storia Letteraria d'Italia*, 3rd ed., Milano 1973.

Jarro, (ed.) *Lettere inedite di Lorenzo Magalotti, Francesco Redi, Alessandro Marchetti e Andrea Moniglia a Carlo Dati*, Firenze 1889.

Jensen, Michael, *Bibliographia Nicolai Stenonis*, Mørke, Denmark 1986.

Jensen, N.P., *Medlemmerne i Hellig Trefoldigheds Laug udi Det Danske Kompagni eller Det Kjøbenhavnske Skydeselskab og Danske Broderskab, 1447–1901*, Kjøbenhavn 1904.

Johannesson, Gösta, Skåne, Halland og Blekinge, *Danmarks Historie uden for Danmark*, København 1981.

Johnsen, A., Die Geschichte einer kristallmorphologischen Erkenntnis, *Sitzungsberichte der Preussischen Akademie der Wissenschaften. Physisch-mathematische Klasse*, Berlin 1932, 404–415.

Jørgensen, A.D., *Peter Schumacher Griffenfeld*, vol. 1–2, København 1893/1894.

Jørgensen, A.D., *Nils Stensen. Et Mindeskrift.* 2nd enlarged ed. by Scherz, Gustav København 1958.

Jorink, Eric, Modus politicus vivendi: Nicolaus Steno and the Dutch (Swammerdam, Spinoza and other friends). In Andrault and Laerke (eds.), *Steno and the Philosophers*, Leiden: Brill, 2018,13–44.

Kardel, Troels. A specimen of observations upon the muscles: taken from that noble anatomist Nicholas Steno. In: Poulsen, Jacob E. and Snorrason, Egill (eds.) *Nicolaus Steno, 1638–1686, a Reconsideration by Danish Scientists*, Gentofte 1986, pp. 97–134, www.book.google.com

Kardel, Troels, ed., Steno – Life, Science, Philosophy, *Acta Historica Scientiarum Naturalium et Medicinalium*, 1994; 42, 1–159 Copenhagen, e REX.

Kardel, Troels, Function and structure in early modern biomechanics: four episodes and a dialogue between Stensen and Borelli on two chief muscular systems, *Acta Anatomica* 159, 61–70., Basel 1997.

Kardel, Troels, Prelude to two dissertations by Johann Bernoulli, *Transactions American Philosophical Society*, 87, Part 3, Philadelphia 1997. **OA**.

Kardel, Troels, Nicolaus Steno's New Myology 1667: rather than muscle, the motor fibre should be called animal's organ of movement, *Nuncius, Journal of the History of Science* 23(1), 37–64, Firenze 2008.

Kardel, Troels, Prompters of Steno's geological principles: generation of stones in living beings, glossopetrae and molding. In: Rosenberg, G., *Revolution in Geology ...*, 2009, 127–134.

Kardel, Troels, *On deducing the unobservable: Nicolaus Steno's seminal illustrations in Earthscience and biology tell about a method to describe concealed mechanisms and solidtransformations*. https://gsa.confex.com/gsa/2015AM/webprogram/Paper260257.html.

Kardel, T. and P. Maquet, Anatomy of a reindeer dissected in Copenhagen in 1672 by Niels Stensen as reported by Thomas Bartholin. *Rangifer* 2012; 32 (1), 49–55.

Keele, Kenneth D., Niels Stensen and the neurophysiosology of pain. In: Scherz, Gustav (ed.) *Steno and Brain Research*, Analecta Medico-Historica, Oxford 1968; 3, 225–231.

Kermit, Hans, *Niels Stensen. The Scientist Who Was Beatified*. Leominster, UK, Gracewings 2003.

Kircher, Heinrich, *Nord-Stern, Führer zur Seeligkeit durch drey Kräfftige Wirckungen Beten, Psallieren, Lesen*, Köln 1680.

Kobell, Franz von, Geschichte der Mineralogie von 1650–1860, *Geschichte der Wissenschaften, in Deutschland, Neuere Zeit*, vol. II, München 1864.

Koch, E.F., *Oluf Borck, en litterærhistorisk-biografisk Skildring*, København 1866.

Koch, L, *Jesuiten-Lexikon. Die Gesellschaft Jesu einst und jetzt*, Paderborn 1934, Photogr. reprint, Löwen-Heverlee 1962.

Koehler, P.J., Bruyn, G.W. and Pearce, J.M.S., *Neurological Eponyms*, Oxford University Press, Oxford 2000.

Kornerup, Bjørn, *Frederiksborg Statsskoles Historie, 1630–1830*, København 1933.

Kornerup Bjørn, Biskop Hans Poulsen Resen. In: *Studier over Kirke og Skolehistorie i det 16. og 17. Aarhundrede*, vol. 1–2 (vol 2, Vello Helk, ed.), København 1928–1968.

Kornerup, Bjørn (ed.), Visitatsbog for Lundestift 1611–1637. In: Carlquist, G. (ed.) *Lunds stifts Herdaminne från reformationen til nyaste tid på uppdrag av Lunds stifts domkapitel*, Ser. 1: Urkunden och aktstycken, vol. 1., Lund 1943.

Kornerup, Bjørn, Til Lunde stifts kirkehistorie i det 17. aarhundrede, *Vetenskaps-societeten i Lund, Årsbok 1943*, pp. 1–40, Lund 1944.

Kragelund, Aage, Den humanistiske renæssance og antikken – 15 portrætter, Chapter 12. *Niels Stensen*, pp. 226–243, København 1976.

Krebs, Maria, Maria Elisabeth von Rantzau (1625?–1706). Gründerin des Annuntiaten-Klosters in Hildesheim. Ein Leben am Rande großer Weltgeschichte, *Die Diözese Hildesheim in Vergangenheit und Gegenwart. Zeitschrift des Vereins für Heimatkunde im Bistum Hildesheim*, Hildesheim 1976: 44, 45–154.

Krey, J.B., *Andenken an die hiesigen Gelehrnten aus den drei lezten Jahrhunderten*, 1. Stück, Neue veränd. Ausgabe, Rostock 1814.

Kyrre, Hans and Langkilde, H.P., Byens skole. *Københavns Kommunale Skoles Historie*, København 1926.

Laros, M., Pascal, *Lexicon für Theologie und Kirche*. 2nd ed., Freiburg 1930–1939, vol. 7, pp. 992–993.

Larsen, Knud, Stensens forhold til filosofi og religion, *Kirkehistoriske Samlinger*, VI, R., vol. 2, København 1938, pp. 511–553.

Laursen, L. (ed.), *Kancelliets Brevbøger 1627–1629 vedrørende Danmarks indre Forhold*, København 1929.

Leibniz, Gottfried Wilhelm, *Sämtliche Schriften und Briefe*. Hrsg. von der Preussischen Akademie der Wissenschaften. 1. R.: Allgemeiner politischer und historischer Briefwechsel, vol. 2–3, Leipzig 1927–1938. 2. R.: Philosophischer Briefwechsel, vol. 1: 1663–1685, Darmstadt 1926. **Leibniz, SSB**.

Leibniz, Gottfried Wilhelm, *Textes inédits d'après les manuscripts de la Bibliothèque provinciale de Hanovre*, publ. and annotated by Grua, Gaston, vol. 1, Paris 1948.

Leikola, Anto, Francesco Redi as a pioneer of experimental biology, *Lychnos 1977/78*, Uppsala 1978, pp. 115–122.

Le Maguét, Paul-Émile, *Le Monde Médical Parisien sous le Grand Roi*. Thèse, Paris. Macon 1899.

Lemoine, Henri, *Manuel d'histoire de Paris*. Paris 1924.

Lesky, Erna, Die Entdeckung der Funktion des Säugetierovars durch Nicolaus Stensen. In: Scherz, Gustav (ed.) *Steno and Brain Research* Analecta Medico-Historica, Oxford, 1968; 3, pp. 235–251.

Lesky, Erna, Ovum uteri expositus. Eine Harvey-Stensen-Interpretation, *Sudhoffs Archiv für Geschichte der Medizin und der Naturwissenschaften*, 7. Beiheft, Wiesbaden 1966, 105–114.

Lévy-Valensi, J., *La médecine et les médecins français au XVII⁰ siècle*, Paris 1933.

Lexicon für Theologie und Kirche, eds. Höfer, J., Rahner, K., 2nd ed., vol. 1–10, Freiburg 1957–1967. **LTHK**.

Lindeboom, Gerrit A., *Geschiedenis van de medische Wetenschap in Nederland*, Bussum 1972.

Lindeboom, Gerrit A., Dog and Frog. Physiological experiments at Leiden during the seventeenth century. In: Lunsingh Scheurleer, Th.H and Posthumus Meyjes, G.H.M., *Leiden University in the Seventeenth Century. An Exchange of Learning*. Leiden 1975, pp. 279–293.

Lister, Martin, 1673 A description of certain stones figured like plants, and by some observing men esteemed to be plants petrified. *Phil. Trans.* London, 1673; 8, 6181–619. (doi:10.1098/rstl.1673.0061)

Livorno e Pisa: due città e un territorio nella politico dei Medici. Il Giardino dei Semplice, Pisa 1980.

Lorenzen, Vilhelm, Metropolitanskolen, 1537–1938, *Metropolitanbogen*, København 1939, 5–176.

Lovén, Pehr, *Dissertatio gradualis de Gothungia ("Giönge-Härad")*, Londini Gothorum (Lund) 1745, reprint Malmö 1934.

Lyell, Charles, *Principles of Geology*, vol. 1–3, London 1830–1833.

Maar, Vilhelm, Om Opdagelsen af Ductus vitello-intestinalis, *Oversigt over Det kongelige Videnskabernes Selskabs Forhandlinger*, København 1908; 5, 233–265.

Maar, Vilhelm, Et blad af Domus Anatomicas Historie, *Thomas Bartholin. Mindeskrift*, København 1916, 14–19.

Maar, Vilhelm, (ed.), *Mindeskrift for Oluf Borch paa 300 Aarsdagen for hans Fødsel*, København 1926.

Magalotti, Lorenzo, *Scritti di corte e di mondo. A cura di Enrico Falqui*, Roma 1945.

Malcolm, Noel, *The Library of Henry Oldenburg* (eBLJ 2005, Article 7, 1–55).

Malloch, Archibald, *Finch and Baines. A Seventeenth Century Friendship*, Cambridge 1917.

Malpighi, Marcello, *Opera posthuma … quibus præfixa est Ejusdem VITA à seipso scripta*, London 1697.

(Malpighi, Marcello), *The Correspondence of Marcello Malpighi*. Edited by Adelmann, Howard, vol. 1–5, Ithaca, NY/London 1965.

Manni, Dominco Maria, *Vita del litteratissimo Monsig. Niccolò Stenone di Danimarca. Vescovo di Titiopolis e Vicario Apostolico*. Firenze 1775.

Mansa, F.V., *Bidrag til Folkesygdommenes og Sundhedspleiens Historie i Danmark fra de ældste Tider til Begyndelsen af det attende Aarhundrede*, København 1873.

Manfredini, M., De Iasio, S. and Lucchetti, E., The plague in Parma: Outbreak and effects of a crisis. *International Journal of Anthropology*, 2002; 17, 41–57 [SpringerLink].

Marquard, Emil, *Kjøbenhavns Borgere 1659*, København 1920.

Marquard, Emil (ed.), *Kancelliets Brevbøger 1627–1629 vedrørende Danmarks indre Forhold, i Uddrag*, København 1919.

Marquard, Emil (ed.), *Kongelige Kammerregnskaber fra Frederik III's og Christian V's Tid*, København 1918.

Massai, Ferdinando, *Lo Stravizzo della Crusca del 12. September, 1666 e l'origine del "Bacco in Toscana di Francesco Redi"*, Rocca S. Casciano 1916.

Matzen, Henning, *Kjøbenhavns Universitets Retshistorie 1479–1879 efter Konsistoriums Opfordring*, Kjøbenhavn 1879.

May, Margaret Tallmadge, On the passage of yolk into the intestines of the chick (De vitelli in intestina pulli transitu). Nicolaus Steno, *Journal of the History of Medicine and Allied Sciences*, 1950; 5, 119–143.

Mazzolini, Renato, Schemes and models of the thinking machine. In: Corsi, Pietro (ed.) *The Enchanted Loom*, New York/Oxford 1991, pp. 68–143.

McKeon, Robert M., Une lettre de Melchisédech Thévenot sur les débuts de l'Académie des Sciences, *Revue d'Histoire des Sciences et de Leur Application*, Paris 1965; 18, 1–6.

Meschini, Franco A., *Neurofisiologia Cartesiana*, Firenze 1998.

Meinsma, K.O., *Spinoza und sein Kreis*, Berlin 1909.

Meisen, Valdemar and Larsen, Knud (eds.), *Stenoniana*, vol. 1 [of 1], København 1933.

Meli, Domenico Bertoloni, The collaboration between anatomists and mathematicians in the midseventeenth century with a study of images as experiments and Galileo's role in Steno's myology. *Early Science and Medicine*, 2008; 13(6), 665–709.

Middleton, W.E. Knowles, *The Experimenters: A Study of the Accademia del Cimento*, Baltimore–London 1971.

Miniati, Stefano, *Nicholas Steno's Challenge for Truth. Reconciling science and faith*. Milan, FrancoAngeli 2009.

Moe, Harald, When Steno brought new esteem to glands. In: Poulsen, Jacob E. and Snorrason, Egill (eds.) *Nicolaus Steno, 1638–1686, A Reconsideration by Danish Scientists*, Gentofte 1986, pp. 51–92.

Moe, Harald, *Niels Stensen, en Billedbiografi*, København 1986.

Moe, Harald, *Nicolaus Steno, An illustrated Biography*, Copenhagen 1994.

Moe, Harald and Bojsen-Møller, Finn The fine structure of the lateral nasal gland (Steno's gland) of the rat, *Journal of Ultrastructure Research* 1971; 36, 127–148.

Molhuysen, P.C., *Bronnen tot de geschiedenis der Leidsche universiteit*, vol. 3 (1647–1682), s'Gravenhage 1918.

Molhuysen, P.C. and Blok, P.J. (eds.) *Nieuw Nederlandsch biographisch Woordenboek*, vol. 1–10, Leiden 1911–1937. **NNBW**.

Møller-Christensen, Vilhelm, Stensen som kongelig anatom i København, *Dansk medicinhistorisk Årbog*, København 1973, 171–197.

Mollerus, Johannes, *Cimbria Literata*, vol. 2, Copenhagen 1744.

(Monrad, Johan), *Etatsraad og Landkommisær Johan Monrads Selvbiografi (1638–1692)*, edited by S. Birket Smith, København 1888.

Morello, Nicoletta, "De glossopetris dissertation": tThe Demonstration by Fabio Colonna of the true nature of fossils, *Archives Internationales d'Historie des Sciences*, Wiesbaden, 1981; 31 [106], 63–71.

Morello, Nicoletta, *La nascita della paleontologia nel seicento: Colonna, Stenone e Scilla*, Milano 1979.

Mosca, Gioseppe, *Vita di Lucantonio Porzio*, Napoli 1765.

Mouy, Paul, *Le développement de la Physique Cartésienne 1646–1712*, Paris 1934.

Müller, Johannes, Über den glatten Hai des Aristoteles und über die Verschiedenheiten unter den Haifischen und Röchen in der Entwicklung des Eies, *Abhandlungen der Königlichen Akademie der Wissenschaften zu Berlin 1840*, Berlin 1842, 187–257.

Newman, John Henry, *"Apologia Pro Vita Sua" Being a History of His Religious Opinions* (with an Introduction and notes by Svaglic, Martin J.), Oxford 1967.

Nielsen, Oluf, (ed.) *Kjøbenhavns Diplomatarium. Samling af Dokumenter, Breve og andre Kilder til Oplysning om Københavns ældre Forhold før 1728*, vol. 1–8, *og Register*, København 1872– 1887. **Kbh. Dipl.**

Nielsen, Oluf, Uddrag af St. Nikolaj Kirkes Begravelsesprotokol. Et Bidrag til den kjøbenhavnske Genealogi i det 17. Aarhundrede, *Personalhistorisk Tidsskrift* København 1880; 1, 192–223.

Nielsen, Oluf, *Kjøbenhavns Historie og Beskrivelse*, vol. 1: Kjøbenhavn i Middelalderen, Kjøbenhavn 1877.

Nordström, Johan, Swammerdamiana, *Lychnos 1954–1955*, Uppsala 1955, 21–65.

Nordström, Johan, Antonio Magliabechi och Nicolaus Steno. Ur Magliabechis brev till Jacob Gronovius, *Lychnos 1962*, Uppsala 1963, 1–41.

Norvin, William, *Københavns Universitet i Reformationens og Orthodoxiens Tidsalder*, vol. 1–2, København 1937–1940.

Nyrop, C., *Meddelelser om Dansk Guldsmedekunst*. Ved Kjøbenhavns Guldsmedelavs Jubilæum d. 7. November 1885, Kjøbenhavn 1885.

Ødum, Hilmar, Niels Stensens geologiske syn og videnskabelige tankesæt. Ved 300-aaret for hans fødsel, *Naturens Verden*, København 1938; 22, 49–60.

(Oldenburg, Henry), *The Correspondence of Henry Oldenburg*. Edited and translated by Hall, A. Rupert and Hall, Marie Boas, vol. 1 ff., Madison–Milwaukee 1965 ff.

Olden-Jørgensen, Sebastian, Nicholas Steno and René Descartes: a Cartesian perspective on Steno's scientific development. In *Rosenberg, G.A., Revolution in Geology*, 2009, 149–157.

Olden-Jørgensen, Sebastian, Jesuits, Women, Money or Natural Theology? Nicolas Steno's Conversion to Catholicism in 1667. In Andrault and Laerke (eds.), *Steno and the Philosophers*, Leiden: Brill, 2018, 45–6.

de Oliveira, Viviane Bastos, Simone Peres Carneiro, Liliam Fernandes de Oliveira. "Reliability of biceps femoris and semitendinosus muscle architecture measurements obtained with ultra-sonography", *Res. Biomed. Eng.* vol.32 no.4 Rio de Janeiro. http://dx.doi.org/10.1590/2446-4740.04115, 2016.

Olrik, Hans, Borchs Kollegiums Historie i de første 40 aar, 1689–1728, *Festskrift i Anledning af Borchs Kollegiums 200-Aars Jubilæum*, Kjøbenhavn 1889, 1–162.

Omaggio a Leopoldo de'Medici. Parte I: Disegni, Catalogo della mostra a cura di Anna Forlani Tempesti e Anna Maria Petrioli Tofani. Saggi di Paola Barocchi e Gloria Chiarini di Anna. Parte II: Ritrattini. Introduzione et catalogo della mostra a cura di Silvia Meloni Trkulja (Gabinetto disegna e stampe degli Uffizi, vol. 44), Firenze 1976.

Otto, Alfred, *Jesuitterne og Kirken i Danmark i det 17. og 18. Aarhundrede*, København 1940 (42 pp).

Panum, P.L., Bidrag til Kundskab om vort medicinske Fakultets Historie med Hensyn til dets Betydning for Naturvidenskabernes og Lægevidenskabens Udvikling i Danmark fra Frederik den 3[dies] Tronbestigelse indtil Frederik den 5tes Død, *Indbydelsesskrift til Kjøbenhavns Universitets Aarsfest til Erindring om Kirkens Reformation*, Kjøbenhavn 1880.

Pappas, George P., Deanna S. Asakawa, Scott L. Delp, Felix E. Zajac, John E. Drace, "Nonuniform shortening in the biceps brachii during elbow flexion", *Journal of Applied Physiology*, 2002 Vol. 92 no. 6, 2381–2389 DOI:10.1152/japplphysiol.00843.2001.

Partington, J.R., *A History of Chemistry*, vol. 1–4, (1961), reprint London 1969.

Pastor, Ludwig von, *Geschichte der Päpste seit dem Ausgang des Mittelalters*, vol. 12, 14^I and 14^{II}, Freiburg 1927–1930.

Paulli, Simon, *Flora Danica, Det er: Dansk Urtebog*. Published with an Introduction and comments by Lange, Johan and Møller-Christensen, Vilhelm, København 1971.

Payne, L.M., Wilson, Leonard G. and Hartley, Harold, William Croone, *Notes and Records of the Royal Society of London*, London 1960; 15, 211–219.

Perrini, Paolo, Lanzini, Giuseppe and Parenti, Giuliano F. Niels Stensen (1638–1686): scientist, neuroanatomist, and saint, *Neurosurgery* 2010; 67, 3–9.

Personalhistorisk Tidsskrift. Samfundet for Dansk Genealogi og Personalhistorie, København 1880 ff. **PHT**.

Petersen, Carl S., Fra Folkevandringstiden indtil Holberg. In: Petersen, Carl S. and Andersen Vilhelm (eds.) *Illustreret Dansk Litteraturhistorie*, vol. 1, København 1929.

Petersen, Julius, *Bartholinerne og Kredsen om dem*, Kjøbenhavn 1898.

Petersen, N.M., *Bidrag til den Danske Litteraturs Historie, Bd. 3: Det lærde Tidsrum 1560–1710*, København 1855/1856.

Petrocchi, Massimo, Roma nel Seicento, *Storia di Roma*, vol. 14, Bologna 1970.

Pfeiffer, Dieter Ildefons, Stensens erste geologische Schrift "De Thermis", Scherz, *Dissertations on Steno as Geologist*, 1971, 232–236.

Pflieger, Michael, *Priesterlich Existens*, Innsbruck/Wien/München 1953.

Philipsen, H.P., Ductus parotideus stenonianus, *Tandlægebladet*, København 1960; 64, 223.

Pieper, Anton, *Die Propaganda-Congregation und die nordischen Missionen im siebzehnten Jahrhundert*. Köln 1886.

Pieraccini, Gaetano, *La stirpe de'Medici di Cafaggiolo*, vol. 2, Firenze 1925.

Poëte, Marcel, *Une vie de cité. Paris de la naissance à nos jours*, vol. 1–3, Paris 1924–1931.

Popper, Karl R., *Conjectures and Refutations*, London 1963, 5th revised ed. 1989.

Porter, Jan H., The Bartholins. A seventeenth century family Study, *Medicinskhistorisk Årsbok*, Stockholm 1964, 13 p.

Poynter, F.N.L., Nicolaus Steno and the Royal Society of London. In: Scherz, Gustav *Steno and Brain Research*, Analecta Medico-Historica, Oxford, 1968; 3, pp. 273–280.

Preller, C.S. du Riche, *Italian Mountain Geology, Parts 1–2*, London 1918.

Préposiet, Jean, Bibliographie Spinoziste, *Travaux du Centre de Documentation et de Bibliographie Philosophiques de Besançon*, vol. 1, Paris 1973.

Ramsing, H.U., *Københavns Ejendomme 1377–1728. Oversigt over Skøder og Adkomster. I. Østre Kvarter*, København 1943.

Räss, Andreas, *Die Convertiten seit der Reformation nach ihrem Leben und aus ihren Schriften dargestellt*, Freiburg 1875.

Raven, Charles E., *John Ray, Naturalist. His Life and Works*. Cambridge 1942.

Redi, Francesco, *Opera*, vol. 1–9, Milano 1809–1811.

Redlich, Oswald, *Weltmacht des Barocks. Österreich in der Zeit Kaiser Leopolds*, 4th ed., Vienna 1961.

Resen, Petrus Johannes, *Bibliotheca*, Hafniæ 1685.

Reumont, Alfred von, *Geschichte Toscanas seit dem Ende des florentinischen Freistaats, Bd. 1: Die Medici Jahren 1530–1737, Gotha 1876.

Richa, Giuseppe, *Notizie istoriche delle chiese fiorentine divisi ne' suoi Quartini*, vol. 1–10, Firenze 1754–1762, reprint Rome 1972.

Richter, Wilhelm, *Studien und Quellen zu Paderborner Geschichte*, vol. 1, Paderborn 1893.

Rodolico, Francesco, *L'Esplorazione naturalistica dell' Appenino*, Firenze 1963.

Rodolico, Francesco, *La Toscana descritta dai naturalisti del settecento*, Firenze 1945.

Rodolico, Francesco, *Die Florentinische Landschaft* with 52 photos taken by the author, Firenze 1959.

Rodolico, Francesco, Niels Stensen, founder of the geology of Tuscany, Scherz, *Dissertations on Steno as Geologist*, 1971, 237–243.

Rome, Remacle, Nicolas Sténon et la Royal Society of London, *Osiris*, Brugis 1956; 12, 244–268.

Rome, Remacle, Nicolas Sténon Paleóntologiste, *Nicolaus Steno and his Indice*, Copenhagen 1958, 93–98.

Romé de l'Isle, Jean Baptiste Louis de, *Cristallographie, ou Description des formes propres à tous les corps du Règne minéral, dans l'état de Combinaison saline, pierreuse ou métallique*, 2nd ed. Paris 1783.

Roos, Anna Marie, ed., *The Correspondence of Dr. Martin Lister (1639–1712)*, Volume One: 1662–1677, Leiden 2015.

Rørdam, Holger, *Kjøbenhavns Universitets Historie fra 1537 til 1621*. Vol. 3, 1588–1621, Kjøbenhavn 1877.

Rørdam, Holger (ed.), *Monumenta Historiæ Danicae. Historiske Kildeskrifter og Bearbejdelser af dansk Historie især i det 16. Aarhundrede*, 1. R, Bd. 1., København 1873.

Rørdam, Holger, *De danske og norske Studenters Deltagelse i Kjøbenhavns Forsvar mod Karl Gustav*, Kjøbenhavn 1855.

Rosa, Stefano de, Un inventario naturalistico inedito di Niccolò Stenone, *Atti della Fondazione Giorgio Ronchi*, 1985; 40(3), 349–357.

Rosa, Stefano de, *Niccolò Stenone nella Firenze e nell'Europa del suo tempo*. Firenze, 1986.

Rosa, Stefano de, Niccolò Stenone e la politica culturale Medicea: i suoi rapporti con lo studio Pisano. In: *Il Futuro dell'Uomo*, 1987; 4 (1–2), 83–96 (Proceedings, N. Stenone, due giornate di studio, Biblioteca Medicea Laurenziana).

Rosenberg, Gary D. (ed.), *The Revolution in Geology from the Renaissance to the Enlightenment*, The Geological Society of America, Boulder, Colorado (*Memoir* 203), 2009.

Rosenfield, Leonora Cohen, *From Beast-machine to Man-Machine. Animal Soul in French Letters from Descartes to La Mettre* (1940). New and enlarged ed., New York 1968.

Rossi, Paolo, La fisica spinoziana e la fisica moderna, *Spinoza nel terzo centenario della sua nascita*. Publ. a cura della Facoltà di filosofia dell'università cattolica del Sacro Cuore, Rivista di filosofia neo-scolastica, special supplement to vol 25, August 1933, Milan 1934, 117–131.

Rothschuh K.E., Descartes, Stensen und der Discours sur l'anatomie du cerveau (1665). In: Scherz, Gustav (ed.) *Steno and Brain Research*, Analecta Medico-Historica, Oxford 1968; 3, pp. 49–57.

Rovere, Maxime, *Le Clan Spinoza*, Paris, Flammarion 2017.

Rudwick, Martin J.S., *The Meaning of Fossils. Episodes in the History of Palaeontology*, London–New York 1972.

Schafranovski, J.J., Die kristallographischen Entdeckungen N. Stensens, Scherz, *Dissertations on Steno as Geologist*, 1971, 244–259.

Schepelern, H.D., *Museum Wormianum. Dets Forudsætninger og Tilblivelse*, Højbjerg 1971.

Scherz, Gustav, Da Stensen var i Paris, *Fund og Forskning*, København 1969; 16, 43–52.

Scherz, Gustav, Danmarks Stensen-manuskript, *Fund og Forskning*, København 1958/1959; 5–6, 19–33.

Scherz, Gustav, Niels Stensen und Galileo Galilei, *Saggi su Galileo Galilei. Comitato nazionale per le manifestazioni celebrative del IV centenario della nascita di Galileo Galilei*, Firenze 1967, 1–65.

Scherz, Gustav, Niels Stensen und Antonio Magliabechi, *Sudhofs Archiv* 1961; 45, 23–33.

Scherz, Gustav, Niels Stensens Reisen, Scherz, *Dissertations on Steno as Geologist*, 1971, 9–133. **Reisen**.

Scherz, Gustav, (ed.) Steno and Brain Research in the Seventeenth Century, *Analecta Medico-Historica*, vol. 3, Oxford 1968.

Scherz, Gustav, Vom Wege Niels Stensens. Beiträge zu seiner naturwissenschaftlichen Entwicklung. Dissertation, *Acta Historica Naturalium et Medicinalium*, vol. 14, Kopenhagen 1956.

Scherz, Gustav, Niels Stensen's first dissertation, *Journal of the History of Medicine and Allied Sciences*, 1960; 15, 247–264.

Scherz, Gustav, The Indice of Nicolaus Steno, *Nicolaus Steno and his Indice*, Copenhagen 1958, 189–199. **Indice**.

Scherz, Gustav, ed., Steno. Geological paper, *Acta Historica Naturalium et Medicinalium*, vol. 20., Odense 1969. **GP**.

Scherz, Gustav, Niels Stensen's geological work, *Steno. Geological Papers*, Odense 1969, 11–47.

Scherz, Gustav (ed.), Dissertations on Steno as geologist, *Acta Historica Naturalium et Medicinalium*, vol. 34, Odense 1971.

Scherz, Gustav, Viviani, Galileis letzter Schüler, Stensens Freund, Scherz, *Dissertations on Steno as Geologist*, 1971, 260–292.

Scherz, Gustav, To berømte Grotter i Alperne, *Naturens Verden*, København 1950; 34, 324–332.

Scherz, Gustav, Niels Stensens Smaragdreise, *Centaurus*, København 1955; 4, 51–57.

Scherz, Gustav, Pionier der Wissenschaft. Niels Stensen in seinen Schriften, *Acta Historica Naturalium et Medicinalium*, vol. 17. Copenhagen 1963. **Pionier**.

Scherz, Gustav, Stensens englische Verbindungen, *Stenoniana Catholica* 1955; 1, 79–81.

Scherz, Gustav, Niels Stensens København, *Historiske Meddelelser om København*, Årbog 1962, 177–216.

Scherz, Gustav, Briefe aus der Bartholinerzeit, *Centaurus*, København 1961; 7, 157–196.

Scherz, Gustav, Neue Stensenbriefe, *Centaurus*, 1967; 12, 167–181.

Scherz, Gustav, Gespräche zwischen Leibniz und Stensen, *Akten des Internationalen Leibniz-Kongresses*, Hannover, 14–19 November, 1966, *Studia Leibnitiana Supplementa*, vol. 5: Geschichte der Philosophie, Wiesbaden 1971, 81–104.

Scherz, Gustav, Niels Stensen und Ferdinand von Fürstenberg, *Theologie und Glaube*, Paderborn 1958; 48, 401–422.

Scherz, Gustav, Niels Stensens Handschrift, *Stenoniana Catholica*, København 1961; 7, 36–42.

Schierbeek, A., *Jan Swammerdam. Zijn leven en werken*, Lochum 1947. [Also: *Jan Swammerdam. His life and works*, Amsterdam 1967].

Schiller, J. and Théodoridès, J., Sténon et les milieux scientifiques parisiens. In: Scherz, Gustav (ed.) *Steno and Brain Research*, Analecta Medico-Historica, Oxford 1968; 3, 155–170.

Schlegel, Johann Heinrich, *Samlung zu dänischen Geschichte, Münzkenntnis, Oekonomie und Sprache*, vol. 2, 1, Copenhagen 1774.

Schneer, Cecil J., Steno on Crystals and the Corpuscular Hypothesis, *Dissertations on Steno as Geologist, Acta Historica Naturalium et Medicinalium*, 1971; 34, 293–307 Odense.

Schneer, Cecil J. (ed.), *Toward a History of Geology*, Proceedings of Interdisciplinary Conference on the History of Geology, Cambridge, Mass/London 1969.

Scilla, Agostina, *La vana speculazione disinganata dal senso*, Napoli 1670.

Serpell, C., Elba, J. *The Tuscan Archipelago* (Traveller's Guide), London 1977.

Shackelford, Jole, *A philosophical Path for Paracelsian Medicine – Ideas, Intellectual Context and Influence of Petrus Severinus (1540/2–1602), Copenhagen, 2004, 264.*

Sibbald, Sir Robert, *Memoirs*, edited Paget Hett, Francis, Oxford 1932.

Skippon, Philip, An Account made of a Journal Made Thro' Part of the Low-Countries, Germany, Italy and France, *A Collection of Voyages and Travels*, vol. 6, London 1732, 359–736.

Slottved, Ejvind, *Lærestole og lærere ved Københavns Universitet 1537–1977*, København 1978.

(Sørensen, Peder), Petrus Severinus of hans "Idea medicinæ philosophiæ". En dansk paracelsist. Idea medicinæ translated by Skov, Hans. Introduction and comments by Bastholm, E. *Acta Historica Naturalium et Medicinalium*, vol. 32, Odense 1979.

(Smid or Smith), *Henrik Smiths Lægebog*. Publ. with an epilogue by Anna-Elisabeth Brade, København 1976.

Sobiech, Frank, *Herz, Gott, Kreuz. Die Spiritualität des Anatomen, Geologen und Bischofs Dr. med. Niels Stensen (1638–86)*, Münster 2004.

Sobiech, Frank, *Radius in manu Dei. Das medizinische Ethos in Leben, Werk und Rezeption des Anatomen Niels Stensen*, Münster 2012.

Sobiech, Frank, *Ethos, Bioethics, and Sexual Ethics in Work and Reception of the Anatomist Niels Stensen (1638–1686)*, Springer Nature, Switzerland, (Philosophy and Medicine, v. 117), 2016.

(Sophie von Hannover), *Briefwechsel der Herzogin Sophie von Hannover mit ihrem Bruder, dem Kurfürsten Karl Ludwig von der Pfalz, und des Letzeren mit seiner Schwägerin, der Pfalzgräfin Anna*. Publ. by Bodemann, Eduard. New edition 1885, Osnabrück 1966.

Sorø, Klostret-Skolen-Akademiet gennem Tiderne. Skrevet af gamle Soranere. Vol. 1: Tiden før 1737, København 1924.

Spanheim, Friedrich, *Operum, tomus tertius. Qui complectitur Theologica scripta omnia, exegetico-didactico elenctica*, Lugduni Batavorum 1703.

Spinoza, Baruch, *Briefwechsel*, edited with an Introduction, comments and register by Gebhardt, Carl, Leipzig 1914.

Spinoza, Baruch, *Opera*, edited by Gebhardt, Carl, Heidelberg 1925.

Springer, Stewart, It began with a shark, *Dissertations on Steno as Geologist*, Odense 1971, 308–319.

Spruit, Leen and Totaro, Pina, The Vatican Manuscript of Spinoza's Ethica, *Brill's Texts and Sources in Intellectual History*, 205 (1) 2011.

Stenoniana Catholica. *Berichte über Leben, Bedeutung und Prozeß Niels Stensens, des Anatomen, Geologen, Bischofs und Diener Gottes*, Edited by Scherz, Gustav, Vol. 1–7, Copenhagen 1955–1961.

(Stensen, Niels), *Nicolai Stenonis Opera Philosophica*, edited by Maar, Vilhelm, vol. 1–2, Copenhagen 1910. **OPH**.

(Stensen, Niels), *Niccolò Stenone Opere Scientifiche. Traduzione integrale dai testi originali*, edited by Casella Luciano and Coturri Enrico, vol. 1–2, Firenze 1986.

(Stensen, Niels), *Nicolai Stenonis Opera Theologica cum prooemiis ac notis germanice scriptis*, edited by Larsen Knud and Scherz Gustav, vol. 1–2, Hafniæ 1944. **OTH**.

(Stensen, Niels), *Nicolai Stenonis Epistolae et epistolae ad eum datae quas cum prooemio ac notis germanice scriptis*, edited by Scherz Gustav, adjuvant Ræder, Joanne, vol. 1–2, Hafniae/Fribourgi 1952. **EP**.

(Stensen, Niels), Extrait de la dissertation de Nicolas Sténon sur les corps solides qui se trouvent continues naturellement dans d'autres corps solides. *Collection académique. Partie etrangère*, tome 4, Dijon 1757, 377–414.

(Stensen, Niels), *Niels Stensen's korrespondance i dansk oversættelse*, edited by Hansen, Harriet M., vol. 1–2, København 1987.

(Stensen, Niels), Niels Stensen's Chaos-manuscript Copenhagen, 1659. Complete edition with Introduction, notes and commentary by Ziggelaar, August, *Acta Historica Naturalium et Medicinalium*, vol. 44. Copenhagen 1997, 520 pp. **Chaos 'ed. Z'** eREX.

Stensen, Niels, *Disputatio physica De Thermis*. A cura di Gustav Scherz, presentazione di Gustav Scherz e Enrico Coturri, Montecatini 1966.

(Stensen, Niels), Indice di cose naturali, forse dettato da Niccolò Stenone. In: Scherz, Gustav *Vom Wege* ... Kopenhagen 1956, 141–217, also: *Nicolaus Steno and his Indice*, Copenhagen 1958, 201–207.

(Stensen, Niels), *Anatomical Observations of the Glands of the Eye and Their New Vessels, Thereby Revealing the True Source of Tears by Nicolaus Steno* (with a preface and notes by Gotfredsen, Edvard), Copenhagen 1951.

(Stensen, Niels), *Nicolaus Steno's Lecture on the Anatomy of the Brain.* Intr. by Scherz, Gustav, Copenhagen 1965. A facsimile edition of 1669 edition is available at http://gallica.bnf.fr/ark:/12148/bpt6k106685b/f1.image.pagination.

(Stensen, Niels), *Nicolaus Stenos Foredrag om Hjernens Anatomi.* Translated by Maar, Vilhelm, København 1903. New edition with an epilogue by Kardel, Troels, København 1997.

Stensen, Niels, *Discours sur l'anatomie du cerveau*, edited and commented by Raphaële Andrault. Paris, Classiques Garnier 2009.

(Stensen, Niels), *Nicolas Sténon, Oeuvres choisies.* Paris: Les belles lettres 2010. Letters and the Prooemium translated by Olesen, Birger Munk.

(Stensen, Niels), *Niels Steensens (Stenonis) Værker i Oversættelse.* Edited by Christensen, R.E., Hansen, A and Larsen Knud, vol. 1 [of 1] (Klassisk dansk Medicin, Bd. 2), København 1939. **NSVO**.

(Stensen, Niels), Steno on Hydrocephalus. *Journal of the History of Neurosciences* 1993; 2, 179–202, introduction and commentary by Kardel, Troels, ibid. 171–178.

(Stensen, Niels), *Steno on Muscles, Stensen's New Structure of the Muscles and Heart and Specimen of Elements of Myology, Transactions American Philosophical Society* 84(1), Philadelphia 1994, 2nd print 1995, www.books.google.com

(Stensen, Niels), Steno. Geological Papers. Edited by Gustav Scherz, translated by Pollock, Alex J. *Acta Historica Scientiarum Naturalium et Medicinalium*, vol. 20, Odense 1969.

(Stensen, Niels), *Nicolaus Steno. Foreløbig Meddelelse til en Afhandling om faste Legemer, der findes naturligt indlejrede i andre faste Legemer.* Tranlated by Krogh, August and Maar, Vilhelm, København 1910.

(Stensen, Niels), *Das Feste im Festen. Vorläufer einer Abhandlung über Festes, das in der Natur in anderem Festen eingeschlossen ist.* Translated by Mietleitner, Karl. Revised with an introduction and explanations by Scherz, Gustav (*Ostwalds Klassiker der exakten Wissenschaften, Neue Folge*, vol. 3), Frankfurt am Main 1967.

(Stensen, Niels), *The Prodromus of Nicolaus Steno's Dissertation Concerning a Solid Body Enclosed by Process of Nature within a Solid.* An English Version with an introduction and explanatory notes by Winter, John Garrett, a Foreword by Hobbs, William H. (Facsimile of University of Michigan Humanistic Studies, Vol. XI, pt. 2, 1916.) Introduction by White, George W. (*Contributions to the History of Geology*, vol. 4), New York–London 1968.

(Stensen, Niels), "Preface to a Demonstration in the Copenhagen Anatomical theater in the Year 1673", and Jacobæus, Holger: "Niels Stensen's Anatomical Demonstration no. XVI" and other texts see Kardel, *Steno, Life, Science, Philosophy, Acta Historica Naturalium et Medicinalium*, vol. 42, 159 pp. Copenhagen 1994.

Sticher, Bernhard, Leibniz' Beitrag zur Theorie der Erde, *Sudhofs Archiv*, Wiesbaden 1967; 51, 244–259.

Sticher, Bernhard, Naturam cognosci per analogiam. Das Prinzip der Analogie in der Naturforschung bei Leibniz, *Akten des internazionalen Leibniz-Kongresses* Hannover, 1966 [Studia Leibnitiana Supplementa, Bd. 5], Bd. 2: Mathematik-Naturwissenschaften, Wiesbaden 1969, 176–196.

Stillman, John Maxson, *The Story of Alchemy and Early Chemistry.* New ed., New York 1960.

Storia d'Italia, II, 1–2: Dalla caduta del'Impero romano al secolo XVIII, Torino 1974.

Struik, Dirk J., *The land of Stevin and Huygens. A Sketch of Science and Technology in the Dutch Republic During the Golden Century* (translated from "Het Land van Stevin en Huygens", 3rd rev. ed. 1979), Dordrecht 1981.

Stybe, Svend Erik, *Universitet og Åndsliv i 500 år.* København 1979.

Sundbo, Arne, *Frederikssunds og købstaden Slangerups Historie. I.: Tiden indtil 1809*, Hillerød 1931.

Swammerdam, Johannis, *Miraculum Naturae sive Uteri muliebris fabrica.* Lugduni Batavorum 1672.

Targioni-Tozzetti, Giovanni, Catalogo delle produzioni naturali che si conservana nella Galleria Imperiale di Firenze distesto dal Dottor Giovanni Targioni-Tozzetti. Scherz, Gustav, *Vom Wege Niels Stensens*, 1956, 218–229. Also: Nicolaus Steno and his Indice, Copenhagen 1958, 278–289.

Teresa de Jesu, *Sämtliche Schriften von der hl. Theresa von Jesu*, Bd. 1: *Leben von ihr selbst beschrieben*. Translated by Aloysius, P., München 1933.

Tertsch, Hermann, *Das Geheimnis der Kristallwelt. Roman einer Wissenschaft*, Wien 1947.

Tertsch, Hermann, Niels Stensen und die Kristallographie, *Nicolaus Steno and his Indice*, Copenhagen 1958, 120–139.

Thijssen-Schroute, C. Louise, *Nederlands Cartesianisme* Verhandelingen der Koninklijke Nederlandse Akademie van Wetenschappen. Afd. Letterkunde. N. R. 9, Amsterdam 1954.

Thomsen, Elsebeth, Niels Stensen-Steno in the world of collections and museums, Rosenberg, *Revolution in Geology*, 2009, 75–92.

Tolmer, L., *Une page d'histoire des sciences 1661–1669. Vingt-deux Lettres inédites d'André de Graindorge à P.D. Huët* Mémoires de l'Académie des sciences, arts et belles-lettres de Caen, n.s. 10, 1941, 245–337. The Graindorge-letters are web accessible from www.kb.dk/rex.

Totaro, Pina, "Ho certi amici in Ollandi": Stensen and Spinoza – science versus faith, *Analecta Romana Instituti Danici*, suppl. XXXI, Rome 2002, 27–38.

Ueberweg, Friedrich, *Grundriss der Geschichte der Philosophie*, 2nd edition fully revised by Frischeisen-Köhler, Max and Moog Willy, vol. III: Die Philosophie der Neuzeit bis zum Ende des XVIII. Jahrhunderts, Berlin 1924.

Universitetsprogram og Mindeskrift over Jørgen Eilersen, død 12. Oktober 1686, Hafniae 1686.

Vaupell, Otto, *Rigskansler Grev Griffenfeld. Et Bidrag til Nordens historie i det 17. Hundredaar*, vol. 1–2, Kjøbenhavn 1880–1882.

Vai, Gian Battista and Caldwell, W. Glen E. (eds.), The origin of geology in Italy, *Geological Society of America*, Special Paper 411, Boulder, Colorado 2006.

Vai, Gian Battista, The scientific revolution and Nicholas Steno's twofold conversion, in Rosenberg, G.A., ed., *The Revolution in Geology from the Renaissance to Enlightenment*, 2009, 187 ff.

Veibel, Stig, *Kemien i Danmark*, I: *Kemiens Historie i Danmark*, København 1939.

Vida, Tivadar, Niels Stensens Ungarnreise im Jahre 1669, *Centaurus* 1984; 27, 167–172.

Vijver, C. van der, *Geschiedkundige Beschrijving der Stad Amsterdam*, vol. 3, Amsterdam 1848.

Vugs, Joseph Gerard, *Leven en Werk van Niels Stensen (1638–1686)*. With the Dutch translation of the "Discours sur l'anatomie du cerveau". Thesis, Leiden 1968.

Watson, Richard A., *The Downfall of Cartesianism 1673–1712. A Study of Epistemiological Issues in Late 17th Century Cartesianism*. The Hague 1966.

Willius, Fredrick A., An unusually early description of the so-called Tetralogy of Fallot, *Proceedings of the Staff Meetings of the Mayo Clinic* 1948; 23(14): 316–320.

Winslow, Jaques-Benigne, *L'Autobiographie*, publ. by Maar, Vilhelm. Paris-Copenhagen 1912.

Winslow, Jaques-Benigne, *Exposition anatomique de la structure du corps humain*. Paris 1732.

Woodward, John, *Essay toward a natural history of the earth,* London 1695.

Wulff, D.H., Biografiske Optegnelser over Biskop Frands Thestrup og hans Familie, *Kirkehistoriske Samlinger*, 5. R. Bd. 5, København 1909–1911, 1–25.

Yamada, Toshihiro, Hooke-Steno relations reconsidered: Reassessing the role of Ole Borch and Robert Boyle, *Rosenberg, Revolution in Geology*, 2009, 107–126.

Yamada, Toshihiro, *The Transformation of Geocosmos: Perception of Earth from Descartes toLeibniz with chapters on Kircher, Varenius, Hooke, Spinoza and Steno*. Tokyo, Keiso 2017.ISBN 978-4326148295

Ziggelaar, August, The age of Earth in Niels Stensen's geology., In: Rosenberg, *Revolution in Geology* …, 2009, 135–142.

Zittel, Karl Alfred von, *Geschichte der Geologie und Paläontologie bis Ende des 19. Jahrhunderts*, München–Leipzig, 1899.

Index of Persons

© Springer-Verlag GmbH Germany, part of Springer Nature 2018
T. Kardel and P. Maquet, *Nicolaus Steno*,
https://doi.org/10.1007/978-3-662-55047-2

Printed in the United States
By Bookmasters